Engineering Principles of Mechanical Vibration

4th Edition

Douglas D. Reynolds, Ph.D.

Cover Pictures
Tacoma Narrows November 7, 1940, Bridge Collapse
Pictures extracted from an Internet YouTube video of the bridge collapse.

Pictures extracted by:
Douglas D. Reynolds, Ph.D.

Order this book online at www.trafford.com
or email orders@trafford.com

Most Trafford titles are also available at major online book retailers.

Print information available on the last page.

ISBN: 978-1-4907-1437-0 (sc)
ISBN: 978-1-4907-1436-3 (e)

Trafford rev. 06/09/2017

Trafford PUBLISHING® www.trafford.com
North America & international
toll-free: 1 888 232 4444 (USA & Canada)
fax: 812 355 4082

TABLE OF CONTENTS

PREFACE

The overall contents of Chapters 1 through 11 in the 4th edition of this textbook are essentially the same as their contents in the 3rd edition. However, these chapters were extensively edited and some chapters reorganized to correct typos and other errors and to more clearly present the vibration concepts and math principles covered in these chapters. Important equations in the developments of vibration concepts and math principles and the answers to EXAMPLE problems are "boxed" throughout this textbook. Chapter 12 is new to the 4th edition.

Many mechanical and structural systems are designed from a statics and strength of materials viewpoint. They take into account design concepts that include, but are not limited to:

1. Stress-strain relations;
2. Structural analyses that include compressive, tensile and shear forces and bending moments;
3. Yield and endurance limits associate with static and slowly cyclic system loading;
4. Structural deformations associated with applied structural forces and moments; and
5. Design factors of safety.

These and other design concepts treat mechanical and structural systems as if they are exposed only to static or slowly varying system forces and moments.

Problems often arise when mechanical and structural systems are exposed to dynamic forces and moments that rapidly vary as a function of time. These excitation forces and moments can vary sinusoidally with time. They can be comprised of many simultaneously superimposed sinusoidal components with different excitation frequencies. They can be transient or random in nature. When these forces and moments act on mechanical and structural systems and the right conditions exist, system resonances can occur. These are situations where the dynamic motions and related system stresses can be many times greater than those predicted by a static and strength of materials design approach. These dynamic motions and related stresses can result in unacceptable performances of mechanical and structure systems and in certain cases can result in catastrophic system failures.

Mechanical vibration is a study of mechanical and structural systems that are exposed to time-varying dynamic forces and moments. This study involves applying known laws of physics and related principles to develop system equations that describe the motions and responses of mechanical and structural systems when they are not in static equilibrium. In these analyses, the system forces and moments are set equal to appropriate mass times acceleration terms or mass moment of inertia times angular acceleration terms that will result in a system of ordinary or partial differential equations. These equations must then be appropriately solved to determine how the mechanical or structural system will respond to applied system initial conditions and externally applied forces and moments.

Effective mechanical vibration analyses of mechanical and structural systems require the use of several math and engineering modeling disciplines. These include:

1. A good understanding of how differential equations can be used to describe the responses of real mechanical and structural systems to initial conditions and externally applied time-varying forces and moments;
2. An ability to conceptually visualize how mechanical and structural system components go together and to develop realistic schematic representations of the system and related system components based on this visualization;
3. An ability to diagram how forces and moments act on mechanical and structural system components by means

of free-body diagrams based on the schematic representations of the system developed in (2);

4. An ability to develop appropriate dynamic differential equations of motion from the free-body diagrams developed in (3) and then to properly develop the solutions to these equations of motion;
5. An understanding of the concept of system resonances and how these resonances can be highly detrimental to mechanical and structural systems;
6. An ability to interpret the solutions to the equations of motion developed in (4) to determine whether or not it predicts a realistic response of the mechanical or structural system to its initial conditions and externally applied time-varying forces and moments; and
7. An ability to apply the results of (4) through (6) to understanding and resolving real-world vibration problems.

Undergraduate or graduate students who enroll in an introductory course in mechanical vibration should have had prerequisite courses that cover many of the above engineering modeling disciplines. They should have had courses in statics and rigid-body dynamics where they learn how to draw free-body diagrams and do static and dynamic vector force and moment analyses. They should have had courses in calculus and ordinary differential equations that give them the math background necessary for an introductory course in mechanical vibration.

Even though students have often taken the appropriate prerequisite courses before enrolling in an introductory course in mechanical vibration, they are often unable to initially pull together the above engineering modeling disciplines to conduct effective vibration analyses of mechanical and structural systems. They have been taught how to draw free-body diagrams, but they have difficulty in applying this knowledge to mechanical systems comprised of coupled mass, spring and damper elements. They have been taught how to solve ordinary differential equations, but they have difficulty in visualizing how these equations can describe the motions of real physical systems. Concepts associated with system resonances are new to them, and it takes time and practice for them to fully understand how system resonances can negatively and sometimes catastrophically affect the responses of mechanical and structural systems. These are barriers that must be effectively addressed before a student can begin to develop a basic knowledge of mechanical vibration.

Another major barrier to understanding mechanical vibration is how students initially approach problem solving. They often use a visualization or perception approach to problem solving. That is, they attempt to first visualize or perceive how they think a mechanical or structural system will respond to applied system forces and moments. Then, they attempt to solve the vibration problem before they fully understand how the different components of the system interact. This approach will not work in mechanical vibration analyses. Students must learn to develop a disciplined system approach to problem solving based on the above engineering modeling disciplines. They must learn how to:

1. Visualize how the components of a mechanical or structural system interact and draw appropriate schematic and related free-body diagrams;
2. Develop the appropriate equations of motion based on the free-body diagrams;
3. Trust the analyses and math principles associated with setting up and solving a system of differential equations;
4. Interpret the system responses that these equations predict to determine how a specific mechanical or structural system will respond to sets of initial system conditions and dynamically applied system forces and moments; and
5. Apply the system response information to solve real world vibration problems.

Currently, several good textbooks exist that address many of the areas associated with mechanical vibration. Several of these textbooks are listed in the General Reference section at the end of this textbook. Some are identified as intermediate and advanced level mechanical vibration textbooks and mechanical vibration handbooks. Several are identified as introductory textbooks for mechanical vibration.

I believe that many of the textbooks that are identified as introductory mechanical vibration textbooks fail in one or more important areas. They often assume students who enroll in an introductory course in mechanical vibration have a better knowledge of the technical contents of the prerequisite courses than they actually do. Students may only be briefly exposed to the required technical concepts in these course. It may take extensive exposures to these concepts before students fully understand them. As a result, introductory mechanical vibration textbooks often omit the full developments of important math and analytical concepts assumed to have been covered in prerequisite courses. They also abbreviate analytical and mathematical developments used to describe important vibration principles, concepts and modeling protocols by leaving out important steps in these developments. This usually makes it more difficult for students to fully understand vibration principles, concepts and modeling protocols presented in these textbooks.

Introductory textbooks often assume students have a better understanding of differential equation solution techniques and of how these equations can describe the vibration responses of real physical systems than they actually do. Therefore, even though they present solutions to system differential equations of motion, they often do not effectively explain how these solutions are obtained.

Some introductory mechanical vibration textbooks do not have adequate numbers and varieties of system examples necessary to strengthen a student's understanding of basic vibration concepts, principles and modeling protocols presented in the textbooks. In addition, they treat the discipline of mechanical vibrations as one of applied mathematics. They fail to clearly demonstrate the link between the applied math covered in the textbooks and how the math can be used to address real world vibration concepts and problems.

It is easy to criticize the work of other authors and perhaps others may find it appropriate to present similar criticisms to this textbook. However, using 43 years of teaching, research and consulting experience in the area of mechanical vibration, I have attempted to write an introductory mechanical vibration textbook that addresses many of the perceived shortcomings associated with other mechanical vibration textbooks. This textbook has complete analytical and mathematical developments of the vibration principles, concepts and modeling protocols that are presented. It does not presume students have an adequate understanding of important materials that are traditionally covered in prerequisite courses, nor does it assume students fully understand the developments and related solution techniques associated with the many systems of differential equations that are presented. All important vibration principles, concepts and modeling protocols are fully developed and all differential equation development protocols and related solution methods are fully explained. This textbook has a large number and variety of vibration system examples that are presented in each chapter to help students better understand the vibration principles, concepts and modeling protocols that are presented. Where appropriate, this textbook identifies the relationships between the math and vibration principles, concepts and modeling protocols that are discussed and their use to solve real world vibration problems.

The contents of the 12 chapters in the 4th edition of this textbook include:

Chapter 1 – Basic Concepts: simple harmonic motion; complex vector operations; periodic and complex periodic signals; random and deterministic signals; degrees-of-freedom; mass, spring and damping elements; equations of motion development techniques; and vibration analyses techniques.

Chapter 2 – 1-DOF Vibration Systems - Unforced Vibration: unforced vibration with and without viscous damping; logarithmic decrement; and unforced vibration with Coulomb and structural damping.

Chapter 3 – 1-DOF Vibration Systems - Response to Harmonic Excitation: forced response of systems with and without viscous damping; forced response of systems with non-viscous and structural damping; response of vibration systems attached to moving structural bases; critical speed of a rotating disc on a shaft; and forced response of systems to complex periodic excitation.

Chapter 4 – 2-DOF Vibration Systems – Response to Harmonic Excitation: unforced vibration with no damping; coordinate coupling; forced response of systems with viscous damping; tuned absorbers; and machine mounted on a flexible structure.

Chapter 5 – n-DOF Vibration Systems – Matrix Eq's and Solution Methods: work and energy relations; generalized coordinates; principle of virtual work; d'Alembert's principle; Hamilton's equation; Lagrange's equation; influence coefficients; solutions to matrix equations of motion; orthogonality of modal vectors; semi-definite systems; and modal analysis of systems with and without damping.

Chapter 6 – Vibration Systems – Response to Non-Periodic Excitation: shock excitation – classical solutions; design considerations for systems exposed to shock excitation; Fourier transforms; Laplace transforms; inverse Laplace transforms; and convolution integral.

Chapter 7 – Vibration Systems - Random Excitation: mean value, mean squared value and variance; probability functions; expected values and moments; correlation functions; power spectral density functions; relations between correlation and power spectral density functions; relations between Fourier transforms and power spectral density functions, and coherence function.

Chapter 8 – Vibration of Continuous Systems: transverse vibration of a string; forced vibration of a string; longitudinal vibration of a bar; force longitudinal vibration of a bar; transverse vibration of a beam; transverse vibration of thin membranes; and transverse vibration of thin plates.

Chapter 9 - Vibration Measurements: vibration transducers, piezoelectric accelerometer characteristics; shock and overload protection of accelerometers; accelerometer mountings; accelerometer cables; accelerometer calibration; and frequency analyzers.

Chapter 10 - Digital Processing of Vibration Signals: relation between the time and frequency domains, digital quantization and sampling of analog signals, aliasing and anti-aliasing filters, dynamic signal analysis and FFT time domain windows, spectral functions and units, averaging, and digital-to-analog conversion.

Chapter 11 - Analytical and Experimental Modal Analyses: formulation of the eigenvalue problem using the dynamical matrix and symmetric eigenvalue solution methods, measurement techniques and protocols associated with experimental modal analysis, methods for processing and analyzing experimental modal analysis data, and forced response of n-DOF vibration systems using analytical modal analysis techniques.

Chapter 12 - Building Mechanical Equipment Vibration Isolation: vibration isolation system design criteria and guidelines, commercial vibration isolation mounts and structural bases, vibration isolation of mechanical systems, investigation and causes of mechanical equipment vibratos problems, measurement of mechanical equipment vibration, and specifications related to and the field inspection of mechanical equipment vibration isolation systems.

I have used this textbook for one senior level undergraduate course and two graduate courses that address topics in the field of mechanical vibration. I cover Chapters 1 - 4, 12 in the senior level undergraduate course. This course is a lecture course. The two graduate course are independent courses with the undergraduate course being a pre-requisite for both courses. I review Chapters 3 and 4 and cover Chapters 5, 6 and 8 in the first graduate course. This course is a lecture course. I cover Chapters 4, 7, 9 - 11 in the second graduate course. This course is a lecture course with four laboratory exercises. The laboratory exercises address topics covered in Chapters 9 and 11.

Douglas D. Reynolds, Ph.D.

ACKNOWLEDGEMENTS

The writing and publishing of this textbook has been a process that has covered a period of nearly 20 years. It represents a compilation of materials that I have developed and used in my 43 year teaching career that started at The University of Texas at Austin, progressed to the University of Pittsburgh, and will conclude at the University of Nevada, Las Vegas.

I am grateful for the continued support and feedback of my undergraduate and graduate students. Their constructive comments, criticisms and suggestions have played an important role in determining and organizing the materials contained in this textbook.

I owe a great debt of gratitude to Dr. Peter Baade. He is the reason I entered the university teaching and research profession. He has continued to be a life-long friend and mentor who has strongly supported and encouraged me in the writing of this textbook.

The writing of any textbook often involves collaborative efforts with other trusted colleges. This has been the case in the writing of this textbook. The writing of Chapters 9 and 10 was a collaborative effort with Dr. Thomas Lagö, a trusted friend and business associate. The writing of Chapter 11 was a collaborative effort with Mr. George Ladkany, an exceptionally talented former graduate student. I am grateful for the contributions these individuals made in the writing of these chapters. I could not have written them without their assistance. I believe these chapters contain information in the areas of vibration measurements, digital signal processing of vibration signals, and analytical and experimental modal analyses that is not contained or easily found in other vibration textbooks or commercially available literature.

Finally, I want to express my heart-felt appreciations for the continued support, encouragement and patience of my wife, Linda, who has been my lifelong companion and supporter for 48 years. I owe a huge debt for her continued support during the many evenings over the years that I have spent writing, editing and re-editing the many parts of this textbook when I know she wanted me to be actively engaged in other activities around our home and in our yard.

CHAPTER 1
VIBRATION SYSTEMS BASIC CONCEPTS

1.1 Components of a Vibration System

Mass and *elasticity* are necessary to support vibration. Physical systems that support vibration consist of one or more mass elements that are connected together with elastic or spring elements. *Mass elements* are associated with *kinetic energy*. This energy is related to the motion of the mass elements of a system. *Elastic* or *spring elements* are associated with *potential energy*. This energy is related to the positions of the mass elements of a system relative to specified reference positions and energy stored within the elastic elements of the system. When a system vibrates, the energy in the system alternately changes between kinetic and potential energy. Theoretically, in the absence of any mechanism to remove energy from a system, once the system begins to vibrate, it will vibrate forever. However, this rarely occurs.

The third element associated with vibration is *damping*. *Damping* is a mechanism that removes energy from a vibrating system, usually in the form of heat. When damping is present, if energy is not added to the system to keep it in motion, damping will eventually cause the motion of the system to stop.

Figure 1.1 shows a schematic representation of a one-degree-of-freedom vibration system. The system is conceptualized as a mass element M that is connected to the ground by means of a spring or other type of elastic element K and a damping element C.

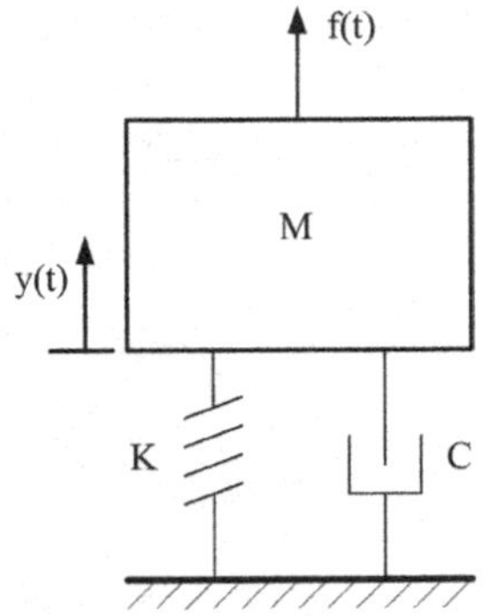

Figure 1.1 One-Degree-of-Freedom Vibrating System

1.2 Simple Harmonic Motion

Motion associated with the mechanical system in Figure 1.1 is oscillatory in nature. Such motion is called *simple harmonic motion*. In *simple harmonic motion*, mass M starts from an undisturbed or equilibrium position, moves to a position of maximum displacement in one direction, moves back through its equilibrium position to a position of maximum displacement in the opposite direction, and finally moves back to its equilibrium position (Figure 1.2). This motion can be described by either a sine or cosine function:

$$y(t) = Y_P \sin \omega t \tag{1.1}$$

where:

Y_P = maximum displacement amp. (in. or m)
ω = angular frequency of oscillation (radians/s)
t = time (s).

The *angular frequency* ω can be expressed:

$$\boxed{\omega = 2\pi f \qquad \text{or} \qquad f = \frac{\omega}{2\pi}} \tag{1.2}$$

where:

π = 3.141593
f = frequency (cycles/s or Hz).

The frequency f represents the number of complete cycles of oscillations an object makes in one second. For example, if an object undergoes ten complete cycles of oscillation in one second, it has a frequency of 10 cycles/s or 10 Hz. The period of oscillation T (s) is given by:

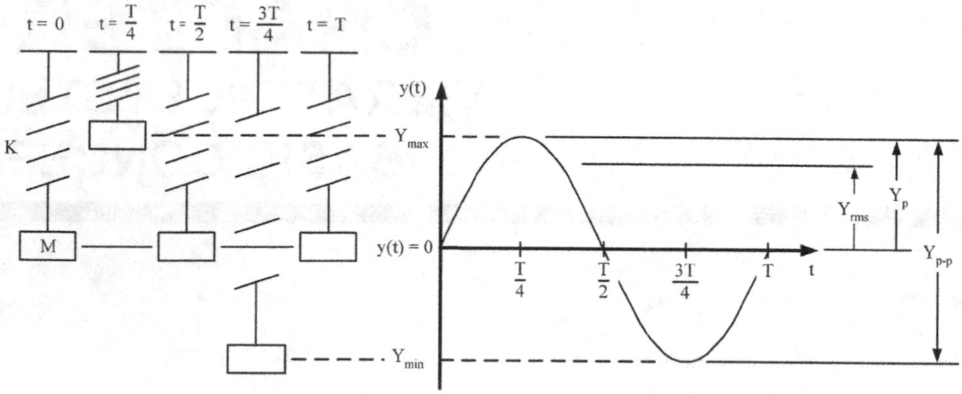

Figure 1.2 Harmonic Oscillation

$$\boxed{T = \frac{1}{f}.} \tag{1.3}$$

The *period of oscillation* represents the time it takes an object to complete one cycle of oscillation.

The amplitude of harmonic motion can also be specified by the *peak-to-peak* displacement $Y_{P\text{-}P}$:

$$Y_{p-p} = Y_{max} - Y_{min} \quad \text{or} \quad Y_{p-p} = 2\,Y_{max} \tag{1.4}$$

and the *root-mean-squared (rms)* displacement, Y_{rms}:

$$\boxed{Y_{rms} = \sqrt{\frac{1}{T}\int_0^T y(t)^2\,dt}\ .} \tag{1.5}$$

1.3 Complex Vector Representation of Harmonic Motion

Simple harmonic motion can also be represented by a *complex rotation phasor* (Figure 1.3). For this case, y(t) is written as a complex vector $\overline{y}(t)$ or:

$$\boxed{\overline{y}(t) = Y\,e^{j\omega t}} \tag{1.6}$$

where:

$$e^{j\omega t} = \cos\omega t + j\sin\omega t\ . \tag{1.7}$$

The equation:

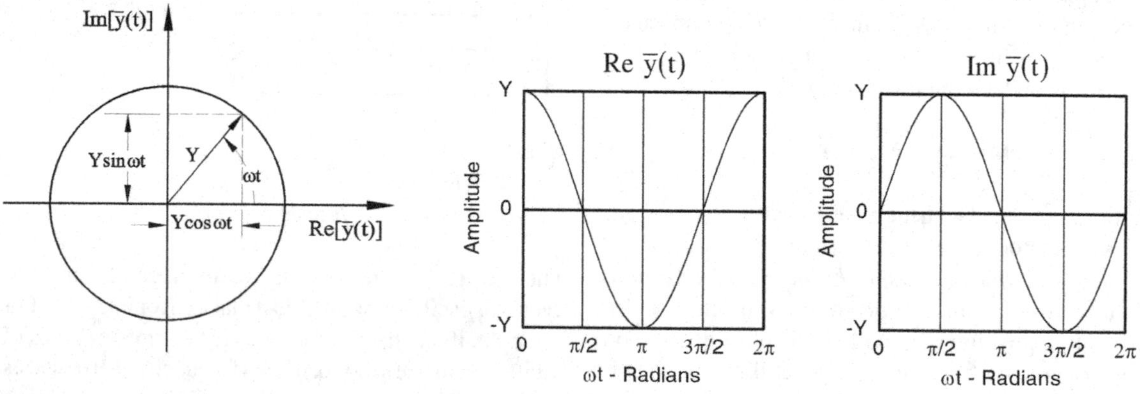

Figure 1.3 Vectorial Representation of Simple Harmonic Motion

$$e^{\pm j\theta} = \cos\theta \pm j\sin\theta \tag{1.8}$$

is called *Euler's equation*. The real part of $\bar{y}(t)$ is written:

$$\text{Re}\left[\bar{y}(t)\right] = Y\cos\omega t \tag{1.9}$$

and the imaginary part is written:

$$\text{Im}\left[\bar{y}(t)\right] = Y\sin\omega t\,. \tag{1.10}$$

A *harmonic function* given as $y(t) = Y\cos\omega t$ can be written as $y(t)=\text{Re}[Ye^{j\omega t}]$. Similarly, $y(t) = Y\sin\omega t$ can be expressed as $y(t) = \text{Im}[Ye^{j\omega t}]$. Harmonic motion is an oscillatory motion, and all physical quantities such as displacement, velocity and acceleration are real-valued quantities. Therefore, these variables can be represented as a function of the real part of a rotating phasor. The use of complex vectors and phasors greatly simplifies the manipulation and solving of the differential equations that describe the motions of vibration systems.

1.4 Complex Vector Operations

Complex phasors are used to describe harmonic signals and harmonic motion throughout this book. Therefore, some of the operations associated with complex numbers must be known and understood. A complex number $\bar{A}$ can be expressed:

$$\bar{A} = a_1 + ja_2 \tag{1.11}$$

where a_1 is the real part and a_2 is the imaginary part of the complex vector (Figure 1.4). $\bar{A}$ can also be expressed in polar form as:

$$\bar{A} = |A|\,e^{j\phi} \tag{1.12}$$

where:

$$|A| = \sqrt{a_1^2 + a_2^2} \qquad \text{and} \qquad \phi = \tan^{-1}\left[\frac{a_2}{a_1}\right]. \tag{1.13}$$

The complex conjugate of $\bar{A}$ is designated $\bar{A}^*$ and is written:

$$\bar{A}^* = a_1 - ja_2 \qquad \text{or} \qquad \bar{A}^* = |A|\,e^{-j\phi} \tag{1.14}$$

where $|A|$ and ϕ are specified by equation (1.13).

If another complex number $\bar{B}$ is designated:

$$\bar{B} = b_1 + jb_2 \qquad \text{or} \qquad \bar{B} = |B|\,e^{j\beta} \tag{1.15}$$

where:

$$|B| = \sqrt{b_1^2 + b_2^2} \qquad \text{and} \qquad \beta = \tan^{-1}\left[\frac{b_2}{b_1}\right], \tag{1.16}$$

$\bar{A} \pm \bar{B}$ is given by:

$$\bar{A} \pm \bar{B} = (a_1 \pm b_1) + j(a_2 \pm b_2)\,. \tag{1.17}$$

$\bar{A} \times \bar{B}$ is given by:

$$\bar{A} \times \bar{B} = |A| \times |B|\,e^{j(\phi+\beta)}\,. \tag{1.18}$$

$\bar{A}/\bar{B}$ is given by:

$$\frac{\bar{A}}{\bar{B}} = \frac{|A|\,e^{j\phi}}{|B|\,e^{j\beta}}\,. \tag{1.19}$$

Equation (1.19) can also be written:

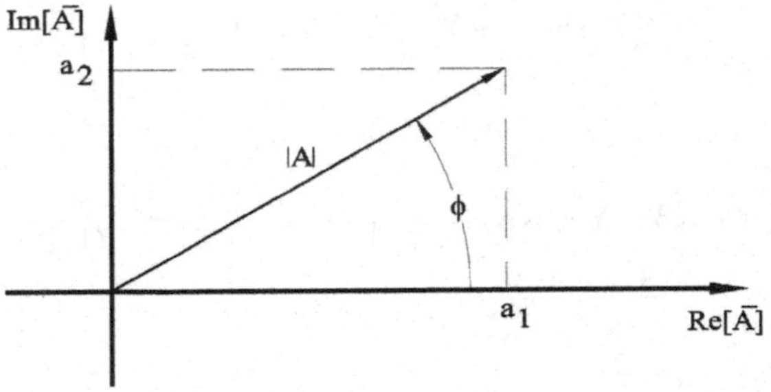

Figure 1.4 Complex Vector

$$\boxed{\frac{\bar{A}}{\bar{B}} = \frac{|A|}{|B|}\, e^{j(\phi-\beta)}\ .} \tag{1.20}$$

EXAMPLE 1.1

Given: $\bar{A} = 1 + j3$ and $\bar{B} = 3 - j2$. Determine:
(a) $\bar{A}+\bar{B}$,
(b) $\bar{A}\times\bar{B}$,
(c) $\bar{A}/\bar{B}$ and
(d) the complex conjugates of $\bar{A}$ and $\bar{B}$.

SOLUTION

(a) $\bar{A}+\bar{B}$ is:

$$\bar{A}+\bar{B}=(1+j3)+(3-j2) \quad \text{or} \quad \bar{A}+\bar{B}=4+j1. \tag{1.1a}$$

$\bar{A}+\bar{B}$ can also be written in polar form:

$$\bar{A}+\bar{B}=\sqrt{4^2+1^2}\, e^{j\theta} \qquad \text{where} \qquad \theta=\tan^{-1}\left[\frac{1}{4}\right] \tag{1.1b}$$

or:

$$\boxed{\bar{A}+\bar{B}=4.12\, e^{j0.245}\ .} \tag{1.1.c}$$

The unit on θ is radians.

(b) To determine $\bar{A}\times\bar{B}$, first write $\bar{A}$ and $\bar{B}$ in polar form:

$$\bar{A}=3.16\, e^{j1.249} \qquad \text{and} \qquad \bar{B}=3.61\, e^{-j0.588}\ . \tag{1.1.d}$$

Therefore, $\bar{A}\times\bar{B}$ is:

$$\bar{A}\times\bar{B}=3.16\times3.61\, e^{j(1.249-0.588)} \tag{1.1.e}$$

or:

$$\boxed{\bar{A}\times\bar{B}=11.41\, e^{j0.661}\ .} \tag{1.1.f}$$

(c) $\bar{A}/\bar{B}$ is:

$$\boxed{\frac{\bar{A}}{\bar{B}}=\frac{3.16\, e^{j1.249}}{3.61\, e^{-j0.588}} \qquad \text{or} \qquad \frac{\bar{A}}{\bar{B}}=0.875\, e^{j1.837}\ .} \tag{1.1.g}$$

(d) The complex conjugates of $\bar{A}$ and $\bar{B}$ are:

$$\boxed{\bar{A}^*=1-j3 \qquad \text{or} \qquad \bar{A}^*=3.16\, e^{-j1.249}} \tag{1.1h}$$

$$\boxed{\bar{B}^*=3+j2 \qquad \text{or} \qquad \bar{B}^*=3.61\, e^{j0.588}\ .} \tag{1.1i}$$

1.5 Addition of Harmonic Signals

The sum of two harmonic signals with the same frequency but different phases will result in a harmonic signal with the same frequency. Let:

$$y_1(t)=Y_1\cos(\omega t+\alpha_1) \tag{1.21}$$

and:

$$y_2(t)=Y_2\cos(\omega t+\alpha_2)\ . \tag{1.22}$$

The sum of $y_1(t)$ plus $y_2(t)$ is:

$$y(t)=y_1(t)=Y_1\cos(\omega t+\alpha_1)+Y_2\cos(\omega t+\alpha_2)\ . \tag{1.23}$$

Using the identity, cos(a + b) = cos a cos b - sin a sin b, the above equation becomes:

$$y(t)=(Y_1\cos\alpha_1+Y_2\cos\alpha_2)\cos\omega t - (Y_1\sin\alpha_1+Y_2\sin\alpha_2)\sin\omega t\ . \tag{1.24}$$

If:

$$Y\cos\beta=Y_1\cos\alpha_1+Y_2\cos\alpha_2 \tag{1.25}$$

and:

$$Y\sin\beta=Y_1\sin\alpha_1+Y_2\sin\alpha_2 \tag{1.26}$$

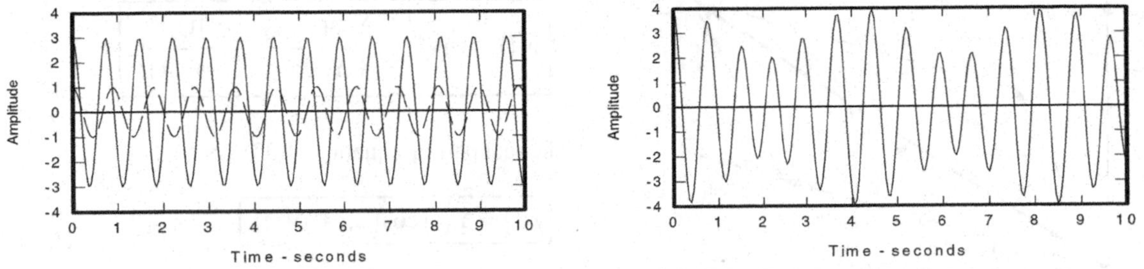

$y_1(t) = 3\cos 8.5t \quad y_2(t) = 1\cos 7.0t$ $\qquad$ $y(t) = 3\cos 8.5t + 1\cos 7.0t$

(a) Sum of Two Simple Harmonic Signals Resulting in Beats

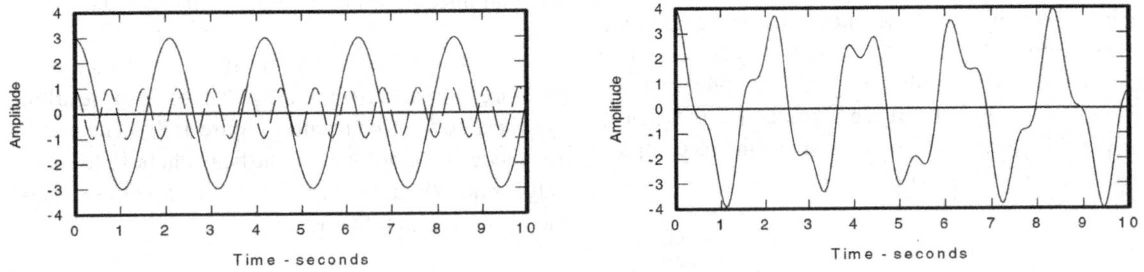

$y_1(t) = 3\cos 3.0t \quad y_2(t) = 1\cos 8.3t$ $\qquad$ $y(t) = 3\cos 3.0t + 1\cos 8.3t$

(b) Sum of Two Simple Harmonic Signals Resulting in an Almost Periodic Signal

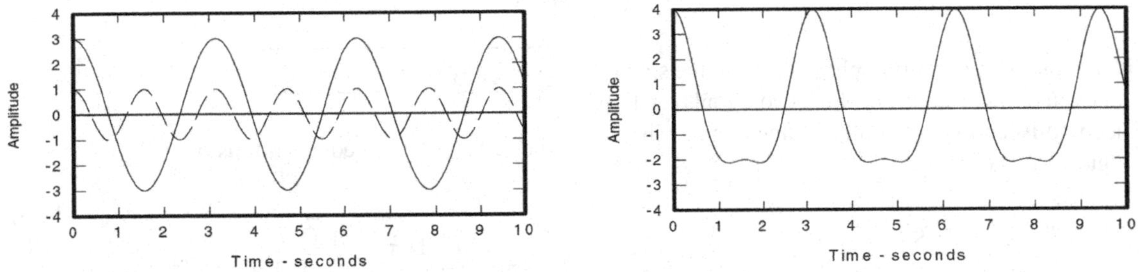

$y_1(t) = 3\cos 2t \quad y_2(t) = 1\cos 4t$ $\qquad$ $y(t) = 3\cos 2t + 1\cos 4t$

(c) Sum of Two Simple Harmonic Signals Resulting in a Complex Periodic Signal

Figure 1.5 Addition of Harmonic Signals

then y(t) can be written:

$$\boxed{y(t) = Y\cos(\omega t + \beta)} \tag{1.27}$$

where:

$$\boxed{Y = \sqrt{Y_1^2 + Y_2^2 + 2\,Y_1\,Y_2\cos(\alpha_1 - \alpha_2)}} \tag{1.28}$$

and:

$$\boxed{\beta = \tan^{-1}\left[\frac{Y_1\sin\alpha_1 + Y_2\sin\alpha_2}{Y_1\cos\alpha_1 + Y_2\cos\alpha_2}\right]}. \tag{1.29}$$

Equations (1.27) through (1.29) indicate even though the frequency of the resulting signal is the same as the

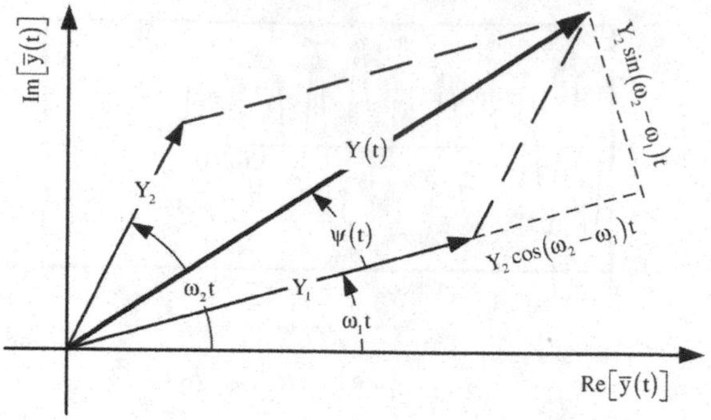

Figure 1.6 Vector Addition of Harmonic Signals

two initial signals, the phase and amplitude are different.

Two harmonic signals with different frequencies, when added together, can result in a signal with a phenomenon referred to as beating (Figure 1.5a), in a signal that is non-periodic (Figure 1.5b), or a signal that is complex periodic (Figure 1.5c). Let:

$$y_1(t) = Y_1 \cos \omega_1 t \tag{1.30}$$

and:

$$y_2(t) = Y_2 \cos \omega_2 t. \tag{1.31}$$

With no loss of generality, a phase between the signals is not included. The addition of the two signals can best be accomplished by vectorially adding them as shown in Figure 1.6. Therefore:

$$\bar{y}(t) = Y_1 e^{j\omega_1 t} + Y_2 e^{j\omega_2 t} \tag{1.32}$$

or::

$$\boxed{\bar{y}(t) = Y(t)\, e^{j(\omega_1 t + \psi(t))}} \tag{1.33}$$

where:

$$\boxed{Y(t) = \sqrt{Y_1^2 + Y_2^2 + 2\, Y_1\, Y_2 \cos(\omega_2 - \omega_1)t}} \tag{1.34}$$

and:

$$\boxed{\psi(t) = \tan^{-1}\left[\frac{Y_2 \sin((\omega_2 - \omega_1)t)}{Y_1 + Y_2 \cos((\omega_2 - \omega_1)t)}\right].} \tag{1.35}$$

The real part of equation (1.33) is:

$$\boxed{y(t) = Y(t) \cos(\omega_1 t + \psi(t)).} \tag{1.36}$$

If the ratio of the frequencies of two signals that are added together is an integer, the resulting signal will be a *complex periodic signal* (Figure 1.5c). If the ratio is not an integer, a non-periodic signal results (Figure 1.5b).

If the difference between the two frequencies $(\omega_2 - \omega_1)$ is small, a rhythmic pulsation in the resulting signal called *beating* occurs (Figure 1.5a). The period, frequency and amplitude of the beat can be determined as follows. When the $\cos(\omega_2 - \omega_1)t$ in equations (1.34) and (1.35) is equal to one:

$$\tau_1 = \frac{2n\,\pi}{\omega_2 - \omega_1} \qquad n = 0, 1, 2, \ldots \quad (\text{for } \omega_2 > \omega_1) \tag{1.37}$$

and:

$$\boxed{y(\tau_1) = (Y_1 + Y_2) \cos \omega_1 \tau_1} \tag{1.38}$$

When $\cos(\omega_2 - \omega_1)t$ equals minus one:

$$\tau_2 = \frac{(2n+1)\,\pi}{\omega_2 - \omega_1} \qquad n = 0, 1, 2, \ldots \quad (\text{for } \omega_2 > \omega_1) \tag{1.39}$$

and:

$$\boxed{y(\tau_2) = (Y_1 - Y_2) \cos \omega_1 \tau_2} \tag{1.40}$$

The period τ of the beat is:

$$\tau = 2(\tau_2 - \tau_1) \tag{1.41}$$

or:

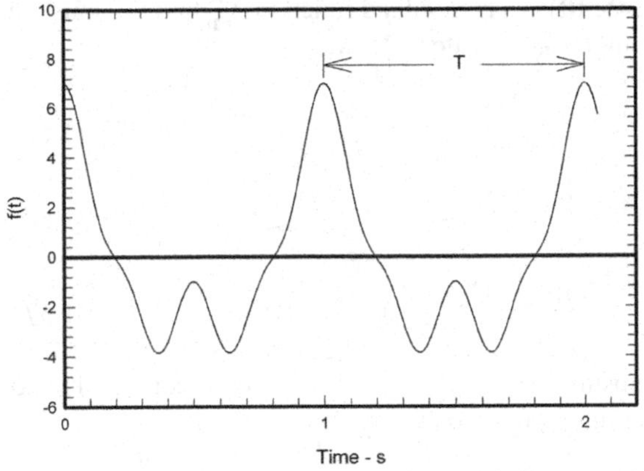

Figure 1.7 Complex Periodic Signal

$$T = \frac{2\pi}{\omega_2 - \omega_1} \quad \left(\text{for } \omega_2 > \omega_1\right) \tag{1.42}$$

and the frequency of the beat is:

$$f = \frac{\omega_2 - \omega_1}{2\pi}. \tag{1.43}$$

1.6 Complex Periodic Signals

For *complex periodic excitation*, a vibration system is excited by a signal composed of a series of sinusoidal components where the higher frequencies are integral multiples of the frequency of the lowest frequency component. Figure 1.7 shows a complex periodic signal that repeats itself over a period of time T. Using complex rotating phasors, the signal indicated in this figure can be expressed as a series of complex counter-rotating phasors (Figure 1.8) where:

$$f(t) = \sum_{n=-\infty}^{\infty} C_n \, e^{jn\omega_o t} \tag{1.44}$$

and where n is an integer that varies from $-\infty$ to $+\infty$ and:

$$\omega_o = \frac{2\pi}{T}. \tag{1.45}$$

ω_o is the fundamental frequency of the series of rotating phasors. Equation (1.44) is commonly referred to as the

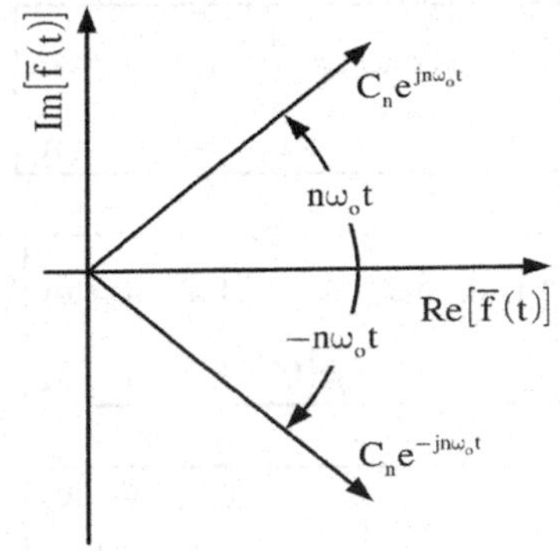

Figure 1.8 Counter-Rotating Phasors

complex Fourier series equation.

The counter-rotating phasors indicated in equation (1.44) are complex conjugates. Therefore, when they are vectorially added, the imaginary parts cancel, leaving real components that have amplitudes that are twice the value of the amplitudes of the real components of the individual phasors.

To evaluate the expression for C_n in equation (1.44), multiply both sides of the equation by $e^{-jm\omega_o t}$ and then integrate both sides with respect to time t over one period T. This gives:

$$\int_0^T f(t)\, e^{-jm\omega_o t}\, dt = \sum_{n=-\infty}^{\infty} C_n \int_0^T e^{jn\omega_o t}\, e^{-jm\omega_o t}\, dt. \tag{1.46}$$

Substituting Euler's equation into the right side of equation (1.46), expanding and then collecting terms yields:

$$\int_0^T f(t)\, e^{-jm\omega_o t}\, dt = \sum_{n=-\infty}^{\infty} C_n \int_0^T \left[\begin{pmatrix} \cos n\omega_o t \cos m\omega_o t + \\ \sin n\omega_o t \sin m\omega_o t \end{pmatrix} + j \begin{pmatrix} \sin n\omega_o t \cos m\omega_o t - \\ \sin m\omega_o t \cos n\omega_o t \end{pmatrix} \right] dt. \tag{1.47}$$

The sine and cosine terms are orthogonal functions that have the following properties:

$$\int_0^T \cos n\omega_o t \cos m\omega_o t = \frac{T}{2} \quad \text{for} \quad n = m$$
$$= 0 \quad \text{for} \quad n \neq m \tag{1.48}$$

$$\int_0^T \sin n\omega_o t \sin m\omega_o t = \frac{T}{2} \quad \text{for} \quad n = m$$
$$= 0 \quad \text{for} \quad n \neq m \tag{1.49}$$

$$\int_0^T \sin n\omega_o t \cos m\omega_o t = 0 \quad \text{for all values of m and n.} \tag{1.50}$$

Equation (1.47) only has a non-zero value for the case where m = n. Therefore, it reduces to:

$$C_n = \frac{1}{T}\int_0^T f(t)\, e^{-jn\omega_o t}\, dt. \tag{1.51}$$

The limits of integration for equation (1.51) do not need to be from 0 to T. They can be from -T/2 to T/2 or any other set of limits, so long as the total interval between the limits equals T.

The complex Fourier series described by equation (1.44) results in pairs of counter-rotating phasors that, when vectorially added together, result in real-valued functions. It is possible to derive these functions from equation (1.44). First, expand equation (1.44) to the form:

$$f(t) = \sum_{n=-\infty}^{-1} C_n\, e^{jn\omega_o t} + C_0 + \sum_{n=1}^{\infty} C_n\, e^{jn\omega_o t}. \tag{1.52}$$

Equation (1.52) can be rewritten:

$$f(t) = C_0 + \sum_{n=1}^{\infty}\left(C_n\, e^{jn\omega_o t} + C_{-n}\, e^{-jn\omega_o t}\right). \tag{1.53}$$

Substituting Euler's equation into equation (1.53) and rearranging the terms gives:

$$f(t) = C_0 + \sum_{n=1}^{\infty}\begin{bmatrix}(C_n + C_{-n})\cos n\omega_o t + \\ j(C_n - C_{-n})\sin n\omega_o t\end{bmatrix}. \tag{1.54}$$

Since f(t) is a real valued function, C_n and C_{-n} must be complex conjugates. Define:

$$a_o = C_o \tag{1.55}$$

$$a_n = C_n + C_{-n} \tag{1.56}$$

$$b_n = j(C_n - C_{-n}) \tag{1.57}$$

Substituting equation (1.55) into equation (1.51) and noting n = 0 yields for a_o:

$$a_0 = \frac{1}{T}\int_0^T f(t)\, dt. \tag{1.58}$$

Substituting equation (1.51) into equation (1.56) yields:

$$a_n = \frac{1}{T}\int_0^T f(t)\left(e^{-jn\omega_o t} + e^{jn\omega_o t}\right) dt. \tag{1.59}$$

Noting that:

$$e^{-jn\omega_o t} + e^{jn\omega_o t} = 2\cos n\omega_o t, \tag{1.60}$$

equation (1.59) becomes:

$$a_n = \frac{2}{T}\int_0^T f(t)\cos n\omega_o t\, dt. \tag{1.61}$$

Substituting equation (1.51) into equation (1.57) yields:

$$b_n = \frac{j}{T}\int_0^T f(t)\left(e^{-jn\omega_o t} - e^{jn\omega_o t}\right) dt. \tag{1.62}$$

Noting that:

$$e^{-jn\omega_o t} - e^{jn\omega_o t} = -2j\sin n\omega_o t, \tag{1.63}$$

equation (1.62) becomes:

$$b_n = \frac{2}{T}\int_0^T f(t)\sin n\omega_o t\, dt. \tag{1.64}$$

Finally, the *real Fourier series* equation can be written:

$$f(t) = a_0 + \sum_{n=1}^{\infty}\left(a_n \cos n\omega_o t + b_n \sin n\omega_o t\right) \tag{1.65}$$

where the coefficients a_o, a_n and b_n are given by equations (1.58), (1.61), and (1.64), respectively.

Equation (1.65) can be written as a cosine series plus a phase angle. Let:

$$a_n = F_n \cos\phi_n \quad \text{and} \quad b_n = -F_n \sin\phi_n \,. \tag{1.66}$$

Substituting equation (1.66) into equation (1.65) yields:

$$f(t) = a_o + \sum_{n=1}^{\infty} F_n \begin{pmatrix} \cos\phi_n \cos n\omega_o t + \\ \sin\phi_n \sin n\omega_o t \end{pmatrix} \tag{1.67}$$

Noting that:

$$\cos\phi_n \cos n\omega_o t + \sin\phi_n \sin n\omega_o t = \cos(n\omega_o t - \phi_n), \tag{1.68}$$

equation (1.67) can be written:

$$f(t) = a_0 + \sum_{n=1}^{\infty} F_n \cos(n\omega_o t - \phi_n), \tag{1.69}$$

where:

$$F_n = \sqrt{a_n^2 + b_n^2} \quad \text{and} \quad \phi_n = \tan^{-1}\left(\frac{b_n}{a_n}\right). \tag{1.70}$$

Following are some simple rules that are useful when evaluating the coefficients for equation (1.65):

1. If the area above the time axis equals the area below the time axis, $a_o = 0$ [Figure 1.9(a) and 1.9(b)].

2. If the periodic function is an even function where $f(t) = f(-t)$, $b_n = 0$ [Figure 1.9(a)].

3. If the periodic function is an odd function where $f(t) = -f(-t)$, $a_n = 0$ [Figure 1.9(b)].

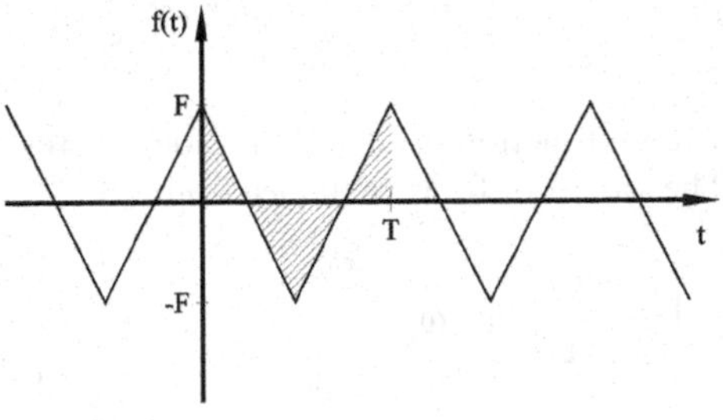

(a) Even Function with $a_o = 0$

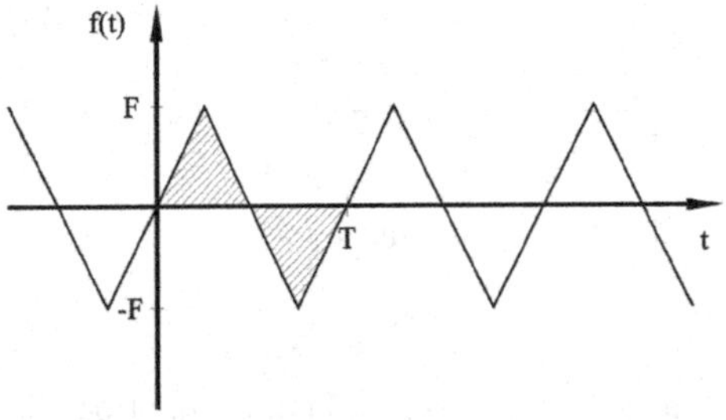

(b) Odd Function with $a_o = 0$

Figure 1.9 Even and Odd Functions

EXAMPLE 1.2

Determine the real Fourier series of the saw tooth function in Figure 1.10. Determine the amplitudes of the first six terms of the series.

SOLUTION

a_o is obtained from equation (1.58):

$$a_0 = \frac{1}{T}\int_0^T \left(F\frac{t}{T}\right)dt \quad \text{or} \quad a_0 = \frac{F}{2T^2}\,t^2\Big|_0^T . \tag{1.2a}$$

Evaluating the above equation for the stated limits of integration yields:

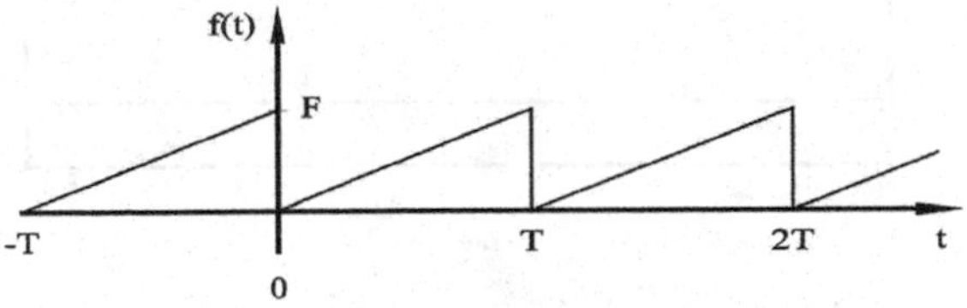

Figure 1.10 Saw Tooth Function

$$a_0 = \frac{F}{2}. \tag{1.2b}$$

The saw tooth function is an odd function. Therefore, $a_n = 0$. The values of b_n are obtained from:

$$b_n = \frac{2}{T}\int_0^T \left(F\frac{t}{T}\right)\sin n\omega_o t\, dt \tag{1.2c}$$

or:

$$b_n = \frac{2F}{T^2}\left[\begin{array}{l} -\dfrac{t}{n\omega_o}\cos n\omega_o t + \\ \dfrac{1}{(n\omega_o)^2}\sin n\omega_o t \end{array}\right]_0^T. \tag{1.2d}$$

Noting that $\omega_o = 2\pi/T$ and evaluating equation (1,2d) over the limits of 0 to T yields:

$$b_n = -\frac{F}{n\pi}. \tag{1.2e}$$

Therefore, the real Fourier series of the saw tooth function in Figure 1.11 can be expressed:

$$f(t) = \frac{F}{2} - \sum_{n=1}^{\infty}\frac{F}{n\pi}\sin n\omega_o t. \tag{1.2f}$$

The amplitudes of the first six terms of the series specified by equation (1.2f) are:

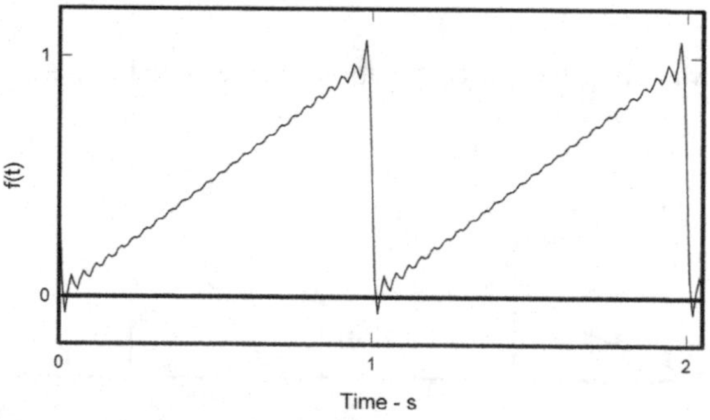

Figure 1.11 Fourier Series Representation of Saw Tooth Function - 26 Terms in Series

$a_o = 0.5\ F$

$b_1 = 0.318\ F;\ b_2 = 0.159\ F;\ b_3 = 0.106\ F;$

$b_4 = 0.079\ F;\ b_5 = 0.064\ F;\ b_6 = 0.053\ F.$

Figure 1.11 shows a construction of the saw tooth function from its real Fourier series components where T = 1 s and F = 1.

When f(t) has a deterministic form, equations (1.58), (1.61), (1.64) and (1.65) can be used to obtain the Fourier series expression for f(t). However, carrying out the integrations specified by equations (1.58), (1.61), and (1.64) can become involved if f(t) does not have a simple form. When f(t) is more complicated, the Fourier coefficients a_0, a_n and b_n can be obtained using a numerical integration procedure.

Referring to Figure 1.12, let $t_1, t_2, \ldots, t_N$ be a number of N equidistant points over the period T of the function that correspond to the values of f(t) given by:

$$f_1 = f(t_1), f_2 = f(t_2), \quad \ldots, f_N = f(t_N) \tag{1.71}$$

where:

$$T = N\,\Delta t \tag{1.72}$$

The values for a_0, a_n, and b_n can then be obtained from:

$$a_0 = \sum_{i=1}^{N}(f_i\,\Delta t) \tag{1.73}$$

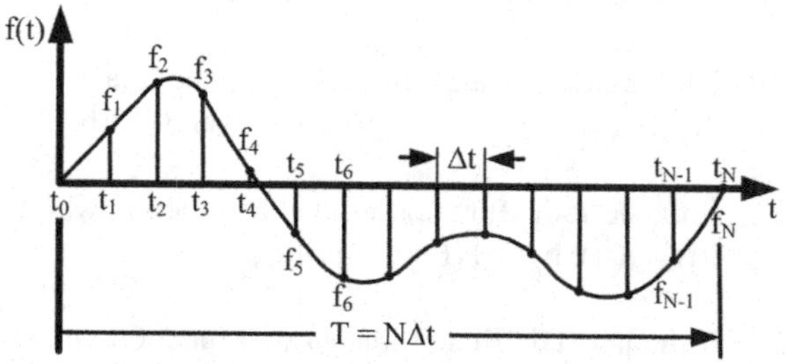

Figure 1.12 Values of a Periodic Function f(t) at discrete points: $t_1, t_2, \ldots, t_N$

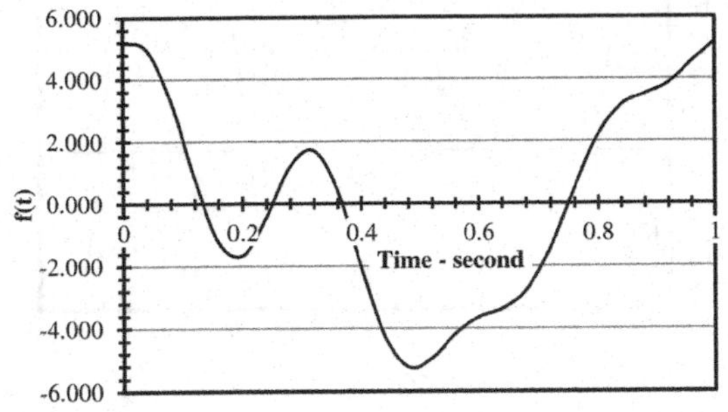

Figure 1.1 Complex Periodic Signal for EXAMPLE 1.3

$$\boxed{a_n = 2\sum_{i=1}^{N}\left[f_i \cos\left(2n\pi f_o \frac{t_i}{T}\right)\Delta t\right]_n} \tag{1.74}$$

$$\boxed{b_n = 2\sum_{i=1}^{N}\left[f_i \sin\left(2n\pi f_o \frac{t_i}{T}\right)\Delta t\right]_n} \tag{1.75}$$

where $f_o = \omega_o/(2\pi)$ is the fundamental frequency of f(t) in the unit Hz.

EXAMPLE 1.3

Figure 1.13 shows a plot of a complex periodic signal f(t). Table 1.1 shows the values of f(t) over the signal period T of 1 second. Determine the values of the first five terms of the Fourier series, and write the Fourier series expression for the function.

SOLUTION

The MATLAB® script in Appendix 1B was used to compute the Fourier coefficients associated with the first five terms of the Fourier series. The results are:

```
fo =

   1

azero =

  -0.0074

coef_a_number =

   1   2   3   4   5

a_coeficients =

  3.9893  -0.0006   1.2094   0.0141   0.0116

coef_b_number =

   1   2   3   4   5

b_coeficients =

  0.0100  -1.4858   0.0111   1.1019  -0.0085
```

The Fourier series expression (rounding to 2 significant figures) for y(t) is:

$$\boxed{\begin{aligned} f(t) = 4.0\cos(2\pi f_o t) + 1.2\cos(6\pi f_o t) - \\ 1.5\sin(4\pi f_o t) + 1.1\sin(8\pi f_o t) \end{aligned}} \tag{1.3a}$$

where $f_o = 1$ Hz.

1.7 Relationship between Displacement, Velocity, and Acceleration Signals

Specific relations exist between the displacement, velocity and acceleration signals when they are harmonic signals. Let the *displacement* y(t) be described by a cosine function:

Table 3.1 f(t) Values for EXAMPLE 1.3

n	0	1	2	3	4	5	6	7	8	9	10	11	12	13	14	15	16	17
Time - sec.	0	0.02	0.04	0.06	0.08	0.1	0.12	0.14	0.16	0.18	0.2	0.22	0.24	0.26	0.28	0.3	0.32	0.34
f(t)	5.200	5.241	4.955	4.301	3.309	2.085	0.792	-0.380	-1.252	-1.695	-1.663	-1.199	-0.435	0.435	1.198	1.663	1.696	1.252
n	18	19	20	21	22	23	24	25	26	27	28	29	30	31	32	33	34	35
Time - sec.	0.36	0.38	0.4	0.42	0.44	0.46	0.48	0.5	0.52	0.54	0.56	0.58	0.6	0.62	0.64	0.66	0.68	0.7
f(t)	0.381	-0.791	-2.085	-3.309	-4.301	-4.955	-5.241	-5.200	-4.927	-4.543	-4.159	-3.852	-3.645	-3.510	-3.377	-3.158	-2.777	-2.194
n	36	37	38	39	40	41	42	43	44	45	46	47	48	49	50			
Time - sec.	0.72	0.74	0.76	0.78	0.8	0.82	0.84	0.86	0.88	0.9	0.92	0.94	0.96	0.98	1			
f(t)	-1.412	-0.489	0.487	1.411	2.193	2.777	3.157	3.376	3.510	3.645	3.852	4.159	4.543	4.927	5.200			

$$\boxed{y(t) = Y\cos\omega t} \tag{1.76}$$

The related *velocity* is the derivative of displacement with respect to time:

$$v(t) = \frac{d\,y(t)}{dt} \tag{1.77}$$

or:

$$\boxed{v(t) = -\omega Y\sin\omega t\,.} \tag{1.78}$$

The derivative of y(t) with respect to t is often written:

$$\dot{y}(t) = \frac{d\,y(t)}{dt}\,. \tag{1.79}$$

The velocity can be expressed:

$$\boxed{\dot{y}(t) = \omega Y\cos\left(\omega t + \frac{\pi}{2}\right)} \tag{1.80}$$

which indicates velocity leads displacement by a phase of 90°.

Acceleration is the derivative of velocity with respect to time, and it is the second derivative of displacement with respect to time:

$$a(t) = \frac{d^2\,y(t)}{dt^2} \tag{1.81}$$

or:

$$\boxed{a(t) = -\omega^2 Y\cos\omega t\,.} \tag{1.82}$$

The second derivative of y(t) with respect to t is often written:

$$\ddot{y}(t) = \frac{d^2\,y(t)}{dt^2}\,. \tag{1.83}$$

The acceleration can be expressed:

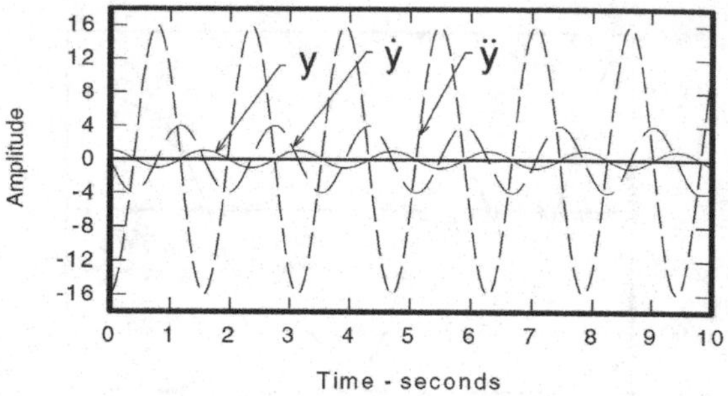

Figure 1.14 Amplitude and Phase Relationships between Displacement, Velocity and Acceleration

$$\boxed{\ddot{y}(t) = \omega^2 Y\cos(\omega t + \pi)} \tag{1.84}$$

which indicates acceleration leads velocity by a phase of 90° and the displacement by a phase of 180°. Figure 1.14 shows a graphical representation of the relations between displacement, velocity and acceleration when they are describe by sinusoidal terms.

The above expressions can also be obtained using rotating phasors. Let the displacement be expressed:

$$\boxed{y(t) = Y\,e^{j\omega t}\,.} \tag{1.85}$$

The velocity and acceleration are:

$$\dot{y}(t) = j\omega Y\,e^{j\omega t} \tag{1.86}$$

and:

$$\ddot{y}(t) = -\omega^2 Y\,e^{j\omega t}\,. \tag{1.87}$$

(Note: $j = \sqrt{-1}$ and $j^2 = -1$) Noting:

$$j = e^{j\left(\frac{\pi}{2}\right)} \quad \text{and} \quad -1 = e^{j\pi}\,, \tag{1.88}$$

equations (1.84) and (1.85) can be written:

$$\boxed{\dot{y}(t) = \omega Y\,e^{j\left(\omega t + \frac{\pi}{2}\right)}} \tag{1.89}$$

$$\boxed{\ddot{y}(t) = \omega^2 Y\,e^{j(\omega t + \pi)}\,.} \tag{1.90}$$

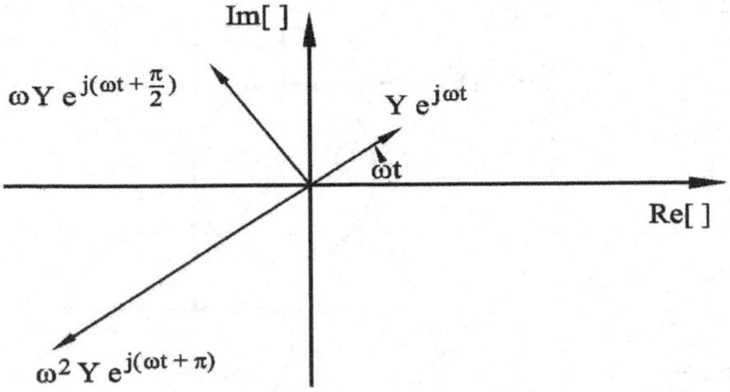

Figure 1.15 Complex Vector Representation of the Amplitude and Phase Relationships between Displacement, Velocity and Acceleration

The complex displacement, velocity and acceleration phasors are graphically shown in the complex plane in Figure 1.15. The real parts of equations (1.85), (1.86) and (1.90) are the same as equations (1.76), (1.80) and (1.84), respectively.

1.8 Deterministic and Random Vibration Signals

There are two basic classes of signals that can excite vibration systems. They are *deterministic* and *random* signals (Figure 1.16). *Deterministic signals* are those signals that can be expressed by explicit math functions. They can be separated into *periodic* and *non-periodic* signals. The response of linear vibration systems to multiple simultaneous deterministic signals can be described by using the *principle of superposition* to sum the responses of the system to the individual signals. *Random* signals cannot be expressed by explicit math functions. Therefore, they must be described by statistical functions. Random signals can be separated into *stationary* and *non-stationary* signals.

Periodic signals repeat themselves over specified time intervals of T, such that f(t) = f(t + nT), where $n = 1, 2, ..., n$. *Harmonic* (sinusoidal) and *complex periodic* (Fourier series) signals are two types of periodic signals that have been discussed in preceding sections of this chapter.

Non-periodic signals can be separated into *almost periodic* and *transient signals*. *Almost periodic signals* are very similar to complex periodic signals. All of the higher harmonics that make up a complex periodic signal are integral multiples of the fundamental or lowest frequency component of the signal. In an almost periodic signal there will be at least one and possibly more higher frequency components in the signal that will not be an integral multiple of the lowest frequency component of the signal. *Transient signals* occur for a specified period of time (may be a short or long time) and then disappear. Transient signals can be described with Fourier and Laplace transforms.

Many signals cannot be described by explicit math functions. For example, if an identical experiment is repeated many times and the measured outcomes are always different, the process is a random process. *Random signals* are described in terms of probability statements and statistical functions. Random signals can be separated into *stationary* and *non-stationary signals*. A *stationary signal* occurs when an ensemble of multiple sample lengths of an overall data record have identical statistical properties as other ensembles of data from the same data record. If this is not the case, the process is *non-stationary*. If all of the individual data records

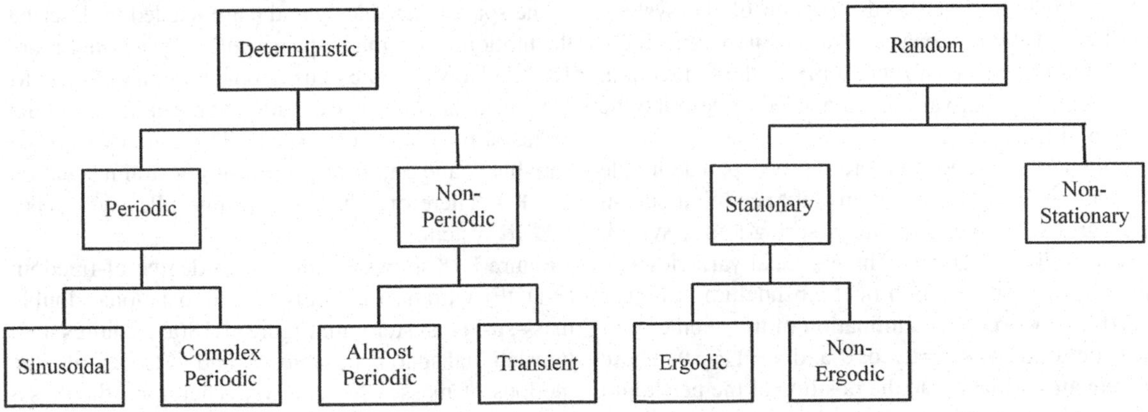

Figure 1.16 Classifications of Deterministic and Random Signals

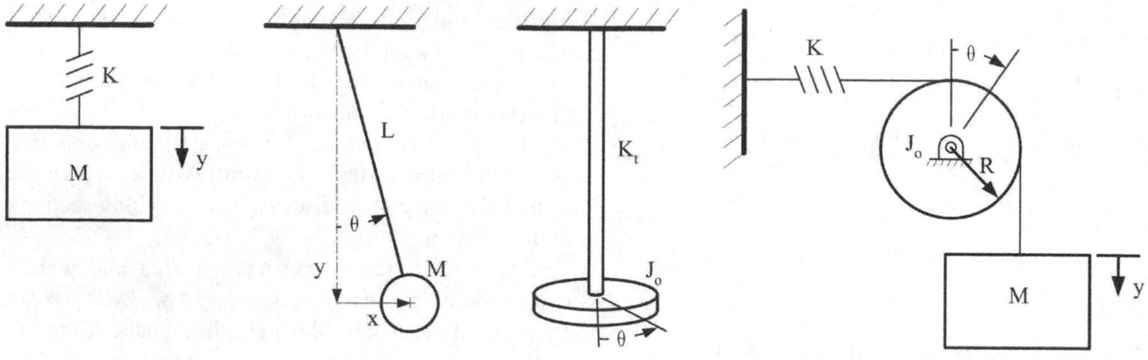

(a) Mass-Spring System (b) Pendulum (c) Torsional Pendulum (d) Mass-Spring-Pulley System

Figure 1.17 Systems with One Degree of Freedom

within multiple ensembles of data from an overall data record have identical statistical properties, the process described by the overall data record is *ergodic*. For a process to be *ergodic*, it first must be *stationary*. A process may be *stationary*, but not *ergodic*.

1.9 Degrees of Freedom (DOF)

The number of independent spacial variables necessary to describe the location of all the mass elements of a vibration system determines the number of *degrees of freedom* (DOF) of the system. If one spacial variable is needed, the system has one degree of freedom. If three independent spatial variables are necessary, the system has three degrees of freedom. Figure 1.17 shows examples of one-degree-of-freedom (1-DOF) vibration systems. If the rectilinear mass-spring system in Figure 1.17(a) is constrained to move in only the vertical direction, it takes only one translational spatial variable y to completely describe the motion of the system. Similarly, if the torsional pendulum shown in Figure 1.17(c) is restrained to rotate only in the θ-direction, one angular spatial variable is needed to describe the motion of the system.

On the surface, it appears more than one spatial variable is needed to describe the motion of the pendulum shown in Figures 1.16(b) and the mass-spring-pulley system shown in Figure 1.17(d). Three spacial variables x, y and θ describe the motion of the pendulum in Figure 1.17(b). However, an examination of the geometry of the system indicates $y = L\cos\theta$ and $x = L\sin\theta$. These two equations that relate the position of the pendulum in terms of the rectilinear spatial variables x and y to the rotational spatial variable θ are called *geometric constraint equations*. Therefore, even though there are three spatial variables that can be used to describe the motion of the pendulum, they all are not independent. In the case of the pendulum, there is only one independent spatial variable θ. This gives rise to a more complete rule that can be used to determine the *number of degrees of freedom* (DOF) associated with a vibration system:

> *The number of degrees of freedom associated with a vibration system is equal to the number of spatial variables that describe the motion of the system minus the number of corresponding geometric constraint equations that describe the geometric relations between these variables.*

In the case of the pendulum, there are three spatial variables and two geometric constraint equations. Therefore, the pendulum is a 1-DOF system.

The spatial variables y and θ are needed to describe the motion of the mass-spring-pulley system in Figure 1.17(d). However, the translational motion y of mass M is equal to the rotational motion θ of the pulley times the radius of the pulley, or $y = R\theta$. There are two spacial variables y and θ and one geometric constraint equation $y = R\theta$. Therefore, the mass-spring-pulley system is a 1-DOF system.

Figure 1.18 shows examples two-degree-of-freedom (2-DOF) vibration systems. The translational double mass-spring system in Figure 1.18(a) requires two translational spatial variables y_1 and y_2 to describe the motions of masses M_1 and M_2. Therefore, the mass-spring system is a 2-DOF system.

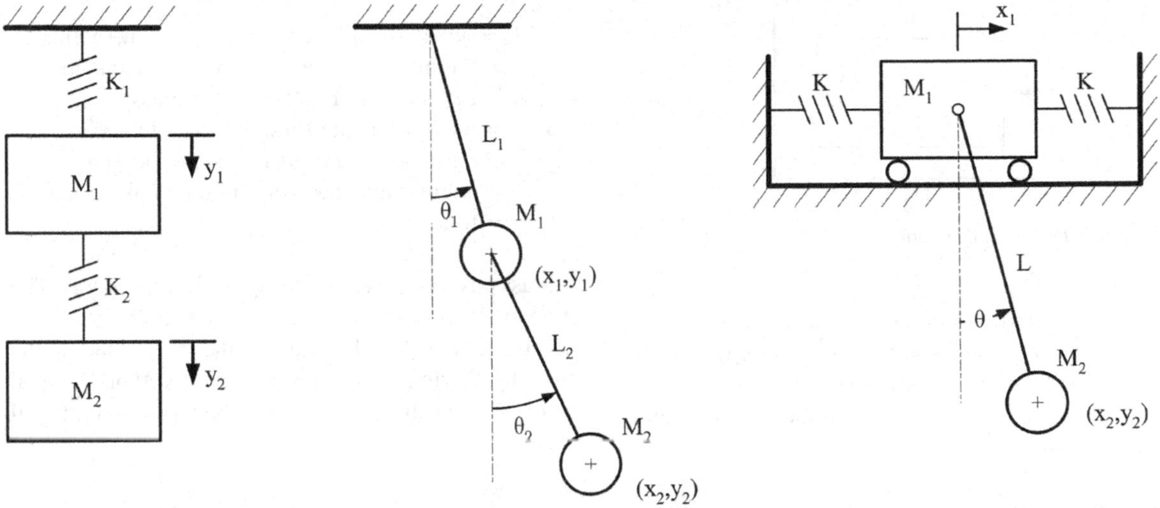

(a) Double Mass-Spring System (b) Double Pendulum (c) Mass-Spring-Pendulum System

Figure 1.18 Systems with Two Degrees of Freedom

For the double pendulum in Figure 1.18(b), there are four translational spatial variables x_1, y_1, x_2 and y_2 and two rotational spatial variables θ_1 and θ_2 that describe the motion of the system. However, there are four geometric constraint equations that relate x_1, y_1, x_2 and y_2 to θ_1 and θ_2. The six spatial variables and four geometric constraint equations indicate the double pendulum is a 2-DOF system.

The motion of the mass-spring-pendulum system in Figure 1.18(c) can be expressed in terms of three translational spatial variables x_1, x_2 and y_2 and one rotational spatial variable θ. However, there are two geometric constraint equations that relate x_2 and y_2 to x_1 and θ. Therefore, this system is a 2-DOF system.

Masses associated with vibrating systems potentially have six degrees of freedom: three rectilinear and three rotational. However, in many cases, because of constraints placed on a vibration system, the number of degrees of freedom needed to describe the motion of the system can be reduced. In some cases, it can be reduced to one spatial variable. When this is done, four important system characteristics can be examined. They include:

- The concept of resonance associated with a vibration system;
- The response of a vibration system to external excitation;
- The role damping plays in the motion of the system; and
- The nature of the force a vibration system transmits to the base or foundation to which it is attached.

An unforced vibrating system oscillates at its resonance frequency when it is given an initial displacement and/or velocity. When a vibrating system is excited at its resonance frequency by an externally applied force, the amplitude of the motion of the system can become very large. Damping is often necessary to control the response of a vibrating system in the region near and at resonance. It is often desirable in many vibration systems to control and/or minimize the amount of vibration energy that is transmitted into the structure to which the vibrating system is attached.

1.10 Elements of a Vibration System

Mass Elements

Vibration systems contain *mass elements*. From *Newton's second law*, the time rate of change of momentum of a mass that has translational motion is equal to the sum of the forces ΣF acting on the mass (Figure 1.19):

$$\sum F = \frac{d\left[M\left(v - v_r\right)\right]}{dt} \tag{1.91}$$

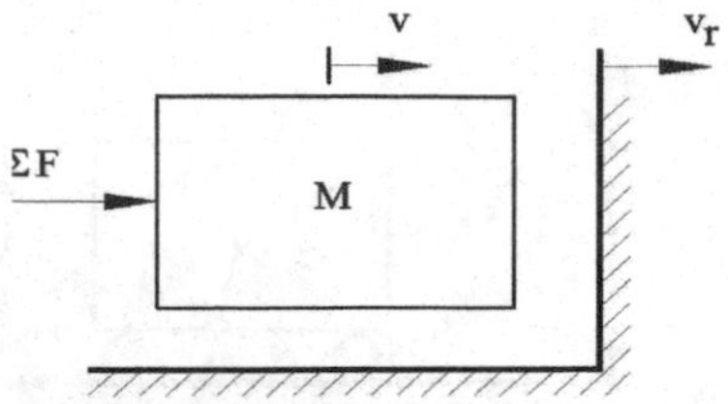

Figure 1.19 Mass Element

where:
ΣF = force (lb_f or N) acting on the mass
M = value of mass M (lb_f - s^2/in. or kg)
v = velocity (in./s or m/s) of mass M
v_r = velocity (in./s or m/s) of the inertial reference
t = time (s).

The inertial reference point is generally assumed to be fixed. Therefore, v_r is normally equal to zero. Also, the value of mass M is usually assumed to be constant. Therefore, equation (1.91) becomes:

$$\sum F = M \frac{dv}{dt} . \tag{1.92}$$

Noting:

$$v = \frac{dx}{dt} \tag{1.93}$$

where x equals the displacement (in. or m) of mass M, equation (1.91) can also be written:

$$\boxed{\sum F = M \frac{d^2 x}{dt^2} \quad \text{or} \quad \sum F = M \ddot{x} .} \tag{1.94}$$

Equation (1.94) indicates, when mass M is not in static equilibrium where $\Sigma F = 0$, the sum of the forces acting on the mass is equal to the value of the mass times its resulting acceleration.

If a mass has rotational motion, the sum of the moments acting on it equals the mass moment of inertia of the mass times its angular acceleration:

$$\boxed{\sum M_a = J_a \frac{d^2 \theta}{dt^2} \quad \text{or} \quad \sum M_a = J_a \ddot{\theta}} \tag{1.95}$$

where:
ΣM_a = sum of the moment (lb_f - in. or N-m) acting on the mass relative to a point a associated with the center of rotation of the mass
J_a = mass moment of inertial (lb_f - in. - s^2 or kg-m^2) of the mass relative to point a
θ = angular displacement (radians) of the mass
t = time (s)

Mass possesses *inertial storage* or *kinetic energy*. The amount of kinetic energy a particular mass possesses can be obtained by determining the work done on the mass by moving it a specified distance x. Work is equal to force times distance. This can be written in integral form as:

$$W = \int_0^x F \, dx . \tag{1.96}$$

where:
W = work (lb_f - in. or N-m)
F = force (lb_f or N)
x = displacement (in. or m).

Substituting equation (1.92) into equation (1.96) yields:

$$W = \int_0^x M \frac{dv}{dt} dx . \tag{1.97}$$

Rearranging the terms in the integral and changing the limits of integration from x to v, equation (1.97) becomes:

$$W = \int_0^v M \, v \, dv . \tag{1.98}$$

Carrying out the prescribed integration gives:

$$W = \frac{1}{2} M v^2 . \tag{1.99}$$

Equation (1.99) is referred to as the *kinetic energy* of the mass and is usually written:

$$\boxed{T = \frac{1}{2} M v^2 \quad \text{or} \quad T = \frac{1}{2} M \dot{x}^2} \tag{1.100}$$

where:

T = kinetic energy (lb_f - in. or N-m)
M = mass (lb_f - s^2/in. or kg)
x = displacement (in. or m) of the mass
v = velocity (in./s or m/s) of the mass.

If a mass has rotational motion instead of translational motion, the kinetic energy associated with the mass is given by:

$$\boxed{T = \frac{1}{2} J_0 \omega^2 \quad \text{or} \quad T = \frac{1}{2} J_0 \dot{\theta}^2} \tag{1.101}$$

where:

T = kinetic energy (lb_f - in. or N-m)
J_o = mass moment of inertia (lb_f - in. - s^2 or kg-m^2) of the mass relative to its center of gravity
θ = angular displacement (radians) of the mass
ω = angular velocity (radian/s) of the mass.

The total kinetic energy of a system that is comprised of several mass elements that are coupled together can be calculated by summing the kinetic energies of each individual mass element. This process can be used to determine the equivalent mass or the equivalent mass moment of inertia of a system. Some examples will be used to illustrate this process.

EXAMPLE 1.4

Determine the equivalent mass moment of inertia of the mass-spring-pulley system n Figure 1.17(d) relative to the rotational variable θ of the pulley.

SOLUTION

The total kinetic energy of the mass-spring-pulley system is:

$$T = \frac{1}{2} J_o \dot{\theta}^2 + \frac{1}{2} M \dot{y}^2 . \tag{1.4a}$$

The constraint equation that relates θ and y is:

$$y = R\,\theta . \tag{1.4b}$$

Substituting equation (1.4b) into equation (1.4a) yields:

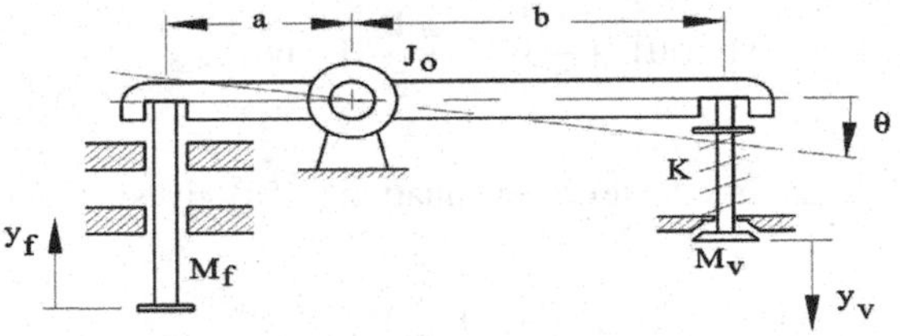

Figure 1.20 Rocker Arm

$$T = \frac{1}{2} J_o \dot{\theta}^2 + \frac{1}{2} M \left(R\,\dot{\theta}\right)^2 . \tag{1.4c}$$

Rearranging equation (1.4c) gives:

$$T = \frac{1}{2}\left(J_o + M R^2\right)\dot{\theta}^2 . \tag{1.4d}$$

The equivalent mass moment of inertia is:

$$\boxed{J_{eq} = J_o + M R^2 .} \tag{1.4e}$$

EXAMPLE 1.5

Determine the equivalent mass moment of inertia of the cam-follower, rocker arm and valve of the system shown in Figure 1.20 relative to the rotation angle θ of the rocker arm.

SOLUTION

The total kinetic energy of the cam-follower, rocker arm and valve is:

$$T = \frac{1}{2} M_f \dot{y}_f^2 + \frac{1}{2} J_o \dot{\theta}^2 + \frac{1}{2} M_v \dot{y}_v^2 . \tag{1.5a}$$

y_f in terms of θ is given by:

$$y_f = a\,\theta . \tag{1.5b}$$

y_v in terms of θ is given by:

$$y_v = b\,\theta . \tag{1.5c}$$

Substituting equations (1.5b) and (1.5c) into equation (1.5a) yields:

$$T = \frac{1}{2} M_f \left(a\dot{\theta}\right)^2 + \frac{1}{2} J_o \dot{\theta}^2 + \frac{1}{2} M_v \left(b\dot{\theta}\right)^2 . \tag{1.5d}$$

Rearranging the terms in equation (1.5d) gives:

$$T = \frac{1}{2}\left(J_o + M_f a^2 + M_v b^2\right)\dot{\theta}^2 . \tag{1.5e}$$

The equivalent mass moment of inertia is:

$$\boxed{J_{eq} = J_o + M_f a^2 + M_v b^2 .} \tag{1.5f}$$

Spring Elements

Vibration systems contain *spring or elastic elements.* The force acting through a translational spring element (Figure 1.21) is:

$$\boxed{f_s = K x .} \tag{1.102}$$

where:

f_s = spring force (lb_f or N)
K = stiffness coefficient (lb_f/in. or N/m) of the spring element
x_1 = displacement (in. or m)
x_2 = displacement (in. or m).

When $x = x_2 - x_1$ is the displacement across the spring, equation (1.102) becomes:

$$f_s = K\left(x_2 - x_1\right) \tag{1.103}$$

When the spring element is a torsional spring element (Figure 1.22), the moment acting on the spring is:

$$\boxed{M_s = K_t \theta} \tag{1.104}$$

where:

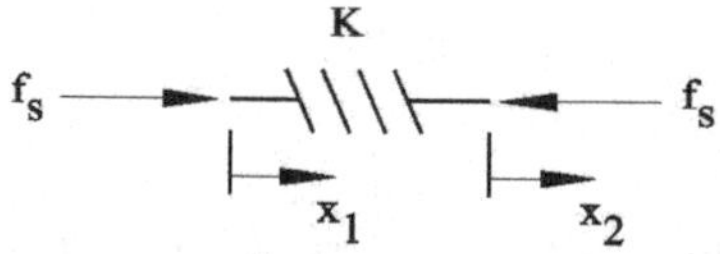

Figure 1.21 Translational Spring Element

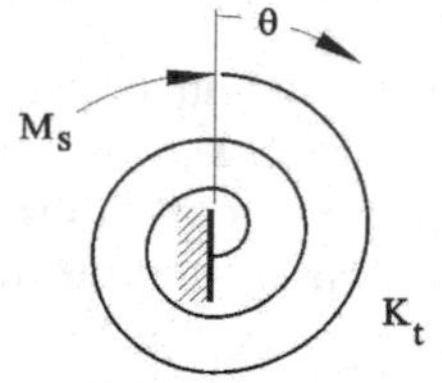

Figure 1.22 Torsional Spring Element

M_s = spring moment (lb_f - in. or N-m)
K_t = torsional stiffness coefficient (lb_f - in. or N•m) of the spring element
θ = angular displacement (radians)

A spring possesses potential energy. The amount of potential energy present in a spring element can be found by determining the work done on the spring as it is compressed or expanded a distance x. Substituting equation (1.102) into equation (1.96) yields:

$$W = \int_0^x K x \, dx . \tag{1.105}$$

Carrying out the prescribed operations yields:

$$W = \frac{1}{2} K x^2 . \tag{1.106}$$

For the case of a spring element, the work specified by equation (1.106) is the potential energy of the spring element. Therefore, equation (1.106) can be written:

$$\boxed{U = \frac{1}{2} K x^2} \tag{1.107}$$

where U is the potential (lb_f - in. or N•m). If the spring element is a torsional spring element, the *potential energy* associated with the spring element is:

$$\boxed{U = \frac{1}{2} K_t \theta^2 .} \tag{1.108}$$

The total potential energy associated with several spring or elastic elements in a system can be obtained by summing the potential energies of each individual element. This process can be used to determine the equivalent stiffness of a system.

EXAMPLE 1.6

Determine the equivalent stiffness of the spring in the mass-spring-pulley system of Figure 1.17(d) in terms of the rotation angle θ of the pulley.

SOLUTION

The deflection of the spring for a rotation angle θ is:

$$y = R\,\theta\,. \tag{1.6a}$$

The potential energy associated with the spring is:

$$U = \frac{1}{2} K\,y^2 \quad \text{or} \quad U = \frac{1}{2} K\left(R\,\theta\right)^2. \tag{1.6b}$$

The potential energy associated with the spring in terms of θ is:

$$U = \frac{1}{2} K R^2\,\theta^2\,. \tag{1.6c}$$

Therefore, the equivalent stiffness of the mass-spring-pulley system in terms of the rotation angle, θ, of the pulley is:

$$\boxed{K_{eq} = K R^2\,.} \tag{1.6d}$$

It may often be necessary to support a mass with several spring elements. If the mass is assumed to vibrate in only one direction, the spring elements are attached to the mass in parallel as shown in Figure 1.23. When the spring elements shown in Figure 1.23 are deflected a distance y downwards, they will exert a resisting or restoring force in the upward direction. If y is defined as being positive downward, the sum of the forces at the point where the force F is applied to the springs is:

$$F - K_1\,y - K_2\,y = 0 \quad \text{or} \quad F = \left(K_1 + K_2\right)y\,. \tag{1.109}$$

Since the deflection of both spring elements is y, equation (1.109) can be written:

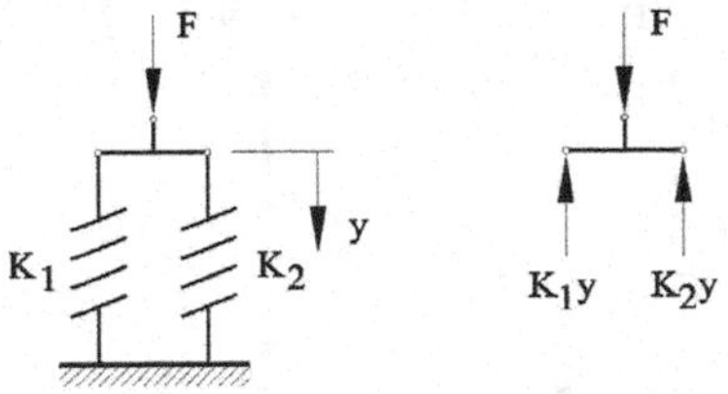

Figure 1.23 Spring Elements in Parallel

$$F = K_{eq}\,y \tag{1.110}$$

where:

$$K_{eq} = K_1 + K_2\,. \tag{1.111}$$

When the number of springs attached in parallel to a mass that vibrates in only one direction is n, the equivalent stiffness K_{eq} of the n spring elements is:

$$\boxed{K_{eq} = \sum_{i=1}^{n} K_i\,.} \tag{1.112}$$

The stiffness coefficient of the spring element(s) used to support the mass of a vibration system can be expressed:

$$\boxed{K = \frac{We}{\delta}} \tag{1.113}$$

where:

K = stiffness coefficient (lb_f/in. or N/m) of the spring
We = Weight of mass (lb_f or N) supported by the spring
δ = static displacement (in. or m) of the spring.

When more than one spring element is used in parallel to support the mass of a vibration system, the sum of the stiffness coefficients of the individual spring elements must add up to K. For example, if four spring elements, each with a stiffness coefficient K_e, are located such that each spring element supports 1/4 of the total weight of a vibration system, K_e is given by:

$$K_e = \frac{K}{4} \tag{1.114}$$

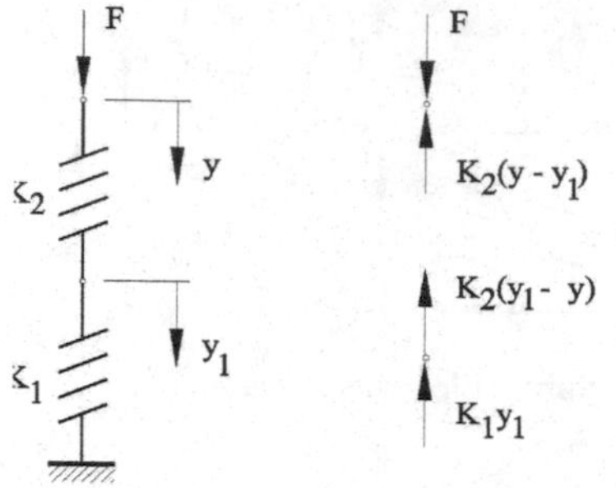

Figure 1.24 Spring Elements in Series

Situations often exist where spring elements may be connected to each other in series, as shown in Figure 1.24. If y is defined as being positive downwards, the sum of the forces at the point where the spring elements are attached together is:

$$-K_2\left(y_1-y\right)-K_1\,y_1=0 \tag{1.115}$$

or solving equation (1.115) in terms of y_1:

$$y_1=\frac{K_2\,y}{K_1+K_2}. \tag{1.116}$$

Summing the forces at the point where the force F is applied to spring element K_2 yields:

$$F-K_2\left(y-y_1\right)=0 \quad \text{or} \quad F=K_2\left(y-y_1\right). \tag{1.117}$$

Substituting equation (1.116) for y_1 into equation (1.117) and simplifying gives:

$$F=\left(\frac{K_1\,K_2}{K_1+K_2}\right)y. \tag{1.118}$$

Equation (1.118) can be written in the format of equation (1.110) where:

$$K_{eq}=\frac{K_1\,K_2}{K_1+K_2}. \tag{1.119}$$

Rearranging equation (1.119), K_{eq} is:

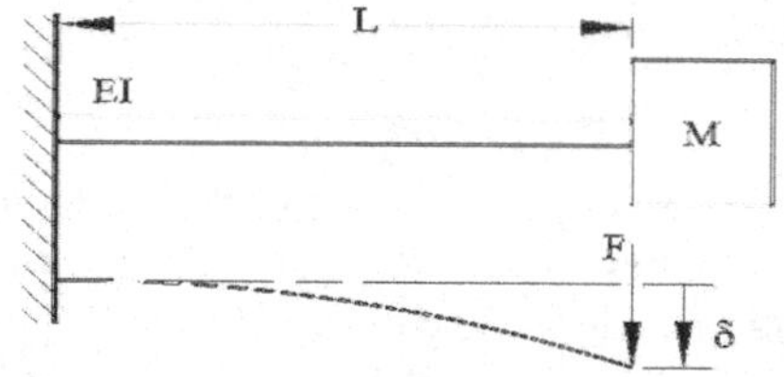

Figure 1.25 Mass at the End of a Slender Beam

$$\frac{1}{K_{eq}}=\frac{1}{K_1}+\frac{1}{K_2}. \tag{1.120}$$

When there are n springs connected together in series, the equivalent stiffness K_{eq} of the spring elements is given by:

$$\boxed{\frac{1}{K_{eq}}=\sum_{i=1}^{n}\left(\frac{1}{K_i}\right).} \tag{1.121}$$

EXAMPLE 1.7

Determine the equivalent stiffness coefficient of a slender beam of length L with a mass attached at the end of the beam (Figure 1.25).

SOLUTION

If the mass vibrates in the direction perpendicular to the beam, the deflection δ of the end of the beam due to a force F at the end is:

$$\delta=\frac{F\,L^3}{3\,EI}. \tag{1.7a}$$

where:

- δ = deflection (in. or m)
- E = Young's modulus (lb_f/in.^{2} or N/m^2)
- I = area moment (in.^{4} or m^4)
- F = force (lb_f or N)
- L = length of the beam (in. or m).

The equivalent spring coefficient K_{eq} for the beam is:

$$\boxed{K_{eq}=\frac{F}{\delta} \quad \text{or} \quad K_{eq}=\frac{3\,EI}{L^3}.} \tag{1.7b}$$

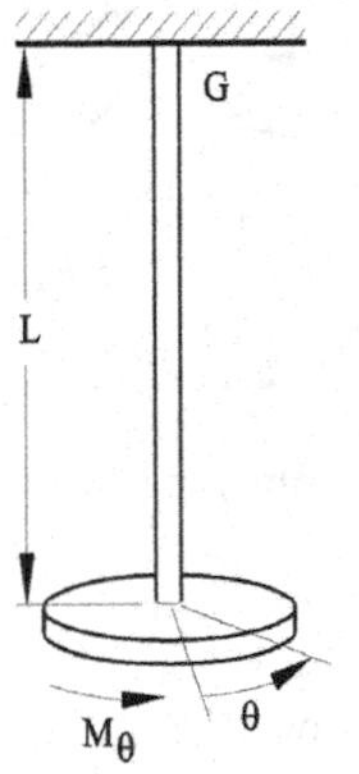

Figure 1.26 Disc at the End of a Slender Shaft

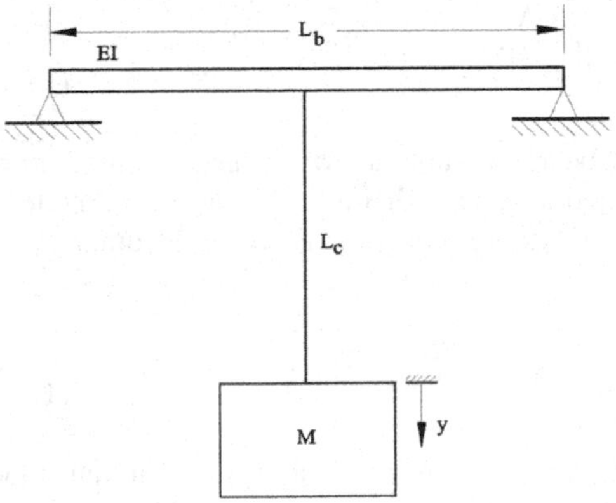

Figure 1.27 Mass Suspended from a Cable Attached to the Center of a Simply Supported Slender Beam

----

EXAMPLE 1.8

Determine the equivalent torsional stiffness coefficient of a slender shaft of length L with a disc attached to the end of the shaft (Figure 1.26). Assume the shaft is constrained to move only in the θ direction.

SOLUTION

The angular deflection, δ_θ, associated with a moment, M_θ, applied to the end of the shaft at the point where the disc is attached to the shaft is:

$$\delta_\theta = \frac{32 M_\theta L}{\pi G d^4} \tag{1.8a}$$

where:

δ_θ = angular deflection (radians)
G = shear modulus ($lb_f/in.^2$ or N/m^2)
M_θ = moment (lb_f-in. or N-m)
d = diameter (in. or m) of the shaft
L = length of the beam (in. or m).

The equivalent torsional stiffness coefficient, K_{teq}, is:

$$\boxed{K_{teq} = \frac{M_\theta}{\delta_\theta} \quad \text{or} \quad K_{teq} = \frac{\pi G d^4}{32 L}.} \tag{1.8b}$$

EXAMPLE 1.9

The system shown in Figure 1.27 represents a block and tackle suspended from the center of a simply supported beam . The length of the beam is L_b and of the cable is L_c. (a) Determine the equivalent stiffness coefficients of the simply supported beam with the cable attached to its center and of the cable from which the mass is suspended. (b) Determine the equivalent stiffness coefficient of both the beam and cable.

SOLUTION

(a) The static deflection of a simply support beam with a point force applied to its center is:

$$\delta_b = \frac{F L_b^3}{48 EI}. \tag{1.9a}$$

The equivalent stiffness coefficient associated with the beam is:

$$K_{eq(b)} = \frac{48 EI}{L_b^3}. \tag{1.9b}$$

The static deflection at the end of the cable is:

$$\delta_c = \frac{F L_c}{EA}. \tag{1.9c}$$

where A is the cross-section area of the cable (in.^{2} or m^2). The equivalent stiffness coefficient associated with the cable is:

$$K_{eq(c)} = \frac{EA}{L_c}. \tag{1.9d}$$

(b) The beam and cable are two spring elements that are connected in series (Figure 1.27). Therefore, the total equivalent spring coefficient is obtained from:

$$\frac{1}{K_{eq(T)}} = \frac{1}{K_{eq(b)}} + \frac{1}{K_{eq(c)}}. \tag{1.9e}$$

Substituting equations (1.9b) and (1.9d) into equation (1.9e) and simplifying yields:

$$K_{eq(T)} = \frac{K_{eq(b)}\, K_{eq(c)}}{K_{eq(b)} + K_{eq(c)}} \tag{1.9f}$$

or:

$$\boxed{K_{eq(T)} = \frac{\left(\frac{48\,EI}{L_b^3}\right)\left(\frac{EA}{L_c}\right)}{\left(\frac{48\,EI}{L_b^3}\right) + \left(\frac{EA}{L_c}\right)}.} \tag{1.9g}$$

Damping Elements

Vibration systems often contain *damping elements.* When the damping element is a viscous damping element, the force acting through a translational damper (Figure 1.28) is:

$$\boxed{f_d = C\,v \quad \text{or} \quad f_d = C\,\dot{x}.} \tag{1.122}$$

where:

f_d = damping force (lb_f or N)
C = viscous damping coefficient (lb_f - s/in. or N-s/m) of the damping element
x_1 = displacement (in. or m)

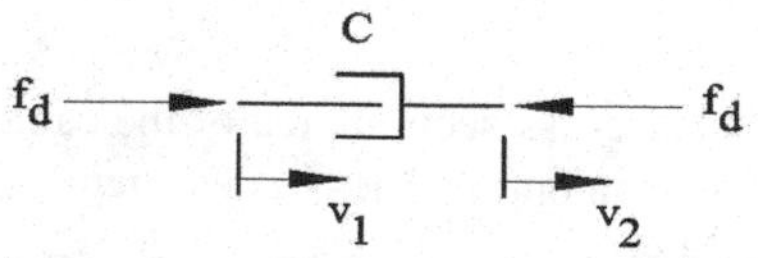

Figure 1.28 Translational Damping Element

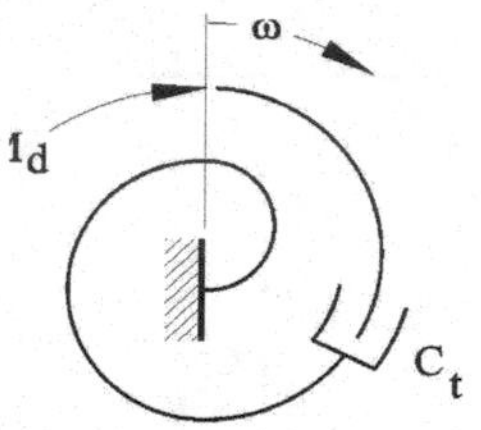

Figure 1.29 Torsional Damping Element

x_2 = displacement (in. or m)
v_1 = velocity (in./s or m/s)
v_2 = velocity (in./s or m/s).

If $v = v_2 - v_1$ is the velocity across the damping element, equation (1.122) can be written:

$$f_d = C\,(v_2 - v_1) \quad \text{or} \quad f_d = C\,(\dot{x}_2 - \dot{x}_1) \tag{1.123}$$

When the viscous damping element is a torsional damping element (Figure 1.29), the moment acting on the damping element is:

$$\boxed{M_d = C_t\,\omega \quad \text{or} \quad M_d = C_t\,\dot{\theta}} \tag{1.124}$$

where:

M_d = damping moment (lb_f - in. or N-m)
C_t = torsional damping coefficient (lb_f - in. - s or N-m-s) of the viscous damping element
θ = angular displacement (radians)
ω = angular velocity (radians/s).

Damping elements do not possess energy like mass and spring elements. They dissipate or remove energy from a vibration system, usually in the form of heat. The *power dissipated* by a damping element as it moves over a specified distance is given by:

$$D = \int_0^v F\,dv. \tag{1.125}$$

Substituting equation (1.122) for viscous damping into equation (1.125) and carrying out the prescribed integration yields:

$$\boxed{D = \frac{1}{2} C\, v^2} \tag{1.126}$$

where:
D = power (lb_f-in./s or J/s)
C = viscous damping coefficient (lb_f-s/in. or N-s/m) of the damping element
v = velocity (in./s or m/s)

When the damping element is a torsional viscous damping element, power is given by:

$$\boxed{D = \frac{1}{2} C_t\, \omega^2 \quad \text{or} \quad D = \frac{1}{2} C_t\, \dot{\theta}^2 .} \tag{1.127}$$

Like spring elements, damping elements can be connected together in parallel or in series. The relationships associated with springs elements connected together in parallel or in series also apply to damping elements. When n damping elements are connected together in parallel, the equivalent damping coefficient is:

$$C_{eq} = \sum_{i=1}^{n} C_i . \tag{1.128}$$

When n damping elements are connected together in series, the equivalent damping coefficient is:

$$\frac{1}{C_{eq}} = \sum_{i=1}^{n} \left(\frac{1}{C_i} \right). \tag{1.129}$$

A simple viscous damping element consists of two parallel plates that are separated by a distance d and that have a fluid with viscosity μ between them (Figure 1.30). If the top plate moves at a velocity v relative to the bottom plate, the damping force f_d is:

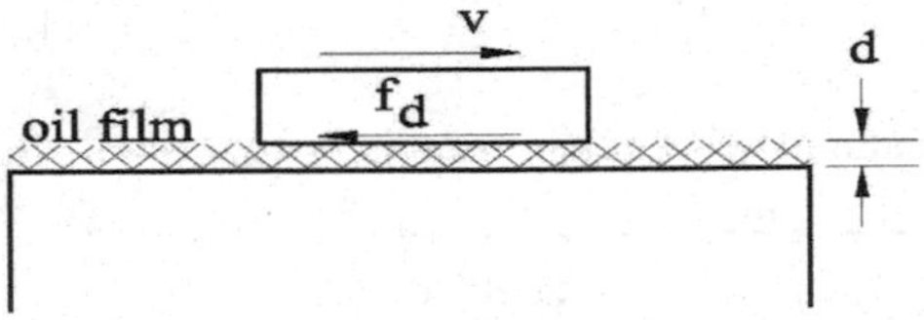

Figure 1.30 Simple Viscous Dampe

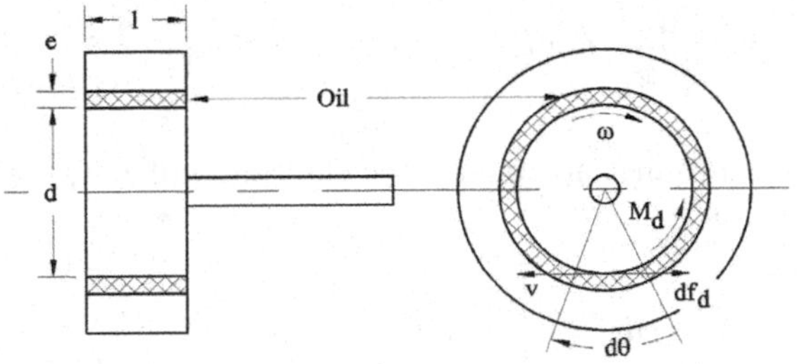

Figure 1.31 Simple Torsional Viscous Damper

$$f_d = \frac{\mu A}{d}\, v \tag{1.130}$$

where:
f_d = damping force (lb_f or N)
μ = coefficient of viscosity (lb_f-s/in.^{2} or N-s/m^2)
A = area of top plate (in.^{2} or m^2)
d = distance between plates (in. or m)
v = velocity (in./s or m/s).

Therefore, the damping constant C is:

$$\boxed{C = \frac{f_d}{v} \quad \text{or} \quad C = \frac{\mu A}{d}.} \tag{1.131}$$

This model for viscous damping is linear for a wide range of velocities. Therefore, the damping constant C is relatively insensitive to velocity.

Figure 1.31 shows a simple torsional viscous damping element that consists of a circular cylinder inside a circular ring. The center lines of the cylinder and ring are concentric. The clearance annulus of thickness e between the ring and cylinder is filled with a viscous fluid. The incremental damping force df_d acting on the circumferential surface of the ring is:

$$df_d = \frac{\mu dA}{e}\, v . \tag{1.132}$$

Noting that:

$$dA = l\, r\, d\theta \quad \text{and} \quad v = r\, \omega \tag{1.133}$$

where r = d/2, equation (1.132) becomes:

$$df_d = \frac{\mu r^2 l}{e} \omega d\theta . \quad (1.134)$$

The incremental damping moment acting on the circular ring is:

$$dM_d = r\, df_d = \frac{\mu r^3 l}{e} \omega d\theta . \quad (1.135)$$

The total damping moment is obtained by integrating equation (1.135) with respect to θ from $\theta = 0$ to $\theta = \pi$, or:

$$M_d = \frac{\mu r^3 l}{e} \omega \int_0^{2\pi} d\theta = \frac{2\pi \mu r^3 l}{e} \omega . \quad (1.136)$$

The torsional damping constant for this system is:

$$\boxed{C_t = \frac{M_d}{\omega} \quad \text{or} \quad C_t = \frac{2\pi \mu r^3 l}{e} .} \quad (1.137)$$

Figure 1.32 shows a viscous damping element that is similar to that used in an automobile shock absorber. When the piston in the figure is moved downwards with a velocity v, the viscous fluid is forced upwards through the holes in the piston. The average velocity of the fluid through the holes is U and the pressure drop across the piston is Δp. From laminar flow theory in pipes, Δp for a single hole is:

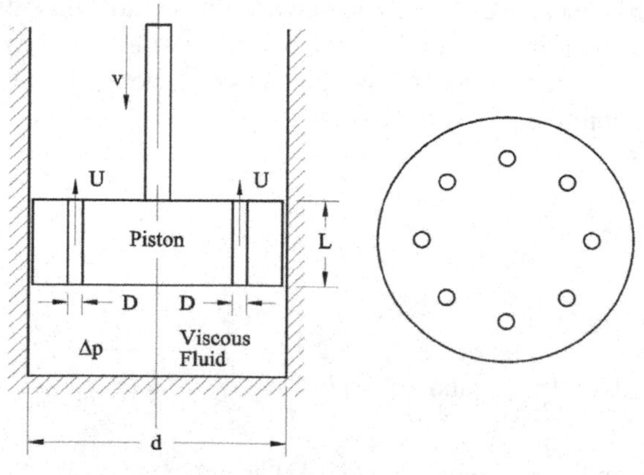

Figure 1.32 Shock Absorper Type Viscous Damper

$$\Delta p = \rho \left(\frac{L}{D} \right) U^2 f \quad (1.138)$$

where ρ is the density of the fluid and f is the friction factor. f is given by:

$$f = \frac{64 \mu}{U D \rho} . \quad (1.139)$$

Substituting equation (1.139) into equation (1.138), Δp for n holes is:

$$\Delta p_n = \frac{64 L \mu}{n D^2} U . \quad (1.140)$$

The average velocity U for laminar flow through a single hole is:

$$U = \frac{1}{2} \left(\frac{d}{D} \right)^2 v . \quad (1.141)$$

Substituting equation (1.141) into equation (1.140) yields:

$$\Delta p_n = \frac{64 L \mu}{2n D^2} \left(\frac{d}{D} \right)^2 v . \quad (1.142)$$

The force acting on the piston (assuming small holes) is:

$$f_d = \frac{\pi d^2}{4} \Delta p . \quad (1.143)$$

Substituting equation (1.142) into equation (1.143) and simplifying yields:

$$f_d = \frac{8\pi L \mu}{n} \left(\frac{d}{D} \right)^4 v . \quad (1.144)$$

The damping coefficient for this system is:

$$\boxed{C = \frac{f_d}{v} \quad \text{or} \quad C = \frac{8\pi L \mu}{n} \left(\frac{d}{D} \right)^4 .} \quad (1.145)$$

1.11 Methods for Developing Equations of Motion

Newton's Second Law

Many vibration systems experience both rotational and translational motion. Figure 1.33 represents this combined motion in a two-dimensional plane. Newton's second law can be used to sum the forces and moments acting on the system. The general mass shown in Figure 1.33 is constrained to move in the x-y plane. Therefore, the mass has translational motion in the x and y directions, and it rotates about point A in the x-y plane. Let x and y be positive to the right and up, respectively, and let θ be positive in the counterclockwise direction. θ represents the angular position, $\dot{\theta}$ the angular velocity and $\ddot{\theta}$ the angular acceleration of the mass relative to point A [Figure 1.33(a)]. The translational accelerations $\ddot{x}$ and $\ddot{y}$ of the elemental mass dm at point B are the same as those at point A. The normal and tangential accelerations associated with the rotation of dm about A are $r\dot{\theta}^2$ and $r\ddot{\theta}$, respectively. ΣF represents the sum of all the non-inertial forces acting on the mass, and ΣM represents the sum of all the non-inertial moments acting on the mass. From Newton's second law, the elemental forces dFx and dF_y acting on the mass element dm are equal to dm times its accelerations in the x and y directions, respectively (Figure 1.33(b)):

$$dF_x = dm\left(\ddot{x} - r\dot{\theta}^2\cos\theta - r\ddot{\theta}\sin\theta\right) \tag{1.146}$$

and:

$$dF_y = dm\left(\ddot{y} - r\dot{\theta}^2\sin\theta + r\ddot{\theta}\cos\theta\right). \tag{1.147}$$

Noting that $x_b = r\cos\theta$ and $y_b = r\sin\theta$ and integrating the above equations over the entire mass yields:

$$\sum F_x = \ddot{x}\int_m dm - \dot{\theta}^2\int_m x_b\,dm - \ddot{\theta}\int_m y_b\,dm \tag{1.148}$$

and:

$$\boxed{\sum F_y = \ddot{y}\int_m dm - \dot{\theta}^2\int_m y_b\,dm + \ddot{\theta}\int_m x_b\,dm} \tag{1.149}$$

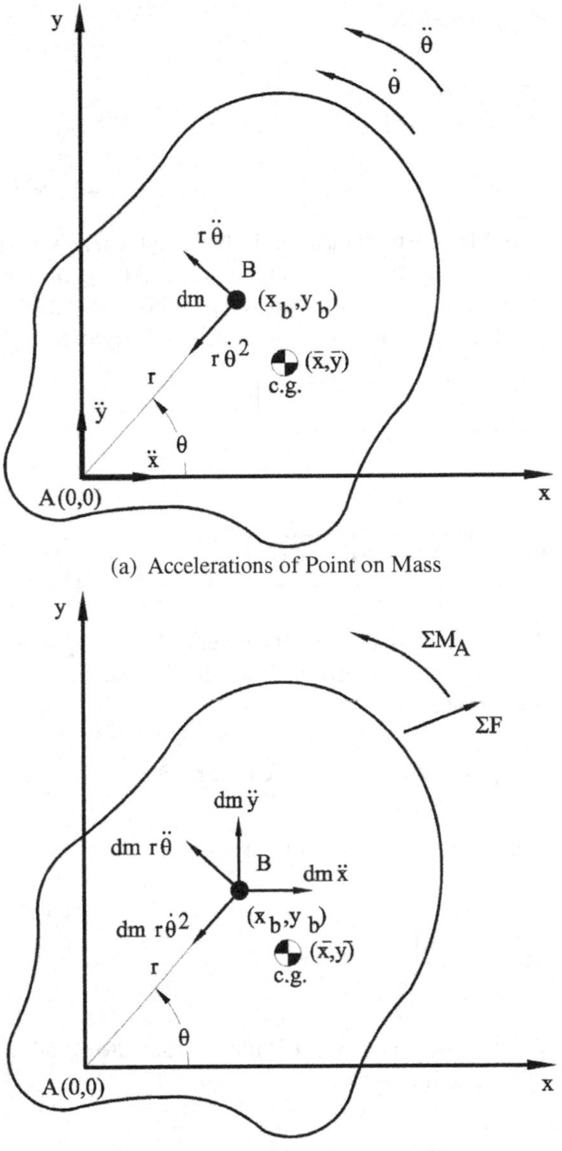

(a) Acceleration of Point on Mass

(b) Forces and Moments Acting on Mass

Figure 1.33 Equations of Motion Using Newton's Second Law

where ΣF_x and ΣF_y are the sum of the non-inertial forces in the x and y directions, respectively, acting on the entire mass. Note that:

$$\int_m dm = M \tag{1.150}$$

$$\int_m y_b \, dm = \bar{y} M \tag{1.151}$$

$$\int_m x_b \, dm = \bar{x} M \tag{1.152}$$

where M is the total mass of the body and where $\bar{x}$ and $\bar{y}$ are the distances between point A and the center of gravity of the mass in the x and y directions, respectively. Therefore, equations (1.148) and (1.149) become:

$$\boxed{\sum F_x = M\left(\ddot{x} - \bar{x}\,\dot{\theta}^2 - \bar{y}\,\ddot{\theta}\right)} \tag{1.153}$$

and:

$$\boxed{\sum F_y = M\left(\ddot{y} - \bar{y}\,\dot{\theta}^2 + \bar{x}\,\ddot{\theta}\right).} \tag{1.154}$$

The effect of an elemental moment dM_A acting about point A on dm is determined by summing the moments about point A:

$$dM_A = x_b \, dm \, \ddot{y} - y_b \, dm \, \ddot{x} + r \, dm \, r \, \ddot{\theta} \tag{1.155}$$

or integrating over the entire mass:

$$\sum M_A = \ddot{y} \int_m x_b \, dm - \ddot{x} \int_m y_b \, dm + \ddot{\theta} \int_m r^2 \, dm \,. \tag{1.156}$$

Substituting equations (1.151) and (1.152) into equation (1.136) and noting:

$$J_o = \int_m r^2 \, dm \tag{1.157}$$

where J_o is the mass moment of inertia of the mass in the x-y plane about the center of gravity of the mass yields:

$$\boxed{\sum M_A = J_o \, \ddot{\theta} - \bar{y} \, M \, \ddot{x} + \bar{x} \, M \, \ddot{y} \,.} \tag{1.158}$$

Normally, when the vibration characteristics of a system are examined, the forces and moments acting on a mass are summed relative to the center of gravity of the mass. When this is done, $\bar{y}$ and $\bar{x}$ equal zero, and equations (1.153), (1.154) and (1.158) become:

$$\boxed{\sum F_x = M \, \ddot{x}} \tag{1.159}$$

$$\boxed{\sum F_y = M \, \ddot{y}} \tag{1.160}$$

$$\boxed{\sum M_{c.g.} = J_o \, \ddot{\theta} \,.} \tag{1.161}$$

d'Alembert's Principle

When Newton's second law is used to obtain the equations of motion of a system that has both translational and rotational motion in a plane, the positive directions of the translational accelerations of the center of gravity of the mass of the system and the rotational acceleration about the point of rotation must be clearly specified. The corresponding signs of the inertial components in the system equations of motion must also be correctly identified. If the positive directions of the system coordinates in Figure 1.43 are changed, the corresponding signs associated with the terms in equations (1.153), (1.154) and (1.158) will also change. d'Alembert's principle can be used to more easily address these sign changes.

d'Alembert's principle states;

Every state of motion of a system can be considered at any instant to be in a state of static equilibrium if the inertial forces and moments are properly taken into consideration.

An *inertia force* is defined as the product of a mass times the acceleration of the center of gravity of the mass. The inertia force is shown acting in the negative direction through the center of gravity of the mass on a free-body diagram. An *inertia moment* is defined as the product of the mass moment of inertia of a mass relative to its center of gravity times the rotational acceleration of the mass about its center of gravity. The inertia moment is shown acting in the negative direction about the center of gravity of the mass on a free-body diagram.

Figure 1.34 shows the system with translational and rotational motion that was examined using Newton's second law in the previous section. As before, let x and y be positive to the right and up, respectively, and let θ be

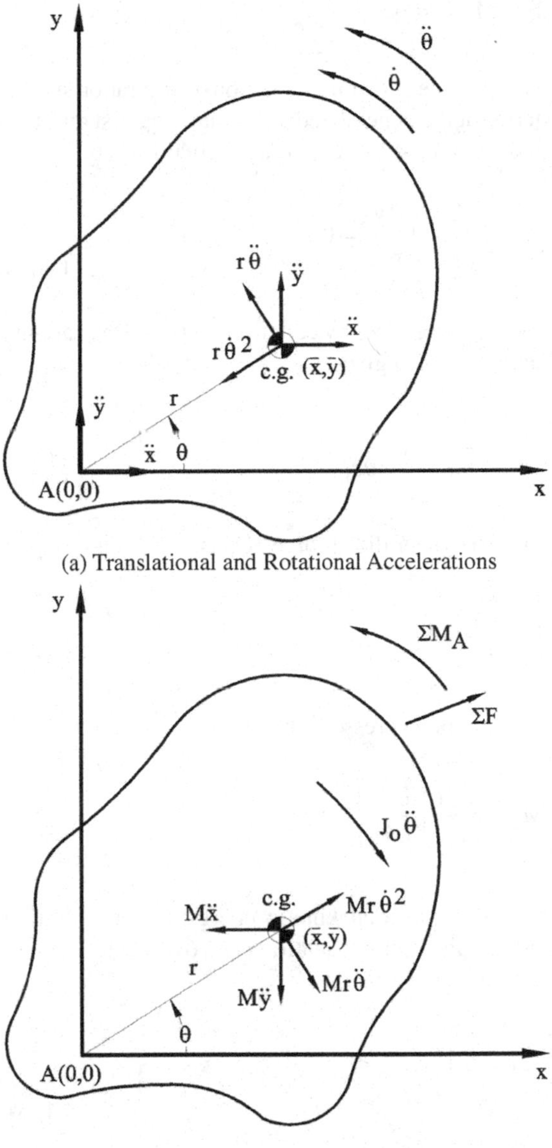

Figure 1.34 Equations of Motion Using d' Alembert's Principle

positive in the counterclockwise direction. Figure 1.34(a) shows the translational and rotational accelerations of the center of gravity of the system. Figure 1.34(b) shows the non-inertial forces and moments acting on the system, the inertia forces acting through the center of gravity of the system, and the inertial moment acting about the center of gravity of the system. Summing the forces in the x and y direction and setting the results equal to zero yields:

$$\sum F_x - M\,\ddot{x} + M\,r\dot{\theta}^2\cos\theta + M\,r\ddot{\theta}\sin\theta = 0 \tag{1.162}$$

and:

$$\sum F_y - M\,\ddot{y} + M\,r\dot{\theta}^2\sin\theta - M\,r\ddot{\theta}\cos\theta = 0\,. \tag{1.163}$$

Noting that:

$$r\cos\theta = \bar{x} \tag{1.164}$$

and:

$$r\sin\theta = \bar{y}\,, \tag{1.165}$$

equations (1.162) and (1.163) become:

$$\boxed{\sum F_x = M\,\ddot{x} - M\,\bar{x}\dot{\theta}^2 - M\,\bar{y}\ddot{\theta}} \tag{1.166}$$

and:

$$\boxed{\sum F_y = M\,\ddot{y} - M\,\bar{y}\dot{\theta}^2 + M\,\bar{x}\ddot{\theta}\,.} \tag{1.167}$$

Summing the moments about A yields:

$$\sum M_A - J_o\,\ddot{\theta} + M\,\ddot{x}\,r\sin\theta - M\,\ddot{y}\,r\cos\theta = 0\,. \tag{1.168}$$

Substituting equations (1.164) and (1.165) into equation (1.168) and simplifying yields:

$$\boxed{\sum M_A = J_o\,\ddot{\theta} - \bar{y}\,M\,\ddot{x} + \bar{x}\,M\,\ddot{y}\,.} \tag{1.169}$$

Equations (1.166), (1.167) and (1.169) are the same as equations (1.153), (1.154) and (1.158).

Energy Equation

If a vibrating system has only mass and spring elements and non-inertial forces and moments do not act on the system, the system is a conservative system. The sum of the kinetic and potential energies of the system is constant and is equal to the initial kinetic and/or potential energy that is imparted to the system. Therefore:

$$T + U = \text{const}. \tag{1.170}$$

For systems with one degree of freedom, when the system is a conservative system, the time rate of the total energy of the system is equal to zero, or:

$$\boxed{\frac{d(T+U)}{dt} = 0.} \tag{1.171}$$

Therefore, to obtain the equation of motion for a 1-DOF system using the energy method, it is necessary to add the kinetic and potential energy expressions for the system, take the derivative of the resulting expression with respect to time, and set the results equal to zero.

Lagrange's Equation

The energy method can only be used to obtain the equation of motion for 1-DOF vibration systems with no damping elements and with no non-inertial forces and moments acting on the system. For vibration systems with more than one degree of freedom, that have damping elements, and they have non-inertial forces and moments acting on them, another energy method must be used to obtain the equations of motion for the system. The following analysis will only consider motion in the y direction. It can be expanded to consider motion in the x and z directions, as well as rotational directions.

In many instances, the coordinates used to describe the motion of a vibration system can be expressed in terms of a set of generalized coordinates q_i where q_i are any set of independent coordinates that can be used to describe the motion of the system. For example, the coordinates y_j can be expressed:

$$y_j = y_j(q_i) \quad \text{where} \quad j = 1, 2, \ldots, n \text{ and } i = 1, 2, \ldots, n. \tag{1.172}$$

For motion in the y direction, d'Alembert's principle can be expressed:

$$F_j - M_j\,\ddot{y}_j = 0. \tag{1.173}$$

To develop the equations of motion of a vibration system in terms of the generalized coordinates q_i, first multiply equation (1.173) by $\partial y_j / \partial q_i$ to obtain:

$$\left[F_j - M_j\,\ddot{y}_j\right]\frac{\partial y_j}{\partial q_i} = 0 \tag{1.174}$$

where y_j is specified by equation (1.172). Rearranging equation (1.174) gives:

$$M_j\,\ddot{y}_j\,\frac{\partial y}{\partial q_i} = F_j\,\frac{\partial y_j}{\partial q_i}. \tag{1.175}$$

Define the generalized forces Q_i as:

$$F_j\,\frac{\partial y_j}{\partial q_i} = Q_i. \tag{1.176}$$

$M_j\,\ddot{y}_j$ can be expressed as:

$$M_j\,\ddot{y}_j = \frac{d}{dt}\left(\frac{\partial T}{\partial \dot{y}_j}\right) \tag{1.177}$$

where T is the total kinetic energy of the vibration system. Substituting equation (1.176) and (1.177) into equation (1.175) yields:

$$\frac{d}{dt}\left(\frac{\partial T}{\partial \dot{y}_j}\right)\frac{\partial y_j}{\partial q_i} = Q_i. \tag{1.178}$$

Use the following relationship:

$$\frac{d}{dt}\left(\frac{\partial T}{\partial \dot{y}_j}\frac{\partial y_j}{\partial q_i}\right) = \frac{d}{dt}\left(\frac{\partial T}{\partial \dot{y}}\right)\frac{\partial y_j}{\partial q_i} + \frac{\partial T}{\partial \dot{y}_j}\frac{\partial \dot{y}_j}{\partial q_i}. \tag{1.179}$$

Rearrange equation (1.179) and substitute the results into equation (1.178) to obtain:

$$\frac{d}{dt}\left(\frac{\partial T}{\partial \dot{y}_j}\frac{\partial y_j}{\partial q_i}\right)-\frac{\partial T}{\partial \dot{y}_j}\frac{\partial \dot{y}_j}{\partial q_i}=Q_i\,. \tag{1.180}$$

Next, take the derivative of y with respect to t. Since y is a function of q_i, the chain rule must be used:

$$\frac{\partial y_j}{\partial t}=\frac{\partial y_j}{\partial q_i}\frac{\partial q_i}{\partial t} \quad \text{or} \quad \dot{y}_j=\frac{\partial y_j}{\partial q_i}\dot{q}_i. \tag{1.181}$$

Taking the partial derivative of equation (1.181) with respect to $\dot{q}_i$ yields:

$$\frac{\partial \dot{y}_j}{\partial \dot{q}_i}=\frac{\partial y_j}{\partial q_i}. \tag{1.182}$$

Substituting equation (1.182) into equation (1.180) and simplifying gives:

$$\frac{d}{dt}\left(\frac{\partial T}{\partial \dot{q}_i}\right)-\frac{\partial T}{\partial q_i}=Q_i\,. \tag{1.183}$$

Equation (1.188) is referred to as *Lagrange's equation*. If the system has i = n generalized coordinates, the system has n degrees of freedom and n equations of motion.

There are two types of generalized forces that act on a vibratsystem. One is associated with the restoring forces of spring and damping elements of a vibration system. The generalized forces associated with the spring elements are:

$$Q_i=-\frac{\partial U}{\partial q_i} \tag{1.184}$$

where U is the total potential energy associated with the system. The generalized forces associated with the damping elements are:

$$Q_i=-\frac{\partial D}{\partial \dot{q}_i} \tag{1.185}$$

where D is the total power dissipation from the system. Substituting equations (1.184) and (1.185) into equation (1.183) yields:

$$\boxed{\frac{d}{dt}\left(\frac{\partial T}{\partial \dot{q}_i}\right)-\frac{\partial T}{\partial q_i}+\frac{\partial D}{\partial \dot{q}_i}+\frac{\partial U}{\partial q_i}=Q_i\,.} \tag{1.186}$$

The other type of generalized forces that can act on a vibration system is externally applied forces. Before the generalized forces can be stated, a Cartesian coordinate system must be defined such that:

$$\boxed{x_j=x_j(q_i) \quad y_j=y_j(q_i) \quad z_j=z_j(q_i)} \tag{1.187}$$

where i = 1, 2, ..., m and j = 1, 2, ..., n. i represents the number of generalized coordinates necessary to describe the motion of a vibration system. j represents the number of mass elements associated with the system. If $F_{x(j)}$, $F_{y(j)}$ and $F_{z(j)}$ represent the external translational forces in the x, y and z directions, respectively, that act on the jth mass element, the generalized forces associated with the externally applied forces are given by:

$$\boxed{Q_i=\sum_{j=1}^{n}\left(F_{x(j)}\frac{\partial x_j}{\partial q_i}+F_{y(j)}\frac{\partial y_j}{\partial q_i}+F_{z(j)}\frac{\partial z_j}{\partial q_i}\right).} \tag{1.188}$$

The generalized coordinates q_i for this development are for translational motions. Therefore, Q_i in equation (1.188) has units of force. Equation (1.188) can be substituted into equation (1.186) for Q_i when a vibration system has externally applied forces. A more robust development of Lagrange's equation is presented in Chapter 5.

1.12 Units

The constitutive equations that have been developed in this chapter were developed for rectilinear and rotational motions. These equations are of fundamental importance in developing the equations that describe the motions of vibration systems. The coefficients and parameters associated with these equations must have the proper units. *English or U.S. units* and *SI units* are used in this book. The *International System of Units (SI)* is the modernized version of the metric system of units. Either the ft-lb_f-s or in.-lb_f-s system can be used for *English units*. When ft (feet) is used, a body falling under the influence of gravity has an acceleration of gravity g equal to 32.2 ft/s^2. When in. (inch) is used,

Table 1.2 Basic Units

Quantity	SI Units				English Units*
	Name	Symbol	In Terms of SI Base Units	In Terms of Other SI Units	
Length, distance	meters	m			in.
Area	square meters	m^2			$in.^2$
Volume	cubic meters	m^3			$in.^3$
Time	second	s			s
Mass	kilogram	kg			$lb_f\text{-}s^2/in.$
Speed, velocity	meters per second	m/s			in./s
Acceleration	meters per second squared	m/s^2			$in./s^2$
Mass density	kilograms per meter cubed	kg/m^3			$lb_f\text{-}s^2/in.^4$
Force	Newton	N	$mkgs^{-2}$		lb_f
Pressure, stress	Pascal	Pa	$m^{-1}kgs^{-2}$	N/m^2	$lb_f/in.^2$
Energy, work	Joule	J	m^2kgs^{-2}	Nm	$in.\text{-}lb_f$
Power	Watt	W	m^2kgs^{-3}	J/s	$in.\text{-}lb_f/s$
Moment	Newton-meter	Nm	m^2kgs^{-2}		$in.\text{-}lb_f$
Frequency	Hertz	Hz	s^{-1}		Hz
Plane angle	radian	rad			rad (degree)
Angular velocity	radian per second	rad/s			rad/s
Angular acceleration	radian per second squared	rad/s^2			rad/s^2

*ft (foot) can be interchanged with in. (inch).

Table 1.3 Prefixes for Multiples and Submultiples of SI Units

Multiple	Prefix	Symbol	Submultiple	Prefix	Symbol
10	deca	dc	10^{-1}	deci	d
10^2	hecto	h	10^{-2}	centi	c
10^3	kilo	k	10^{-3}	milli	m
10^6	mega	M	10^{-6}	micro	μ
10^9	giga	G	10^{-9}	nano	n
10^{12}	tera	T	10^{-12}	pico	p
			10^{-15}	femto	f
			10^{-18}	atto	a

Table 1.4 Units for Parameters of Rectilinear and Rotational Vibration Systems

Parameter	Rectilinear Systems			Rotational Systems		
	Symbol	English Units	SI Units	Symbol	English Units	SI Units
Time	t	s	s	t	s	s
Displacement	x, y, etc	in.	m	θ, ϕ, etc.	rad	rad
Velocity		in./s	m/s		rad/s	rad/s
Acceleration		in./s^2	m/s^2		rad/s^2	rad/s^2
Mass, Mass moment of inertia	M	lb$_f$-s^2/in.	kg	J	in.-lb$_f$-s^2	kgm
Force, moment	F	lb$_f$	N	M, M_o, etc.	in.-lb$_f$	Nm
Momentum		lb$_f$-s	kgm/s		in.-lb$_f$-s	kgm^2/s
Impulse		lb$_f$-s	Ns		in.-lb$_f$-s	Nms
Kinetic energy	T	in.-lb$_f$	J	T	in.-lb$_f$	J
Potential engergy	U	in.-lb$_f$	J	U	in.-lb$_f$	J
Work	W	in.-lb$_f$	J	W	in.-lb$_f$	J
Power	D, P, , etc.	in.-lb$_f$/s	W	D, P, , etc.	in.-lb$_f$/s	W
Stiffness coefficient	K	lb$_f$/in.	N/m	K_t	in.-lb$_f$	Nm
Damping coefficient	C	lb$_f$-s/in.	Ns/m	C_t	in.-lb$_f$-s	Nms
Angular frequency	ω	rad/s	rad/s	ω	rad/s	rad/s
Circular frequency	f	Hz	Hz	f	Hz	Hz
Period of oscillation	T	s	s	T	s	s

the body has an acceleration of gravity g equal to 386 in/s^2. The in-lb$_f$-s system of English units is normally used for the analysis of vibration systems in the United States. The acceleration of gravity in SI units is 9.81 m/s^2. Tables 1.2 and 1.3 list examples of basic units. Table 1.4 lists the symbols and units associated with common parameters of rectilinear and rotational systems.

Often when English units are used, mass is given in terms of weight with the unit of lb$_f$ or in terms of mass with the unit of lb$_m$. When this is done, it is necessary to convert the units of weight or of mass to the units of lb$_f$-s^2/in. Use the units:

We = weight of a mass (lb$_f$)
m' = value of mass (lb$_m$)
g = acceleration of gravity (in./s^2)
g_c = gravitational constant [(lb$_m$-in.)/(lb$_f$-s^2)].

where

$$g = 386 \frac{\text{in.}}{\text{s}^2} \quad \text{and} \quad g_c = 386 \frac{\text{lb}_m - \text{in.}}{\text{lb}_f - \text{s}^2}. \tag{1.189}$$

When the mass of the body is given in the units of lb$_m$, the unit can be converted to (lb$_f$-s^2)/in. by using the equation:

$$M = \frac{m'}{g_c} \frac{\text{lb}_f - \text{s}^2}{\text{in.}}. \tag{1.190}$$

When the weight of a mass is given in the unit lb$_f$, the unit can be converted to (lb$_f$-s^2)/in. by using the equation:

$$We = M\,g \quad \text{or} \quad M = \frac{We}{g} \frac{\text{lb}_f - \text{s}^2}{\text{in.}}. \tag{1.191}$$

1.13 Vibration Analyses Procedures

Analyses of vibration systems often appear to be confusing and difficult. This confusion can be minimized by following a prescribed sequence of steps when analyzing vibration systems.

- *Define the objectives of the analysis*: Most analyses cannot be successfully completed because the analyses objectives are not properly defined. Therefore, the first step in analyzing a vibration system is to properly define the objectives of the analysis. Here are some questions that should be considered when starting the analysis of a vibration system:

  When a vibration system failure is being examined, the following questions should be considered. What is broken, what has failed, and what is not working properly? What must be examined to determine why the vibration system has failed or is not working properly? What existing information is available to help determine why the vibration system has failed or is not working properly? What additional information is needed to determine why the vibration system has failed or is not working right?

  When a future vibration system design is being considered, the following questions should be considered. What is the vibration system being designed to do? What are the operation characteristics and requirements of the vibration system? What are other important characteristics or requirements of the vibration system?

  Relative to the failure or the new design of a vibration system, the following questions should be considered. What are the physical and other constraints that must be considered when examining or designing the vibration system? What parameters must be specified and used to properly examine or design the vibration system?

- *Define the objectives of the vibration analysis*: After the above and other related questions have been answered, the vibration analysis objectives should be formulated and clearly defined. The following should be considered when defining the vibration analysis objectives:

1. Specify the design requirements of the vibration system.
2. Identify the vibration system information that is available.
3. Identify the vibration system information that must be obtained.
4. Identify the constraints associated with the vibration system.
5. Identify the resources that are available to analyze the vibration system.
6. Specify time and resource constraints that are associated with the vibration analysis.
7. Consider whether an analytical, experimental, or a combination of analytical and experimental approach must be used for the vibration analysis.
8. Specify the accuracy that is required for the vibration analysis.

When conducting an analysis of a vibration system, the following steps are important.

- *Obtain or develop an accurate representation of the vibration system being analyzed*: This representation can be a three-dimensional drawing, a two-dimensional drawing, a picture, or other similar representation of the system.

- *From the accurate representation of the vibration system, develop a schematic drawing of the system*: Separate the system into individual mass elements. These mass elements may have translational motion, rotational motion, or both. Show the spring and damping elements that connect the individual mass elements.

- *Specify the positive coordinate system(s) and the related spatial variables associated with the mass elements in the schematic drawing*: The positive coordinate system(s) may or may not be a traditional right-handed coordinate system. Identify all of the related translational and rotational spatial variables that are associated with the motions of each mass element. Make sure the specified positive motions of all of the mass elements are consistent with the overall positive coordinate system. Generally (but not always) the spatial variables will be associated with the movements of the centers-of-gravity of the individual mass elements.

- *Draw the free-body diagrams associated with all of the system mass elements*: It will be necessary to specify the number of degrees of freedom that will be associated with each mass element and with the overall system. To do this, identify all the spatial variables necessary to properly describe the motions of each mass element and the geometric constraint equations that relate these spatial variables. Show all of the spring and damping forces and moments that act on the individual mass elements as restoring forces and moments that act in the appropriate negative directions. Show all of the externally applied and internally generated forces and moments that act on the individual mass elements as forces and moments that act in the appropriate positive directions. Do not attempt to anticipate the actual directions these forces and moments will act. Their directions will change with time over cycles of oscillation. If the above procedures are followed, the actual directions of the restoring and externally and internally applied forces and moments will be properly specified by the solution to the system equations of motion.

- *Develop the equations of motion for the vibration system*: These equations will be coupled second order differential equations of motion if the system has more than 1-DOF. There will be one equation of motion associated with each system degree-of-freedom. If the vibration system only has one degree of freedom, there will be one equation of motion. A n-DOF system will have n coupled equations of motion.

- *Solve the vibration system equations of motion*: Use appropriate differential equation solution techniques to solve the vibration system equations of motion.

- *Interpret the results of the solution to the vibration system equations of motion*: This is the final step in any vibration system analysis. One of the most common mistakes in analyzing vibration systems is to intuitively anticipate the system response before employing the above analysis procedures. Most often intuitive anticipation results in an incorrect assessment of the system response. This sometimes makes formulating and defining the system analysis objectives more difficult. However, after the vibration system has been properly defined, formulated and analyzed, applying one's intuition to determine whether or not the system solution makes sense is necessary. Does the system response correspond to what has been physically observed? Is it realistic? Does it make sense? If it is not possible to answer these or other related questions from physical observations of similar systems that have been observe or analyzed, it may be necessary to design and conduct appropriate experiments to validate the analytical system response. If the results of the experimental analysis of the vibration system do not reasonably correlate with the results of the analytical analysis, it may be necessary to revise the schematic representation of the vibration system and repeat the vibration 0analysis process. Often the modelling of vibration systems and related experimental analyses are part of an iterative process that must be repeated to achieve acceptable results.

APPENDIX 1A - CONVERSION OF UNITS

LENGTH

1 ft = 12 in.	1 in. = 0.8333 ft
1 yd = 3 ft	1 ft = 0.3333 yd
1 in. = 25.4 mm	1 mm = 0.03937 in.
1 ft = 0.3048 m	1 m = 3.2808 ft
1 yd = 0.9144 m	1 m = 1.0936 m
1 mile = 1.6093 km	1 km = 0.6214 mile
1 mph = 88 fpm	1 fpm = 00.01136 mph
1 mph = 1.4667 fps	1 fps = 0.6182 mph
1 mph = 0.44704 m/s	1 m/s = 2.2369 mph
1 fpm = 0.00508 m/s	1 m/s = 196.8504 fpm
1 fps = 0.3048 m/s	1 m/s = 3.2808 fps
1 mph = 1.6093 km/h	1 km/h = 0.6214 mph

AREA

1 ft^2 = 144 $in.^2$	1 $in.^2$ = 0.006944 ft^2
1 yd^2 = 9 ft^2	1 ft^2 = 0.1111 yd^2
1 acre = 43,560 ft^2	
1 acre = 0.001562 $mile^2$	
1 $in.^2$ = 0.00064516 m^2	1 m^2 = 1,550.0031 $in.^2$
1 ft^2 = 0.09290304 m^2	1 m^2 = 10.76391 ft^2
1 yd^2 = 0.836127 m^2	1 m^2 =1.19599 yd^2
1 acre = 4046.86 m^2	1 m^2 = 0.0002471 acre
1 $mile^2$ = 2.58999 km^2	1 km^2 = 0.3861 $mile^2$
1 a (are) = 10^2 m^2	1 m^2 = 10^{-2} a
1 ha (hectare) = 10^4 m^2	1 m^2 =10^{-4} ha
1 acre = 0.40468 ha	1 ha = 2.47109 ha

VOLUME

1 ft^3 = 1,723 $in.^3$	1 $in.^3$ = 0.0005787 ft^3
1 yd^3 = 27 ft^3	1 ft^3 = 0.037037 yd^3
1 ft^3 = 0.0283168 m^3	1 m^3 = 35.3147 ft^3
1 ft^3 = 28.3168 litres	1 litres = 0.03531 ft^3
1 US pint = 0.4732 litre	1 litre = 2.1133 US pints
1 US gal = 3.7853 litres	1 litre = 0.26418 US gal
1 litre = 10^{-3} m	1 m = 10^3 litres

ANGLES

2π rad = 360 degrees	
1 deg = 0.0174533 rad	1 rad = 57.295754 deg
1 minute = 2.90888 x 10^{-4} rad	
1 second = 4.84814 x 10^{-6} rad	
1 cycle = 2π rad	
1 rpm = 0.166667 cycle/s	1 cycle/s = 60 rpm
1 rpm = 0.104720 rad/s	1 rad/s = 9.54909 rpm
1 cycle/s = 6.28318 rad/s	1 rad/s = 0.159155 cycle/s

MASS

16 oz = 1 lb	1 lb = 0.0625 oz
1 oz = 28.3495 g	1 g = 0.035274 ox
1 lb = 0.45359237 kg	1 kg = 2.2046 lb
1 long ton = 1.01604 Mg	1 Mg = 0.9842 long ton
1 lb/ft^3 = 16.0185 kg/m^3	1 kg/m^3 = 0.06243 lb/ft^3
1 g/cm^3 = 10^3 kg/m^3	1 kg/m^3 = 10^{-3} g/cm^3
1 slug = 32.17 lb	1 lb = 0.03108 slug
1 lb/ft^2 = 0.006944 lb/$in.^2$	1 lb/$in.^2$ = 144 lb/ft^2
1 lb/ft^3 = 0.0005785 lb/$in.^3$	1 lb/$in.^3$ = 1,728 lb/ft^3
1 lb/ft^3 = 16.02 kg/m^3	1 kg/m^3 = 0.06243 lb/ft^3

FORCE

1 N = 1 kg-m/s^2	
1 Pa = 1 N/m^2	
1 μbar = 1 dyne/cm^2	
1 lb_f = 4.44822 N	1 N = 0.22481 lb_f
1 pdl (poundal) = 32.17 lb_f	1 lb_f = 0.031085 pdl
1 pdl = 0.138255 N	1 N = 7.2330 pdl
1 lb_f/$in.^2$ = 144 lb_f/ft^2	1 lb_f/ft^2 = 0.006945 lb_f/$in.^2$
1 N = 10^5 dynes	1 dyne = 10^{-5} N
1 lb_f/$in.^2$ = 6894.76 Pa	1 Pa = 0.000145038 1 lb_f/$in.^2$
1 lb_f/ft^2 = 47.88 Pa	1 Pa = 0.020885 lb_f/ft^2
1 dyne/cm^2 = 10^{-1} Pa	1 Pa = 10 dynes/cm^2
1 bar = 10^5 Pa	1 Pa = 10^{-5} bar
1 mm H_2O = 9.80665 Pa	1 Pa = 0.10197 mm H_2O
1 in. H_2O = 249.089 Pa	1 Pa = 0.0040146 in. H_2O
1 mm Hg = 133.322 Pa	1 Pa = 0.007501 mm Hg
1 atm = 101.325 Pa	1 Pa = 0.0098692 atm

ENERGY

1 J = 1 N-m = 1 W-s	
1 ft-lb_f = 1.355818 J	1 J = 737562 ft-lb_f
1 in.-lb_f = 0.1129848 J	1 J = 8.850744 in.-lb_f
1 ft-pdl = 0.04214 J	1 J = 23.7304 ft-pdl
1 kW-hr = 3.6 MJ	1 MJ = 0.2778 kW-hr
1 BTU = 1.055056 kJ	1 kJ = 0.947817 BTU
1 J = 10^7 ergs	1 erg = 10^{-7} J

POWER

1 W = 1 J/s = 1 N-m/s	
1 ft-lb_f/s = 1.355818 W	1 W = 0.737562 ft-lb_f/s
1 in.-lb_f/s = 0.1129848 W	1 W = 8.850744 1 in.-lb_f/s
1 BTU/hr = 0.293071 W	1 W =3.41214 BTU/hr
1 hp = 0.7457 kW	1 kW = 1.34102 hp
1 hp = 550 ft-lb_f/s	1 ft-lb_f/s = 0.0018182 hp

AREA MOMENT OF INTERIA

1 in.^{4} = 41.6231 x 10^{-8} m^4

1 ft^4 = 86.3097 x 10^{-4} m^4

1 m^4 = 240.251 x 10^4 in.^{4}

1 m^4 = 115.862 ft^4

MASS MOMENT OF INTERIA

1 in.-lb$_f$-s^2 = 0.1129848 m^2-kg

1 m^2-kg = 8.850744 in.-lb$_f$-s^2

SPRING CONSTANT

1 lb$_f$/in. = 175.1268 N/m

1 lb$_f$/ft = 14.5939 N/m

1 in.-lb$_f$/rad = 0.1129848 N-m/rad

1 N/m = 5.71017 x 10^{-3} lb$_f$/in.

1 N/m = 68.5221 x 10^{-3} lb$_f$/ft

1 N-m/rad = 8.850744 10^{-3} in.-lb$_f$/rad

1 N-m/rad = 0.737562 ft-lb$_f$/rad

DAMPING CONSTANT

1 lb$_f$-s/in. = 175.1268 N-s/m

1 in.-lb$_f$-s/rad = 0.1129848 N-m-s/rad

1 N-s/m = 5.71017 x 10^{-3} lb$_f$-s/in.

1 N-m-s/rad = 8.850744 10^{-3} in.-lb$_f$-s/rad

APPENDIX 1B - MATLAB® Script for EXAMPLE 1.3

```
%% MATLAB Script for EXAMPLE 1.3
%%Calculates the Fourier Series Coefficients for a
%%complex periodic signal

clear all
close all
clc

m = 51; %Number of terms in the time and response arrays
delta_time=0.02; % Time interval between array values

%%time array
time=[0 0.02 0.04 0.06 0.08 0.10 0.12 0.14 0.16 0.18 0.20 0.22 0.24 ...
   0.26 0.28 0.30 0.32 0.34 0.36 0.38 0.40 0.42 0.44 0.46 0.48 ...
   0.50 0.52 0.54 0.56 0.58 0.60 0.62 0.64 0.66 0.68 0.70 0.72 0.74 ...
   0.76 0.78 0.80 0.82 0.84 0.86 0.88 0.90 0.92 0.94 0.96 0.98 1.00];

%%Response f(t) array
response=[5.200 5.241 4.955 4.301 3.309 2.085 0.792 -0.380 -1.252 -1.699...
   -1.663 -1.199 -0.435 0.435 1.198 1.663 1.696 1.252 0.381 -0.791 ...
   -2.085 -3.309 -4.301 -4.955 -5.241 -5.200 -4.927 -4.543 -4.159 ...
   -3.852 -3.645 -3.510 -3.377 -3.158 -2.777 -2.194 -1.412 -0.489 ...
   0.487 1.411 2.193 2.777 3.157 3.376 3.150 3.645 3.852 4.159 4.543 ...
   4.927 5.200];

%%frequency = 1 Hz
period=(m-1)*delta_time; %Period of complex periodic signal
fo=1/period

%%Calculate the a0 Fourier series coefficient
azero = 0;
for z=2:1:51 %z=2 is t1; z=51 is tN
  azero=azero+response(z)*delta_time;
end

%%Calculate the an Fourier series coefficients
for n=1:5
   a = 0;
 for z=2:51 %z=2 is t1; z=51 is tN
    a=a+(response(z)*cos(2*n*pi*fo*time(z)/period))*delta_time;
 end
   coef_a_number(n)=n;a_coeficients(n)=2*a;
end
```

```
%%Calculate the bn Fourier series coefficients
for n=1:5
   b = 0;
  for z=2:51 %z=2 is t1; z=51 is tN
      b = b+response(z)*sin(2*n*pi*fo*time(z)/period)*delta_time;
  end
    coef_b_number(n)=n;b_coeficients(n) = 2*b;
end

%%Print out the results
azero
coef_a_number
a_coeficients
coef_b_number
b_coeficients
```

PROBLEMS - CHAPTER 1

1. A harmonic signal is expressed as:
$x(t) = 0.6 \cos(18\pi t - \pi/6)$,
where x(t) is measured in inches, t in seconds and the phase angle in radians. Determine:
(a) the frequency in rad/s and in Hz,
(b) the period of motion,
(c) the amplitude of maximum displacement, velocity and acceleration,
(d) the amplitude of displacement, velocity and acceleration at t = 0 s, and
(e) the amplitude of displacement, velocity and acceleration at t = 0.2 s.

2. A harmonic signal is described by the equation:
$x(t) = X \cos(100t + \beta)$.
The initial displacement is x(0) = 0.15 in. and the initial velocity is $\dot{x}(0) = 50$ in./s. Determine the constants X and β.

3. Use the algebraic method to determine the sum of the harmonic signals:
$x_1(t) = 2 \sin(\omega t + \pi/3)$
$x_2(t) = 3 \cos(\omega t - 2\pi/3)$.
Express the results in the form:
$x(t) = X \cos(\omega t + \beta)$.
Check the results graphically.

4. Two harmonic signals are described by the equations:
$x_1(t) = X\cos \omega t$
$x_2(t) = (X + \delta) \cos (\omega + \varepsilon)t$
where $\delta << X$ and $\varepsilon << \omega$. Determine the sum of the two harmonic signals. If beating occurs, write the equation for the beat amplitude and specify the beat frequency.

5. Express the following complex numbers in polar form:
(a) = 3 + j4
(b) = -4 + j3
(c) = (6 + j6)/(5 + j7)
(d) = (6 - j6) x (4 + j7).

6. Determine the Fourier series equation of the wave form shown in Figure P1.1.

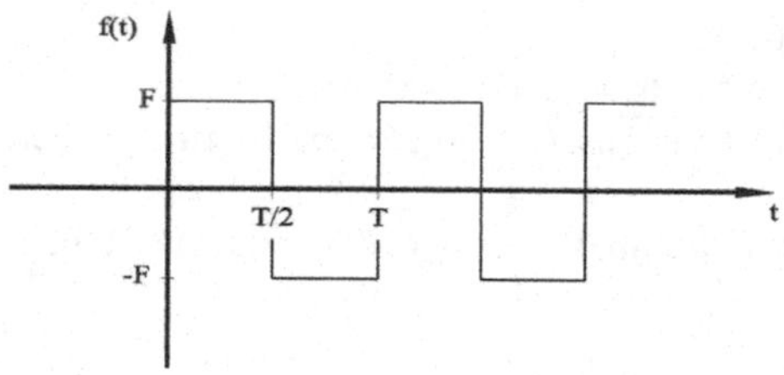

Figure P1.1

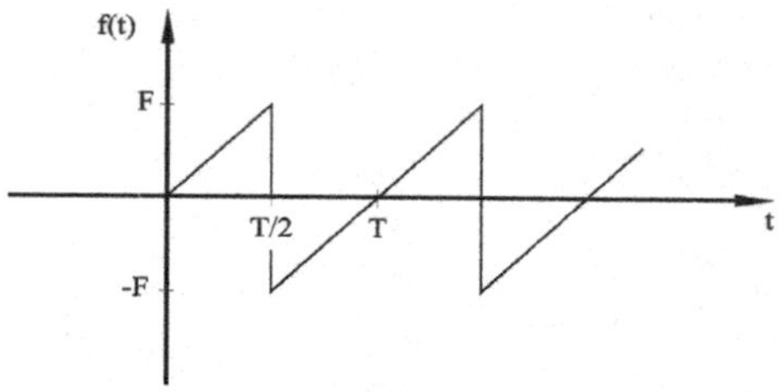

Figure P1.2

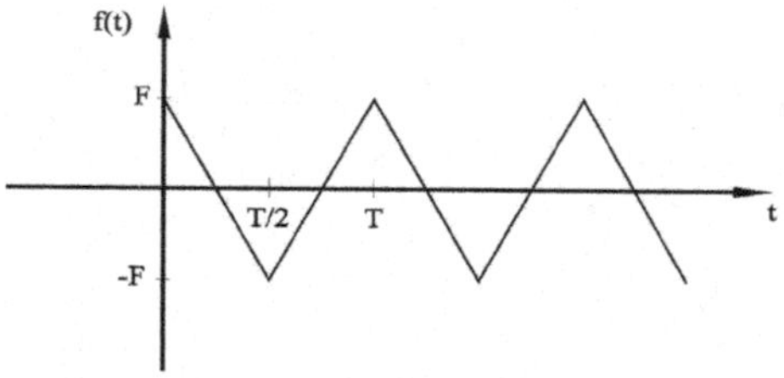

Figure P1.3

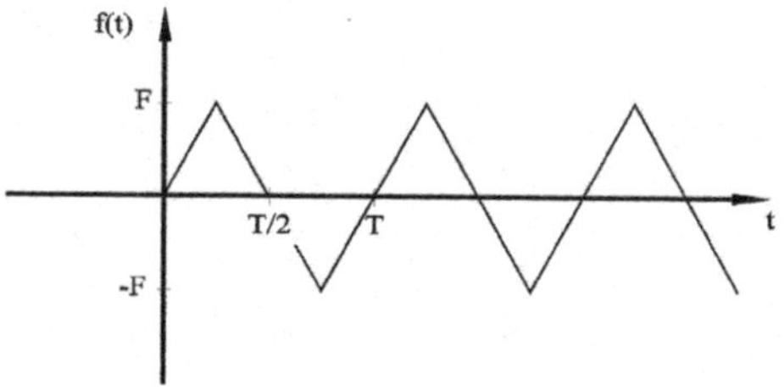

Figure P1.4

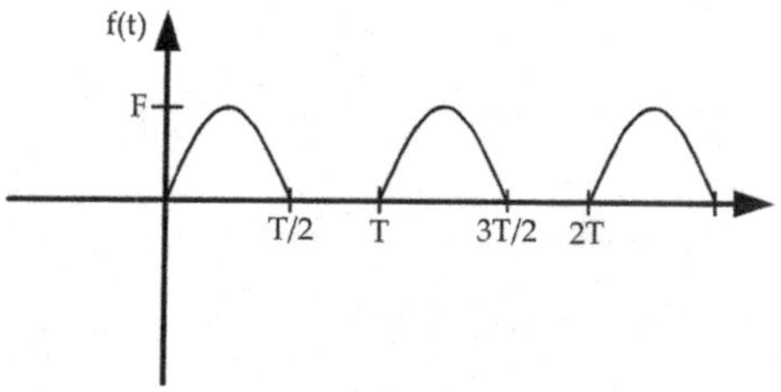

Figure P1.5

7. Determine the Fourier series equation of the wave form shown in Figure P1.2.

8. Determine the Fourier series equation of the wave form shown in Figure P1.3.

9. Determine the Fourier series equation of the wave form shown in Figure P1.4

10. Determine the Fourier series equation of the wave form shown in Figure P1.5. The wave form in Figure P1.5 is given by:

$$f(t) = F\sin(2\pi t) \quad \text{for} \quad 0 \le t \le T/2$$
$$f(t) = 0 \quad \text{for} \quad T/2 < t \le T.$$

11. The origin of the square wave form in Figure P1.1 is shifted by a value of t = -T/4 so that the equation of the wave form over one period is given by:

$$f(t) = F_o \quad \text{for} \quad 0 \le t \le T/4$$
$$f(t) = -F_o \quad \text{for} \quad T/4 < t \le 3T/4$$
$$f(t) = F_o \quad \text{for} \quad 3T/4 < t \le T.$$

Determine the Fourier series equation of the wave form. How does this equation differ relative to the equation obtained for Problem 6?

12. A complex periodic signal is shown in Figure P1.6. The signal has a period of 1 s. The values of the signal are shown in Table P1. Determine the values forf the first five terms of the Fourier series. Write the Fourier series in a forms similar to equation 1.3a in EXAMPLE 3.1..

13. Find the equivalent mass moment of inertia of the system shown in Figure P1.7 relative to the spatial variable θ.

14. Find the equivalent mass of the system shown in Figure P1.7 relative to the spatial variable x_1.

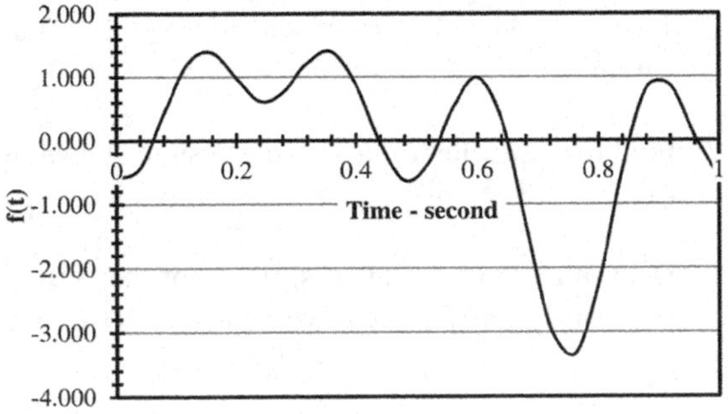

Figure P1.6

15. Find the equivalent mass of the system shown in Figure P1.7 relative to the spatial variable x_2.

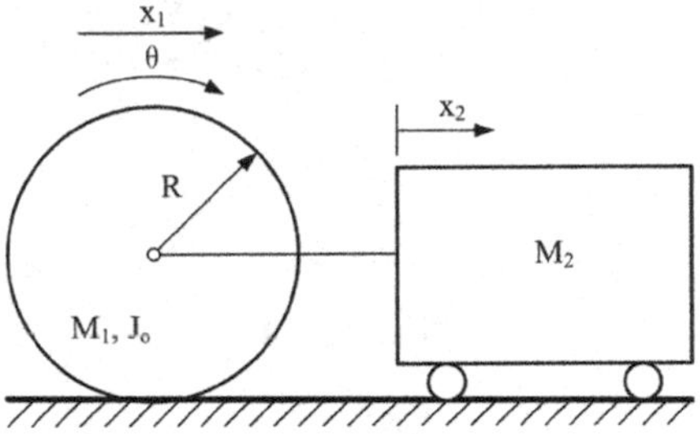

Figure P1.7

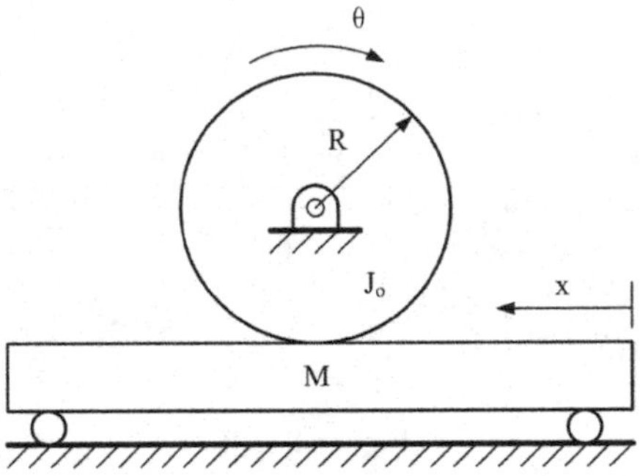

Figure P1.8

Table P1

n	0	1	2	3	4	5	6	7	8	9	10	11	12	13	14	15	16	17
Time - sec.	0	0.02	0.04	0.06	0.08	0.1	0.12	0.14	0.16	0.18	0.2	0.22	0.24	0.26	0.28	0.3	0.32	0.34
f(t)	-0.600	-0.633	-0.435	-0.053	0.420	0.877	1.222	1.394	1.380	1.217	0.979	0.753	0.618	0.618	0.753	0.979	1.217	1.380
n	18	19	20	21	22	23	24	25	26	27	28	29	30	31	32	33	34	35
Time - sec.	0.36	0.38	0.4	0.42	0.44	0.46	0.48	0.5	0.52	0.54	0.56	0.58	0.6	0.62	0.64	0.66	0.68	0.7
f(t)	1.394	1.222	0.877	0.420	-0.053	-0.434	-0.633	-0.600	-0.345	0.064	0.511	0.860	0.988	0.812	0.316	-0.445	-1.352	-2.244
n	36	37	38	39	40	41	42	43	44	45	46	47	48	49	50			
Time - sec.	0.72	0.74	0.76	0.78	0.8	0.82	0.84	0.86	0.88	0.9	0.92	0.94	0.96	0.98	1			
f(t)	-2.955	-3.349	-3.349	-2.956	-2.245	-1.353	-0.447	0.315	0.812	0.988	0.861	0.511	0.064	-0.344	-0.600			

16. Find the equivalent mass moment of inertia of the system shown in Figure P1.8 relative to the spatial variable θ.

17. Find the equivalent mass of the system shown in Figure P1.8 relative to the spatial variable x.

18. Determine the equivalent stiffness of the system shown in Figure P1.9 relative to the spatial variable y.

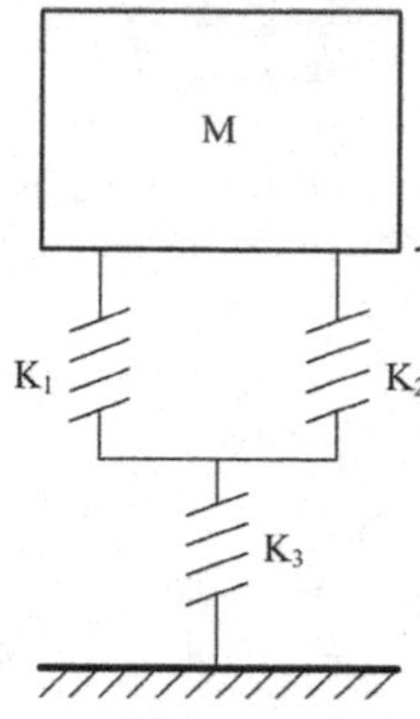

Figure P1.9

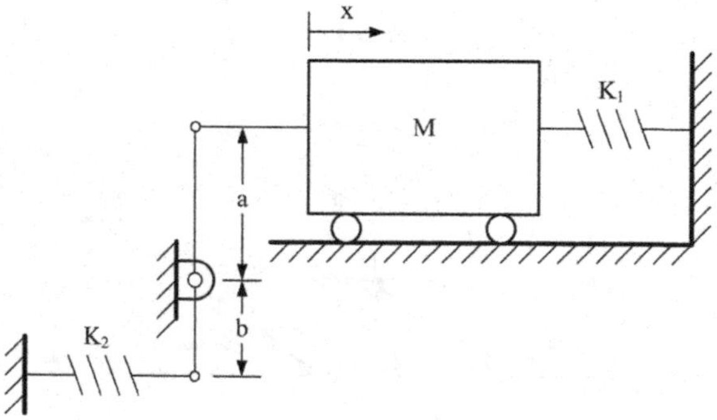

Figure P1.10

19. Determine the equivalent stiffness of the system shown in Figure P1.10 relative to the spatial variable x.

20. Determine the equivalent stiffness of the system shown in Figure P1.11 relative to the spatial variable θ.

21. The system in Figure P1.12 is a cantilever beam with a mass on a cable suspended from the end of the beam. Determine the equivalent stiffness of the system in terms of the spatial variable y.

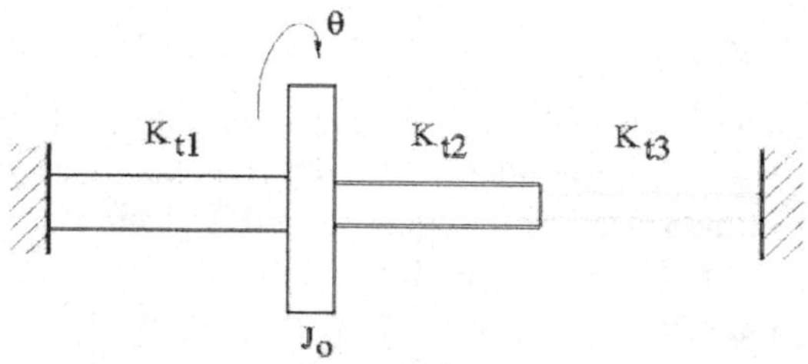

Figure P1.11

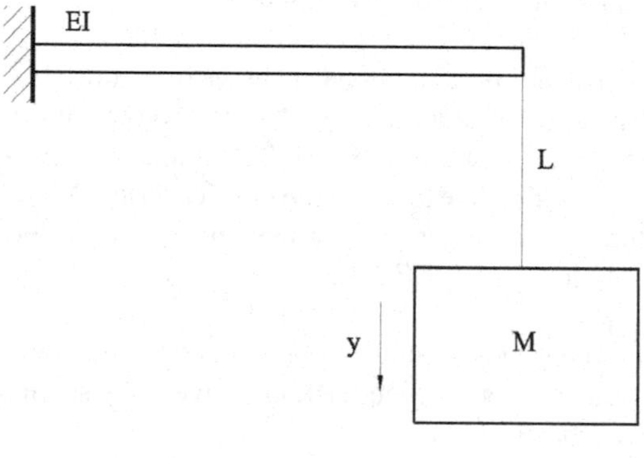

Figure P1.12

CHAPTER 2
1-DOF VIBRATION SYSTEMS UNFORCED VIBRATION

2.1 Introduction

Many physical or mechanical systems possess characteristics that support vibration motion similar to that discussed in Chapter 1. These systems usually consist of one or more mass elements that are connected together with spring and damping elements. Some of these systems may be very complicated and difficult to model. However, they often can be approximated as systems that consist of mass, spring and damping elements.

2.2 Unforced Vibration with No Damping

A simplest vibration system that consists of a mass element M suspended from a spring element K is schematically represented in Figure 2.1. The system is constrained to move only in the y direction. The positive direction of motion for mass M is shown to be downward. Newton's second law is used to develop the equation of motion.

When mass M is attached to the unstretched spring element K in Figure 2.2(a), the spring deflects a distance δ downward in the positive y direction. This causes the spring to exert a static force on M that is equal to the weight of M and that acts in the negative y direction. Figure 2.3(a) shows a free-body diagram associated with the static deflection of M. Summing the static forces yields:

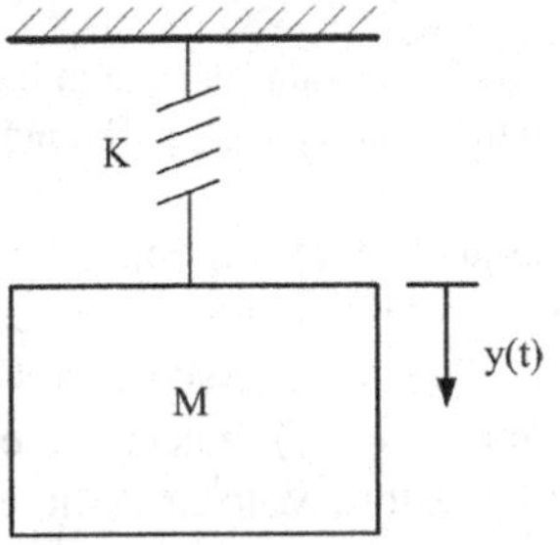

Figure 2.1 1-DOF Vibration System with Mass and Spring Elements

$$\sum F = 0 = Mg - K\delta \tag{2.1}$$

or:

$$Mg = K\delta\,. \tag{2.2}$$

The weight of M equals the stiffness coefficient of spring K times its static deflection δ.

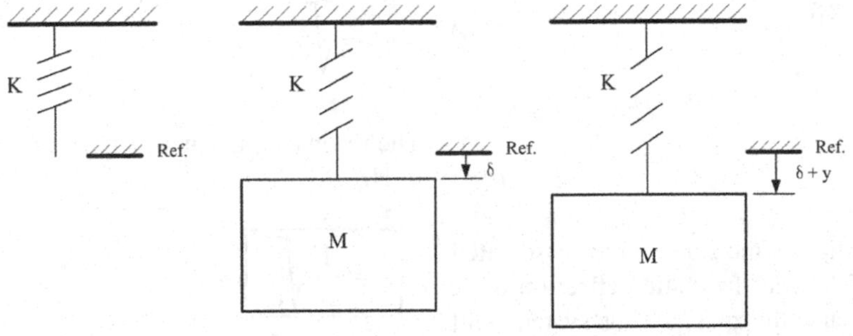

Figure 2.2 1-DOF Vibration System with Mass and Spring Elements: (a) Unstreched Spring, (b) Spring Displaced by Static Weight of Mass M, and (c) Spring Displaced by Static Weight and Dynamic Motion of Mass M

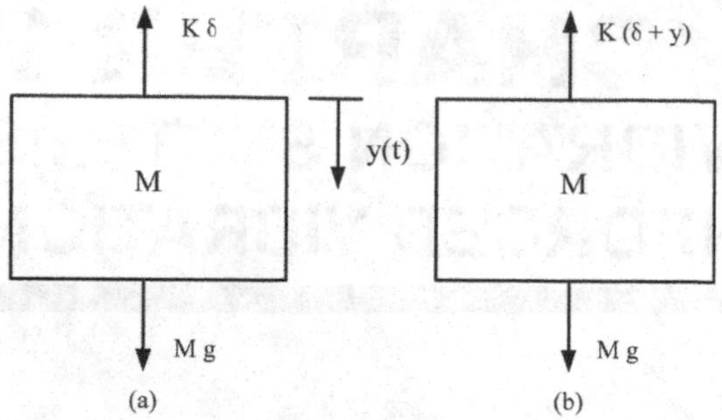

Figure 2.3 Free Body Diagrams of System in Figure 2.2: (a) Static Equation and (b) Dynamic Equation

The dynamic equation of motion is obtained by showing M with an additional downward displacement y [Figure 2.2(c)]. The force associated with K that acts on M is equal to K(δ + y). It acts in the negative y direction to pull or restore M to its static equilibrium position. Therefore, *the spring force associated with K and, subsequently, all spring forces that act on system masses are often called restoring forces and are shown acting in a direction opposite the assigned positive directions of motion of the masses.*

The mass-spring system is no longer in static equilibrium. The restoring force associated with K is greater than the force associated with the weight of M. Therefore, when the mass is released, the spring will pull it back toward its static equilibrium position. For this case, the sum of the forces acting on M is given by [Figure 2.3(b)]:

$$\sum F = M\ddot{y} \quad \text{or} \quad M\ddot{y} = Mg - K\left(Y + \delta\right). \tag{2.3}$$

Rearranging the terms in equation (2.3) yields:

$$M\ddot{y} + Ky = Mg - K\delta. \tag{2.4}$$

Since Mg = Kδ, equation (2.4) becomes:

$$M\ddot{y} + Ky = 0. \tag{2.5}$$

Equation (2.5) indicates the static forces associated with the weight of M and the static deflection of K cancel each other, and the mass-spring system will oscillate about its static equilibrium position. When analyzing lumped-parameter vibration systems, static force terms within the system normally cancel each other. Therefore, *the reference positions associated with the motions of mass elements in vibration systems are their static equilibrium positions, and only the dynamic terms associated with system spring and damping elements are considered when developing the system equations of motion.*

To find the solution to equation (2.5), first rewrite it in the form:

$$\boxed{\ddot{y} + \omega_n^2 y = 0} \tag{2.6}$$

where:

$$\boxed{\omega_n = \sqrt{\frac{K}{M}}} \tag{2.7}$$

is called the *resonance frequency* of the system and has units of rad/s. ω_n is the frequency at which the system will naturally oscillate after it has been disturbed from its static equilibrium position by an initial displacement and/or velocity.

Equation (2.6) is the general form of the equation of motion for all 1-DOF vibration systems without damping that are not acted upon by an external or internal time-varying force. Systems with more complicated geometries associated with their mass and spring elements will have more complicated expressions for M and K in equations (2.5) and (2.7). However, these expressions will resolve to single numeric values represented by M and K in equations (2.5) and (2.7).

The resonance frequency ω_n can also be written in terms of the static deflection of the mass element. Substituting equation (2.2) into equation (2.7) yields:

$$\omega_n = \sqrt{\frac{g}{\delta}}. \tag{2.8}$$

The resonance frequency f_n can be written with units of Hz:

$$\boxed{f_n = \frac{1}{2\pi}\sqrt{\frac{g}{\delta}}.} \tag{2.9}$$

Equation (2.9) is an important equation. In some applications where a vibrating system can be treated

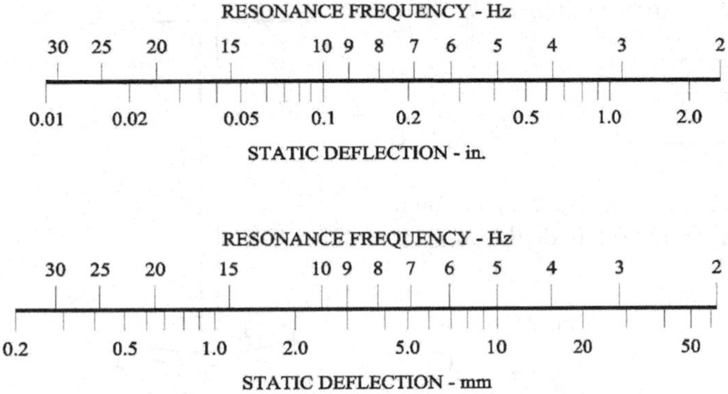

Figure 2.4 Relationship between Static Deflection δ and Resonance Frequency f_n

as a single mass-spring system, it may not be possible to determine the value of the mass or the stiffness of the spring, but it may be possible to measure the static deflection of the system elastic element. When this is the case, the resonance frequency of the system can be obtained from equation (2.9). Figure 2.4 is a nomogram for equation (2.9).

The solution to equation (2.6) can be obtained by first assuming a solution of the form:

$$y(t) = Y e^{st}. \tag{2.10}$$

Substituting equation (2.10) into equation (2.6) and carrying out the prescribed operations yields:

$$\left(s^2 + \omega_n^2\right) Y e^{st} = 0 \quad \text{or} \quad s^2 + \omega_n^2 = 0. \tag{2.11}$$

Therefore, s is given by:

$$s = \pm j\omega_n. \tag{2.12}$$

Substituting the expressions for s into equation (2.10) yields:

$$y(t) = Y_1 e^{j\omega_n t} + Y_2 e^{-j\omega_n t}. \tag{2.13}$$

Using Euler's equation for the complex exponential terms in equation (2.13) and rearranging the terms gives:

$$y(t) = (Y_1 + Y_2)\cos\omega_n t + j(Y_1 - Y_2)\sin\omega_n t. \tag{2.14}$$

The response of the 1-DOF vibration system in Figure 2.1 is real-valued. For this to occur, Y_1 and Y_2 must be complex conjugates where:

$$Y_1 = a + jb \quad \text{and} \quad Y_2 = a - jb. \tag{2.15}$$

Therefore, y(t) becomes:

$$y(t) = 2a\cos\omega_n t - 2b\sin\omega_n t. \tag{2.16}$$

Since 2a and 2b are arbitrary constants that are determined by the initial conditions associated with the system described by equation (2.6), y(t) can be written::

$$y(t) = A\cos\omega_n t + B\sin\omega_n t. \tag{2.17}$$

A and B are evaluated by applying the initial conditions: $y(0) = y_o$ and $\dot{y}(0) = v_o$. Taking the derivative of equation (2.17) with respect to t yields:

$$\dot{y}(t) = -\omega_n A\sin\omega_n t + \omega_n B\cos\omega_n t. \tag{2.18}$$

Applying the initial conditions to equations (2.17) and (2.18) gives:

$$A = y_o \tag{2.19}$$

and:

$$B = \frac{v_o}{\omega_n}. \tag{2.20}$$

Therefore, equation (2.17) can be written:

$$\boxed{y(t) = y_o \cos\omega_n t + \frac{v_o}{\omega_n}\sin\omega_n t\,.} \tag{2.21}$$

Sometimes, it is desirable to write equation (2.21) in terms of one trigonometric function. To do this, first let:

$$y_o = Y\cos\phi \qquad \text{and} \qquad \frac{v_o}{\omega_n} = -Y\sin\phi\,. \tag{2.22}$$

Substituting these expressions into equation (2.21) yields:

$$y(t) = Y(\cos\phi\cos\omega_n t - \sin\phi\sin\omega_n t) \tag{2.23}$$

which reduces to:

$$\boxed{y(t) = Y\cos(\omega_n t + \phi).} \tag{2.24}$$

Y is obtained by squaring both sides of the expressions in equation (2.22) and then adding them to obtain:

$$Y^2(\sin^2\phi + \cos^2\phi) = \left(\frac{v_o}{\omega_n}\right)^2 + y_o^2. \tag{2.25}$$

Noting $\sin^2\phi + \cos^2\phi = 1$,Y becomes:

$$\boxed{Y = \sqrt{y_o^2 + \left(\frac{v_o}{\omega_n}\right)^2}\,.} \tag{2.26}$$

ϕ is found by dividing the expressions in equation (2.22) and noting $\tan^{-1}(-\phi) = -\tan^{-1}(\phi)$ to obtain:

$$\boxed{\tan\phi = -\frac{v_o}{y_o\,\omega_n} \quad \text{or} \quad \phi = -\tan^{-1}\left(\frac{v_o}{y_o\,\omega_n}\right).} \tag{2.27}$$

EXAMPLE 2.1

A 1-DOF vibration system similar to the one shown in Figure 2.1 has the following characteristics: We = 40 lb_f

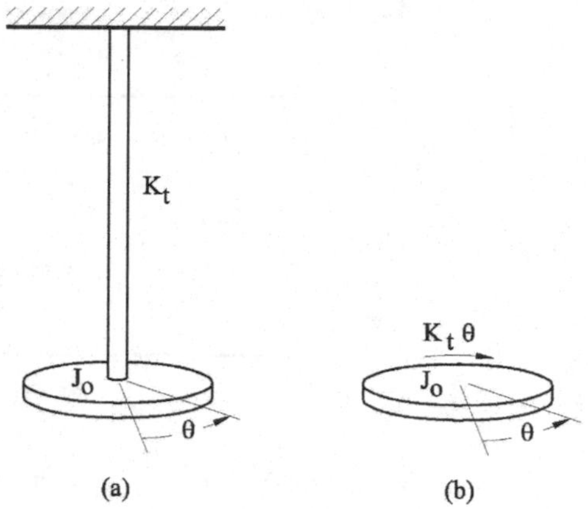

Figure 2.5 1-DOF Torsional Pendulum

and K = 50 lb_f/in. Determine the resonance frequency of the system and determine the response of the system if it has an initial displacement of 0.15 in. and an initial velocity of 2 in./s.

SOLUTION

To determine the resonance frequency, the mass must be converted to the proper units. From equation (1.196):

$$M = \frac{40}{386} = 0.1035\,\frac{lb_f - s^2}{in.}\,. \tag{2.1a}$$

From equation (2.7):

$$\omega_n = \sqrt{\frac{50}{0.1035}} = 21.98\,\frac{rad}{s} \tag{2.1b}$$

and:

$$f_n = \frac{21.98}{2\pi} = 3.50\ \text{Hz}\,. \tag{2.1c}$$

From equation (2.26):

$$Y = \sqrt{0.15^2 + \left(\frac{2}{21.98}\right)^2} = 0.175\ \text{in.} \tag{2.1d}$$

From equation (2.27):

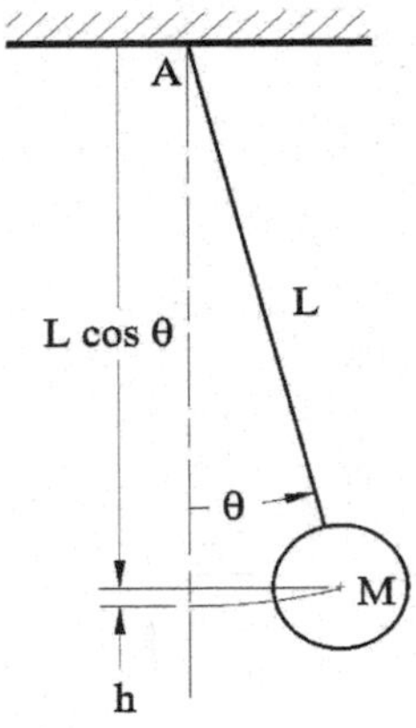

Figure 2.6 1-DOF Pendulum

$$\phi = \tan\left(-\frac{2}{0.15\times 21.98}\right) = -0.545 \text{ rad}. \tag{2.1e}$$

Therefore, the response of the system is:

$$\boxed{y(t) = 0.175\cos(21.98\,t - 0.545).} \tag{2.1f}$$

EXAMPLE 2.2

Use Newton's second law to derive the equation of motion of the torsional pendulum in Figure 2.5(a). Write the expression for the resonance frequency.

SOLUTION

This system has only rotational motion. Therefore, the equation of motion is obtained by summing the moments acting on the pendulum mass using equation (1.166). A free-body diagram of the torsional pendulum is shown in Figure 2.5(b). If the pendulum is given a positive rotation in the counterclockwise direction, the torsional spring K_t will tend to rotate the pendulum in the clockwise direction back to its equilibrium position. Therefore, the restoring moment $K_t\theta$ is shown acting in the negative θ direction in Figure 2.5(b). Summing the moments in Figure 2.5(b) yields:

$$\sum M_{c.g} = J_o\ddot{\theta} \quad \text{or} \quad J_o\ddot{\theta} = -K_t\,\theta. \tag{2.2a}$$

Therefore:

$$\boxed{J_o\ddot{\theta} + K_t\theta = 0.} \tag{2.2b}$$

The resonance frequency is given by:

$$\boxed{\omega_n = \sqrt{\frac{K_t}{J_o}}.} \tag{2.2c}$$

EXAMPLE 2.3

Use the energy method to derive the equation of motion of the pendulum shown in Figure 2.6. Write the expression for the resonance frequency.

SOLUTION

It is first necessary to determine the kinetic and potential energies associated with the pendulum. The pendulum has only rotational motion about point A. Therefore, the kinetic energy of the pendulum is equal to:

$$T = \frac{1}{2}J_A\,\dot{\theta}^2 \tag{2.3a}$$

where, using the parallel axis theorem, $J_A = J_o + ML^2$. J_o is the mass moment of inertia of the pendulum mass about its center of gravity. For pendulums, $J_o << ML^2$. Therefore, the kinetic energy becomes:

$$T = \frac{1}{2}ML^2\,\dot{\theta}^2. \tag{2.3b}$$

In this problem the change in potential energy is associated with a change in height h of the pendulum as it moves an angle θ. From Figure 2.6:

$$U = Mgh. \tag{2.3c}$$

U is positive since a positive motion of θ results in an increase in potential energy associated with an increase in the height of the pendulum mass. h is given by:

$$h = L - L\cos\theta \qquad \text{or} \qquad h = L(1-\cos\theta). \tag{2.3d}$$

Substituting equation (2.2d) into equation (2.2c) yields:

$$U = MgL(1-\cos\theta). \tag{2.3e}$$

The sum of the potential plus kinetic energies is:

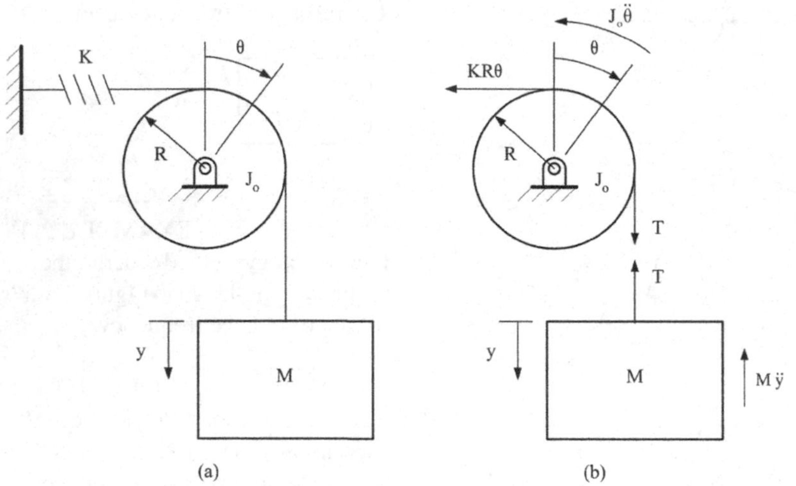

Figure 2.7 1-DOF Vibration System with a Spring Element and Rotational and Translational Mass Elements

$$T + U = \frac{1}{2} ML^2 \dot{\theta}^2 + MgL\left(1 - \cos\theta\right). \tag{2.3f}$$

Taking the derivative of equation (2.3f) with respect to time t and setting the results equal to zero yields:

$$ML^2 \dot{\theta}\ddot{\theta} + MgL \sin\theta\, \dot{\theta} = 0 \tag{2.3g}$$

or:

$$ML^2 \ddot{\theta} + MgL \sin\theta = 0 \tag{2.3h}$$

When θ is assumed to be very small, sin θ ≈ θ, and the equation (2.3h) can be written:

$$\boxed{ML^2 \ddot{\theta} + MgL\,\theta = 0 \qquad \text{or} \qquad \ddot{\theta} + \frac{g}{L}\theta = 0.} \tag{2.3i}$$

The expression for the resonance frequency is:

$$\boxed{\omega_n = \sqrt{\frac{g}{L}}.} \tag{2.3j}$$

EXAMPLE 2.4

Use d'Alembert's principle to determine the equation of motion of the 1-DOF vibration system shown in Figure 2.7(a). Write the equation in terms of the rotational variable θ. Write the expression for the resonance frequency.

SOLUTION

Figure 2.7(b) shows the free-body diagrams of the forces and moments acting on the pulley and the forces acting on the mass. The pulley is pinned at its center and only has rotational motion. The mass is constrained to move only in the vertical direction. The spatial variables θ and y are needed to describe the motion of the system. However, there is one constraint equation:

$$y = R\theta. \tag{2.4a}$$

Therefore, the system is a 1-DOF system. Figure 2.7(b) shows the free-body diagrams of the pulley and mass. The rotation of the pulley is specified to be positive in the clockwise direction. Therefore, the motion of the mass is specified to be positive in the downward direction. Shown acting on the pulley are the restoring spring force KRθ, the negative inertia moment $J_o\ddot{\theta}$, and the tension force T. Shown acting on the mass are the negative inertia force $M\ddot{y}$ and the tension force T. The cable from the pulley to the mass is a two-force member. Therefore, the tension forces on the pulley and the mass are shown acting in opposite directions.

Summing the moments on the pulley relative to its center of rotation and setting the results equal to zero yields:

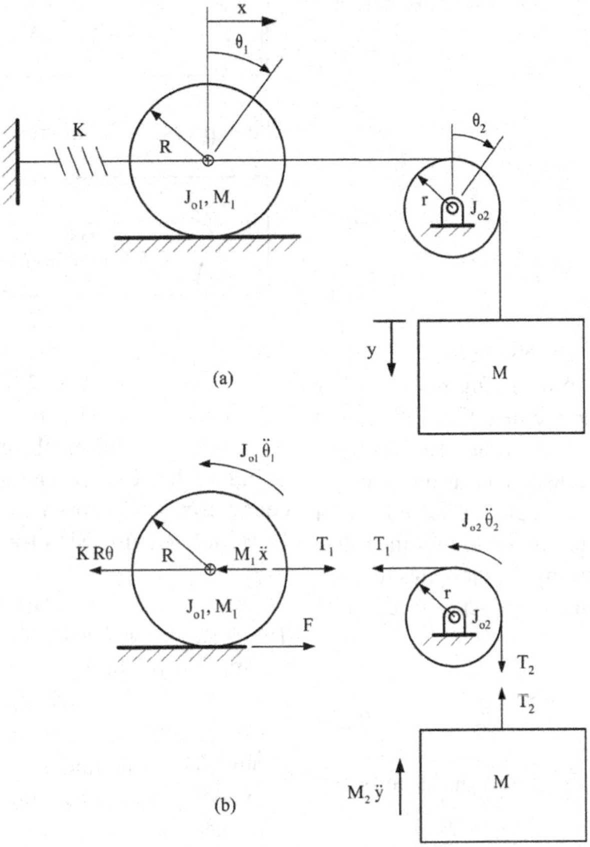

Figure 2.8 1-DOF Vibration System with a Spring Element and Rolling, Rotational and Translational Mass Elements

$$-J_o\,\ddot{\theta} - KR^2\,\theta + R\,T = 0\,. \tag{2.4b}$$

Summing the forces acting in the vertical direction on the mass and setting the results equal to zero gives:

$$-T - M\,\ddot{y} = 0\,. \tag{2.4c}$$

Substituting equations (2.4a) and (2.4c) into equation (2.4b) and rearranging the terms yields:

$$\boxed{\left(J_o + MR^2\right)\ddot{\theta} + KR^2\,\theta = 0\,.} \tag{2.4d}$$

The expression for the resonance frequency is:

$$\boxed{\omega_n = \sqrt{\frac{KR^2}{J_o + MR^2}}\,.} \tag{2.4e}$$

EXAMPLE 2.5

Use d'Alembert's principle to develop the equation of motion of the system shown in Figure 2.8(a). Develop the equation in terms of the rotation variable of the roller θ_1. Assume there is no slipping at the point of contact between the roller and the flat surface on which it rolls. Write the expression for the resonance frequency.

SOLUTION

The roller has both translational and rotational motion; the pulley only has rotational motion; and the mass only has translational motion. Figure 2.8(a) indicates that four spatial variables x, θ_1, θ_2 and y are needed to describe the motion of the system. However, there are three constraint equations:

$$x = R\theta_1 \tag{2.5a}$$

$$R\theta_1 = r\theta_2 \qquad \text{or} \qquad \theta_2 = \frac{R}{r}\theta_1 \tag{2.5b}$$

$$y = R\theta_1 \qquad \text{and} \qquad y = r\theta_2 \,. \tag{2.5c}$$

Therefore, the system is a 1-DOF system.

The inertia moment and force acting on the roller are shown acting in the negative directions about and through the center of mass of the roller, respectively. The roller has an instantaneous center at the point of contact A between the roller and the surface on which it rolls. Therefore, the equation of motion for the roller can best be obtained by summing the moments relative to point A. Summing the moments and setting the results equal to zero yields:

$$-J_{o1}\,\ddot{\theta}_1 - M_1 R\,\ddot{x} - KR^2\,\theta_1 + T_1 R = 0 \,. \tag{2.5d}$$

Summing the moments on the pulley and setting the results equal to zero gives:

$$-rT_1 - J_{o2}\,\ddot{\theta}_2 + rT_2 = 0 \,. \tag{2.5e}$$

Summing the forces on the mass in the y direction and setting the results equal to zero yields:

$$-T_2 - M_2\,\ddot{y} = 0 \,. \tag{2.5f}$$

Substituting equations (2.5b), (2.5c) and (2.5f) into equation (2.5e) and solving for T_1 gives:

$$T_1 = -J_{o2}\frac{R}{r^2}\,\ddot{\theta}_1 - M_2 R\,\theta_1 \,. \tag{2.5g}$$

Substituting equations (2.5a) and (2.5g) into equation (2.5d) and rearranging the terms yields:

$$\left[J_{o1} + J_{o2}\frac{R^2}{r^2} + \left(M_1 + M_2\right)R^2\right]\ddot{\theta}_1 + KR^2\,\theta_1 = 0 \,. \tag{2.5h}$$

The expression for the resonance frequency is:

$$\omega_n = \sqrt{\frac{KR^2}{J_{o1} + J_{o2}\dfrac{R^2}{r^2} + \left(M_1 + M_2\right)R^2}} \,. \tag{2.5i}$$

EXAMPLE 2.6

A cylinder of mass M and of radius r rolls without slipping on a concave cylindrical surface of radius R, as shown in Figure 2.9. Use the energy method to determine the equation of motion for small oscillations θ about the point C. Write the expression for the resonance frequency.

SOLUTION

Two spatial variables θ and ϕ are needed to describe the motion of the cylinder. There are two constraint equations for this system. One relates θ and ϕ, and the other relates $\dot{\theta}$ and $\dot{\phi}$. This system is a 1-DOF system. The constraint equation that relates θ and ϕ is obtained by noting that the arc C'B must equal the arc CB. Therefore,

$$R\theta = r\phi \qquad \text{or} \qquad \phi = \frac{R}{r}\theta \,. \tag{2.6a}$$

The constraint equation that relates $\dot{\theta}$ and $\dot{\phi}$ is obtained by noting there are two instantaneous centers in this system: one at point A and the other at point B. The translational velocity v at the center of the cylinder associated with rotation about point A must equal the translational velocity v associated with rotation about point B. Therefore,

$$v = r\dot{\phi} = \left(R - r\right)\dot{\theta} \qquad \text{or} \qquad \dot{\phi} = \frac{R - r}{r}\,\dot{\theta} \,. \tag{2.6b}$$

The potential energy of the system is associated with the increase in height h of the system as it rolls an angle θ. Therefore:

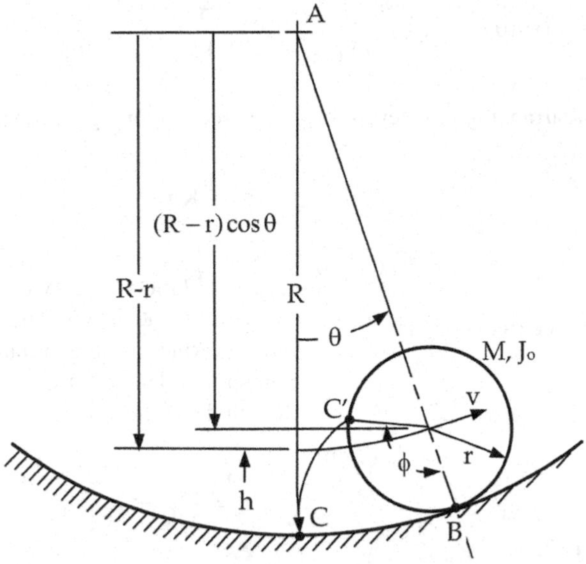

Figure 2.9 1-DOF Vibration System Consisting of a Cylindrical Roller Rolling on a Concave Cylindrical Surface

$$U = mgh \tag{2.6c}$$

where:

$$h = (R - r)(1 - \cos\theta). \tag{2.6d}$$

Substituting equation (2.6d) into equation (2.6c) yields:

$$U = Mg(R - r)(1 - \cos\theta). \tag{2.6e}$$

Because the cylinder rotates about point B, it has both translational and rotational kinetic energy. The translational kinetic energy can be written:

$$T_{trans} = \frac{1}{2} M v^2 \quad \text{or} \quad T_{trans} = \frac{1}{2} M (R - r)^2 \dot{\theta}^2. \tag{2.6f}$$

The rotational kinetic energy relative to point B is:

$$T_{rot} = \frac{1}{2} J_o \dot{\phi}^2. \tag{2.6g}$$

The mass moment of inertia J_o of a cylinder relative to its center of gravity is given by:

$$J_o = \frac{1}{2} M r^2. \tag{2.6h}$$

Substituting equations (2.6b) and (2.6h) into equation (2.6g) yields:

$$T_{rot} = \frac{1}{4} M (R - r)^2 \dot{\theta}^2. \tag{2.6i}$$

The total kinetic energy is obtained by adding equations (2.6f) and (2.6i), or:

$$T = \frac{3}{4} M (R - r)^2 \dot{\theta}^2. \tag{2.6j}$$

The total energy of the system is obtained by adding equations (2.6e) and (2.6j), or:

$$T + U = \frac{3}{4} M (R - r)^2 \dot{\theta}^2 + Mg(R - r)(1 - \cos\theta). \tag{2.6k}$$

Taking the derivative of equation (2.6k) with respect to time yields:

$$\frac{3}{2}M(R-r)^2\,\dot{\theta}\,\ddot{\theta}+Mg(R-r)\sin\theta\,\dot{\theta}=0\,. \tag{2.6l}$$

Noting that $\sin\,\theta \approx \theta$ and rearranging the terms of equation (2.6l) gives:

$$\boxed{\ddot{\theta}+\frac{2g}{3(R-r)}\,\theta=0\,.} \tag{2.6m}$$

The expression for the resonance frequency is:

$$\boxed{\omega_n=\sqrt{\frac{2g}{3(R-r)}}\,.} \tag{2.6n}$$

EXAMPLE 2.7

In most mass-spring vibration systems, it is assumed that the value of the mass of the spring is much less than the value of the mass which is supported by the spring. For most cases, this assumption is valid. However, there may be some systems for which this assumption is not true. Use the energy method to develop the equation of a mass-spring vibration system for which the mass of the spring cannot be neglected. Write the expression for the resonance frequency.

SOLUTION

Figure 2.10 schematically shows a mass-spring vibration system in which the mass of the spring is not negligible. For this case, it is assumed that the deflection of the spring along its length L is linear and that the displacement y_s at any point along the length of the spring is given by:

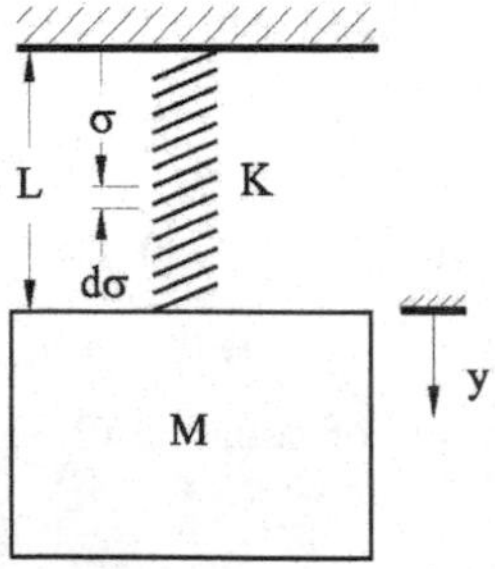

Figure 2.10 Equivalent Mass of a Spring

$$y_s=\frac{\sigma}{L}\,y\,. \tag{2.7a}$$

The potential energy associated with the system is:

$$U=\frac{1}{2}K\,y^2\,. \tag{2.7b}$$

The total kinetic energy associated with the system is the sum of the kinetic energy relative to the mass M plus the kinetic energy relative to the effective mass of the spring. The kinetic energy dT_s of an incremental length $d\sigma$ of the spring is:

$$dT_s=\frac{1}{2}(\rho\,d\sigma)\left(\frac{\sigma}{L}\,\dot{y}\right)^2 \tag{2.7c}$$

where ρ is the mass per unit length of the spring. The total kinetic energy associated with the spring is obtained by integrating equation (2.7e) over the length of the spring. Therefore:

$$T_s=\frac{1}{2}\rho\left(\frac{\dot{y}}{L}\right)^2\int_0^L \sigma^2\,d\sigma \tag{2.7d}$$

or:

$$T_s=\frac{1}{6}\rho L\,\dot{y}^2\,. \tag{2.7e}$$

The total kinetic energy of the system is:

$$T=\frac{1}{2}M\,\dot{y}^2+\frac{1}{6}\rho L\,\dot{y}^2 \tag{2.7f}$$

or:

$$T=\frac{1}{2}\left(M+\frac{\rho L}{3}\right)\dot{y}^2\,. \tag{2.7g}$$

The total energy in the system is:

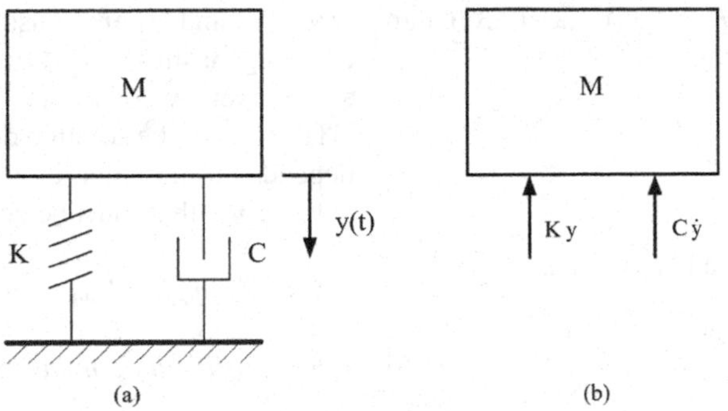

Figure 2.11 1-DOF Vibration System with Viscous Damping

$$T + U = \frac{1}{2}\left(M + \frac{\rho L}{3}\right)\dot{y}^2 + \frac{1}{2}K\,y^2 . \tag{2.7h}$$

Taking the derivative of equation (2.7h) with respect to time t and setting the results equal to zero yields:

$$\boxed{\left(M + \frac{\rho L}{3}\right)\ddot{y}^2 + K\,y = 0 .} \tag{2.7i}$$

The expression for the resonance frequency is:

$$\boxed{\omega_n = \sqrt{\frac{K}{M + \frac{\rho L}{3}}} .} \tag{2.7j}$$

2.3 Unforced Vibration with Viscous Damping

Most mechanical systems have some form of damping present. Damping dissipates energy in a vibration system until, in the absence of an external or internal force to replace it, the vibration will stop.

Figure 2.11(a) shows a schematic representation of a system with mass, damping and spring elements. Damping can be present in many different forms. Viscous damping is initially chosen because it results in a linear differential equation of motion.

Figure 2.11(b) shows a free-body diagram of the system. The y direction is shown to be positive downward. The force associated with spring element K is shown as a *restoring force* acting in the negative y direction. The damping element C acts to resist the motion of M. Therefore, the damping force is shown as a *restoring force* acting in the negative direction. Summing the forces acting on M yields:

$$\sum F = M\ddot{y} \quad \text{or} \quad M\ddot{y} = -C\dot{y} - Ky. \tag{2.28}$$

Therefore:

$$M\ddot{y} + C\dot{y} + Ky = 0 . \tag{2.29}$$

Before solving equation (2.29), it is desirable to rewrite it in a different form. First, divide the equation by M to get:

$$\ddot{y} + \frac{C}{M}\dot{y} + \frac{K}{M}y = 0 . \tag{2.30}$$

Equation (2.7) indicates $K/M = \omega_n^2$. Define the *critical damping coefficient* C_c as:

$$\boxed{C_c = 2\sqrt{KM}} \tag{2.31}$$

and the *damping ratio* ξ as:

$$\boxed{\xi = \frac{C}{C_c} .} \tag{2.32}$$

C/M can be written:

$$\frac{C}{M} = \frac{2C}{2\sqrt{KM}}\sqrt{\frac{K}{M}} . \tag{2.33}$$

Substituting equations (2.7), (2.31) and (2.32) into equation (2.33) yields:

$$\frac{C}{M} = 2\,\xi\,\omega_n\,. \tag{2.34}$$

Therefore, equation (2.29) can be written:

$$\boxed{\ddot{y} + 2\,\xi\,\omega_n\,\dot{y} + \omega_n^2\,y = 0\,.} \tag{2.35}$$

Equation (2.35) is the general form of the equation of motion for all 1-DOF vibration systems with mass, spring and viscous damping elements that are not acted upon by an external or internal force. Systems with more complicated geometries associated with their mass, spring and damping elements will have more complicated expressions for M, C and K. However, these complicated expressions will resolve to single numeric values represented by M, C and K in equations (2.7), (2.29), (2.31), (2.32) and (2.34).

The solution to equation (2.35) is obtained by substituting equation (2.10) for y(t) into it and carrying out the prescribed operations to obtain:

$$\left(s^2 + 2\,\xi\,\omega_n\,s + \omega_n^2\right) Ye^{st} = 0 \tag{2.36}$$

or:

$$s^2 + 2\,\xi\,\omega_n\,s + \omega_n^2 = 0\,. \tag{2.37}$$

The quadratic equation can be used to solve for s_1 and s_2 in equation (2.37). Therefore:

$$s_{1,2} = -\xi\omega_n \pm \sqrt{(\xi\omega_n)^2 - \omega_n^2} \tag{2.38}$$

or:

$$s_{1,2} = \left(-\xi \pm \sqrt{\xi^2 - 1}\right)\omega_n\,. \tag{2.39}$$

The response of a system with damped free oscillation can be written:

$$y(t) = Y_1\,e^{s_1 t} + Y_2\,e^{s_2 t}\,. \tag{2.40}$$

where Y_1 and Y_2 are constants that are determined by applying the initial conditions of the system and s_1 and s_2 are given by equations (2.38) and (2.39).

The form of the solution described by equation (2.40) depends on the value of ξ. There are three ranges of values for ξ that must be considered:

- $\xi < 1$, *damping less than critical damping*;
- $\xi = 1$, *critical damping*; and
- $\xi > 1$, *damping greater than critical damping*.

Damping less than critical - $\xi < 1$: For this case the roots of equation (2.39) are complex with a real and imaginary part, or:

$$s_{1,2} = \left(-\xi \pm j\sqrt{1-\xi^2}\right)\omega_n\,. \tag{2.41}$$

Therefore, the response of the system is given by:

$$y(t) = Y_1\,e^{\left(-\xi + j\sqrt{1-\xi^2}\right)\omega_n t} + Y_2\,e^{\left(-\xi - j\sqrt{1-\xi^2}\right)\omega_n t} \tag{2.42}$$

or:

$$y(t) = \left(Y_1\,e^{j\sqrt{1-\xi^2}\,\omega_n t} + Y_2\,e^{-j\sqrt{1-\xi^2}\,\omega_n t}\right)e^{-\xi\omega_n t}\,. \tag{2.43}$$

Equation (2.43) indicates there are two parts to the solution. The complex exponential terms in the brackets indicate the system is oscillatory. The negative exponential term outside of the brackets indicates the amplitudes of oscillation decrease with increasing time. Note also that the frequency of oscillation is not ω_n. The presence of viscous damping alters the frequency at which the system vibrates. For systems with viscous damping in which the damping ratio is less than one, the *damped resonance frequency* ω_d is:

$$\boxed{\omega_d = \sqrt{1-\xi^2}\,\omega_n\,.} \tag{2.44}$$

Using Euler's equation and equation (2.44) for ω_d and carrying out the necessary algebraic operations, equation (2.43) can be written:

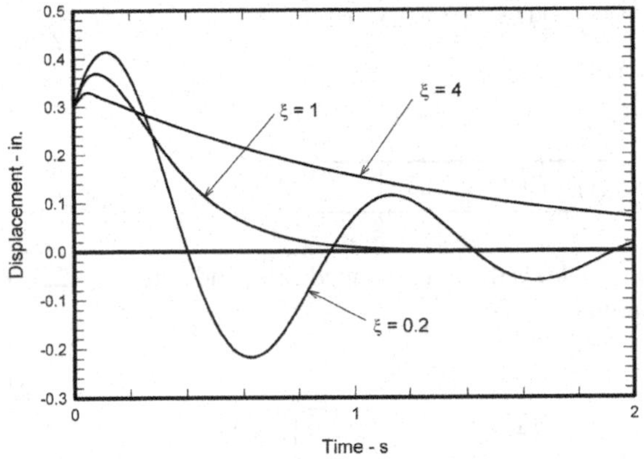

Figure 2.12 Response of a 1-DOF Vibration System with Viscous Damping to an Initial Displacement and an Initial Velocity

$$y(t) = \left(A \cos\omega_d t + B \sin\omega_d t\right) e^{-\xi\omega_n t} \ . \tag{2.45}$$

Applying the initial conditions $y(0) = y_o$ and $\dot{y}(0) = v_o$ yields for A and B:

$$A = y_o \qquad \text{and} \qquad B = \frac{v_o + \xi\omega_n\, y_o}{\omega_d} \ . \tag{2.46}$$

In a manner similar to that used to derive equation (2.24), equation (2.45) can be written:

$$\boxed{y(t) = Y\, e^{-\xi\omega_n t} \cos\left(\omega_d t + \phi\right)} \tag{2.47}$$

where:

$$\boxed{Y = \sqrt{y_o^2 + \left(\frac{v_o + \xi\omega_n\, y_o}{\omega_d}\right)^2}} \tag{2.48}$$

and noting $\tan^{-1}(-\phi) = -\tan^{-1}(\phi)$:

$$\boxed{\phi = -\tan^{-1}\left(\frac{v_o + \xi\omega_n\, y_o}{y_o\, \omega_d}\right).} \tag{2.49}$$

Figure 2.12 shows the response of y(t) as a function of time for a system with less than critical damping.

Critical Damping - $\xi = 1$: For the case of critical damping, the roots of equation (2.38) are real and equal, or:

$$s_{1,2} = -\omega_n \ . \tag{2.50}$$

Therefore, the response of the system is written:

$$y(t) = Y_1\, e^{-\omega_n t} + Y_2\, t\, e^{-\omega_n t} \ . \tag{2.51}$$

The term t must be added to the second term of equation (2.51) to be able to apply the initial conditions to solve for Y_1 and Y_2. Applying the initial conditions y_o and v_o yields:

$$\boxed{y(t) = \left[y_o + \left(v_o + y_o\, \omega_n\right) t\right] e^{-\omega_n t} \ .} \tag{2.52}$$

A critically damped system will return to its static equilibrium position after an initial disturbance in the shortest period of time without the response changing sign. Critical damping also signifies the transition of the response of the system from being oscillatory to being non-oscillatory. Figure 2.12 shows the response of a critically damped system.

Damping greater than critical - $\xi > 1$: For the case of a system with greater than critically damping system, the roots of equation (2.39) are real and negative, or:

$$s_{1,2} = \left(-\xi \pm \sqrt{\xi^2 - 1}\right) \omega_n \ . \tag{2.53}$$

Therefore, the response of the system becomes:

$$\boxed{y(t) = Y_1\, e^{\left(-\xi - \sqrt{\xi^2 - 1}\right)\omega_n t} + Y_2\, e^{\left(-\xi + \sqrt{\xi^2 - 1}\right)\omega_n t} \ .} \tag{2.54}$$

The above equation indicates the response of the system will be exponentially decaying and non-oscillatory. Applying the initial displacement y_o and velocity v_o yields for Y_1 and Y_2:

$$\boxed{Y_1 = -\frac{v_o + y_o\left(\xi - \sqrt{\xi^2 - 1}\right)\omega_n}{2\sqrt{\xi^2 - 1}\, \omega_n}} \tag{2.55}$$

and:

$$Y_2 = \frac{v_o + y_o\left(\xi + \sqrt{\xi^2 - 1}\right)\omega_n}{2\sqrt{\xi^2 - 1}\,\omega_n}. \tag{2.56}$$

Figure 2.12 shows the response of an overdamped system as a function of time.

EXAMPLE 2.8

A 1-DOF vibration system similar to the one shown in Figure 2.11 is given an initial displacement of 0.3 in. and an initial velocity of 2 in./s. The system has a resonance frequency of 6.283 rad/s. Determine the equations of motion when: (a) $\xi = 0.2$, (b) $\xi = 1.0$, and (c) $\xi = 4.0$. Compare the three cases by plotting the displacement as a function of time.

SOLUTION

(a) The motion for $\xi = 0.2$ is oscillatory and is described by equation (2.47). The damped resonance frequency is:

$$\omega_d = \sqrt{1 - 0.2^2} \times 6.283 = 6.156\,\frac{\text{rad}}{\text{s}}. \tag{2.8a}$$

The amplitude of Y is:

$$Y = \sqrt{0.3^2 + \left(\frac{2 + 0.2 \times 6.283 \times 0.3}{6.156}\right)} = 0.489 \text{ in.} \tag{2.8b}$$

and the amplitude of the phase angle ϕ is:

$$\phi = \tan^{-1}\left(-\frac{2 + 0.2 \times 6.283 \times 0.3}{0.3 \times 6.156}\right) = -0.910 \text{ rad.} \tag{2.8c}$$

The equation of motion is:

$$y(t) = 0.589\, e^{-1.257t} \cos(6.156t - 0.920). \tag{2.8d}$$

(b) $\xi = 1.0$ is the critically damped case, and equation (2.52) applies. Therefore,

$$y(t) = \left[0.3 + (2 + 0.3 \times 6)\,t\right] e^{-6.283t} \tag{2.8e}$$

or:

$$y(t) = \left[0.3 + 3.885\,t\right] e^{-6.283t}. \tag{2.8f}$$

(c) $\xi = 4.0$ is the over damped case, and equation (2.54) applies.

$$Y_1 = -\frac{2 + 0.3 \times \left(4 - \sqrt{4^2 - 1}\right) \times 6.283}{2\sqrt{4^2 - 1} \times 6.283} = -0.046 \text{ in.} \tag{2.8g}$$

and:

$$Y_2 = \frac{2 + 0.3 \times \left(4 + \sqrt{4^2 - 1}\right) \times 6.283}{2\sqrt{4^2 - 1} \times 6.283} = 0.346 \text{ in.} \tag{2.8h}$$

Therefore:

$$y(t) = -0.046\, e^{-49.465t} + 0.346\, e^{-0.798t}. \tag{2.8i}$$

The equations of motion for the above three cases are plotted in Figure 2.12.

EXAMPLE 2.9

A 1-DOF vibration system similar to the one in Figure 2.11 has the following characteristics: We = 40 lb_f, K = 50 lb_f/in., and C = 0.85 lb_f-s/in. (a) Determine the resonance frequency, the damping ratio, and the damped resonance frequency of the system. (b) Write the expression for the response of the system to an initial displacement of 0 in and an initial velocity of 4 in./s.

SOLUTION

(a) To find the resonance frequency, we must first convert to the proper mass units. From equation (1.194):

$$M = \frac{40}{386} = 0.1035\,\frac{lb_f - s^2}{\text{in.}}. \tag{2.9a}$$

From equation (2.7):

$$\omega_n = \sqrt{\frac{50}{0.1035}} = 21.98\ \frac{\text{rad}}{\text{s}} \tag{2.9b}$$

and:

$$f_n = \frac{21.98}{2\pi} = 3.50\ \text{Hz}\,. \tag{2.9c}$$

The damping ratio is obtained from equation (2.32):

$$\xi = \frac{0.85}{2\sqrt{50 \times 0.1035}} = 0.187\,. \tag{2.9d}$$

Since $\xi < 1$, the system is underdamped. Therefore, the motion will be oscillatory and will decrease with increasing time. Equation (2.44) is used to determine the damped resonance frequency. Therefore:

$$\omega_d = \sqrt{1 - 0.187^2} \times 21.98 = 21.59\ \frac{\text{rad}}{\text{s}} \tag{2.9e}$$

and:

$$f_d = \frac{21.59}{2\pi} = 3.44\ \text{Hz}\,. \tag{2.9f}$$

(b) Since the system is underdamped, equations (2.48) and (2.49) are used to determine the response of the system to the initial conditions:

$$Y = \frac{4}{21.59} \quad \text{or} \quad Y = 0.185\ \text{in.} \tag{2.9g}$$

and:

$$\phi = \tan^{-1}(-\infty) \quad \text{or} \quad \phi = -\frac{\pi}{2}\ \text{rad}\,. \tag{2.9h}$$

Therefore, the response of the system is:

$$\boxed{y(t) = 0.185\, e^{-4.04t} \cos\left(21.59t - \frac{\pi}{2}\right).} \tag{2.9i}$$

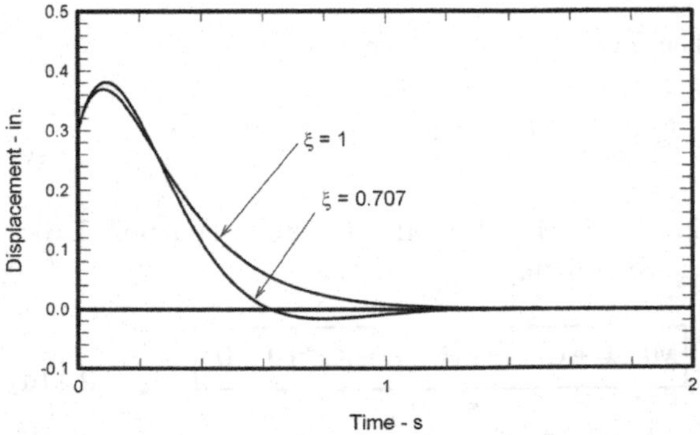

Figure 2.13 Response of a 1-DOF Vibration System with Viscous Damping at and near Critical Damping

Damping is often added to a mechanical system to control the vibration response of the system to an initial input. An example is the use of shock absorbers on an automobile. When a car hits a bump, without shock absorbers, it will continue to oscillate for an unacceptably long period of time after it passes over the bump. This presents a serious stability problem for the automobile, and it will be unsafe. Typically, the shock absorbers are selected so, when they are combined with the suspension system, the car will cease to oscillate in the quickest possible time after passing over a bump. Initially, it appears the damping coefficient of the shock absorbers should be selected such that the damping ratio $\xi = 1$. However, Figure 2.13 shows a damping ratio of $\xi = 0.707$ may be more desirable. Even though the car may have a slight negative response after returning to the equilibrium position, it nonetheless returns to the equilibrium position in a faster time than if the damping ratio was $\xi = 1$.

EXAMPLE 2.10

Determine the expression for the response of mass M_2 in Figure 2.14 when it is impacted by mass M_1 from a height of h. Assume the collision is a perfect inelastic collision and that $\xi < 1$.

SOLUTION

This problem can be solved by treating the system described by M_2, C and K as if it were given an initial

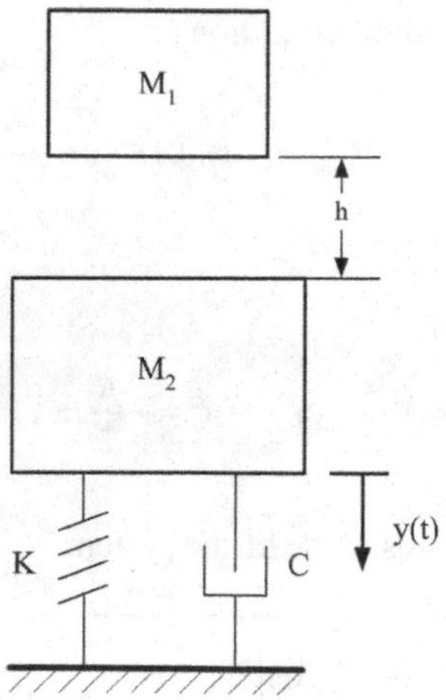

Figure 2.14 1-DOF Vibration System Where Mass M_1 Impacts Mass M_2 from a Height h

velocity v_o at time t = 0. First, the velocity v_1 of mass M_1 at the time of impact must be determined. The potential energy of mass M_1 relative to mass M_2, before it is released, is equal to the kinetic energy of mass M_1 when it impacts mass M_2. Therefore:

$$M_1 g h = \frac{1}{2} M_1 v_1^2 \quad \text{or} \quad v_1 = \sqrt{2gh}\,. \tag{2.10a}$$

The initial velocity v_o of masses M_1 and M_2 after impact is given by:

$$M_1 v_1 = (M_1 + M_2) v_o \quad \text{or} \quad v_o = \frac{M_1}{M_1 + M_2} v_1\,. \tag{2.10b}$$

The initial displacement x_o of masses M_1 and M_2 after impact is equal to zero. Since the system has less than critical damping, the response is given by equation (2.47) for the case where:

$$Y = \frac{v_o}{\omega_d} \quad \text{and} \quad \phi = -\frac{\pi}{2} \tag{2.10c}$$

and where:

$$\omega_d = \sqrt{1-\xi^2}\,\omega_n\,. \tag{2.10d}$$

Therefore, the response of the system after mass M_1 impacts mass M_2 is:

$$y(t) = \frac{v_o}{\omega_d} e^{-\xi\omega_n t} \sin\omega_d t\,. \tag{2.10e}$$

Substituting equations (2.10a) and (2.10b) into equation (2.10e) yields:

$$\boxed{y(t) = \frac{M_1 \sqrt{2gh}}{(M_1 + M_2)\,\omega_d} e^{-\xi\omega_n t} \sin\omega_d t\,.} \tag{2.10f}$$

EXAMPLE 2.11

Use Newton's method to determine the equation of motion of the pendulum 1-DOF vibrationsystem shown in Figure 2.15(a). Assume θ is very small. Write the expressions for the resonance frequency and the damping ratio.

SOLUTION

The free-body diagram for the system is shown in Figure 2.15(b). Since θ is very small:

$$\sin\theta \approx \theta \quad \text{and} \quad \cos\theta \approx 1\,. \tag{2.11a}$$

To determine the equation of motion, sum the moments about point A. Therefore:

$$\sum M_A = J_A\,\ddot{\theta} \quad \text{or} \quad J_A\,\ddot{\theta} = -MgL\,\theta - Ka^2\,\theta - Ca^2\,\dot{\theta}. \tag{2.11b}$$

Therfore:

$$J_A\,\ddot{\theta} + Ca^2\,\dot{\theta} + \left(MgL + Ka^2\right)\theta = 0\,. \tag{2.11c}$$

Since $J_o << ML^2$, $J_A = ML^2$. Therefore, equation (2.10c) can be written:

$$\boxed{ML^2\,\ddot{\theta} + Ca^2\,\dot{\theta} + \left(MgL + Ka^2\right)\theta = 0\,.} \tag{2.11d}$$

The expression for the resonance frequency is:

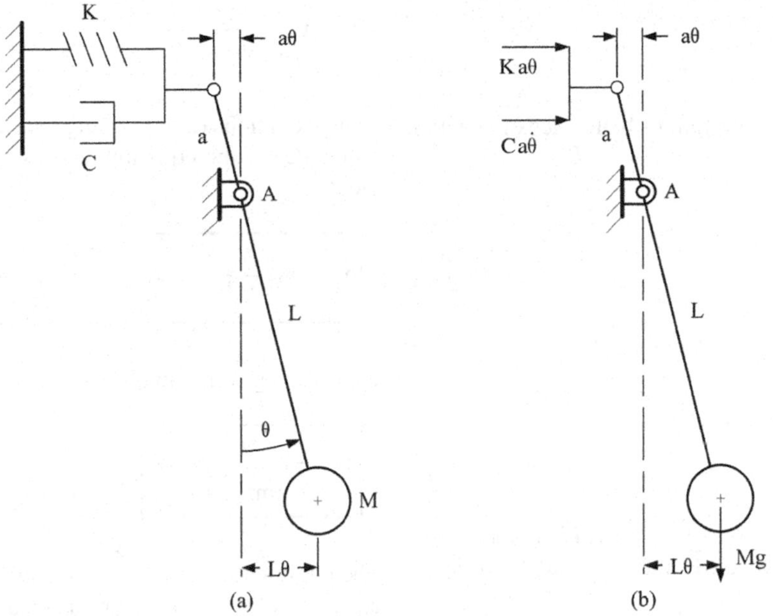

Figure 2.15 1-DOF Pendulum Vibration System with Spring and Damping Elements

$$\boxed{\omega_n = \sqrt{\frac{g}{L} + \frac{Ka^2}{ML^2}}\,.} \tag{2.11e}$$

The expression for the damping ratio is:

$$\boxed{\xi = \frac{Ca^2}{2\sqrt{ML^2\left(MgL + Ka^2\right)}}\,.} \tag{2.11f}$$

2.4 Logarithmic Decrement

Most vibration systems have some viscous damping present. However, in some cases, it may not be possible to determine the amount of damping present in the system by analytical methods. Many vibration systems can be approximated as a 1-DOF system with a low level of viscous damping. When this is the case, the response of the system can be measured with respect to time (Figure 2.16). Then, the ratio of peak response values over consecutive cycles of oscillation can be obtained.

When $\xi < 1$, the response of a 1-DOF vibration system with viscous damping is given by equation (2.47). The ratio of two consecutive peak values over one cycles of oscillation is:

$$\frac{Y_1}{Y_2} = \frac{Y\,e^{-\xi\omega_n t_1}}{Y\,e^{-\xi\omega_n t_2}} \quad \text{or} \quad \frac{Y_1}{Y_2} = e^{\xi\omega_n (t_2 - t_1)}\,. \tag{2.57}$$

Since $t_2 - t_1$ represents the time interval between two consecutive peak response values over one cycle:

$$t_2 - t_1 = \frac{2\pi}{\omega_d}\,. \tag{2.58}$$

Substituting equation (2.58) into equation (2.57) and noting that $\omega_d = \sqrt{1-\xi^2}\,\omega_n$ yields:

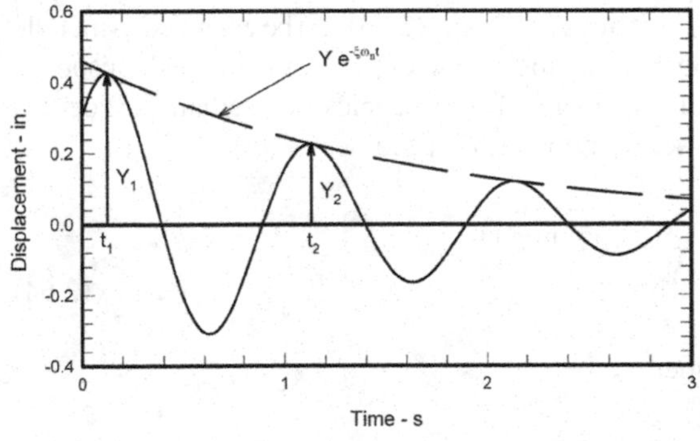

Figure 2.16 Logarithmic Decrement

$$\frac{Y_1}{Y_2} = e^{\left(2\pi\xi/\sqrt{1-\xi^2}\right)}. \tag{2.59}$$

Taking the natural logarithm of both sides of equation (2.59) yields:

$$\ln\left(\frac{Y_1}{Y_2}\right) = \frac{2\pi\xi}{\sqrt{1-\xi^2}}. \tag{2.60}$$

Define the left side of equation (2.60) as:

$$\boxed{\alpha = \ln\left(\frac{Y_1}{Y_2}\right)} \tag{2.61}$$

where α is referred to as the *logarithmic decrement.* Substituting equation (2.61) into equation (2.60) yields:

$$\alpha = \frac{2\pi\xi}{\sqrt{1-\xi^2}}. \tag{2.62}$$

Rearranging equation (2.62) yields:

$$\boxed{\xi = \frac{\alpha}{\sqrt{(2\pi)^2 + \alpha^2}}}. \tag{2.63}$$

The damping ratio ξ is obtained by first measuring the response of the system being examined, as shown in Figure 2.16. The logarithmic decrement α is then obtained using equation (2.61). Finally, the damping ratio ξ is obtained by substituting the value of α into equation 2.63).

Equations (2.62) and (2.63) can be extended to include the natural logarithm of the ratio of peak response values separated by m cycles of oscillation. For this case, equation (2.57) becomes:

$$\frac{Y_1}{Y_m} = e^{\xi\omega_n(t_m - t_1)} \tag{2.64}$$

where:

$$t_m - t_1 = \frac{2m\pi}{\omega_d}. \tag{2.65}$$

Using equations (2.59), taking the natural logarithm of both sides of the equation, and combining the results yield:

$$\boxed{\alpha_m = \ln\left(\frac{Y_1}{Y_m}\right) \quad \text{and} \quad \alpha_m = \frac{2m\pi\xi}{\sqrt{1-\xi^2}}}. \tag{2.66}$$

Rearranging equation (2.66) gives:

$$\boxed{\xi = \frac{\alpha_m}{\sqrt{(2m\pi)^2 + \alpha_m^2}}}. \tag{2.67}$$

The comments that apply to equation (2.63) apply to equation (2.67). However, instead of obtaining the logarithmic decrement α from the ration of peak response values separated by one cycle, as indicated by equation (2.262), the logarithmic decrement α_m is obtained from the ration of peak response values separated by m cycle, as indicated by equation (2.66). In most cases, this results in greater accuracy in the calculation of the damping ratio ξ.

EXAMPLE 2.12

A 1-DOF vibration system similar to the one shown in Figure 2.11 consists of a mass which weighs 7 lb_f, a spring element with a stiffness coefficient of $K = 23$ lb_f/in, and a damping element with an unknown damping coefficient. The system is set into vibration and the amplitude of vibration decreases to a value of 0.15 of the initial amplitude of vibration after 6 cycles of oscillation. Determine the damping ratio ξ of the system and the damping coefficient C of the damping element.

SOLUTION

The logarithmic decrement a_m is:

$$\alpha_m = \ln\left(\frac{1}{0.15}\right) \quad \text{or} \quad \alpha_m = 1.90. \tag{2.12a}$$

Therefore, the damping ratio is:

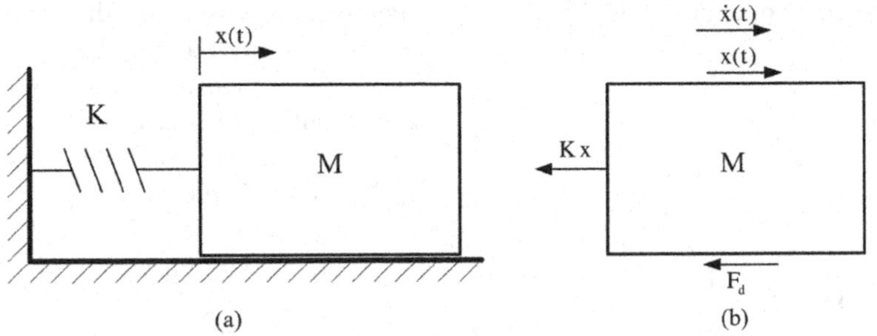

Figure 2.17 1-DOF Vibration System with Coulomb Damping

(2.12b)

$$\xi = \frac{1.90}{\sqrt{(2\times 6\times \pi)^2 + 1.90^2}} \quad \text{or} \quad \xi = 0.05 .$$

The damping coefficient is obtained from equation (2.34). Therefore:

(2.12c)

$$\boxed{C = 2\times 0.05 \times \sqrt{\frac{23\times 7}{386}} \quad \text{or} \quad C = 0.065\,\frac{lb_f - s}{in.}.}$$

2.5 Unforced Vibration with Coulomb Damping

Figure 2.17(a) shows a 1-DOF vibration system with Coulomb damping. The mass M is in contact with a surface where there is no lubricating medium between the mass and surface. *Coulomb damping* results from the friction force generated by the mass sliding over the surface. The friction force is equal to the force associated with the product of the weight of the mass acting on the surface times the surface friction coefficient. When the mass is starting from a "stopped" position, the friction coefficient is the surface *static friction coefficient*, μ_s. When the mass is moving over the surface, the friction coefficient is the surface *dynamic friction coefficient* μ.

Figure 2.17(b) shows a free-body diagram of a 1-DOF vibration system with Coulomb damping. F_d is the friction force. Summing the forces in the x direction yields:

$$\sum F_x = M\ddot{x} \quad \text{or} \quad M\ddot{x} = -F_d - K\,x. \tag{2.68}$$

Therefore:

$$M\ddot{x} + K\,x = -F_d\,. \tag{2.69}$$

The friction force can be represented as acting in a direction opposite the direction of the velocity of mass M. It is independent of the amplitudes of the velocity and displacement of the mass. The friction force F_d can be represented by:

$$F_d = Mg\,\mu\,\text{sgn}(\dot{x}) \quad \text{where} \quad \text{sgn}(\dot{x}) = \frac{\dot{x}}{|\dot{x}|}. \tag{2.70}$$

Therefore, equation (2.69) is written:

$$M\ddot{x} + K\,x = -Mg\,\mu \quad \text{when} \quad \dot{x} > 0 \tag{2.71}$$

$$M\ddot{x} + K\,x = Mg\,\mu \quad \text{when} \quad \dot{x} < 0. \tag{2.72}$$

To solve for x in equations (2.71) and (2.72), the equations must be solved for the respective half cycles corresponding to $\dot{x} > 0$ and $\dot{x} < 0$. $\dot{x} = 0$ when mass M is at the end of a half cycle and before beginning a new half cycle.

Assume mass M is given an initial positive displacement x_o and then released ($\dot{x} = 0$). Over the first half cycle $\dot{x}$ will be negative. Therefore, equation (2.72) must be solved for x. There are two parts to the solution. One is the complimentary solution associated with the initial conditions, which is given by equation (2.17). The second part is the particular solution, which is associated with the friction forcing F_d. Since F_d is a constant, assume a particular solution of the form:

$$x(t)_p = X_p\,. \tag{2.73}$$

Substituting equation (2.73) into equation (2.72) and

carrying out the prescribed operations yields:

$$X_p = \frac{F_d}{K}. \tag{2.74}$$

The total solution is:

$$x(t) = A\cos\omega_n t + B\sin\omega_n t + \frac{F_d}{K}. \tag{2.75}$$

To solve for A and B, apply the initial conditions $x(0) = x_o$ and $\dot{x}(0) = 0$ to equation (2.75). Taking the derivative of equation (2.75) with respect to time yields:

$$\dot{x}(t) = -A\omega_n \sin\omega_n t + B\omega_n \cos\omega_n t. \tag{2.76}$$

Applying the initial conditions to equations (2.74) and (2.75) gives:

$$A = x_o - \frac{F_d}{K} \quad \text{and} \quad B = 0. \tag{2.77}$$

Therefore, the displacement over the first half cycle is:

$$x(t) = \left(x_o - \frac{Mg\,\mu}{K}\right)\cos\omega_n t + \frac{Mg\,\mu}{K} \tag{2.78}$$

and the related velocity is:

$$\dot{x}(t) = -\omega_n\left(x_o - \frac{Mg\,\mu}{K}\right)\sin\omega_n t. \tag{2.79}$$

where ω_n is given by equation (2.7).

The first half cycle continues until the velocity equals zero. The velocity equals zero for the first time when:

$$\omega_n t_1 = \pi \quad \text{or} \quad t_1 = \frac{\pi}{\omega_n}. \tag{2.80}$$

The displacement at t_1 is:

$$x(t_1) = -\left(x_o - \frac{2\,Mg\,\mu}{K}\right). \tag{2.81}$$

For the response over the second half cycle, the velocity is positive and equation (2.71) must be solved. The initial displacement at t_1 for the beginning of the second half cycle is given by equation (2.81), and the related velocity is zero. The solution to equation (2.71) for the second half cycle is:

$$x(t) = \left(x_o - \frac{3\,Mg\,\mu}{K}\right)\cos\omega_n t - \frac{Mg\,\mu}{K} \tag{2.82}$$

and the velocity is:

$$\dot{x}(t) = -\left(x_o - \frac{3\,Mg\,\mu}{K}\right)\sin\omega_n t. \tag{2.83}$$

The second time the velocity equals zero is at the end of the second half cycle where:

$$\omega_n t_2 = 2\pi \quad \text{or} \quad t_2 = \frac{2\pi}{\omega_n}. \tag{2.84}$$

The displacement at the end of the second half cycle is:

$$x(t_2) = \left(x_o - \frac{4\,Mg\,\mu}{K}\right). \tag{2.85}$$

A continuation of this process of sequentially solving equations (2.71) and (2.72) over successive half cycles of oscillation results in the expression:

$$\boxed{x(t) = \left[x_o - (2n-1)\frac{Mg\,\mu}{K}\right]\cos\omega_n t + (-1)^{(n+1)}\frac{Mg\,\mu}{K}} \tag{2.86}$$

where n is an integer that begins at 1 and is increased by one at the end of each half cycle of oscillation. At the end of a specified number of n half cycles, time t_n is given by:

$$\boxed{t_n = \frac{n\pi}{\omega_n}.} \tag{2.87}$$

Substituting equation (2.87) into the cosine term in equation (2.86) yields for the value of the cosine term at the end of n half cycles:

$$\cos(\omega_n t_n) = (-1)^n . \tag{2.88}$$

Substituting equation (2.88) into equation (2.86) and simplifying yields for the value of $x(t_n)$ at the end of n half cycles of oscillation:

$$\boxed{x(t_n) = (-1)^n \left(x_o - 2n \frac{Mg\,\mu}{K} \right).} \tag{2.89}$$

The displacement amplitude decreases by 2Mgμ/K per half cycle and by 4Mgμ/K per full cycle of oscillation.

Figure 2.18 shows the vibration response of an unforced 1-DOF system with Coulomb damping. Mass M in Figure 2.17 will continue to oscillate as long as the restoring force associated with spring K at the end of each half cycle of oscillation is greater than the static friction force associated with mass M. When the restoring force associated with the spring at the end of a given half cycle is not greater than the static friction force, mass M will cease to oscillate. This occurs when:

$$\left(x_o - 2N \frac{Mg\,\mu}{K} \right) K < Mg\,\mu_s \tag{2.90}$$

where N is the number of the half cycle of oscillation at which motion ceases. Rearranging equation (2.90):

$$\boxed{N > \frac{K x_o}{2 Mg\,\mu} - \frac{\mu_s}{2\mu} .} \tag{2.91}$$

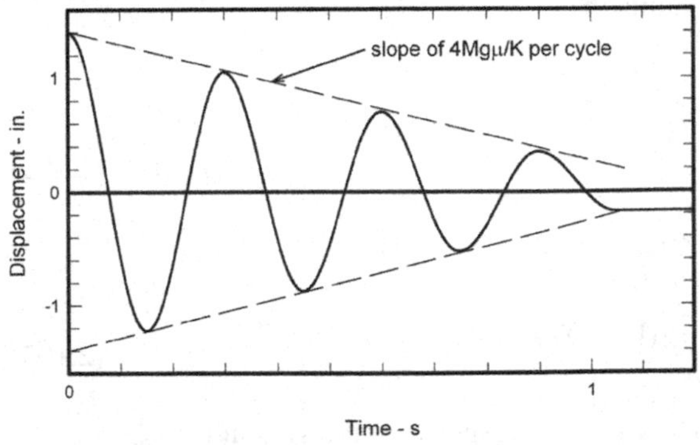

Figure 2.18 Response of a 1-DOF Vibration System with Coulomb Damping

The number obtained for the expression on the right side of equation (2.91) will most often not be an integer. Therefore, N will be the next higher integer associated with this number. The displacement value for $x(t_N)$ after N half cycles of oscillation is obtained by letting n in equation (2.89) equal N.

EXAMPLE 2.13

Let the weight of mass M in Figure 2.17 be 700 lb_f and the stiffness coefficient of spring K be 800 lb_f/in. Let $\mu_s = 0.35$ and $\mu = 0.1$. Determine: (a) the decay per cycle, (b) the number of half cycles before oscillation stops and (c) the displacement amplitudes of the mass when oscillation ceases for the initial conditions of x_o = 1.4 in. and $v_o = 0$.

SOLUTION

(a) From equation (2.89), the decay per cycle is:

$$\boxed{\frac{4 \times 700 \times 0.1}{800} = 0.35 \frac{\text{in.}}{\text{cycle}} .} \tag{2.13a}$$

(b) N is obtained from equation (2.91):

$$N > \frac{800 \times 1.4}{2 \times 700 \times 0.1} - \frac{0.35}{2 \times 0.1} \tag{2.13b}$$

or:

$$\boxed{N > 6.25 .} \tag{2.13c}$$

Therefore, N = 7, or the mass will stop oscillating after 7 half cycles.

(c) The displacement amplitude of the mass after 7 half cycles is obtained from equation (2.87).

$$x(t_7) = (-1)^7 \left(1.4 - 2 \times 7 \times \frac{700 \times 0.1}{800} \right). \tag{2.13d}$$

or:

$$\boxed{x(t_7) = -0.175 \text{ in.}} \tag{2.13e}$$

Figure 2.18 shows a plot of the response of the system in this example.

2.6 Unforced Vibration with Structural Damping

All mechanical systems possess mechanisms that dissipate energy. This is true even with those systems that do not contain damping elements. There are a large variety of structural materials that exhibit a stress-strain relationship that is characterized by a hysteresis loop when they are subjected to cyclic stresses below their elastic limits. The energy that is dissipated per cycle is associated with internal friction within the material matrix and is proportional to the area within the hysteresis loop. Therefore, this type of damping is referred to as *hysteresis or structural damping*. For most structural materials, the internal friction is independent of the rate of strain and, consequently, frequency. It is proportional to displacement over a wide frequency range. Therefore, the damping force is proportional to the elastic or spring force. However, since energy is dissipated, it must be in phase with velocity.

If harmonic excitation is assumed, the effects of structural damping can be incorporated in a complex stiffness coefficient $K(1 + j\gamma)$ where γ is called the *structural damping factor*. Therefore, the equation of motion of a 1-DOF vibration system with structural damping can be written:

$$M\ddot{y} + K(1 + j\gamma)y = 0 . \tag{2.92}$$

Since harmonic motion is assumed, y(t) can be written:

$$y(t) = Y\, e^{j\omega_n t} \tag{2.93}$$

and:

$$\dot{y}(t) = j\omega_n\, Y\, e^{j\omega_n t} . \tag{2.94}$$

Equating equations (2.93) and (2.94) yields:

$$\dot{y}(t) = j\omega_n\, y(t) \qquad \text{or} \qquad y(t) = \frac{\dot{y}(t)}{j\omega_n} . \tag{2.95}$$

Substituting equation (2.95) into equation (2.92) and rearranging the terms yields:

$$M\ddot{y} + C_{eq}\,\dot{y} + Ky = 0 . \tag{2.96}$$

where:

$$C_{eq} = \frac{\gamma K}{\omega_n} . \tag{2.97}$$

Dividing equation (2.96) by M and simplifying yields:

$$\ddot{y} + \gamma\,\omega_n\,\dot{y} + \omega_n^2\, y = 0 . \tag{2.98}$$

The solution to equation (2.98) can be obtained by substituting the expression $y(t) = Y e^{st}$ into it and carrying out the prescribed operations to obtain:

$$\left(s^2 + \gamma\,\omega_n\, s + \omega_n^2\right) Y\, e^{st} = 0 . \tag{2.99}$$

or:

$$s^2 + \gamma\,\omega_n\, s + \omega_n^2 = 0 . \tag{2.100}$$

Normally, the damping associated with structural damping is very small, such that $\gamma \leq 0.1$. Therefore, the expressions for s_1 and s_2 are:

$$s_{1,2} = \left(-\frac{\gamma}{2} \pm j\sqrt{1 - \left(\frac{\gamma}{2}\right)^2}\,\right)\omega_n . \tag{2.101}$$

Noting that:

$$\sqrt{1 - \left(\frac{\gamma}{2}\right)^2} \approx 1 , \tag{2.102}$$

equation (2.93) can be written:

$$y(t) = \left(Y_1\, e^{j\omega_n t} + Y_2\, e^{-j\omega_n t}\right) e^{-(\gamma\omega_n/2)t} . \tag{2.103}$$

Therefore, the response of the system is:

$$y(t) = \left(Y_1\, e^{j\omega_n t} + Y_2\, e^{-j\omega_n t}\right) e^{-(\gamma\omega_n/2)t}\,. \tag{2.104}$$

Equation (2.103) can be written:

$$\boxed{y(t) = Y\, e^{-(\gamma\omega_n/2)t} \cos\left(\omega_n t + \phi\right)} \tag{2.105}$$

where:

$$\boxed{Y = \sqrt{y_o^2 + \left(\frac{v_o}{\omega_n} + \frac{\gamma\, y_o}{2}\right)^2}} \tag{2.106}$$

and:

$$\boxed{\phi = \tan^{-1}\left[-\left(\frac{v_o}{y_o\,\omega_n} + \frac{\gamma}{2}\right)\right].} \tag{2.107}$$

The relation between the damping ratio, ξ, and the structural damping factor γ is obtained by substituting equation (2.96) into equation (2.32), or:

$$\xi = \frac{\gamma K}{2\omega_n \sqrt{KM}}\,. \tag{2.108}$$

Noting that $K/M = \omega_n^2$ and simplifying yields:

$$\boxed{\xi = \frac{\gamma}{2} \qquad \text{or} \qquad \gamma = 2\xi\,.} \tag{2.109}$$

PROBLEMS - CHAPTER 2

1. Use d'Alembert's principle to develop the equation of motion of the system shown in Figure P2.1. Write the equation in terms of θ. Assume there is no slipping between the roller and surface. Write the expression for the resonance frequency.

2. Use the energy method to develop the equation of motion of the system shown in Figure P2.1 in terms of θ. Assume there is no slipping between the roller and surface. Write the expression for the resonance frequency.

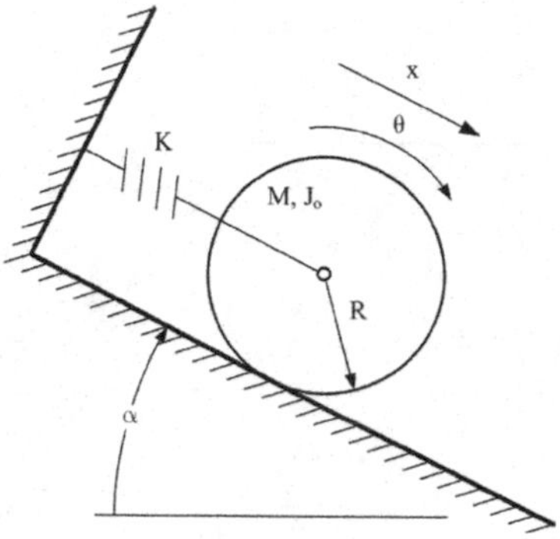

Figure P2.1

3. Use Newton's method to develop the equation of motion of the system shown in Figure P2.2 in terms of θ. Assume there is no slipping between the roller and surface. Write the expression for the resonance frequency.

4. Use the d'Alembert's principle to develop the equation of motion of the system shown in Figure P2.2 in terms of θ. Assume there is no slipping between the roller and surface. Write the expression for the resonance frequency.

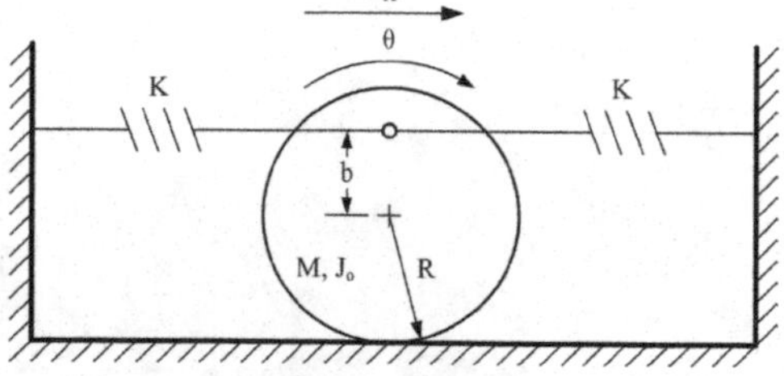

Figure P2.2

5. Use Newton's method to develop the equation of motion of the system shown in Figure P2.3 in terms of θ. Write the expression for the resonance frequency.

6. Use d'Alembert's principle to develop the equation of motion of the system shown in Figure P2.3 in terms of θ. Write the expression for the resonance frequency.

7. Use the energy method to develop the equation of motion of the system shown in Figure P2.3 in terms of y_2. Write the expression for the resonance frequency.

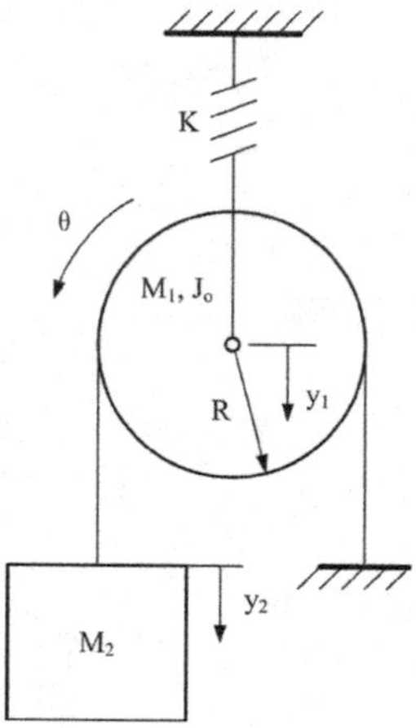

Figure P2.3

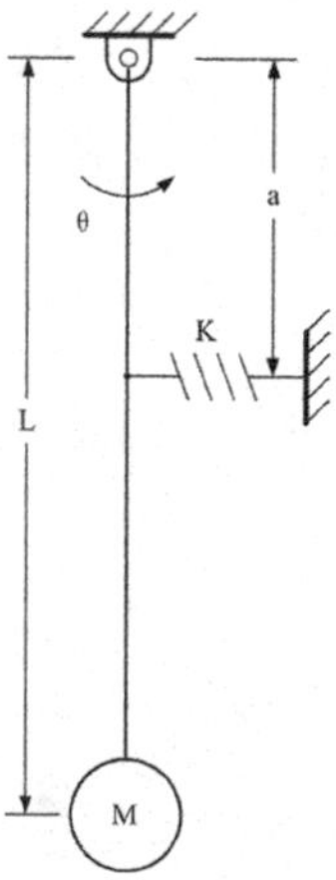

Figure P2.4

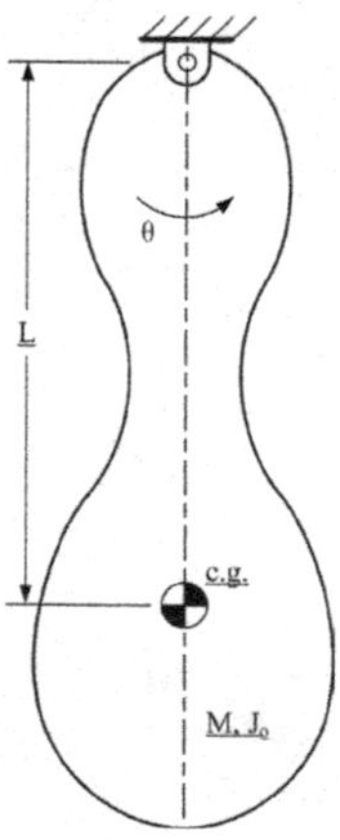

Figure P2.5

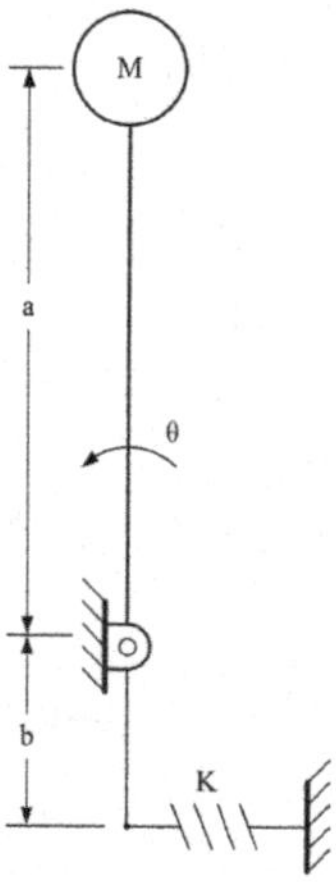

Figure P2.6

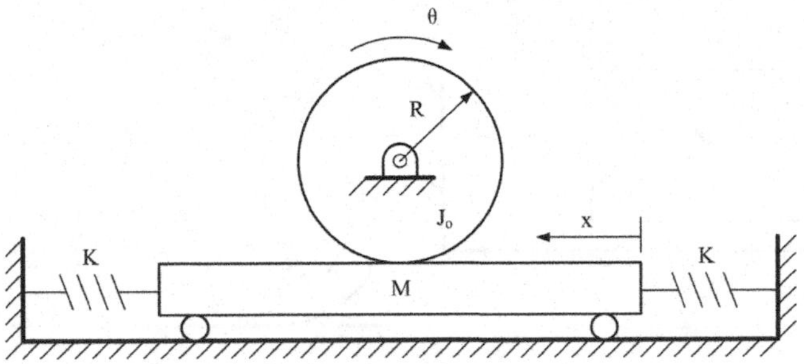

Figure P2.7

8. Use Newton's method to develop the equation of motion of the pendulum system shown in Figure P2.4 in terms of θ. Write the expression for the resonance frequency.

9. Use the energy method to develop the equation of motion of the pendulum system shown in Figure P2.4 in terms of θ. Write the expression for the resonance frequency.

10. Use d'Alembert's principle to develop the equation of motion of the compound pendulum shown in Figure P2.5 in terms of θ. Write the expression for the resonance frequency.

11. Use the energy method to develop the equation of motion of the compound pendulum shown in Figure P2.5 in terms of θ. Write the expression for the resonance frequency.

12. Use d'Alembert's principle to develop the equation of motion of the inverted pendulum system shown in Figure 2.6 in terms of θ. Write the expression for the resonance frequency.

13. Use the energy method to develop the equation of motion of the inverted pendulum system shown in Figure 2.6 in terms of θ. Write the expression for the resonance frequency.

14. Use d'Alembert's principle to develop the equation of motion of the system shown in Figure P2.7 in terms of θ. Assume no slipping between the roller and the plate. Write the expression for the resonance frequency.

15. Use the energy method to develop the equation of motion of the system shown in Figure P2.7 in terms of θ. Assume no slipping between the roller and the plate. Write the expression for the resonance frequency.

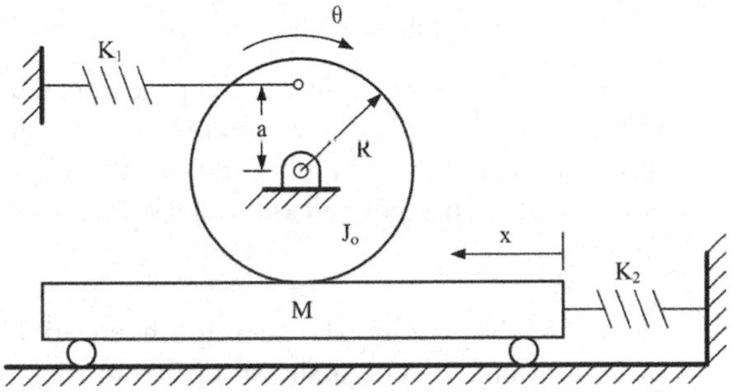

Figure P2.8

16. Use d'Alembert's principle to develop the equation of motion of the system shown in Figure P2.8 in terms of θ. Assume no slipping between the roller and the plate. Write the expression for the resonance frequency.

17. Use the energy method to develop the equation of motion of the system shown in Figure P2.8 in terms of θ. Assume no slipping between the roller and the plate. Write the expression for the resonance frequency.

18. Use d'Alembert's principle to develop the equation of motion of the system shown in Figure P2.9 in terms of x. Write the expression for the resonance frequency.

19. Use the energy method to develop the equation of motion of the system shown in Figure P2.9 in terms of x. Write the expression for the resonance frequency.

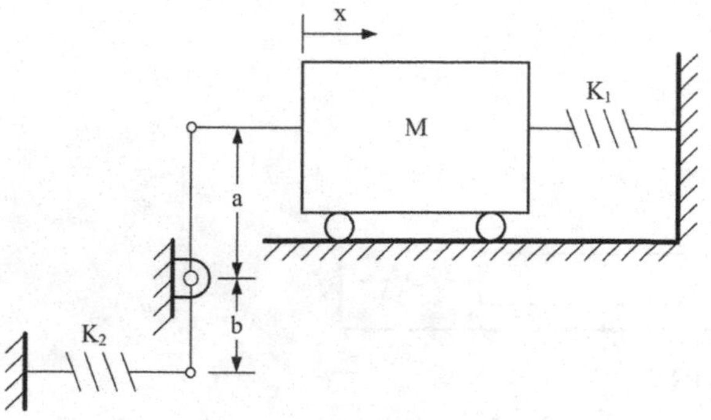

Figure P2.9

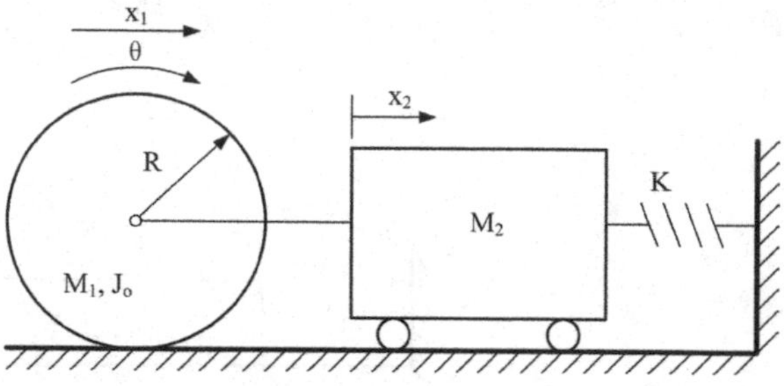

Figure P2.10

20. Use d'Alembert's principle to develop the equation of motion of the system shown in Figure P2.10 in terms of θ. Assume no slipping between the roller and the plate. Write the expression for the resonance frequency.

21. Use d'Alembert's principle to develop the equation of motion of the system shown in Figure P2.10 in terms of x_1. Assume no slipping between the roller and the plate. Write the expression for the resonance frequency.

22. Use d'Alembert's principle to develop the equation of motion of the system shown in Figure P2.11 in terms of θ_1. Assume no slipping between the roller and the plate. Write the expression for the resonance frequency.

23. Use the energy method to develop the equation of motion of the system shown in Figure P2.11 in terms of θ_1. Assume no slipping between the roller and the plate. Write the expression for the resonance frequency.

24. Use Newton's method to develop the equation of motion of the system shown in Figure P2.12 in terms of θ. Write the expression for the resonance frequency.

25. An automobile that weighs 1,820 kg has its weight equally distributed over four wheels. Using a bumper jack, the car is jacked up over one wheel.

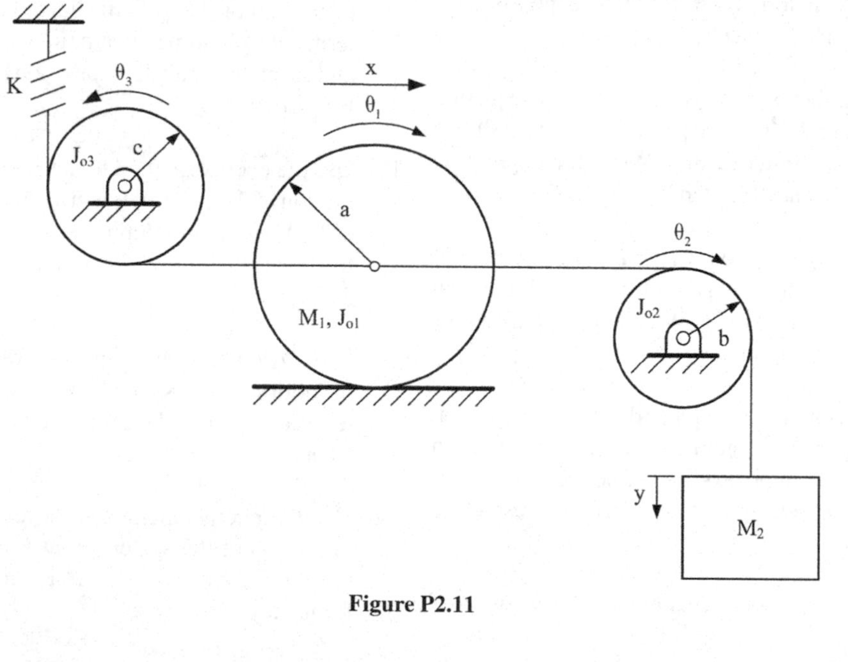

Figure P2.11

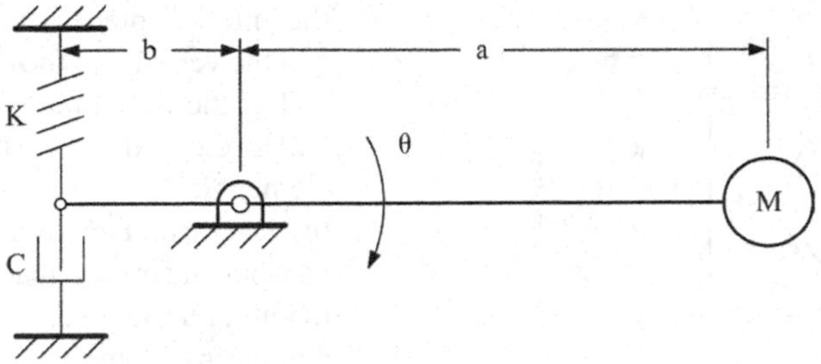

Figure P2.12

The jack has to move the bumper up 30.5 cm to get the wheel barely off the ground. (a) Find the resonance frequency of vibration in Hz of the automobile on its springs. (b) Find the stiffness coefficient of a single spring element. (c) What must the damping coefficient C of a single shock absorber be to achieve critical damping for the automobile?

26. A machine that weighs 60 lb_f is mounted on spring and damping elements in a manner similar to that shown in Figure 2.11. The combined stiffness of the spring elements is 120 lb_f/in. and the total damping is 20 lb_f-s/in. Determine the motion y(t) of the system when:
(a) $y(0) = 1$ in. and $\dot{y}(0) = 0$,
(b) $y(0) = 0$ and $\dot{y}(0) = 10\,\text{in./s}$,
(c) $y(0) = 1$ in. and $\dot{y}(0) = 10\,\text{in./s}$.

27. You are asked to design a braking system for railroad cars. A schematic of such a system is shown in Figure P2.13. The mass of the braking system relative to the mass of a railroad car is small and, therefore, can be neglected.
(a) Derive the equation of motion of the railroad car and braking system for the case after the car impacts the brake with an initial velocity v_o.
(b) State the assumed initial conditions after the car impacts the brake and solve the equation of motion from part (a) for the case where critical damping is assumed.
(c) Derive the expression that gives the maximum displacement of the car after impact in terms of the other appropriate system parameters.

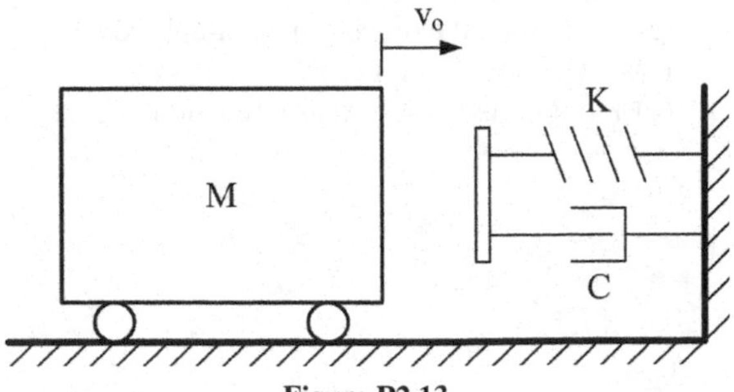

Figure P2.13

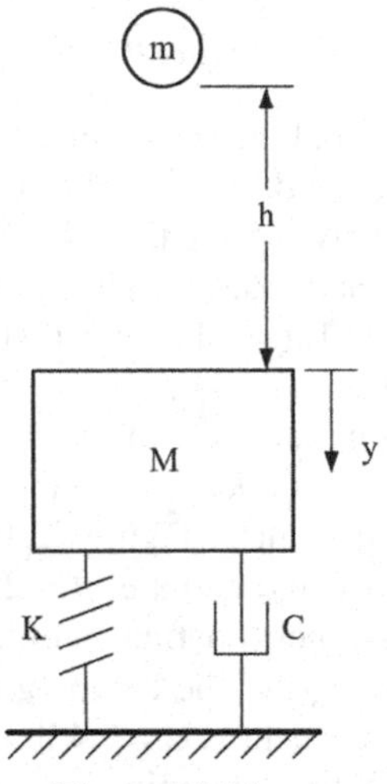

Figure P2.14

28. A falling steel sphere impacts the mass in Figure P2.14 with a velocity v_1. An elastic impact is assumed between the sphere and mass M. The momentum of the sphere at the point of impact is mv_1 where $v_1 = \sqrt{2gh}$. The momentum of mass M after the impact with the spherec is Mv_2. If K/M = 8.0 (1/s^2), C/M = 6.0 (1/s), h = 16.1 ft, and m/M = 0.1, determine the response y(t) after the impact of the sphere. Plot y(t) as a function of t.

29. A model of a mechanical baseball throwing machine is shown in Figure P2.15. The ball is

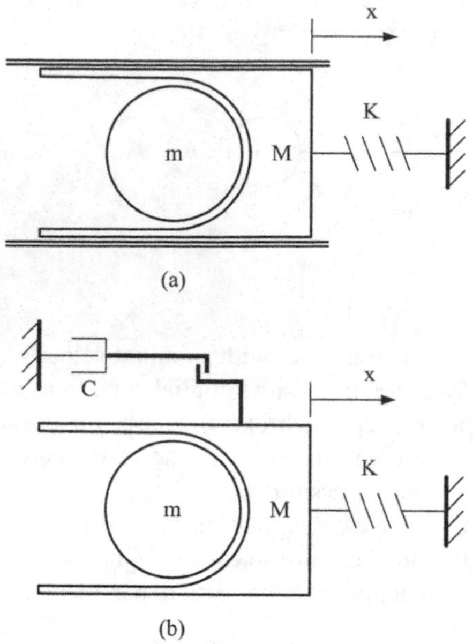

Figure P2.15

thrown by pulling the block M and the ball m in the positive x direction a distance x_o, resulting in a compressive spring force Kx_o being applied the the block and baseball. The block is then released [Figure P2.15(a)]. For the first quarter cycle of motion, the ball is in contact with the block. After the first quarter cycle of motion [when x(t) = 0], the baseball is released and the block M engages a damping element (Figure P2.15(b)]. The weight of the block and ball are M = 2 lb_f and m = 0.25 lb_f. The stiffness coefficient of the spring element is K = 85 lb_f/in. The damping coefficient of the damping element is C = 0.4 lb_f-s/in.

(a) Neglecting damping, derive the equation of motion of the mechanical baseball throwing machine for the first quarter cycle.

(b) Assuming an initial displacement of x_o and an initial velocity equal to zero, obtain the solution for the equation developed in part (a). What should the initial displacement of the block and ball be so that the velocity of the ball will be 1,200 in./s at the end of the first quarter cycle of motion when the ball is released and the block engages the damping element?

(c) Derive the equation of motion of the mechanical baseball throwing machine for the motion after the first quarter cycle when the block engages the dampering element.

(d) Determine the general solution for the equation developed in part (c). Write the expressions for ω_n, ω_d and ξ.

(e) Applying the appropriate initial conditions, write the expression for x(t) after the first quarter cycle.

(f) Plot x(t) from the expression developed in parts (b) and (e).

30. A 1-DOF vibration system with viscous damping has the following characteristics: M = 12 lb_f, K = 35 lb_f/in., and C = 0.1 lb_f-s/in. Determine the logarithmic decrement α and the ratio of the amplitudes of any two consecutive cycles.

31. The ratio of the amplitudes of oscillation of two consecutive cycles is 0.85. Determine the logarithmic decrement and the damping ratio.

32. A system similar to the one shown in Figure 2.17 has Coulomb damping. The mass weighs 500 lb_f, the stiffness coefficient of the springelement is element600 lb_f/in., the static coefficient of friction μ_s is 0.25, and the dynamic coefficient of friction μ is 0.05. The initial conditions are: x_o = 1.5 in. and v_o = 0.

(a) Determine the decay per cycle.

(b) Determine the number of half cycles before oscillation stops.

(c) Determine the displacement amplitude of the mass when oscillation ceases.

(d) Plot the response x(t) as a function of t.

CHAPTER 3
1-DOF VIBRATION SYSTEMS RESPONSE TO HARMONIC EXCITATION

3.1 Introduction

Vibrating systems are often excited by externally applied or internally generated forces that vary with time. Externally applied forces can have deterministic or random signal characteristic like those identified in Figure 1.15 in Chapter 1. These forces can be applied directly to the mass of a system or to the mass through spring and damping elements that connect it to a vibrating structure. Internally generated forces are normally associated with unbalanced rotating and reciprocating components in machinery. These forces most often are internally applied directly to the machine mass.

This chapter addresses 1-DOF vibration systems that are excited by a harmonic force. Equations of motion are developed for systems that have mass, damping and spring elements. Viscous damping is initially chosen for damping elements because it results in linear differential equations of motion. However, other types of damping elements will be discussed. Principles associated with discussions in this chapter form an introductory basis for analyses of vibration systems.

3.2 Equation of Motion

Figures 3.1(a) and (b) show schematic and free body diagrams of a 1-DOF vibration system with viscous damping and with a force f(t) applied to mass M. The system is constrained to move only in the y direction, and y(t) is shown as positive downward. When M is given a positive displacement in the y direction, the restoring forces associated with K and C are shown acting in the negative y direction.

The force f(t) in Figure 3.1 can have deterministic, random or both types of signal characteristics. These signal characteristics do not enter into the development of the system equation of motion. However, they determine the solution technique that must be used to solve the system equation of motion to obtain the expression for the system response. Since f(t) varies with time, it is shown as acting in the positive y direction.

Summing the forces associated with K, C and f(t) that act on M yields:

$$\sum F_y = M\,\ddot{y} \quad \text{or} \quad M\,\ddot{y} = f(t) - Ky - C\dot{y}. \tag{3.1}$$

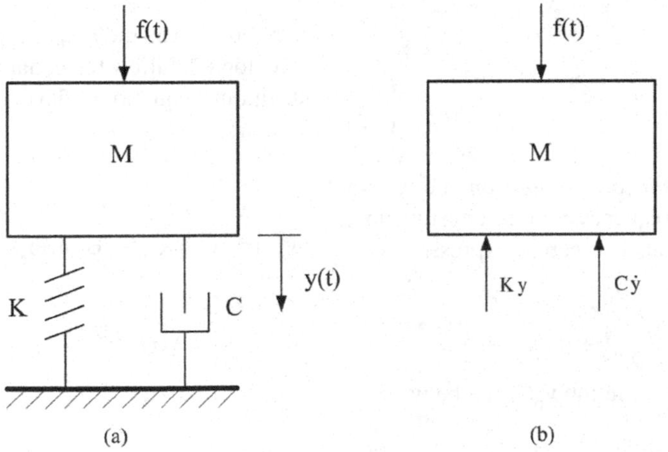

Figure 3.1 1-DOF Vibration System with Mass, Spring and Damping Elements Excited by an Externally Applied Force

Rearranging the terms:

$$M\ddot{y} + C\dot{y} + Ky = f(t). \tag{3.2}$$

There are two parts to the solution for equation (3.2). They are the complementary and particular solutions. Therefore, the total solution is written:

$$y(t) = y_c(t) + y_p(t). \tag{3.3}$$

$y_c(t)$ is the *complementary solution* of equation (3.2). It is the response of the system to its initial conditions. $y_p(t)$ is the *particular solution* of equation (3.2). It is the response of the system to the forcing function f(t).

3.3 Response of a 1-DOF Vibration System without Damping

Displacement Response - Total Response

For the case when the system has no damping, equation (3.2) reduces to:

$$M\ddot{y} + Ky = f(t). \tag{3.4}$$

Equation (2.17) in Section 2.2 gives the expression for $y_c(t)$:

$$\boxed{y_c(t) = A\cos\omega_n t + B\sin\omega_n t} \tag{3.5}$$

where:

$$\boxed{\omega_n = \sqrt{\frac{K}{M}}.} \tag{3.6}$$

For the case of a *harmonic forcing function*, the system will vibrate at the same frequency as the forcing function. The harmonic forcing function can be expressed:

$$f(t) = F\,e^{j\omega t}. \tag{3.7}$$

Therefore, the particular solution $y_p(t)$ can be written:

$$y_p(t) = \bar{Y}\,e^{j\omega t} \quad \text{where} \quad \bar{Y} = Y\,e^{-j\phi}. \tag{3.8}$$

Substituting equation (3.8) into equation (3.4) and carrying out the prescribed operations yields:

$$\left(-\omega^2 M + K\right)\bar{Y}\,e^{j\omega t} = F\,e^{j\omega t} \tag{3.9}$$

or rearranging terms:

$$\bar{Y} = \frac{F}{K - M\omega^2}. \tag{3.10}$$

$\bar{Y}$ can be written:

$$\bar{Y} = \frac{F/K}{1 - \dfrac{\omega^2}{\omega_n^2}}. \tag{3.11}$$

The particular solution can be written in terms of the real part of $e^{j\omega t}$, or:

$$\boxed{y_p(t) = \left(\frac{F/K}{1 - r^2}\right)\cos\omega t} \tag{3.12}$$

where:

$$\boxed{r = \frac{\omega}{\omega_n}.} \tag{3.13}$$

r is referred to as the *frequency ratio*.

The total solution for equation (3.4) is obtained by substituting equations (3.5) and (3.12) into equation (3.3), or:

$$y(t) = A\cos\omega_n t + B\sin\omega_n t + \left(\frac{F/K}{1 - r^2}\right)\cos\omega t. \tag{3.14}$$

The velocity $\dot{y}(t)$, is:

$$\dot{y}(t) = -\omega_n A \sin\omega_n t + \omega_n B \cos\omega_n t + \left(\frac{-\omega F/K}{1-r^2}\right)\sin\omega t. \tag{3.15}$$

Applying the initial conditions, $y(0) = y_o$ and $\dot{y}(0) = v_o$, yields:

$$A = y_o - \frac{F/K}{1-r^2} \quad \text{and} \quad B = \frac{v_o}{\omega_n}. \tag{3.16}$$

Substituting equation (3.16) into equation (3.14) gives:

$$y(t) = \left(y_o - \frac{F/K}{1-r^2}\right)\cos\omega_n t + \frac{v_o}{\omega_n}\sin\omega_n t + \left(\frac{F/K}{1-r^2}\right)\cos\omega t. \tag{3.17}$$

Equation (3.17) can be written in the form:

$$\boxed{y(t) = y_o \cos\omega_n t + \frac{v_o}{\omega_n}\sin\omega_n t + \frac{F}{K}\left(\frac{\cos\omega t - \cos\omega_n t}{1-r^2}\right).} \tag{3.18}$$

If F = 0, equation (3.18) reduces to equation (2.21) in Section 2.2.

Relative to equation (3.18), there are three frequency regions of interest: $r < 1$, $r > 1$ and $r = 1$.

Frequency ratio r < 1: For $r < 1$, the value of $(1 - r^2)$ is positive. Therefore, the response for y(t) is given by equation (3.18). When $r << 1$, equation (3.18) becomes:

$$\boxed{y(t) = y_o \cos\omega_n t + \frac{v_o}{\omega_n}\sin\omega_n t + \frac{F}{K}(\cos\omega t - \cos\omega_n t).} \tag{3.19}$$

Frequency ratio r > 1: For $r > 1$, the value of $(1 - r^2)$ in equation (3.18) is negative. Therefore, the response for y(t) can be written:

$$y(t) = y_o \cos\omega_n t + \frac{v_o}{\omega_n}\sin\omega_n t - \frac{F}{K}\left(\frac{\cos\omega t - \cos\omega_n t}{r^2-1}\right). \tag{3.20}$$

If $r >> 1$, the value of (r^2-1) becomes r^2 and equation (3.20) becomes:

$$y(t) = y_o \cos\omega_n t + \frac{v_o}{\omega_n}\sin\omega_n t - \frac{F}{K}\left(\frac{\cos\omega t - \cos\omega_n t}{r^2}\right). \tag{3.21}$$

or:

$$\boxed{y(t) = y_o \cos\omega_n t + \frac{v_o}{\omega_n}\sin\omega_n t - \frac{F}{M\omega^2}(\cos\omega t - \cos\omega_n t).} \tag{3.22}$$

When $\omega >> F/M$, equation (3.22) becomes:

$$y(t) = y_o \cos\omega_n t + \frac{v_o}{\omega_n}\sin\omega_n t. \tag{3.23}$$

Frequency ratio r = 1: When $r = 1$, $\omega = \omega_n$. Therefore, the last term in equation (3.18) becomes indefinite. Both the numerator and denominator equal zero. L'Hopital's rule must be used to determine the limit of the last term of equation (3.18) as ω approaches ω_n:

$$\lim_{\omega\to\omega_n}\left(\frac{\cos\omega t - \cos\omega_n t}{1-\dfrac{\omega^2}{\omega_n^2}}\right) = \lim_{\omega\to\omega_n}\left(\frac{\dfrac{d}{d\omega}(\cos\omega t - \cos\omega_n t)}{\dfrac{d}{d\omega}\left(1-\dfrac{\omega^2}{\omega_n^2}\right)}\right) \tag{3.24}$$

or:

$$\lim_{\omega\to\omega_n}\left(\frac{\cos\omega t-\cos\omega_n t}{1-\dfrac{\omega^2}{\omega_n^2}}\right)=\lim_{\omega\to\omega_n}\left(\frac{-t\sin\omega t}{-\dfrac{2\omega}{\omega_n^2}}\right). \tag{3.25}$$

Therefore, when r = 1, equation (3.18) becomes:

$$\boxed{y(t)=y_o\cos\omega_n t+\frac{v_o}{\omega_n}\sin\omega_n t+\frac{F\,t}{2\sqrt{KM}}\sin\omega_n t.} \tag{3.26}$$

Equation (3.26) indicates when $\omega=\omega_n$ the response y(t) of the system increases linearly with time. As t becomes very large, the response also becomes very large.

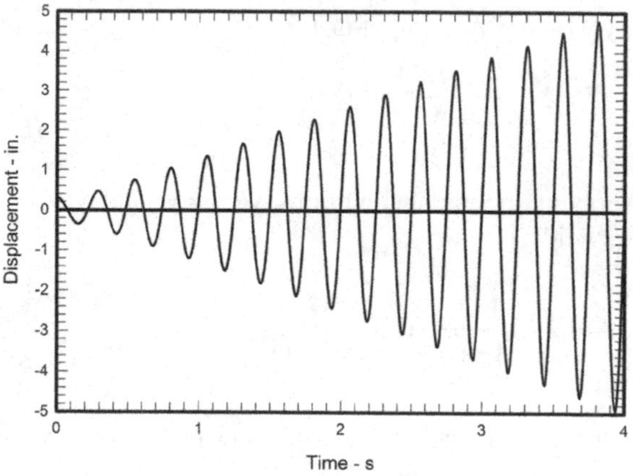

Figure 3.2 Response of a 1-DOF Vibration System without Damping when Excited at Its Resonance Frequency

EXAMPLE 3.1

A 1-DOF vibration system similar to Figure 3.1 without damping that is excited at resonance has the following properties: F = 100 lb_f, K = 1,000 lb_f/in., ω_n = 25.132 rad/s, y_o = 0.3 in. and v_o = 0. Write the expression for the response y(t) and plot y(t) as a function of t.

SOLUTION

The expression for the response y(t) is:

$$y(t)=0.3\cos(25.132\,t)+\frac{100\times 25.132\,t}{2\times 1{,}000}\sin(25.132\,t). \tag{3.1a}$$

or:

$$\boxed{y(t)=0.3\cos(25.132\,t)+1.257\,t\sin(25.132\,t).} \tag{3.1b}$$

Figure 3.2 shows a plot of equation (3.1b)

Displacement Response - Forced Response Only

Often when examining the response of a 1-DOF mass-spring vibration system to a harmonic force, the initial conditions are assumed to be zero. When this is the case, only the particular solution is considered. Therefore, the response y(t) is given by equation (3.12). The expression for $\bar{Y}$ in equation (3.11) can be written:

$$\bar{Y}=\frac{F/K}{\left|1-r^2\right|}e^{-j\phi} \tag{3.27}$$

where:

$$\phi=0 \quad \text{when} \quad r<1$$

$$\phi=\pi \quad \text{when} \quad r>1. \tag{3.28}$$

The response y(t) can then be written:

$$\boxed{y(t)=\frac{F/K}{\left|1-r^2\right|}e^{j(\omega t-\phi)}.} \tag{3.39}$$

Equation (3.29) is the response of all 1-DOF vibration systems without damping that are excited by a harmonic forcing function.

It is desirable to plot the value of Y as a function of frequency ratio r. To do this, Y is written:

$$Y=\frac{F/K}{\left|1-r^2\right|} \tag{3.30}$$

or:

$$\boxed{\frac{Y}{F/K} = \frac{1}{\left|1 - r^2\right|}} \tag{3.31}$$

where Y/(F/K) is referred to as the *frequency response function*. Figure 3.3 shows a plot of Y/(F/K) as a function of the frequency ratio r for a 1-DOF vibration system without damping. At low frequencies where r << 1, Y becomes:

$$Y = \frac{F}{K}. \tag{3.32}$$

At high frequencies where r >> 1, Y becomes:

$$Y = \frac{F}{M\,\omega^2}. \tag{3.33}$$

The value of Y decreases as a function of ω^2. Y can be further reduced by decreasing F, increasing M, or both.

When r = 1, equation (3.29) implies the value of Y goes to infinity. However, referring to equation (3.1b), it takes a finite period of time for Y to become very large. The time for this to occur is a function of the ratio $F/\sqrt{KM}$. When $\sqrt{KM} >> F$, it may take a long time for the system response to become dangerously large. If the system is excited at resonance for only a short period of time and then the frequency is shifted to below or above the resonance frequency, even though the system response may be noticeable, it will not be dangerously noticeable. As will be shown later, when damping is present, even if the system is excited at resonance for a long period of time, the value of Y will reach a limiting value. This occurs when the energy directed into the system by the exciting force equals the energy removed from the system by damping.

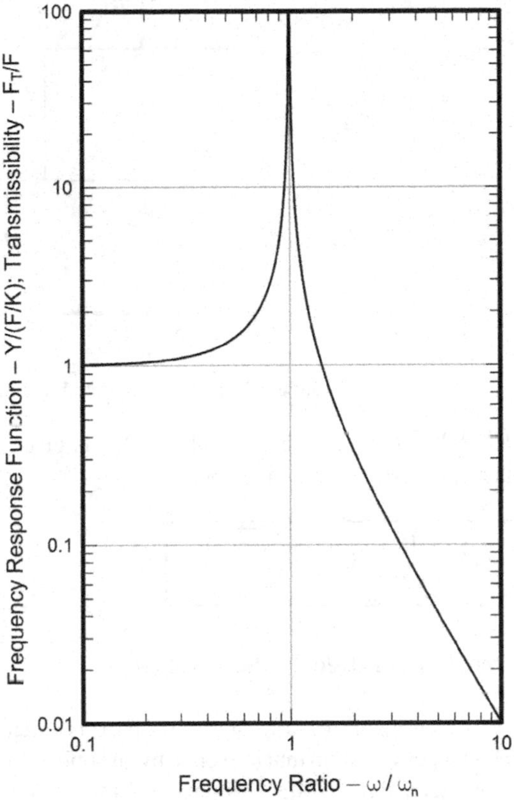

Figure 3.3 Amplitudes of the Frequency Response Function and Transmissibility for a 1-DOF Vibration System with No0 Damping

EXAMPLE 3.2

Determine the expressions for obtaining the value of r in equation (3.31) in terms of the value for KY/F.

SOLUTION

Rearranging equation (3.31):

$$\left|1 - r^2\right| = \frac{F}{KY}. \tag{3.2a}$$

When KY/F = 1, r = 0 or:

$$r^2 - 1 = 1. \tag{3.2b}$$

Therefore:

$$\boxed{r = \sqrt{2} = 1.414.} \tag{3.2c}$$

When KY/F < 1:

$$\boxed{r^2 - 1 = \frac{F}{KY} \quad \text{or} \quad r = \sqrt{1 + \frac{F}{KY}}.} \tag{3.2d}$$

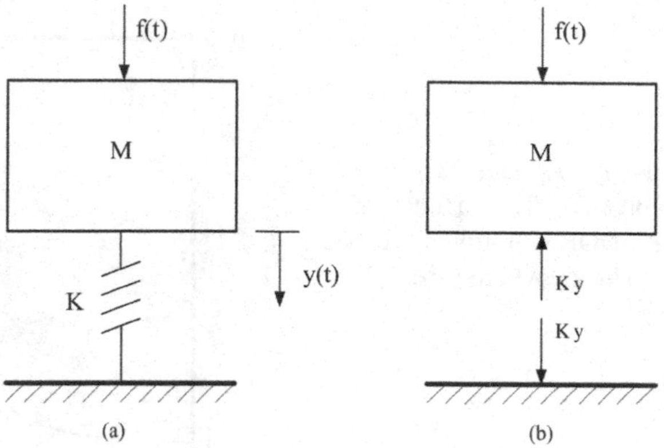

Figure 3.4 Force Transmited to a Structural Base where Damping Is Not Present

When KY/F > 1, r has two values. One is given by equation (3.2d) and the other by:

$$\boxed{1 - r^2 = \frac{F}{KY} \quad \text{or} \quad r = \sqrt{1 - \frac{F}{KY}}.} \tag{3.2e}$$

Force Transmitted to the Base

The force transmitted to a structural base is often of interest when a system that is excited by an applied force is examined [Figure 3.4(a)]. Figure 3.4(b) shows a free-body diagram of the force transmitted to the base. It is:

$$f_T(t) = K\, y(t). \tag{3.34}$$

Substituting equation (3.29) into equation (3.34) yields:

$$f_T(t) = \frac{F}{|1 - r^2|} e^{j(\omega t - \phi)}. \tag{3.35}$$

The value of the force F_T transmitted to the base is:

$$F_T = \frac{F}{|1 - r^2|}. \tag{3.36}$$

The ratio of the transmitted force F_T divided by the applied force F is:

$$\boxed{\frac{F_T}{F} = \frac{1}{|1 - r^2|}.} \tag{3.37}$$

F_T/F is referred to as the *transmissibility*. Equation (3.37) is the same as equation (3.31). Therefore, Figure 3.3 also shows a plot of equation (3.37). As with the frequency response ratio, the transmissibility has three frequency regions. At low frequencies where r << 1, equation (3.37) becomes:

$$F_T = F. \tag{3.38}$$

At frequencies where r >> 1, equation (3.37) becomes:

$$F_T = \frac{F\,\omega_n^2}{\omega^2} \quad \text{or} \quad F_T = \frac{F\,K}{M\,\omega^2}. \tag{3.39}$$

The value of F_T decreases as a function of ω^2. The value of F_T can be reduced by increasing M, decreasing K or both. Changes in $|\bar{F}_T|$ must be coordinated with changes in Y.

3.4 Response of a 1-DOF Vibration System with Viscous Damping

Displacement Response - Total Response

When a 1-DOF vibration system has a viscous damping element, equation (3.2) must be solved. If the damping is less than critical damping, the complimentary solution is given by equation (2.45) in Section 2.3, or:

$$y_c(t) = (A \cos \omega_d t + B \sin \omega_d t)\, e^{-\xi \omega_n t}. \tag{3.40}$$

To obtain the particular solution, equations (3.7) and

(3.8) are substituted into equation (3.2) to obtain:

$$\left(-M\omega^2 + jC\omega + K\right)\bar{Y}\, e^{j\omega t} = F\, e^{j\omega t}\,. \tag{3.41}$$

Rearranging the terms yields:

$$\bar{Y} = \frac{F}{K - M\omega^2 + jC\omega}\,. \tag{3.42}$$

Dividing the numerator and denominator by K and noting:

$$\frac{M}{K} = \frac{1}{\omega_n^2} \quad \text{and} \quad \frac{C}{K} = 2\xi\frac{1}{\omega_n}\,. \tag{3.43}$$

gives:

$$\bar{Y} = \frac{F/K}{1 - \dfrac{\omega^2}{\omega_n^2} + j2\xi\dfrac{\omega}{\omega_n}}\,. \tag{3.44}$$

Substituting equation (3.13) for the frequency ratio r into equation (3.44) yields:

$$\bar{Y} = \frac{F/K}{1 - r^2 + j2\xi r}\,. \tag{3.45}$$

The denominator of equation (3.45) can be written:

$$1 - r^2 + j2\xi r = \sqrt{\left(1 - r^2\right)^2 + \left(2\xi r\right)^2}\; e^{j\phi} \tag{3.46}$$

where:

$$\boxed{\phi = \tan^{-1}\left(\frac{2\xi r}{1 - r^2}\right).} \tag{3.47}$$

Therefore, equation (3.45) can be written:

$$\bar{Y} = \frac{(F/K)\, e^{-j\phi}}{\sqrt{\left(1 - r^2\right)^2 + \left(2\xi r\right)^2}}\,. \tag{3.48}$$

The particular solution is obtained by substituting equation (3.48) into equation (3.8):

$$\boxed{y_p(t) = \frac{F/K}{\sqrt{\left(1 - r^2\right)^2 + \left(2\xi r\right)^2}}\, e^{j(\omega t - \phi)}\,.} \tag{3.49}$$

Writing the response in terms of the real part of $e^{j(\omega t - \phi)}$:

$$y_p(t) = \frac{F/K}{\sqrt{\left(1 - r^2\right)^2 + \left(2\xi r\right)^2}}\cos(\omega t - \phi)\,. \tag{3.50}$$

The total solution is the sum of the complimentary and particular solutions, or:

$$\boxed{\begin{aligned} y(t) = {} & \left(A\cos\omega_d t + B\sin\omega_d t\right) e^{-\xi\omega_n t} + \\ & \frac{F/K}{\sqrt{\left(1 - r^2\right)^2 + \left(2\xi r\right)^2}}\cos(\omega t - \phi)\,. \end{aligned}} \tag{3.51}$$

A and B are obtained by applying the initial conditions to equation (3.51).

Attempting to solve for A and B in equation (3.51) can be an involved process. When forced vibration of systems with viscous damping is examined, the response of the system to its initial conditions is normally ignored. It is assumed damping will eventually cause the complimentary part of the solution to go to zero.

Displacement Response - Forced Response Only

Referring to equation (3.49), the response of a 1-DOF vibration system with viscous damping to an applied harmonic force is:

$$\boxed{y(t) = \frac{F/K}{\sqrt{\left(1 - r^2\right)^2 + \left(2\xi r\right)^2}}\, e^{j(\omega t - \phi)}\,.} \tag{3.52}$$

where ϕ is given by equation (3.47), and r is given by equation (3.13).

The response of a 1-DOF vibration system with viscous damping can be described in terms of the *frequency response function*. Referring to equation (3.45), equation (3.52) can be written:

$$y(t) = \bar{Y}\, e^{j\omega t} = \frac{F}{K}\left[\frac{1}{1-r^2+j2\xi r}\right] e^{j\omega t}. \tag{3.53}$$

The *frequency response function* (*or complex frequency response function*) $\bar{H}(\omega)$ of the system is defined as:

$$\bar{H}(\omega) = \frac{\bar{Y}}{F/K} \tag{3.54}$$

where:

$$\bar{H}(\omega) = \left[\frac{1}{1-r^2+j2\xi r}\right]. \tag{3.55}$$

Therefore, equation (3.53) can be written:

$$y(t) = \frac{F}{K}\bar{H}(\omega)\, e^{j\omega t}. \tag{3.56}$$

Equations (3.53) and (3.56) indicate the displacement can be represented as a complex quantity that has both real and imaginary components.

These components of the frequency response function $\bar{H}(\omega)$ can be obtained by multiplying the numerator and denominator of equation (3.55) by the complex conjugate of the denominator and simplifying, or:

$$\bar{H}(\omega) = \frac{1-r^2}{\left(1-r^2\right)^2+\left(2\xi r\right)^2} - \frac{j2\xi r}{\left(1-r^2\right)^2+\left(2\xi r\right)^2}. \tag{3.57}$$

Equation (3.57) indicates $\bar{H}(\omega)$ has one component:

$$\boxed{\mathrm{Re}\left[\bar{H}(\omega)\right] = \frac{1-r^2}{\left(1-r^2\right)^2+\left(2\xi r\right)^2}} \tag{3.58}$$

that is *in phase* with the applied force and a second component:

$$\boxed{\mathrm{Im}\left[\bar{H}(\omega)\right] = \frac{-j2\xi r}{\left(1-r^2\right)^2+\left(2\xi r\right)^2}} \tag{3.59}$$

that has a phase lag of $\pi/2$ radians relative to the applied force. The imaginary component is said to be in *quadrature* with the applied force.

The frequency response function can be written:

$$\boxed{\bar{H}(\omega) = \left|\bar{H}(\omega)\right| e^{j\phi}} \tag{3.60}$$

where:

$$\left|\bar{H}(\omega)\right| = \sqrt{\mathrm{Re}^2\left[\bar{H}(\omega)\right] + \mathrm{Im}^2\left[\bar{H}(\omega)\right]}. \tag{3.61}$$

and:

$$\phi = \tan^{-1}\left[\frac{\mathrm{Im}\left[\bar{H}(\omega)\right]}{\mathrm{Re}\left[\bar{H}(\omega)\right]}\right]. \tag{3.62}$$

Substituting equations (3.58) and (3.59) into equations (3.61) and (3.62) and simplifying yields:

$$\boxed{\left|\bar{H}(\omega)\right| = \frac{1}{\sqrt{\left(1-r^2\right)^2+\left(2\xi r\right)^2}}} \tag{3.63}$$

and noting $\tan(-\phi) = -\tan(\phi)$:

$$\boxed{\phi = -\tan^{-1}\left(\frac{2\xi r}{1-r^2}\right).} \tag{3.64}$$

ϕ is the *phase lag* associated with y(t) relative to the applied force f(t).

The response y(t) is obtained by substituting equation (3.60) into (3.56):

$$\boxed{y(t) = \frac{F}{K}\left|\bar{H}(\omega)\right| e^{j(\omega t+\phi)}.} \tag{3.65}$$

Substituting equation (3.63) into equation (3.65) gives:

$$\boxed{y(t) = \frac{F/K}{\sqrt{\left(1-r^2\right)^2+\left(2\xi r\right)^2}}\, e^{j(\omega t+\phi)}} \tag{3.66}$$

where ϕ is given by equation (3.64) and where:

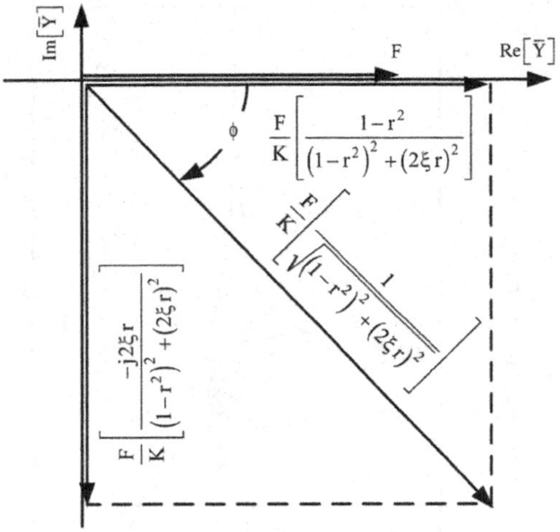

Figure 3.5 Real and Imaginary Components of the Frequency Respnse Function Relative to the Excitation Force F

$$\boxed{Y = \frac{F/K}{\sqrt{(1-r^2)^2+(2\xi r)^2}}} \tag{3.67}$$

Referring to the negative sign in equation (3.64) and the lack of one in equation (3.47), equation (3.66) is the same as equation (3.52). *Equations (3.65) through (3.67) represent the response of all 1-DOF vibration systems with viscous damping to an applied harmonic force.*

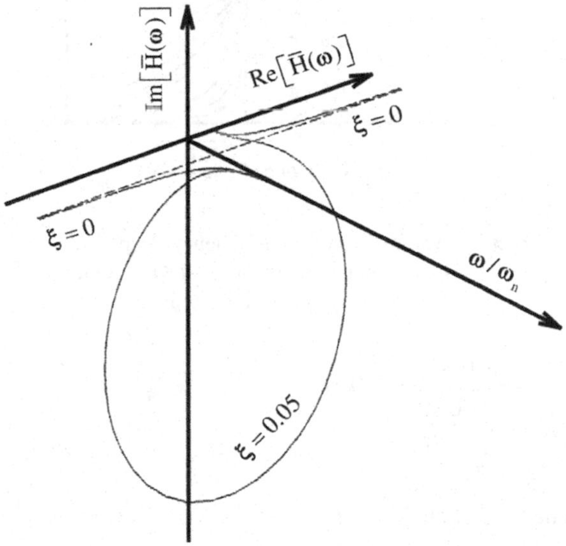

Figure 3.6 Three-Dimensional Plot of the Frequency Response Function for a Vibration System with Viscous Damping

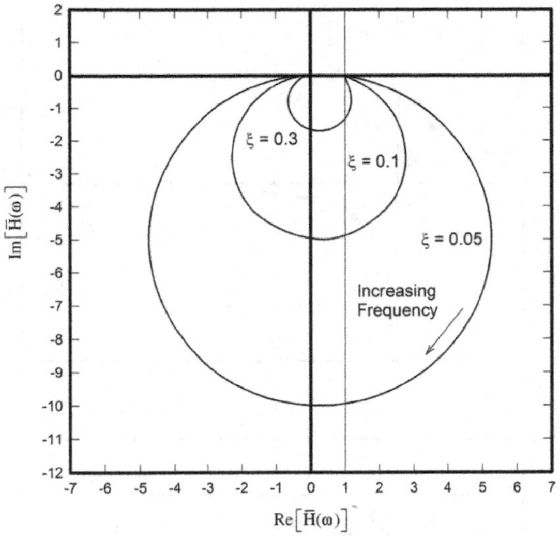

Figure 3.7 Nyquist Plots of the Real and Imaginary Components of the Frequency Response Function for a 1-DOF Vibration System with Viscous Damping for Selected Values of the Damping Ratio ξ

Figure 3.5 shows a plot of the real and complex components of the displacement vector and of the displacement vector relative to the force vector in the complex plane. Figure 3.6 shows a three-dimensional plot of the frequency response function $\bar{H}(\omega)$ as a function of the frequency ratio ω/ω_n. The dashed line represents the case where the damping ratio ξ equals zero. For this case, the curve solely lies in the real plane represented by Re[$\bar{H}(\omega)$] and ω/ω_n. Figure 3.7 shows projections of the three-dimensional plot of the frequency response function onto the complex plane represented by Re[$\bar{H}(\omega)$] and Im[$\bar{H}(\omega)$] for three different values of the damping ratio ξ. This plot is called a *Nyquist plot*. Figure 3.8 shows projections of the frequency response function onto the {Re[$\bar{H}(\omega)$], ω/ω_n} and {Im[$\bar{H}(\omega)$], ω/ω_n} planes, respectively. Figure 3.9 shows plots of the values of the frequency response function $|\bar{H}(\omega)|$ and phase ϕ as a function of the frequency ratio ω/ω_n.

EXAMPLE 3.3

A 1-DOF vibration system similar to the one shown in

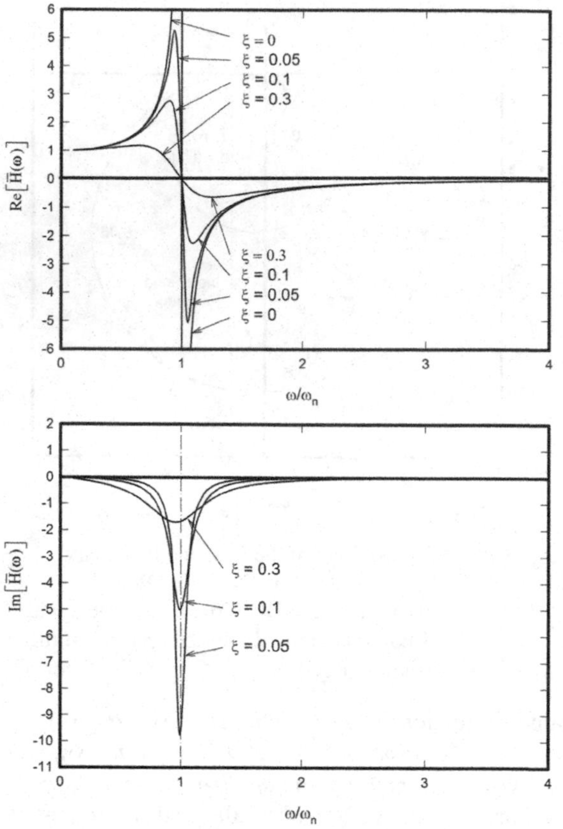

Figure 3.8 Real and Imaginary Components of the Frequency Response Function for a 1-DOF Vibration System with Viscous Damping

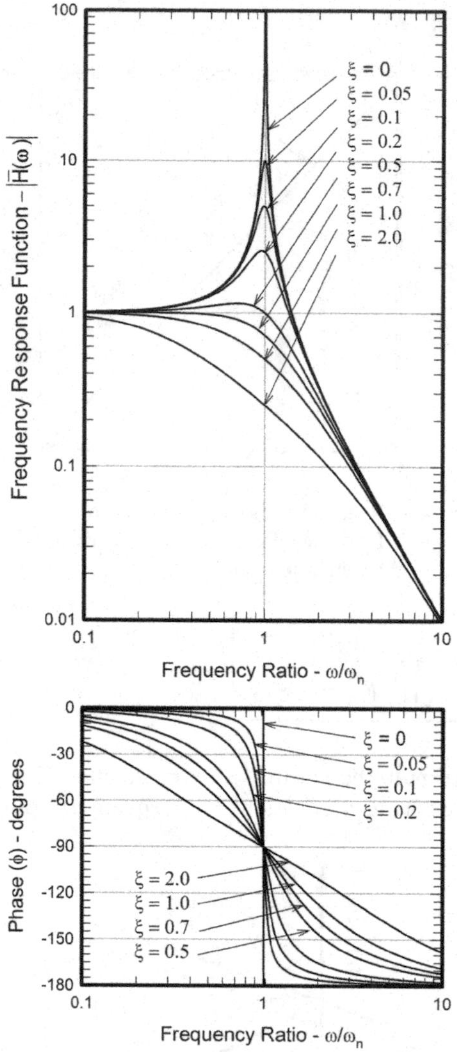

Figure 3.9 Amplitudes of the Frequency Response Function and Phase for a 1-DOF Vibration System with Viscouse Damping

Figure 3.1 has the following characteristics: We = 40 lb_f, K = 50 lb_f/in. and C = 0.85 lb_f-s/in. The system is excited by an external force with a value of F = 10 lb_f and a frequency $\omega = 15$ rad/s. Determine: (a) the value of the harmonic response function $\left|\bar{H}(\omega)\right|$; (b) the phase angle ϕ; and (c) the displacement value Y.

SOLUTION

(a) The resonance frequency is:

$$\omega_n = \sqrt{\frac{50}{40/386}} = 22.0\ \frac{\text{rad}}{\text{s}}. \tag{3.3a}$$

The damping ratio is:

$$\xi = \frac{0.85}{2\sqrt{\frac{40 \times 50}{386}}} = 0.187. \tag{3.3b}$$

The frequency ratio is:

$$\frac{\omega}{\omega_n} = \frac{15}{22} = 0.682 \tag{3.3c}$$

The value of the frequency response function is obtained from equation (3.63) or:

$$\left|\bar{H}(\omega)\right| = \frac{1}{\sqrt{\left(1-0.682^2\right)^2 + \left(2\times 0.187\times 0.682\right)^2}} = 1.69\,. \tag{3.3d}$$

(b) The phase lag is obtained from equation (3.64), or:

$$\phi = -\tan^{-1}\left(\frac{2\times 0.187\times 0.682}{1-0.682^2}\right) = -0.445 \text{ rad}\,. \tag{3.3e}$$

(c) The displacement value Y is obtained from equation (3.67), or:

$$Y = \frac{10}{50}\times 1.69 = 0.338 \text{ in.} \tag{3.3f}$$

Therefore, the response of the system is:

$$\boxed{y(t) = 0.338\, e^{j(15t-0.445)}} \tag{3.3g}$$

or:

$$\boxed{y(t) = 0.338\cos(15t-0.445)\,.} \tag{3.3h}$$

3.5 Characteristic of 1-DOF Vibration Systems with Viscous Damping

An examination of Figure 3.9 indicates when $\xi < 1$ the maximum value of the frequency response function does not occur at ω_n. It occurs at some value less than ω_n. The frequency ω at which the frequency response function is a maximum can be obtained by taking the derivative of equation (3.63) with respect to ω and setting the results equal to zero. Rewrite equation (3.63) in the form:

$$\left|\bar{H}(\omega)\right| = \frac{\omega_n^2}{\sqrt{\left(\omega_n^2-\omega^2\right)^2 + 4\xi^2\omega_n^2\omega^2}}\,. \tag{3.68}$$

Taking the derivative of equation (3.68) with respect to ω yields:

$$\frac{d\left|\bar{H}(\omega)\right|}{d\omega} = \frac{-\dfrac{\omega_n^2}{2}\dfrac{d}{d\omega}\left[\left(\omega_n^2-\omega^2\right)^2 + \left(2\xi\omega_n\omega\right)^2\right]}{\left[\left(\omega_n^2-\omega^2\right)^2 + \left(2\xi\omega_n\omega\right)^2\right]^{3/2}}\,. \tag{3.69}$$

Equation (3.69) equals zero when the numerator equals zero. Therefore:

$$\frac{d}{d\omega}\left[\left(\omega_n^2-\omega^2\right)^2 + \left(2\xi\omega_n\omega\right)^2\right] = 0 \tag{3.70}$$

or:

$$-4\omega\left(\omega_n^2-\omega^2\right) + 8\xi^2\omega_n^2\omega = 0\,. \tag{3.71}$$

Simplifying equation (3.71), the frequency ω_{max} at which the value of the harmonic response function $\left|\bar{H}(\omega)\right|$ is a maximum is:

$$\boxed{\omega_{max} = \omega_n\sqrt{1-2\xi^2}\,.} \tag{3.72}$$

The expression within the square root symbol must be real-valued. Therefore, equation (3.72) only has a real value when $\xi < 0.707$.

When $\xi < 0.707$, the maximum value for $\left|\bar{H}(\omega)\right|$ is obtained by substituting equation (3.72) into equation (3.68) and simplifying:

$$\boxed{\left|\bar{H}(\omega)\right|_{max} = \frac{1}{2\xi\sqrt{1-\xi^2}}\,.} \tag{3.73}$$

When $\xi < 0.3$, the maximum value for $\left|\bar{H}(\omega)\right|$ occurs approximately at $\omega/\omega_n = 1$. For this case, the maximum value of $\left|\bar{H}(\omega)\right|$ is:

$$\boxed{\left|\bar{H}(\omega_n)\right| \approx \frac{1}{2\xi}\,.} \tag{3.74}$$

An examination of Figure 3.8 indicates Re[$\bar{H}(\omega)$] always equals zero when $\omega = \omega_n$. Furthermore, when $\xi < 0.707$, Re[$\bar{H}(\omega)$] has a maxima and minima. The frequencies at which these maxima and minima occur are obtained by taking the derivative of equation (3.58) with respect to ω and setting the results equal to zero. Rewrite equation (3.58) in the form:

$$\mathrm{Re}\left[\bar{H}(\omega)\right] = \frac{\omega_n^2\left(\omega_n^2 - \omega^2\right)}{\left(\omega_n^2 - \omega^2\right)^2 + 4\xi^2\,\omega_n^2\,\omega^2}. \tag{3.75}$$

Taking the derivative of equation (3.75) with respect to ω yields:

$$\frac{d\,\mathrm{Re}\left[\bar{H}(\omega)\right]}{d\omega} = \omega_n^2 \left[\frac{-2\omega}{\left(\omega_n^2 - \omega^2\right)^2 + 4\xi^2\,\omega_n^2\,\omega^2} + \frac{-\left(\omega_n^2 - \omega^2\right)\begin{bmatrix} -4\omega\left(\omega_n^2 - \omega^2\right) + \\ 8\xi^2\,\omega_n^2\,\omega \end{bmatrix}}{\left[\left(\omega_n^2 - \omega^2\right)^2 + 4\xi^2\,\omega_n^2\,\omega^2\right]^2} \right] \tag{3.76}$$

or:

$$\frac{d\,\mathrm{Re}\left[\bar{H}(\omega)\right]}{d\omega} = \omega_n^2 \frac{\begin{bmatrix} -2\omega\left[\left(\omega_n^2 - \omega^2\right)^2 + 4\xi^2\,\omega_n^2\,\omega^2\right] - \\ \left(\omega_n^2 - \omega^2\right)\begin{bmatrix} -4\omega\left(\omega_n^2 - \omega^2\right) + \\ 8\xi^2\,\omega_n^2\,\omega \end{bmatrix} \end{bmatrix}}{\left[\left(\omega_n^2 - \omega^2\right)^2 + 4\xi^2\,\omega_n^2\,\omega^2\right]^2}. \tag{3.77}$$

Equation (3.77) equals zero when its numerator equals zero. Setting the numerator of equation (3.77) equal to zero and simplifying yields:

$$\left(\omega_n^2 - \omega^2\right)^2 - 4\xi^2\,\omega_n^4 = 0 \tag{3.78}$$

or:

$$\left(\omega_n^2 - \omega^2\right)^2 = 4\xi^2\,\omega_n^4. \tag{3.79}$$

Taking the square root of both sides of equation (3.79) yields:

$$\omega_n^2 - \omega^2 = \pm 2\xi\,\omega_n^2. \tag{3.80}$$

To find the frequency ω_1 for the maxima, use the equation:

$$\omega_n^2 - \omega^2 = 2\xi\,\omega_n^2. \tag{3.81}$$

Rearranging equation (3.80) for ω_1 yields:

$$\boxed{\omega_1 = \omega_n\sqrt{1 - 2\xi}\,.} \tag{3.82}$$

To find the frequency ω_2 for the minima, use the equation:

$$\omega_n^2 - \omega^2 = -2\xi\,\omega_n^2. \tag{3.83}$$

Rearranging equation (3.82) for ω_2 yields:

$$\boxed{\omega_2 = \omega_n\sqrt{1 + 2\xi}\,.} \tag{3.84}$$

Values for the maxima and minima associated with Re[$\bar{H}(\omega)$] are obtained by substituting equations (3.82) and (3.84) into equation (3.63):

$$\left|\bar{H}(\omega_1)\right| = \frac{1}{2\sqrt{2}\,\xi\sqrt{1-\xi}}. \tag{3.85}$$

$$\left|\bar{H}(\omega_2)\right| = \frac{1}{2\sqrt{2}\,\xi\sqrt{1+\xi}}. \tag{3.86}$$

When $\xi < 0.1$, equations (3.85) and (3.86) reduce to:

$$\left|\bar{H}(\omega_1)\right| \approx \left|\bar{H}(\omega_2)\right| \approx \frac{1}{2\sqrt{2}\,\xi}. \tag{3.87}$$

Power is proportional to $\left|\bar{H}(\omega)\right|^2$. When $\xi < 0.1$, the half-power are obtained by dividing equation (3.87)

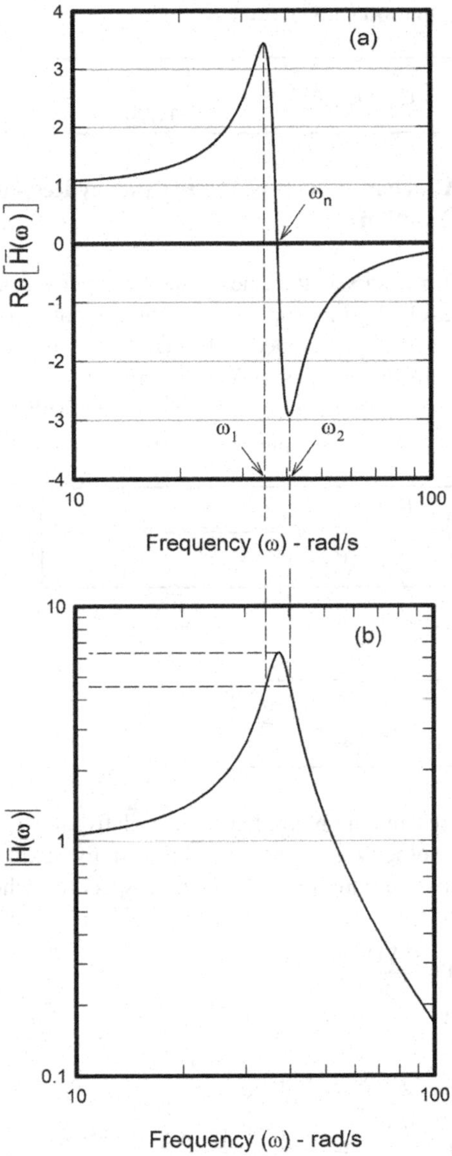

Figure 3.10 Half Power Points of a 1-DOF Vibration

squared by equation (3.74) squared:

$$\frac{|\bar{H}(\omega_1)|^2}{|\bar{H}(\omega_n)|^2} \approx \frac{|\bar{H}(\omega_2)|^2}{|\bar{H}(\omega_n)|^2} \approx \frac{1}{2} \tag{3.88}$$

The value specified by equation (3.88) is referred to as the *half power point(s)* (Figure 3.10). The frequency bandwidth $\Delta\omega$ between the two half power points is:

$$\Delta\omega = \omega_2 - \omega_1 \,. \tag{3.89}$$

where ω_1 and ω_2 are given by equations (3.82) and (3.84). Substituting these equations into equation (3.89) yields:

$$\boxed{\Delta\omega = \omega_n\left(\sqrt{1+2\xi} - \sqrt{1-2\xi}\right).} \tag{3.90}$$

The square root terms in equation (3.91) can be expanded in a Taylor series expansion. When $\xi < 0.1$, the second and higher order terms in the Taylor series expansion can be neglected. Therefore:

$$\omega_2 - \omega_1 \approx \omega_n\left[\left(1+\frac{1}{2}(2\xi)\right) - \left(1+\frac{1}{2}(-2\xi)\right)\right] \tag{3.91}$$

and:

$$\boxed{\xi \approx \frac{1}{2}\frac{\omega_2 - \omega_1}{\omega_n}.} \tag{3.92}$$

Equation (3.92) has the general restriction that $\xi < 0.1$.

ξ can be determined solely from ω_1 and ω_2. Dividing equation (3.84) by equation (3.82), and rearranging the terms yields:

$$\boxed{\xi = \frac{1}{2}\frac{\left(\frac{\omega_2}{\omega_1}\right)^2 - 1}{\left(\frac{\omega_2}{\omega_1}\right)^2 + 1}.} \tag{3.93}$$

Equation (3.93) does not have the restriction that $\xi < 0.1$.

When examining systems with viscous damping, a term referred to as the *Q or quality factor* is defined such that at the resonance frequency $Q = |\bar{H}(\omega_n)|$ where $|\bar{H}(\omega_n)|$ is given by equation (3.74). A high Q factor $Q > 10$ implies light damping, while a low Q factor $Q < 0.5$ implies significant damping.

When $r > 3$ and $\xi < 0.3$, equations (3.63) and (3.67) reduce to:

$$\left|\bar{H}(\omega)\right| \approx \frac{1}{r^2} \quad \text{and} \quad Y \approx \frac{F}{M\omega^2}. \tag{3.94}$$

This indicates when $\omega > 3\omega_n$ and $\xi < 0.3$ the value of Y decreases as a function of ω^2. Y can be further reduced by decreasing the value of F, increasing the value of M, or both. When the value of M is increased, noting that $\omega_n = \sqrt{K/M}$, the value of K must also be increased to maintain the same or achieve a lower value for ω_n.

EXAMPLE 3.4

A 1-DOF vibration system has the following properties: ω_n = 37.7 rad/s, ω_1 = 34.6 rad/s, and ω_2 = 40.6 rad/s. Figure 3.10 shows a plot of Re[$\bar{H}(\omega)$] and $\left|\bar{H}(\omega)\right|$ as a function of ω for the system.

(a) Determine the value of the damping ratio using equation (3.93).
(b) Determine the value of the damping ratio using equation (3.94).
(c) Determine the values of the frequency response function at the resonance frequency and at the half power points.

SOLUTION

(a) From equation (3.92):

$$\xi = \frac{1}{2}\frac{40.6 - 34.6}{37.7} = 0.080. \tag{3.4a}$$

(b) From Equation (3.93):

$$\xi = \frac{1}{2}\frac{\left(\frac{40.6}{34.6}\right)^2 - 1}{\left(\frac{40.6}{34.6}\right)^2 + 1} = 0.079. \tag{3.4b}$$

(c) The value of $\left|\bar{H}(\omega)\right|$ at the resonance frequency is obtained from equation (3.84), or:

$$\left|\bar{H}(\omega_n)\right| = \frac{1}{2 \times 0.079} = 6.26. \tag{3.4c}$$

The value of $\left|\bar{H}(\omega)\right|$ at the half power points is obtained from equation (3.87), or:

$$\left|\bar{H}(\omega_1)\right| = \left|\bar{H}(\omega_2)\right| = \frac{1}{2\sqrt{2} \times 0.079} = 4.42. \tag{3.4d}$$

3.6 Alternate Forms of the Frequency Response Function

In the previous section, the frequency response function was developed in terms of the displacement response. It can also be developed in terms of the velocity and acceleration response. With regard the frequency response function associated with the displacement response, substitute equation (3.68) into equation (3.65):

$$y(t) = \frac{F}{K}\frac{\omega_n^2}{\sqrt{\left(\omega_n^2 - \omega^2\right)^2 + 4\xi^2\,\omega_n^2\,\omega^2}}\, e^{j(\omega t + \phi_d)} \tag{3.95}$$

where:

$$\phi_d = -\tan^{-1}\left(\frac{2\xi\,\omega_n\,\omega}{\omega_n^2 - \omega^2}\right). \tag{3.96}$$

ϕ_d represents the phase lag of y(t) relative to f(t).

The velocity response is obtained by taking the derivative of equation (3.95) with respect to t where:

$$v(t) = \frac{d\,y(t)}{dt} \tag{3.97}$$

or:

$$v(t) = \frac{F}{K}\frac{j\omega\,\omega_n^2}{\sqrt{\left(\omega_n^2 - \omega^2\right)^2 + 4\xi^2\,\omega_n^2\,\omega^2}}\, e^{j(\omega t + \phi_d)} \tag{3.98}$$

v(t) can be written:

$$v(t) = \frac{F}{K}\frac{\omega\,\omega_n^2}{\sqrt{\left(\omega_n^2 - \omega^2\right)^2 + 4\xi^2\,\omega_n^2\,\omega^2}}\, e^{j(\omega t + \phi_v)} \tag{3.99}$$

where:

$$\phi_v = \phi_d + \frac{\pi}{2}. \tag{3.100}$$

ϕ_v represents the phase lag of v(t) relative to f(t).

The acceleration response is obtained by taking the second derivative of equation (3.95) with respect to t where:

$$a(t) = \frac{d^2\, y(t)}{dt^2} \tag{3.101}$$

or:

$$a(t) = \frac{F}{K} \frac{-\omega^2\, \omega_n^2}{\sqrt{\left(\omega_n^2 - \omega^2\right)^2 + 4\xi^2\, \omega_n^2\, \omega^2}}\, e^{j(\omega t + \phi_d)} \tag{3.102}$$

a(t) can be written:

$$a(t) = \frac{F}{K} \frac{\omega^2\, \omega_n^2}{\sqrt{\left(\omega_n^2 - \omega^2\right)^2 + 4\xi^2\, \omega_n^2\, \omega^2}}\, e^{j(\omega t + \phi_a)} \tag{3.103}$$

where:

$$\phi_a = \phi_d + \pi. \tag{3.104}$$

ϕ_a represents the phase lag of a(t) relative to f(t).

ISO Standard 7626-1, *Vibration and Shock - Experimental Determination of Mechanical Mobility - Part 1: Basic Definitions and Transducers*, indicates there are six related frequency response functions that can be used to describe the response of a vibration system. They are:

Dynamic compliance (Y/F - m/N)
Mobility ($\mathcal{Y}$ = V/F - m/N-s)*
Accelerance (A/F - m/N-s^2 or kg^{-1})
Dynamic stiffness (F/Y - N/m)
Impedance (Z = F/V - N-s/m)
Effective mass (F/A - N-s^2/m or kg).

Note: $\mathcal{Y}$ is used to denote mobility, whereas Y is used to denote displacement.

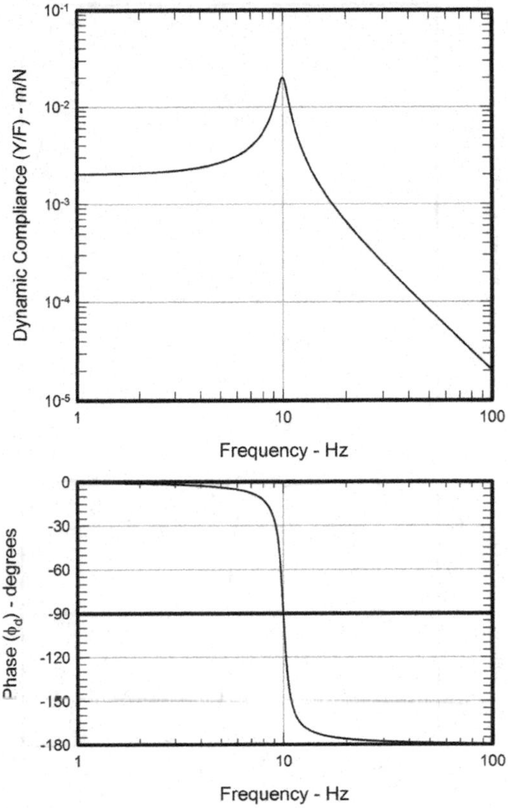

Figure 3.11 Dynamic Compliance Response for a 1-DOF Vibration System

It is easier to measure the response divided by the applied force. The values of the dynamic compliance, mobility, and accelerance associated with equations (3.95), (3.199) and (3.103) can be written:

$$\left|\frac{Y(\omega)}{F}\right| = \frac{1}{K} \frac{\omega_n^2}{\sqrt{\left(\omega_n^2 - \omega^2\right)^2 + 4\xi^2\, \omega_n^2\, \omega^2}} \tag{3.105}$$

$$\left|\frac{V(\omega)}{F}\right| = \frac{1}{K} \frac{\omega\, \omega_n^2}{\sqrt{\left(\omega_n^2 - \omega^2\right)^2 + 4\xi^2\, \omega_n^2\, \omega^2}} \tag{3.106}$$

$$\left|\frac{A(\omega)}{F}\right| = \frac{1}{K} \frac{\omega^2\, \omega_n^2}{\sqrt{\left(\omega_n^2 - \omega^2\right)^2 + 4\xi^2\, \omega_n^2\, \omega^2}}. \tag{3.107}$$

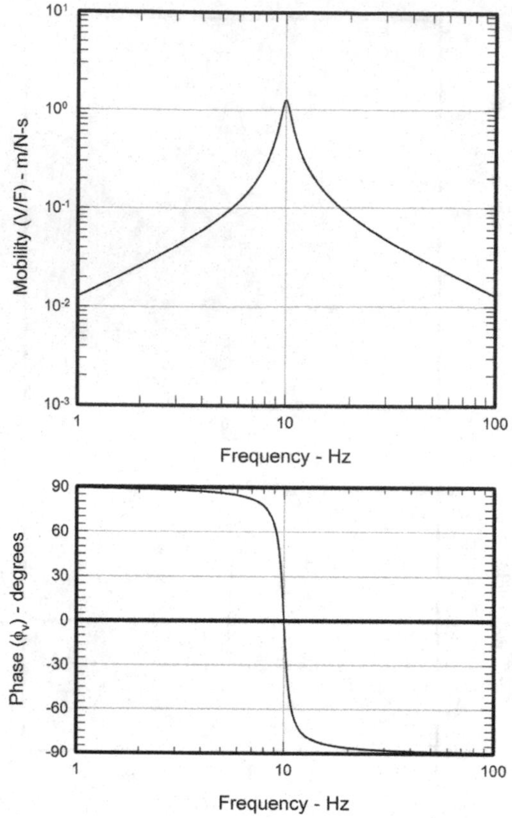

Figure 3.12 Mobility Response for a 1-DOF Vibration System

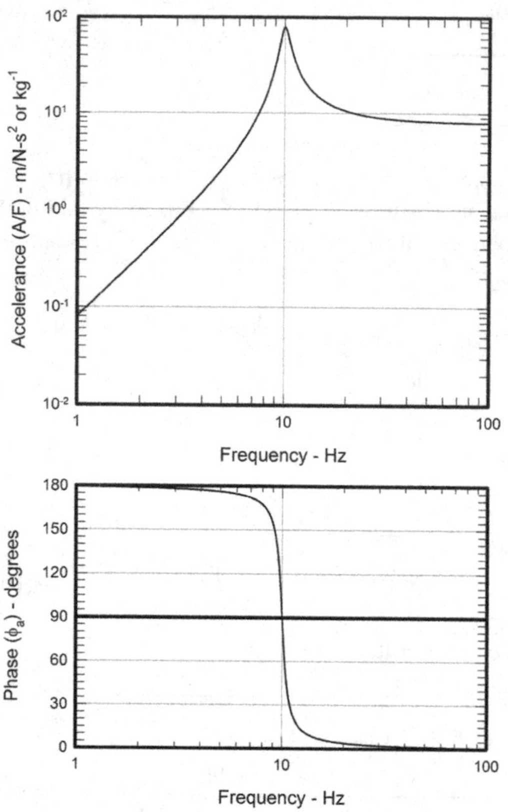

Figure 3.13 Accelerance Response for a 1-DOF Vibration System

The corresponding phases are given by equations (3.96), (3.100), and (3.104).

EXAMPLE 3.5

A 1-DOF vibration system has the following properties: $K = 500$ N/m, $f_n = 10$ Hz, and $\xi = 0.05$.
(a) Plot the dynamic compliance, mobility, and accelerance of the system as a function of frequency from a frequency of 10 Hz to 100 Hz.
(b) Plot the associated phases.

SOLUTION

Figure 3.11 shows a plot of the dynamic compliance and associated phase. Figure 3.12 shows a plot of the mobility and associated phase. Figure 3.13 shows a plot of the accelerance and associated phase.

3.7 Force Transmitted to a Structural Base of a 1-DOF Vibration Systems with Viscous Damping

Figure 3.14(b) shows a free-body diagram of the force transmitted to a structural base for a 1-DOF vibration system with viscous damping. The transmitted force is:

$$f_T(t) = K\,y(t) + C\,\dot{y}(t)\,. \tag{3.108}$$

Substituting equations (3.66) into equation (3.108) and carrying out the prescribed operations yields:

$$f_T(t) = \frac{F\left(1 + j\dfrac{C}{K}\omega\right)}{\sqrt{\left(1-r^2\right)^2 + \left(2\xi r\right)^2}}\, e^{j(\omega t + \phi)} \tag{3.109}$$

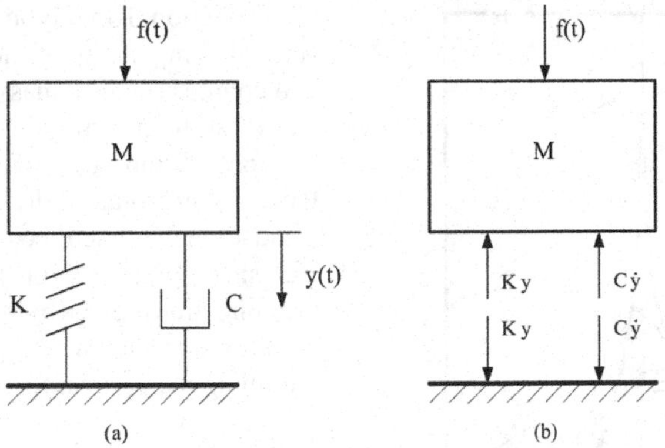

Figure 3.14 F orce Transmitted to a Structural Base where Damping Is Present

where ϕ is given by equation (3.64) and $C/K = 2\xi r$. Therefore, equation (3.109) can be written:

$$f_T(t) = \frac{F(1 + j\,2\xi r)}{\sqrt{(1-r^2)^2 + (2\xi r)^2}}\, e^{j(\omega t + \phi)} . \tag{3.110}$$

The numerator in equation (3.110) can be written:

$$F(1 + j\,2\xi r) = F\sqrt{1 + (2\xi r)^2}\; e^{j\theta} \tag{3.111}$$

where:

$$\boxed{\theta = \tan^{-1}(2\xi r).} \tag{3.112}$$

Therefore, equation (3.110) can be written:

$$f_T(t) = \frac{F\sqrt{1 + (2\xi r)^2}}{\sqrt{(1-r^2)^2 + (2\xi r)^2}}\, e^{j(\omega t + \phi + \theta)} . \tag{3.113}$$

The expression for the transmitted force is:

$$\bar{F}_T = \frac{F\sqrt{1 + (2\xi r)^2}}{\sqrt{(1-r^2)^2 + (2\xi r)^2}}\, e^{j(\phi + \theta)} . \tag{3.114}$$

The *force transmissibility* is defined as:

$$\boxed{\frac{\bar{F}_T}{F} = \frac{\sqrt{1 + (2\xi r)^2}}{\sqrt{(1-r^2)^2 + (2\xi r)^2}}\, e^{j(\phi + \theta)}} \tag{3.115}$$

where the value of the transmissibility is:

$$\boxed{\left|\frac{\bar{F}_T}{F}\right| = \frac{\sqrt{1 + (2\xi r)^2}}{\sqrt{(1-r^2)^2 + (2\xi r)^2}} .} \tag{3.116}$$

The transmitted force lags the applied force by a phase of $\phi + \theta$ where ϕ is given by equation (3.64) and θ is given by equation (3.112). Figure 3.15 shows plots of the transmissibility and phase for a 1-DOF vibration system with viscous damping.

When $r > 3$ and $\xi < 0.3$, equation 3.116 becomes:

$$\boxed{\left|\bar{F}_T\right| = \frac{F\,\omega_n \sqrt{\omega_n^2 + (2\xi\omega)^2}}{\omega^2} .} \tag{3.117}$$

When $r > 3$, the value of $\left|\bar{F}_T\right|$ decreases as a function of ω^2. $\left|\bar{F}_T\right|$ can be further reduced by decreasing the value of ω_n, ξ or both. Noting that $\omega_n = \sqrt{K/M}$, the value of ω_n can be reduced by increasing M, decreasing K or both. Changes in $\left|\bar{F}_T\right|$ must be coordinated with changes in Y.

3.8 Vibration Systems with Viscous Damping and Rotating and Reciprocating Unbalance

Many vibration systems have unbalanced rotating machine components. These include blowers, turbines,

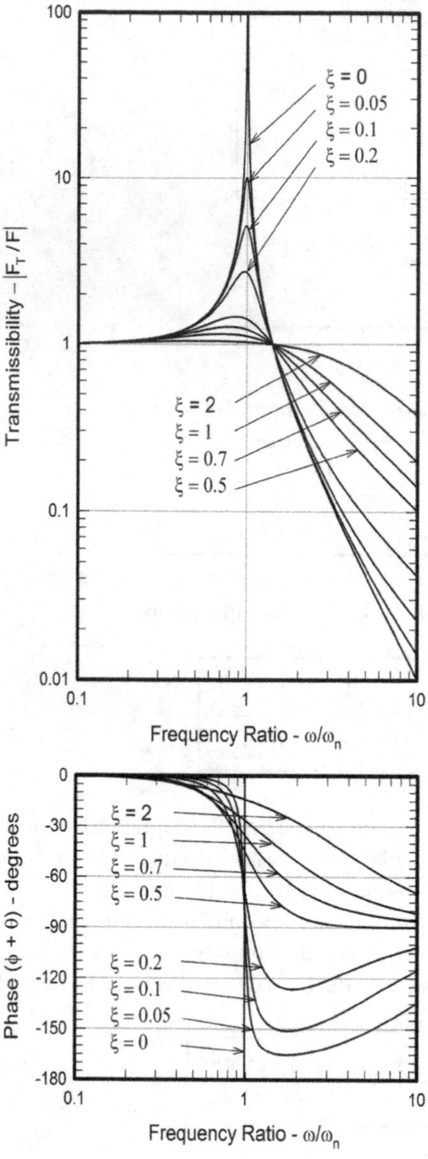

Figure 3.15 Amplitudes of Transmissibility and Phase for a 1-DOF System with Viscous Damping

pumps, fans, electric motors, etc. This unbalance can be represented as a mass m rotating with an eccentricity e that is specified as the product me. A schematic representation of a vibration system with mass M that has within it two counter rotating masses m/2 that rotate about points with a radius of rotation e is shown in Figure 3.16(a).

The system mass M is constrained to move only in the y direction. To analytically facilitate this motion, the vertical components of the forces associated with the two counter rotating masses m/2 combine to produce an overall force acting in the vertical y direction. The horizontal components of their forces cancel each other. If masses m/2 rotate with a constant angular velocities of ω and $-\omega$, their related accelerations in the y direction are $e\,\omega^2 \sin\omega t$ [Figure 3.16(a)], and the combined force in the y direction is given by $me\,\omega^2 \sin\omega t$ [Figure 3.16(b)]. This force is shown acting in the positive y direction Summing the vertical forces acting on M yields:

$$\sum F_y = M\ddot{y} \quad \text{or} \quad M\ddot{y} = me\,\omega^2 \sin\omega t - Ky - C\dot{y}. \tag{3.118}$$

Therefore:

$$M\ddot{y} + C\dot{y} + Ky = me\,\omega^2 \sin\omega t\,. \tag{3.119}$$

If the equivalent force F_{eq} is defined as:

$$F_{eq} = me\,\omega^2, \tag{3.120}$$

equation (3.118) can be written:

$$M\ddot{y} + C\dot{y} + Ky = F_{eq} \sin\omega t\,. \tag{3.121}$$

Noting that $\sin\omega t$ can be represented by $e^{j\omega t}$, equation (3.121) has the same form as equations (3.2) and (3.7). Therefore, noting $F_{eq} = me\,\omega^2$ the solution is of the form:

$$y(t) = \frac{me\,\omega^2}{K}\left|\bar{H}(\omega)\right| e^{j(\omega t + \phi)} \tag{3.122}$$

where ϕ is given by equation (3.64). If y(t) is multiplied and divided by M, equation (3.123) becomes:

$$\boxed{y(t) = \frac{me}{M}\frac{\omega^2}{\omega_n^2}\left|\bar{H}(\omega)\right| e^{j(\omega t + \phi)}\,.} \tag{3.123}$$

The value of Y is given by:

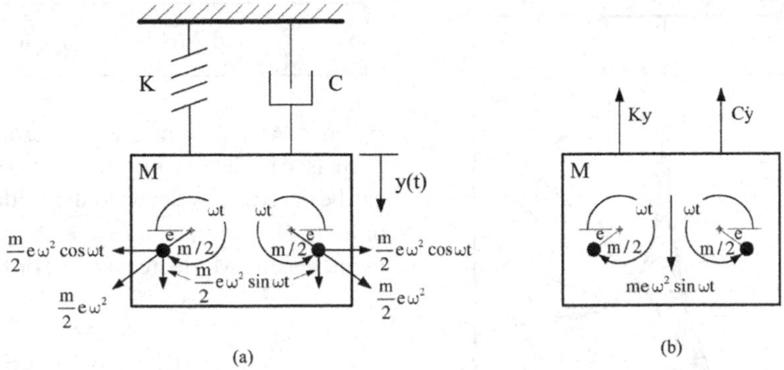

Figure 3.16 1-DOF Vibration System with Damping that Has a Rotating Unbalance

$$\boxed{Y = \frac{me}{M}\frac{\omega^2}{\omega_n^2}\left|\bar{H}(\omega)\right|} \tag{3.124}$$

or rearranging the terms:

$$\boxed{\frac{YM}{me} = \frac{\omega^2}{\omega_n^2}\left|\bar{H}(\omega)\right| \quad \text{or} \quad \frac{YM}{me} = r^2\left|\bar{H}(\omega)\right|.} \tag{3.125}$$

Equation (3.125) is a non-dimensional form of the response and is plotted in Figure 3.17 for several values of the damping ratio ξ.

The force F_T transmitted to the base is obtained by substituting equation (3.120) into equation (3.114) for F:

$$\bar{F}_T = \frac{me\,\omega^2\sqrt{1+(2\xi r)^2}}{\sqrt{\left(1-r^2\right)^2+(2\xi r)^2}}\,e^{j(\phi+\theta)} \tag{3.126}$$

where ϕ is given by equation (3.64) and θ is given by equation (3.112). Dividing and multiplying equation (3.127) by ω_n and defining:

$$\boxed{F_n = me\,\omega_n^2} \tag{3.127}$$

yields:

$$\boxed{\frac{\bar{F}_T}{F_n} = \frac{r^2\sqrt{1+(2\xi r)^2}}{\sqrt{\left(1-r^2\right)^2+(2\xi r)^2}}\,e^{j(\phi+\theta)}.} \tag{3.128}$$

Equation (3.128) can be written:

$$\frac{\bar{F}_T}{F_n} = r^2\left|\frac{\bar{F}_T}{F_n}\right| e^{j(\phi+\theta)} \tag{3.129}$$

where $|F_T/F|$ is given by equation (3.116) and:

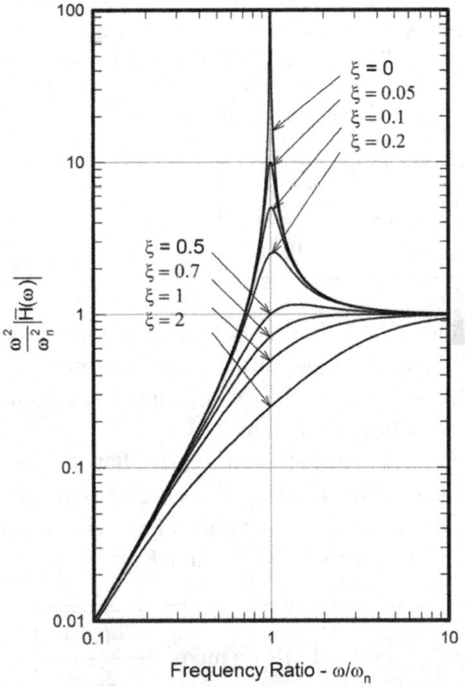

Figure 3.17 Harmonic Response Function Times the Frequency Ratio Squared for a 1-DOF Vibration System with Viscous Damping that Has a Rotating or Reciprocating Unbalance

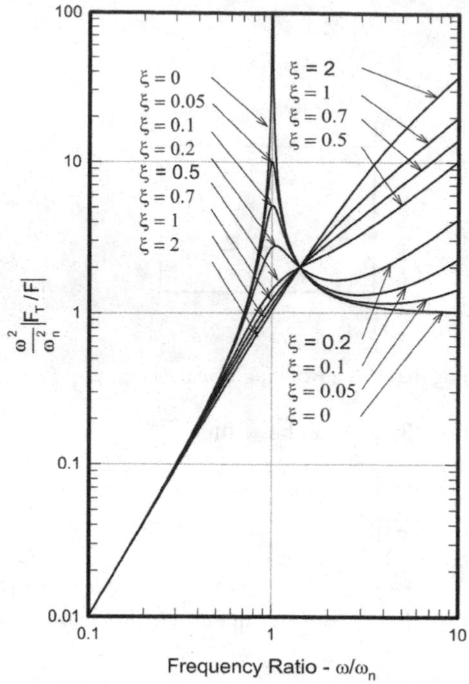

Figure 3.18 Transmissibility Times the Frequency Ratio Squared for a 1-DOF Vibration System with Viscous Damping that Hasa Rotatingo or Reciprocating Unbalance

$$\left|\frac{\bar{F}_T}{F_n}\right| = \frac{r^2\sqrt{1+(2\xi r)^2}}{\sqrt{(1-r^2)^2+(2\xi r)^2}}. \tag{3.130}$$

Equation (3.130) is a non-dimensional equation for determining the value of the force transmitted to the base for a system with a rotating unbalance. Equation (3.130) is plotted in Figure 3.18.

When the speed of rotation ω is significantly less than the resonance frequency ω_n, $F_{eq} = me\,\omega^2$ is low, and the values of vibration and the transmitted force are low.

When r = 1, the values of Y and $|F_T|$ are:

$$Y = \frac{me}{M}\frac{1}{2\xi} \quad \text{and} \quad \left|\bar{F}_T\right| = me\omega_n^2\,\frac{\sqrt{1+(2\xi)^2}}{2\xi}. \tag{3.131}$$

When r > 3 and $\xi < 0.3$, the values of Y and $|F_T|$ are:

$$Y = \frac{me}{M} \quad \text{and} \quad \left|\bar{F}_T\right| = me\,\omega_n\sqrt{\omega_n^2+(2\xi\omega)^2}. \tag{3.132}$$

When $\xi \neq 0$, $|\bar{F}_T|$ increase as ω increases. Y can be decreased by decreasing me, increasing M, or both. $|\bar{F}_T|$ can be decrease by decreasing me, decreasing ω_n and ξ, or both. Noting that $\omega_n = \sqrt{K/M}$, ω_n can be reduced by increasing M, decreasing K, or both.

EXAMPLE 3.6

A counter-rotating eccentric mass exciter was used to produce forced vibration of a 1-DOF vibration system as shown in Figure 3.16. By varying the speed of rotation, the vibration amplitude at the resonance frequency was found to be 0.55 in. As the speed of rotation of the eccentric masses was increased considerably beyond the resonance frequency, the value of vibration approached a fixed value of 0.07 in. Determine the damping ratio ξ for the vibration system.

SOLUTION

Equation (3.131) is used to obtain the response at ω_n:

$$Y = \frac{me}{M}\frac{1}{2\xi} \quad \text{or} \quad \xi = \frac{me}{M}\frac{1}{2Y}. \tag{3.6a}$$

When $\omega \gg \omega_n$, Y becomes:

$$Y = \frac{me}{M} \quad \text{or} \quad \frac{me}{M} = 0.07 \text{ in.} \tag{3.6b}$$

Substituting equation (3.5b) into the equation (3.5a) yields:

$$\xi = 0.07 \times \frac{1}{2\times 0.55} = 0.064. \tag{3.6c}$$

Vibration systems often have unbalanced reciprocating components, such as internal combustion engines and reciprocating compressors and pumps. The discussion associated with rotating unbalance can be extended to include reciprocating unbalance. Figure 3.19 shows a schematic drawing of a simple reciprocating mechanism.

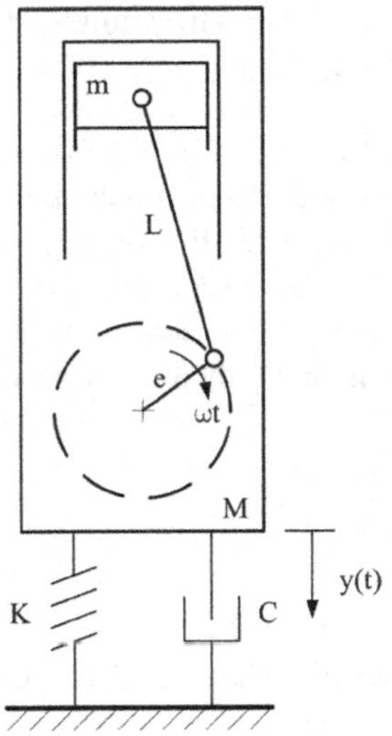

Figure 3.19 1-DOF Vibration System with Damping that Has a Reciprocating Unbalance

There are two types of unbalance present in this type of mechanism. The first is the rotating unbalance associated with the crank shaft, crank pin and part of the connecting rod. The other is the reciprocating unbalance associated with the motion of the piston, the wrist pin and the other part of the connecting rod.

With regard to single cylinder mechanisms, nearly all of the rotating unbalance and part of the reciprocating unbalance can be reasonably well balanced. However, it is normally not possible to completely balance all of the reciprocating unbalance. Define the unbalanced mass m as the mass of the parts of the piston, wrist pin and connecting rod that cannot be balanced. The acceleration a_p of the reciprocating unbalanced mass m is given by:

$$a_p = e\omega^2\left(\sin\omega t + \frac{e}{L}\sin 2\omega t\right) \tag{3.133}$$

where e is the crank radius, L is the length of the connecting rod, and ω is the rotational speed of the crank shaft. If e/L is small, the first harmonic term $(e/L)\sin(2\omega t)$ can be neglected. Therefore, the equivalent excitation force F_{eq} equals the unbalanced mass m times the acceleration a_p of the piston:

$$F_{eq} = me\,\omega^2 \tag{3.134}$$

which is the same as equation (3.119) for a system with a rotating unbalance. Therefore, the discussions associated with a vibration system with a rotating unbalance also apply to a system with a reciprocating unbalance.

EXAMPLE 3.7

A reciprocating air compressor which weighs 1,600 lb_f is to be operated at a speed of 1,800 rpm. The unbalanced reciprocating parts weigh 25 lb_f and the rotating parts are well balanced. The crank radius is 4 in. The type of isolators that are to be used have a damping ratio of $\xi = 0.05$ associated with them. Four spring isolation mounts, one at each corner of the compressor mounting base, are to be used to vibration isolate the compressor from its foundation.

(a) Determine the stiffness coefficient K of each spring isolation mount if only 10% of the unbalanced force is to be transmitted to the compressor's foundation. Assume the weight of the compressor is evenly distributed amount the four mounts.

(b) Determine the value of the transmitted force.

(c) If the operating speed of the compressor is increased to 2,400 rpm, what will be the value of the force transmitted to the base?

SOLUTION

(a) From equation (3.116), the frequency ratio r that is required for a transmissibility Tr = 0.1 is found to be r = 3.23. The operating speed of the compressor in rad/s is:

$$\omega = \frac{2\pi \times 1{,}800}{60} = 188.5\,\frac{\text{rad}}{\text{s}}. \tag{3.7a}$$

Therefore, the required resonance frequency is:

$$\omega_n = \frac{188.5}{3.23} = 58.6\,\frac{\text{rad}}{\text{s}}. \tag{3.7b}$$

The total stiffness of the spring isolation mounts is:

$$K = M\,\omega_n^2 \tag{3.7c}$$

or:

$$K = \frac{1{,}600}{386} \times 58.6^2 = 14{,}234 \frac{\text{lb}_f}{\text{in.}} . \tag{3.7d}$$

The stiffness of each spring isolation mount is:

$$\boxed{\frac{K}{4} = \frac{14{,}234}{4} = 3{,}559 \frac{\text{lb}_f}{\text{in.}} .} \tag{3.7 e}$$

(b) The value of the transmitted force is given by equation (3.130), or:

$$\boxed{\left|\bar{F}_T\right| = \frac{25}{386} \times 4 \times 188.5^2 \times 0.1 = 920.5 \text{ lb}_f .} \tag{3.7f}$$

(c) If the speed of the compressor is increased to 2,400 rpm, the angular velocity w becomes:

$$\omega = \frac{2\pi \times 2{,}400}{60} = 251.3 \frac{\text{rad}}{\text{s}} . \tag{3.7g}$$

Therefore, the frequency ratio is:

$$r = \frac{251.3}{58.6} = 4.3 . \tag{3.7h}$$

Using equation (3.116), the transmissibility for a frequency ratio of $r = 4.3$ is:

$$\left|\frac{\bar{F}_T}{F}\right| = 0.062 . \tag{3.7i}$$

The value of the transmitted force is obtained from equation (3.130), or:

$$\boxed{\left|\bar{F}_T\right| = \frac{25}{386} \times 4 \times 251.3^2 \times 0.062 = 1{,}014 \text{ lb}_f .} \tag{3.7j}$$

Therefore, even though the transmissibility is decreased from 10% down to 6.2% when the speed of the compressor is increased from 1,800 rpm to 2,400 rpm, the value of the transmitted force is increased from 920.5 lb_f to 1,014 lb_f, an increase of nearly 94 lb_f. This increase will continue to get larger as the speed of the compressor is increased.

3.9 Response of 1-DOF Vibration Systems with Structural Damping

Many vibration isolation systems use metal springs or other similar resilient elements that have structural damping. Therefore, 1-DOF vibration systems with structural damping are examined. The equation of motion for a system with structural damping where γ is the structural damping factor can be written [from equation (2.92)]:

$$M\ddot{y} + K(1 + j\gamma)y = f(t) . \tag{3.135}$$

When $f(t) = Fe^{j\omega t}$, $y(t) = \bar{Y}e^{j\omega t}$ where $\bar{Y} = Ye^{j\phi}$, using the results of Section 2.6, equation (3.135) can be written:

$$M\ddot{y} + \frac{\gamma K}{\omega}\dot{y} + Ky = f(t) . \tag{3.136}$$

Substituting the expressions for y(t) and f(t) into equation (3.136), carrying out the prescribed operations, and rearranging the terms yields:

$$\bar{Y} = \frac{F}{K - M\omega^2 + j\gamma K} \tag{3.137}$$

or:

$$\frac{\bar{Y}}{F/K} = \bar{H}(\omega) \quad \text{where} \quad \bar{H}(\omega) = \frac{1}{1 - r^2 + j\gamma} . \tag{3.138}$$

r is the frequency ratio given by equation (3.13). Multiplying the numerator and denominator of equation (3.138) by the complex conjugate of the denominator yields:

$$\bar{H}(\omega) = \frac{1 - r^2 - j\gamma}{\left(1 - r^2\right)^2 + \gamma^2} \tag{3.139}$$

The real and imaginary parts of the frequency response function are:

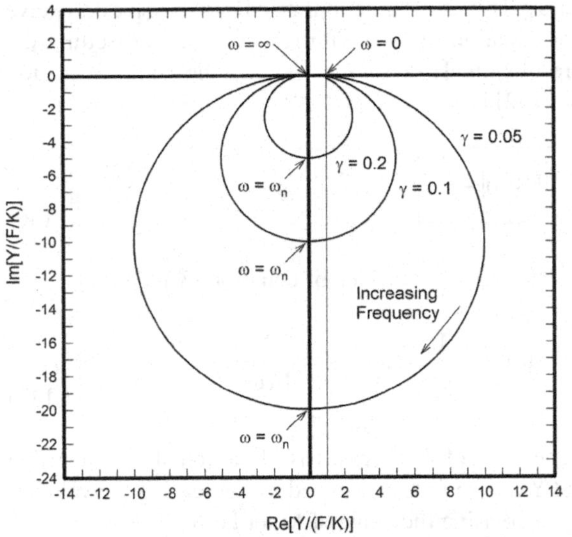

Figure 3.20 Nyquist Plots of the Frequency Response Function for a 1-DOF Vibration System with Structural Damping for Selected Values of the Structural Damping Factor γ

$$\boxed{\mathrm{Re}\left[\bar{\mathrm{H}}(\omega)\right]=\frac{1-r^2}{\left(1-r^2\right)^2+\gamma^2}} \tag{3.140}$$

and:

$$\boxed{\mathrm{Im}\left[\bar{\mathrm{H}}(\omega)\right]=\frac{-j\gamma}{\left(1-r^2\right)^2+\gamma^2}} \tag{3.141}$$

Figure 3.20 shows Nyquist plots of a system with structural damping for several different values of γ. Figure 3.21 shows the corresponding plots of the real and imaginary values of the frequency response function as a function of the frequency ratio r for several different values of γ.

The value and corresponding phase associated with the frequency response function are:

$$\boxed{\left|\bar{\mathrm{H}}(\omega)\right|=\frac{1}{\sqrt{\left(1-r^2\right)^2+\gamma^2}}} \tag{3.142}$$

and noting tan(-ϕ) = -tan(ϕ):

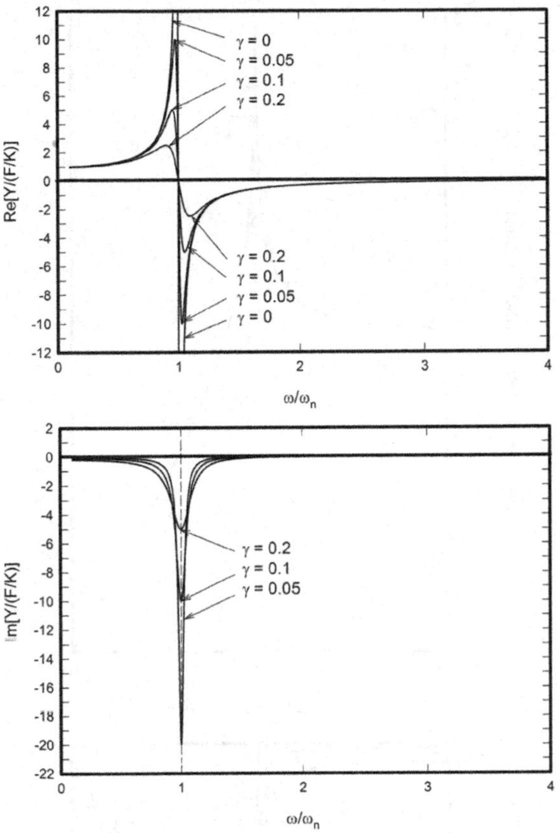

Figure 3.21 Real and Imaginary Components of the Frequency Response Function for a 1-DOF Vibration System with Structural Damping

$$\boxed{\phi=-\tan^{-1}\left(\frac{\gamma}{1-r^2}\right).} \tag{3.143}$$

y(t) for a system with structural damping can be written:

$$\boxed{y(t)=\frac{F}{K}\left|\bar{\mathrm{H}}(\omega)\right|e^{j(\omega t+\phi)}} \tag{3.144}$$

where:

$$\boxed{Y=\frac{F/K}{\sqrt{\left(1-r^2\right)^2+\gamma^2}}} \tag{3.145}$$

and ϕ is given by equation (3.143). Figure 3.22 shows

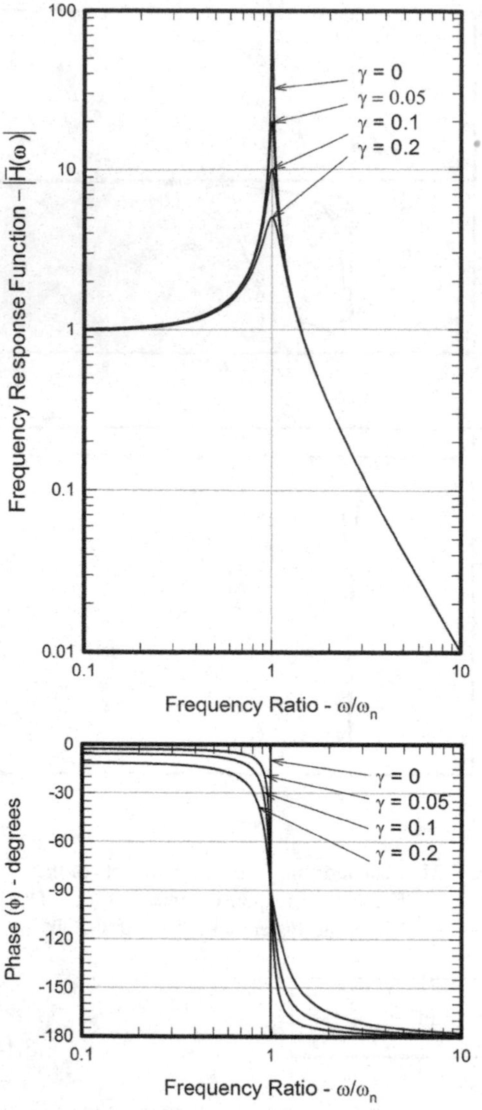

Figure 3.22 Amplitudes of the Frequency Response Function and Phase for a 1-DOF Vibration System with Structural Damping

the values of the frequency response function and the corresponding phase for different values of γ.

3.10 Characteristics of A 1-DOF Vibration System with Structural Damping

An examination of Figure 3.22 indicates the maximum value of the frequency response function always occurs at ω_n. This can also be confirmed by taking the derivative of equation (3.142) with respect to ω and setting the results equal to zero. When $\omega = \omega_n$, the value of equation (3.142) is:

$$\boxed{\left|\bar{H}(\omega)\right| = \frac{1}{\gamma}.} \tag{3.146}$$

When r >3, the value of equation (3.142) is:

$$\left|\bar{H}(\omega_n\right| = \frac{1}{r^2} \quad \text{and} \quad Y = \frac{F}{M\omega^2}. \tag{3.147}$$

The value of Y decreases as a function of ω^2. The value of Y can be further reduced by decreasing the value of F, increasing the value of M, or both.

The maxima and minima of the real part of the frequency response function (Figure 3.21) is obtained by taking the derivative of equation (3.140) with respect to ω and setting the results equal to zero. Using the steps associated with equations (3.75) - (3.84) yields for ω_1 and ω_2:

$$\omega_1 = \omega_n\sqrt{1-\gamma} \tag{3.148}$$

and:

$$\omega_2 = \omega_n\sqrt{1+\gamma}\,. \tag{3.149}$$

As was the case with viscous damping, ω_1 and ω_2 are the half power points. Substituting equations (3.148) and (3.149) into equation (3.89) and simplifying yields:

$$\boxed{\gamma \approx \frac{\omega_2 - \omega_1}{\omega_n}.} \tag{3.150}$$

Dividing equation (3.149) by equation (3.148) and simplifying gives:

$$\boxed{\gamma = \frac{\left(\frac{\omega_2}{\omega_1}\right)^2 - 1}{\left(\frac{\omega_2}{\omega_1}\right)^2 + 1}.} \tag{3.151}$$

As was shown in Section 2.6:

$$\xi = \frac{\gamma}{2} \quad \text{or} \quad \gamma = 2\xi . \tag{2.152}$$

3.11 Force Transmitted to the Base of a 1-DOF Vibration System with Structural Damping

The expression for the force transmitted to the base for a vibration system with structural damping is:

$$f_T(t) = K\, y(t) + \frac{\gamma K}{\omega}\, \dot{y}(t). \tag{3.153}$$

Substituting equations (3.144) and (3.145) into equation (3.155) and simplifying yields:

$$f_T(t) = \frac{F\sqrt{1+\gamma^2}}{\sqrt{\left(1-r^2\right)^2+\gamma^2}}\, e^{j(\omega t+\phi+\theta)} \tag{3.154}$$

where ϕ is given by equation (3.143) and:

$$\theta = \tan^{-1}(\gamma). \tag{3.155}$$

The transmissibility is given by:

$$\left|\frac{\bar{F}_T}{F}\right| = \frac{\sqrt{1+\gamma^2}}{\sqrt{\left(1-r^2\right)^2+\gamma^2}} . \tag{3.156}$$

Figure 3.23 shows plots of the transmissibility and corresponding phase for several values of γ.

When $r > 3$, equation (3.156) becomes:

$$\left|\bar{F}_T\right| = \frac{F}{r^2} \quad \text{or} \quad \left|\bar{F}_T\right| = \frac{FK}{M\omega^2}. \tag{3.157}$$

The value of $\left|\bar{F}_T\right|$ decreases as a function of ω^2. The value of $\left|\bar{F}_T\right|$ can be further reduced by decreasing the values of F and ω_n. Noting that $\omega_n = \sqrt{K/M}$, when M is decreased per the discussion associated with equation (3.147), K must be decreased proportionately more to achieve a decrease in the value of ω_n.

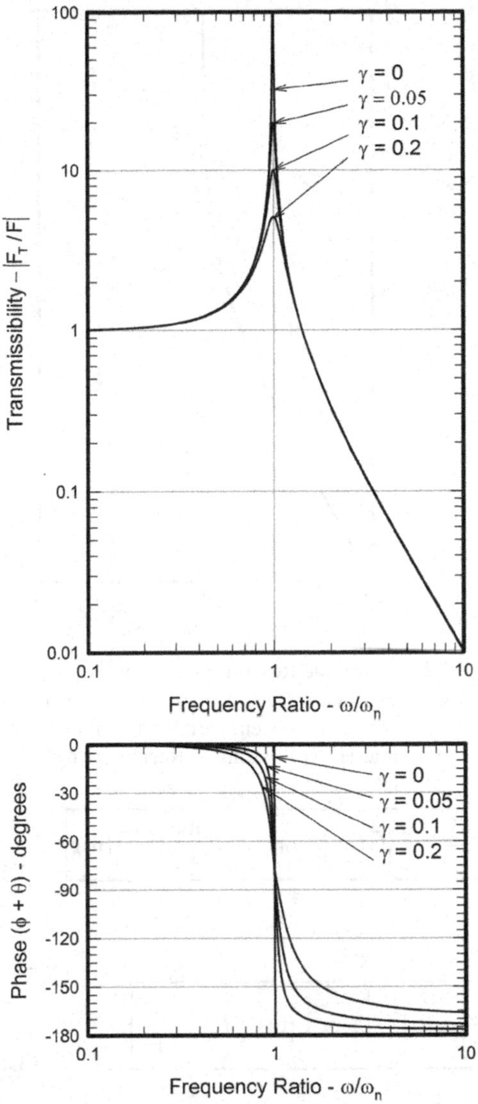

Figure 3.23 Amplitudes of Transmissibility and Phase for a 1-DOF Vibration System with Structural Damping

3.12 Vibration Systems with Structural Damping and Rotating and Reciprocating Unbalance

For systems with structural damping that have rotating and reciprocating unbalance:

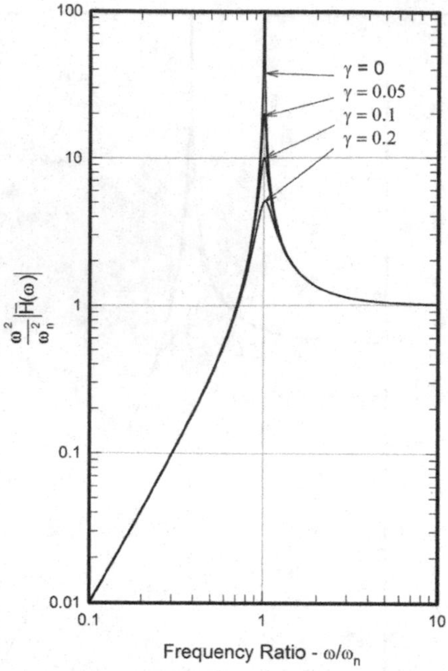

Figure 3.24 Harmonic Response Function Times the Frequency Ratio Squared for a 1-DOF Vibration System with Structural Damping that Has a Rotating or Reciprocating

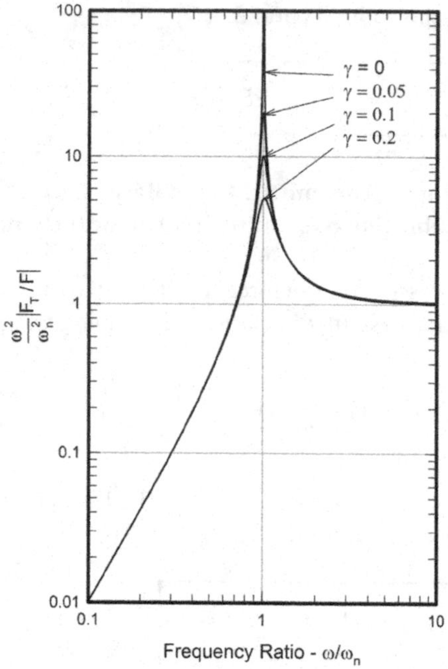

Figure 3.25 Transmissibility Times the Frequency Ratio Squared for a 1-DOF Vibration System with Structural Damping that Has a Rotating or Reciprocating Unbalance

$$\boxed{\frac{YM}{me} = r^2\left|\bar{H}(\omega)\right| \quad \text{or} \quad Y = \frac{me}{M}r^2\left|\bar{H}(\omega)\right|} \tag{3.158}$$

and:

$$\boxed{\left|\frac{\bar{F}_T}{F_n}\right| = r^2\left|\frac{\bar{F}_T}{F}\right| \quad \text{or} \quad \left|\bar{F}_T\right| = F_n r^2\left|\frac{\bar{F}_T}{F}\right|} \tag{3.159}$$

where $|\bar{H}(\omega)|$ is given by equation (3.142) and $|\bar{F}_T/F|$ is given by equation (3.158). Figures (3.24) and (3.25) show plots of equations (3.159) and (3.160), respectively, for different values of γ.

When $r > 3$, equations (3.158) and (3.159) become:

$$\boxed{Y = \frac{me}{M} \quad \text{and} \quad \left|\bar{F}_T\right| = me\,\omega_n^2 \quad \text{or} \quad \left|\bar{F}_T\right| = \frac{meK}{M}.} \tag{3.160}$$

The value of Y is reduced by decreasing me, increasing M, or both. The value of $|\bar{F}_T|$ is reduced by decreasing the values me, ω_n, or both. Noting that $\omega_n = \sqrt{K/M}$, when M is decreased to reduce the value of Y, K must be decreased proportionately more to achieve a reduction in the value of ω_n.

3.13 Response of a 1-DOF Vibration System with Non-Viscous Damping

Other types of damping that can exist in vibration systems include Coulomb damping and velocity-squared damping. The mass and spring elements of systems with these types of damping behave in a linear manner.

The overall damping associated with Coulomb or velocity-squared damping is usually very small. Therefore, it is possible to develop an approximate expression for this non-viscous damping element. Consider a change in energy ΔE associated with a damping force F_d that is dissipated over one cycle of oscillation. For this case, ΔE can be expressed:

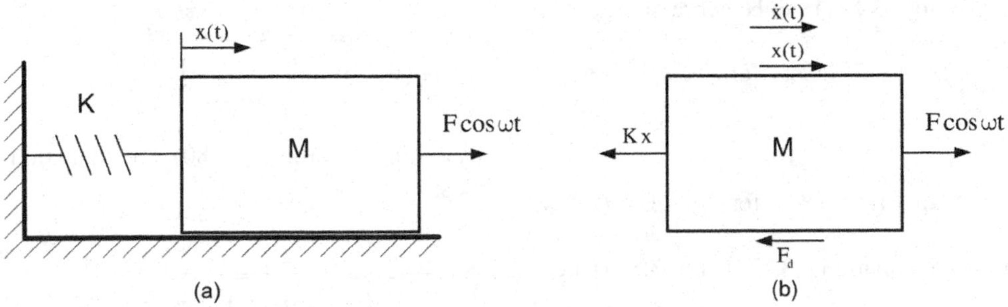

Figure 3.26 Force Vibration of 1-DOF Vibration System with Coulomb Damping

$$\Delta E = \int_{cycle} F_d \, dy \quad \text{or} \quad \Delta E = \int_0^T F_d \frac{dy}{dt} dt. \tag{3.161}$$

T is the period of oscillation. For the case where viscous damping is assumed:

$$F_d = -C\frac{dy}{dt}. \tag{3.162}$$

Substituting equation (3.162) into equation (3.161) yields:

$$\Delta E = -\int_0^T C\left(\frac{dy}{dt}\right)^2 dt. \tag{3.163}$$

If harmonic motion is assumed:

$$y(t) = Y\cos(\omega t - \phi) \tag{3.164}$$

and:

$$\frac{dy}{dt} = -\omega Y \sin(\omega t - \phi). \tag{3.165}$$

Substituting equation (3.165) into equation (3.163) yields:

$$\Delta E = \omega^2 Y^2 C \int_0^T \sin^2(\omega t - \phi)\, dt \tag{3.166}$$

or:

$$\Delta E = \frac{\omega^2 Y^2 C T}{2}. \tag{3.167}$$

Noting that $T = 2\pi/\omega$, ΔE becomes:

$$\Delta E = C\omega\pi Y^2. \tag{3.168}$$

If the non-viscous damping present in a 1-DOF vibration system is small, an *equivalent damping coefficient* C_{eq} can be defined such that:

$$C_{eq} = \frac{\Delta E}{\omega\pi Y^2}. \tag{3.169}$$

Substituting equation (3.161) into equation (3.169) yields:

$$C_{eq} = \frac{1}{\omega\pi Y^2}\int_0^T F_d \frac{dy}{dt} dt. \tag{3.170}$$

Systems with Coulomb Damping

Unforced vibration of a 1-DOF vibration system with Coulomb damping was examined in Section 2.5. The equation of motion for this system when acted upon by a harmonic force $F\cos(\omega t)$ can be written (Figure 3.26):

$$M\ddot{x} + Kx = F_d + F\cos\omega t \tag{3.171}$$

where:

$$F_d = -Mg\mu\, \mathrm{sgn}(\dot{x}). \tag{3.172}$$

Therefore, equation (3.171) can be expressed:

$$M\ddot{x} + Kx = -Mg\mu + F\cos\omega t \quad \text{for} \quad \dot{x} > 0 \quad (3.173)$$

and:

$$M\ddot{x} + Kx = Mg\mu + F\cos\omega t \quad \text{for} \quad \dot{x} < 0. \quad (3.174)$$

The solutions to equations (3.173) and (3.174) are:

$$x(t) = -\frac{Mg\mu}{K} + X\cos\omega t \quad \text{for} \quad \dot{x} > 0 \quad (3.175)$$

and:

$$x(t) = \frac{Mg\mu}{K} + X\cos\omega t \quad \text{for} \quad \dot{x} < 0. \quad (3.176)$$

The corresponding velocity is:

$$\frac{dx}{dt} = -\omega X \sin\omega t. \quad (3.177)$$

Substituting equations (3.175) through (3.177) into equation (3.170) yields:

$$C_{eq} = \frac{1}{\pi\omega X^2}\left[\begin{array}{l}\int_0^{T/2}(-Mg\mu)(-\omega X\sin\omega t)\,dt + \\ \int_{T/2}^{T} Mg\mu(-\omega X\sin\omega t)\,dt\end{array}\right] \quad (3.178)$$

or:

$$C_{eq} = \frac{Mg\mu}{\pi X}\left[\int_0^{T/2}\sin\omega t\,dt - \int_{T/2}^{T}\sin\omega t\,dt\right]. \quad (3.179)$$

Carrying out the above integrations and simplifying gives:

$$C_{eq} = \frac{4\,Mg\mu}{\pi\omega X}. \quad (3.180)$$

The response of a 1-DOF vibration system with viscous damping when acted upon by a harmonic force is given by equation (3.67). The response X can be written:

$$X = \frac{F}{\sqrt{\left(K - M\omega^2\right)^2 + \left(C_{eq}\omega\right)^2}}. \quad (3.181)$$

Substituting equation (3.180) into equation (3.181) yields:

$$X = \frac{F}{\sqrt{\left(K - M\omega^2\right)^2 + \left(\frac{4\,Mg\mu}{\pi X}\right)^2}}. \quad (3.182)$$

Since X appears on both sides of the equal sign in equation (3.182), the equation must be rearranged to solve for X. Therefore:

$$X = \frac{\sqrt{F^2 - \left(\frac{4\,Mg\mu}{\pi}\right)^2}}{K - M\omega^2}. \quad (3.183)$$

Dividing the numerator and denominator by K and simplifying yields:

$$\boxed{X = \frac{\frac{F}{K}\sqrt{1 - \left(\frac{4\,Mg\mu}{F\pi}\right)^2}}{1 - \frac{\omega^2}{\omega_n^2}}}. \quad (3.184)$$

Two observations can be made from equation (3.184). First, for vibration to occur, X must be real-valued. Therefore, for forced vibration to exist:

$$F > \frac{4\,Mg\mu}{\pi}. \quad (3.185)$$

Second, when Coulomb damping is present and the system is excited at its resonance frequency, the value of X goes to infinity.

Equation (3.184) can be written:

$$\boxed{\frac{X}{\frac{F}{K}\sqrt{1 - \left(\frac{4\,Mg\mu}{F\pi}\right)^2}} = \frac{1}{\left|1 - \frac{\omega^2}{\omega_n^2}\right|}e^{j\phi}} \quad (3.186)$$

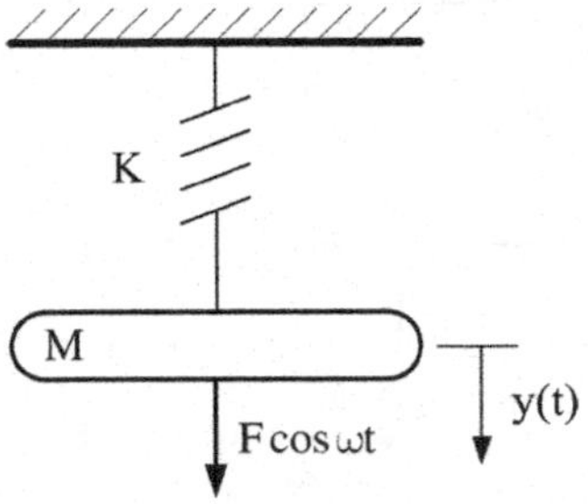

Figure 3.27 System with Velocity Squared Damping

where $\phi = 0$ when $\omega < \omega_n$, and $\phi = -\pi$ when $\omega > \omega_n$. A plot of the value term on the right side of equation (3.186) is shown in Figure 3.3.

Systems with Velocity Squared Damping

Objects that move in fluids such as water or air at moderate velocities are resisted by a damping force that is proportional to their velocity squared. If a fluid is forced through an orifice, the resulting damping force is often assumed to be proportional to velocity squared.

Figure 3.27 shows a relatively thin plate that is attached to a resilient element. As the plate vibrates, it experiences a drag force D that can be expressed:

$$D = \frac{C_d \rho S}{2}\dot{y}^2 \tag{3.187}$$

where C_d is the drag coefficient, ρ is the density of the fluid, and S is the surface area of the plate. The damping force acts in the direction opposite the velocity of the plate. Therefore, F_d can be written:

$$F_d = -\frac{C_d \rho S}{2}|\dot{y}|\,\dot{y} \tag{3.188}$$

or:

$$F_d = -\frac{C_d \rho S}{2}\dot{y}^2 \quad \text{for} \quad \dot{y} > 0 \tag{3.189}$$

and:

$$F_d = \frac{C_d \rho S}{2}\dot{y}^2 \quad \text{for} \quad \dot{y} < 0. \tag{3.190}$$

The equation of motion of the system in Figure 3.27 is:

$$M\ddot{y} + \frac{C_d \rho S}{2}|\dot{y}|\,\dot{y} + Ky = F\cos\omega t. \tag{3.191}$$

When $C_d\rho S/2$ is small, the response of the system can be assumed to be:

$$y(t) = Y\cos\omega t. \tag{3.192}$$

Therefore, the velocity is:

$$\frac{dy}{dt} = -\omega Y\sin\omega t. \tag{3.193}$$

Substituting the appropriate values for force and velocity into equation (3.170) yields:

$$C_{eq} = \frac{1}{\pi\omega Y^2}\left[-\int_0^{T/2}\left(\frac{C_d\rho S}{2}\dot{y}^2\right)\dot{y}\,dt + \int_{T/2}^{T}\left(\frac{C_d\rho S}{2}\dot{y}^2\right)\dot{y}\,dt\right] \tag{3.194}$$

or:

$$C_{eq} = \frac{C_d\rho SY\omega^2}{2\pi}\left[\int_0^{T/2}\sin^3\omega t\,dt - \int_{T/2}^{T}\sin^3\omega t\,dt\right]. \tag{3.195}$$

Carrying out the above integrations and simplifying gives:

$$C_{eq} = \frac{4}{3\pi}C_d\rho SY\omega. \tag{3.196}$$

Substituting equation (3.196) into equation (3.181) yields:

$$Y = \frac{F}{\sqrt{\left(K - M\omega^2\right)^2 + \left(\frac{4}{3\pi}C_d\rho SY\omega^2\right)^2}}. \tag{3.197}$$

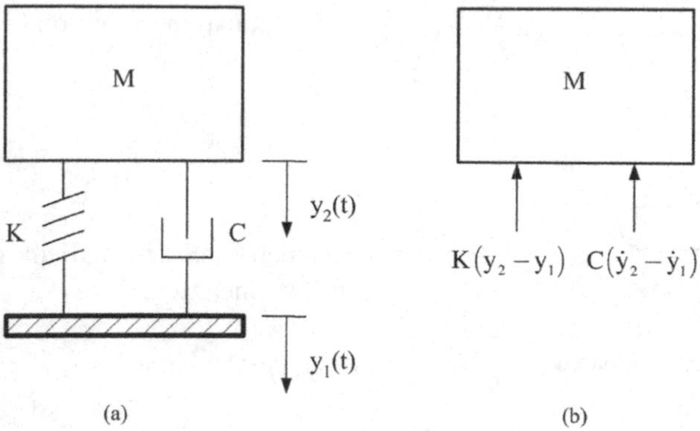

Figure 3.28 1-DOF System with Viscous Damping Attached to a Moving Base

Y is obtained by rearranging equation (3.197) to get:

$$Y = \frac{\sqrt{-\left(k - M\omega^2\right)^2 + \sqrt{\left(k - M\omega^2\right)^4 + 4F\left(\frac{4}{3\pi} C_d \rho S \omega^2\right)^2}}}{\frac{4\sqrt{2}}{3\pi} C_d \rho S \omega^2}. \tag{3.198}$$

3.14 1-DOF Vibrating System with Viscous Damping Attached to a Moving Base

The mass of a vibration system can be excited by motion of the base or support structure to which the system is attached. Figure 3.28(a) shows a schematic representation of this type of system. Both the motion of the mass $y_2(t)$ and the base $y_1(t)$ must be considered. The positive directions for both $y_1(t)$ and $y_2(t)$ are shown downward.

Forces are applied to mass M through a spring element K and a damping element C that connect it to the base. These forces are shown as restoring forces that act in the negative y direction [Figure 2.28(b)]. The spring force is associated with the relative displacement $(y_2 - y_1)$ across the spring element. The damping force is associated with the relative velocity $(\dot{y}_2 - \dot{y}_1)$ across the damping element. The first term within each bracket is always referenced to the motion of the mass to which the end of the spring or damping element is attached. Summing the forces acting mass M yields:

$$\sum F_y = M\ddot{y}_2 \text{ or } M\ddot{y}_2 = -K(y_2 - y_1) - C(\dot{y}_2 - \dot{y}_1). \tag{3.199}$$

Therefore:

$$M\ddot{y}_2 + C\dot{y}_2 + Ky_2 = C\dot{y}_1 + Ky_1 . \tag{3.200}$$

$C\dot{y}_1 + Ky_1$ are shown on the right side of the equal sign because they are the forces associated with the motion of the base that are acting on C and K.

To determine the response $y_2(t)$ of mass M to a harmonic input displacement $y_1(t)$ and velocity $\dot{y}_1(t)$ let:

$$y_1(t) = Y_1\, e^{j\omega t} \quad \text{and} \quad \dot{y}_1(t) = j\omega\, Y_1\, e^{j\omega t}. \tag{3.201}$$

The response of the system can be written:

$$y_2(t) = \overline{Y}_2\, e^{j\omega t} . \tag{3.202}$$

Substituting equations (3.201) and (3.202) into equation (3.200) and carrying out the prescribed operations yields:

$$\frac{\overline{Y}_2}{Y_1} = \frac{K + jC\omega}{\left(K - M\omega^2\right) + jC\omega}. \tag{3.203}$$

Dividing the numerator and denominator by K and simplifying gives:

$$\frac{\bar{Y}_2}{Y_1} = \frac{1 + j2\xi \dfrac{\omega}{\omega_n}}{\left(1 - \dfrac{\omega^2}{\omega_n^2}\right) + j2\xi \dfrac{\omega}{\omega_n}}. \tag{3.204}$$

Noting $r = \omega/\omega_n$, equation (3.204) can be written:

$$\boxed{\frac{\bar{Y}_2}{Y_1} = \frac{\sqrt{1+(2\xi r)^2}}{\sqrt{(1-r^2)^2 + (2\xi r)^2}} e^{j(\phi_1+\phi_2)}} \tag{3.205}$$

where:

$$\boxed{\phi_1 = -\tan^{-1}\left(\frac{2\xi r}{1-r^2}\right)} \tag{3.206}$$

and:

$$\boxed{\phi_2 = \tan^{-1}(2\xi r).} \tag{3.207}$$

The response $y_2(t)$ of mass M can be written:

$$\boxed{y_2(t) = Y_1 \left|\frac{\bar{Y}_2}{Y_1}\right| e^{j(\omega t+\phi_1+\phi_2)}} \tag{3.208}$$

where:

$$\boxed{\left|\frac{\bar{Y}_2}{Y_1}\right| = \frac{\sqrt{1+(2\xi r)^2}}{\sqrt{(1-r^2)^2 + (2\xi r)^2}}.} \tag{3.209}$$

The expression for $\left|\bar{Y}_2 / Y_1\right|$ in equation (3.209) is the same as the expression for $\left|F_T / F\right|$ in equation (3.116). The expressions for $\phi_1 + \phi_2$ in equation (3.205) are the same as the expressions for $\theta + \phi$ in equation (3.115). Therefore, the plots in Figure 3.15 apply to $\left|\bar{Y}_2 / Y_1\right|$ and $\phi_1 + \phi_2$.

When r = 1, the value of $\left|\bar{Y}_2\right|$ is given by:

$$\left|\bar{Y}_2\right| = \frac{Y_1\sqrt{1+(2\xi r)^2}}{2\xi}. \tag{3.210}$$

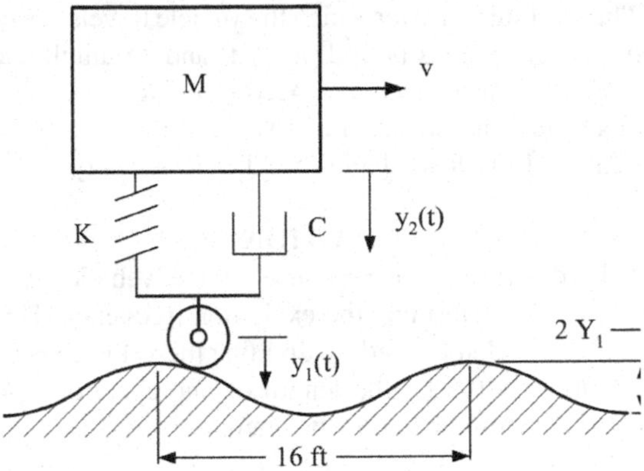

Figure 3.29 Schematic Representation of a 1-DOF Vehicle Moving over a Rough Road

When r > 3 and $\xi < 0.3$ the value of $\left|\bar{Y}_2\right|$ is given by:

$$\left|\bar{Y}_2\right| \approx \frac{Y_1\,\omega_n\sqrt{\omega_n^2 + (2\xi\omega)^2}}{\omega^2}. \tag{3.211}$$

For this case, the value of $\left|\bar{Y}_2\right|$ decreases as ω increases. For a specified value of Y_1, $\left|\bar{Y}_2\right|$ can be further reduced by decreasing the value of ω_n, ξ, or both. Noting $\omega_n = \sqrt{K/M}$, the value of $\left|\bar{Y}_2\right|$ can be reduced by increasing M, decreasing K, or both, so long as the resulting value of ω_n is less than $\omega/3$ where ω is the excitation frequency. When $\xi = 0$, equation (3.211) becomes:

$$\left|\bar{Y}_2\right| \approx \frac{Y_1\,\omega_n^2}{\omega^2} \quad \text{or} \quad \left|\bar{Y}_2\right| \approx \frac{Y_1 K}{M\omega^2}. \tag{3.212}$$

EXAMPLE 3.8

A motor vehicle is a complex system that has many degrees of freedom. Figure 3.29 represents a first approximation of a vehicle traveling over a rough road. It is assumed: the vehicle is constrained to move only in the vertical or y direction; the stiffness coeficient of the tires is infinite, that is, roughness of the road surface is directly transmitted to the suspension system of the vehicle; and the tires do not leave the road surface.

The vehicle weighs 1.00 ton when fully loaded and 0.25 ton when empty. The stiffness coeficient for the total vehicle suspension system is 2,000 lb_f/in. The damping ratio ξ is 0.5 when the vehicle is fully loaded.

The road surface over which the vehicle travels varies sinusoidally with a period of 16 ft and an amplitude value of Y_1 in. The vehicle speed is 60 mph.

Determine the displacement value of the vehicle (a) when it is fully loaded and (b) when it is empty.

SOLUTION

(a) To determine the response of the vehicle, it is necessary to determine the excitation frequency. This frequency is determined by dividing the vehicle speed with units of ft/s by the length of one cycle of road roughness with the unit of ft. Therefore:

$$f = \frac{60\left(\frac{\text{mile}}{\text{hr}}\right)\times 5{,}280\left(\frac{\text{ft}}{\text{mile}}\right)}{3{,}600\left(\frac{\text{s}}{\text{hr}}\right)}\times\frac{1}{16(\text{ft})} = 5.5\ \text{Hz}. \tag{3.8a}$$

The resonance frequency of the vehicle when it is fully loaded is:

$$f_n = \frac{1}{2\pi}\sqrt{\frac{2{,}000\times 386}{2{,}000}} = 3.13\ \text{Hz}. \tag{3.8b}$$

Therefore, the frequency ratio r when the vehicle is fully loaded is:

$$r = \frac{5.5}{3.13} = 1.76. \tag{3.8c}$$

Substituting the values of r = 1.76 and ξ = 0.5 into equation (3.209) yields:

$$\left|\frac{\bar{Y}_2}{Y_1}\right| = \frac{\sqrt{1+(2\times 0.5\times 1.76)^2}}{\sqrt{\left(1-1.74^2\right)^2+(2\times 0.5\times 1.76)^2}} = 0.74. \tag{3.8d}$$

Therefore:

$$\boxed{\left|\bar{Y}_2\right| = 0.74\ Y_1.} \tag{3.8e}$$

(b) When the vehicle is empty, the excitation frequency will be the same, but the resonance frequency and the damping ratio will change. The resonance frequency of the empty vehicle is:

$$f_n = \frac{1}{2\pi}\sqrt{\frac{2{,}000\times 386}{500}} = 6.25\ \text{Hz}. \tag{3.8f}$$

From part (a), the damping constant for the fully loaded vehicle is:

$$C = 0.5\times 2\sqrt{\frac{2{,}000}{386}\times 2{,}000} = 101.8\ \frac{\text{lb}_f - \text{s}}{\text{in.}}. \tag{3.8g}$$

The damping ratio for the empty vehicle is:

$$\xi = \frac{101.8}{2\sqrt{\frac{500}{386}\times 2{,}000}} = 1.0. \tag{3.8h}$$

The frequency ratio for the empty vehicle is:

$$r = \frac{5.5}{6.25} = 0.88. \tag{3.8i}$$

Substituting the values of r = 0.88 and ξ = 1.0 into equation (3.209) yields:

$$\left|\frac{\bar{Y}_2}{Y_1}\right| = \frac{\sqrt{1+(2\times 1.0\times 0.88)^2}}{\sqrt{\left(1-0.88^2\right)^2+(2\times 1.0\times 0.88)^2}} = 1.14. \tag{3.8j}$$

Therefore:

$$\boxed{\left|\bar{Y}_2\right| = 1.14\ Y_1.} \tag{3.8k}$$

Figure 3.28 can represents a 1-DOF mounting arrangement of sensitive equipment with mass M that is placed on a vibrating structure or base. For this case, the acceleration of M and the force transmitted to M are of interest. The motion of the base is given by equation (3.201), and the response of M is given by equation (3.202). Therefore, the acceleration of mass M is:

$$\ddot{y}_2(t) = -\omega^2 \, \bar{Y}_2 \, e^{j\omega t} . \tag{3.213}$$

The value of the acceleration of M can be written:

$$\left|\ddot{\bar{Y}}_2\right| = -\omega^2 \left|\bar{Y}_2\right| . \tag{3.214}$$

Substituting equations (3.208) and (3.209) for $\left|\bar{Y}_2\right|$ into equation (3.214) yields:

$$\ddot{\bar{Y}}_2 = -\omega^2 \, Y_1 \frac{\sqrt{1+(2\xi r)^2}}{\sqrt{\left(1-r^2\right)^2+(2\xi r)^2}} e^{j(\phi_1+\phi_2)} \tag{3.215}$$

or:

$$\ddot{\bar{Y}}_2 = \omega^2 \, Y_1 \frac{\sqrt{1+(2\xi r)^2}}{\sqrt{\left(1-r^2\right)^2+(2\xi r)^2}} e^{j(\phi_1+\phi_2+\pi)} . \tag{3.216}$$

The value of the acceleration of M is:

$$\left|\ddot{\bar{Y}}_2\right| = \omega^2 \, Y_1 \frac{\sqrt{1+(2\xi r)^2}}{\sqrt{\left(1-r^2\right)^2+(2\xi r)^2}} . \tag{3.217}$$

Multiplying and dividing equation (3.217) by ω_n^2 and rearranging the terms yields:

$$\boxed{\left|\frac{\ddot{\bar{Y}}_2}{\omega_n^2 \, Y_1}\right| = r^2 \frac{\sqrt{1+(2\xi r)^2}}{\sqrt{\left(1-r^2\right)^2+(2\xi r)^2}} .} \tag{3.218}$$

Equation (3.218) is the dame as equation (3.130) and is plotted in Figure 3.18.

When $r > 3$ and $\xi < 0.3$, $\left|\ddot{\bar{Y}}_2\right|$ reduces to:

$$\boxed{\left|\ddot{\bar{Y}}_2\right| \approx Y_1 \, \omega_n \sqrt{\omega_n^2+(2\xi\omega)^2} .} \tag{3.219}$$

For this case, $\left|\ddot{\bar{Y}}_2\right|$ increases as ω increases. For a specified value of Y_1, $\left|\ddot{\bar{Y}}_2\right|$ can be reduced by decreasing ω_n, ξ or both. Noting $\omega_n = \sqrt{K/M}$, ω_n can be reduced by increasing M, decreasing K, or both, so long as the resulting value of ω_n is less than $\omega/3$ where ω is the excitation frequency. When $\xi = 0$, $\left|\ddot{\bar{Y}}_2\right|$ becomes:

$$\boxed{\left|\ddot{\bar{Y}}_2\right| = Y_1 \, \omega_n^2 \quad \text{or} \quad \left|\ddot{\bar{Y}}_2\right| = \frac{Y_1 K}{M} .} \tag{3.220}$$

The expression for the force $f_T(t)$ transmitted to M is given by:

$$f_T(t) = M \, \ddot{y}(t) \quad \text{or} \quad \left|\bar{F}\right|_T = -\omega^2 M \left|\bar{Y}_2\right| . \tag{3.221}$$

Substituting equation (3.209) for $\left|\bar{Y}_2\right|$ into equation (3.221) yields:

$$F_T = -\omega^2 M \, Y_1 \frac{\sqrt{1+(2\xi r)^2}}{\sqrt{\left(1-r^2\right)^2+(2\xi r)^2}} e^{j(\phi_1+\phi_2)} \tag{3.222}$$

or:

$$\bar{F}_T = \omega^2 M \, Y_1 \frac{\sqrt{1+(2\xi r)^2}}{\sqrt{\left(1-r^2\right)^2+(2\xi r)^2}} e^{j(\phi_1+\phi_2+\pi)} . \tag{3.223}$$

Multiplying and dividing equation (3.223) by K and rearranging the terms yields:

$$\boxed{\left|\frac{\bar{F}_T}{K \, Y_1}\right| = r^2 \frac{\sqrt{1+(2\xi r)^2}}{\sqrt{\left(1-r^2\right)^2+(2\xi r)^2}} .} \tag{2.224}$$

Equation (3.224) is the same as equation (3.130) and is plotted in Figure 3.18.

When $r > 3$ and $\xi < 0.3$, $\left|\bar{F}_T\right|$ reduces to:

$$\boxed{\left|\bar{F}_T\right| \approx Y_1 \sqrt{K M} \sqrt{\omega_n^2+(2\xi\omega)^2} .} \tag{3.225}$$

For this case, $\left|\bar{F}_T\right|$ increases as ω increases. For a specified value of Y_1, $\left|\bar{F}_T\right|$ can be reduced by reducing $\sqrt{K M}$, ξ or both, so long as the resulting value of ω_n is less than $\omega/3$ where ω is the excitation frequency. When $\xi = 0$, $\left|\bar{F}_T\right|$ becomes:

$$\left|\bar{F}_T\right| = Y_1 K. \tag{3.226}$$

For a specified value of Y_1, $\left|\bar{F}_T\right|$ can be reduced by decreasing K. As previously stated, the value of $\left|\ddot{\bar{Y}}_2\right|$ can be reduced by increasing M, decreasing K, or both, so long as the resulting value of ω_n is less than $\omega/3$ where ω is the excitation frequency.

3.15 1-DOF Vibration System with Structural Damping Attached to a Moving Base

When damping present in a base-excited vibration system is structural damping and the excitation is harmonic, equation (3.200) is written:

$$M\ddot{y}_2 + \frac{\gamma K}{\omega}\dot{y}_2 + Ky_2 = \frac{\gamma K}{\omega}\dot{y}_1 + Ky_1 . \tag{3.227}$$

Substituting equations (3.201) and (3.202) into equation (3.227) and simplifying yields:

$$\frac{\bar{Y}_2}{Y_1} = \frac{\sqrt{1+\gamma^2}}{\sqrt{\left(1-r^2\right)^2+\gamma^2}} e^{j(\phi_1+\phi_2)} \tag{3.228}$$

where:

$$\phi_1 = -\tan^{-1}\left(\frac{\gamma}{1-r^2}\right) \tag{3.229}$$

and:

$$\phi_2 = \tan^{-1}(\gamma). \tag{3.230}$$

The value of $\left|\bar{Y}_2 / Y_1\right|$ is:

$$\left|\frac{\bar{Y}_2}{Y_1}\right| = \frac{\sqrt{1+\gamma^2}}{\sqrt{\left(1-r^2\right)^2+\gamma^2}}. \tag{3.231}$$

The response of a 1-DOF vibration system with structural damping to harmonic excitation of its base is given by:

$$y_2(t) = Y_1 \left|\frac{\bar{Y}_2}{Y_1}\right| e^{j(\omega t+\phi_1+\phi_2)}. \tag{3.232}$$

$\left|\bar{Y}_2 / Y_1\right|$ in equation (3.231) is the same as $\left|\bar{F}_T / F\right|$ in equation (3.158). $\phi_1 + \phi_2$ in equation (3.232) is the same as $\theta + \phi$ in equation (3.156). Therefore, Figure 3.23 also shows plots of $\left|\bar{Y}_2 / Y_1\right|$ and $\phi_1 + \phi_2$ as a function of the frequency ratio r for several values of the structural damping factor γ.

When $r > 3$ and noting γ is normally less than 0.1, $\left|\bar{Y}_2\right|$ is given by:

$$\left|\bar{Y}_2\right| = \frac{Y_1 \omega_n^2}{\omega^2} \quad \text{or} \quad \left|\bar{Y}_2\right| = \frac{Y_1 K}{M\omega^2}. \tag{2.233}$$

For this case, the value of $\left|\bar{Y}_2\right|$ decreases as as a function of ω^2. For a specified value of Y_1, $\left|\bar{Y}_2\right|$ can be further reduced by increasing M, decreasing K, or both, so long as the resulting value of ω_n is less than $\omega/3$ where ω is the excitation frequency.

The value of the acceleration of M of a base-excited system with structural damping is:

$$\left|\ddot{\bar{Y}}_2\right| = \omega^2\, Y_1 \frac{\sqrt{1+\gamma^2}}{\sqrt{\left(1-r^2\right)^2+\gamma^2}}. \tag{3.234}$$

Multiplying and dividing equation (3.234) by ω_n^2 and rearranging the terms yields:

$$\left|\frac{\ddot{\bar{Y}}_2}{\omega_n^2\, Y_1}\right| = r^2 \frac{\sqrt{1+\gamma^2}}{\sqrt{\left(1-r^2\right)^2+\gamma^2}}. \tag{3.235}$$

Equation (3.235) is the same as equation (3.159) and is plotted in Figure 3.25.

When $r > 3$ and noting γ is normally less than 0.1, equation (3.235) reduces to:

$$\left|\ddot{\bar{Y}}_2\right| = Y_1\, \omega_n^2 \quad \text{or} \quad \left|\ddot{\bar{Y}}_2\right| = \frac{Y_1 K}{M}. \tag{3.236}$$

For a specified value of Y_1, $\left|\ddot{\bar{Y}}_2\right|$ can be reduced by increasing M, decreasing K, or both, so long as the resulting value of ω_n is less than $\omega/3$ where ω is the excitation frequency.

The force transmitted to M of a base-excited system with structural damping is:

$$\bar{F}_T = \omega^2 M\, Y_1 \frac{\sqrt{1+\gamma^2}}{\sqrt{\left(1-r^2\right)^2+\gamma^2}}\, e^{j(\phi_1+\phi_2+\pi)} . \tag{3.237}$$

Multiplying and dividing equation (3.237) by K and rearranging the terms yields:

$$\boxed{\left|\frac{\bar{F}_T}{K\, Y_1}\right| = r^2 \frac{\sqrt{1+\gamma^2}}{\sqrt{\left(1-r^2\right)^2+\gamma^2}} .} \tag{3.238}$$

Equation (3.238) is the same as equation (3.159) and is plotted in Figure 3.25.

When $r > 3$ and noting γ is normally less than 0.1, equation (3.237) reduces to::

$$\boxed{\left|\bar{F}_T\right| = K\, Y_1 .} \tag{3.239}$$

For a specified value of Y_1, $\left|\bar{F}_T\right|$ can be reduced by decreasing K. As previously stated, the value of $\left|\ddot{\bar{Y}}_2\right|$ can be reduced by increasing M, decreasing K, or both, so long as the resulting value of ω_n is less than $\omega/3$ where ω is the excitation frequency.

3.16 Critical Speed of a Rotating Disk on a Shaft

Vibration associated with a rotating unbalance was examined in Sections 3.8 and 3.13. For these cases, the rotating unbalance was treated as a rotating eccentric mass m that rotated about a fixed center relative to a larger mass M (Figure 3.16).

Another vibration system with a rotating unbalance is a rotating shaft that supports an unbalanced disc. This system is schematically shown in Figure 3.30(a). Even though the disc is shown at the middle of the shaft in this figure, it can be located at any point along the shaft. The bearing supports at the two ends of the shaft are assumed to be rigid in the transverse direction and to allow the shaft to behave as a beam that is simply supported at its two ends.

A *critical speed* occurs when the speed of rotation of the shaft and disc equals one of the transverse resonance frequencies associated with the shaft and disk. Since the shaft has distributed mass and elasticity along its length, it has more than one degree of freedom and, consequently, more than one transverse resonant frequency. Normally, only the first resonant frequency is of interest.

The anticipated displacement of a rotating disc of mass M in the x-y plane is shown in Figure 3.30(b). O is the center of rotation. P is the geometric center of the disc. G is the center of gravity (or mass center) of the disc.

The shaft is treated as having negligible mass relative to the mass M of the disc. The damping associated with the shaft and disc is treated as a viscous damping element C, which is related to the transverse velocities of the disc in the x and y directions. The stiffness associated

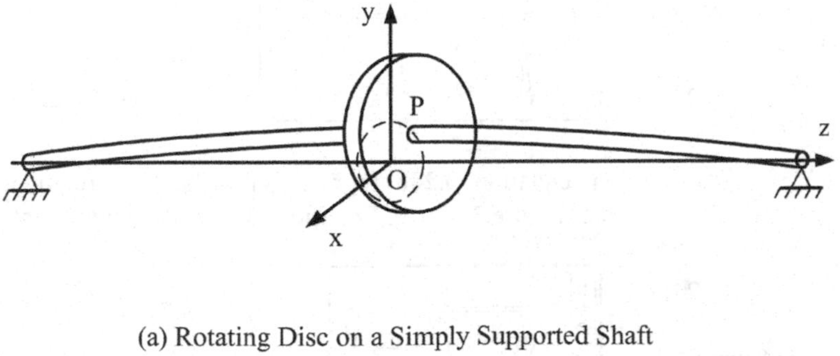

(a) Rotating Disc on a Simply Supported Shaft

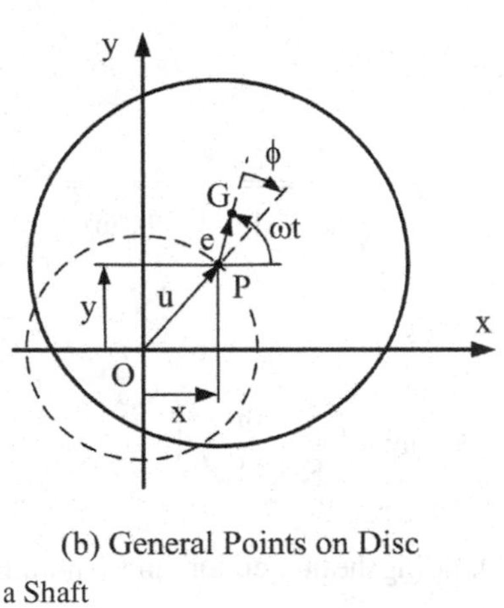

(b) General Points on Disc

Figure 3.30 Critical Speed of a Rotating Disk on a Shaft with the shaft is treated as a transverse spring element K, which is related to the transverse motions of the disc in the x and y directions. (See Section 1.10 for a method of calculating the lateral stiffness of a beam.)

Resolving the forces acting on the disc in the x and y directions yields:

$$M\frac{d^2(x+e\cos\omega t)}{dt^2}=-Kx-C\dot{x} \tag{3.240}$$

and:

$$M\frac{d^2(y+e\sin\omega t)}{dt^2}=-Ky-C\dot{y} \tag{3.241}$$

where e is the distance from the geometric center P to the mass center G of the disc. Carrying out the derivative operations on the left side of the equal sign and rearranging the above equations gives:

$$M\ddot{x}+C\dot{x}+Kx=Me\,\omega^2\cos\omega t \tag{3.242}$$

and:

$$M\ddot{y}+C\dot{y}+Ky=Me\,\omega^2\sin\omega t. \tag{3.243}$$

Solving the above equations yields:

$$x(t)=\frac{Me\,\omega^2}{\sqrt{(K-M\omega^2)^2+(C\omega)^2}}\cos(\omega t-\phi) \tag{3.244}$$

and:

$$y(t)=\frac{Me\,\omega^2}{\sqrt{(K-M\omega^2)^2+(C\omega)^2}}\sin(\omega t-\phi) \tag{3.245}$$

where:

$$\phi=\tan^{-1}\left(\frac{C\omega}{K-M\omega^2}\right). \tag{3.246}$$

Dividing the numerators and denominators of equations (3.244) through (3.246) by K and simplifying gives:

$$x(t)=\frac{e\dfrac{\omega^2}{\omega_n^2}}{\sqrt{\left(1-\dfrac{\omega^2}{\omega_n^2}\right)^2+\left(2\xi\dfrac{\omega}{\omega_n}\right)^2}}\cos(\omega t-\phi) \tag{3.247}$$

$$y(t)=\frac{e\dfrac{\omega^2}{\omega_n^2}}{\sqrt{\left(1-\dfrac{\omega^2}{\omega_n^2}\right)^2+\left(2\xi\dfrac{\omega}{\omega_n}\right)^2}}\sin(\omega t-\phi) \tag{3.248}$$

and:

$$\phi=\tan^{-1}\left(\frac{2\xi\dfrac{\omega}{\omega_n}}{1-\dfrac{\omega^2}{\omega_n^2}}\right). \tag{3.249}$$

Equations (3.247) - (3.249) indicate the motions x(t) and y(t) are equal in magnitude, have the same frequency, and have a phase of $\pi/2$ radians relative to each other. Therefore, their vector sum forms a circle of radius u(t) where:

$$u(t)=\sqrt{x^2(t)+y^2(t)} \tag{3.250}$$

where:

$$\boxed{|u|=\frac{e\dfrac{\omega^2}{\omega_n^2}}{\sqrt{\left(1-\dfrac{\omega^2}{\omega_n^2}\right)^2+\left(2\xi\dfrac{\omega}{\omega_n}\right)^2}}.} \tag{3.251}$$

ϕ in equation (3.248) represents the phase of u relative to e [Figure 3.30(b)]. Equation (3.250) can be written:

$$\boxed{\left|\frac{u}{e}\right|=\frac{r^2}{\sqrt{(1-r^2)^2+(2\xi r)^2}}} \tag{3.252}$$

where $r = \omega/\omega_n$. Equation (2,252) is the same as equation (3.126) and is plotted in Figure 3.17.

Some interesting observations can be made from the above results. Referring to Figure 3.17, when the rotational speed is much less than the critical speed $(r \ll 1)$, $|u|$ is very small compared to e and the mass center G is pointed away from the center of rotation O ($\phi = 0$ radians). When the rotational speed is much greater than the critical speed ($r \gg 1$), $|u|$ equals e and the mass center G coincides with the center of rotation O ($\phi = \pi$ radians). When the rotational speed equals the critical speed, the shaft and disc are at resonance. The value of $|u|$ is then limited by the amount of damping or other physical constraints present in the system. e and u have a phase relation of $\pi/2$ radians.

3.17 Response of 1-DOF Vibration Systems to a Complex Periodic Forcing Function

Complex periodic signals were discussed in Section 1.6 of Chapter 1. The equation of motion of a 1-DOF vibration system similar to the one in Figure 3.1 that is excited by a complex periodic forcing function is:

$$M\ddot{y} + c\dot{y} + Ky = a_0 + \sum_{n=1}^{\infty}\left(a_n \cos n\omega_o t + b_n \sin n\omega_o t\right). \tag{3.253}$$

$$f(t) = a_0 + \sum_{n=1}^{\infty}\left(a_n \cos n\omega_o t + b_n \sin n\omega_o t\right) \tag{3.254}$$

is the Fourier series representation of the complex periodic forcing function.

The total response of the system can be obtained by using the principle of superposition. First determine the response of the system to each individual component of the Fourier series representation of the forcing function. The response of the system to a_0 is:

$$y_0(t) = \frac{a_0}{K}. \tag{3.255}$$

The response of the system to each of the cosine and sine terms is:

$$y_{c(n)}(t) = \frac{a_n / K}{\sqrt{\left(1 - r_n^2\right)^2 + \left(2\xi r_n\right)^2}} \cos\left(n\omega_o t + \phi_n\right) \tag{3.256}$$

and:

$$y_{s(n)}(t) = \frac{b_n / K}{\sqrt{\left(1 - r_n^2\right)^2 + \left(2\xi r_n\right)^2}} \sin\left(n\omega_o t + \phi_n\right) \tag{3.257}$$

where:

$$\phi_n = -\tan^{-1}\left(\frac{2\xi r_n}{1 - r_n^2}\right) \tag{3.258}$$

and:

$$r_n = \frac{n\omega_o}{\omega_n}. \tag{3.259}$$

The total response of the system is:

$$y(t) = y_0(t) + \sum_{n=1}^{\infty}\left[y_{c(n)}(t) + y_{s(n)}(t)\right] \tag{3.260}$$

or:

$$y(t) = \frac{a_0}{K} + \sum_{n=1}^{\infty}\left[\frac{a_n \cos\left(n\omega_o t + \phi_n\right) + b_n \sin\left(n\omega_o t + \phi_n\right)}{K\sqrt{\left(1 - r_n^2\right)^2 + \left(2\xi r_n\right)^2}}\right]. \tag{3.261}$$

By combining the sine and cosine terms, equation (3.261) can be expressed:

$$\boxed{y(t) = \frac{a_0}{K} + \sum_{n=1}^{\infty}\left[\frac{C_n \cos\left(n\omega_o t + \phi_n + \beta_n\right)}{K\sqrt{\left(1 - r_n^2\right)^2 + \left(2\xi r_n\right)^2}}\right]} \tag{3.262}$$

where:

$$\boxed{C_n = \sqrt{a_n^2 + b_n^2}} \tag{3.263}$$

and:

$$\boxed{\beta_n = -\tan^{-1}\left(\frac{b_n}{a_n}\right).} \tag{3.264}$$

EXAMPLE 3.9

A 1-DOF vibration system similar to the one in Figue 3.1 is excited by the saw tooth forcing function described in EXAMPLE 1.2 in Chapter 1. The system has the following properties: M = 500 lb_f, K = 5,108 lb_f/in., and $\xi = 0.05$. Determine the response of the system when F = 200 lb_f and T = 40 ms.

SOLUTION

From EXAMPLE 1.2,

$$f(t) = \frac{F}{2} - \sum_{n=1}^{\infty} \frac{F}{n\pi} \sin(n\omega_o t). \tag{3.9a}$$

The resonance frequency of the system is:

$$\omega_n = \sqrt{\frac{5{,}108 \times 386}{500}} = 62.8 \frac{\text{rad}}{\text{s}}. \tag{3.9b}$$

ω_o is:

$$\omega_o = \frac{2\pi}{0.040} = 157 \frac{\text{rad}}{\text{s}}. \tag{3.9c}$$

The frequency ratio is:

$$r_n = \frac{157\ n}{62.8} = 2.5\ n. \tag{3.9d}$$

Substituting the appropriate values into equation (3.261) and noting:

$$a_o = \frac{F}{2} \quad \text{and} \quad b_n = \frac{F}{n\pi} \tag{3.9e}$$

yields:

$$y(t) = \frac{F}{2K} - \sum_{n=1}^{\infty}\left[\frac{F \sin(n\omega_o t + \phi_n)}{n\pi K \sqrt{(1-r_n^2)^2 + (2\xi r_n)^2}}\right]. \tag{3.9f}$$

Substituting the appropriate numerical values into equation (3.9f) gives:

$$\boxed{y(t) = 0.0196 - \sum_{n=1}^{\infty}\left[\frac{0.0392 \sin(n\omega_o t + \phi_n)}{n\pi \sqrt{\left[1-(2.5\ n)^2\right]^2 + (0.25\ n)^2}}\right]} \tag{3.9g}$$

where:

$$\boxed{\phi_n = -\tan^{-1}\left(\frac{0.25\ n}{1-(2.5\ n)^2}\right).} \tag{3.9h}$$

PROBLEMS - CHAPTER 3

1. Use d'Alembert's principle to develop the equation of motion of the system shown in Figure P3.1 in terms of θ(t). Assume no slipping between roller and surface. Write the expressions for ω_n, θ(t) and the phase φ between f(t) and θ(t).

2. Use Newton's method to develop the equation of motion of the system shown in Figure P3.1 in terms of $x_1(t)$. Assume no slipping between roller and surface. Write the expressions for ω_n, θ(t) and the phase φ between f(t) and $x_1(t)$.

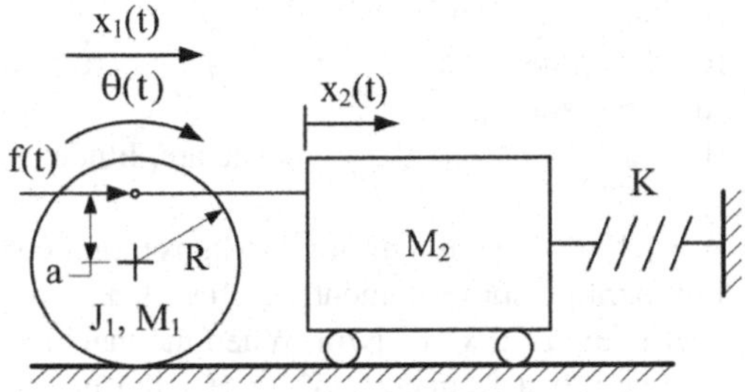

Figure P3.1

3. Use d'Alembert's principle to develop the equation of motion of the system shown in Figure P3.2 in terms of θ(t). Assume no slipping between roller and platform. Write the expressions for ω_n, θ(t) and the phase φ between m(t) and θ(t).

4. Use Newton's method to develop the equation of motion of the system shown in Figure P3.2 in terms of x(t). Assume no slipping between roller and platform. Write the expressions for ω_n, x(t) and the phase φ between m(t) and x(t).

5. Use d'Alembert's principle to develop the equation of motion of the system shown in Figure P3.3 in terms of θ(t). Write the expressions for ω_n, θ(t) and the phase φ between m(t) and θ(t).

6. Use Newton's method to develop the equation of motion of the system shown in Figure P3.3 in terms of $x_2(t)$. Write the expressions for ω_n, x(t) and the phase φ between m(t) and $x_2(t)$.

7. Use d'Alembert's principle to develop the equation of motion of the system shown in Figure P3.4 in terms of x(t). Write the expressions for ω_n, x(t) and the phase φ between m(t) and x(t).

8. Use d'Alembert's principle to develop the equation of motion of the system shown in Figure P3.5 in terms of θ(t). Assume no slipping between the roller and platform. Write the expressions for ω_n, θ(t) and the phase φ between m(t) and θ(t).

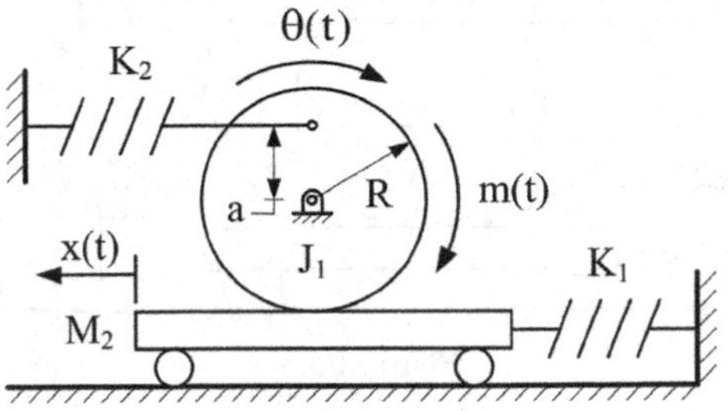

Figure P3.2

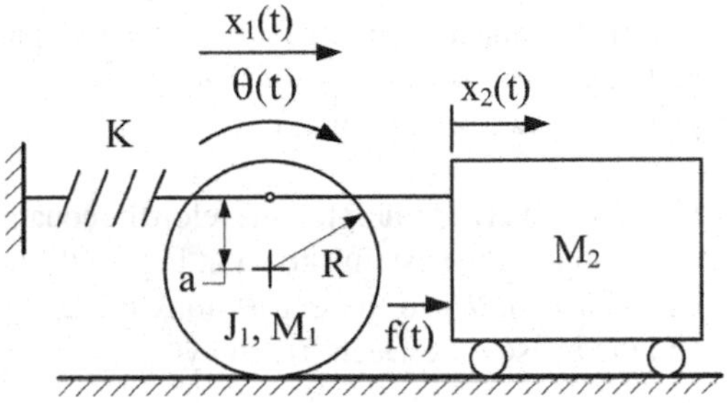

Figure P3.3

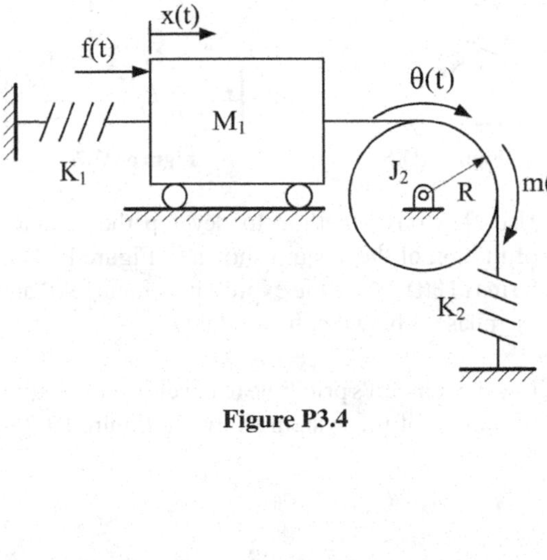

Figure P3.4

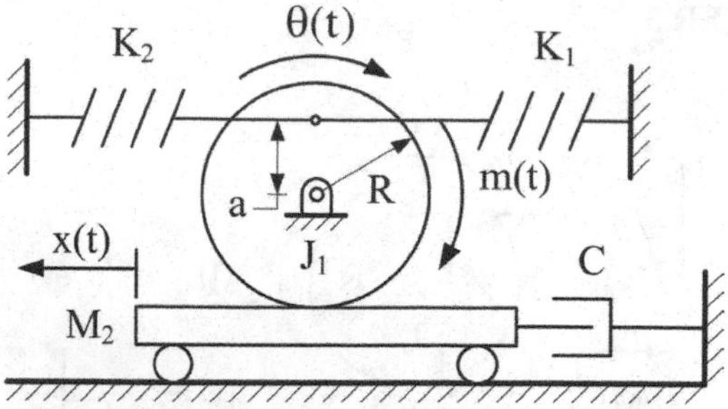

Figure P3.5

9. Use Newton's method to develop the equation of motion of the system shown in Figure P3.5 in terms of x(t). Assume no slipping between the roller and platform. Write the expressions for ω_n, x(t) and the phase ϕ between m(t) and x(t).

10. Use d'Alembert's principle to develop the equation of motion of the system shown in Figure P.3.6 in terms of y(t). Write the expressions for ω_n, y(t) and the phase ϕ between f(t) and y(t).

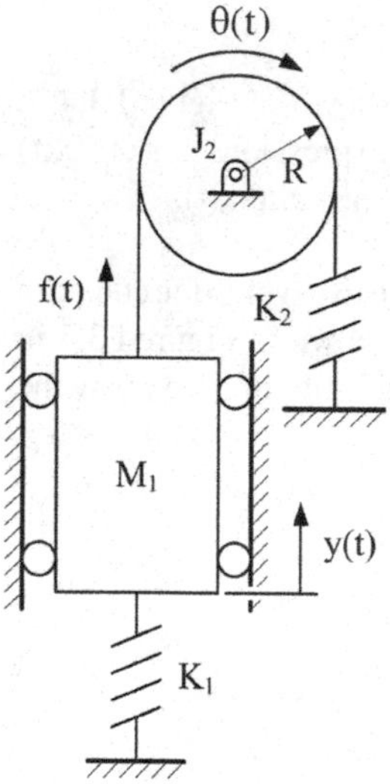

Figure P3.6

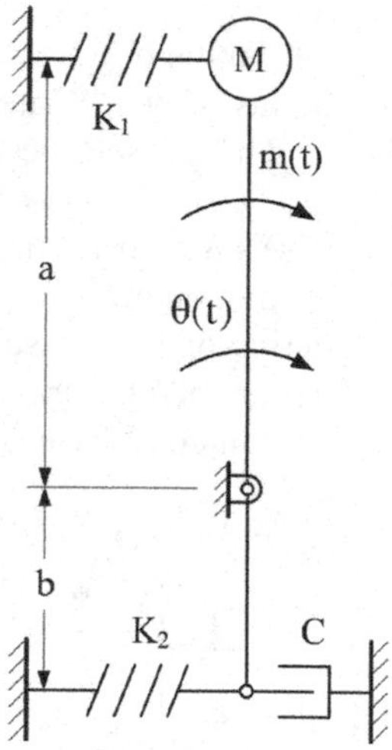

Figure P3.7

11. Use Newtons's method to develop the equation of motion of the system shown in Figure P.3.6 in terms of θ(t). Write the expressions for ω_n, θ(t) and the phase ϕ between f(t) and θ(t).

12. Use d'Alembert's principle to develop the equation of motion of the system shown in Figure P3.7 in terms of θ(t). Write the expressions for ω_n, θ(t) and the phase ϕ between m(t) and θ(t).

13. A 1-DOF vibration system similar to the one shown in Figure 3.1 has the following properties: M = 500 lb_f, ω_n = 15 rad/s, and ξ = 0.20. The external force applied to the mass has the following properties: F = 12 lb_f and f = 1.5 Hz. Determine:
(a) The stiffness coefficient K of the spirng element;
(b) The damping coefficient C of the damping element;
(c) The damped resonance frequency ω_d;
(d) The amplitude $|\bar{H}(\omega)|$ of the frequency response function;
(e) The phase angle ϕ of the frequency response function; and
(f) The amplitude of Y.

14. A machine that has a mass of 45 kg is mounted on four parallel sping elements, each with a stiffness coefficient of 2 x 10^5 N/m. When the machine is operated at a frequency of 32 Hz, the machine has a vibration amplitude of 1.5 mm. What is the amplitude of the excitation force for the machine?

15. A machine that can be represented by Figure 3.1 has a mass of 110 kg and is mounted on an elastic foundation that has a stiffness coefficient of 2 x 10^6 N/m. The machine operates at a frequency of 150 rad/s and has an excitation force with an amplitude of 1,500 N. The response Y of the machine is 0.0019 m. What is the damping ratio ξ of the foundation?

16. A vibration system similar to the one shown in Figure 3.1 has the following properties: We = 24.0 lb_f, K = 3,967 lb_f/in., and C = 0.994 lb_f-s/in.
(a) Determine the damped and undamped resonance frequencies in rad/s and Hz, the damping ratio ξ, and the amplitude of Y when the forcing frequency equals two times the undamped resonance frequency of the system.
(b) Repeat part (a) when C = 0.0497 lb_f-s/in.

17. A motor is mounted on spring isolation mounts that have negligible damping. The motor and spring isolation mounts can be represented by Figure 3.1. The motor weighs 150 lb_f and operates at a

rotational speed of 1,800 rpm. The desired force transmissibility $|\bar{F}_T / F|$ to the base of the isolation system is 0.1. What is the required combined stiffness coefficient K of the sping isolation mounts?

18. The equation of motion of a vibration system is given as:

$$\left(m+\frac{J}{a^2}\right)\ddot{x}+c\dot{x}+5kx=\frac{M_o}{a}\cos\omega t$$

where m = 10 kg; J = 0.1 $kg \cdot m^2$; a = 10 cm; k = 1.6 x 10^5 N/m; c = 640 $N \cdot s / m$; M_o = 100 $N \cdot m$; and ω = 180 rad/s.
(a) Write the expression for the response x(t) and phase φ of the system.
(b) Determine the values for X and φ when the system is excited at the frequency ω = 180 rad/s.

19. A compressor that has a mass of 136 kg is placed on the floor of a building. The deflection of the floor caused by the weight of the compressor was measured to be approximately 0.076 mm.
(a) If the compressor operates at a rotational speed of 3,600 rpm, will a vibration problem exist? Justify your answer. The system can be represented by Figure 3.1.
(b) It has been determined that a force transmission reduction of 94 percent (Transmissibility = 0.06) is desired. If the a frequency ratio r = 5 is used for the design criteria, determine the resonance frequency of the vibration isolation system and the values of the combined stiffness coefficient K of the spring isolation mounts and the damping ratio ξ associated with the damping elements. Assume viscous damping.

20. A vertical single-stage compressor that has a mass of 300 kg is mounted on an isolation system that has the following properties: K = 50 N/mm and ξ = 0.17. The value of the unbalanced mass of the compressor is 10 kg, and the piston stroke is 0.2 m. The compressor is operated at a rotational speed of 400 rpm. The system is represented by Figure 3.19. Determine:
(a) The amplitude Y and phase φ of the vertical motion of the compressor;
(b) The force transmissibility $|\bar{F}_T / F|$ associated with the compressor isolation system; and
(c) The amplitude of the transmitted force $|\bar{F}_T|$.

21. A motor that has a mass of 80 kg is mounted on a vibration isolation system designed to have a force transmissibility $|\bar{F}_T / F|$ of 0.08. The system has negligible damping, and the motor operates at a rotational speed of 1,800 rpm. The system can be represented by Figure 3.1.
(a) Determine the value of the combined stiffness coefficient K of the vibration isolation system.(b) What will the force transmissibility $|\bar{F}_T / F|$ be if the motor is operated at a rotational speed of 1,150 rpm?

22. A vibration isolation system similar to the one shown in Figure 3.1 has the following properties: We = 3.36 lb_f, K = 20.0 lb_f/in., and C = 0.2 lb_f-s/in. Determine the amplitude of vibration Y and the phase φ when the system is excited by the driving force f(t) = 0.75 cos ωt when:
(a) $\omega = \omega_n$; (b) ω = 6.0 rad/s; and
(c) ω = 120.0 rad/s.

23. A machine that can be represented by Figure 3.16 weighs 115 kg and has a rotor mass of 22 kg with an eccentricity of 0.5 mm. The operating speed of the machine is 600 rpm. The machine is mounted on spring isolation mounts with a combined stiffness coefficient K = 8,700 N/m. Damping is negligible.
(a) Determine the amplitude of vibration Y of the machine.
(b) What is combined damping ratio x of the damping elements that must be added to the compressor mount so that the resulting amplitude of vibration Y is one-half of the value in (a) without changing the undamped resonance frequency of the system.
(c) What is the combined damping coefficient C of the damping elements.

24. A vibration exciter with counter-rotating unbalanced masses similar to the one shown in Figure 3.16 is placed on a flexible floor system. The operating frequency of this device is varied from a very low frequency to a very high frequency.

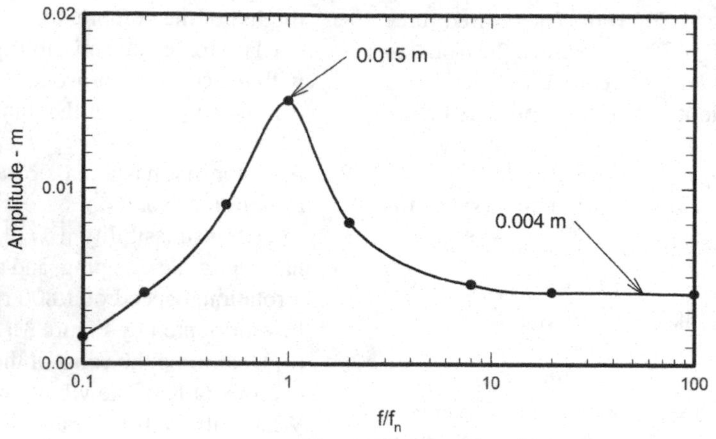

Figure P3.8

The measured amplitude of vibration is show in Figure P3.8. Determine the damping ratio ξ of the floor system.

25. A vertical single-stage air compressor that weighs 1,000 lb_f is mounted on spring isolation mounts with a combined stiffness coefficient K = 1,000 lb_f/in. and damping elements with a damping ratio $\xi = 0.2$. The rotating parts of the compressor are well balanced and the equivalent reciprocating parts weigh 40 lb_f. The compressor stroke is 8 in. Determine the amplitude of vibration Y, the phase angle $\phi + \theta$ with respect to the excitation force, the forc transmissibility $|\bar{F}_T / F|$, and the force $|\bar{F}_T|$ transmitted to the foundation to which the compressor is mounted when the compressor is operated at: (a) a speed of 200 rpm and (b) a speed of 600 rpm.

26. The wind direction is measured by a device called a wind vane (Figure P3.9). V is the wind velocity, J is the mass moment of inertia of the wind vane about the pivot. θ represents the angle of the wind vane relative to the true wind direction. C_t is the torsional damping coefficient of the viscous pivot. L represents the side force on the tail of the vane. From wing theory:

$$L = B\rho V^2 \theta$$

where B is a constant, ρ is the density of air, and V is the wind velocity.

(a) Derive the equation of motion for the wind vane.

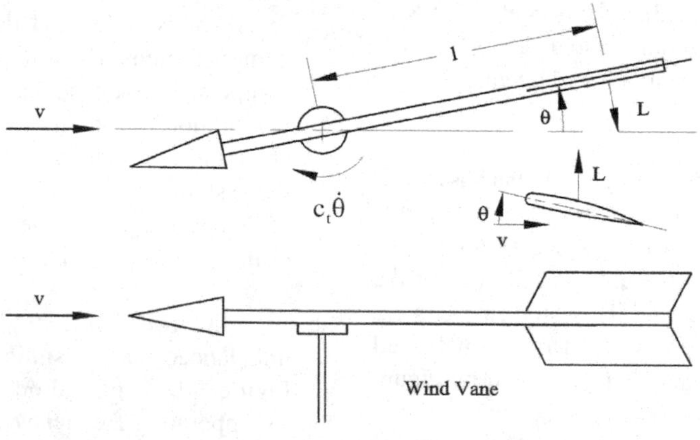

Figure P3.9

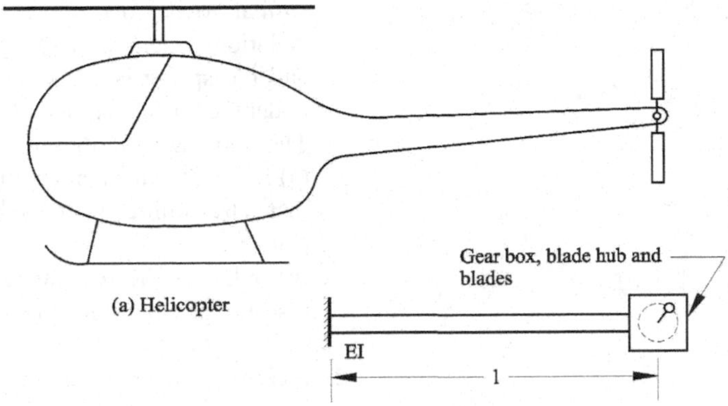

(a) Helicopter

(b) Schematic Representation of of the Tail Section

Figure P3.10

(b) Obtain the solution for the general response of the system. (c) Determine the response of the wind vane for the cases where damping is less than, equal to, and greater than critical damping. (d) Sketch the response of the wind vane for very high and very low wind velocities. Explain the reason for the two responses.

27. Figure P3.10(a) shows a sketch of a helicopter. As a preliminary analysis to see what would happen if one of the anti-torque blades suddenly came off in flight, a simplified model of the anti-torque beam as shown in Figure P3.10(b) is made. The equivalent spring coefficient of the beam is $K = 3EI/l^3$. The amplitude of the mass at the end of the beam results in a resonance frequency of $\omega_n = 50$ rad/s. Each blade weighs 24.1 lb_f. If one blade comes off, the remaining blade can be considered as a point mass acting at a radius of 8 in. from the hub. The weight of the gear box, blade hub, and two blades is 100 lb_f. The blades rotate at 1,000 rpm, and an equivalent damping ratio $\xi = 0.1$ can be assumed for the beam.
(a) Determine the forced response y(t) of the gear box, blade hub and blades when one blade comes off. Note that when one blade comes off, the mass at the end of the beam changes, and the corresponding resonance frequency of the system may also change.
(b) In the situation described above, is it possible for a transient value for y(t) to exist when the blade comes off? Defend your answer.

28. A rotating machine in which the rotor is mounted symmetrically on a shaft has an annular clearance of 0.9 mm between the stator and rotor. The mass of the rotor is 39 kg and has an unbalance of 2 x 10^{-3} kg-m. The length of the shaft is 0.36 m, and the shaft can be assumed to be simply supported as shown in Figure 3.30. If the operating range of the rotor is between 300 and 9,000 rpm, specify the shaft diameter such that its dynamic deflection u is less than 0.2 mm.

29. A circular disc that rotates at a speed of 1,500 rpm is mounted on the center of a shaft that is 0.8 m long and 2 cm in diameter. The disc has a mass of 18 kg and has a mass center which is 4 mm from its geometric center. Assume a damping ratio $\xi = 0.02$ and that the shaft is simply supported as shown in Figure 3.30.
(a) Determine the dynamic deflection u of the shaft and the position of the mass center of the shaft relative to the geometric center and the center of rotation.
(b) At what speed will the shaft be in resonance?
(c) Repeat (a) and (b) for a shaft with a diameter of 3 cm.

30. In order to decrease the vibration motion transmitted to an instrument panel in an aircraft,

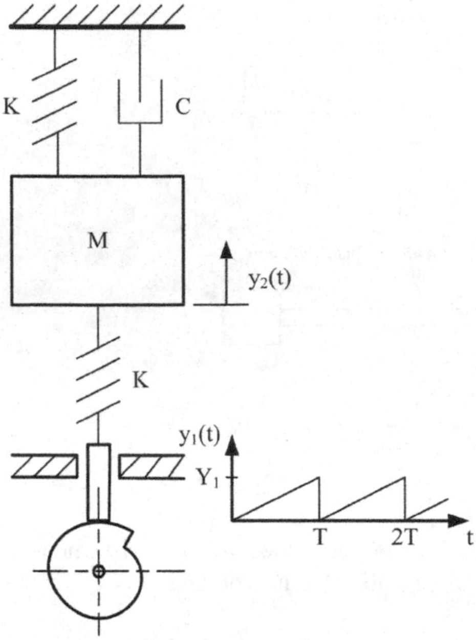

Figure P3.11

the instrument package was mounted on spring isolation mounts. If the spriing isolation mounts, which have structural damping, deflect 0.125 in. under an instrument weight of 50 lb_f, what is the percentage of vibration motion that is transmitted to the instrument packaage when the frequency of vibration is 2,000 rpm. At what frequency in rpm will the vibration motion transmitted to the instrument package be maximum?

31. A sensitive instrument package is isolated from its vibrating base with a vibration isolation system similar to the one shown in Figure 3.28. The isolation system has a damping ratio of $\xi = 0.01$, and the spring isolation mounts deflects 0.2 in. under the 85 lb_f weight of the instrument package. The frequency of vibration of the base is 128 Hz.
(a) Determine the percent of the vibration motion that is transmitted from the base to the instrument package.
(b) What is the resonance frequency f_n of the instrument vibration isolation system?

32. A cam with a sawtooth motion actuates the vibration system shown in Figure P3.11. The total cam lift Y_1 is 1 in. and the cam rotation speed is 60 rpm. The mass weighs 38.6 lb_f. The stiffness coefficient of the two spring elements is 20 lb_f/in., and the damping coefficient of the damping element is 1.0 lb_f-s/in. Determine the expression for the response $y_2(t)$ for the system.

33. (a) Develop the equation that describes the motion of the trailer shown in Figure P3.12 in terms of θ(t). Assume point A moves in a straight line with a velocity V (miles/s) and point B follows the contour of the road. The bumps can be approximated by:

$$y(x) = Y\sin\left(\frac{2\pi}{a}x\right) \quad \text{for} \quad 0 \le x \le \frac{a}{2}$$

$$= 0 \qquad\qquad \text{for} \quad \frac{a}{2} < x \le a.$$

(b) Derive the Fourier series solution for the equation developed in (a).

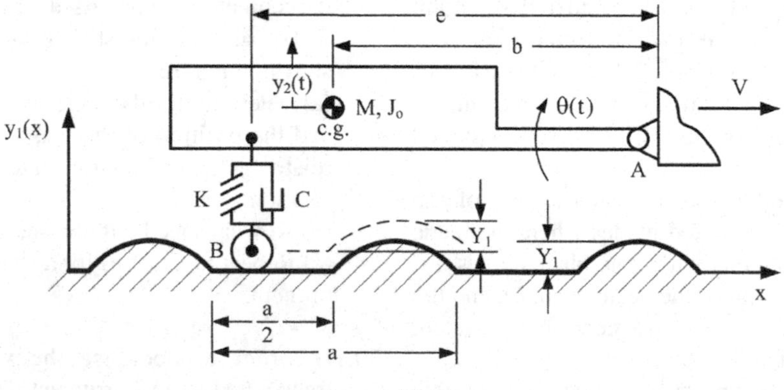

Figure P3.12

(c) What statements can be made with respect to the values of K and C relative to the specified values of M and J_o such that the trailer motion associated with the bumps can be minimized?

34. Use d'Alembert's principle to develop the equation of motion of the system shown in Figure P3.13 in terms of $\theta_1(t)$. Write the expressions for ω_n, $\theta_1(t)$ and the phase ϕ between f(t) and $\theta_1(t)$.

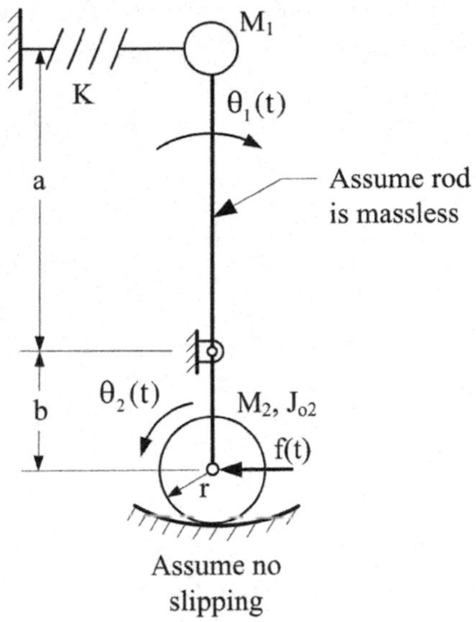

Figure P3.13

CHAPTER 4
2-DOF VIBRATION SYSTEMS RESPONSE TO HARMONIC EXCITATION

4.1 Introduction

Many vibration systems require multiple spatial variables not linked together by geometric constraint equations to describe the motions of the system mass(es). These systems can consist of a single mass supported by spring and damping elements that has translational, rotational or both types of motion in more than one direction or plane. They can consist of numerous masses connected together by spring and damping elements that have translational and rotational motions in multiple directions and planes.

A vibration system that requires numerous spatial variables to describe the motions of its masses is referred to as a *n-DOF vibration system.* n is the number of degrees of freedom associated with the motions of the system masses. A system of n coupled second order differential equations is required to describe the motions of the system masses. These equations must be simultaneously solved to determine the system resonance frequencies and the related motions of the system masses.

The number of degrees of freedom associated with a vibration system is determined by the number of spatial variables needed to describe the motions of the system masses minus the number of geometric constraint equations associated with these variables. For example, if three spatial variables with no geometric constraint equations are needed, the vibration system is a 3-DOF vibration system. One second order differential equation is required for each spatial variable. Therefore, a 3-DOF vibration system requires three coupled second order differential equations to describe the motions of the system masses.

This chapter will discuss 2-DOF vibration systems that require the use of two spatial variables (with no geometric constraint equations) to describe the motions of the system mass(es). The modeling and differential equation solution principles and techniques presented in this chapter can be extended to the analysis of more complicated n-DOF vibration systems.

4.2 Equations of Motion for Unforced Vibration Systems without Damping

A 2-DOF vibration system with two mass elements is shown in Figure 4.1(a). The two mass elements M_1 and M_2 are supported and connected together by three spring elements K_1, K_2 and K_3. The system is constrained to move only in the y direction.

Newton's method can be used to develop the system equations of motion. Figure 4.1(b) shows the free-body diagrams associated with M_1 and M_2. The only forces acting on the mass elements are the restoring forces associated with the spring elements. Per the procedures discussed in Chapters 2 and 3, all restoring forces are shown acting in the negative y direction.

The restoring forces associated with spring element K_2 that connects masses M_1 and M_2 are both shown acting in the negative y direction. The restoring force acting on M_1 is labeled $K_2(y_1 - y_2)$, and the restoring force acting on M_2 is labeled $K_2(y_2 - y_1)$. The first y term within each relative displacement bracket is referenced to the mass on which the respective restoring force acts. Therefore, even though the two restoring force are schematically show acting in the negative y direction, the symbol expression of one is the negative of the other.

Summing the forces that act on M_1 in the y direction yields:

$$\sum F_y = M_1\ddot{y}_1 \quad \text{or} \quad M_1\ddot{y}_1 = -K_1 y_1 - K_2\left(y_1 - y_2\right). \tag{4.1}$$

Therefore:

$$M_1\ddot{y}_1 + \left(K_1 + K_2\right)y_1 - K_2 y_2 = 0. \tag{4.2}$$

In a similar fashion, the sum of the forces acting on mass M_2 in the y direction is:

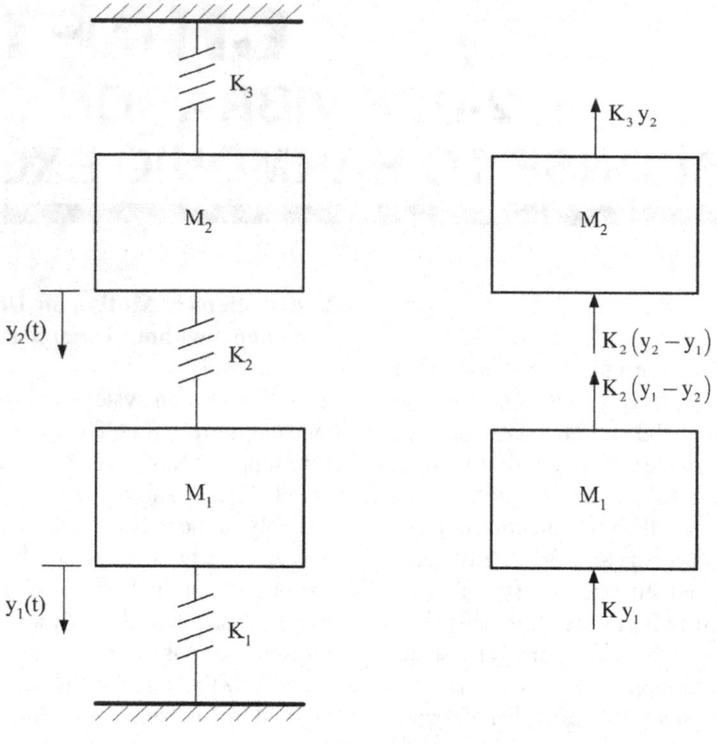

Figure 4.1 2-DOF Vibration System with Mass and Spring Elemrnts

$$\sum F_y = M_2\ddot{y}_2 \quad \text{or} \quad M_2\ddot{y}_2 = -K_3y_2 - K_2\left(y_2 - y_1\right). \tag{4.3}$$

Therefore:

$$M_2\ddot{y}_2 + \left(K_2 + K_3\right)y_2 - K_2y_1 = 0\,. \tag{4.4}$$

Equations (4.2) and (4.4) can be written in matrix form:

$$\boxed{\begin{bmatrix} M_1 & 0 \\ 0 & M_2 \end{bmatrix}\begin{Bmatrix} \ddot{y}_1 \\ \ddot{y}_2 \end{Bmatrix} + \begin{bmatrix} K_1 + K_2 & -K_2 \\ -K_2 & K_2 + K_3 \end{bmatrix}\begin{Bmatrix} y_1 \\ y_2 \end{Bmatrix} = \begin{Bmatrix} 0 \\ 0 \end{Bmatrix}.} \tag{4.5}$$

Vibration System Resonance Frequencies and Mode Vectors

The matrix equation of a vibration system without damping can be written in the form:

$$\left[M\right]\ddot{\bar{y}} + \left[K\right]\bar{y} = \bar{0} \tag{4.6}$$

where:

$[M]$ = the *mass matrix*;
$[K]$ = the *stiffness matrix*;
$\ddot{\bar{y}}$ = the *acceleration vector*; and
$\bar{y}$ = the *displacement vector*.

When harmonic motion is assumed, the displacement vector can be written:

$$\bar{y}\left(t\right) = \bar{Y}\, e^{j\omega t} \tag{4.7}$$

where for a 2-DOF vibration system:

$$\bar{y}\left(t\right) = \begin{Bmatrix} y_1\left(t\right) \\ y_2\left(t\right) \end{Bmatrix} \quad \text{and} \quad \bar{Y} = \begin{Bmatrix} Y_1 \\ Y_2 \end{Bmatrix}. \tag{4.8}$$

Therefore, the acceleration vector can be written:

$$\ddot{\bar{y}}(t) = -\omega^2\ \bar{Y}\ e^{j\omega t} \tag{4.9}$$

or:

$$\begin{Bmatrix} \ddot{y}_1(t) \\ \ddot{y}_2(t) \end{Bmatrix} = -\omega^2 \begin{Bmatrix} Y_1 \\ Y_2 \end{Bmatrix} e^{j\omega t}\,. \tag{4.10}$$

Substituting equations (4.7) and (4.9) into equation (4.6) yields:

$$\left[-\omega^2\,[M]+[K]\right]\bar{Y}\ e^{j\omega t} = \bar{0} \tag{4.11}$$

or:

$$\left[[K]-\omega^2\,[M]\right]\bar{Y} = \bar{0}\,. \tag{4.12}$$

Equation (4.12) is satisfied when the displacement vector equals zero or when the determinant of the matrix equals zero. Setting the displacement vector equal to zero results in the trivial solution. Therefore, equation (4.12) is satisfied when:

$$\boxed{\left|[K]-\omega^2\,[M]\right| = \bar{0}.} \tag{4.13}$$

Equation (4.13) is solved to obtain the values for ω. The values of ω that satisfy equation (4.13) are the *resonance frequencies* of the system. The number of resonance frequencies that are obtained by solving equation (4.13) will equal the number of degrees of freedom of the vibration system that is being analyzed. There will be two resonance frequencies for the 2-DOF vibration systems that are analyzed in this chapter.

EXAMPLE 4.1

Assume the system shown in Figure 4.1 has the following properties: $M_1 = M_2 = M$ and $K_1 = K_2 = K_3 = K$. Determine the expressions for the two resonance frequencies.

SOLUTION

For this problem, equation (4.5) reduces to:

$$\begin{bmatrix} M & 0 \\ 0 & M \end{bmatrix} \begin{Bmatrix} \ddot{y}_1 \\ \ddot{y}_2 \end{Bmatrix} + \begin{bmatrix} 2K & -K \\ -K & 2K \end{bmatrix} \begin{Bmatrix} y_1 \\ y_2 \end{Bmatrix} = \begin{Bmatrix} 0 \\ 0 \end{Bmatrix} \tag{4.1a}$$

and equation (4.13) becomes:

$$\left| \begin{bmatrix} 2K & -K \\ -K & 2K \end{bmatrix} - \omega^2 \begin{bmatrix} M & 0 \\ 0 & M \end{bmatrix} \right| = 0\,. \tag{4.1b}$$

Simplifying equation (4.1b) yields:

$$\begin{vmatrix} 2K - M\omega^2 & -K \\ -K & 2K - M\omega^2 \end{vmatrix} = 0\,. \tag{4.1c}$$

Expanding equation (4.1c) gives:

$$\left(2K - M\omega^2\right)\left(2K - M\omega^2\right) - K^2 = 0 \tag{4.1d}$$

or:

$$M^2\omega^4 - 4KM\omega^2 + 3K^2 = 0\,. \tag{4.1e}$$

Dividing equation (4.1e) by M^2 yields:

$$\omega^4 - \frac{4K}{M}\omega^2 + \frac{3K^2}{M^2} = 0\,. \tag{4.1f}$$

Equation (4.1f) is called the characteristic frequency equation. The values associated with ω^2 are obtained from the quadratic equations, or:

$$\boxed{\omega_{1,2}^2 = \frac{1}{2}\left[\frac{4K}{M} \mp \sqrt{\frac{16K^2}{M^2} - \frac{12K^2}{M^2}}\right].} \tag{4.1g}$$

Simplifying equation (4.1g) yields:

$$\omega_{1,2}^2 = \frac{1}{2}\left[\frac{4K}{M} \mp \frac{2K}{M}\right] \tag{4.1h}$$

or:

$$\boxed{\omega_1 = \sqrt{\frac{K}{M}} \quad \text{and} \quad \omega_2 = \sqrt{\frac{3K}{M}}.} \tag{4.1i}$$

ω_1 and ω_2 are the two resonance frequencies of the 2-DOF vibration system shown in Figure 4.1 when $M_1 = M_2 = M$ and $K_1 = K_2 = K_3 = K$.

The motions of masses M_1 and M_2 relative to each other at each of the resonance frequencies is of interest. Relative to the system in EXAMPLE 4.1, this information can be obtained by first expanding equation (4.12) for the case where $M_1 = M_2 = M$ and $K_1 = K_2 = K_3 = K$. Equation (4.12) can be written:

$$\begin{bmatrix} 2K - M\omega^2 & -K \\ -K & 2K - M\omega^2 \end{bmatrix} \begin{Bmatrix} Y_1 \\ Y_2 \end{Bmatrix} = \begin{Bmatrix} 0 \\ 0 \end{Bmatrix}. \tag{4.14}$$

Equation (4.14) results in two equations, or:

$$\left(2K - M\omega^2\right) Y_1 - KY_2 = 0 \tag{4.15}$$

$$-KY_1 + \left(2K - M\omega^2\right) Y_2 = 0. \tag{4.16}$$

Equations (4.15) and (4.16) are not independent equations when they are solved for the cases when $\omega = \omega_1$ or $\omega = \omega_2$. Therefore, it is not possible to solve for both Y_1 and Y_2 when $\omega = \omega_1$ or $\omega = \omega_2$. However, Y_1 can be obtained in terms of Y_2, or Y_2 can be obtained in terms of Y_1. Use equation (4.15) to solve for Y_2 in terms of Y_1. Therefore:

$$\frac{Y_2}{Y_1} = \frac{2K - M\omega^2}{K} \tag{4.17}$$

or:

$$\frac{Y_2}{Y_1} = 2 - \frac{M}{K}\omega^2. \tag{4.18}$$

Solving for Y_2/Y_1 for ω_1 yields:

$$\frac{Y_{21}}{Y_{11}} = 2 - \frac{M}{K}\omega_1^2. \tag{4.19}$$

Substituting equation (4.1i) into equation (4.20) and simplifying gives:

$$\frac{Y_{21}}{Y_{11}} = 1. \tag{4.20}$$

Solving for Y_2/Y_1 for ω_2 yields:

$$\frac{Y_{22}}{Y_{12}} = -1. \tag{4.21}$$

Usually the displacement vectors $\bar{Y}^{(1)}$ and $\bar{Y}^{(2)}$ are normalized relative to Y_1. Therefore, for the system in EXAMPLE 4.1:

$$\boxed{\frac{\bar{Y}^{(1)}}{Y_{11}} = \begin{Bmatrix} Y_{11}/Y_{11} \\ Y_{21}/Y_{11} \end{Bmatrix} = \begin{Bmatrix} 1 \\ 1 \end{Bmatrix}} \tag{4.22}$$

and:

$$\boxed{\frac{\bar{Y}^{(2)}}{Y_{12}} = \begin{Bmatrix} Y_{12}/Y_{12} \\ Y_{22}/Y_{12} \end{Bmatrix} = \begin{Bmatrix} 1 \\ -1 \end{Bmatrix}.} \tag{4.23}$$

Equations (4.22) and (4.23) are referred to as the *normalized mode vectors* associated with the resonance frequencies ω_1 and ω_2, respectfully. They indicate that at the first resonance frequency, masses M_1 and M_2 vibrate in phase with equal amplitudes. At the second resonance frequency, masses M_1 and M_2 vibrate 180 deg out-of-phase with equal amplitudes.

Vibration System Response to Initial Conditions

The solution to equation (4.6) associated with initial conditions is obtained by separating the spatial and time components of the system response. Therefore:

$$\bar{y}(t) = \bar{Y}\, T(t) \tag{4.24}$$

and:

$$\ddot{\bar{y}}(t) = \bar{Y}\,\ddot{T}(t). \tag{4.25}$$

Therefore, equation (4.6) becomes:

$$[M]\bar{Y}\,\ddot{T}(t) + [K]\bar{Y}\,T(t) = \bar{0} \tag{4.26}$$

Rearranging equation (4.26) yields:

$$-\frac{\ddot{T}(t)}{T(t)} = \frac{[K]\bar{Y}}{[M]\bar{Y}} = \omega^2 . \tag{4.27}$$

The two sides of equation (4.27) are separated by letting both sides of the equation equal ω^2. Therefore, equation (4.27) is separated into:

$$\ddot{T}(t) + \omega^2\,T(t) = 0 \tag{4.28}$$

and:

$$\big[[K] - \omega^2\,[M]\big]\,\bar{Y} = \bar{0}. \tag{4.29}$$

The solution to equation (4.28) for the resonance frequency ω_i is:

$$T_i(t) = C_i\,\cos(\omega_i t + \phi_i). \tag{4.30}$$

As before, equation (4.29) is solved to obtain the values for ω_i and $\bar{Y}^{(i)}/Y_{1i}$.

The total solution to equation (4.6) associated with initial conditions is obtained by first writing equation (4.24):

$$\bar{y}(t) = \sum_{i=1}^{n} \frac{\bar{Y}^{(i)}}{Y_{1i}}\,Y_{1i}\,T_i(t). \tag{4.31}$$

Substitute equation (4.30) into equation (4.31):

$$\bar{y}(t) = \sum_{i=1}^{n} \frac{\bar{Y}^{(i)}}{Y_{1i}}\,Y_{1i}\,C_i\,\cos(\omega_i t + \phi_i). \tag{4.32}$$

$\bar{Y}^{(i)}/Y_{1i}$ are the *normalized mode vectors*, and ω_i are the corresponding *resonance frequencies*. The products $Y_{1i}\,C_i$ are constants and can be replaced by another constant where:

$$Y_{1i}\,C_i = A_i\,. \tag{4.33}$$

Therefore, equation (4.32) can be written:

$$\bar{y}(t) = \sum_{i=1}^{n} \frac{\bar{Y}^{(i)}}{Y_{1i}}\,A_i\,\cos(\omega_i t + \phi_i). \tag{4.34}$$

For a 2-DOF vibration system, equation (4.34) can be written:

$$y_1(t) = \frac{Y_{11}}{Y_{11}}\,A_1\cos(\omega_1 t + \phi_1) + \frac{Y_{12}}{Y_{12}}\,A_2\cos(\omega_2 t + \phi_2) \tag{4.35}$$

$$y_2(t) = \frac{Y_{21}}{Y_{11}}\,A_1\cos(\omega_1 t + \phi_1) + \frac{Y_{22}}{Y_{12}}\,A_2\cos(\omega_2 t + \phi_2). \tag{4.36}$$

A_1, A_2, ϕ_1 and ϕ_2 are constants that are obtained by applying the initial conditions $y_1(0)$, $\dot{y}_1(0)$, $y_2(0)$ and $\dot{y}_2(0)$.

The general initial condition equations are:

$$\boxed{\bar{y}(0) = \sum_{i=1}^{n} \frac{\bar{Y}^{(i)}}{Y_{1i}}\,A_i\,\cos\phi_i} \tag{4.37}$$

$$\boxed{\dot{\bar{y}}(0) = -\sum_{i=1}^{n} \frac{\bar{Y}^{(i)}}{Y_{1i}}\,A_i\,\omega_i\,\sin\phi_i\,.} \tag{4.38}$$

For a 2-DOF vibration system:

$$\boxed{y_1(0) = \frac{Y_{11}}{Y_{11}}\,A_1\cos\phi_1 + \frac{Y_{12}}{Y_{12}}\,A_2\cos\phi_2} \tag{4.39}$$

$$\boxed{y_2(0) = \frac{Y_{21}}{Y_{11}}\,A_1\cos\phi_1 + \frac{Y_{22}}{Y_{12}}\,A_2\cos\phi_2} \tag{4.40}$$

$$\dot{y}_1(0) = -\frac{Y_{11}}{Y_{11}} A_1\, \omega_1 \sin\phi_1 - \frac{Y_{12}}{Y_{12}} A_2\, \omega_2 \sin\phi_2 \tag{4.41}$$

$$\dot{y}_2(0) = -\frac{Y_{21}}{Y_{11}} A_1\, \omega_1 \sin\phi_1 - \frac{Y_{22}}{Y_{12}} A_2\, \omega_2 \sin\phi_2\,. \tag{4.42}$$

There are four equations and four unknowns.

EXAMPLE 4.2

Determine the response of the two mass elements in EXAMPLE 4.1 for the initial conditions:

$y_1(0) = 0,\ y_2(0) = 1,\ \dot{y}_1(0) = 0$ and $\dot{y}_2(0) = 0$.

SOLUTION

The responses of the two mass elements are given by equations (4.35) and (4.36). From EXAMPLE 4.1:

$$\omega_1 = \sqrt{\frac{K}{M}} \quad \text{and} \quad \omega_2 = \sqrt{\frac{3K}{M}}\,. \tag{4.2a}$$

From equations (4.22) and (4.23):

$$\frac{\bar{Y}^{(1)}}{Y_{11}} = \begin{Bmatrix} Y_{11}/Y_{11} \\ Y_{21}/Y_{11} \end{Bmatrix} = \begin{Bmatrix} 1 \\ 1 \end{Bmatrix} \tag{4.2b}$$

$$\frac{\bar{Y}^{(2)}}{Y_{12}} = \begin{Bmatrix} Y_{12}/Y_{12} \\ Y_{22}/Y_{12} \end{Bmatrix} = \begin{Bmatrix} 1 \\ -1 \end{Bmatrix}. \tag{4.2c}$$

Substituting equations (4.2a) through (4.2c) into equations (4.39) through (4.42) and applying the initial conditions yield:

$$0 = A_1 \cos\phi_1 + A_2 \cos\phi_2 \tag{4.2d}$$

$$1 = A_1 \cos\phi_1 - A_2 \cos\phi_2 \tag{4.2e}$$

$$0 = -A_1 \sqrt{\frac{K}{M}} \sin\phi_1 - A_2 \sqrt{\frac{3K}{M}} \sin\phi_2 \tag{4.2f}$$

$$0 = -A_1 \sqrt{\frac{K}{M}} \sin\phi_1 + A_2 \sqrt{\frac{3k}{M}} \sin\phi_2\,. \tag{4.2g}$$

Adding equations (4.2f) and (4.2g) yields:

$$0 = -2\, A_1 \sqrt{\frac{K}{M}} \sin\phi_1\,. \tag{4.2h}$$

A_1 cannot equal zero. Therefore, $\sin\phi_1 = 0$ or $\phi_1 = 0$. Applying this result to either equation (4.2f) or (4.2g) yields $\phi_2 = 0$. Equations (4.2d) and (4.2e) become:

$$0 = A_1 + A_2 \tag{4.2i}$$

$$1 = A_1 - A_2\,. \tag{4.2j}$$

Solving equations (4.2i) and (4.2j) for A_1 and A_2 yields:

$$A_1 = \frac{1}{2} \quad \text{and} \quad A_2 = -\frac{1}{2}. \tag{4.2k}$$

Therefore, the responses of masses M_1 and M_2 to the initial conditions are:

$$y_1(t) = \frac{1}{2}\cos\sqrt{\frac{K}{M}}\, t - \frac{1}{2}\cos\sqrt{\frac{3K}{M}}\, t \tag{4.2l}$$

$$y_2(t) = \frac{1}{2}\cos\sqrt{\frac{K}{M}}\, t + \frac{1}{2}\cos\sqrt{\frac{3K}{M}}\, t\,. \tag{4.2m}$$

Figure 4.2 shows the responses for $y_1(t)$) and $y_2(t)$ for the case where $\omega_1 = 6.28$ rad/s and $\omega_2 = 10.88$ rad/s.

Vibration Systems with Two Degrees of Freedom

EXAMPLE 4.3

Use d'Alembert's principle to develop the equations of motion for the double pendulum shown in Figure 4.3.

SOLUTION

This problem can be solved by summing the moments relative to point A and then relative to mass M_1. There are six spatial variables and four geometric constraint

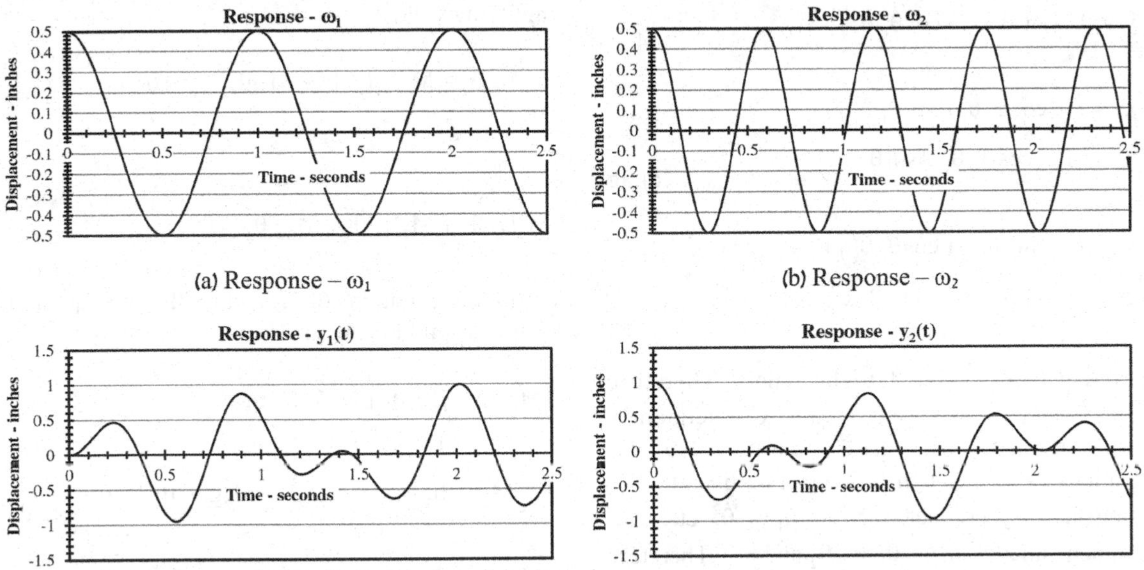

(a) Response – ω_1 (b) Response – ω_2

(c) Response – $y_1(t)$ (d) Response – $y_2(t)$

Figure 4.2 Response $y_1(t)$ and $y_2(t)$ for EXAMPLE 4.2

equations for the double pendulum in Figure 4.2. Therefore, the double pendulum is a 2-DOF vibration system that will require two second-order differential equations to describe its motion.

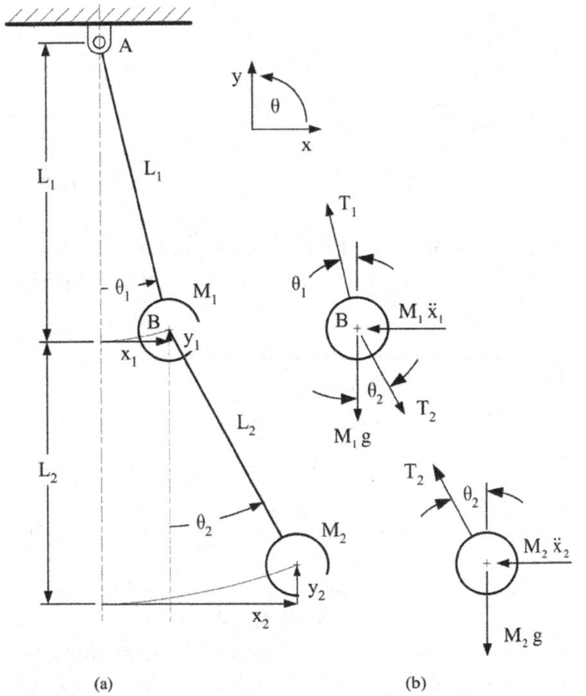

Figure 4.3 2-DOF Double Pendulum Vibration System

The spatial variables are x_1, y_1, θ_1, x_2, y_2 and θ_2. The constraint equations are:

$$x_1 = L_1 \sin\theta_1 \tag{4.3a}$$

$$y_1 = L_1 \left(1 - \cos\theta_1\right) \tag{4.3b}$$

$$x_2 = L_1 \sin\theta_1 + L_2 \sin\theta_2 \tag{4.3c}$$

$$y_2 = L_1 \left(1 - \cos\theta_1\right) + L_2 \left(1 - \cos\theta_2\right). \tag{4.3d}$$

Equations (4.3a) through (4.3d) indicate that θ_1 and θ_2 are the spatial variables that should be used to describe the motion of the system. When d'Alembert's principle is used, it is first necessary to determine the accelerations of the two pendulum masses. Taking the second derivatives of equations (4.3a) through (4.3d) with respect to time yields:

$$\ddot{x}_1 = L_1 \left(\cos\theta_1 \, \ddot{\theta}_1 - \sin\theta_1 \, \dot{\theta}_1^2\right) \tag{4.3e}$$

$$\ddot{y}_1 = L_1\left(\sin\theta_1\,\ddot{\theta}_1 + \cos\theta_1\,\dot{\theta}_1^2\right) \tag{4.3f}$$

$$\ddot{x}_2 = L_1\left(\cos\theta_1\,\ddot{\theta}_1 - \sin\theta_1\,\dot{\theta}_1^2\right) + L_2\left(\cos\theta_2\,\ddot{\theta}_2 - \sin\theta_2\,\dot{\theta}_2^2\right) \tag{4.3g}$$

$$\ddot{y}_2 = L_1\left(\sin\theta_1\,\ddot{\theta}_1 + \cos\theta_1\,\dot{\theta}_1^2\right) + L_2\left(\sin\theta_2\,\ddot{\theta}_2 + \cos\theta_2\,\dot{\theta}_2^2\right). \tag{4.3h}$$

θ_1 and θ_2 are assumed to be small. Therefore, $\dot{\theta}_1, \ddot{\theta}_1, \dot{\theta}_2$ and $\ddot{\theta}_2$ are also small. Consequently, $\sin\theta_1 \approx \theta_1$, $\sin\theta_2 \approx \theta_2$, and $\cos\theta_1 \approx \cos\theta_2 \approx 1$. Equations (4.3e) through (4.3h) can be linearized by noting that any product, such as $\theta_1\,\ddot{\theta}_1, \dot{\theta}_1^2$, etc., will be much smaller than $\dot{\theta}_1, \ddot{\theta}_1, \dot{\theta}_2$ and $\ddot{\theta}_2$. Therefore, equations (4.3e) through (4.3f) reduce to:

$$\ddot{x}_1 = L_1\ddot{\theta}_1 \tag{4.3i}$$

$$\ddot{y}_1 = 0 \tag{4.3j}$$

$$\ddot{x}_2 = L_1\ddot{\theta}_1 + L_2\ddot{\theta}_2 \tag{4.3k}$$

$$\ddot{y}_2 = 0. \tag{4.3l}$$

Figure 4.2 shows the free-body diagrams associated with both pendulum masses. Note that the tension forces T_1 and T_2 pass through the centers of the pendulum masses. Summing the moments relative to point A yields:

$$\sum M_A = 0 = -M_1 g\,L_1\theta_1 - M_1\,L_1\ddot{x}_1 - M_2 g\left(L_1\theta_1 + L_2\theta_2\right) - M_2\left(L_1 + L_2\right)\ddot{x}_2\,. \tag{4.3m}$$

Summing the moments relative to point B on mass M_1 gives:

$$\sum M_B = 0 = -M_2 g\,L_2\theta_2 - M_2\,L_2\ddot{x}_2\,. \tag{4.3n}$$

Substituting equation (4.3k) into equation (4.3n) and simplifying yields:

$$M_2 L_2^2\,\ddot{\theta}_2 + M_2 L_1 L_2\,\ddot{\theta}_1 + M_2 g L_2\,\theta_2 = 0 \tag{4.3o}$$

or:

$$M_2 L_2\,\ddot{\theta}_2 + M_2 L_1\,\ddot{\theta}_1 + M_2 g\,\theta_2 = 0\,. \tag{4.3p}$$

Substituting equations (4.3i) and (4.3k) into equation (4.3m) and simplifying gives:

$$\left(M_1 + M_2\right)L_1^2\,\ddot{\theta}_1 + M_2 L_1 L_2\,\ddot{\theta}_2 + \left(M_1 + M_2\right)g L_1\,\theta_1 + \left[M_2 L_2^2\,\ddot{\theta}_2 + M_2 L_1 L_2\,\ddot{\theta}_1 + M_2 g L_2\,\theta_2\right] = 0\,. \tag{4.3q}$$

The term in the brackets in equation (4.3q) is equation (4.3o), which equals zero. Therefore, equation (4.3q) reduces to:

$$\left(M_1 + M_2\right)L_1^2\,\ddot{\theta}_1 + M_2 L_1 L_2\,\ddot{\theta}_2 + \left(M_1 + M_2\right)g L_1\,\theta_1 = 0 \tag{4.3r}$$

or:

$$\left(M_1 + M_2\right)L_1\,\ddot{\theta}_1 + M_2 L_2\,\ddot{\theta}_2 + \left(M_1 + M_2\right)g\,\theta_1 = 0\,. \tag{4.3s}$$

Equations (4.3p) and (4.3s) are the equations of motion for the double pendulum. They can be written in matrix form, or:

$$\begin{bmatrix} \left(M_1 + M_2\right)L_1 & M_2 L_2 \\ M_2 L_1 & M_2 L_2 \end{bmatrix} \begin{Bmatrix} \ddot{\theta}_1 \\ \ddot{\theta}_2 \end{Bmatrix} + \begin{bmatrix} \left(M_1 + M_2\right)g & 0 \\ 0 & M_2 g \end{bmatrix} \begin{Bmatrix} \theta_1 \\ \theta_2 \end{Bmatrix} = \begin{Bmatrix} 0 \\ 0 \end{Bmatrix}. \tag{4.3t}$$

EXAMPLE 4.4

Use Newton's method to develop the equations of motion of the 2-DOF torsional vibration system shown in Figure 4.4(a).

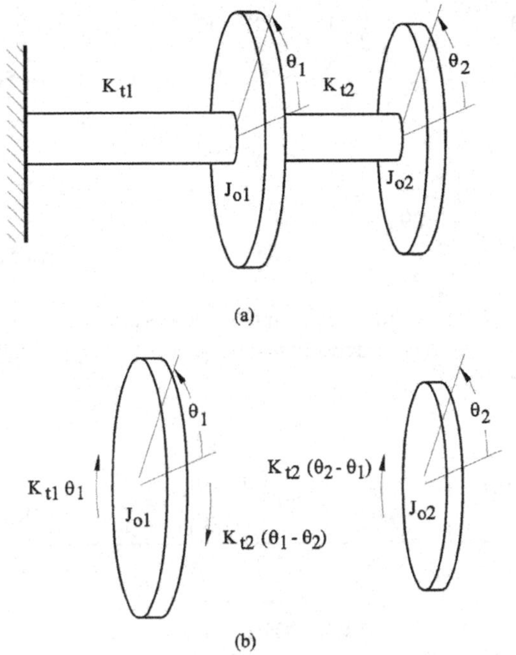

Figure 4.4 2-DOF Torsional Vibration System

SOLUTION

The free-body diagrams of the torsional disks are shown in Figure 4.4(b). Summing the moments on disk 1 yields:

$$\sum M_1 = J_{o1}\,\ddot{\theta}_1 \text{ or } J_{o1}\,\ddot{\theta}_1 = -K_{t1}\,\theta_1 - K_{t2}\,(\theta_1-\theta_2). \tag{4.4a}$$

Therefore:

$$J_{o1}\,\ddot{\theta}_1 + (K_{t1}+K_{t2})\,\theta_1 - K_{t2}\,\theta_2 = 0. \tag{4.4b}$$

Summing the moments on disk 2 gives:

$$\sum M_2 = J_{o2}\,\ddot{\theta}_2 \text{ or } J_{o2}\,\ddot{\theta}_2 = -K_{t2}\,(\theta_2-\theta_1). \tag{4.4c}$$

Therefore:

$$J_{o2}\,\ddot{\theta}_2 + K_{t2}\,\theta_2 - K_{t2}\,\theta_1 = 0. \tag{4.4d}$$

Equations (4.4b) and (4.4d) can be written in matrix form, or:

$$\boxed{\begin{bmatrix} J_{o1} & 0 \\ 0 & J_{o2} \end{bmatrix}\begin{Bmatrix} \ddot{\theta}_1 \\ \ddot{\theta}_2 \end{Bmatrix} + \begin{bmatrix} K_{t1}+K_{t2} & -K_{t2} \\ -K_{t2} & K_{t2} \end{bmatrix}\begin{Bmatrix} \theta_1 \\ \theta_2 \end{Bmatrix} = \begin{Bmatrix} 0 \\ 0 \end{Bmatrix}.} \tag{4.4e}$$

EXAMPLE 4.5

In EXAMPLE 4.4, $J_{o2} = 0.5\,J_{o1}$ and $K_{t2} = 0.8\,K_{t1}$. Determine the expressions for the resonance frequencies and the corresponding mode vectors.

SOLUTION

Equation (4.4e) becomes:

$$\begin{bmatrix} J_{o1} & 0 \\ 0 & 0.5\,J_{o1} \end{bmatrix}\begin{Bmatrix} \ddot{\theta}_1 \\ \ddot{\theta}_2 \end{Bmatrix} + \begin{bmatrix} 1.8\,K_{t1} & -0.8\,K_{t1} \\ -0.8\,K_{t1} & 0.8\,K_{t1} \end{bmatrix}\begin{Bmatrix} \theta_1 \\ \theta_2 \end{Bmatrix} = \begin{Bmatrix} 0 \\ 0 \end{Bmatrix}. \tag{4.5a}$$

For harmonic motion, equation (4.5a) becomes:

$$\begin{bmatrix} 1.8\,K_{t1}-J_{o1}\,\omega^2 & -0.8\,K_{t1} \\ -0.8\,K_{t1} & 0.8\,K_{t1}-0.5\,J_{o1}\,\omega^2 \end{bmatrix}\begin{Bmatrix} \theta_1 \\ \theta_2 \end{Bmatrix} = 0. \tag{4.5b}$$

The determinant of equation (4.5b) is:

$$\begin{vmatrix} 1.8\,K_{t1}-J_{o1}\,\omega^2 & -0.8\,K_{t1} \\ -0.8\,K_{t1} & 0.8\,K_{t1}-0.5\,J_{o1}\,\omega^2 \end{vmatrix} = 0 \tag{4.5c}$$

Solving the determinant yields:

$$0.5\,J_{01}^2\,\omega^4 - 1.7\,J_{o1}K_{t1}\,\omega^2 + 0.8\,K_{t1}^2 = 0 \tag{4.5d}$$

or:

$$\omega^4 - 3.4\,\frac{K_{t1}}{J_{o1}}\,\omega^2 + 1.6\,\frac{K_{t1}^2}{J_{o1}^2} = 0. \tag{4.5e}$$

Solving equation (4.5e) for $\omega_{1,2}^2$ gives:

$$\omega_{1,2}^2=\frac{1}{2}\left[\frac{3.4\ K_{t1}}{J_{o1}}\mp\sqrt{\left(\frac{3.4\ K_{t1}}{J_{o1}}\right)^2-\frac{6.4\ K_{t1}^2}{J_{o1}^2}}\right] \tag{4.5f}$$

or:

$$\omega_{1,2}^2=\frac{1}{2}\left[\frac{3.4\ K_{t1}}{J_{o1}}\mp\frac{2.27\ K_{t1}}{J_{o1}}\right]. \tag{4.5g}$$

Therefore:

$$\boxed{\omega_1=\sqrt{\frac{0.565\ K_{t1}}{J_{o1}}}\quad\text{and}\quad\omega_2=\sqrt{\frac{2.835\ K_{t1}}{J_{o1}}}.} \tag{4.5h}$$

Use the first equation in equation (4.5b) to obtain the mode vectors. The first equation is:

$$\left(1.8\ K_{t1}-J_{o1}\ \omega^2\right)\theta_1-0.8\ K_{t1}=0 \tag{4.5i}$$

or:

$$\frac{\theta_2}{\theta_1}=\frac{1.8\ K_{t1}-J_{o1}\ \omega^2}{0.8\ K_{t1}}. \tag{4.5j}$$

Simplifying equation (4.5j) yields:

$$\frac{\theta_2}{\theta_1}=2.25-1.25\frac{J_{o1}}{K_{t1}}\ \omega^2. \tag{4.5k}$$

For the first resonance frequency is:

$$\frac{\theta_{21}}{\theta_{11}}=2.25-1.25\frac{J_{o1}}{K_{t1}}\left(\frac{0.565\ K_{t1}}{J_{o1}}\right) \tag{4.5l}$$

or:

$$\frac{\theta_{21}}{\theta_{11}}=1.54. \tag{4.5m}$$

For the second resonance frequency is:

$$\frac{\theta_{22}}{\theta_{12}}=2.25-1.25\frac{J_{o1}}{K_{t1}}\left(\frac{2.835\ K_{t1}}{J_{o1}}\right) \tag{4.5n}$$

or:

$$\frac{\theta_{22}}{\theta_{12}}=-1.29. \tag{4.5o}$$

Therefore, the normalized mode vectors associated with the first and second resonance frequencies, respectively, are:

$$\boxed{\begin{Bmatrix}\theta_{11}/\theta_{11}\\ \theta_{21}/\theta_{11}\end{Bmatrix}=\begin{Bmatrix}1\\ 1.54\end{Bmatrix}\ \text{and}\ \begin{Bmatrix}\theta_{12}/\theta_{12}\\ \theta_{22}/\theta_{12}\end{Bmatrix}=\begin{Bmatrix}1\\ -1.29\end{Bmatrix}.} \tag{4.5p}$$

EXAMPLE 4.6

Use Lagrange's equation to determine the equations of motion of the 2-DOF vibration system shown in Figure 4.1.

SOLUTION

Since y_1 and y_2 are independent coordinates, they can represent the generalized coordinates associated with the system where:

$$q_1=y_1 \quad\text{and}\quad q_2=y_2. \tag{4.6a}$$

The total kinetic energy for the system is:

$$T=\frac{1}{2}M_1\dot{y}_1^2+\frac{1}{2}M_2\dot{y}_2^2. \tag{4.6b}$$

The total potential energy for the system is:

$$U=\frac{1}{2}K_1y_1^2+\frac{1}{2}K_2\left(y_1-y_2\right)^2+\frac{1}{2}K_3y_2^2. \tag{4.6c}$$

There are no damping or externally applied forces.

Taking the derivatives of equation (4.6b) with respect to $\dot{y}_1$ and $\dot{y}_2$ yields:

$$\frac{\partial T}{\partial \dot{y}_1} = M_1 \dot{y}_1 \qquad (4.6d)$$

$$\frac{\partial T}{\partial \dot{y}_2} = M_1 \dot{y}_2 \,. \qquad (4.6e)$$

Taking the derivatives of equation (4.6c) with respect to y_1 and y_2 gives:

$$\frac{\partial U}{\partial y_1} = K_1 y_1 + K_2 \left(y_1 - y_2 \right) \qquad (4.6f)$$

$$\frac{\partial U}{\partial y_2} = -K_2 \left(y_1 - y_2 \right) + K_3 y_2 \,. \qquad (4.6g)$$

Substituting equations (4.6d) and (4.6f) into equation (1.185) in Chapter 1 and simplifying yields:

$$M_1 \ddot{y}_1 + \left(K_1 + K_2 \right) y_1 - K_2 y_2 = 0 \,. \qquad (4.6h)$$

Substituting equations (4.6e) and (4.6f) into equation (1.185) gives:

$$M_2 \ddot{y}_2 + \left(K_2 + K_3 \right) y_2 - K_2 y_1 = 0 \,. \qquad (4.6i)$$

Equations (4.6h) and (4.6i) can be written in matrix form, or:

$$\begin{bmatrix} M_1 & 0 \\ 0 & M_2 \end{bmatrix} \begin{Bmatrix} \ddot{y}_1 \\ \ddot{y}_2 \end{Bmatrix} + \begin{bmatrix} K_1 + K_2 & -K_2 \\ -K_2 & K_2 + K_3 \end{bmatrix} \begin{Bmatrix} y_1 \\ y_2 \end{Bmatrix} = \begin{Bmatrix} 0 \\ 0 \end{Bmatrix} . \qquad (4.6j)$$

Equation (4.6j) is the same as equation (4.5).

EXAMPLE 4.7

Use d'Alembert's principle to develop the equations of motion of the roller-pulley-mass system in Figure 4.5(a). Write the equations of motion in terms of θ_1 and θ_2.

SOLUTION

Figure 4.5(b) shows the free-body diagrams of the roller-pulley-mass system. There are four spatial variables: θ_1, x, θ_2 and y. There are two geometric constraint equations:

$$x = R\theta_1 \qquad \text{and} \qquad y = r\theta_2 \,. \qquad (4.7a)$$

Therefore, the system is a 2-DOF vibration system. Sum the moments acting on the roller relative to point A:

$$\sum M_A = 0 = -K_1 R^2 \,\theta_1 - K_2 R \left(R\theta_1 - r\theta_2 \right) - J_{o1} \,\ddot{\theta}_1 - M_1 R^2 \,\ddot{\theta}_2 \qquad (4.7b)$$

or:

$$\left(J_{o1} + M_1 R^2 \right) \ddot{\theta}_1 + \left(K_1 R^2 + K_2 R^2 \right) \theta_1 - K_2 R\, r\, \theta_2 = 0 \,. \qquad (4.7c)$$

Sum the moments acting on the pulley relative to its center of rotation, B:

$$\sum M_B = 0 = -K_2 r \left(r\theta_2 - R\theta_1 \right) - J_{o2} \,\ddot{\theta}_2 + r\,T \,. \qquad (4.7d)$$

Finally, sum the forces acting on mass M_2 in the y direction, or:

$$\sum F_y = 0 = -T - M_2 r \,\ddot{\theta}_2 \qquad (4.7e)$$

or:

$$T = -M_2 r \,\ddot{\theta}_2 \,. \qquad (4.7f)$$

Substituting equation (4.7f) into equation (4.7d) and simplifying yields

$$\left(J_{o2} + M_2 r^2 \right) \ddot{\theta}_2 + K_2 r^2 \,\theta_2 - K_2 R\, r\, \theta_1 = 0 \,. \qquad (4.7g)$$

Equations (4.7c) and (4.7g) are the equations of motion of the 2-DOF vibration system in Figure 4.5. These equations can be written in matrix form, or:

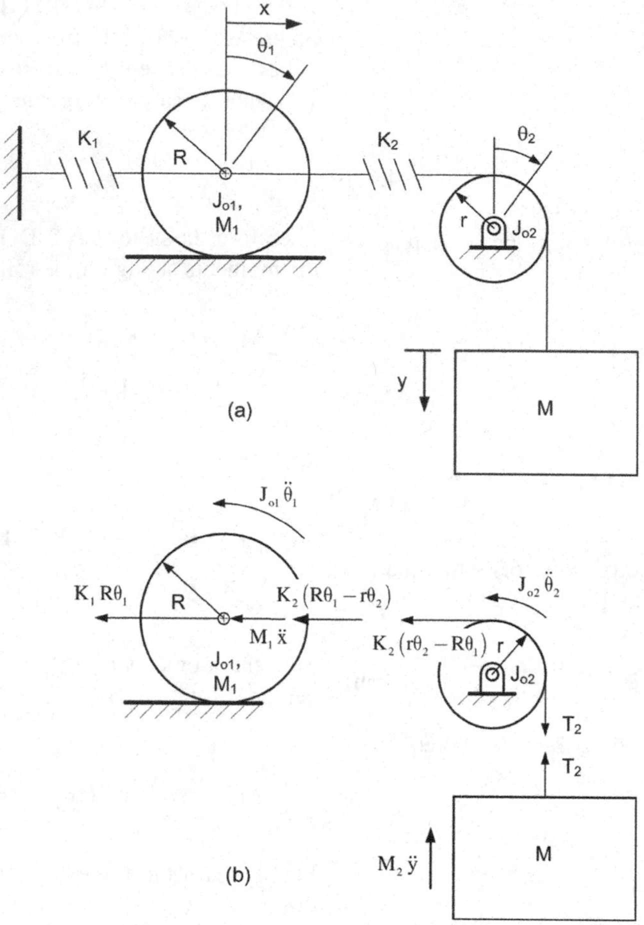

Figure 4.5 2-DOF Vibration System with Spring Elements and Rolling, Rotational and Translational Mass Elements

$$\begin{bmatrix} J_{o1}+M_1R^2 & 0 \\ 0 & J_{o2}+M_2r^2 \end{bmatrix}\begin{Bmatrix} \ddot{\theta}_1 \\ \ddot{\theta}_2 \end{Bmatrix}+ \begin{bmatrix} K_1R^2+K_2R^2 & -K_2R\,r \\ -K_2R\,r & K_2r^2 \end{bmatrix}\begin{Bmatrix} \theta_1 \\ \theta_2 \end{Bmatrix}=\begin{Bmatrix} 0 \\ 0 \end{Bmatrix}. \tag{4.7h}$$

EXAMPLE 4.8

Use Lagrange's equation to develop the equations of motion of the roller-pulley-mass system in Figure 4.5(a). Write the equations of motion in terms of θ_1 and θ_2.

SOLUTION

The total kinetic energy of the system is:

$$T=\frac{1}{2}J_{o1}\,\dot{\theta}_1^2+\frac{1}{2}M_1\,\dot{x}^2+\frac{1}{2}J_{o2}\,\dot{\theta}_2^2+\frac{1}{2}M_2\,\dot{y}^2\,. \tag{4.8a}$$

Substituting equation (4.7a) into equation (4.8a) and simplifying yields:

$$T=\frac{1}{2}\left(J_{o1}+M_1R^2\right)\dot{\theta}_1^2+\frac{1}{2}\left(J_{o2}+M_2r^2\right)\dot{\theta}_2^2\,. \tag{4.8b}$$

The total potential energy is:

$$U=\frac{1}{2}K_1R^2\,\theta_1^2+\frac{1}{2}K_2\left(R\theta_1-r\theta_2\right)^2. \tag{4.8c}$$

Taking the derivatives of equation (4.8b) with respect to $\dot{\theta}_1$ and $\dot{\theta}_2$ yields:

$$\frac{\partial T}{\partial \dot{\theta}_1} = \left(J_{o1} + M_1 R^2\right)\dot{\theta}_1 \tag{4.8d}$$

$$\frac{\partial T}{\partial \dot{\theta}_2} = \left(J_{o2} + M_2 r^2\right)\dot{\theta}_2 \,. \tag{4.8e}$$

Taking the derivatives of equation (4.8c) with respect to θ_1 and θ_2 gives:

$$\frac{\partial U}{\partial \theta_1} - K_1 R^2\, \theta_1 + K_2\left(R\theta_1 - r\theta_2\right)R \tag{4.8f}$$

or:

$$\frac{\partial U}{\partial \theta_1} = \left(K_1 R^2 + K_2 R^2\right)\theta_1 - K_2 R\, r\, \theta_2 \tag{4.8g}$$

and:

$$\frac{\partial U}{\partial \theta_2} = K_2\left(R\theta_1 - r\theta_2\right)(-r) \tag{4.8h}$$

or:

$$\frac{\partial U}{\partial \theta_2} = -K_2 R\, r\, \theta_1 + K_2 r^2\, \theta_2 \,. \tag{4.8i}$$

Substituting equations (4.8d) and (4.8g) into equation (1.185) yields:

$$\left(J_{o1} + M_1 R^2\right)\ddot{\theta}_1 + \left(K_1 R^2 + K_2 R^2\right)\theta_1 - K_2 R\, r\, \theta_2 = 0 \,. \tag{4.8j}$$

Substituting equations (4.8e) and (4.8i) into equation (1.185) gives:

$$\left(J_{o2} + M_2 r^2\right)\ddot{\theta}_2 + K_2 r^2\, \theta_2 - K_2 R\, r\, \theta_1 = 0 \,. \tag{4.8k}$$

Equations (4.8j) and (4.8k) can be written in matrix form, or:

$$\begin{bmatrix} J_{o1} + M_1 R^2 & 0 \\ 0 & J_{o2} + M_2 r^2 \end{bmatrix} \begin{Bmatrix} \ddot{\theta}_1 \\ \ddot{\theta}_2 \end{Bmatrix} + \begin{bmatrix} K_1 R^2 + K_2 R^2 & -K_2 R\, r \\ -K_2 R\, r & K_2 r^2 \end{bmatrix} \begin{Bmatrix} \theta_1 \\ \theta_2 \end{Bmatrix} = \begin{Bmatrix} 0 \\ 0 \end{Bmatrix} . \tag{4.8l}$$

4.3 Coordinate Coupling

Normally, spatial variables are selected to describe the motions of masses in a vibration system relative to coordinate systems centered on the centers of gravity of thc masscs. Howcver, this is not always the case. Spatial variables can be referenced to coordinate systems centered on instant centers and centers of rotation, and they can be referenced to coordinate systems centered on arbitrary points on or off of system masses. Figure 4.6(a) shows a schematic representation of an automobile. Figure 4.6(b) shows a 2-DOF representation of the automobile where the automobile is allowed to move vertically in the y direction and rotate in the x-y plane. Any number of spatial variable and coordinate system combinations can be used to describe the motion of the automobile. They include:

1. y of the c.g. and θ about the c.g.;
2. y_1 at point A and y_2 at point B;
3. y_1 at point A and θ about point A;
4. y_2 at point B and θ about point B; or
5. y and θ relative to any other point on or off of mass M.

h_1 and h_2 are not system coordinates; they are vertical displacement inputs to the tires. Any combination of the coordinates shown in Figure 4.5(b) [(y, θ), (y_1, y_2), (y_1, θ), or (y_2, θ)] can be used as a set of coordinates to describe the motion of the automobile.

EXAMPLE 4.9

Use Newton's method to develop the equations of motion of the automobile shown in Figure 4.6 relative to its center of gravity. Assume h_1 and h_2 equal zero.

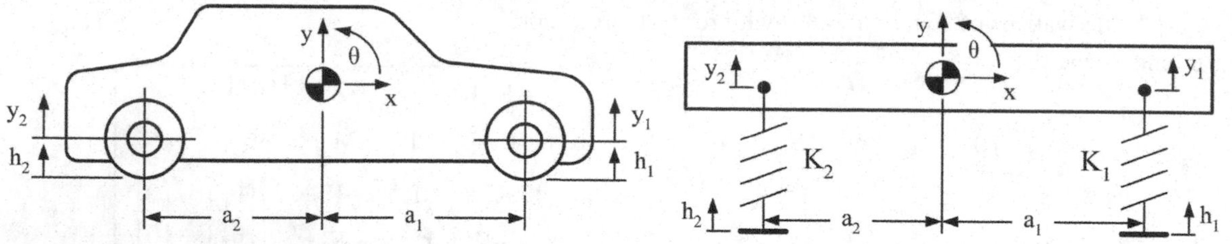

(a) Schematic Representation of an Automobile (b) 2-DOF Representation of an Automobile

Figure 4.6 2-DOF Vibration Model of an Automobile

SOLUTION

Figure 4.7 shows the free-body diagram of the automobile when its motion is referenced to its center of gravity. To obtain the motions of points A and B, first give the c.g. a positive displacement y upward and then a positive rotation θ in the counter clockwise direction. Assume the upward displacement y of the c.g. is sufficiently large so the overall displacement $y - a_2\theta$ at point B is positive. Therefore, the restoring force $K_2(y - a_2\theta)$ at point B can be shown acting in the negative y direction.

Summing the forces in the y directions yields:

$$\sum F_y = M\ddot{y} \quad \text{or} \quad M\ddot{y} = -K_1\left(y + a_1\theta\right) - K_2\left(y - a_2\theta\right). \tag{4.9a}$$

Therefore:

$$M\,\ddot{y} + \left(K_1 + K_2\right)y + \left(K_1a_1 - K_2a_2\right)\theta = 0. \tag{4.9b}$$

Summing the moments relative to the c.g. gives:

$$\sum M_{c.g.} = J_o\,\theta \quad \text{or}$$

$$J_o\,\theta = -K_1a_1\left(y + a_1\theta\right) + K_2a_2\left(y - a_2\theta\right). \tag{4.9c}$$

Therefore:

$$J_o\,\ddot{\theta} + \left(K_1a_1^2 + K_2a_2^2\right)\theta + \left(K_1a_1 - K_2a_2\right)y = 0. \tag{4.9d}$$

Equations (4.9b) and (4.9d) can be written in matrix form, or:

$$\begin{bmatrix} M & 0 \\ 0 & J_o \end{bmatrix}\begin{Bmatrix} \ddot{y} \\ \ddot{\theta} \end{Bmatrix} + \begin{bmatrix} K_1 + K_2 & K_1a_1 - K_2a_2 \\ K_1a_1 - K_2a_2 & K_1a_1^2 + K_2a_2^2 \end{bmatrix}\begin{Bmatrix} y \\ \theta \end{Bmatrix} = \begin{Bmatrix} 0 \\ 0 \end{Bmatrix}. \tag{4.9e}$$

EXAMPLE 4.10

Use d'Alembert's principle to develop the equations of motion of the automobile shown in Figure 4.6 relative to point A. Assume h_1 and h_2 equal zero.

SOLUTION

Figure 4.8 shows the free-body diagram of the automobile when its motion is referenced to point A. To obtain the motions of points A and B and the center of gravity, first give point A a positive displacement y upward and then a positive rotation θ in the counter clockwise direction. Assume the upward displacement y at point A is sufficiently large so the displacements $y - a_1\theta$ at the c.g. and $y - (a_1 + a_2)\theta$ at point B are

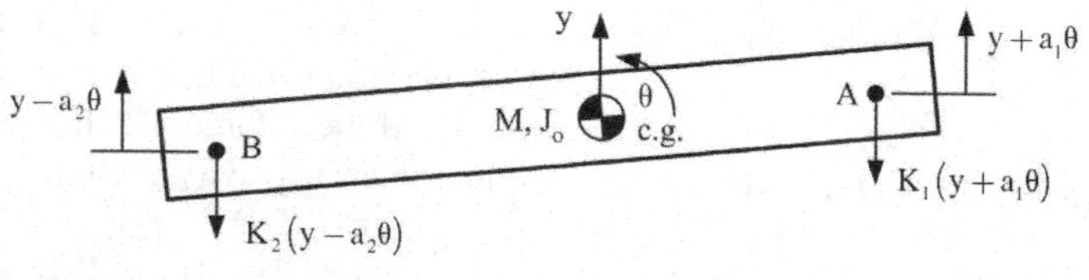

Figure 4.7 2-DOF Representation of an Automobile - Motion Relative to c.g

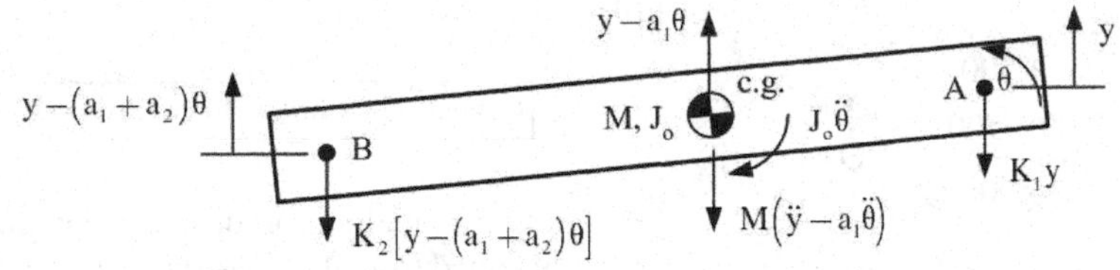

Figure 4.8 2-DOF Representation of an Automobile - Motion Relative to Point A

positive. Therefore, the inertial force $M(\ddot{y} - a_1\ddot{\theta})$ at the c.g. and the restoring force $K_2[y - (a_1 + a_2)\theta]$ at point B can be shown acting in the negative y direction.

Summing the forces in the y direction yields:

$$\sum F_y = 0 = -M(\ddot{y} - a_1\ddot{\theta}) - K_1 y - K_2[y - (a_1 + a_2)\theta] \quad (4.10a)$$

or:

$$M\ddot{y} - Ma_1\ddot{\theta} + (K_1 + K_2)y - K_2(a_1 + a_2)\theta = 0. \quad (4.10b)$$

Summing the moments in the θ direction relative to point A gives:

$$\sum M_A = 0 = -J_o\ddot{\theta} + Ma_1(\ddot{y} - a_1\ddot{\theta}) + K_2(a_1 + a_2)[y - (a_1 + a_2)\theta] \quad (4.10c)$$

or:

$$(J_o + Ma_1^2)\ddot{\theta} - Ma_1\ddot{y} + K_2(a_1 + a_2)^2\theta - K_2(a_1 + a_2)y = 0. \quad (4.10d)$$

Equations (4.10b) and (4.10d) can be written in matrix form, or:

$$\begin{bmatrix} M & -Ma_1 \\ -Ma_1 & J_o + Ma_1^2 \end{bmatrix} \begin{Bmatrix} \ddot{y} \\ \ddot{\theta} \end{Bmatrix} + \begin{bmatrix} K_1 + K_2 & -K_2(a_1 + a_2) \\ -K_2(a_1 + a_2) & K_2(a_1 + a_2)^2 \end{bmatrix} \begin{Bmatrix} y \\ \theta \end{Bmatrix} = \begin{Bmatrix} 0 \\ 0 \end{Bmatrix}. \quad (4.10e)$$

EXAMPLE 4.11

The automobile in EXAMPLES 4.9 and 4.10 has the following properties: $M = 3{,}000\ lb_f$, $J_o = 20{,}000\ lb_f\text{-}s/in.$, $K_1 = 400\ lb_f/in.$, $K_2 = 300\ lb_f/in.$, $a_1 = 55$ in., and $a_2 = 65$ in. Determine the resonance frequencies and the corresponding mode vectors of the automobile using the equations developed in EXAMPLE 4.9.

SOLUTION

Noting that:

$$y(t) = Y\, e^{j\omega t} \quad \text{and} \quad \theta(t) = \Theta\, e^{j\omega t} \quad (4.11a)$$

equation (4.9e) becomes:

$$\begin{bmatrix} K_1 + K_2 - M\omega^2 & K_1 a_1 - K_2 a_2 \\ K_1 a_1 - K_2 a_2 & K_1 a_1^2 + K_2 a_2^2 - J_o\omega^2 \end{bmatrix} \begin{Bmatrix} Y \\ \Theta \end{Bmatrix} = \begin{Bmatrix} 0 \\ 0 \end{Bmatrix}. \quad (4.11b)$$

Noting that:

$$M = \frac{3{,}000}{386} = 7.772\, \frac{lb_f - s^2}{in.} \quad (4.11c)$$

and substituting the appropriate system variables into equation (4.11b) and simplifying yields:

$$\begin{bmatrix} 700-7.772\,\omega^2 & 2{,}500 \\ 2{,}500 & 2{,}477{,}500- \\ & 20{,}000\,\omega^2 \end{bmatrix} \begin{Bmatrix} Y \\ \Theta \end{Bmatrix} = \begin{Bmatrix} 0 \\ 0 \end{Bmatrix}. \tag{4.11d}$$

Factoring 1,000 out of the matrix in equation (4.11d) gives:

$$\begin{bmatrix} 0.7-0.007772\,\omega^2 & 2.5 \\ 2.5 & 2{,}477.5- \\ & 20\,\omega^2 \end{bmatrix} \begin{Bmatrix} Y \\ \Theta \end{Bmatrix} = \begin{Bmatrix} 0 \\ 0 \end{Bmatrix}. \tag{4.11e}$$

Setting the determinant of equation (4.11e) equal to zero yields:

$$\left(0.7-0.007772\,\omega^2\right)\left(2{,}477.5-20\,\omega^2\right)- 2.5^2 = 0 \tag{4.11f}$$

or:

$$0.155\,\omega^4 - 33.255\,\omega^2 - 1{,}728 = 0\,. \tag{4.11g}$$

Divide equation (4.11g) by 0.115 to obtain:

$$\omega^4 - 214.5\,\omega^2 + 11{,}148.3 = 0\,. \tag{4.11h}$$

$\omega_{1,2}^2$ is obtained from the quadratic equation, or:

$$\omega_{1,2}^2 = 0.5\left[214.6 \mp \sqrt{214.6^2 - 4\times 11{,}148.4}\right] \tag{4.11i}$$

or:

$$\omega_{1,2}^2 = 107.3 \mp 19.1\,. \tag{4.11j}$$

Therefore:

$$\boxed{\omega_1 = 9.4\,\frac{\text{rad}}{\text{s}} \quad \text{and} \quad \omega_2 = 11.2\,\frac{\text{rad}}{\text{s}}} \tag{4.11k}$$

or:

$$\boxed{f_1 = 1.5\text{ Hz} \quad \text{and} \quad f_2 = 1.8\text{ Hz}\,.} \tag{4.11l}$$

The equation for determining the mode vector is obtained from the first equation in equation (4.11b), or:

$$\frac{\Theta}{Y} = \frac{M\omega^2 - \left(K_1 + K_2\right)}{K_1 a_1 - K_2 a_2}\,. \tag{4.11m}$$

Substituting the system variables into equation (4.11m) yields:

$$\frac{\Theta}{Y} = \frac{7.772\,\omega^2 - 700}{2{,}500}\,. \tag{4.11n}$$

Substituting ω_1 and ω_2 into equation (4.11n) and simplifying gives:

$$\frac{\Theta_1}{Y_1} = -5.80\times 10^{-3} \text{ and } \frac{\Theta_2}{Y_2} = 1.13\times 10^{-1}\,. \tag{4.11o}$$

The corresponding mode vectors are:

$$\boxed{\begin{Bmatrix} Y_1/Y_1 \\ \Theta_1/Y_1 \end{Bmatrix} = \begin{Bmatrix} 1 \\ -5.80\times 10^{-3} \end{Bmatrix}} \tag{4.11p}$$

$$\boxed{\begin{Bmatrix} Y_2/Y_2 \\ \Theta_2/Y_2 \end{Bmatrix} = \begin{Bmatrix} 1 \\ 1.13\times 10^{-1} \end{Bmatrix}.} \tag{4.11q}$$

EXAMPLE 4.12

Determine the resonance frequencies and the corresponding mode vectors of the automobile in EXAMPLE 4.11 using the equations developed in EXAMPLE 4.10.

SOLUTION

Noting that:

$$y(t) = Y\,e^{j\omega t} \quad \text{and} \quad \theta(t) = \Theta\,e^{j\omega t} \tag{4.12a}$$

equation (4.10e) becomes:

(4.12b)

$$\begin{bmatrix} K_1 + K_2 - M\omega^2 & Ma_1\omega^2 - K_2(a_1 + a_2) \\ Ma_1\omega^2 - K_2(a_1 + a_2) & K_2(a_1 + a_2)^2 - (J_o + Ma_1^2)\omega^2 \end{bmatrix} \begin{Bmatrix} Y \\ \Theta \end{Bmatrix} = \begin{Bmatrix} 0 \\ 0 \end{Bmatrix}.$$

Noting that:

$$M = \frac{3{,}000}{386} = 7.772\ \frac{lb_f - s^2}{in.} \quad (4.12c)$$

and substituting the appropriate system variables into equation (4.12b) and simplifying yields:

(4.12d)

$$\begin{bmatrix} 700 - 7.772\,\omega^2 & 427.46\,\omega^2 - 36{,}000 \\ 427.46\,\omega^2 - 36{,}000 & 4{,}320{,}000 - 43{,}510.36\,\omega^2 \end{bmatrix} \begin{Bmatrix} Y \\ \Theta \end{Bmatrix} = \begin{Bmatrix} 0 \\ 0 \end{Bmatrix}.$$

Factoring 1,000 out of the matrix in equation (4.12d) gives:

(4.12e)

$$\begin{bmatrix} 0.7 - 0.007772\,\omega^2 & 0.427\,\omega^2 - 36 \\ 0.427\,\omega^2 - 36 & 4{,}320 - 43.510\,\omega^2 \end{bmatrix} \begin{Bmatrix} Y \\ \Theta \end{Bmatrix} = \begin{Bmatrix} 0 \\ 0 \end{Bmatrix}.$$

Setting the determinant of equation (4.12e) equal to zero yields:

$$\left(0.7 - 0.007772\,\omega^2\right)\left(4{,}320 - 43.510\,\omega^2\right) - \left(0.427\,\omega^2 - 36\right)^2 = 0 \quad (4.12f)$$

or:

$$0.155\,\omega^4 - 33.255\,\omega^2 - 1{,}728 = 0\,. \quad (4.12g)$$

Equation (4.12g) is the same as equation (4.11g). Therefore:

$$\boxed{\omega_1 = 9.4\ \frac{rad}{s} \quad \text{and} \quad \omega_2 = 11.2\ \frac{rad}{s}} \quad (4.12h)$$

or:

$$\boxed{f_1 = 1.5\text{ Hz} \quad \text{and} \quad f_2 = 1.8\text{ Hz}\,.} \quad (4.12i)$$

The equation for determining the mode vector is obtained from the first equation in equation (4.12b), or:

$$\frac{\Theta}{Y} = \frac{M\omega^2 - (K_1 + K_2)}{Ma_1\omega^2 - K_2(a_1 + a_2)}\,. \quad (4.12j)$$

Substituting the system variables into equation (4.12j) yields:

$$\frac{\Theta}{Y} = \frac{0.07772\,\omega^2 - 0.7}{0.427\,\omega^2 - 36}\,. \quad (4.12k)$$

Substituting ω_1 and ω_2 into equation (4.12k) and simplifying gives:

$$\frac{\Theta_1}{Y_1} = -8.53 \times 10^{-3} \text{ and } \frac{\Theta_2}{Y_2} = 1.57 \times 10^{-2}\,. \quad (4.12l)$$

The corresponding mode vectors are:

$$\boxed{\begin{Bmatrix} Y_1/Y_1 \\ \Theta_1/Y_1 \end{Bmatrix} = \begin{Bmatrix} 1 \\ -8.53 \times 10^{-3} \end{Bmatrix}} \quad (4.12m)$$

$$\boxed{\begin{Bmatrix} Y_2/Y_2 \\ \Theta_2/Y_2 \end{Bmatrix} = \begin{Bmatrix} 1 \\ 1.566 \times 10^{-2} \end{Bmatrix}.} \quad (4.12n)$$

Some observations can be made relative to EXAMPLES 4.9 through 4.12. First, examine the coordinate coupling. Relative to equation (4.9e) in EXAMPLE 4.9, the off-diagonal terms in the mass matrix are zero while the off-diagonal terms in the stiffness matrix are nonzero. With respect to the spatial coordinates that were selected for EXAMPLE 4.9, the system is coupled through the stiffness matrix. Therefore, the system is *statically coupled.* An examination of equation (4.10e) in EXAMPLE 4.10 indicates the off-diagonal terms in both the mass and stiffness matrices are nonzero. With respect to the spatial coordinates that were selected for EXAMPLE 4.10, the system is coupled through the mass matrix (*dynamic coupling*) and the stiffness matrix (*static coupling*). It is possible to selected spatial coordinates where the automobile in Figure 4.6 is neither statically nor dynamically coupled. Such coordinates are referred to as *orthogonal coordinates*. Orthonormal coordinates will be discussed in Chapter 5.

EXAMPLES 4.11 and 4.12 indicate that, irrespective of the spatial coordinates that are selected to model the automobile, the resonance frequencies are the same. However, the resulting mode vectors are a function of the spatial coordinates that are used to model the system.

Because the automobile has both translational and rotational motion, there is a node point associated with zero y motion about which the automobile rotates. The distance a_{node} between the vibration reference point on the automobile and the node point can be found using the equation:

$$Y + a_{node}\,\Theta = 0 \tag{4.43}$$

or:

$$a_{node} = -\frac{Y}{\Theta}. \tag{4.44}$$

$a_{node(1)} = 172.4$ in. and $a_{node(2)} = -8.9$ in. in EXAMPLE 4.12. The positive value of $a_{node(1)}$ indicates the node point is in the positive direction to the right of the c.g. on the vibration reference line. The negative value indicates the node point is in the negative direction to the left of the c.g. on the vibration reference line. Figure 4.9 shows the vibration modes of the automobile (associated with the two resonance frequencies) relative to the spatial variables that were selected for EXAMPLE 4.10.

$a_{node(1)} = 117.2$ in. and $a_{node(2)} = -8.9$ in. in EXAMPLE 4.12. These dimensions are referenced to point B. Figure 4.10 shows the vibration modes of the automobile (associated with the two resonance frequencies) relative to the spatial variables that were selected for EXAMPLE 4.12.

If $a_1 = 55$ in. is added to $a_{node(1)}$ for EXAMPLE 4.12, the value for $a_{node(1)}$ for EXAMPLE 4.11 is obtained. Similarly, if $a_1 = 55$ in. is subtracted from $a_{node(2)}$ for EXAMPLE 4.12, the value for $a_{node(2)}$ for EXAMPLE 4.11 is obtained. This indicates the overall vibration motion of the automobile is independent of the reference point and related spatial variables used to model the automobile vibration.

The following general observations can be made:

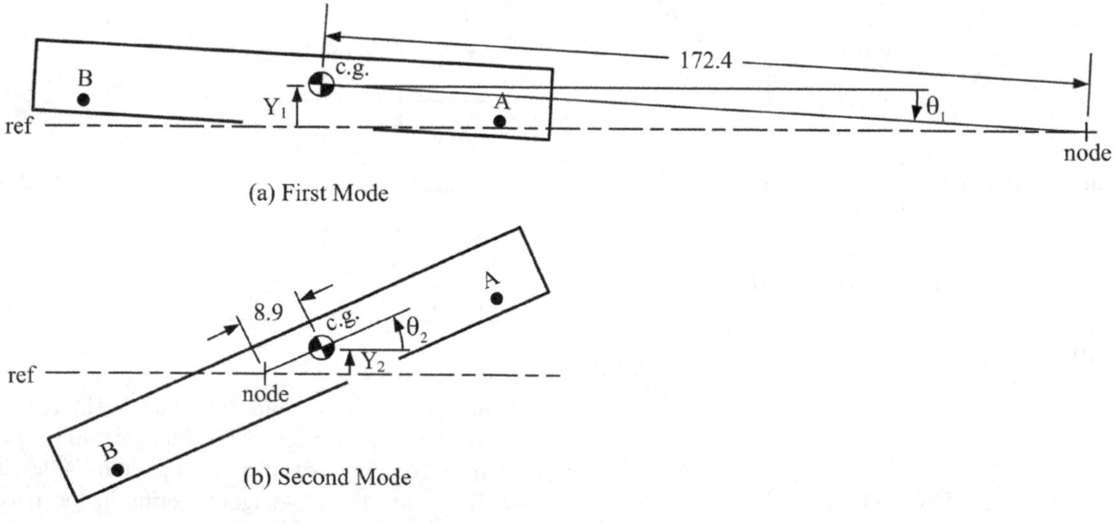

Figure 4.9 Modes of Vibration for Automobile in EXAMPLE 4.10

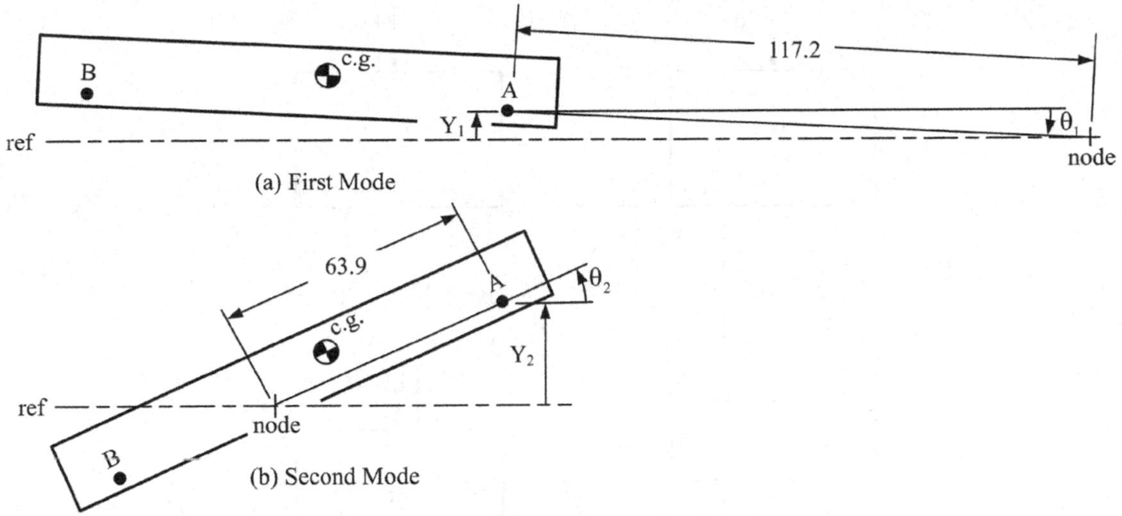

Figure 4.10 Modes of Vibration for Automobile in EXAMPLE 4.11

- The resonance frequencies of a vibration system are independent of the spatial variables and related coordinate system(s) used to describe the motion of the system. They are a function of the physical properties of a vibration system

- The overall motion of a vibration system is independent of the spatial variables and related coordinate system(s) used to describe the motion of the system. The overall motion of a vibration system is a function of the physical properties of the system. However, the mode vectors associated with the respective resonance frequencies of a vibration system are a function of the spatial variables and related coordinate system(s) used to describe the motion of the system. The selection of the spatial variables and related coordinate system(s) used to model a vibration system is arbitrary and is often based on convenience.

- A system is *statically coupled* if the off-diagonal terms in the stiffness matrix are nonzero. The system is *dynamically coupled* if the off-diagonal terms in the mass or damping matrix are nonzero.

- Static and dynamic coupling are not physical properties of a vibration system. Static and dynamic coupling are math matrix properties associated with the selection of the spatial variables and related coordinate system(s) used to model a vibration system. It is possible to select spatial variables associated with a specific coordinate system that will result in a vibration system being neither statically nor dynamically coupled. Such coordinates are referred to as *orthonormal coordinates*.

4.4 Equations of Motion for Forced Vibration Systems with Damping

Figure 4.11(a) shows a 2-DOF vibration system with viscous damping. Figure 4.11(b) shows the free-body diagram of the system. Summing the forces on masses M_1 and M_2 in the y direction yields:

$$\sum F_y = M_1\ddot{y}_1 \text{ or } M_1\ddot{y}_1 = -K_1y_1 - K_2(y_1 - y_2) - C_1\dot{y}_1 - C_2(\dot{y}_1 - \dot{y}_2) + f_1(t) \tag{4.45}$$

and:

$$\sum F_y = M_2\ddot{y}_2 \text{ or } M_2\ddot{y}_2 = -K_2(y_2 - y_1) - C_2(\dot{y}_2 - \dot{y}_1) + f_2(t). \tag{4.46}$$

Therefore:

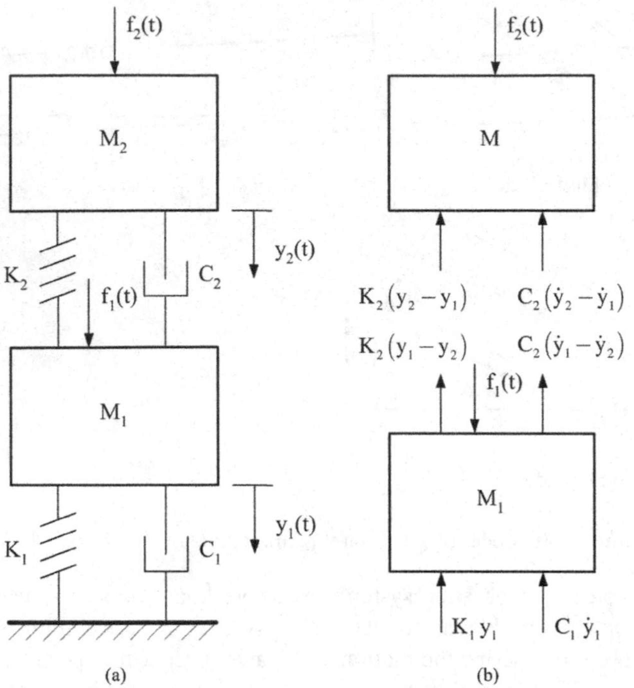

Figure 4.11 2-DOF Vibration System with Mass, Spring and Damp Elements

$$M_1\,\ddot{y}_1 + (C_1 + C_2)\dot{y}_1 + (K_1 + K_2)y_1 - C_2\,\dot{y}_2 - K_2\,y_2 = f_1(t) \tag{4.47}$$

and:

$$M_2\,\ddot{y}_2 + C_2\,\dot{y}_2 + K_2\,y_1 - C_2\,\dot{y}_1 - K_2\,y_1 = f_2(t). \tag{4.48}$$

Equations (4.47) and (4.48) can be written in matrix form, or:

$$\boxed{\begin{bmatrix} M_1 & 0 \\ 0 & M_2 \end{bmatrix}\begin{Bmatrix} \ddot{y}_1 \\ \ddot{y}_2 \end{Bmatrix} + \begin{bmatrix} C_1 + C_2 & -C_2 \\ -C_2 & C_2 \end{bmatrix}\begin{Bmatrix} \dot{y}_1 \\ \dot{y}_2 \end{Bmatrix} + \begin{bmatrix} K_1 + K_2 & -K_2 \\ -K_2 & K_2 \end{bmatrix}\begin{Bmatrix} y_1 \\ y_2 \end{Bmatrix} = \begin{Bmatrix} f_1(t) \\ f_2(t) \end{Bmatrix}.} \tag{4.49}$$

When $f_1(t)$ and $f_2(t)$ are:

$$f_1(t) = F_1\,e^{j\omega t} \quad \text{and} \quad f_2(t) = F_2\,e^{j\omega t} \tag{4.50}$$

then:

$$y_1(t) = \bar{Y}_1\,e^{j\omega t} \quad \text{where} \quad \bar{Y}_1 = Y_1\,e^{j\phi_1} \tag{4.51}$$

and:

$$y_2(t) = \bar{Y}_2\,e^{j\omega t} \quad \text{where} \quad \bar{Y}_2 = Y_2\,e^{j\phi_2}\,. \tag{4.52}$$

Substituting equations (4.50) through (4.52) into equation (4.49) and simplifying yields:

$$\begin{bmatrix} K_1 + K_2 - M_1\omega^2 + j(C_1 + C_2)\omega & -K_2 - jC_2\omega \\ -K_2 - jC_2\omega & K_2 - M_2\omega^2 + jC_2\omega \end{bmatrix}\begin{Bmatrix} \bar{Y}_1 \\ \bar{Y}_2 \end{Bmatrix} = \begin{Bmatrix} F_1 \\ F_2 \end{Bmatrix}. \tag{4.53}$$

$\bar{Y}_1$ and $\bar{Y}_2$ can be obtained by using Cramer's rule.

EXAMPLE 4.13

(a) Determine the responses for $\bar{Y}_1$ and $\bar{Y}_2$ associated

with equation (4.53) when $C_1 = C_2 = 0$ and $F_2 = 0$.
(b) Determine the resonance frequencies associated with the system.

SOLUTION

(a) Using *Cramer's rule*:

$$\bar{Y}_1 = \frac{\begin{vmatrix} F_1 & -K_2 \\ 0 & K_2 - M_2\,\omega^2 \end{vmatrix}}{\begin{vmatrix} K_1 + K_2 - M_1\,\omega^2 & -K_2 \\ -K_2 & K_2 - M_2\,\omega^2 \end{vmatrix}} \tag{4.13a}$$

and:

$$\bar{Y}_2 = \frac{\begin{vmatrix} K_1 + K_2 - M_1\,\omega^2 & F_1 \\ -K_2 & 0 \end{vmatrix}}{\begin{vmatrix} K_1 + K_2 - M_1\,\omega^2 & -K_2 \\ -K_2 & K_2 - M_2\,\omega^2 \end{vmatrix}}. \tag{4.13b}$$

Expanding the determinants in the numerator and denominator of equations (4.13a) and (4.13b) and simplifying yields:

$$\bar{Y}_1 = \frac{F_1\left(K_2 - M_2\,\omega^2\right)}{M_1 M_2\,\omega^4 - \begin{bmatrix} M_1 K_2 + \\ M_2\left(K_1 + K_2\right) \end{bmatrix}\omega^2 + K_1 K_2} \tag{4.13c}$$

and:

$$\bar{Y}_2 = \frac{F_1 K_2}{M_1 M_2\,\omega^4 - \begin{bmatrix} M_1 K_2 + \\ M_2\left(K_1 + K_2\right) \end{bmatrix}\omega^2 + K_1 K_2}. \tag{4.13d}$$

Y_1 and Y_2 are given by:

$$Y_1 = \frac{F_1\left|K_2 - M_2\,\omega^2\right|}{\left|M_1 M_2\,\omega^4 - \begin{bmatrix} M_1 K_2 + \\ M_2\left(K_1 + K_2\right) \end{bmatrix}\omega^2 + K_1 K_2\right|} \tag{4.13e}$$

and:

$$Y_2 = \frac{F_1 K_2}{\left|M_1 M_2\,\omega^4 - \begin{bmatrix} M_1 K_2 + \\ M_2\left(K_1 + K_2\right) \end{bmatrix}\omega^2 + K_1 K_2\right|}. \tag{4.13f}$$

Dividing the numerator and denominator of equations (4.13e) and (4.13f) by $M_1 M_2$ and simplifying yields:

$$Y_1 = \frac{\dfrac{F_1}{M_1}\left|\,\omega_2^2 - \omega^2\,\right|}{\left|\omega^4 - \left[\omega_1^2 + \omega_2^2\left(1 + \dfrac{M_2}{M_1}\right)\right]\omega^2 + \omega_1^2\omega_2^2\right|} \tag{4.13g}$$

and:

$$Y_2 = \frac{\dfrac{F_1}{M_1}\omega_2^2}{\left|\omega^4 - \left[\omega_1^2 + \omega_2^2\left(1 + \dfrac{M_2}{M_1}\right)\right]\omega^2 + \omega_1^2\omega_2^2\right|} \tag{4.13h}$$

where:

$$\omega_1 = \sqrt{\frac{K_1}{M_1}} \quad \text{and} \quad \omega_2 = \sqrt{\frac{K_2}{M_2}}. \tag{4.13i}$$

It is desirable to write equations (4.13g) and (4.13h) in non-dimensional form. This can be accomplished by dividing the numerator and denominator of equations (4.13g) and (4.13h) by ω_1^4 and simplifying. Therefore:

$$\frac{Y_1}{F_1/K_1} = \frac{\left|\dfrac{\omega_2^2}{\omega_1^2} - \dfrac{\omega^2}{\omega_1^2}\right|}{\left|\dfrac{\omega^4}{\omega_1^4} - \left[1 + \dfrac{\omega_2^2}{\omega_1^2}\left(1 + \dfrac{M_2}{M_1}\right)\right]\dfrac{\omega^2}{\omega_1^2} + \dfrac{\omega_2^2}{\omega_1^2}\right|} \tag{4.13j}$$

and:

$$\frac{Y_2}{F_1/K_1} = \frac{\dfrac{\omega_2^2}{\omega_1^2}}{\left|\dfrac{\omega^4}{\omega_1^4} - \left[1 + \dfrac{\omega_2^2}{\omega_1^2}\left(1 + \dfrac{M_2}{M_1}\right)\right]\dfrac{\omega^2}{\omega_1^2} + \dfrac{\omega_2^2}{\omega_1^2}\right|}. \tag{4.13k}$$

Noting that:

$$\frac{M_2}{M_1} = \frac{\omega_1^2}{\omega_2^2}\frac{K_2}{K_1} \tag{4.13l}$$

equations (4.13j) and (4.13l) can be written:

$$\frac{Y_1}{F_1/K_1} = \frac{\left|\dfrac{\omega_2^2}{\omega_1^2} - \dfrac{\omega^2}{\omega_1^2}\right|}{\left|\dfrac{\omega^4}{\omega_1^4} - \left[1 + \dfrac{K_2}{K_1} + \dfrac{\omega_2^2}{\omega_1^2}\right]\dfrac{\omega^2}{\omega_1^2} + \dfrac{\omega_2^2}{\omega_1^2}\right|} \tag{4.13m}$$

and:

$$\frac{Y_2}{F_1/K_1} = \frac{\dfrac{\omega_2^2}{\omega_1^2}}{\left|\dfrac{\omega^4}{\omega_1^4} - \left[1 + \dfrac{K_2}{K_1} + \dfrac{\omega_2^2}{\omega_1^2}\right]\dfrac{\omega^2}{\omega_1^2} + \dfrac{\omega_2^2}{\omega_1^2}\right|}. \tag{4.13n}$$

(b) The resonance frequencies are obtained by letting the denominator of equations (4.13m) and (4.13n) equal zero. If the resonance frequencies are written in terms of M_2/M_1, then:

$$\omega_{n(1,2)}^2 = \frac{\omega_1^2}{2}\left\{\begin{array}{l}\left[1 + \dfrac{\omega_2^2}{\omega_1^2}\left(1 + \dfrac{M_2}{M_1}\right)\right] \mp \\ \sqrt{\left[1 + \dfrac{\omega_2^2}{\omega_1^2}\left(1 + \dfrac{M_2}{M_1}\right)\right]^2 - 4\dfrac{\omega_2^2}{\omega_1^2}}\end{array}\right\}. \tag{4.13o}$$

If the resonance frequencies are written in terms of K_2/K_1, then:

$$\omega_{n(1,2)}^2 = \frac{\omega_1^2}{2}\left\{\begin{array}{l}\left[1 + \dfrac{K_2}{K_1} + \dfrac{\omega_2^2}{\omega_1^2}\right] \mp \\ \sqrt{\left[1 + \dfrac{K_2}{K_1} + \dfrac{\omega_2^2}{\omega_1^2}\right]^2 - 4\dfrac{\omega_2^2}{\omega_1^2}}\end{array}\right\}. \tag{4.13p}$$

EXAMPLE 4.14

(a) Determine the responses for $\bar{Y}_1$ and $\bar{Y}_2$ associated with equation (4.53) when $C_1 = C_2 = 0$ and $F_1 = 0$.
(b) Determine the resonance frequencies associated with the system.

SOLUTION

(a) Using *Cramer's rule*:

$$\bar{Y}_1 = \frac{\begin{vmatrix} 0 & -K_2 \\ F_2 & K_2 - M_2\,\omega^2 \end{vmatrix}}{\begin{vmatrix} K_1 + K_2 - M_1\,\omega^2 & -K_2 \\ -K_2 & K_2 - M_2\,\omega^2 \end{vmatrix}} \tag{4.14a}$$

and:

$$\bar{Y}_2 = \frac{\begin{vmatrix} K_1 + K_2 - M_1\,\omega^2 & 0 \\ -K_2 & F_2 \end{vmatrix}}{\begin{vmatrix} K_1 + K_2 - M_1\,\omega^2 & -K_2 \\ -K_2 & K_2 - M_2\,\omega^2 \end{vmatrix}}. \tag{4.14b}$$

Expanding the determinants in the numerator and denominator of equations (4.14a) and (4.14b) and simplifying yields:

$$\bar{Y}_1 = \frac{F_2\,K_2}{M_1 M_2\,\omega^4 - \left[\begin{array}{l} M_1 K_2 + \\ M_2\left(K_1 + K_2\right)\end{array}\right]\omega^2 + K_1 K_2} \tag{4.14c}$$

and:

$$\bar{Y}_2 = \frac{F_2\left(K_1 + K_2 - M_1\,\omega_2\right)}{M_1M_2\,\omega^4 - \left[\begin{matrix} M_1K_2 + \\ M_2\left(K_1 + K_2\right)\end{matrix}\right]\omega^2 + K_1K_2} \tag{4.14d}$$

Y_1 and Y_2 are given by:

$$\bar{Y}_1 = \frac{F_2\,K_2}{\left|M_1M_2\,\omega^4 - \left[\begin{matrix} M_1K_2 + \\ M_2\left(K_1 + K_2\right)\end{matrix}\right]\omega^2 + K_1K_2\right|} \tag{4.14e}$$

and:

$$\bar{Y}_2 = \frac{F_2\left|K_1 + K_2 - M_1\,\omega^2\right|}{\left|M_1M_2\,\omega^4 - \left[\begin{matrix} M_1K_2 + \\ M_2\left(K_1 + K_2\right)\end{matrix}\right]\omega^2 + K_1K_2\right|}. \tag{4.14f}$$

Dividing the numerator and denominator of equations (4.14e) and (4.14f) by M_1M_2 and simplifying yields:

$$Y_1 = \frac{\dfrac{F_2}{M_1}\omega_2^2}{\left|\omega^4 - \left[\omega_1^2 + \omega_2^2\left(1 + \dfrac{M_2}{M_1}\right)\right]\omega^2 + \omega_1^2\omega_2^2\right|} \tag{4.14g}$$

and:

$$Y_2 = \frac{\dfrac{F_2}{M_2}\left|\omega_1^2 + \omega_2^2\dfrac{M_2}{M_1} - \omega^2\right|}{\left|\omega^4 - \left[\omega_1^2 + \omega_2^2\left(1 + \dfrac{M_2}{M_1}\right)\right]\omega^2 + \omega_1^2\omega_2^2\right|} \tag{4.14h}$$

where:

$$\omega_1 = \sqrt{\frac{K_1}{M_1}} \quad \text{and} \quad \omega_2 = \sqrt{\frac{K_2}{M_2}}. \tag{4.14i}$$

It is desirable to write equations (4.14g) and (4.14h) in non-dimensional form. This can be done by dividing the numerator and denominator of the two equations by ω_2^4 and simplifying. Therefore:

$$\frac{Y_1}{F_2/K_2} = \frac{\dfrac{K_2}{K_1}\dfrac{\omega_1^2}{\omega_2^2}}{\left|\dfrac{\omega^4}{\omega_2^4} - \left[1 + \dfrac{\omega_1^2}{\omega_2^2} + \dfrac{M_2}{M_1}\right]\dfrac{\omega^2}{\omega_2^2} + \dfrac{\omega_1^2}{\omega_2^2}\right|} \tag{4.14j}$$

and:

$$\frac{Y_2}{F_2/K_2} = \frac{\left|\dfrac{\omega_1^2}{\omega_2^2} + \dfrac{M_2}{M_1} - \dfrac{\omega^2}{\omega_2^2}\right|}{\left|\dfrac{\omega^4}{\omega_2^4} - \left[1 + \dfrac{\omega_1^2}{\omega_2^2} + \dfrac{M_2}{M_1}\right]\dfrac{\omega^2}{\omega_2^2} + \dfrac{\omega_1^2}{\omega_2^2}\right|}. \tag{4.14k}$$

Noting that:

$$\frac{M_2}{M_1} = \frac{\omega_1^2}{\omega_2^2}\frac{K_2}{K_1} \tag{4.14l}$$

equations (4.14j) and (4.14k) can be written:

$$\frac{Y_1}{F_2/K_2} = \frac{\dfrac{K_2}{K_1}\dfrac{\omega_1^2}{\omega_2^2}}{\left|\dfrac{\omega^4}{\omega_2^4} - \left[1 + \dfrac{\omega_1^2}{\omega_2^2}\left(1 + \dfrac{K_2}{K_1}\right)\right]\dfrac{\omega^2}{\omega_2^2} + \dfrac{\omega_1^2}{\omega_2^2}\right|} \tag{4.14m}$$

and:

$$\frac{Y_2}{F_2/K_2} = \frac{\left|\dfrac{\omega_1^2}{\omega_2^2}\left(1 + \dfrac{K_2}{K_1}\right) - \dfrac{\omega^2}{\omega_2^2}\right|}{\left|\dfrac{\omega^4}{\omega_2^4} - \left[1 + \dfrac{\omega_1^2}{\omega_2^2}\left(1 + \dfrac{K_2}{K_1}\right)\right]\dfrac{\omega^2}{\omega_2^2} + \dfrac{\omega_1^2}{\omega_2^2}\right|}. \tag{4.14n}$$

(b) The resonance frequencies are obtained by letting the denominator of equations (4.14m) and (4.14n) equal zero. If the resonance frequencies are written in terms of M_2/M_1, then:

$$\omega^2_{n(1,2)} = \frac{\omega_2^2}{2}\left\{\left[1+\frac{\omega_1^2}{\omega_2^2}+\frac{M_2}{M_1}\right] \mp \sqrt{\left[1+\frac{\omega_1^2}{\omega_2^2}+\frac{M_2}{M_1}\right]^2 - 4\frac{\omega_1^2}{\omega_2^2}}\right\}. \quad (4.14o)$$

If the resonance frequencies are written in terms of K_2/K_1, then:

$$\omega^2_{n(1,2)} = \frac{\omega_2^2}{2}\left\{\left[1+\frac{\omega_1^2}{\omega_2^2}\left(1+\frac{K_2}{K_1}\right)\right] \mp \sqrt{\left[1+\frac{\omega_1^2}{\omega_2^2}\left(1+\frac{K_2}{K_1}\right)\right]^2 - 4\frac{\omega_1^2}{\omega_2^2}}\right\}. \quad (4.14p)$$

EXAMPLE 4.15

Use Lagrange's equation to determine the equations of motion for the system shown in Figure 4.12.

SOLUTION

Let x_1 represent the motion of mass M_1 in the x direction, and let x_2 and y_2 represent the motion of mass M_2 in the x and y directions. Mass M_2 has two constraint equations:

$$x_2 = x_1 + L\sin\theta \quad (4.15a)$$

$$y_2 = L(1-\cos\theta). \quad (4.15b)$$

Since x_2 and y_2 can be expressed in terms of x_1 and θ, use the generalized coordinates x_1 and θ to describe the motion of the system.

The total kinetic energy of the system is:

$$T = \frac{1}{2}M_1\dot{x}_1^2 + \frac{1}{2}M_2\dot{x}_2^2 + \frac{1}{2}M_2\dot{y}_2^2 \quad (4.15c)$$

or:

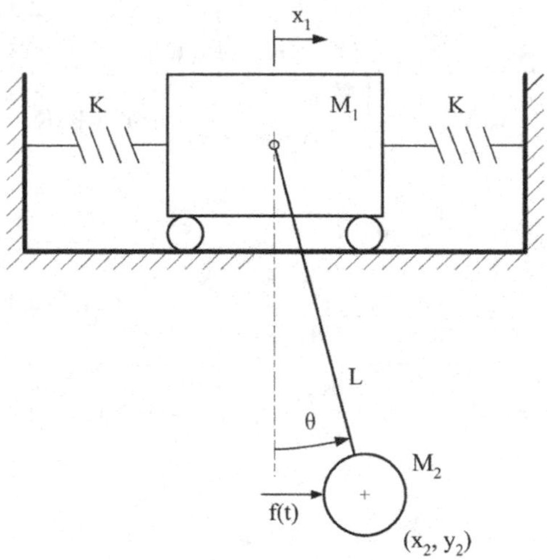

Figure 4.12 2-DOF Mass-Spring-Pendulum Syste

$$T = \frac{1}{2}M_1\dot{x}_1^2 + \frac{1}{2}M_2\left(\dot{x}_1 + L\cos\theta\,\dot{\theta}\right)^2 + \frac{1}{2}M_2 L^2\sin^2\theta\,\dot{\theta}^2. \quad (4.15d)$$

If θ is assumed to be small, $\cos\theta \approx 1$ and $\sin\theta \approx \theta$. An examination of equation (4.15d) indicates the kinetic energy of mass M_2 in the y directions is much smaller than the other kinetic energy terms and can be neglected. Therefore, equation (4.15d) becomes:

$$T = \frac{1}{2}M_1\dot{x}_1^2 + \frac{1}{2}M_2\left(\dot{x}_1 + L\dot{\theta}\right)^2. \quad (4.15e)$$

The total potential energy of the system is:

$$U = \frac{1}{2}K x_1^2 + \frac{1}{2}K x_1^2 + mgL(1-\cos\theta). \quad (4.15f)$$

Taking the derivative of equation (4.15e) with respect to $\dot{x}_1$ and $\dot{\theta}$ yields:

$$\frac{\partial T}{\partial \dot{x}_1} = M_1\dot{x}_1 + M_2\left(\dot{x}_1 + L\dot{\theta}\right). \quad (4.15g)$$

$$\frac{\partial T}{\partial \dot{\theta}} = M_2 L\left(\dot{x}_1 + L\dot{\theta}\right). \quad (4.15h)$$

Taking the derivative of equation (4.15f) with respect to x_1 and θ gives:

$$\frac{\partial U}{\partial x_1} = 2K\,x_1 \tag{4.15i}$$

$$\frac{\partial U}{\partial \theta} = mg\,L\sin\theta \approx mg\,L\,\theta\,. \tag{4.15j}$$

The externally applied force acts only in the x direction. Therefore, equation (1.193) in Chapter 1 reduces to:

$$Q_i = \sum_{j=1}^{2} F_{x(j)}\,\frac{\partial x_j}{\partial q_i} \tag{4.15k}$$

where $q_1 = x_1$ and $q_2 = \theta$. The generalized force associated with mass M_1 is:

$$Q_1 = F_{x(1)}\,\frac{\partial x_1}{\partial x_1} + F_{x(2)}\,\frac{\partial x_2}{\partial x_1} \tag{4.15l}$$

where x_2 is given by equation (4.15a) and:

$$F_{x(1)} = 0 \qquad \text{and} \qquad F_{x(2)} = f(t)\,. \tag{4.15m}$$

Substituting x_1 and equation (4.15a) for x_2 into equation (4.15l) and carrying out the prescribed operations yields:

$$Q_1 = f(t)\,. \tag{4.15n}$$

The generalized force associated with mass M_2 is:

$$Q_2 = F_{x(1)}\,\frac{\partial x_1}{\partial \theta} + F_{x(2)}\,\frac{\partial x_2}{\partial \theta} \tag{4.15o}$$

or:

$$Q_2 = L\,f(t)\,. \tag{4.15p}$$

Taking the derivative of equation (4.15g) with respect to t and adding to the results equations (4.15i) and (4.15n) yields:

$$M_1\,\ddot{x}_1 + M_2\left(\ddot{x}_1 + L\ddot{\theta}\right) + 2K\,x_1 = f(t)\,. \tag{4.15q}$$

or:

$$\left(M_1 + M_2\right)\ddot{x}_1 + M_2\,L\ddot{\theta} + 2K\,x_1 = f(t)\,. \tag{4.15r}$$

Taking the derivative of equation (4.15h) with respect to t and adding to the results equations (4.15j) and (4.15p) gives:

$$M_2\,L\left(\ddot{x}_1 + L\ddot{\theta}\right) + M_2 g\,L\,\theta = L\,f(t) \tag{4.15s}$$

or:

$$M_2\,L\ddot{x}_1 + M_2\,L^2\,\ddot{\theta} + M_2 g\,L\,\theta = L\,f(t) \tag{4.15t}$$

Writing equations (4.15r) and (4.15t) in matrix form yields:

$$\boxed{\begin{bmatrix} M_1 + M_2 & M_2\,L \\ M_2\,L & M_2\,L^2 \end{bmatrix}\begin{Bmatrix} \ddot{x}_1 \\ \ddot{\theta} \end{Bmatrix} + \begin{bmatrix} 2K & 0 \\ 0 & M_2 g L \end{bmatrix}\begin{Bmatrix} x_1 \\ \theta \end{Bmatrix} = \begin{Bmatrix} f(t) \\ L\,f(t) \end{Bmatrix}.} \tag{4.15u}$$

EXAMPLE 4.16

Use Lagrange's equation to develop the equations of motion of the automobile in Figure 4.6 when the suspension system has both spring elements and shock absorbers (damping elements).

SOLUTION

Figure 4.13 shows a schematic representation of the automobile. The total kinetic energy is:

$$T = \frac{1}{2}M\,\dot{y}^2 + \frac{1}{2}J_o\,\dot{\theta}^2\,. \tag{4.16a}$$

The total potential energy is:

$$U = \frac{1}{2}K_1\left(y_1 - h_1\right)^2 + \frac{1}{2}K_2\left(y_2 - h_2\right)^2 \tag{4.16b}$$

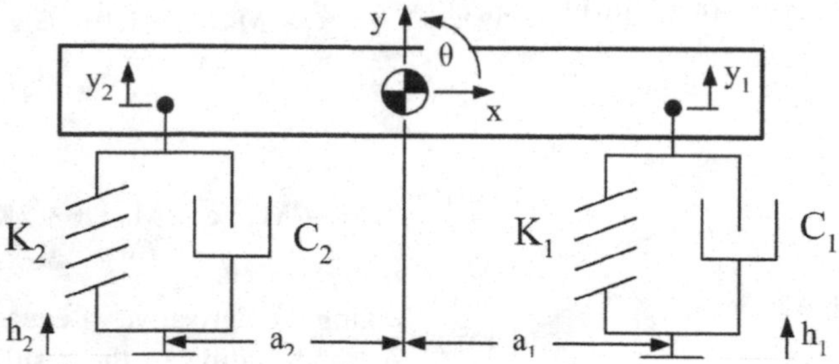

Figure 4.13 2-DOF Representation of the Automobile in Figure 4.5 with Spring Elements and Damping Elements (Shock Absorbers)

where:

$$y_1 = y + a_1\,\theta \tag{4.16c}$$

and:

$$y_2 = y - a_2\,\theta\,. \tag{4.16d}$$

Therefore, the potential energy expression can be written:

$$U = \frac{1}{2}K_1\left(y + a_1\,\theta - h_1\right)^2 + \frac{1}{2}K_2\left(y - a_2\,\theta - h_2\right)^2. \tag{4.16e}$$

The total power dissipation is:

$$D = \frac{1}{2}C_1\left(\dot{y}_1 - \dot{h}_1\right)^2 + \frac{1}{2}C_2\left(\dot{y}_2 - \dot{h}_2\right)^2 \tag{4.16f}$$

where:

$$\dot{y}_1 = \dot{y} + a_1\,\dot{\theta} \tag{4.16g}$$

and:

$$\dot{y}_2 = \dot{y} - a_2\,\dot{\theta}\,. \tag{4.16h}$$

Therefore, the power dissipation expression can be written:

$$D = \frac{1}{2}C_1\left(\dot{y} + a_1\,\dot{\theta} - \dot{h}_1\right)^2 + \frac{1}{2}C_2\left(\dot{y} - a_2\,\dot{\theta} - \dot{h}_2\right)^2. \tag{4.16i}$$

Taking the derivative of equation (4.16a) with respect to $\dot{y}$ and $\dot{\theta}$ yields:

$$\frac{\partial T}{\partial \dot{y}} = M\,\dot{y} \tag{4.16j}$$

and:

$$\frac{\partial T}{\partial \dot{\theta}} = J_o\,\dot{\theta}\,. \tag{4.16k}$$

Taking the derivative of equation (4.16e) with respect to y and θ gives:

$$\frac{\partial U}{\partial y} = K_1\left(y + a_1\,\theta - h_1\right) + K_2\left(y - a_2\,\theta - h_2\right) \tag{4.16l}$$

and:

$$\frac{\partial U}{\partial \theta} = K_1 a_1\left(y + a_1\,\theta - h_1\right) - a_2 K_2\left(y - a_2\,\theta - h_2\right). \tag{4.16m}$$

Taking the derivative of equation (4.16i) with respect to $\dot{y}$ and $\dot{\theta}$ yields:

$$\frac{\partial D}{\partial \dot{y}} = C_1\left(\dot{y} + a_1\,\dot{\theta} - \dot{h}_1\right) + C_2\left(\dot{y} - a_2\,\dot{\theta} - \dot{h}_2\right) \tag{4.16n}$$

and:

$$\frac{\partial D}{\partial \dot{\theta}} = C_1 a_1 \left(\dot{y} + a_1 \dot{\theta} - \dot{h}_1\right) - a_2 C_2 \left(\dot{y} - a_2 \dot{\theta} - \dot{h}_2\right). \tag{4.16o}$$

Taking the derivative of equation (4.16j), adding to the results equations (4.16l) and (4.16n), and collecting terms yields:

$$\begin{aligned} &M\ddot{y} + (C_1 + C_2)\dot{y} + (K_1 + K_2)y + \\ &\quad (C_1 a_1 - C_2 a_2)\dot{\theta} + (K_1 a_1 - K_2 a_2)\theta = C_1 \dot{h}_1 + \\ &\quad C_2 \dot{h}_2 + K_1 h_1 + K_2 h_2 \end{aligned} \tag{4.16p}$$

Taking the derivative of equation (4.16k), adding to the results equations (4.16m) and (4.16o), and collecting terms gives:

$$\begin{aligned} &J_o \ddot{\theta} + \left(C_1 a_1^2 + C_2 a_2^2\right)\dot{\theta} + \left(K_1 a_1^2 + K_2 a_2^2\right)\theta + \\ &\quad (C_1 a_1 - C_2 a_2)\dot{y} + (K_1 a_1 - K_2 a_2)y = C_1 a_1 \dot{h}_1 - \\ &\quad C_2 a_2 \dot{h}_2 + K_1 a_1 h_1 - K_2 a_2 h_2 . \end{aligned} \tag{4.16q}$$

Equations (4.16p) and (4.16q) in matrix form are:

$$\begin{aligned} &\begin{bmatrix} M & 0 \\ 0 & J_o \end{bmatrix} \begin{Bmatrix} \ddot{y} \\ \ddot{\theta} \end{Bmatrix} + \\ &\begin{bmatrix} C_1 + C_2 & C_1 a_1 - C_2 a_2 \\ C_1 a_1 - C_2 a_2 & C_1 a_1^2 + C_2 a_2^2 \end{bmatrix} \begin{Bmatrix} \dot{y} \\ \dot{\theta} \end{Bmatrix} + \\ &\begin{bmatrix} K_1 + K_2 & K_1 a_1 - K_2 a_2 \\ K_1 a_1 - K_2 a_2 & K_1 a_1^2 + K_2 a_2^2 \end{bmatrix} \begin{Bmatrix} y \\ \theta \end{Bmatrix} = \\ &\begin{Bmatrix} C_1 \dot{h}_1 + C_2 \dot{h}_2 + K_1 h_1 + K_2 h_2 \\ C_1 a_1 \dot{h}_1 - C_2 a_2 \dot{h}_2 + K_1 a_1 h_1 - K_2 a_2 h_2 \end{Bmatrix}. \end{aligned} \tag{4.16r}$$

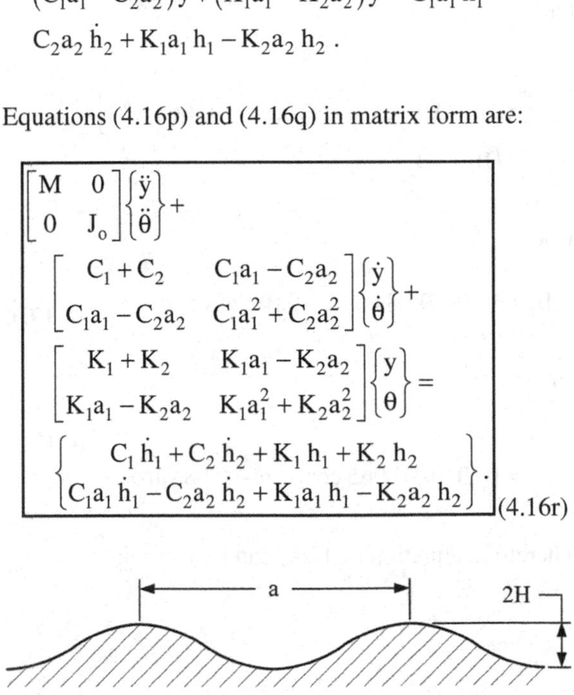

Figure 4.14 Road Profile for EXAMPLE 4.1

EXAMPLE 4.17

The automobile in EXAMPLES 4.16 has the following properties: $M = 3{,}000\ lb_f$, $J_o = 20{,}000\ lb_f$-s/in., $K_1 = 400\ lb_f$/in., $K_2 = 300\ lb_f$/in., $C_1 = 25\ lb_f$-s/in., $C_2 = 20\ lb_f$-s/in., $a_1 = 55$ in., and $a_2 = 65$ in. The automobile travels over a road that has a contour specified by:

$$h(x) = H \cos\left(\frac{2\pi}{a} x\right) \tag{4.17a}$$

where H and a are specified in Figure 4.14. If a = 16 ft. determine the expressions for and plot Y/H and θ/H as a function of automobile speed in mph.

SOLUTION

Substituting the values for the system parameters into equations (4.16p) and (4.16q) yields:

$$\begin{aligned} &7.772\ddot{y} + 45\dot{y} + 700y + 75\dot{\theta} + 2{,}500\theta = \\ &\quad 25\dot{h}_1 + 20\dot{h}_2 + 400h_1 + 300h_2 \end{aligned} \tag{4.17b}$$

or:

$$\begin{aligned} &\ddot{y} + 5.79\dot{y} + 90.07y + 9.65\dot{\theta} + 321.67\theta = \\ &\quad 3.22\dot{h}_1 + 2.57\dot{h}_2 + 51.47h_1 + 38.60h_2 \end{aligned} \tag{4.17c}$$

and:

$$\begin{aligned} &20{,}000\ddot{\theta} + 160{,}125\dot{\theta} + 2{,}477{,}500\theta + \\ &\quad 75\dot{\theta} + 2{,}500y = 1{,}375\dot{h}_1 - 1{,}300\dot{h}_2 + \\ &\quad 22{,}000h_1 - 19{,}500h_2 \end{aligned} \tag{4.17d}$$

or:

$$\begin{aligned} &\ddot{\theta} + 8.01\dot{\theta} + 123.88\theta + 0.0038\dot{y} + 0.125y = \\ &\quad 0.069\dot{h}_1 - 0.065\dot{h}_2 + 1.10h_1 - 0.98h_2 \end{aligned} \tag{4.17e}$$

Writing equations (4.17c) and (4.17e) in matrix form yields:

$$\begin{bmatrix} 1 & 0 \\ 0 & 1 \end{bmatrix} \begin{Bmatrix} \ddot{y} \\ \ddot{\theta} \end{Bmatrix} + \begin{bmatrix} 5.79 & 9.65 \\ 0.0038 & 8.01 \end{bmatrix} \begin{Bmatrix} \dot{y} \\ \dot{\theta} \end{Bmatrix} + \begin{bmatrix} 90.7 & 321.67 \\ 0.125 & 123.88 \end{bmatrix} \begin{Bmatrix} y \\ \theta \end{Bmatrix} = \begin{Bmatrix} 3.22\,\dot{h}_1 + 2.57\,\dot{h}_2 + 51.47\,h_1 + 38.60\,h_2 \\ 0.069\,\dot{h}_1 - 0.065\,\dot{h}_2 + 1.10\,h_1 - 0.98\,h_2 \end{Bmatrix}. \tag{4.17f}$$

$h_1(x)$ is given by equation (4.17a), and $h_2(x)$ is given by:

$$h_2(x) = H\cos\left(\frac{2\pi}{a}(x + a_1 + a_2)\right). \tag{4.17g}$$

Noting that x = v t, $h_1(t)$ and $h_2(t)$ can be expressed:

$$h_1(t) = H\cos\omega t \quad \text{and} \quad h_2(t) = H\cos(\omega t + \alpha) \tag{4.17h}$$

where:

$$\omega = \frac{2\pi}{a} v \quad \text{and} \quad \alpha = \frac{2\pi}{a}(a_1 + a_2). \tag{4.17i}$$

v is the automobile speed in ft/s. It is desirable to represent $h_1(t)$ and $h_2(t)$ as complex exponential, or:

$$h_1(t) = H\,e^{j\omega t} \quad \text{and} \quad h_2(t) = H\,e^{j(\omega t + \alpha)}. \tag{4.17j}$$

Since $h_1(t)$ and $h_2(t)$ are represented as complex exponential, y(t) and θ(t) can be written:

$$y(t) = \bar{Y}\,e^{j\omega t} \quad \text{where} \quad \bar{Y} = Y\,e^{j\phi_1} \tag{4.17k}$$

and:

$$\theta(t) = \bar{\Theta}\,e^{j\omega t} \quad \text{where} \quad \bar{\Theta} = \Theta\,e^{j\phi_2}. \tag{4.17l}$$

Substituting equations (4.17j) through (4.17l) into equation (4.17f) and simplifying yields:

$$\begin{bmatrix} 90.067 - \omega^2 + j5.79\omega & 321.67 + j9.65\omega \\ 0.125 + j0.0038\omega & 123.88 - \omega^2 + j8.01\omega \end{bmatrix} \begin{Bmatrix} \bar{Y} \\ \bar{\Theta} \end{Bmatrix} = \begin{Bmatrix} H\left[51.47 + 38.60\,e^{j\alpha} + j\omega\left(3.22 + 2.57\,e^{j\alpha}\right)\right] \\ H\left[1.10 - 0.98\,e^{j\alpha} + j\omega\left(0.069 - 0.065\,e^{j\alpha}\right)\right] \end{Bmatrix}. \tag{4.17m}$$

Noting that :

$$H\left[51.47 + 38.60\,e^{j\alpha} + j\omega\left(3.22 + 2.57\,e^{j\alpha}\right)\right] = H\left(A_1 + jA_2\right) \tag{4.17n}$$

where:

$$A_1 = 51.44 + 38.60\cos\alpha - 2.57\,\omega\sin\alpha \tag{4.17o}$$

$$A_2 = \left(3.22 + 2.57\cos\alpha\right)\omega + 38.6\sin\alpha \tag{4.17p}$$

and:

$$H\left[1.10 - 0.98\,e^{j\alpha} + j\omega\left(0.069 - 0.065\,e^{j\alpha}\right)\right] = H\left(B_1 + jB_2\right) \tag{4.17q}$$

where:

$$B_1 = 1.10 - 0.98\cos\alpha + 0.065\,\omega\sin\alpha \tag{4.17r}$$

and:

$$B_2 = \left(0.069 - 0.065\cos\alpha\right)\omega - 0.98\sin\alpha. \tag{4.17s}$$

Therefore, equation (4.17m) can be written:

$$\begin{bmatrix} 90.067-\omega^2+ & 321.67+j9.65\omega \\ j5.79\omega & \\ 0.125+j0.0038\omega & 123.88-\omega^2+ \\ & j8.01\omega \end{bmatrix} \begin{Bmatrix} \bar{Y} \\ \bar{\Theta} \end{Bmatrix} = \begin{Bmatrix} H\left(A_1+jA_2\right) \\ H\left(B_1+jB_2\right) \end{Bmatrix}. \tag{4.17t}$$

$\bar{Y}/H$ and $\bar{\Theta}/H$ are obtained by using Cramer's rule. Therefore:

$$\frac{\bar{Y}}{H} = \frac{\begin{bmatrix} A_1\left(123.9-\omega^2\right)+\left(9.7B_2-8.0A_2\right)\omega- \\ 321.7B_1 \end{bmatrix} + j\begin{bmatrix} A_2\left(123.9-\omega^2\right)+\left(8.0A_1-9.7B_1\right)\omega- \\ 321.7B_2 \end{bmatrix}}{\omega^4-260.3\omega^2+11{,}116.8+ j\left(1436.3\omega-13.8\omega^3\right)} \tag{4.17u}$$

and:

$$\frac{\bar{\Theta}}{H} = \frac{\begin{bmatrix} B_1\left(90.1-\omega^2\right)+ \\ \left(0.0038A_2-5.8B_2\right)\omega-0.125A_1 \end{bmatrix} + j\begin{bmatrix} B_2\left(90.1-\omega^2\right)+ \\ \left(5.8B_1-0.0038A_1\right)\omega-0.125A_2 \end{bmatrix}}{\omega^4-260.3\omega^2+11{,}116.8+ j\left(1436.3\omega-13.8\omega^3\right)}. \tag{4.17v}$$

The amplitudes of $\left|\bar{Y}/H\right|$ and $\left|\bar{\Theta}/H\right|$ are:

$$\left|\frac{\bar{Y}}{H}\right| = \frac{\sqrt{\begin{bmatrix} A_1\left(123.9-\omega^2\right)+ \\ \left(9.7B_2-8.0A_2\right)\omega-321.7B_1 \end{bmatrix}^2 + \begin{bmatrix} A_2\left(123.9-\omega^2\right)+ \\ \left(8.0A_1-9.7B_1\right)\omega-321.7B_2 \end{bmatrix}^2}}{\sqrt{\left(\omega^4-260.3\omega^2+11{,}116.8\right)^2+ \left(1436.3\omega-13.8\omega^3\right)^2}} \tag{4.17w}$$

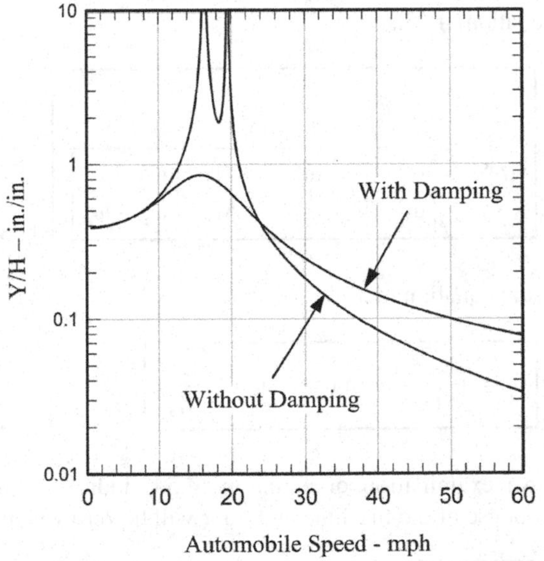

Figure 4.15 Vibration Response of Automobile in EXAMPLE 4.17 - $\left|\bar{Y}/H\right|$

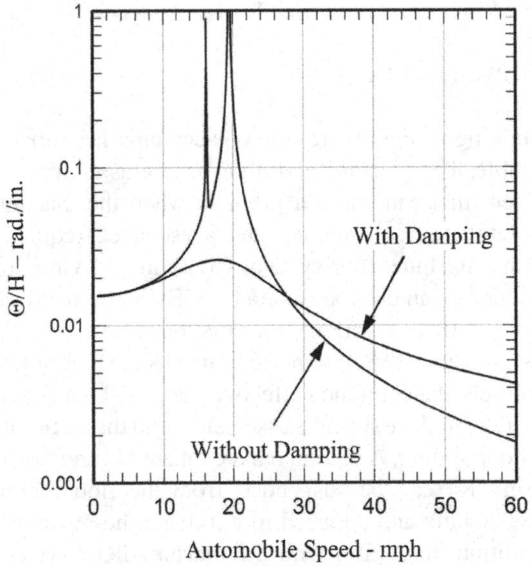

Figure 4.16 Vibration Response of Automobile in EXAMPLE 4.17 - $\left|\bar{\Theta}/H\right|$

and:

$$\left|\frac{\bar{\Theta}}{H}\right| = \frac{\sqrt{\begin{bmatrix} B_1\left(90.1-\omega^2\right)+ \\ \left(0.0038\,A_2-5.8\,B_2\right)\omega-0.125\,A_1 \end{bmatrix}^2 + \begin{bmatrix} B_2\left(90.1-\omega^2\right)+ \\ \left(5.8\,B_1-0.0038\,A_1\right)\omega-0.125\,A_2 \end{bmatrix}^2}}{\sqrt{\left(\omega^4-260.3\,\omega^2+11{,}116.8\right)^2+\left(1436.3\,\omega-13.8\,\omega^3\right)^2}}. \tag{4.17x}$$

ω and α are specified by equation (4.17i).

Equations (4.17w) and (4.17x) are plotted in Figures 4.15 and 4.16 as a function of automobile speed in mph. It should be noted that:

$$v(\text{mph}) = 0.682\,v(\text{ft/s}). \tag{4.17y}$$

$\left|\bar{Y}/H\right|$ and $\left|\bar{\Theta}/H\right|$ are plotted with the damping specified in this example and with zero damping. The necessity of shock absorbers (damping elements) on automobiles is easily understood from this example.

4.5 Vibration Dynamic (Tuned) Absorber

Displacement Response

Vibrating machines are often placed on a floor that is flexible. The machine and floor behave as a vibration system that can cause a problem when the machine operates at a frequency at or near a resonance frequency of the machine-floor system. Conventional vibration isolation techniques based on a 1-DOF vibration analysis may not always work in this situation.

Mass element M_1 and spring element K_1 in Figure 4.17 represent the combined effective mass of the machine and floor and the stiffness associated with the flexibility of floor system. A second mass element M_2 and spring spring K_2 can be suspended from the floor below the machine and adjusted to eliminate the resonance condition that exists with the machine-floor system. The equations of motion for this system are those that were developed in EXAMPLE 4.13. The response of the machine and floor represented by mass M_1 is given by equation (4.13j):

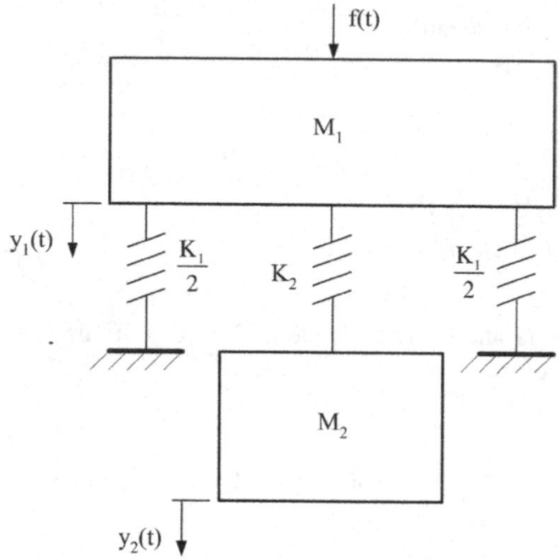

Figure 4.17 Schematic Representation of a Tuned Absorber

$$\frac{Y_1}{F_1/K_1} = \frac{\left|\dfrac{\omega_2^2}{\omega_1^2}-\dfrac{\omega^2}{\omega_1^2}\right|}{\left|\dfrac{\omega^4}{\omega_1^4}-\left[1+\dfrac{\omega_2^2}{\omega_1^2}\left(1+\dfrac{M_2}{M_1}\right)\right]\dfrac{\omega^2}{\omega_1^2}+\dfrac{\omega_2^2}{\omega_1^2}\right|}. \tag{4.54}$$

The response of the suspended mass M_2 is given by equation (4.13k):

$$\frac{Y_2}{F_1/K_1} = \frac{\dfrac{\omega_2^2}{\omega_1^2}}{\left|\dfrac{\omega^4}{\omega_1^4}-\left[1+\dfrac{\omega_2^2}{\omega_1^2}\left(1+\dfrac{M_2}{M_1}\right)\right]\dfrac{\omega^2}{\omega_1^2}+\dfrac{\omega_2^2}{\omega_1^2}\right|}. \tag{4.55}$$

From equation (4.13i):

$$\omega_1 = \sqrt{\frac{K_1}{M_1}} \quad \text{and} \quad \omega_2 = \sqrt{\frac{K_2}{M_2}}. \tag{4.56}$$

An examination of equation (4.54) indicates the response of the machine and floor will be zero when:

$$\omega = \sqrt{\frac{K_2}{M_2}}. \tag{4.57}$$

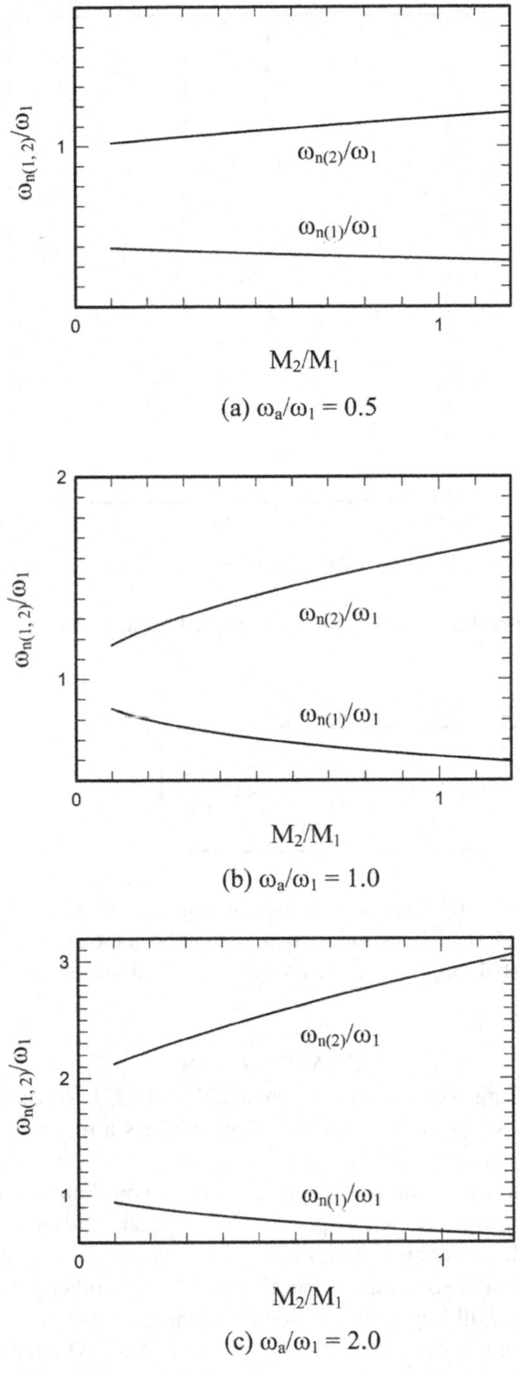

Figure 4.18 Effects of the Ratios M_2/M_1 and ω_a/ω_1 on the Resonance Frequencies of a System with a Tuned Absorber

Define this frequency as the *"tuned" frequency* ω_a where:

$$\omega_a = \sqrt{\frac{K_2}{M_2}} \quad \text{or} \quad \omega_a = \omega_2 \,. \tag{4.58}$$

The values of M_2 and K_2 are selected so the frequency ω_a equals the excitation frequency of the machine.

The system resonance frequencies are obtained by setting the denominator of equations (4.54) and (4.55) equal to zero. The expression for these frequencies is:

$$\frac{\omega^2_{n(1,2)}}{\omega_1^2} = \frac{1}{2}\left[1+\frac{\omega_2^2}{\omega_1^2}\left(1+\frac{M_2}{M_1}\right) \mp \sqrt{\left[1+\frac{\omega_2^2}{\omega_1^2}\left(1+\frac{M_2}{M_1}\right)\right]^2 - 4\frac{\omega_2^2}{\omega_1^2}} \right]. \tag{4.59}$$

Figure 4.18 shows plots of $\omega_{n(1,2)}/\omega_1$ as a function of M_2/M_1 for three different ratios of ω_a/ω_1. The figure indicates the span between the two resonance frequencies increase as the ratios M_2/M_1 and ω_a/ω_1 increase. It is desirable to make this range as large as physical constraints allow.

EXAMPLE 4.18

A large vertical reciprocating compressor was mounted in the center of a first floor mechanical equipment room in a large laboratory building. The compressor operated at a single rotational speed of 1,200 rpm. When the compressor was started for the first time, it was discovered that the operating speed of the compressor was the same as the resonance frequency of the compressor-floor system. Because of the low operating speed (20 Hz) and the size of the compressor, conventional vibration control techniques based on a 1-DOF analysis were considered to be impractical. There was an empty room below the mechanical equipment room in which the compressor was located. Therefore, it was decided to suspend a mass-spring system from the deck area below the floor on which the compressor was mounted. Assume that $M_2/M_1 = 0.8$. Plot the frequency response functions specified by equations (4.54) and (4.55).

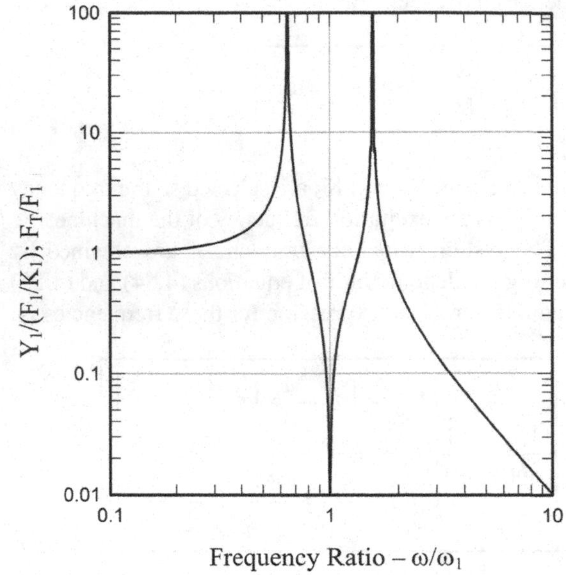

Figure 4.19 Response of Mass M_1 in EXAMPLE 4.18

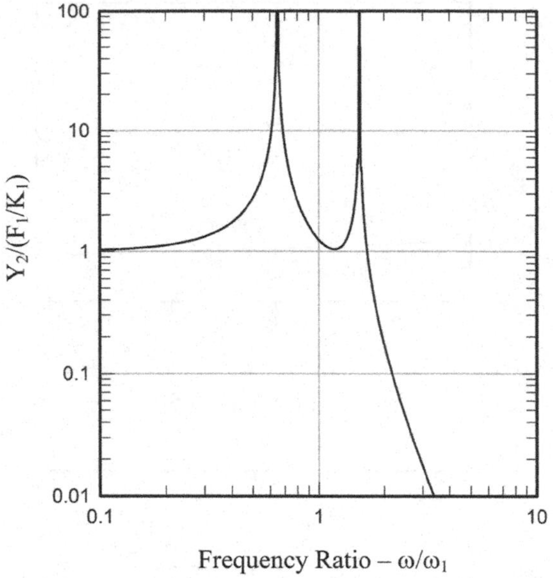

Figure 4.20 Response of Mass M_2 in EXAMPLE 4.18

SOLUTION

For this situation, the operating rotational speed of the compressor equals the resonance frequency of the floor-compressor system. Therefore, ω_a must equal ω_1 where ω_1 is the resonance frequency of the 1-DOF machine-floor system. From equation (4.58), $\omega_1 = \omega_2$, which can be written:

$$\sqrt{\frac{K_2}{M_2}} = \sqrt{\frac{K_1}{M_1}}. \tag{4.18a}$$

Rearranging equation (4.19a) yields:

$$\boxed{K_2 = K_1 \frac{M_2}{M_1} \quad \text{or} \quad K_2 = 0.8\,K_1.} \tag{4.18b}$$

When $\omega_1 = \omega_2$, equations (4.54) and (4.55) reduce to:

$$\boxed{\frac{Y_1}{F_1/K_1} = \frac{\left|1 - \frac{\omega^2}{\omega_1^2}\right|}{\left|\frac{\omega^4}{\omega_1^4} - \left[2 + \frac{M_2}{M_1}\right]\frac{\omega^2}{\omega_1^2} + 1\right|}} \tag{4.18c}$$

and:

$$\boxed{\frac{Y_2}{F_1/K_1} = \frac{1}{\left|\frac{\omega^4}{\omega_1^4} - \left[2 + \frac{M_2}{M_1}\right]\frac{\omega^2}{\omega_1^2} + 1\right|}}. \tag{4.18d}$$

Figures 4.19 and 4.20 show plots of equations (4.18b) and (4.18c). Note in Figure 4.18 that the response Y_1 for mass M_1 approaches zero when $\omega/\omega_1 = 1$ or when $\omega = \omega_a$.

EXAMPLE 4.19

The large vertical compressor in EXAMPLE 4.18 is now mounted in the center of a floor that has a resonance frequency of 10 Hz. The compressor operates at a single speed of 1,200 rpm (20 Hz). Even though the compressor does not operate at a speed that corresponds to the resonance frequency of the compressor-floor system, it still causes vibration problems in other parts of the building. it was decided to suspend a mass-spring system from the deck area below the floor on which the compressor was mounted that has a value of $\omega_a = \omega_2$ that corresponds to the operating speed of the compressor. Assume that $M_2/M_1 = 0.8$. Plot the frequency response functions specified by equations (4.54) and (4.55).

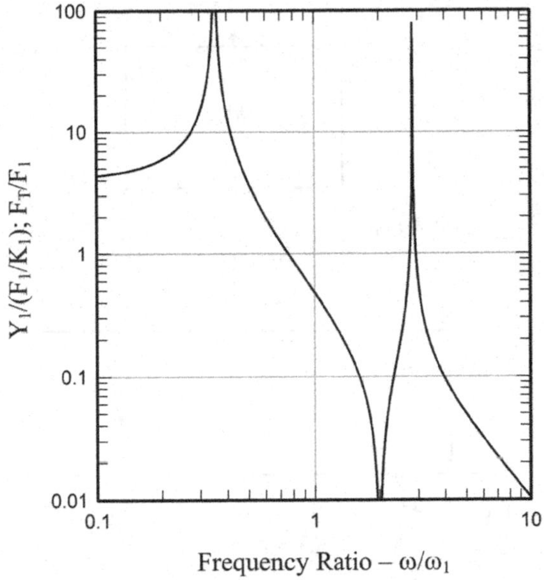

Figure 4.21 Response of Mass M_1 in EXAMPLE 4.19

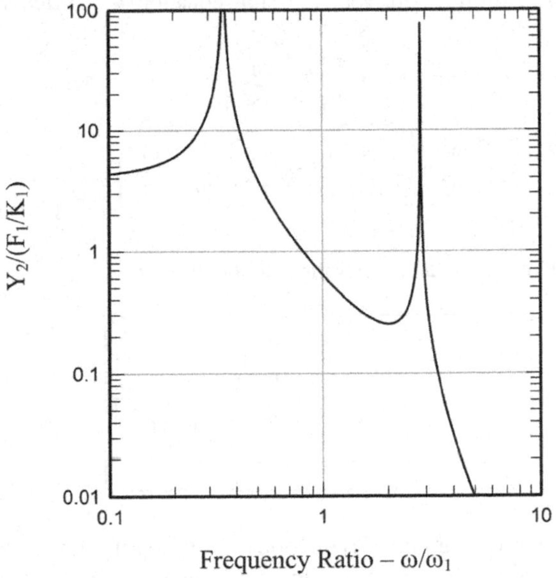

Figure 4.22 Response of Mass M_2 in EXAMPLE 4.19

SOLUTION

The operating speed of the compressor is two times the resonance frequency of the floor-compressor system. Therefore, ω_a must equal $2\omega_1$ and $\omega_2 = 2\,\omega_1$.

$$\sqrt{\frac{K_2}{M_2}} = 2\sqrt{\frac{K_1}{M_1}}\,. \tag{4.19a}$$

Rearranging equation (4.19a) yields:

$$\boxed{K_2 = 4\,K_1\frac{M_2}{M_1} \quad \text{or} \quad K_2 = 3.2\,K_1.} \tag{4.19b}$$

When $\omega_1 = 2\omega_2$, equations (4.54) and (4.55) reduce to:

$$\boxed{\frac{Y_1}{F_1/K_1} = \frac{\left|4 - \dfrac{\omega^2}{\omega_1^2}\right|}{\left|\dfrac{\omega^4}{\omega_1^4} - \left[5 + 4\dfrac{M_2}{M_1}\right]\dfrac{\omega^2}{\omega_1^2} + 4\right|}} \tag{4.19c}$$

and:

$$\boxed{\frac{Y_2}{F_1/K_1} = \frac{4}{\left|\dfrac{\omega^4}{\omega_1^4} - \left[5 + 4\dfrac{M_2}{M_1}\right]\dfrac{\omega^2}{\omega_1^2} + 4\right|}} \tag{4.19d}$$

Figures 4.21 and 4.22 show plots of equations (4.19c) and (4.19d. Note in Figure 4.20 that the response Y_1 for mass M_1 approaches zero when $\omega/\omega_1 = 2$ or when $\omega = \omega_a = 2\,\omega_1$.

When vibration dynamic (tuned) absorbers are used to address a structural vibration problem, M_1 and K_1 are often not known. When this is the case, M_2 and K_2 are selected to yield a value for ω_a that will match the excitation frequency the machine that is to be reduced. Mass M_2 is usually selected to accommodate the physical constraints of the structure that will support it. The selection of M_2 determines the required value for K_2.

Force Transmitted to a Structual Base

The equation for the force transmitted to a structural base is:

$$F_T = K_1\,Y_1\,. \tag{4.60}$$

Substituting equation (4.54) into equation (4.60) yields:

$$F_T = \frac{F_1 \left| \frac{\omega_2^2}{\omega_1^2} - \frac{\omega^2}{\omega_1^2} \right|}{\left| \frac{\omega^4}{\omega_1^4} - \left[1 + \frac{\omega_2^2}{\omega_1^2}\left(1 + \frac{M_2}{M_1}\right)\right] \frac{\omega^2}{\omega_1^2} + \frac{\omega_2^2}{\omega_1^2} \right|}. \tag{4.61}$$

The force transmissibility can be written:

$$\frac{F_T}{F_1} = \frac{\left| \frac{\omega_2^2}{\omega_1^2} - \frac{\omega^2}{\omega_1^2} \right|}{\left| \frac{\omega^4}{\omega_1^4} - \left[1 + \frac{\omega_2^2}{\omega_1^2}\left(1 + \frac{M_2}{M_1}\right)\right] \frac{\omega^2}{\omega_1^2} + \frac{\omega_2^2}{\omega_1^2} \right|} \tag{4.62}$$

which is the same as equation (4.54). Therefore, Figure 4.18 is also a plot of the force transmissibility for EXAMPLE 4.18, and Figure 4.20 is a plot of the force transmissibility for EXAMPLE 4.19.

4.6 Machine Mounted on a Flexible Structure

General System Equations

Vibrating machines are mounted on flexible supporting structures. These systems can be modeled as a 2-DOF vibration system similar to the one shown in Figure 4.23. M_1 and K_1 represent the effective mass and stiffness of the structural system, which is often a flexible floor system. M_2 and K_2 represent the machine mass and the stiffness of the resilient elements that support the machine. The equations of motion for this system are those that were developed in EXAMPLE 4.14. The response of the floor system is given by equation (4.14m):

$$\frac{Y_1}{F_2/K_2} = \frac{\frac{K_2}{K_1}\frac{\omega_1^2}{\omega_2^2}}{\left| \frac{\omega^4}{\omega_2^4} - \left[1 + \frac{\omega_1^2}{\omega_2^2}\left(1 + \frac{K_2}{K_1}\right)\right] \frac{\omega^2}{\omega_2^2} + \frac{\omega_1^2}{\omega_2^2} \right|}. \tag{4.63}$$

The response of the machine is given by equation (4.14n):

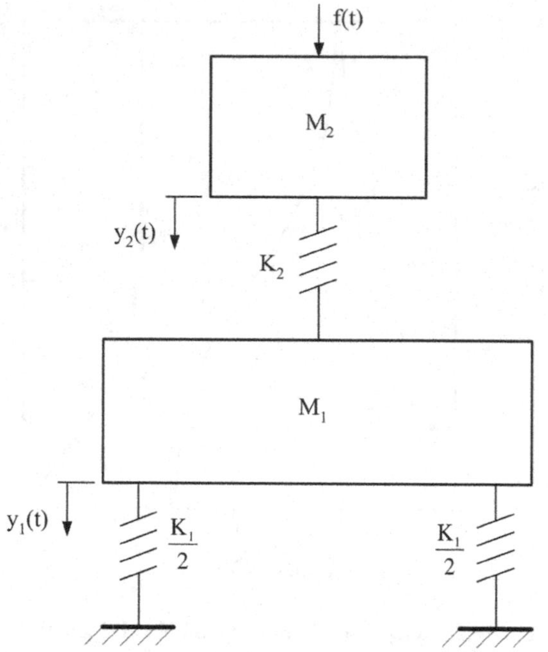

Figure 4.23 Schematic Representation of Machine Mounted on a Flexible Support Structure

$$\frac{Y_2}{F_2/K_2} = \frac{\left| \frac{\omega_1^2}{\omega_2^2}\left(1 + \frac{K_2}{K_1}\right) - \frac{\omega^2}{\omega_2^2} \right|}{\left| \frac{\omega^4}{\omega_2^4} - \left[1 + \frac{\omega_1^2}{\omega_2^2}\left(1 + \frac{K_2}{K_1}\right)\right] \frac{\omega^2}{\omega_2^2} + \frac{\omega_1^2}{\omega_2^2} \right|}. \tag{4.64}$$

From equation (4.14i):

$$\omega_1 = \sqrt{\frac{K_1}{M_1}} \quad \text{and} \quad \omega_2 = \sqrt{\frac{K_2}{M_2}}. \tag{4.65}$$

The system resonance frequencies are given by equation (4.14p), or:

$$\frac{\omega_{n(1,2)}^2}{\omega_2^2} = \frac{1}{2} \left\{ \begin{aligned} &\left[1 + \frac{\omega_1^2}{\omega_2^2}\left(1 + \frac{K_2}{K_1}\right)\right] \mp \\ &\sqrt{\left[1 + \frac{\omega_1^2}{\omega_2^2}\left(1 + \frac{K_2}{K_1}\right)\right]^2 - 4\frac{\omega_1^2}{\omega_2^2}} \end{aligned} \right\}. \tag{4.66}$$

Note that all of the frequency ratios can also be written:

$$\frac{\omega_1^2}{\omega_2^2} = \frac{f_1^2}{f_2^2}\,;\quad \frac{\omega}{\omega_2^2} = \frac{f}{f_2^2}\,;\quad \text{and}\quad \frac{\omega_{n(1,2)}^2}{\omega_2^2} = \frac{f_{n(1,2)}^2}{f_2^2}\,. \tag{4.67}$$

Equations (4.14j), (4.14k) and (4.14o) that are written in terms of M_2/M_1 can also be used to describe the response of the system in Figure 4.22..

The force transmitted from the machine to the flexible structure is given by:

$$F_{T(M)} = K_2\, Y_2\,. \tag{4.68}$$

Substituting equation (4.64) into equation (4.68) and rearranging the terms yields:

$$\boxed{\frac{F_{T(M)}}{F_2} = \frac{\left|\dfrac{\omega_1^2}{\omega_2^2}\left(1+\dfrac{K_2}{K_1}\right) - \dfrac{\omega^2}{\omega_2^2}\right|}{\left|\dfrac{\omega^4}{\omega_2^4} - \left[1+\dfrac{\omega_1^2}{\omega_2^2}\left(1+\dfrac{K_2}{K_1}\right)\right]\dfrac{\omega^2}{\omega_2^2} + \dfrac{\omega_1^2}{\omega_2^2}\right|}}\,. \tag{4.69}$$

The force transmitted from the flexible floor to the structure supporting the floor is given by:

$$F_{T(S)} = K_1\, Y_1\,. \tag{4.70}$$

Substituting equation (4.63) into equation (4.70) and rearranging the terms yields:

$$\boxed{\frac{F_{T(S)}}{F_2} = \frac{\dfrac{\omega_1^2}{\omega_2^2}}{\left|\dfrac{\omega^4}{\omega_1^4} - \left[1+\dfrac{\omega_1^2}{\omega_2^2}\left(1+\dfrac{K_2}{K_1}\right)\right]\dfrac{\omega^2}{\omega_2^2} + \dfrac{\omega_1^2}{\omega_2^2}\right|}}\,. \quad (4.71$$

Value for the Disturbing Force

For most machines, the *disturbing force* F_2 in the above equations usually represents the force associated with the out-of-balance weight of the rotating and reciprocating parts of a machine. The maximum value for out-of-balance forces associated with rotating and reciprocating machines is set by various engineering standards. For rotating and reciprocating machines, F_2 is given by:

$$\boxed{F_2 = (m_{imb}\; e)\, \omega^2} \tag{4.72}$$

where:

- m_{imb} = value of the unbalanced mass (lb_f-s^2/in. or kg);
- e = distance (in. or m) between the centers of rotation and the unbalanced mass; and
- ω = rotational speed (rad/s) of the machine.

If rotational speed is specified in terms of rpm, equation (4.72) becomes:

$$\boxed{F_2 = (m_{imb}\; e)\left(\frac{\pi\, f(rpm)}{30}\right)^2} \tag{4.73}$$

where:

f(rpm) = operating speed (rpm) of the machine.

The values of m_{imb} and e are impossible to individually determine for rotating machines. However, the product $m_{imb}e$ is readily measured in rotating machines. Maximum recommended unbalance values associated with $m_{imb}e$ that apply to general classes of machines are specified in ANSI Standard S2.19, *Standard for Balance Quality of Rotating Rigid Bodies*. Using the information in ANSI Standard S2.19, values for $m_{imb}e$ can be obtained from:

$$\boxed{m_{imb}\; e = \frac{W_{rot}}{g}\,\theta_f} \qquad (4.74\ \text{U.S.})$$

$$\boxed{m_{imb}\; e = M_{rot}\;\theta_f} \qquad (4.74\ \text{SI})$$

where:

- m_{imb} = value of the unbalanced mass (lb_f-s^2/in. or kg);
- e = distance (in. or m) between the center of rotation and the unbalanced mass;
- W_{rot} = weight (lb_f) of the rotating component of the machine;
- M_{rot} = mass (kg) of the rotating component of the machine;
- g = acceleration of gravity (386 in./s^2); and
- θ_f = specific unbalance (lb_f-in./lb_f or kg-m/kg).

θ_f is given by:

$$\boxed{\theta_f = \frac{A}{f(rpm)}} \quad (4.75\ U.S.)$$

$$\boxed{\theta_f = 0.0254\,\frac{A}{f(rpm)}} \quad (4.75\ SI)$$

where:

f(rpm) = operating speed (rpm) of the machine; and
A = constant.

Specified values for A for different classes of HVAC equipment are given in Table 4.1. These values are derived from the information in ANSI Standard S2.19. If the machine being investigated is a fan or blower, maximum recommended balance limits are specified in ARI Guideline G, *Guideline for Mechanical Balance of Fans and Blowers*. These limits are given in Table 4.2 as a function of fan or blower wheel diameter. Corresponding values for $m_{imb}e$ are obtained from:

$$\boxed{m_{imb}\ e = 0.0625\,\frac{me}{g}} \quad (4.76\ U.S.)$$

$$\boxed{m_{imb}\ e = me \times 10^{-6}} \quad (4.76\ SI)$$

where:

m_{imb} = value of the unbalanced mass (lb_f-s^2/in. or kg);
e = distance (in. or m) between the center of rotation and the unbalanced mass;
me = Representative commercial balance limits from Table 4.2 (oz-in. or g-mm); and
g = acceleration of gravity (386 in./s^2 or 9.8 m/s^2).

If the machine being investigated is a centrifugal fan or blower, the potential maximum unbalance is 2 times the respective value listed in Table 4.2.

Table 4.1 Specified Values of A for Groups of Representative Rotating Machines

Rotor Types	A
Drive shafts, parts of crushing machinery, parts of agricultural machinery, individual components of engines, crankshaft drives of engines with six or more cylinders, slurry or dredge pump impeller	6.0
Parts of process plant machines, marine main turbine gears, centrifuge drums, fans, aircraft gas turbine rotors, fly wheels, pump impellers, machine-tool and general machinery parts, normal electrical armatures, individual components of engines under special requirements	2.4
Gas and steam turbines, rigid turbo-generator rotors, rotors, turbo-compressors, machine-tool drives, medium and large electrical armatures with special requirements, small electrical armatures, turbine-driven pumps	1.0

Estimating Effective Structure Mass and Stiffness

The values for M_2 and K_2 in Figure 4.22 are the mass of the machine and the total stiffness coefficient of the resilient element between the machine base and the structural support. These values are easily obtained. However, the values of the effective mass M_f and the effective stiffness K_f of a nonrigid support structure are more difficult to obtain. Realistic estimates of M_f and K_f can be obtained if the structural support system is viewed properly. The following approach is associated with the manner in which loads are applied to building floor and roof systems.

Two fundamental parameters are generally used in the design of a floor or roof system: *live load* and *dead load*. *Live loads* are the weights that must be supported by a floor or roof system that are not associated with the structural weight of the system. Items such as the building occupants, office equipment, wall partitions, carpeting, fans, pumps, chillers, and other mechanical devices and building systems are examples of live loads. The *dead load* is the weight of a floor or roof system associated with the weight of the structural elements that comprise the system.

Structural engineers usually design floor and roof

Table 4.2 **(U.S.)** Representative Commercial Balance Limits

For Propeller Fans		For Blower Wheels	
Fan Diameter (in.)	Amount of Unbalance (oz-in.)	Blower Diameter (in.)	Amount of Unbalance per Plane (oz-in.)
8	0.10	4 (or less)	0.07
9	0.10	6	0.10
10	0.10	7 & 8	0.13
11	0.10	9, 10 & 11	0.15
12	0.10	12	0.25
14	0.10	14 & 15	0.25
16	0.15	16	0.45
18	0.15	18	0.68
20	0.20	20	0.92
22	0.25	22	1.15
24	0.30	24	1.39
26	0.30	26	1.74
28	0.40	28	2.09
30	0.45	30	2.43
36	0.60	32	2.78
42	1.00	34	3.13
48	1.40	36	3.48
54	1.50	38	3.82
60	2.00	40	4.17

Table 4.2 **(SI)** Representative Commercial Balance Limits

For Propeller Fans		For Blower Wheels	
Fan Diameter (mm)	Amount of Unbalance (g-mm)	Blower Diameter (mm)	Amount of Unbalance per Plane (g-mm)
203	72	102 (or less)	50.4
229	72	152	72.0
254	72	178 & 203	93.6
279	72	229, 254 & 279	108.0
305	72	305	180.0
356	72	356 & 381	180.0
406	108	406	324.0
457	108	457	489.6
508	144	508	662.4
559	180	559	828.0
610	216	610	1,000.8
660	216	660	1,252.8
711	288	711	1,504.8
762	324	762	1,749.6
914	432	813	2,001.6
1,067	720	864	2,253.6
1,219	1,008	914	2,505.6
1,372	1,080	965	2,750.4
1,524	1,440	1,016	3,002.4

structural systems to meet the anticipated live loads the floor and roof systems must support. They check to determine whether the resulting live plus dead load total weight produces a static deflection that does not exceed building code requirements and customer specifications. Building codes permit a live load floor deflection up to 1/360 of the length of the floor span and a live plus dead load deflection up to 1/240 of the length. In practice, live load deflections are usually limited to 1/1400 to 1/800 of the length of the span. Roof decks are permitted slightly greater live load deflections. Refer to Table 4.3.

Deflection equations are found in many engineering handbooks for point and distributed loads. The basic expressions for simply supported beams are given below.

Load in center of beam:

$$\boxed{\delta = \frac{F\,L^3}{48\,EI}} \tag{4.77}$$

Load not in the center of the beam:

$$\boxed{\delta = \frac{F\left(a\,L - a^2\right)^2}{3\,EI\,L}} \tag{4.78}$$

Uniform load on beam:

$$\boxed{\delta = \frac{5w\,L^4}{384\,EI}} \tag{4.79}$$

Table 4.3 Values of Allowed Maximum Deflections Used in Structural Design

Deflections as a Function of Span Length, L				
Deck Type	Live Load (max)	Dead Load (max)	Live + Dead Load (max)	Live Load (Typical)
Floor	L /360	L /720	L /240	L /1400 to L /800
Roof	L /240	L /720	L /180	

where:

- F = point load (lb_f or N);
- w = distributed load (lb_f/in. or N/m) acting along the length of the beam
- L = length of beam (in. or m);
- a = distance (in. or m) from the left end of the beam;
- E = modulus of elasticity (Young modulus) of the beam (lb_f/in^2 or N/m^2);
- I = area moment of the beam cross section ($in.^4$ or m^4).

The stiffness term used in Figure 4.23 for the effective stiffness of the support structure can be obtained from the appropriate deflection by recalling [equation (1.113) in Chapter 1] the stiffness of an elastic element is given by dividing the weight supported by the element by the displacement (deflection) of the element:

$$K_1' = \frac{F}{\delta} \quad \text{or} \quad K_1' = \frac{wL}{\delta} \tag{4.80}$$

Applying equation (4.80) to equations (4.77) through (4.79) gives effective stiffness terms for the above beam configurations.

Load in center of beam:

$$K_1' = \frac{48\,EI}{L^3} \tag{4.81}$$

Load not in the center of the beam:

$$\boxed{K_1' = \frac{3\,EIL}{\left(aL - a^2\right)^2}} \tag{4.82}$$

Uniform load on beam:

$$\boxed{K_1' = \frac{384\,EI}{5L^3}} \tag{4.83}$$

Before equations (4.81) - (4.83) can be used, two adjustments must be made to correct for the way the load and the floor span approximated by a beam are modeled. Structural engineers usually design floor and roof systems using anticipated distributed live loads. Typical values for distributed live and dead loads are given in Table 4.4. The vibration analysis in this section assumes a point or lumped mass for calculating resonance frequencies. Therefore, an adjustment must be made, converting from the distributed mass used in the structural design to the lumped mass used in this analysis. The deflection equations used in this section are for simply supported beams. Even though the connections between floor joists and structural columns are close to being simply supported, they are not purely simply supported. Therefore, a correction must be made for this.

The first of two adjustments is made by determining an equivalent mass that, if applied to the center of a beam or floor span, will yield the same point load static deflection in equation (4.77) as the distributed load static deflection in equation (4.79). Setting equation (4.77) equal to equation (4.79), the *equivalent mass of a floor or roof* is given by:

$$\boxed{M_1 = 0.625\,\frac{wL}{g}} \tag{4.84}$$

where the factor 0.625 is the adjustment made as compensation for converting a uniform weight w acting over the length L of a floor span to a single lumped mass M_1.

A second adjustment is made in the floor or roof stiffness values determined by equations (4.81) through (4.83) This adjustment is necessary to account for beam or joist members exhibiting characteristics that lie somewhere between being simply supported and

Table 4.4 (U.S.) Floor and Roof Load Distributions

Design Load Distribution - lb_f/ft^2		
Deck Type	Live Load	Dead Load
Floor		
Construction known	50	
Construction unknown	100	
Composite Floor		
Dense Concrete (150 lb_f/ft^3)		67
Lt. Wt. Concrete (100 lb_f/ft^3)		51
Composite Built-up Roof	20	20-45
Typical Spacing between Joist = 10 ft		

Table 4.4 (SI) Floor and Roof Load Distributions

Design Load Distribution - N/m^2		
Deck Type	Live Load	Dead Load
Floor		
Construction known	2,395	
Construction unknown	4,789	
Composite Floor		
Dense Concrete (2,405 kg/m^3)		3,209
Lt. Wt. Concrete (1603 kg/m^3)		2,443
Composite Built-up Roof	958	958-2,155
Typical Spacing between Joist = 3.05 m		

clamped at their ends. Normally, the ends of floor joist are securely fastened to vertical structural columns. It is usually noted that the vertical columns are elastic, resulting in the joists exhibiting behavior characteristics between being simply supported and clamped. The characteristics more closely resemble being simply supported. The literature provides studies that indicate the resonance frequency of a floor system is predicted by:

$$\boxed{f_1 = 1.57\sqrt{\frac{g\,EI}{w\,L^3}}} \tag{4.85}$$

where f_1, E, I, w, and L are as previously defined. Equating equations (4.85) and (4.81) gives:

$$\boxed{K_1 = 1.267\,K_1^{'}} \tag{4.86}$$

The factor 1.267 is the compensating adjustment made to the stiffness coefficient to correct for the fact a joist is not purely simply supported. Although the mass was adjusted based on the equation for a load in the center of a joist, the stiffness adjustment is better represented by a load not in the center of a joist. Therefore, the expression for the adjusted stiffness is obtained by substituting equation (4.82) into equation (4.86):

$$\boxed{K_1 = 1.267\frac{3\,EI\,L}{\left(aL - a^2\right)^2}}. \tag{4.87}$$

The expressions in equation (4.84) for M_1 and in cquation (4.87) for K_1 are substituted into equation (4.65) to calculate the value for f_1.

Design Procedures

The following is a step-by-step procedure for performing an analysis relative to a vibrating machine on a flexible supporting structure.

1. Determine the required static deflection associated with the resilient elements supporting the machine base. Use equation 2.9 or Figure 2.4 in Chapter 2 to determine the corresponding frequency f_2. It is also possible to start with f_2 and calculate the corresponding static deflection using Figure 2.4.

2. Calculate the isolator stiffness coefficient K_2 by solving equation (4.65) for K_2 using the known mass M_2 (or weight W_2) and the frequency f_2 or the static deflection δ from the previous step.

3. Find the effective mass M_1 of the floor or roof system using equation (4.84). w in equation (4.84) is obtained by multiplying the dead load design load

distribution in the units of lb_f/ft^2 (N/m^2) by the space between the joists in the unit of ft (m). The dead load design load distribution can be obtained from Table 4.3. Note, if the units on w from the above procedure are lb_f/ft, the units must be converted to $lb_f/in.$ by dividing the results by 12:

$$w\left(\frac{lb_f}{in.}\right) = w\left(\frac{lb_f}{ft}\right)\frac{1}{12}. \tag{4.88}$$

4. Determine the value of EI of the floor or roof system for the design live load by using equation (4.79). The value for the design live load static deflection δ in equation (4.79) is obtained from the live load deflection values specified in Table 4.3 as a function of the length of the span. (Estimating EI values based on a conservative (typical) live load deflection is a normally accepted practice among structural engineers.)

For example, rearranging equation (4.79) gives:

$$EI = \frac{5\,w\,L^4}{384\,\delta}. \tag{4.89}$$

Assume from Table 4.3 the design live load deflection δ is L/1000. Substituting this expression for δ into equation (4.89) yields:

$$EI = \frac{5000\,w\,L^3}{384}. \tag{4.90}$$

Note, if the units for E are $lb_f/in.^2$ (N/m^2) and the unit for I is $in.^4$ (m^4) the units for w are $lb_f/in.$ (N/m) and the unit for L is in. (m).

5. Calculate the value of the stiffness coefficient K_1 using equation (4.87). Note, if the units for E are $lb_f/in.^2$ (N/m^2) and the unit for I is in^4 (m^4), the unit for a and L is in. (m).

6. Calculate the frequency of the floor f_1 using equation (4.65) with the values of M_1 and K_1 determined from Steps 3 and 5.

7. Calculate the two resonance frequencies of the machine-support structural system by substituting appropriate values from the previous steps into equation (4.66).

A cursory evaluation can be made once the two resonance frequencies of the system are known by comparing them to the operating frequency (in like units) of the vibrating machine. Typically, there should be at least a 1 Hz difference between the machine's operating frequency and either of the two resonance frequencies. A difference of approximately 1 Hz or less signals the potential for excessive vibration. In any case, the analysis must be carried to conclusion using the following steps.

8. Determine the normalized machine and floor displacements from equations (4.63) and (4.64) based on the specified operating frequency of the machine. [The normalized transmitted forces to the floor and supporting building structure can be found from equations (4.69) and (4.71).]

The results of the previous step gives the percentage of deflection and excitation force that is transmitted from the machine to the floor and related structural elements. These values should be relatively small values for the vibration to be considered acceptable. Finally, calculate the actual dynamic floor displacement and related acceleration values.

9. Calculate the excitation force F_2 from equation (4.73), and then calculate the actual floor and machine displacement values Y_1 and Y_2 by multiplying the corresponding normalized displacements from Step 8 by the value of F_2/K_2.

10. Calculate the floor and equipment velocities using the appropriate displacement values obtained in Step 9 and the operating frequency of the machine in radians/s. Use Figure 1.45 in Chapter 1 to determine the acceptability of the floor vibration and Figure 1.46 to determine the severity of the machine vibration.

EXAMPLE 4.20

A cabinet centrifugal fan unit which operates at 300 rpm and weighs 2,500 lb_f (11,121 N) is to be installed on a floor system that has a span of 20 ft (6.1 m) and constructed of lightweight concrete. The wheel diameter of the centrifugal fan is 38 in. (965 mm). The fan will be positioned 6 ft (1.83 m) from the end of the floor span and will be mounted on spring isolation mounts that have a static deflection of 1 in. (25.4 mm) The floor is designed for a maximum live load of 50 lb_f/ft^2 (2,395 N/m^2). However, the actual live load is only 30 lb_f/ft^2 (1,437 N/m^2). The design live load deflection is 1/1200 of the length of the span. What will an analysis of the vibration show regarding the acceptability or unacceptability of this configuration?

SOLUTION

From equations (2.9) and (1.113):

U.S. Units

$$f_2 = \frac{1}{2\pi}\sqrt{\frac{386}{1}} = 3.1 \text{ Hz} \qquad (4.20a\ U.S.)$$

$$K_2 = \frac{2{,}500}{1} = 2{,}500 \quad lb_f/in. \qquad (4.20b\ U.S.)$$

SI Units

$$f_2 = \frac{1}{2\pi}\sqrt{\frac{9.8}{0.0254}} = 3.1 \text{ Hz} \qquad (4.20a\ SI)$$

$$K_2 = \frac{11{,}121}{0.0254} = 437{,}835 \quad N/m \qquad (4.20b\ SI)$$

The equivalent mass of the floor system is obtained from equation (4.84). The equivalent mass includes the dead load distributed weight and the actual live load distributed weight.

U.S. Units

$$w = (51 + 30) \times 10 = 810 \quad lb_f/ft \qquad (4.20c\ U.S.)$$

$$M_2 = 0.625 \times \frac{810 \times 20}{386} = 26.23 \quad \frac{lb_f - s^2}{in.} \qquad (4.20d\ U.S.)$$

SI Units

$$w = (2{,}395 + 1{,}437) \times 3.05 = 11{,}688 \quad N/m \qquad (4.20c\ SI)$$

$$M_2 = 0.625 \times \frac{11{,}688 \times 6.1}{9.8} = 4{,}547 \quad kg \qquad (4.20d\ SI)$$

EI is obtained from equation (4.89); w in equation (4.89) is the maximum live load weight.

U.S. Units

$$w = \frac{50 \times 10}{12} = 41.67 \quad lb_f/in. \qquad (4.20e\ U.S.)$$

SI Units

$$w = 2{,}395 \times 3.05 = 7{,}305 \quad N/m \qquad (4.20e\ SI)$$

EI is:

U.S. Units

$$EI = \frac{1200 \times 5 \times 41.67 \times (20 \times 12)^3}{384} = 9.0 \times 10^9 \quad lb_f - in^2 \qquad (4.20f\ U.S.)$$

SI Units

$$EI = \frac{1200 \times 5 \times 7{,}305 \times (6.1)^3}{384} = 2.591 \times 10^7 \quad N - m^2 \qquad (4.20f\ SI)$$

K_1 is obtained from equation (4.87).

U.S. Units

$$K_1 = \frac{1.267 \times 3 \times (9.0 \times 10^9) \times (20 \times 12)}{\left[(6 \times 12) \times (20 \times 12) - (6 \times 12)^2\right]^2} = 5.611 \times 10^4 \quad lb_f/in. \qquad (4.20g\ U.S.)$$

SI Units

$$K_1 = \frac{1.267 \times 3 \times (2.591\times 10^7) \times (6.1)}{(1.83 \times 6.1 - 1.83^2)^2} = 9.839 \times 10^6 \quad N/m \qquad (4.20g\ SI)$$

f_1 is obtained from equation (4.65).

U.S. Units

$$f_1 = \frac{1}{2\pi}\sqrt{\frac{5.61\times 10^4}{26.23}} = 7.4 \quad Hz \qquad (4.20h\ U.S.)$$

SI Units

$$f_1 = \frac{1}{2\pi}\sqrt{\frac{9.839\times 10^6}{4,547}} = 7.4 \quad Hz \qquad (4.20h\ SI)$$

The ratios $(f_1/f_2)^2$ and K_1/K_2 are:

$$\frac{f_1^2}{f_2^2} = \frac{7.4^2}{3.1^2} = 5.7 \qquad (4.20i)$$

U.S. Units

$$\frac{K_2}{K_1} = \frac{2,500}{5.61\times 10^4} = 0.0446 \qquad (4.20j\ U.S.)$$

SI Units

$$\frac{K_2}{K_1} = \frac{437,835}{9.839\times 10^6} = 0.0445 \qquad (4.20j\ SI)$$

The two resonance frequency are obtained by substituting the appropriate values into equation (4.66). The resonance frequencies are:

$$f_{n(1)} = 3.1 \quad Hz \quad \text{and} \quad f_{n(2)} = 7.6 \quad Hz. \qquad (4.20k)$$

$Y_2/(F_2/K_2)$ and $Y_1/(F_2/K_2)$ are obtained by substituting the appropriate values into equations (4.63) and (4.64):

$$\frac{Y_2}{F_2/K_2} = 0.612 \quad \text{and} \quad \frac{Y_1}{F_2/K_2} = 0.0468. \qquad (4.20l)$$

The disturbing force is obtained by first determining $m_{imb}e$. $m_{imb}e$ is obtained from equation (4.76) for a centrifugal fan. me is obtained from Table 4.2.

U.S. Units

$$me = 2 \times 3.82 = 7.64 \quad oz-in. \qquad (4.20m\ U.S.)$$

$$m_{imb}\, e = \frac{0.0625 \times 7.64}{386} = 0.00124 \quad lb_f - s^2 \qquad (4.20n\ U.S.)$$

SI Units

$$me = 2 \times 2,750 = 5,500 \quad g-mm \qquad (4.20m\ SI)$$

$$m_{imb}\, e = 5,500 \times 10^{-6} = 0.0055 \quad kg-m \qquad (4.20n\ SI)$$

Substituting this result into equation (4.73) gives:

U.S. Units

$$F_d = 0.00124 \times \left(\frac{300\,\pi}{30}\right)^2 = 1.22 \quad lb_f \qquad (4.20o\ U.S.)$$

SI Units

$$F_d = 0.0055 \times \left(\frac{300\,\pi}{30}\right)^2 = 5.43 \quad N \qquad (4.20o\ SI)$$

Y_2 and Y_1 are obtained from equations (4.63) and (4.64).

U.S. Units

$$Y_2 = 0.612 \times \frac{1.22}{2,500} = 2.99 \times 10^{-4} \quad in. \qquad (4.20p\ U.S.)$$

$$Y_1 = 0.0468 \times \frac{1.22}{2,500} = 2.28 \times 10^{-5} \quad \text{in.} \tag{4.20q U.S.}$$

SI Units

$$Y_2 = 0.612 \times \frac{5.43}{437,835} = 7.59 \times 10^{-6} \quad \text{m} \tag{4.20p SI}$$

$$Y_1 = 0.0477 \times \frac{5.43}{437,835} = 5.85 \times 10^{-7} \quad \text{m} \tag{4.20q SI}$$

The corresponding velocities are:

U.S. Units

$$\boxed{V_2 = 2.99 \times 10^{-4} \text{ x} \left(\frac{300\pi}{30} \right) = 9.39 \times 10^{-3} \quad \text{in./s}} \tag{4.20r U.S.}$$

$$\boxed{V_1 = 2.28 \times 10^{-5} \text{ x} \left(\frac{300\pi}{30} \right) = 7.16 \times 10^{-4} \quad \text{in./s}} \tag{4.20s U.S.}$$

SI Units

$$\boxed{V_2 = 7.59 \times 10^{-6} \text{ x} \left(\frac{300\pi}{30} \right) = 2.38 \times 10^{-4} \quad \text{m/s}} \tag{4.20r SI}$$

$$\boxed{V_1 = 5.85 \times 10^{-7} \text{ x} \left(\frac{300\pi}{30} \right) = 1.84 \times 10^{-5} \quad \text{m/s}} \tag{4.20s SI}$$

Figures 4.24 and 4.25 shows plots of the vibration responses associated with Y_2 and Y_1.

The analysis in EXAMPLE 4.20 indicates the velocity amplitude on the floor is between curves C and D in Figure12.2/12.3. This amplitude is significantly below what is required for any type of human occupancy. With respect to sensitive electronic or optical equipment, the type of equipment expected to be present must be specified before an assessment can be made.

With respect to vibration severity of the fan assembly, the analysis in EXAMPLE 4.20 indicates the velocity amplitude on the fan assembly is in the smooth range in Figure 12.1/12.2. This is what would normally be expected if the fan is properly balanced and supported on resilient elements.

Figures 4.24 and 4.25 show plots of the vibration responses associated with Y_1 and Y_2 for the system examined in EXAMPLE 4.20. Some important observations can be made from this figure. The figures clearly shows the effects of the two resonance frequencies at 3.1 Hz and 7.6 Hz. The rotational speed of the fan must not correspond to either of these two frequencies. The lower resonance frequency is primarily controlled by the value of f_2 [ω_2 - equation (4.65)]. The higher resonance frequency is controlled by the value of f_1

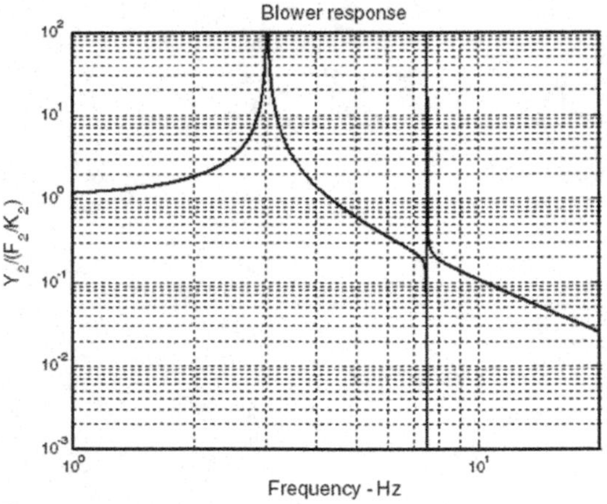

Figure 4.24 Response of Mass M_2 in Example 4.20

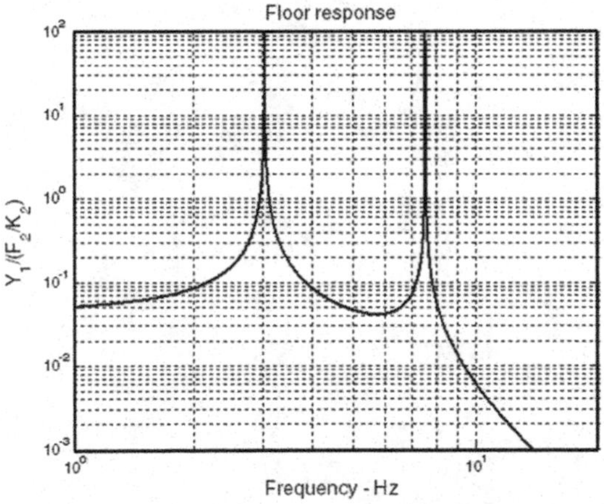

Figure 4.25 Response of Mass M_1 in Example 4.20

[ω_1 - equation (4.65)]. If the static deflection of the resilient elements between the machine and the floor in the example is increased to 2 in., the lower resonance frequency would be reduced to around 2.7 Hz, while the higher resonance frequency would remain relatively unchanged.

Assume the fan in EXAMPLE 4.20 operates at a rotational speed near 454 rpm (7.6 Hz). Because this speed corresponds to the higher resonance frequency of the fan-floor system, the vibration levels will be very high. If only a 1-DOF analysis is used, it might be presumed increasing the static deflection of the spring mountss supporting the fan to 2 in. will result in a significant reduction in the vibration levels associated with the fan. However, the 2-DOF analysis indicates that, even though the lower resonance frequency will be reduced, the higher resonance frequency will remain relatively unchanged. As a result, the high vibration levels will not be reduced. For this case, two solutions may be appropriate:

(1) Change the operating speed of the fan by at least 60 rpm so it no longer operates at a rotational speed that corresponds to the higher resonance frequency.

(2) Increase the stiffness of the floor structure that supports the fan to increase the higher resonance frequency so it no longer corresponds to the operating speed of the fan.

If the operating frequency of a machine with a rotating unbalance corresponds to one of the resonance frequencies associated with a building structure, increasing the static deflection of the resilient elements supporting the machine will not always reduce the vibration levels associated with the machine.

APPENDIX 4A - MATLAB® Script for EXAMPLE 4.20

```
clc, clear all, close all;
% This program prompts the user for the parameters of a blower fan mounted on
% a flexible structure. Then it processes the inputs to plot the system
% response and give specific results for k1, k2, f1, f2, resposne at the given
% rpm, the disturbing force and the displacements of the fan and floor
% along with corresponding velocites.

format compact;

% creates gui prompt input
prompt = {'Enter the weight of the fan(lb):','Enter the static deflection of the spring isolators(in):',...
   'Enter the fan speed (rpm):','Enter the fan unbalance (see Table)(oz-in):'....
   'Enter the live load(lb/ft^2):','Enter the length of floor span(ft):',...
   'Enter the maxiumum live load (lb/ft^2):','Enter the denomitor of the deflection ratio:',...
   'Enter the position of the fan from the nearest end of the floor span (ft):'};

dlg_title = 'Blower Mounted on a Flexible Structure';
num_lines = 1;                  % scales and titles the input
def = {'','','','','','','','',''};     % prompt box
answ = inputdlg(prompt,dlg_title,num_lines,def);

%Print the input values.
weight=str2double(answ(1))
static_deflection=str2double(answ(2))
rpm=str2double(answ(3))
fanub=str2double(answ(4))
live_load=str2double(answ(5))
floor_span=str2double(answ(6))
max_live_load=str2double(answ(7))
lldef=str2double(answ(8))
position=str2double(answ(9))

% Plots the response function Y1/(F2/K2) and Y2/(F2/K2) with respect to
%  frequency (Hz)for a centrifugal fan unit operatins on a lightweight
%  concrete floor system.
% Provide the following information to the function:
% weight: weight of the fan (lb)
% static_deflection: static deflection of the spring isolators (in)
% live_load: actual live load (lb/ft^2)
% floor_span: span of the floor (ft)
% max_live_load: maximum live load for design of floor (lb/ft^2)
% lldef: 1/(live load deflection ratio of the length of the span)
% position: position of fan from the end of the floor span (ft)
% rpm: speed at which fan operates (rpm)

% calculate the resonant frequency of the fan only.
```

```
f2 = (1/(2*pi))*sqrt(386/static_deflection)

% calculate the stiffness coefficient of the fan's spring support
k2 = weight/static_deflection

% calculate the distributed load: multiply dead load design load
% distrubtion in lb/ft^2 by the spacing between the joists in ft
w = (51 + live_load)*10;

% calculate the effective mass of the floor system
M1 = 0.625*(w * floor_span)/386;

% calculate maximum live load weight
w2 = (max_live_load * 10)/12;

% calculate the value of EI of the floor system for the design live load
EI = (5 * w2 * lldef * (floor_span * 12)^3)/384;

% calculate the stiffness coefficient of the floor
% the multiplicative factor is the compensating adjustment made to the
% stiffness coefficient to correct for the fact that a joist is no purely
% simply supported
k1 = (1.267 * 3 * EI * (floor_span * 12))/(((position * 12) * (floor_span * 12) -...
   (position * 12)^2)^2)

% calculate the resonant frequency of the floor system only.
f1 = (1/(2*pi))*sqrt(k1/M1)

% calculates fan imbalance
me = 2*fanub*0.0625/386;

% calculates the disturbing force
Disturbing_Force = me*(rpm*pi()/30)^2

% convert rpm to hz
hzz = rpm/60;

% the response of Y1/(F2/K2) and Y2/(F2/K2) when the fan runs at rpm input
response_1 = ((k2/k1)*(f1^2/f2^2))/abs((hzz^4/f2^4)-(1+(f1^2/f2^2)*(1+(k2/k1)))*...
   (hzz^2/f2^2)+(f1^2/f2^2))
response_2 = abs((f1^2/f2^2)*(1+(k2/k1))-(hzz^2/f2^2))/abs((hzz^4/f2^4)-(1+(f1^2/f2^2)*...
   (1+(k2/k1)))*(hzz^2/f2^2)+(f1^2/f2^2))

% calaculates displacement of machine and floor
Y2 = response_2*Disturbing_Force/k2
Y1 = response_1*Disturbing_Force/k2

% calaculates velocity of machine and floor
```

```
V2 = Y2*rpm*pi/30
V1 = Y1*rpm*pi/30

% generate points in array response1 and response2
T = 1;
for fn = 1:.0001:20
   response1(T) = ((k2/k1)*(f1^2/f2^2))/abs((fn^4/f2^4)-(1+(f1^2/f2^2)*(1+(k2/k1)))*...
      (fn^2/f2^2)+(f1^2/f2^2));
   response2(T) = abs((f1^2/f2^2)*(1+(k2/k1))-(fn^2/f2^2))/abs((fn^4/f2^4)-(1+(f1^2/f2^2)*...
      (1+(k2/k1)))*(fn^2/f2^2)+(f1^2/f2^2));
   T = T+1;
end


%Plote the results for Y1 and Y2.

   figure(1);
   loglog(1:.0001:20,response1,'linewidth',1.5,'color','black')
   grid on
   ylim([0.001,100]);
   xlim([1,20]);
   xlabel('Frequency - Hz','fontsize',12);
   ylabel('Y_1/(F_2/K_2)','fontsize',12);
   title('Floor response','fontsize',12);

   figure(2);
   loglog(1:.0001:20,response2,'linewidth',1.5,'color','black')
   grid on
   ylim([0.001,100]);
   xlim([1,20]);
   xlabel('Frequency - Hz','fontsize',12);
   ylabel('Y_2/(F_2/K_2)','fontsize',12);
   title('Blower response','fontsize',12);
```

PROBLEMS - CHAPTER 4

Note: For all problems involving the development of equations of motion, write the equations in matrix format.

1. Write the equations of motion of the system shown in Figure P4.1, using d'Alembert's principle. Write the equations in terms of θ and y_2.

2. Write the equations of motion of the system shown in Figure P4.1, using Lagrange's equation. Write the equations in terms of θ and y_2.

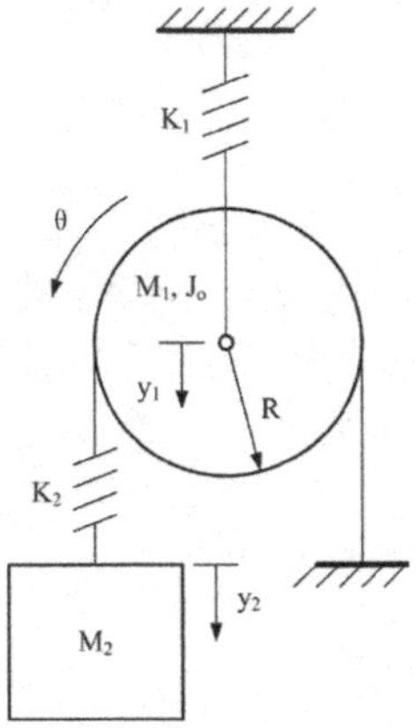

Figure P4.1

3. Write the expressions for the resonance frequencies and corresponding mode vectors for the system shown in Figure P4.1, assuming that:

$$K_1 = K_2 = K;\ M_1 = M_2 = M;\ J_o = 0.5MR^2.$$

Use the results of either Problem 4.1 or 4.2.

4. Write the equations of motion of the system shown in Figure P4.2, using d'Alembert's principle. Write the equations in terms of x_1 and x_2.

5. Write the equations of motion of the system shown in Figure P4.2, using d'Alembert's principle. Write the equations in terms of θ and x_2.

6. Write the equations of motion of the system shown in Figure P4.2, using Lagrange's equation. Write the equations in terms of x_1 and x_2.

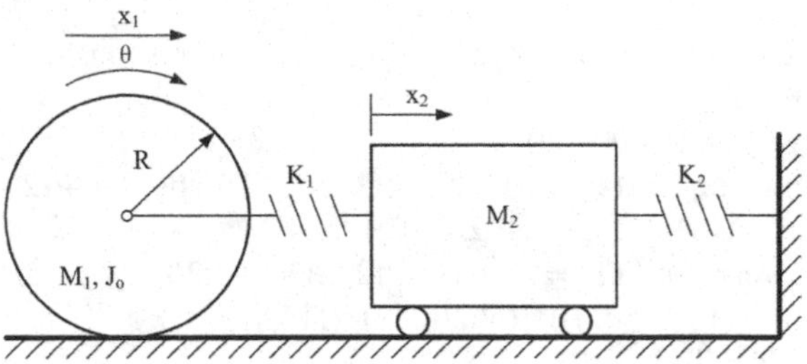

Figure P4.2

7. Write the equations of motion of the system shown in Figure P4.2, using Lagrange's equation. Write the equations in terms of θ and x_2.

8. Write the expressions for the resonance frequencies and corresponding mode vectors for the system shown in Figure P4.2, assuming that:

$$K_1 = K_2 = K \text{ and } M_1 = M_2 = M.$$

Use the results of either Problem 4.4 or 4.6 for the case where the equations are written in terms of x_1 and x_2.

9. Write the expressions for the resonance frequencies and corresponding mode functions for the system shown in Figure P4.2, assuming that:

$$K_1 = K_2 = K;\ M_1 = M_2 = M;\ J_o = 0.5MR^2.$$

Use the results of either Problem 4.5 or 4.66 for the case where the equations are written in terms of θ and x_2.

10. Write the equations of motion of the system shown in Figure P4.3, using d'Alembert's principle. Write the equations in terms of θ_1 and θ_2.

11. Write the equations of motion of the system shown in Figure P4.3, using d'Alembert's principle. Write the equations in terms of x_1 and x_2.

12. Write the equations of motion of the system shown in Figure P4.3, using Lagrange's equation. Write

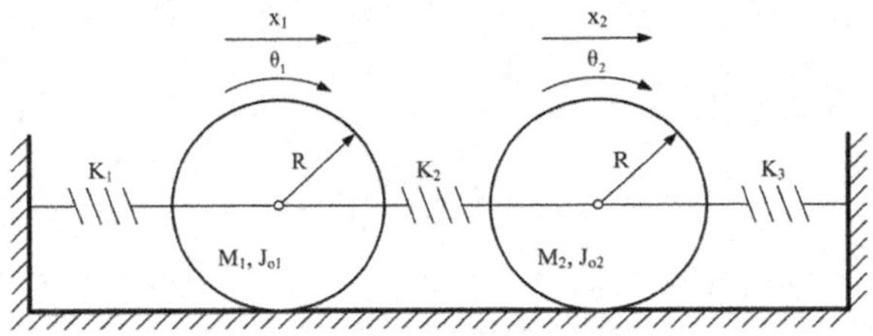

Figure P4.3

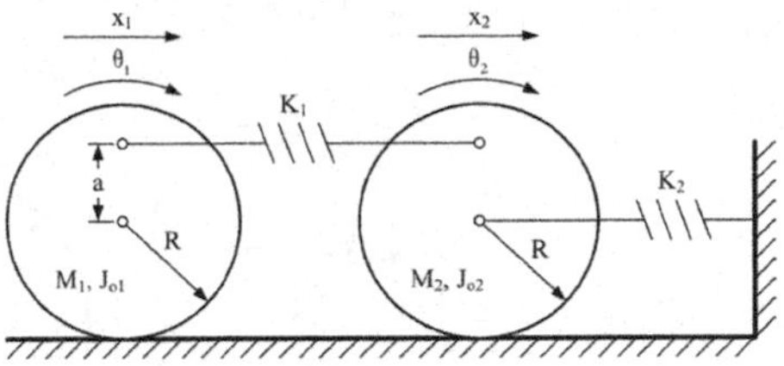

Figure P4.4

the equations in terms of θ_1 and θ_2.

13. Write the equations of motion of the system shown in Figure P4.3, using Lagrange's equation. Write the equations in terms of x_1 and x_2.

14. Write the expressions for the resonance frequencies and corresponding mode vectorss for the system shown in Figure P4.3, assuming that:

$$K_1 = K_2 = K_3 = K;\ M_1 = M_2 = M; \text{ and}$$
$$J_{o1} = J_{02} = 0.5MR^2.$$

Use the results of either Problem 4.10 or 4.12 for the case where the equations are written in terms of θ_1 and θ_2.

15. Write the expressions for the resonance frequencies and corresponding mode vectorss for the system shown in Figure P4.3, assuming that:

$$K_1 = K_2 = K_3 = K;\ M_1 = M_2 = M; \text{ and}$$
$$J_{o1} = J_{02} = 0.5MR^2.$$

Use the results of either Problem 4.11 or 4.13 for the case where the equations are written in terms of x_1 and x_2.

16. Write the equations of motion of the system shown in Figure P4.4, using d'Alembert's principle. Write the equations in terms of θ_1 and θ_2.

17. Write the equations of motion of the system shown in Figure P4.4, using d'Alembert's principle. Write the equations in terms of x_1 and x_2.

18. Write the equations of motion of the system shown in Figure P4.4, using Lagrange's equation. Write the equations in terms of θ_1 and θ_2.

19. Write the equations of motion of the system shown in Figure P4.4, using Lagrange's equation. Write the equations in terms of x_1 and x_2.

20. Write the expressions for the resonance frequencies and corresponding mode vectors for the system shown in Figure P4.4, assuming that:

$$K_1 = K_2 = K;\ M_1 = M_2 = M;\ a = 0.5R;$$
$$J_{o1} = J_{02} = 0.5MR^2; \text{ and } a = 0.5R.$$

Use the results of either Problem 4.16 or 4.18 for the case where the equations are written in terms of θ_1 and θ_2.

21. Write the expressions for the resonance frequencies and corresponding mode vectors for the system shown in Figure P4.4, assuming that:

$$K_1 = K_2 = K;\ M_1 = M_2 = M;\ a = 0.5R;$$
$$J_{o1} = J_{02} = 0.5MR^2; \text{ and } a = 0.5R.$$

Use the results of either Problem 4.17 or 4,19 for the case where the equations are written in terms of x_1 and x_2.

22. Write the equations of motion of the system shown in Figure P4.5, using d'Alembert's principle. Write the equations in terms of θ_1 and y.

23. Write the equations of motion of the system shown in Figure P4.5, using d'Alembert's principle. Write the equations in terms of x and y.

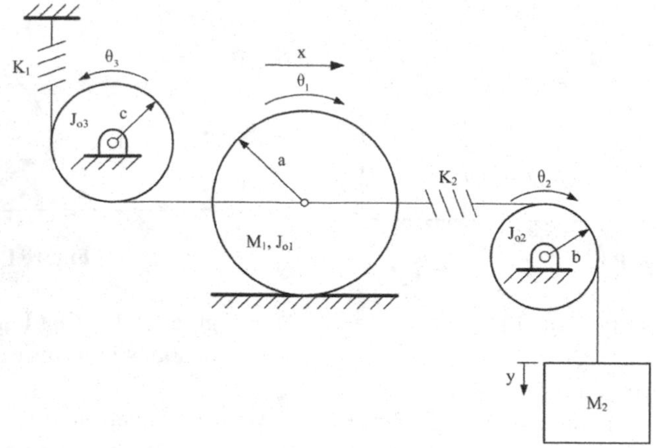

Figure P4.5

24. Write the equations of motion of the system shown in Figure P4.5, using Lagrange's equation. Write the equations in terms of θ_1 and y.

25. Write the equations of motion of the system shown in Figure P4.5, using Lagrange's equation. Write the equations in terms of x and y.

26. Write the expressions for the resonance frequencies and corresponding mode vectors for the system shown in Figure P4.5, assuming that:

$$K_1 = K_2 = K;\ M_1 = M_2 = M;\ J_{o1} = 0.5\ M a^2;$$
$$J_{o2} = J_{o3} = 0.5 J_{01};\ \text{and } b = c = 0.5 a.$$

Use the results of either Problem 4.22 or 4.24 for the case where the equations are written in terms of θ_1 and y.

27. Write the expressions for the resonance frequencies and corresponding mode vectors for the system shown in Figure P4.5, assuming that:

$$K_1 = K_2 = K;\ M_1 = M_2 = M;\ J_{o1} = 0.5\ M a^2;$$
$$J_{o2} = J_{o3} = 0.5 J_{01};\ \text{and } b = c = 0.5 a.$$

Use the results of either Problem 4.23 or 4.25 for the case where the equations are written in terms of x and y.

28. Write the equations of motion of the system shown in Figure P4.6, using d'Alembert's principle. Write the equations in terms of y and θ.

29. Write the equations of motion of the system shown in Figure P4.6, using Lagrange's equation. Write the equations in terms of y and θ.

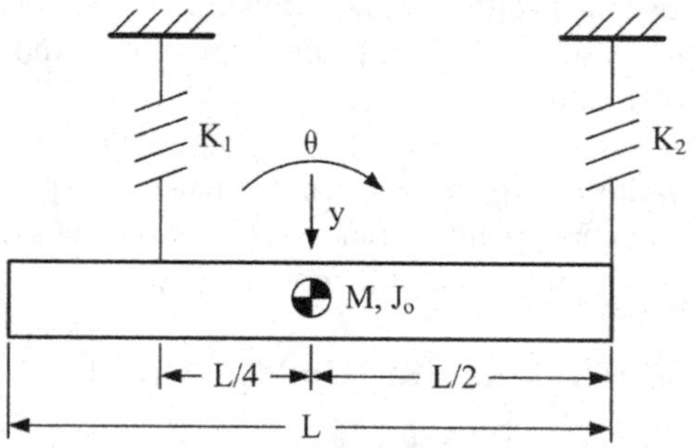

Figure P4.6

30. Write the equations of motion of the system shown in Figure P4.7, using d'Alembert's principle. Write the equations in terms of θ_1 and θ_2.

31. Write the equations of motion of the system shown in Figure P4.7, using Lagrange's equation. Write the equations in terms of θ_1 and θ_2.

32. Write the expressions for the resonance frequencies and corresponding mode vectors for the system shown in Figure P4.7. Use the results of either Problem 4.30 or 4.31.

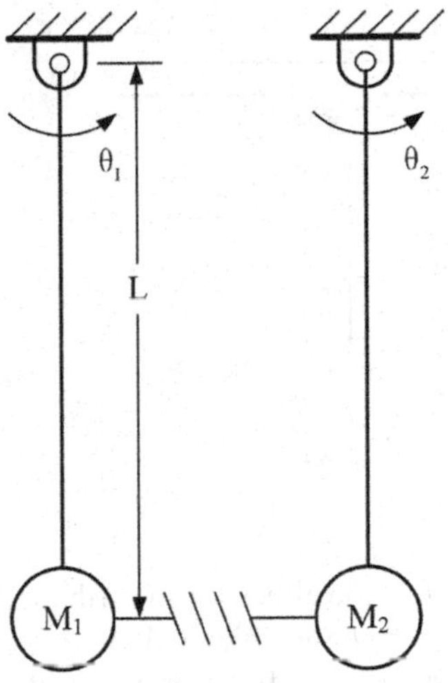

Figure P4.7

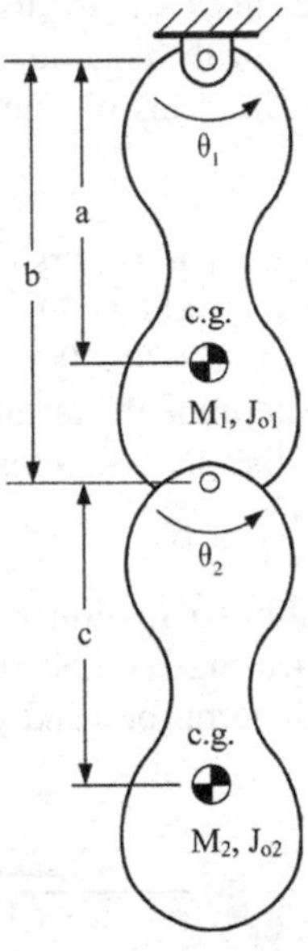

Figure P4.8

33. Write the equations of motion of the system shown in Figure P4.8, using d'Alembert's principle. Write the equations in terms of θ_1 and θ_2.

34. Write the equations of motion of the system shown in Figure P4.8, using Lagrange's equation. Write the equations in terms of θ_1 and θ_2.

35. Write the equations of motion of the system shown

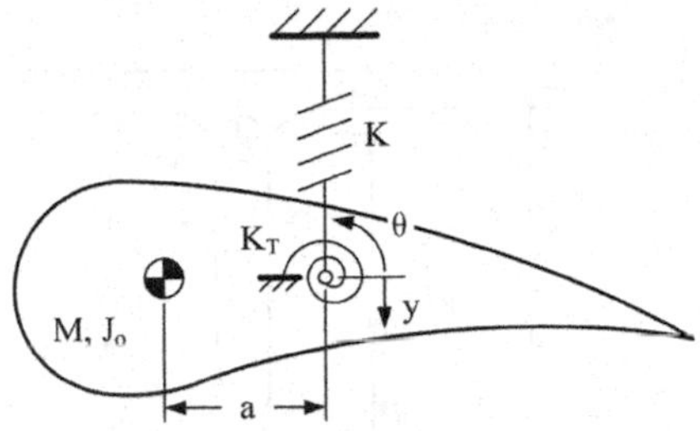

Figure P4.9

in Figure P4.9, using d'Alembert's principle. Write the equations in terms of y and θ.

36. Write the equations of motion of the system shown in Figure P4.9, using Lagrange's equation. Write the equations in terms of y and θ.

37. Write the equations of motion of the system shown in Figure P4.10, using d'Alembert's principle. Write the equations in terms of y and θ.

38. Write the equations of motion of the system shown in Figure P4.10, using Lagrange's equation. Write the equations in terms of y and θ.

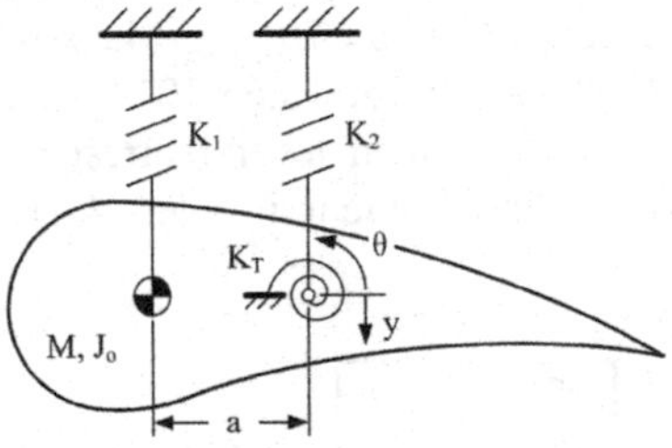

Figure P4.10

39. Write the equations of motion of the system shown in Figure P4.11, using d'Alembert's principle. Write the equations in terms of y and θ.

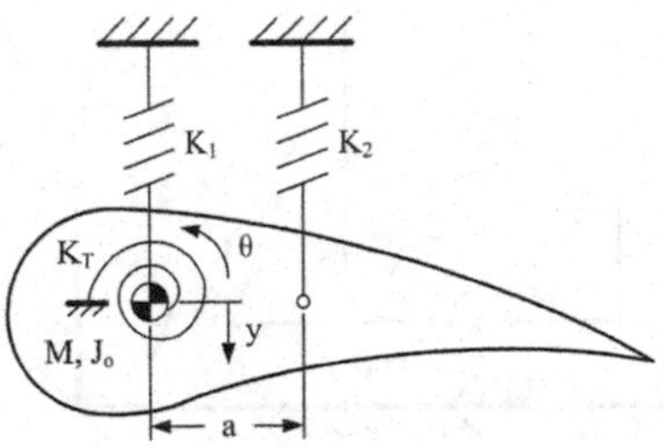

Figure P4.11

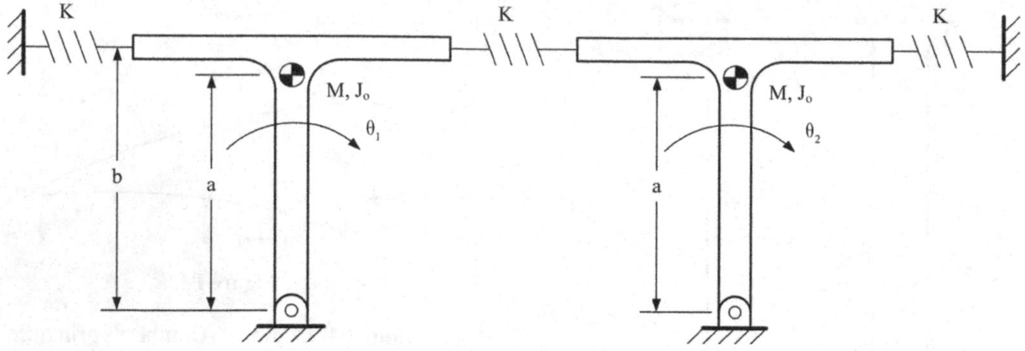

Figure P4.12

40. Write the equations of motion of the system shown in Figure P4.11, using Lagrange's equation. Write the equations in terms of y and θ.

41. A bridge consisting of two rigid sections attached together and to ground by resilient coupling elements of stiffness K can be modeled as shown in Figure P4.12. Determine the equations of motion of the bridge using d'Alembert's principle in terms of θ_1 and θ_2.

42. The values of the parameters of the bridge in Figure P4.12 are: a = 60 ft, b = 70 ft, $J_o = 4.0 \times 10^6$ lb_f-s^2-ft, M = 30,000 lb_f, and K = 2.6×10^5 lb_f/ft. Determine the resonance frequencies and corresponding mode vectors for the bridge in Problem 4.41.

43. To simulate the effects of an earthquake on a rigid building, the base of the building is assumed to be connected to ground through a translational resilient element K and a torsional resilient element K_T. If the ground is assumed to have a translational motion of $x_1(t)$, determine the equations of motion of the building using in terms of y and θ d'Alembert's principle.

44. The values of the parameters of the building in Figure P4.13 are: a = 80 ft, M = 550,000 lb_f, J_o = 5.0×10^7 lb_f-s^2-ft, K = 80,000 lb_f/ft, and K_T = 5.2 x 10^7 lb_f-ft. Determine the resonance frequencies and corresponding mode vectos for the building in Problem 4.43.

45. Write the equations of motion of the system shown in Figure P4.14, using d'Alembert's principle. Write the equations in terms of θ and y_2.

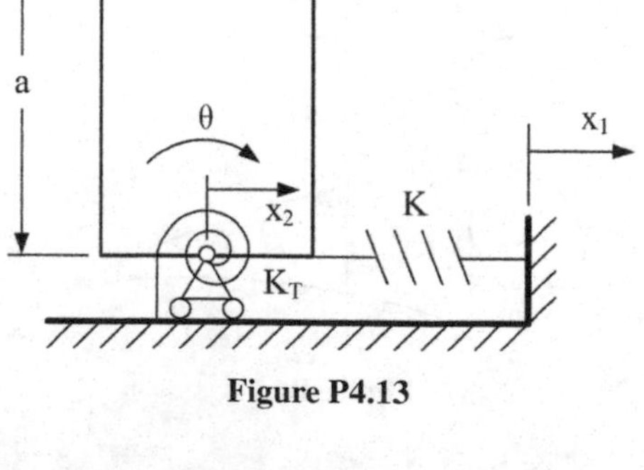

Figure P4.13

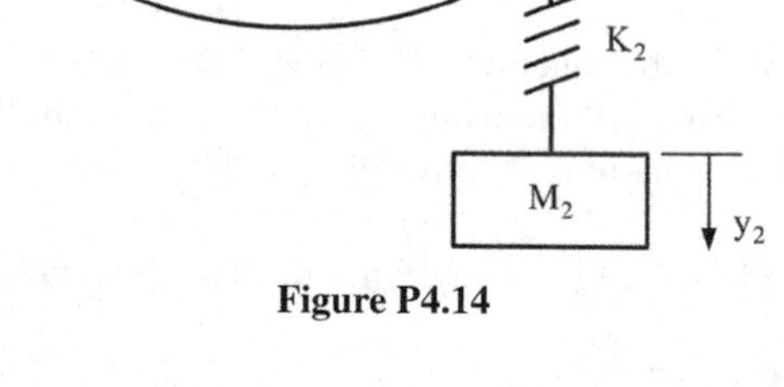

Figure P4.14

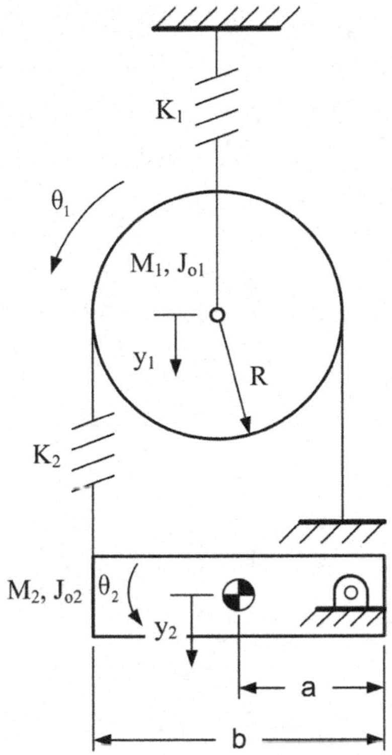

Figure P4.15

46. Write the equations of motion of the system shown in Figure P4.14, using Lagrange's equation. Write the equations in terms of θ and y_2.

47. Write the equations of motion of the system shown in Figure P4.15, using d'Alembert's principle. Write the equations in terms of θ_1 and θ_2.

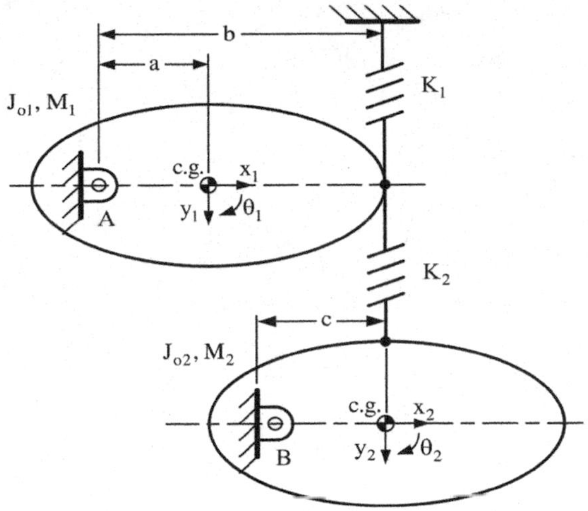

Figure P4.16

48. Write the equations of motion of the system shown in Figure P4.15, using Lagrange's equation. Write the equations in terms of θ_1 and θ_2.

49. Write the equations of motion of the system shown in Figure P4.16, using d'Alembert's principle. Write the equations in terms of θ_1 and θ_2.

50. Write the equations of motion of the system shown in Figure P4.16, using Lagrange's equation. Write the equations in terms of θ_1 and θ_2.

CHAPTER 5
n-DOF VIBRATION SYSTEMS MATRIX EQ'S AND SOLUTION METHODS

5.1 Introduction

Newton's method, d'Alembert's principle and Lagrange's equation were presented in Chapter 1 as methods used to develop the equations of motion for vibration systems. These methods were then used to develop the equations of motion for vibration systems that were examined in Chapters 2 through 4.

Concepts associated with virtual work, d'Alembert's principle, Hamilton's principle, Lagrange's equation and influence coefficients are presented in this chapter. Discussions associated with these methods include the use of relevant coordinate systems and how to combine equations of motion for vibration systems into related matrix equations. Solution methods are then presented for solving these matrix equations to obtain the resonance frequencies and mode vectors for a vibration system.

5.2 Work and Energy Relations

Figure 5.1 shows a mass element m that moves along a path C_1 from position 1 to position 2. Both *non-conservative* and *conservative forces* can act on this mass element. *Non-conservative forces* remove energy from or add energy to a system. For example, they can be damping and friction forces or external forces that act on system masses. Non-conservative forces often cannot be determined based on the positions of mass elements alone nor can they be determined from associated potential energy relationships. *Conservative forces* depend only on the positions of mass elements and can be derived from associated potential energy relationships.

The *incremental work* dW performed on the mass element in Figure 5.1 by a non-conservative *vector force* $\mathbf{F_{nc}}$ over an *incremental vector distance* $\mathbf{dr}$ is given by the dot product:

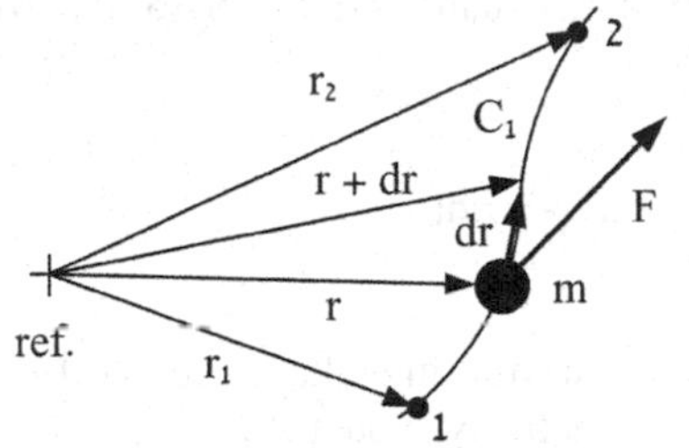

Figure 5.1 Mass Element Moving under the Action of a force **F**

$$dW = \mathbf{F_{nc}} \cdot \mathbf{dr}. \tag{5.1}$$

where the position of the mass element along the path C_1 relative to a reference position is designated by the position vector **r**. The dot product indicates only the component of the force vector that acts in the same direction as the corresponding displacement vector performs work on the mass element. The work performed on the mass element by $\mathbf{F_{nc}}$ as it moves from position 1 to position 2 along path C_1 is:

$$\boxed{W_{12} = \int_{t_1}^{t_2} \mathbf{F_{nc}} \cdot \mathbf{dr}} \tag{5.2}$$

where in Cartesian coordinates:

$$\boxed{\mathbf{F_{nc}} = F_x\,\mathbf{i} + F_y\,\mathbf{j} + F_z\,\mathbf{k}} \tag{5.3}$$

and:

$$\boxed{\mathbf{dr} = dx\,\mathbf{i} + dy\,\mathbf{j} + dz\,\mathbf{k}.} \tag{5.4}$$

i, **j** and **k** are *unit vectors* in the x, y and z directions, respectively.

If the mass of the mass element remains constant over the path from point 1 to 2, Newton's second law states:

$$\mathbf{F}_{nc} = m\frac{d\dot{\mathbf{r}}}{dt}. \tag{5.5}$$

Noting that:

$$\frac{d\mathbf{r}}{dt} = \dot{\mathbf{r}} \quad \text{or} \quad d\mathbf{r} = \dot{\mathbf{r}}\,dt \tag{5.6}$$

and substituting equations (5.5) and (5.6) into equation (5.2) yields:

$$W_{12} = \int_{t_1}^{t_2} m\frac{d\dot{\mathbf{r}}}{dt} \bullet \dot{\mathbf{r}}\,dt \tag{5.7}$$

where t_1 and t_2 are the times that correspond to positions r_1 and r_2, respectively. Note that:

$$\frac{d(\dot{\mathbf{r}} \bullet \dot{\mathbf{r}})}{dt} = \frac{d\dot{\mathbf{r}}}{dt} \bullet \dot{\mathbf{r}} + \dot{\mathbf{r}} \bullet \frac{d\dot{\mathbf{r}}}{dt} \tag{5.8}$$

or:

$$\frac{d\dot{\mathbf{r}}}{dt} \bullet \dot{\mathbf{r}} = \frac{1}{2}\frac{d(\dot{\mathbf{r}} \bullet \dot{\mathbf{r}})}{dt}. \tag{5.9}$$

Therefore, equation (5.7) can be written:

$$W_{12nc} = \frac{1}{2} m \int_{t_1}^{t_2} \frac{d(\dot{\mathbf{r}} \bullet \dot{\mathbf{r}})}{dt} dt \tag{5.10}$$

or:

$$W_{12nc} = \frac{1}{2} m \int_{\dot{r}_1}^{\dot{r}_2} d(\dot{\mathbf{r}} \bullet \dot{\mathbf{r}}) \tag{5.11}$$

where $\dot{r}_1$ and $\dot{r}_2$ are the velocities at positions 1 and 2, respectively. Evaluating the integral in equation (5.10) yields:

$$W_{12nc} = \frac{1}{2} m \left[\dot{\mathbf{r}} \bullet \dot{\mathbf{r}}\right]_{\dot{r}_1}^{\dot{r}_2} \tag{5.12}$$

or:

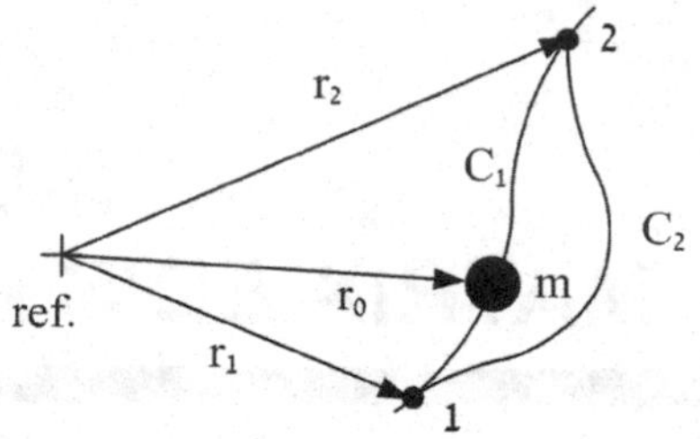

Figure 5.2 Different Paths for Mass Element from Position 1 to Position 2

$$W_{12nc} = \frac{1}{2} m \dot{\mathbf{r}}_2 \bullet \dot{\mathbf{r}}_2 - \frac{1}{2} m \dot{\mathbf{r}}_1 \bullet \dot{\mathbf{r}}_1. \tag{5.13}$$

The kinetic energy at any position can be written:

$$\boxed{T = \frac{1}{2} m \dot{\mathbf{r}} \bullet \dot{\mathbf{r}}} \tag{5.14}$$

Carrying out the dot product operations in equation (5.13) yields:

$$\boxed{W_{12nc} = \frac{1}{2} m \dot{r}_2^2 - \frac{1}{2} m \dot{r}_1^2.} \tag{5.15}$$

Noting equation (5.14), equation (5.15) can be written:

$$\boxed{W_{12nc} = T_2 - T_1.} \tag{5.16}$$

T_1 and T_2 are the *kinetic energies* at positions 1 and 2, respectively. Equation (5.16) indicates the work performed by a non-conservative force in moving mass element m in Figure 5.1 from position 1 to position 2 equals the change in kinetic energy of the mass element between positions 1 and 2. The incremental work dW_{nc} associated with a non-conservative force can be written:

$$\boxed{dW_{12nc} = dT.} \tag{5.17}$$

The work performed on the mass element by a conservative force $\mathbf{F_c}$ in moving it from position 1 to position 2 depends only on the position vectors $\mathbf{r}_1$ and $\mathbf{r}_2$ and is independent of the path taken by the mass element in moving from positions 1 to 2. Referring to Figure 5.2, the work W_{12c} that is performed in moving the mass element from position 1 to position 2 can be expressed:

$$W_{12c} = \left[\int_{r_1}^{r_2} \mathbf{F}_c \bullet \mathbf{dr}\right]_{\text{Path } C_1} = \left[\int_{r_1}^{r_2} \mathbf{F}_c \bullet \mathbf{dr}\right]_{\text{Path } C_2} . \tag{5.18}$$

A conservative force is normally referenced to a reference position $\mathbf{r}_0$. Therefore, equation (5.18) can be written:

$$W_{12c} = \int_{r_1}^{r_0} \mathbf{F}_c \bullet \mathbf{dr} + \int_{r_0}^{r_2} \mathbf{F}_c \bullet \mathbf{dr} \tag{5.19}$$

or:

$$W_{12c} = -\int_{r_0}^{r_1} \mathbf{F}_c \bullet \mathbf{dr} + \int_{r_0}^{r2} \mathbf{F}_c \bullet \mathbf{dr}. \tag{5.20}$$

Define the *potential energy* U_r as the work that is performed by a conservative force in moving a mass element from a reference position r_0 to a position r. The potential energy can be expressed:

$$U_r = \int_{r_0}^{r} \mathbf{F}_c \bullet \mathbf{dr}. \tag{5.21}$$

Therefore, equation (5.20) becomes:

$$\boxed{W_{12c} = U_2 - U_1.} \tag{5.22}$$

Equation (5.22) indicates the work performed by a conservative force in moving the mass element m in Figure 5.2 from position 1 to position 2 is equal to the change in potential energy of the mass element between positions 1 and 2. The incremental work dW_c associated with a conservative force can be written:

$$\boxed{dW_{12c} = dU.} \tag{5.23}$$

Often both non-conservative and conservative forces act on the mass element in Figures 5.1 and 5.2 as it moves from position 1 to position 2. When this occurs, the total incremental work dW_{12} is:

$$dW_{12} = dW_{12nc} + dW_{12c}. \tag{5.24}$$

Define:

$$\boxed{W_{12} = E_2 - E_1 \quad \text{and} \quad dW_{12} = dE.} \tag{5.25}$$

Substituting equations (5.17) and (5.23) into equation (5.24), equation (5.25) can be written:

$$\boxed{dE = dT + dU.} \tag{5.26}$$

When dE, dT and dU are associated with the motion of the mass element from positions 1 to 2, equation (5.26) can be written:

$$(T_2 - T_1) + (U_2 - U_1) = E_2 - E_1 \tag{5.27}$$

or:

$$\boxed{(T_2 + U_2) - (T_1 + U_1) = E_2 - E_1.} \tag{5.28}$$

E denotes the *total energy* of a mass element at a particular position and is the sum of the kinetic and potential energies at that position. Equation (5.28) indicates the work performed by non-conservative forces acting on the mass element is responsible for the change in the total energy of the element when it moves from a position to another. If the non-conservative forces are equal to zero (energy is neither dissipated from or added to the mass element), equation (5.28) becomes:

$$\boxed{(T_2 + U_2) - (T_1 + U_1) = 0 \quad \text{or} \quad T + U = \text{constant.}} \tag{5.29}$$

5.3 Generalized Coordinates

In the preceding section, the motion of a mass element was designated by:

$$\mathbf{r} = x\,\mathbf{i} + y\,\mathbf{j} + z\,\mathbf{k}. \tag{5.30}$$

If there are n mass elements in a system, the position of the elements can be specified by:

$$\mathbf{r}_i = x_i\,\mathbf{i} + y_i\,\mathbf{j} + z_i\,\mathbf{k} \quad \text{where} \quad i = 1, 2, 3, \cdots, n. \tag{5.31}$$

The response of the system is specified when the motions of the mass elements are known as a function of time, or:

$$\mathbf{r}_i = \mathbf{r}_i(t) \tag{5.32}$$

where:

$$x_i = x_i(t), \quad y_i = y_i(t), \quad \text{and} \quad z_i = z_i(t). \tag{5.33}$$

There are situations where all of the Cartesian coordinates are not independent or when they are not the most convenient coordinates to use to describe the motion of the mass elements in a system. For these situations, another set of coordinates can be used to describe the motions of the mass elements. If the system has m degrees of freedom, a set of m *generalized coordinates*, denoted by q_1, q_2, ..., q_m, can be defined such that:

$$\boxed{\begin{aligned} x_1 &= x_1(q_1, q_2, \ldots, q_m) \\ y_1 &= y_1(q_1, q_2, \ldots, q_m) \\ z_1 &= z_1(q_1, q_2, \ldots, q_m) \\ x_2 &= x_2(q_1, q_2, \ldots, q_m) \\ &\vdots \\ z_n &= z_n(q_1, q_2, \ldots, q_m). \end{aligned}} \tag{5.34}$$

5.4 Principle of Virtual Work

The principle of virtual work addresses concepts that will be used in later developments. The *principle of virtual work* can be stated:

> *The total virtual work done by all of the resultant forces acting on a system associated with their respective virtual displacements is equal to zero.*

Therefore, the principle of virtual work applies to a system that is in static equilibrium.

Define δx_1, δy_1, δz_1, δx_2, ..., δz_n to be *imagined infinitesimal arbitrary changes* in the coordinates x_1, y_1, z_1, x_2, ..., z_n of a system. The symbol "δ" represents the *virtual displacement* of a system variable, whereas, the symbol "d" designates the differential of the system variable. Operations that involve δ follow the rules of elementary calculus.

The virtual displacements are not true displacements. They are infinitesimal changes in the system coordinates that arise from imagining the system in a slightly displaced position. Actual displacements occur over a finite period of time in which the system forces and constraint relations may change. However, virtual displacements occur without any change in time, and therefore, are independent of time. The system forces and constraint relations do not change over a virtual displacement.

If the system coordinates are related by the constraint equation:

$$f(x_1, \ldots, x_n, y_1, \ldots, y_n, z_1, \ldots, z_n, t) = C \tag{5.35}$$

then:

$$f\begin{pmatrix} x_1 + \delta x_1, \ldots, y_1 + \delta y_1, \ldots, z_1 + \\ \delta z_1, \ldots, z_n + \delta z_n, t \end{pmatrix} = C. \tag{5.36}$$

Expanding equation (5.36) in a Taylor series expansion and neglecting the higher order terms yields:

$$f(x_1, \ldots, x_n, y_1, \ldots, y_n, z_1, \ldots, z_n, t) + \sum_{i=1}^{n} \left(\frac{\partial f}{\partial x_i}\delta x_i + \frac{\partial f}{\partial y_i}\delta y_i + \frac{\partial f}{\partial z_i}\delta z_i \right) = C \tag{5.37}$$

or:

$$\boxed{\sum_{i=1}^{n} \left(\frac{\partial f}{\partial x_i}\delta x_i + \frac{\partial f}{\partial y_i}\delta y_i + \frac{\partial f}{\partial z_i}\delta z_i \right) = 0.} \tag{5.38}$$

EXAMPLE 5.1

Determine the constraint equation for the simple lever system show in Figure 5.3.

SOLUTION

The geometric constraint relations for the lever system are:

$$y_1 = -a\theta \tag{5.1a}$$

and:

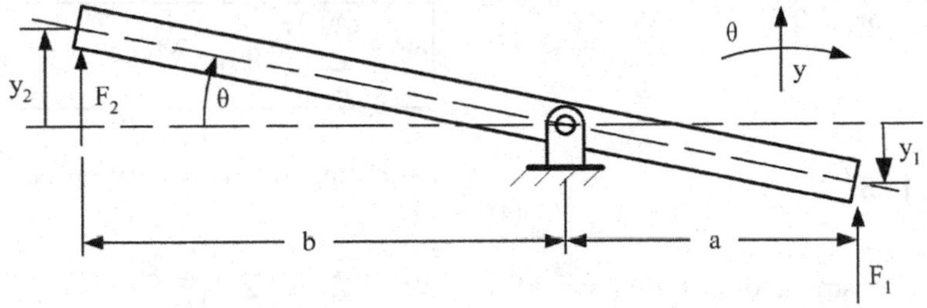

Figure 5.3 Free Body Diagram of a Lever

$$y_2 = b\theta. \quad (5.1b)$$

Substituting equation (5.1b) into equation (5.1a) to eliminate θ yields:

$$a\,y_2 + b\,y_1 = 0. \quad (5.1c)$$

From equation (5.35):

$$f(y_1, y_2) = a\,y_2 + b\,y_1 = 0 \quad (5.1d)$$

where C = 0. Relative to equation (5.36):

$$f(y_1 + \delta y_1, y_2 + \delta y_2) = a(y_2 + \delta y_2) + b(y_1 + \delta y_1) = 0. \quad (5.1e)$$

Since:

$$a\,y_2 + b\,y_1 = 0 \quad (5.1f)$$

then:

$$a\,\delta y_2 + b\,\delta y_1 = 0. \quad (5.1g)$$

EXAMPLE 5.2

Use equation 5.38 to derive equation (5.1g).

SOLUTION

Because the constraint equation is only a function of y, equation (5.38) reduces to:

$$\sum_{i=1}^{2} \frac{\partial f}{\partial y_i} \delta y_i = 0. \quad (5.2a)$$

Substituting equation (5.1d) into equation (5.2a) and carrying out the prescribed operations yields:

$$\boxed{a\,\delta y_2 + b\,\delta y_1 = 0.} \quad (5.2b)$$

When a system is in static equilibrium, the resultant force vector associated with each mass element equals zero. For this case:

$$\mathbf{R}_i = 0 \quad (5.39)$$

where:

$$\mathbf{R}_i = R_x\,\mathbf{i} + R_y\,\mathbf{j} + R_z\,\mathbf{k}. \quad (5.40)$$

$\mathbf{R}_i$ is comprised of two components and can be written:

$$\mathbf{R}_i = \mathbf{F}_i + \mathbf{f}_i. \quad (5.41)$$

$\mathbf{F}_i$ is the resultant of the external forces acting on mass element i, and $\mathbf{f}_i$ is the resultant of the constraint forces acting on mass element i.

If the resultant force vector $\mathbf{R}_i$ acting on a mass element equals zero, the virtual work δW_i associated with the mass element that is performed over the virtual displacement $\delta \mathbf{r}_i$ must equal zero. Therefore:

$$\delta W_i = \mathbf{R}_i \cdot \delta \mathbf{r}_i = 0 \quad (5.42)$$

or:

$$\delta W_i = \mathbf{F}_i \bullet \delta\mathbf{r}_i + \mathbf{f}_i \bullet \delta\mathbf{r}_i = 0 \tag{5.43}$$

where:

$$\delta\mathbf{r}_i = \delta x_i\ \mathbf{i} + \delta y_i\ \mathbf{j} + \delta z_i\ \mathbf{k}\,. \tag{5.44}$$

Normally, the system constraint forces $\mathbf{f}_i$ are assumed to perform zero work. This is because the constraint forces usually do not have force components in the directions of their related displacements. Therefore:

$$\mathbf{f}_i \bullet \delta\mathbf{r}_i = 0 \tag{5.45}$$

and:

$$\delta W_i = \mathbf{F}_i \bullet \delta\mathbf{r}_i = 0. \tag{5.46}$$

If a system has n mass elements:

$$\boxed{\delta W = \sum_{i=1}^{n} \mathbf{F}_i \bullet \delta\mathbf{r}_i = 0.} \tag{5.47}$$

Equation (5.47) is the *equation representation of the principle of virtual work*. Carrying out the dot product in equation (5.47) yields:

$$\delta W = \sum_{i=1}^{n} \left(F_{x(i)}\ \delta x_i + F_{y(i)}\ \delta y_i + F_{z(i)}\ \delta z_i \right) = 0. \tag{5.48}$$

If all of the virtual displacements $\delta\mathbf{r}_i$ are not independent, equation (5.47) cannot be used to derive the equilibrium equations of a system. When this is the case, the virtual displacement $\delta\mathbf{r}_i$ must be expressed in terms of the generalized coordinates of the system. From equation (5.34):

$$\mathbf{r}_i = \mathbf{r}_i\left(q_1, q_2, \ldots, q_m\right) \quad \text{where} \quad i = 1, 2, 3, \cdots, n \tag{5.49}$$

where $q_1, q_2, q_3, \ldots, q_m$ are the independent generalized coordinates. The virtual displacements $\delta\mathbf{r}_i$ are obtained from:

$$\boxed{\delta\mathbf{r}_i = \sum_{j=1}^{m} \frac{\partial \mathbf{r}_i}{\partial q_j}\delta q_j \quad \text{where} \quad i = 1, 2, 3, \cdots, n\,.} \tag{5.50}$$

Substituting equation (5.50) into equation (5.47) yields:

$$\delta W = \sum_{i=1}^{n} \mathbf{F}_i \bullet \sum_{j=1}^{m} \frac{\partial \mathbf{r}_1}{\partial q_j}\ \delta q_j = 0 \tag{5.51}$$

or:

$$\delta W = \sum_{j=1}^{m} \left[\sum_{i=1}^{n} \mathbf{F}_i \bullet \frac{\partial \mathbf{r}_i}{\partial q_j} \right] \delta q_j = 0. \tag{5.52}$$

Equation (5.52) can be written:

$$\boxed{\delta W = \sum_{j=1}^{m} Q_j\ \delta q_j = 0\,.} \tag{5.53}$$

where:

$$\boxed{Q_j = \sum_{i=1}^{n} \mathbf{F}_i \bullet \frac{\partial \mathbf{r}_i}{\partial q_j}\,.} \tag{5.54}$$

Q_j are referred to as *generalized forces*. Because the virtual displacements δq_i are independent and arbitrary, equation (5.53) can only be satisfied when:

$$Q_j = 0 \quad \text{where} \quad j = 1, 2, 3, \cdots, m\,. \tag{5.55}$$

Equation (5.55) are the equilibrium equations.

EXAMPLE 5.3

Use the principle of virtual work to develop the static force equation for the system shown in Figure 5.3.

SOLUTION

Since the forces only act in the y direction, equation (5.48) reduces to:

$$\sum_{i=1}^{2} F_{y(i)}\,\delta y_i = 0 \tag{5.3a}$$

Expanding equation (5.3a) yields:

$$F_1\,\delta y_1 + F_2\,\delta y_2 = 0 \tag{5.3b}$$

Solving equation (5.2b) for δy_2 in terms of δy_1 gives:

$$\delta y_2 = -\frac{b}{a}\,\delta y_1 \tag{5.3c}$$

Substituting equation (5.3c) into equation (5.3b) and rearranging the terms yields:

$$\left(a\,F_1 - b\,F_2\right)\delta y_1 = 0 \tag{5.3d}$$

Since δy_1 is arbitrary:

$$a\,F_1 = b\,F_2 \tag{5.3e}$$

The virtual displacements must obey the system geometric constraint equations. Therefore, from equations (5.1a) and (5.1b):

$$\delta y_1 = -a\,\delta\theta \tag{5.3f}$$

$$\delta y_2 = b\,\delta\theta \tag{5.3g}$$

Substituting equations (5.3f) and (5.3g) into equation (5.3b) yields:

$$\left(b\,F_2 - a\,F_1\right)\delta\theta \tag{5.3h}$$

or:

$$a\,F_1 = b\,F_2 \tag{5.3i}$$

Equations (5.3e) and (5.3i) are the same.

If a system has both translational and rotational motion, equation (5.47) is written:

$$\boxed{\delta W = \sum_{i=1}^{n}\left(\mathbf{F}_i \bullet \delta\mathbf{r}_i + \mathbf{M}_i \bullet \delta\boldsymbol{\theta}_i\right) = 0} \tag{5.56}$$

where $\mathbf{F}_i$ and $\delta\mathbf{r}_i$ are given by equations (5.3) and (5.4) and:

$$\boxed{\mathbf{M}_i = M_{xy(i)}\,\mathbf{i} + M_{xz(i)}\,\mathbf{j} + M_{yz(i)}\,\mathbf{k}} \tag{5.57}$$

and:

$$\boxed{\delta\theta_i = \delta\theta_{xy(i)}\,\mathbf{i} + \delta\theta_{xz(i)}\,\mathbf{j} + \delta\theta_{yz(i)}\,\mathbf{k}.} \tag{5.58}$$

EXAMPLE 5.4

Use the principle of virtual work to develop the static system equation for the crane shown in Figure 5.4(a).

SOLUTION

The two force vectors are [Figure 5.4(b)]:

$$\mathbf{F}_1 = -T\cos\alpha\,\mathbf{i} - T\sin\alpha\,\mathbf{j} \tag{5.4a}$$

and:

$$\mathbf{F}_2 = -mg\,\mathbf{j}. \tag{5.4b}$$

The corresponding displacement vectors associated with each force is:

$$\mathbf{r}_1 = \mathbf{r}_2 = r\cos\theta\,\mathbf{i} + r\sin\theta\,\mathbf{j}. \tag{5.4c}$$

The virtual displacement vectors δr_1 and δr_2 are obtained by taking the derivative of equation (5.4d) and then replacing the d's with δ's. Therefore:

$$\delta\mathbf{r}_1 = \delta\mathbf{r}_2 = -r\sin\theta\,\delta\theta\,\mathbf{i} + r\cos\theta\,\delta\theta\,\mathbf{j}. \tag{5.4d}$$

Substituting equations (5.4a), (5.4b), and (5.4d) into equation (5.47) for the case where n = 2 yields:

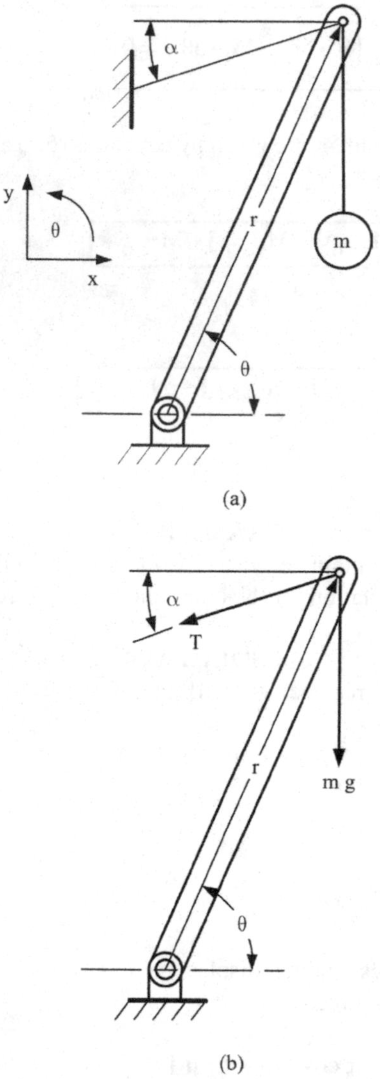

Figure 5.4 Free Body Diagram of a Simple Crane

$$\begin{pmatrix} -T\cos\alpha\,\mathbf{i}- \\ T\sin\alpha\,\mathbf{j} \end{pmatrix} \bullet \begin{pmatrix} -r\sin\theta\,\delta\theta\,\mathbf{i}+ \\ r\cos\theta\,\delta\theta\,\mathbf{j} \end{pmatrix} + (-mg\,\mathbf{j}) \bullet \begin{pmatrix} -r\sin\theta\,\delta\theta\,\mathbf{i}+ \\ r\cos\theta\,\delta\theta\,\mathbf{j} \end{pmatrix} = 0. \tag{5.4e}$$

Carrying out the dot products and simplifying gives:

$$\begin{bmatrix} T r\sin\theta\cos\alpha - T r\cos\theta\sin\alpha - \\ m g r\cos\theta \end{bmatrix} \delta\theta = 0. \tag{5.4f}$$

Since $\delta\theta$ is arbitrary:

$$T r\sin\theta\cos\alpha - T r\cos\theta\sin\alpha - m g r\cos\theta = 0 \tag{5.4g}$$

or:

$$\boxed{T = \frac{mg\cos\theta}{\sin(\theta-\alpha)}}. \tag{5.4h}$$

5.5 d'Alembert's Principle

d'Alembert's principle can be used to extend the discussion of virtual work to the analysis of dynamic systems. *d'Alembert's principle* states:

Every state of motion may be considered at any instant in time to be in a state of equilibrium if the inertia forces are taken into account.

From Newton's second law, the resultant force vector acting on a mass element is equal to the time rate of change of momentum. If mass is assumed to be constrained, Newton's second law can be written:

$$\mathbf{F}_i + \mathbf{f}_i = m_i \frac{d\mathbf{v}_i}{dt} \quad \text{or} \quad \mathbf{F}_i + \mathbf{f}_i = m_i\,\dot{\mathbf{v}}_i \tag{5.59}$$

where $\mathbf{F}_i$ and $\mathbf{f}_i$ are as previously defined, m_i is mass element i, and $\mathbf{v}_i$ is the vector velocity of mass element i. Equation (5.59) can also be written in terms of the displacement vector $\mathbf{r}_i$:

$$\mathbf{F}_i + \mathbf{f}_i = m_i \frac{d^2\mathbf{r}_i}{dt^2} \quad \text{or} \quad \mathbf{F}_i + \mathbf{f}_i = m_i\,\ddot{\mathbf{r}}_i. \tag{5.60}$$

Applying d'Alembert's principle, equation (5.60) can be written:

$$\mathbf{F}_i + \mathbf{f}_i - m_i\,\ddot{\mathbf{r}}_i = 0. \tag{5.61}$$

Using the principle of virtual work, equation (5.61) can be substituted into equation (5.47), or:

$$\sum_{i=1}^{n}\left(\mathbf{F}_i+\mathbf{f}_i-m_i\,\ddot{\mathbf{r}}_i\right)\cdot\delta\mathbf{r}_i=0. \tag{5.62}$$

As was the case in the discussion on virtual work:

$$\mathbf{f}_i\cdot\delta\mathbf{r}=0. \tag{5.63}$$

Therefore, equation (5.62) becomes:

$$\sum_{i=1}^{n}\left(\mathbf{F}_i-m_i\,\ddot{\mathbf{r}}_i\right)\cdot\delta\mathbf{r}_i=0. \tag{5.64}$$

Carrying out the dot product, equation (5.64) can be written:

$$\sum_{i=1}^{n}\left[\begin{array}{l}\left(F_{x(i)}-m_i\,\ddot{x}_i\right)\delta x_i+\left(F_{y(i)}-m_i\,\ddot{y}_i\right)\delta y_i+\\ \left(F_{z(i)}-m_i\,\ddot{z}_i\right)\delta z_i\end{array}\right]=0. \tag{5.65}$$

If a system has both translational and rotational motion, equation (5.63) is written:

$$\sum_{i=1}^{n}\left[\left(\vec{F}_{r(i)}-m_i\,\ddot{\vec{r}}_i\right)\cdot\delta\vec{r}_i+\left(\vec{M}_{r(i)}-J_{o(i)}\,\ddot{\theta}_i\right)\cdot\delta\theta_i\right]=0. \tag{5.66}$$

$J_{o(i)}$ is the mass moment of inertia of mass element i. $J_{o(i)}$ can have different values in the xy, xz and yz planes.

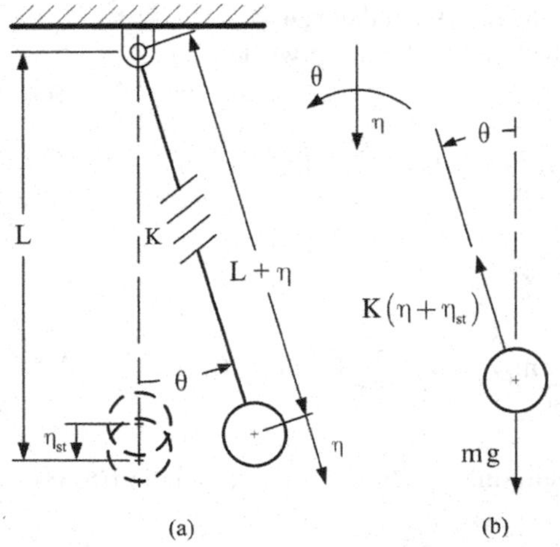

Figure 5.5 2-DOF Pendulum System with a Spring Element

EXAMPLE 5.5

Use d'Alembert's principle to develop the equations of motion of the vibration system shown in Figure 5.5(a).

SOLUTION

For this problem, develop the equations of motion in terms of the forces acting on the pendulum mass in the x and y directions [Figure 5.5(b)]. For this case, equation (5.65) reduces to:

$$\left(F_x-m\ddot{x}\right)\delta x+\left(F_y-m\ddot{y}\right)\delta y=0. \tag{5.5a}$$

F_x and F_y are external forces and are given by:

$$F_x=-K\left(\eta+\eta_{st}\right)\sin\theta \tag{5.5b}$$

and:

$$F_y=-K\left(\eta+\eta_{st}\right)\cos\theta+mg. \tag{5.5c}$$

Note that η_{st} is the static deflection of the spring element due to the pendulum mass (when $\theta = 0$). The corresponding static force equation is:

$$mg=K\,\eta_{st}. \tag{5.5d}$$

The static terms cannot be neglected in this problem because the system is a pendulum.

The geometric constraint equations that relate x, y and θ are:

$$x=\left(L+\eta\right)\sin\theta \tag{5.5e}$$

and:

$$y=\left(L+\eta\right)\cos\theta. \tag{5.5f}$$

The virtual displacements δx and δy are:

$$\delta x = \frac{\partial x}{\partial \eta}\delta\eta + \frac{\partial x}{\partial \theta}\delta\theta = \sin\theta\,\delta\eta + (L+\eta)\cos\theta\,\delta\theta \tag{5.5g}$$

and:

$$\delta y = \frac{\partial y}{\partial \eta}\delta\eta + \frac{\partial y}{\partial \theta}\delta\theta = \cos\theta\,\delta\eta - (L+\eta)\sin\theta\,\delta\theta \tag{5.5h}$$

The velocities $\dot{x}$ and $\dot{y}$ are:

$$\dot{x} = (L+\eta)\cos\theta\,\dot{\theta} + \dot{\eta}\sin\theta \tag{5.5i}$$

and:

$$\dot{y} = -(L+\eta)\sin\theta\,\dot{\theta} + \dot{\eta}\cos\theta. \tag{5.5j}$$

The accelerations $\ddot{x}$ and $\ddot{y}$ are:

$$\ddot{x} = (L+\eta)\cos\theta\,\ddot{\theta} - (L+\eta)\sin\theta\,\dot{\theta}^2 + 2\dot{\eta}\,\dot{\theta}\cos\theta + \ddot{\eta}\sin\theta \tag{5.5k}$$

and:

$$\ddot{y} = -(L+\eta)\sin\theta\,\ddot{\theta} - (L+\eta)\cos\theta\,\dot{\theta}^2 - 2\dot{\eta}\,\dot{\theta}\sin\theta + \ddot{\eta}\cos\theta. \tag{5.5l}$$

Substituting equations (5.5b), (5.5c), (5.5g), (5.5h), (5.5k), and (5.5l) into equation (5.5a) yields:

$$\begin{bmatrix} -K(\eta+\eta_{st})\sin\theta - m(L+\eta)\cos\theta\,\ddot{\theta} + \\ m(L+\eta)\sin\theta\,\dot{\theta}^2 - 2m\,\dot{\eta}\,\dot{\theta}\cos\theta - m\,\ddot{\eta}\sin\theta \end{bmatrix} \times \left[\sin\theta\,\delta\eta + (L+\eta)\cos\theta\,\delta\theta\right] + \begin{bmatrix} -K(\eta+\eta_{st})\cos\theta + mg + m(L+\eta)\sin\theta\,\ddot{\theta} + \\ m(L+\eta)\cos\theta\,\dot{\theta}^2 + 2m\,\dot{\eta}\,\dot{\theta}\sin\theta - m\,\ddot{\eta}\cos\theta \end{bmatrix} \times \left[\cos\theta\,\delta\eta - (L+\eta)\sin\theta\,\delta\theta\right] = 0. \tag{5.5m}$$

Collecting the coefficients of $\delta\eta$ and $\delta\theta$ and noting that $K\eta_{st} = mg$, gives:

$$\begin{bmatrix} m(L+\eta)\dot{\theta}^2 - K\eta - mg(1-\cos\theta) - \\ M\ddot{\eta} \end{bmatrix}\delta\eta + \begin{bmatrix} m(L+\eta)^2\ddot{\theta} + 2m(L+\eta)\dot{\eta}\dot{\theta} + \\ mg(L+\eta)\sin\theta \end{bmatrix}\delta\theta = 0. \tag{5.5n}$$

Since the virtual displacements $\delta\eta$ and $\delta\theta$ are independent and arbitrary, equation (5.5n) is satisfied when:

$$\boxed{M\ddot{\eta} + K\eta - m(L+\eta)\dot{\theta}^2 + mg(1-\cos\theta) = 0} \tag{5.5o}$$

and:

$$\boxed{m(L+\eta)^2\ddot{\theta} + 2m(L+\eta)\dot{\eta}\dot{\theta} + mg(L+\eta)\sin\theta = 0.} \tag{5.5p}$$

5.6 Hamilton's Principle

d'Alembert's principle can be used to formulate Hamilton's principle. *Hamilton's principle* is a variational method in the area of mechanics. It is an integral method that considers the entire motion of a system between two instances in time, which in the following development will be designated t_1 and t_2. Hamilton's principle is invariant with respect to the coordinate system that is used.

Equation (5.64) can be written:

$$\sum_{i=1}^{n}\left(m_i\,\ddot{\mathbf{r}}_i - \mathbf{F}_i\right)\cdot\delta\mathbf{r}_i = 0 \tag{5.67}$$

or:

$$\sum_{i=1}^{n} m_i\,\ddot{\mathbf{r}}_i\cdot\delta\mathbf{r}_i - \sum_{i=1}^{n}\mathbf{F}_i\cdot\delta\mathbf{r}_i = 0. \tag{5.68}$$

Substituting equation (5.46) into equation (5.68) yields:

$$\sum_{i=1}^{n} m_i\,\ddot{\mathbf{r}}_i\cdot\delta\mathbf{r}_i - \delta W = 0. \tag{5.69}$$

Next, consider the identity:

$$\frac{d\left(\dot{\mathbf{r}}_i \bullet \delta \mathbf{r}_i\right)}{dt} = \ddot{\mathbf{r}}_i \bullet \delta \mathbf{r}_i + \dot{\mathbf{r}}_i \bullet \delta \dot{\mathbf{r}}_i. \tag{5.70}$$

Rearranging equation (5.70) yields:

$$\ddot{\mathbf{r}}_i \bullet \delta \mathbf{r}_i = \frac{d\left(\dot{\mathbf{r}}_i \bullet \delta \mathbf{r}_i\right)}{dt} - \dot{\mathbf{r}}_i \bullet \delta \dot{\mathbf{r}}_i. \tag{5.71}$$

Noting that:

$$d\left(\frac{1}{2}x^2\right) = x\,dx \tag{5.72}$$

the second term in equation (5.71) can be written:

$$\dot{\mathbf{r}}_i \bullet \delta \dot{\mathbf{r}}_i = \delta\left(\frac{1}{2}\dot{\mathbf{r}}_i \bullet \dot{\mathbf{r}}_i\right). \tag{5.73}$$

Therefore, equation (5.71) becomes:

$$\ddot{\mathbf{r}}_i \bullet \delta \mathbf{r}_i = \frac{d\left(\dot{\mathbf{r}}_i \bullet \delta \mathbf{r}_i\right)}{dt} - \delta\left(\frac{1}{2}\dot{\mathbf{r}}_i \bullet \dot{\mathbf{r}}_i\right). \tag{5.74}$$

Substituting equation (5.74) into equation (5.69) yields:

$$\sum_{i=1}^{n} m_i \frac{d\left(\dot{\mathbf{r}}_i \bullet \delta \mathbf{r}_i\right)}{dt} - \delta \sum_{i=1}^{n}\left(\frac{1}{2} m_i\, \dot{\mathbf{r}}_i \bullet \dot{\mathbf{r}}_i\right) - \delta W = 0. \tag{5.75}$$

Noting the kinetic energy T is:

$$T = \sum_{i=1}^{n}\left(\frac{1}{2} m_i\, \dot{\mathbf{r}}_i \bullet \dot{\mathbf{r}}_i\right) \tag{5.76}$$

and rearranging the terms in equation (5.75) yields:

$$\delta T + \delta W = \sum_{i=1}^{n} m_i \frac{d\left(\dot{\mathbf{r}}_i \bullet \delta \mathbf{r}_i\right)}{dt}. \tag{5.77}$$

Integrating equation (5.77) with respect to time between the values of t_1 and t_2 gives:

$$\int_{t_1}^{t_2} \delta(T+W)\,dt = \int_{t_1}^{t_2}\left[\sum_{i=1}^{n} m_i \frac{d\left(\dot{\mathbf{r}}_i \bullet \delta \mathbf{r}_i\right)}{dt}\right] dt. \tag{5.78}$$

The summation and integration operations on the right side of equation (5.78) are independent. Therefore, the integration operation can be taken inside of the summation operation to obtain:

$$\int_{t_1}^{t_2} \delta(T+W)\,dt = \sum_{i=1}^{n} \int_{t_1}^{t_2} m_i \frac{d\left(\dot{\mathbf{r}}_i \bullet \delta \mathbf{r}_i\right)}{dt}\,dt \tag{5.79}$$

or:

$$\int_{t_1}^{t_2} \delta(T+W)\,dt = \sum_{i=1}^{n} m_i \left[\dot{\mathbf{r}}_i \bullet \delta \mathbf{r}_i\right]_{t_1}^{t_2}. \tag{5.80}$$

The virtual displacement vector $\delta \mathbf{r}_i$ is arbitrary. With respect to the variational method associated with Hamilton's principle, the values of $\delta \mathbf{r}_i$ are selected such that they equal zero at t_1 and t_2. Therefore, equation (5.80) becomes:

$$\int_{t_1}^{t_2} \delta(T+W)\,dt = 0. \tag{5.81}$$

From equation (5.23), for a conservative system:

$$dW = dU. \tag{5.82}$$

Therefore:

$$\delta W = \delta U. \tag{5.83}$$

The variational operation implied by δ and the integration operation are independent. Therefore , the variational operation can be taken outside of the integral in equation (5.81). Substituting equation (5.83) into equation (5.81) and taking the variational operation outside of the integral yields:

$$\boxed{\delta \int_{t_1}^{t_2} (T + U)\, dt = 0.} \tag{5.84}$$

Equation (5.84) is the mathematical expression of Hamilton's principle.

EXAMPLE 5.6

Use Hamilton's principle to develop the equations of motion for the system shown in Figure 5.5.

SOLUTION

The kinetic energy of the pendulum mass is:

$$T = \frac{1}{2} m \left(\dot{x}^2 + \dot{y}^2 \right). \tag{5.6a}$$

Substituting equation (5.5i) and (5.5j) into equation (5.6a) yields:

$$T = \frac{1}{2} m \begin{Bmatrix} \left[(L+\eta) \cos\theta\, \dot{\theta} + \dot{\eta} \sin\theta \right]^2 + \\ \left[-(L+\eta) \sin\theta\, \dot{\theta} + \dot{\eta} \cos\theta \right]^2 \end{Bmatrix}. \tag{5.6b}$$

Expanding the terms in the brackets in equation (5.6b) and simplifying gives:

$$T = \frac{1}{2} m \left[(L+\eta)^2 \dot{\theta}^2 + \dot{\eta}^2 \right]. \tag{5.6c}$$

The variation of T is:

$$\delta T = \frac{\partial T}{\partial \eta} \delta\eta + \frac{\partial T}{\partial \theta} \delta\theta + \frac{\partial T}{\partial \dot{\eta}} \delta\dot{\eta} + \frac{\partial T}{\partial \dot{\theta}} \delta\dot{\theta}. \tag{5.6d}$$

Substituting equation (5.6c) into equation (5.6d) and carrying out the prescribed operations yields:

$$\delta T = \frac{1}{2} m \begin{bmatrix} 2(L+\eta)\dot{\theta}^2\, \delta\eta + 2\dot{\eta}\, \delta\dot{\eta} + \\ (L+\eta)^2\, 2\dot{\theta}\, \delta\dot{\theta} \end{bmatrix}. \tag{5.6e}$$

The integral of δT with respect to t is:

$$\int_{t_1}^{t_2} \delta T\, dt = \frac{1}{2} m \begin{vmatrix} \int_{t_1}^{t_2} \left[2(L+\eta)\dot{\theta}^2\, \delta\eta \right] dt + \\ \int_{t_1}^{t_2} \left[2\dot{\eta}\, \delta\dot{\eta} \right] dt + \\ \int_{t_1}^{t_2} \left[(L+\eta)^2\, 2\dot{\theta}\, \delta\dot{\theta} \right] dt \end{vmatrix}. \tag{5.6f}$$

To simplify equation (5.6f), integrate the second and third terms on the right hand side of the equations by parts where:

$$\int u\, dv = uv - \int v\, du. \tag{5.6g}$$

Relative to the second term:

$$u = \dot{\eta} \quad \text{and} \quad du = \ddot{\eta}\, dt. \tag{5.6h}$$

$$dv = \delta\dot{\eta}\, dt = \frac{d(\delta\eta)}{dt} dt. \tag{5.6i}$$

$$dv = d(\delta\eta) \quad \text{or} \quad v = \delta\eta. \tag{5.6j}$$

Therefore:

$$2\int_{t_1}^{t_2} \dot{\eta}\, \delta\dot{\eta}\, dt = 2\left\{ \left[\dot{\eta}\, \delta\eta \right]_{t_1}^{t_2} - \int_{t_1}^{t_2} \ddot{\eta}\, \delta\eta\, dt \right\}. \tag{5.6k}$$

Because $\delta\eta$ is arbitrary and equals zero at t_1 and t_2:

$$\left[\dot{\eta}\, \delta\eta \right]_{t_1}^{t_2} = 0 \tag{5.6l}$$

and equation (5.6k) becomes:

$$2\int_{t_1}^{t_2} \dot{\eta}\, \delta\dot{\eta}\, dt = -2\int_{t_1}^{t_2} \ddot{\eta}\, \delta\eta\, dt. \tag{5.6m}$$

Relative to the third term:

$$u = (L+\eta)^2\, \dot{\theta} \quad \text{and}$$
$$du = (L+\eta)^2\, \ddot{\theta}\, dt + 2\dot{\theta}\, (L+\eta)\, \dot{\eta}\, dt. \tag{5.6n}$$

$$dv = \delta\dot{\theta}\,dt = \frac{d(\delta\theta)}{dt}\,dt \tag{5.6o}$$

and:

$$dv = d(\delta\theta) \quad \text{or} \quad v = \delta\theta\,. \tag{5.6p}$$

Therefore:

$$2\int_{t_1}^{t_2}\left[(L+\eta)^2\,\dot{\theta}\,\delta\dot{\theta}\right]dt = 2\left\{\begin{array}{l}\left[(L+\eta)^2\,\dot{\theta}\,\delta\theta\right]_{t_1}^{t_2} - \\ \int_{t_1}^{t_2}\left[\begin{array}{l}(L+\eta)^2\,\ddot{\theta} + \\ \quad 2(L+\eta)\,\dot{\eta}\dot{\theta}\,dt\end{array}\right]\delta\theta\,dt\end{array}\right\}. \tag{5.6q}$$

or:

$$2\int_{t_1}^{t_2}\left[(L+\eta)^2\,\dot{\theta}\,\delta\dot{\theta}\right]dt = -2\int_{t_1}^{t_2}\left[\begin{array}{l}(L+\eta)^2\,\ddot{\theta} + \\ \quad 2(L+\eta)\,\dot{\eta}\dot{\theta}\,dt\end{array}\right]\delta\theta\,dt\,. \tag{5.6r}$$

Substituting equations (5.6m) and (5.6r) into equation (5.6f) and collecting terms yields:

$$\int_{t_1}^{t_2}\delta T\,dt = \int_{t_1}^{t_2}\left\{\begin{array}{l}\left[m(L+\eta)\dot{\theta}^2 - m\ddot{\eta}\right]\delta\eta - \\ \left[\begin{array}{l}m(L+\eta)^2\,\ddot{\theta} + \\ \quad 2m(L+\eta)\dot{\eta}\dot{\theta}\end{array}\right]\delta\theta\end{array}\right\}dt\,. \tag{5.6s}$$

The potential energy is:

$$U = \frac{1}{2}K\eta^2 + mg(L+\eta)(1-\cos\theta)\,. \tag{5.6t}$$

The variation of the potential energy is:

$$\delta U = \frac{\partial U}{\partial \eta}\delta\eta + \frac{\partial U}{\partial \theta}\delta\theta\,. \tag{5.6u}$$

Substituting equation (5.6t) into equation (5.6u) and carrying out the prescribed operations yields:

$$\begin{array}{l}\delta U = K\eta\,\delta\eta + mg(1-\cos\theta)\,\delta\eta + \\ \quad mg(L+\eta)\sin\theta\,\delta\theta\,.\end{array} \tag{5.6v}$$

The integral of dU with respect to time is:

$$\int_{t_1}^{t_2}\delta U\,dt = \int_{t_1}^{t_2}\left\{\begin{array}{l}\left[K\eta + mg(1-\cos\theta)\right]\delta\eta + \\ \quad mg(L+\eta)\sin\theta\,\delta\theta\end{array}\right\}dt\,. \tag{5.6w}$$

Substituting equations (5.6s) and (5.6w) into equation (5.84) and collecting terms yields:

$$\delta\int_{t_1}^{t_2}(T+U)\,dt = \int_{t_1}^{t_2}\left\{\begin{array}{l}\left[\begin{array}{l}m(L+\eta)\dot{\theta}^2 - m\ddot{\eta} - K\eta - \\ \quad mg(1-\cos\theta)\end{array}\right]\delta\eta - \\ \left[\begin{array}{l}m(L+\eta)^2\,\ddot{\theta} + 2m(L+\eta)\dot{\eta}\dot{\theta} + \\ \quad mg(L+\eta)\sin\theta\end{array}\right]\delta\theta\end{array}\right\}dt\,. \tag{5.6x}$$

Since the variational displacements $\delta\eta$ and $\delta\theta$ are arbitrary and independent, equation (5.6x) is only satisfied when:

$$\boxed{m(L+\eta)\dot{\theta}^2 - m\ddot{\eta} - K\eta - mg(1-\cos\theta) = 0} \tag{5.6y}$$

and:

$$\boxed{\begin{array}{l}m(L+\eta)^2\,\ddot{\theta} + 2m(L+\eta)\dot{\eta}\dot{\theta} + \\ \quad mg(L+\eta)\sin\theta = 0\,.\end{array}} \tag{5.6z}$$

5.7 Lagrange's Equation

Hamilton's principle will be used to develop *Langrange's equation*. The following development applies only for holonomic systems. A *holonomic system* is:

A system in which the constraints are functions of

the system coordinates and, if applicable, of time. The constraint equations can be written in the form of equation (5.35).

A *non-holonomic system* has constraint equations that contain velocity expressions.

Let the positions of the mass elements of a system be specified by n independent variables $\mathbf{r}_i$ which are functions of n generalized coordinates q_i and time t. $\mathbf{r}_i$ can be expressed as:

$$\mathbf{r}_i = \mathbf{r}_i\left(q_1, q_2, \ldots, q_m, t\right) \quad \text{where} \quad i = 1, 2, 3, \cdots, n. \tag{5.85}$$

Note that with m generalized coordinates this system has m degrees of freedom. The related velocities are:

$$\dot{\mathbf{r}}_i = \frac{d\mathbf{r}_i}{dt} = \frac{\partial \mathbf{r}_i}{\partial q_1}\dot{q}_1 + \frac{\partial \mathbf{r}_i}{\partial q_2}\dot{q}_2 + \cdots + \frac{\partial \mathbf{r}_i}{\partial q_n}\dot{q}_n + \frac{\partial \mathbf{r}_i}{\partial t}. \tag{5.86}$$

The kinetic energy from equation (5.14) is:

$$T = \frac{1}{2}\sum_{i=1}^{n} m_i\, \dot{\mathbf{r}}_i \bullet \dot{\mathbf{r}}_i. \tag{5.87}$$

Substituting equation (5.86) into equation (5.87) yields:

$$T = \frac{1}{2}\sum_{i=1}^{n} m_i \left[\begin{array}{l} \left(\begin{array}{l} \dfrac{\partial \mathbf{r}_i}{\partial q_1}\dot{q}_1 + \dfrac{\partial \mathbf{r}_i}{\partial q_2}\dot{q}_2 + \\ \cdots + \dfrac{\partial \mathbf{r}_i}{\partial q_n}\dot{q}_n + \dfrac{\partial \mathbf{r}_i}{\partial t} \end{array}\right) \bullet \\ \left(\begin{array}{l} \dfrac{\partial \mathbf{r}_i}{\partial q_1}\dot{q}_1 + \dfrac{\partial \mathbf{r}_i}{\partial q_2}\dot{q}_2 + \cdots \\ + \dfrac{\partial \mathbf{r}_i}{\partial q_n}\dot{q}_n + \dfrac{\partial \mathbf{r}_i}{\partial t} \end{array}\right) \end{array}\right] \tag{5.88}$$

or:

$$T = \frac{1}{2}\sum_{i=1}^{n} m_i \sum_{j=1}^{n}\sum_{k=1}^{n} \left[\begin{array}{c} \dfrac{\partial \mathbf{r}_i}{\partial q_j} \bullet \dfrac{\partial \mathbf{r}_i}{\partial q_k}\dot{q}_j \dot{q}_k + \\ 2\dfrac{\partial \mathbf{r}_i}{\partial t} \bullet \sum_{j=1}^{m} \dfrac{\partial \mathbf{r}_i}{\partial q_j}\dot{q}_j + \\ \dfrac{\partial \mathbf{r}_i}{\partial t} \bullet \dfrac{\partial \mathbf{r}_i}{\partial t} \end{array}\right]. \tag{5.89}$$

Equation (5.89) indicates the kinetic energy is a function of displacements, velocities and time. In the most general case:

$$T = T\left(q_1, q_2, \ldots, q_n, \dot{q}_1, \dot{q}_2, \ldots, \dot{q}_n, t\right). \tag{5.90}$$

The variation of T is:

$$\begin{aligned} \delta T = {} & \frac{\partial T}{\partial q_1}\delta q_1 + \frac{\partial T}{\partial q_2}\delta q_2 + \cdots + \frac{\partial T}{\partial q_n}\delta q_n + \\ & \frac{\partial T}{\partial \dot{q}_1}\delta \dot{q}_1 + \frac{\partial T}{\partial \dot{q}_2}\delta \dot{q}_2 + \cdots + \frac{\partial T}{\partial \dot{q}_n}\delta \dot{q}_n + \frac{\partial T}{\partial t}. \end{aligned} \tag{5.91}$$

Since t is not varied:

$$\frac{\partial T}{\partial t} = 0. \tag{5.92}$$

Therefore, equation (5.91) can be written:

$$\delta T = \sum_{j=1}^{n} \frac{\partial T}{\partial q_j}\delta q_j + \sum_{j=1}^{n} \frac{\partial T}{\partial \dot{q}_j}\delta \dot{q}_j. \tag{5.93}$$

Integrating equation (5.93) with respect to time from t_1 to t_2 yields:

$$\begin{aligned} \int_{t_1}^{t_2} \delta T\, dt &= \int_{t_1}^{t_2} \sum_{j=1}^{n} \frac{\partial T}{\partial q_j}\delta q_j\, dt + \int_{t_1}^{t_2} \sum_{j=1}^{n} \frac{\partial T}{\partial \dot{q}_j}\delta \dot{q}_j\, dt \\ &= \sum_{j=1}^{n} \int_{t_1}^{t_2} \frac{\partial T}{\partial q_j}\delta q_j\, dt + \sum_{j=1}^{n} \int_{t_1}^{t_2} \frac{\partial T}{\partial \dot{q}_j}\delta \dot{q}_j\, dt. \end{aligned} \tag{5.94}$$

With respect to the second integral in equation (5.94):

$$\delta\dot{q}_j = \delta\left(\frac{\partial q_j}{\partial t}\right) = \frac{\partial\,\delta q_j}{\partial t}. \tag{5.95}$$

The second integral can be written:

$$\int_{t_1}^{t_2}\frac{\partial T}{\partial \dot{q}_j}\delta\dot{q}_j\,dt = \int_{t_1}^{t_2}\frac{\partial T}{\partial \dot{q}_j}\frac{\partial\,\delta q_j}{\partial t}\,dt. \tag{5.96}$$

Evaluating equation (5.96) by means of integration by parts where [refer to equation (5.6g) in Example 5.6]:

$$u = \frac{\partial T}{\partial \dot{q}_j} \quad \text{and} \quad du = \frac{d}{dt}\left(\frac{\partial T}{\partial \dot{q}_j}\right)dt \tag{5.97}$$

and:

$$dv = \frac{d\,\delta q_j}{dt}\,dt \quad \text{and} \quad v = \delta q_j. \tag{5.98}$$

Therefore, equation (5.96) becomes:

$$\begin{aligned}\int_{t_1}^{t_2}\frac{\partial T}{\partial \dot{q}_j}\delta\dot{q}_j\,dt &= \int_{t_1}^{t_2}\frac{\partial T}{\partial \dot{q}_j}\frac{\partial\,\delta q_j}{\partial t}\,dt \\ &= \left[\frac{\partial T}{\partial \dot{q}_j}\delta q_j\right]_{t_1}^{t_2} - \int_{t_1}^{t_2}\delta q_j\frac{d}{dt}\left(\frac{\partial T}{\partial \dot{q}_j}\right)dt.\end{aligned} \tag{5.99}$$

δq_j equals zero at t_1 and t_2. Therefore, equation (5.99) becomes:

$$\int_{t_1}^{t_2}\frac{\partial T}{\partial \dot{q}_j}\delta\dot{q}_j\,dt = -\int_{t_1}^{t_2}\delta q_j\frac{d}{dt}\left(\frac{\partial T}{\partial \dot{q}_j}\right)dt. \tag{5.100}$$

Substituting equation (5.100) into equation (5.94) yields:

$$\int_{t_1}^{t_2}\delta T\,dt = \sum_{j=1}^{n}\int_{t_1}^{t_2}\frac{\partial T}{\partial q_j}\delta q_j\,dt - \sum_{j=1}^{n}\int_{t_1}^{t_2}\frac{d}{dt}\left(\frac{\partial T}{\partial \dot{q}_j}\right)\delta q_j\,dt \tag{5.101}$$

or:

$$\int_{t_1}^{t_2}\delta T\,dt = -\sum_{j=1}^{n}\int_{t_1}^{t_2}\left[\frac{d}{dt}\left(\frac{\partial T}{\partial \dot{q}_j}\right) - \frac{\partial T}{\partial q_j}\right]\delta q_j\,dt. \tag{5.102}$$

Next, examine the virtual work. From equation (5.47):

$$\delta W = \sum_{i=1}^{n}\mathbf{F}_i \bullet \delta\mathbf{r}_i. \tag{5.103}$$

Following the development of equations (5.50) through (5.54), equation (5.103) can be written:

$$\delta W = \sum_{j=1}^{n} Q_j\,\delta q_j \tag{5.104}$$

where:

$$Q_j = \sum_{i=1}^{n}\mathbf{F}_i \bullet \frac{\partial \mathbf{r}_i}{\partial q_j}. \tag{5.105}$$

Q_j is referred to as the *generalized force*. Equation (5.104) can be integrated with respect to time from t_1 to t_2 to get:

$$\begin{aligned}\int_{t_1}^{t_2}\delta W\,dt &= \int_{t_1}^{t_2}\sum_{j=1}^{n} Q_j\,\delta q_j\,dt \\ &= \sum_{j=1}^{n}\int_{t_1}^{t_2} Q_j\,\delta q_j\,dt.\end{aligned} \tag{5.106}$$

Substituting equations (5.102) and (5.106) into equation (5.81) yields:

$$\int_{t_1}^{t_2}\delta(T+W)\,dt = -\sum_{j=1}^{n}\int_{t_1}^{t_2}\left[\frac{d}{dt}\left(\frac{\partial T}{\partial \dot{q}_j}\right) - \frac{\partial T}{\partial q_j} - Q_j\right]\delta q_j\,dt = 0. \tag{5.107}$$

The values for δq_j are arbitrary and equal zero at t_1 and t_2. Therefore, equation (5.107) is only satisfied when:

$$\frac{d}{dt}\left(\frac{\partial T}{\partial \dot{q}_j}\right)-\frac{\partial T}{\partial q_j}-Q_j=0 \text{ where } j=1,2,3,\ldots,n. \tag{5.108}$$

The generalized *forces* Q_j are comprised of *conservative forces* $Q_{c(j)}$ and *non-conservative forces* $Q_{nc(j)}$. The conservative forces arise from potential energy relations, and the non-conservative forces arise from either dissipative or external forces. Therefore, Q_j can be written:

$$Q_j = Q_{c(j)} + Q_{nc(j)}. \tag{5.109}$$

For conservative forces, from equations (5.21) and (5.104):

$$\delta W_c = \sum_{j=1}^{n} Q_{c(j)}\,\delta q_j = -\delta U \tag{5.110}$$

where:

$$U = U(q_1, q_2, \ldots, q_n). \tag{5.111}$$

δU is given by:

$$\delta U = \sum_{j=1}^{n} \frac{\partial U}{\partial q_j}\,\delta q_j. \tag{5.112}$$

Substituting equation (5.112) into equation (5.110) yields:

$$\sum_{j=1}^{n} Q_{c(j)}\,\delta q_j = -\sum_{j=1}^{n} \frac{\partial U}{\partial q_j}\,\delta q_j. \tag{5.113}$$

Therefore:

$$Q_{c(j)} = -\frac{\partial U}{\partial q_j}. \tag{5.114}$$

Substituting equations (5.109) and (5.114) into equation (5.108) yields:

$$\boxed{\frac{d}{dt}\left(\frac{\partial T}{\partial \dot{q}_j}\right)-\frac{\partial T}{\partial q_j}+\frac{\partial U}{\partial q_j}=Q_{nc(j)}.} \tag{5.115}$$

Two types of non-conservative forces are considered in this section. They are forces associated with viscous damping and external forces. The non-conservative forces $Q_{ncd(j)}$ associated with damping are:

$$Q_{ncd(j)} = -\frac{\partial D}{\partial \dot{q}_j} \tag{5.116}$$

where D is the *power dissipation function*. The non-conservative force $Q_{nce(j)}$ associated with external forces is given by equation (5.54), or:

$$\boxed{Q_{nce(j)} = \sum_{i=1}^{n} \mathbf{F}_i \bullet \frac{\partial \mathbf{r}_i}{\partial q_j}.} \tag{5.117}$$

Substituting equations (5.116) and (5.117) into equation (5.115) yields:

$$\boxed{\frac{d}{dt}\left(\frac{\partial T}{\partial \dot{q}_j}\right)-\frac{\partial T}{\partial q_j}+\frac{\partial U}{\partial q_j}+\frac{\partial D}{\partial \dot{q}_j}=Q_{nce(j)}.} \tag{5.118}$$

Equations (5.115) and (5.118) are general forms of *Lagrange's equation.*

EXAMPLE 5.7

Use Lagrange's equation to develop the equations of motion of the vibration system shown in Figure 5.5(a).

SOLUTION

The two generalized coordinates for the vibration system are η and θ. Therefore, equation (5.115) can be written:

$$\frac{d}{dt}\left(\frac{\partial T}{\partial \dot{\eta}}\right)-\frac{\partial T}{\partial \eta}+\frac{\partial U}{\partial \eta}=0 \tag{5.7a}$$

and:

$$\frac{d}{dt}\left(\frac{\partial T}{\partial \dot\theta}\right)-\frac{\partial T}{\partial \theta}+\frac{\partial U}{\partial \theta}=0. \tag{5.7b}$$

From equation (5.6c) in Example 5.6, the kinetic energy expression is:

$$T=\frac{1}{2}m\left[(L+\eta)^2\dot\theta^2+\dot\eta^2\right]. \tag{5.7c}$$

From equation (5.6t) in Example 5.6, the potential energy expression is:

$$U=\frac{1}{2}K\eta^2+mg(L+\eta)(1-\cos\theta). \tag{5.7d}$$

Taking the derivatives of equation (5.7c) with respect to η, θ, $\dot\eta$ and $\dot\theta$ yields:

$$\frac{\partial T}{\partial \dot\eta}=m\dot\eta \quad \text{and} \quad \frac{d}{dt}\left[\frac{\partial T}{\partial \dot\eta}\right]=m\ddot\eta \tag{5.7e}$$

$$\frac{\partial T}{\partial \dot\theta}=m(L+\eta)^2\dot\theta \quad \text{and}$$

$$\frac{d}{dt}\left[\frac{\partial T}{\partial \dot\theta}\right]=m(L+\eta)^2\ddot\theta+2m(L+\eta)\dot\eta\dot\theta \tag{5.7f}$$

$$\frac{\partial T}{\partial \eta}=m(L+\eta)\dot\theta^2 \tag{5.7g}$$

$$\frac{\partial T}{\partial \theta}=0. \tag{5.7h}$$

Taking the derivatives of equation (5.7d) with respect to η and θ yields:

$$\frac{\partial U}{\partial \eta}=K\eta+mg(1-\cos\theta) \tag{5.7i}$$

and:

$$\frac{\partial U}{\partial \theta}=mg(L+\eta)\sin\theta. \tag{5.7j}$$

Substituting equations (5.7e), (5.7g), and (5.7i) into equation (5.7a) yields:

$$\boxed{m\ddot\eta+K\eta-m(L+\eta)\dot\theta^2+mg(1-\cos\theta)=0.} \tag{5.7k}$$

Substituting equations (5.7f), (5.7h), and (5.7j) into equation (5.7b) yields:

$$\boxed{m(L+\eta)^2\ddot\theta+2m(L+\eta)\dot\eta\dot\theta+mg(L+\eta)\sin\theta=0.} \tag{5.7l}$$

EXAMPLE 5.8

Use Lagrange's equation to develop the equations of motion of the roller-pendulum system shown in Figure 5.6. Assume no slipping between the roller and surface. Make small amplitude approximations and linearize the equations of motions. Write the equations of motion in matrix form.

SOLUTION

Let the generalized coordinates be $q_1=\theta_1$ and $q_2=\theta_2$. The geometric constraint equations for the system are:

$$x_1=R\theta_1 \tag{5.8a}$$

$$x_2=R\theta_1-\frac{L}{2}\sin\theta_2 \tag{5.8b}$$

$$y_2=-\frac{L}{2}\cos\theta_2 \tag{5.8c}$$

$$x_3=R\theta_1-L\sin\theta_2. \tag{5.8d}$$

The corresponding velocities of the centers of mass are:

$$\dot x_1=R\dot\theta_1 \tag{5.8e}$$

$$\dot x_2=R\dot\theta_1-\frac{L}{2}\cos\theta_2\,\dot\theta_2 \tag{5.8f}$$

$$\dot y_2=\frac{L}{2}\sin\theta_2\,\dot\theta_2. \tag{5.8g}$$

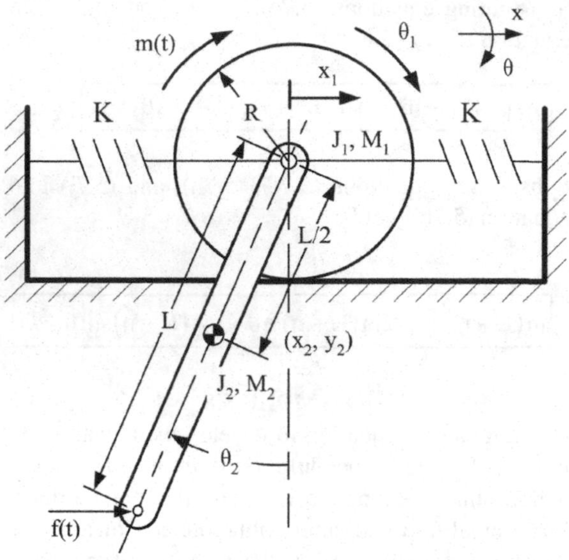

Figure 5.6 2-DOF Roller-Pendulum System with Spring Elements

Making the small amplitude approximations yields:

$$\dot{x}_2 = R\,\dot{\theta}_1 - \frac{L}{2}\,\dot{\theta}_2 \tag{5.8h}$$

and:

$$\dot{y}_2 = 0. \tag{5.8i}$$

The expressions for the kinetic and potential energies are:

$$T = \frac{1}{2} J_1\,\dot{\theta}_1^2 + \frac{1}{2} M_1\,\dot{x}_1^2 + \frac{1}{2} J_2\,\dot{\theta}_2^2 + \frac{1}{2} M_2\,\dot{x}_2^2 + \frac{1}{2} M_2\,\dot{y}_2^2 \tag{5.8j}$$

and:

$$\bullet U = mg\,\frac{L}{2}\left(1-\cos\theta_2\right) + \frac{1}{2} K\,x_1^2 + \frac{1}{2} K\,x_1^2. \tag{5.8k}$$

Substituting equations (5.8e), (5.8h) and (5.8i) into equations (5.8j) and (5.8k) yields:

$$T = \frac{1}{2}\left(J_1 + M_1 R^2\right)\dot{\theta}_1^2 + \frac{1}{2} J_2\,\dot{\theta}_2^2 + \frac{1}{2} M_2\left(R\,\dot{\theta}_1 - \frac{L}{2}\,\dot{\theta}_2\right)^2 \tag{5.8l}$$

and:

$$U = mg\,\frac{L}{2}\left(1-\cos\theta_2\right) + K R^2\,\theta_1^2. \tag{5.8m}$$

The virtual work associated with the roller-pendulum system is:

$$\delta W = m(t)\,\delta\theta_1 + f(t)\,\delta x_3. \tag{5.8n}$$

The virtual displacement δx_3 is:

$$\delta x_3 = R\,\delta\theta_1 - L\,\cos\theta_2\,\delta\theta_2 \tag{5.8o}$$

or:

$$\delta W = m(t)\,\delta\theta_1 + f(t)\,\delta x_3. \tag{5.8p}$$

Substituting equation (5.8p) into equation (5.8n) yields:

$$\delta W = m(t)\,\delta\theta_1 + f(t)\left(R\,\delta\theta_1 - L\,\delta\theta_2\right) \tag{5.8q}$$

or:

$$\delta W = \left[m(t) + R\,f(t)\right]\delta\theta_1 + \left[-L\,f(t)\right]\delta\theta_2. \tag{5.8r}$$

Therefore:

$$Q_{nce(1)} = m(t) + R\,f(t) \tag{5.8s}$$

and:

$$Q_{nce(2)} = -L\,f(t). \tag{5.8t}$$

Taking the derivatives of equation (5.8l) with respect to and simplifying yields:

$$\frac{\partial T}{\partial \dot{\theta}_1} = \left(J_1 + M_1 R^2 + M_2 R^2\right)\dot{\theta}_1 - \frac{M_2 L R}{2}\dot{\theta}_2 \tag{5.8u}$$

and:

$$\frac{\partial T}{\partial \dot{\theta}_2} = \left(J_2 + \frac{M_2 L^2}{4}\right)\dot{\theta}_2 - \frac{M_2 \, L R}{2}\dot{\theta}_1. \tag{5.8v}$$

Taking the derivatives of equation (5.8m) with respect to θ_1 and θ_2 yield:

$$\frac{\partial U}{\partial \theta_1} = 2 K R^2 \, \theta_1 \tag{5.8w}$$

and:

$$\frac{\partial U}{\partial \theta_2} = \frac{m g L}{2}\sin\theta_2 = \frac{m g L}{2}\theta_2. \tag{5.8x}$$

Substituting equations (5.8s) through (5.8x) into equation (5.115) yields:

$$\left(J_1 + M_1 R^2 + M_2 R^2\right)\ddot{\theta}_1 - \frac{M_2 L R}{2}\ddot{\theta}_2 + 2 K R^2 \, \theta_1 = m(t) + R\, f(t) \tag{5.8y}$$

and:

$$\left(J_2 + \frac{M_2 L^2}{4}\right)\ddot{\theta}_2 - \frac{M_2 \, L R}{2}\ddot{\theta}_1 + \frac{M_2 g L}{2}\theta_2 = -L\, f(t). \tag{5.8z}$$

Writing equations (5.8y) and (5.8z) in matrix form gives:

$$\begin{bmatrix} \left(J_1 + M_1 R^2 + M_2 R^2\right) & -\dfrac{M_2 L R}{2} \\ -\dfrac{M_2 L R}{2} & \left(J_2 + \dfrac{M_2 L^2}{4}\right) \end{bmatrix} \begin{Bmatrix} \ddot{\theta}_1 \\ \ddot{\theta}_2 \end{Bmatrix} + \begin{bmatrix} 2 K R^2 & 0 \\ 0 & \dfrac{M_2 g L}{2} \end{bmatrix} \begin{Bmatrix} \theta_1 \\ \theta_2 \end{Bmatrix} = \begin{Bmatrix} m(t) + R f(t) \\ -L f(t) \end{Bmatrix}. \tag{5.8aa}$$

5.8 Influence Coefficients

Modeling methods using influence coefficients can be used to develop the equations of motion for a vibration system. Two types of influence coefficients will be discussed: flexibility influence coefficients and stiffness influence coefficients. Flexibility influence coefficients can be used to develop the equations of motion of vibration systems by treating them as a static system. This is useful when developing approximate vibration models of systems comprised of beam, plate or shell structures. These structures can be divided into discrete mass elements. The deflections at individual mass points associated with forces applied to the other mass points can then be obtained from static load-deflection equations that are commonly found in handbooks.

The *flexibility influence coefficient* a_{ij} is defined as the deflection at point i due to a unit force ($F_j = 1$) at point j (Figure 5.7). Therefore:

$$u_{ij} = a_{ij} \, F_j \tag{5.119}$$

where u_{ij} is the deflection at point i due to a force F_j at point j. The principle of superposition can be used for a linear system. Therefore, the total deflection u_i at point i is obtained by adding all of the individual deflections u_{ij}, or:

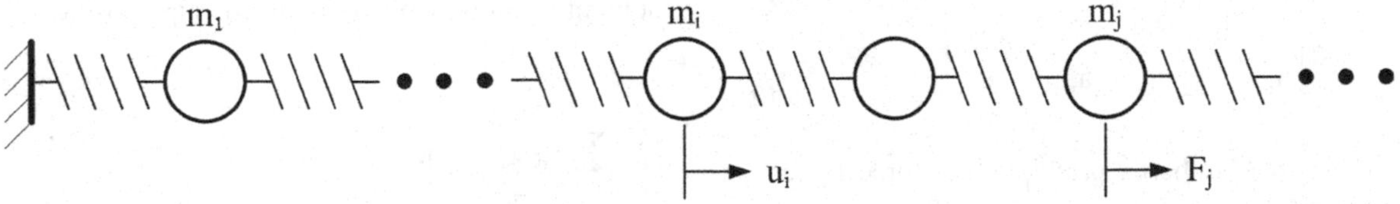

Figure 5.7 n-DOF Vibration System with Mass and Spring Elements

$$u_i = u_{i1} + u_{i2} + \cdots + u_{ii} + \cdots + u_{ij} + \cdots + u_{in} = \sum_{j=1}^{n} u_{ij} \quad \text{where} \quad i = 1, 2, 3, \ldots, n. \tag{5.120}$$

n in equation (5.120) is the number of mass points in the system. Substituting equation (5.119) into equation (5.120) yields:

$$u_i = \sum_{j=1}^{n} a_{ij}\, F_j. \tag{5.121}$$

Maxwell's reciprocity theorem can be used to show that:

$$a_{ij} = a_{ji}. \tag{5.122}$$

Equation (5.121) can be written for the values of i that range from 1 to n in the form:

$$\begin{aligned} u_1 &= a_{11}\, F_1 + a_{12}\, F_2 + \cdots + a_{1n}\, F_n \\ u_2 &= a_{21}\, F_1 + a_{22}\, F_2 + \cdots + a_{2n}\, F_n \\ &\vdots \\ u_n &= a_{n1}\, F_1 + a_{n2}\, F_2 + \cdots + a_{nn}\, F_n . \end{aligned} \tag{5.123}$$

Equation (5.123) can be written in matrix form, or:

$$\{u\} = [a]\{F\} \tag{5.124}$$

where:

$$\{u\} = \begin{Bmatrix} u_1 \\ u_2 \\ \vdots \\ u_n \end{Bmatrix}; \quad \{F\} = \begin{Bmatrix} F_1 \\ F_2 \\ \vdots \\ F_n \end{Bmatrix}; \quad \text{and}$$

$$[a] = \begin{bmatrix} a_{11} & a_{12} & \cdots & a_{1n} \\ a_{21} & a_{22} & & \\ \vdots & & \ddots & \\ a_{n1} & a_{n2} & & a_{nn} \end{bmatrix}. \tag{5.125}$$

Using d'Alembert's principle, the force F_j can be written:

$$F_j = -m_j\, \ddot{u}_j \tag{5.126}$$

and the force vector {F} can be written:

$$\begin{Bmatrix} F_1 \\ F_2 \\ \vdots \\ F_n \end{Bmatrix} = -\begin{bmatrix} m_1 & 0 & \cdots & 0 \\ 0 & m_2 & & 0 \\ \vdots & & \ddots & \vdots \\ 0 & 0 & \cdots & m_n \end{bmatrix} \begin{Bmatrix} \ddot{u}_1 \\ \ddot{u}_2 \\ \vdots \\ \ddot{u}_n \end{Bmatrix} \tag{5.127}$$

or:

$$\{F\} = -[m]\{\ddot{u}\}. \tag{5.128}$$

Substituting equation (5.128) into equation (5.124) and rearranging the terms yields:

$$[a][m]\{\ddot{u}\} + [\,I\,]\{u\} = \{0\} \tag{5.129}$$

where [I] is the unity matrix:

$$[\,I\,] = \begin{bmatrix} 1 & 0 & \cdots & 0 \\ 0 & 1 & & \\ \vdots & & \ddots & \\ 0 & 0 & \cdots & 1 \end{bmatrix}. \tag{5.130}$$

[a] in equation (5.129) is referred to as the *flexibility influence matrix*, and [m] is referred to as the *mass matrix*.

Stiffness influence coefficients can also be used to develop the equations of motion of vibration systems. *The stiffness influence coefficient k_{ij} is defined as the force at point i due to a unit displacement* $(u_j = 1)$ *at point j*. Therefore:

$$F_{ij} = k_{ij}\, u_j \tag{5.131}$$

where F_{ij} is the force at point i due to a displacement u_j at point j. Applying the principle of superposition:

$$F_i = \sum_{j=1}^{n} k_{ij}\, u_j \quad \text{where} \quad i = 1, 2, 3, \ldots, n. \tag{5.132}$$

Equation (5.132) can be written in matrix form:

$$\{F\}=[k]\{u\} \tag{5.133}$$

where:

$$\{u\}=\begin{Bmatrix} u_1 \\ u_2 \\ \vdots \\ u_n \end{Bmatrix}; \quad \{F\}=\begin{Bmatrix} F_1 \\ F_2 \\ \vdots \\ F_n \end{Bmatrix}; \quad \text{and}$$

$$[k]=\begin{bmatrix} k_{11} & k_{12} & \cdots & k_{1n} \\ k_{21} & k_{22} & & \\ \vdots & & \ddots & \\ k_{n1} & k_{n2} & & k_{nn} \end{bmatrix}. \tag{5.134}$$

Substituting equation (5.128) into equation (5.133) and rearranging the terms yields:

$$[m]\{\ddot{u}\}+[k]\{u\}=\{0\}. \tag{5.135}$$

[k] is referred to as the *stiffness matrix*.

Pre-multiplying equation (5.135) by the inverse of the stiffness matrix yields:

$$[k]^{-1}[m]\{\ddot{u}\}+[k]^{-1}[k]\{u\}=[k]^{-1}\{0\}. \tag{5.136}$$

Noting that:

$$[k]^{-1}[k]=[I] \tag{5.137}$$

equation (5.136) becomes:

$$[k]^{-1}[m]\{\ddot{u}\}+[I]\{u\}=\{0\}. \tag{5.138}$$

Comparing equations (5.129) and (5.138), it can be seen that:

$$[a]=[k]^{-1}. \tag{5.139}$$

Conversely:

$$[k]=[a]^{-1}. \tag{5.140}$$

EXAMPLE 5.9

Determine the equations of motion of the 3-DOF vibration system shown in Figure 5.8(a) using flexibility influence coefficients.

SOLUTION

Figure 5.8(e) shows the free-body diagrams of masses 1, 2, and 3 for a 1 N force applied to mass 1. The three static force equations are:

$$1-k_1 a_{11}-k_2(a_{11}-a_{21})=0 \tag{5.9a}$$

$$-k_2(a_{21}-a_{11})-k_3(a_{21}-a_{31})=0 \tag{5.9b}$$

$$-k_3(a_{31}-a_{21})=0. \tag{5.9c}$$

Solving equation (5.9c):

$$a_{31}=a_{21}. \tag{5.9d}$$

Substituting equation (5.9d) into equation (5.9b) and simplifying yields:

$$a_{21}=a_{11}. \tag{5.9e}$$

Substituting equation (5.9e) into equation (5.9a) and solving for a_{11} gives:

$$a_{11}=\frac{1}{k_1}. \tag{5.9f}$$

From equations (5.9d) and (5.9f):

$$a_{21}=\frac{1}{k_1} \quad \text{and} \quad a_{31}=\frac{1}{k_1}. \tag{5.9g}$$

Figure 5.8(f) shows the free-body diagrams of masses 1, 2, and 3 for a 1 N force applied to mass 2. The three static force equations are:

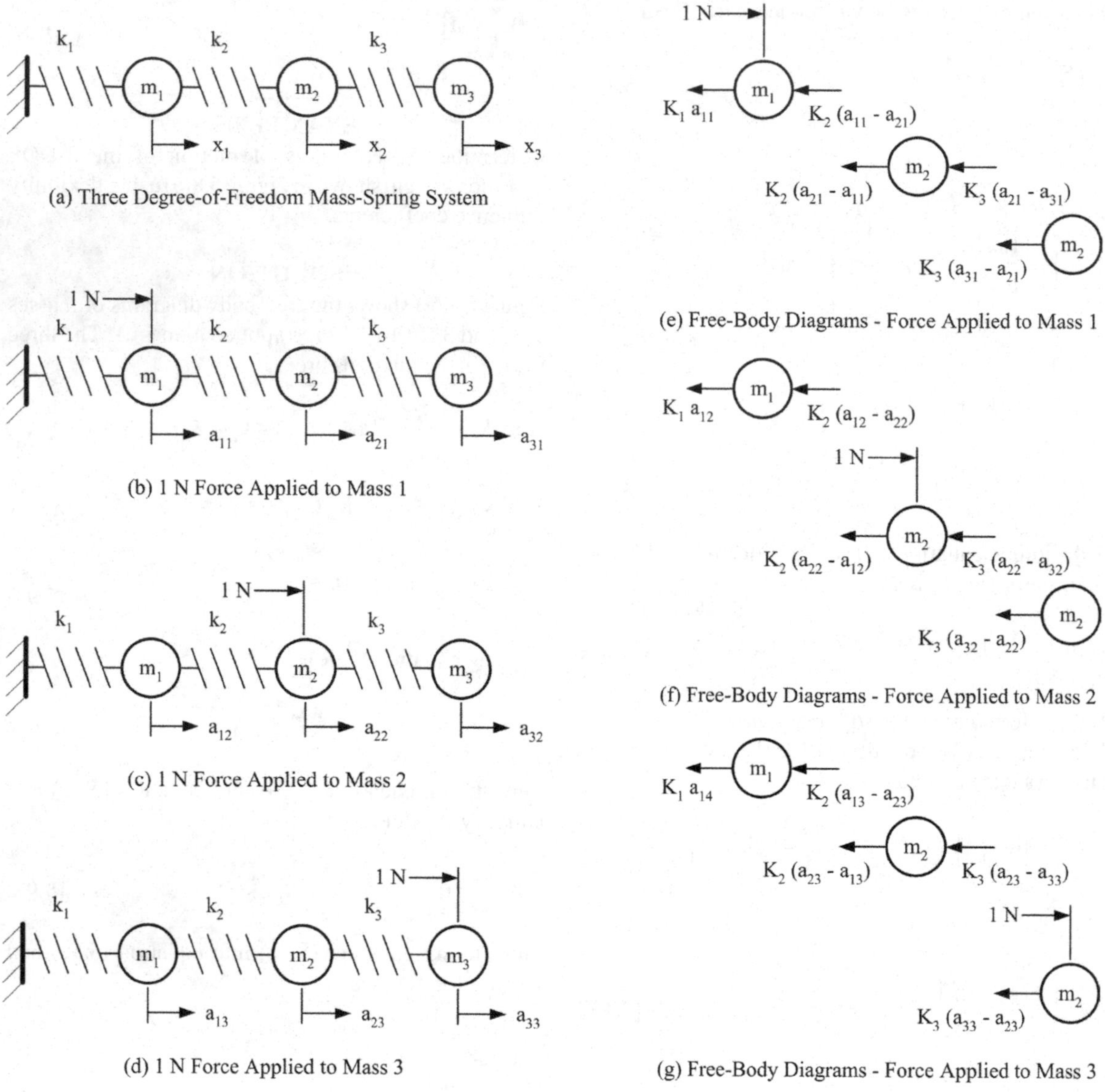

Figure 5.8 3-DOF Vibration System with Mass and Spring Elements

$$-k_1\, a_{12} - k_2\left(a_{12} - a_{22}\right) = 0 \tag{5.9h}$$

$$1 - k_2\left(a_{22} - a_{12}\right) - k_3\left(a_{22} - a_{32}\right) = 0 \tag{5.9i}$$

$$-k_3\left(a_{32} - a_{22}\right) = 0. \tag{5.9j}$$

From equation (5.9j):

$$a_{32} = a_{22}. \tag{5.9k}$$

Substituting equation (5.9k) into equation (5.9i) and solving for a_{22} yields:

$$a_{22} = \frac{1}{k_2} + a_{12}. \tag{5.9l}$$

Noting that $a_{12} = a_{21}$ and substituting equation (5.9g) into equation (5.9l) gives:

$$a_{22} = \frac{1}{k_1} + \frac{1}{k_2}. \quad (5.9m)$$

From equation (5.9g), (5.9k) and (5.9l):

$$a_{12} = \frac{1}{k_1} \quad \text{and} \quad a_{32} = \frac{1}{k_1} + \frac{1}{k_2}. \quad (5.9n)$$

Figure 5.8(g) shows the free-body diagrams of masses 1, 2, and 3 for a 1 N force applied to mass 3. The three static force equations are:

$$-k_1\, a_{13} - k_2\left(a_{13} - a_{23}\right) = 0 \quad (5.9o)$$

$$-k_2\left(a_{23} - a_{13}\right) - k_3\left(a_{23} - a_{33}\right) = 0 \quad (5.9p)$$

$$1 - k_3\left(a_{33} - a_{23}\right) = 0. \quad (5.9q)$$

Noting that $a_{13} = a_{31}$ and $a_{23} = a_{32}$, from equations (5.9g) and (5.9n):

$$a_{13} = \frac{1}{k_1} \quad \text{and} \quad a_{23} = \frac{1}{k_1} + \frac{1}{k_2}. \quad (5.9r)$$

Solving equation (5.9q) for a_{33} yields:

$$a_{33} = \frac{1}{k_1} + \frac{1}{k_2} + \frac{1}{k_3}. \quad (5.9s)$$

The equation of motion for the 3-DOF vibration system in Figure 5.8(a) is:

$$\begin{bmatrix} a_{11} & a_{12} & a_{13} \\ a_{21} & a_{22} & a_{23} \\ a_{31} & a_{32} & a_{33} \end{bmatrix} \begin{bmatrix} m_1 & 0 & 0 \\ 0 & m_2 & 0 \\ 0 & 0 & m_3 \end{bmatrix} \begin{Bmatrix} \ddot{x}_1 \\ \ddot{x}_2 \\ \ddot{x}_3 \end{Bmatrix} + \begin{bmatrix} 1 & 0 & 0 \\ 0 & 1 & 0 \\ 0 & 0 & 1 \end{bmatrix} \begin{Bmatrix} x_1 \\ x_2 \\ x_3 \end{Bmatrix} = \begin{Bmatrix} 0 \\ 0 \\ 0 \end{Bmatrix}. \quad (5.9t)$$

EXAMPLE 5.10

Figure 5.9(a) shows a beam that is clamped at $x = 0$ and free at $x = l$. Figure 5.9(b) shows a 3-DOF approximation to the clamped-free beam in Figure 5.9(a). The beam is divided into three equal masses that are placed a distance of l/3 apart. Use flexibility influence coefficients to develop the equations of motion of the approximate beam in Figure 5.9(b).

SOLUTION

The beam is approximated as a 3-DOF vibration system with three mass elements. Therefore, it will be necessary to calculate nine flexibility influence coefficients. Let y_1, y_2 and y_3 be the downward vertical deflections of mass elements 1, 2, and 3, respectively. The equations that specify the deflection u of a clamped-free beam associated with an applied point force F are:

$$u = \frac{F x^2}{6EI}\left(3a - x\right) \quad \text{from A to B} \quad (5.10a)$$

$$u = \frac{F a^2}{6EI}\left(3x - a\right) \quad \text{from B to C} \quad (5.10b)$$

where E is Young's modulus and I is the area moment of the beam cross-section. a_{11} is obtained by letting $F = 1$, $a = L/3$, and $x = L/3$ in equation (5.10a), or:

$$a_{11} = \frac{L^2}{6EI\,3^2}\left(\frac{3L}{3} - \frac{L}{3}\right) \quad \text{or} \quad a_{11} = \frac{2L^3}{162\,EI}. \quad (5.10c)$$

a_{21} is obtained by letting $F = 1$, $a = L/3$, and $x = 2L/3$ in equation (5.10b), or:

$$a_{21} = \frac{L^2}{6EI\,3^2}\left(\frac{3\times 2L}{3} - \frac{L}{3}\right) \quad \text{or} \quad a_{21} = a_{12} = \frac{5L^3}{162\,EI}. \quad (5.10d)$$

a_{31} is obtained by letting $F = 1$, $a = L/3$, and $x = L$ in equation (5.10b), or:

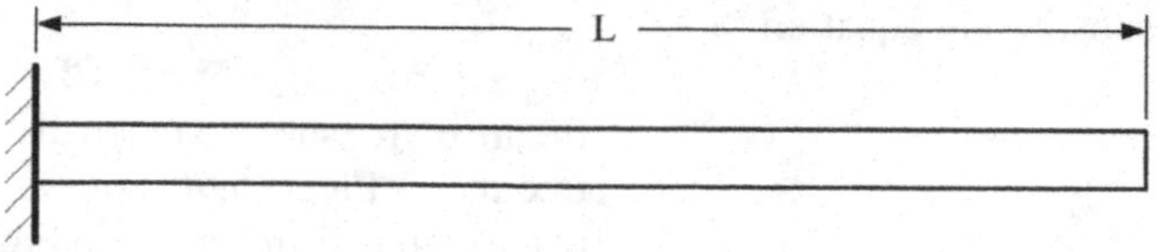

(a) Beam that Is Clamped at x = 0 and Free at x = L

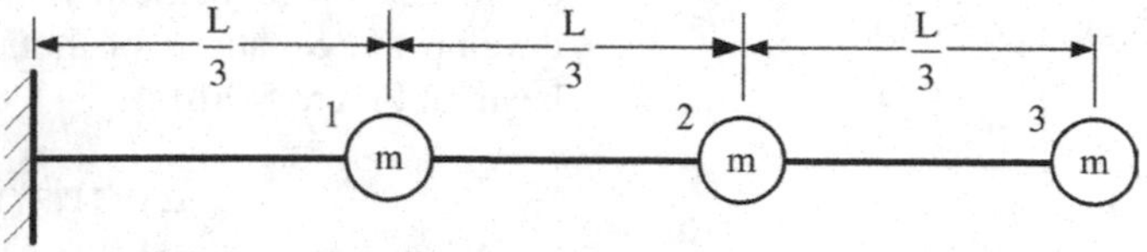

(b) Three-Degree-of-Freedom Approximation of Clamped-Free Beam in (a)

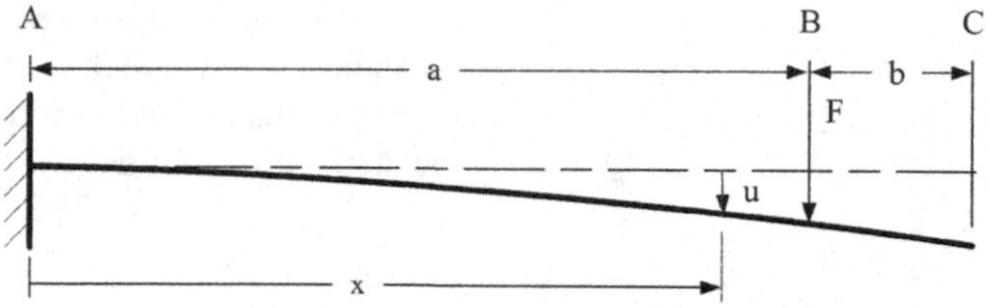

(c) Static Deflection Associated with a Point Force F Applied to a Clamped-Free BeaM

Figure 5.9 3-DOF Approximation of a Beam That is Clamped at One End and Free at the Other End

$$a_{31} = \frac{L^2}{6\,EI\,3^2}\left(3\,L - \frac{L}{3}\right) \text{ or } a_{31} = a_{13} = \frac{8\,L^3}{162\,EI}. \tag{5.10e}$$

In a like fashion:

$$a_{22} = \frac{16\,L^3}{162\,EI};\, a_{32} = a_{23} = \frac{28\,L^3}{162\,EI};\, a_{33} = \frac{54\,L^3}{162\,EI}. \tag{5.10f}$$

Using d'Alembert's principle, equation (5.123) can be written:

$$y_1 = -\frac{2\,L^3}{162\,EI}\,m\,\ddot{y}_1 - \frac{5\,L^3}{162\,EI}\,m\,\ddot{y}_2 - \frac{8\,L^3}{162\,EI}\,m\,\ddot{y}_3 \tag{5.10g}$$

$$y_2 = -\frac{5\,L^3}{162\,EI}\,m\,\ddot{y}_1 - \frac{16\,L^3}{162\,EI}\,m\,\ddot{y}_2 - \frac{28\,L^3}{162\,EI}\,m\,\ddot{y}_3 \tag{5.10h}$$

$$y_2 = -\frac{8\,L^3}{162\,EI}\,m\,\ddot{y}_1 - \frac{16\,L^3}{162\,EI}\,m\,\ddot{y}_2 - \frac{54\,L^3}{162\,EI}\,m\,\ddot{y}_3. \tag{5.10i}$$

Therefore, the equation of motion for the 3-DOF approximation of the clamped-free beam in Figure 5.9(a) is:

$$\boxed{\frac{m\,L^3}{162\,EI}\begin{bmatrix} 2 & 5 & 8 \\ 5 & 16 & 28 \\ 8 & 28 & 54 \end{bmatrix}\begin{Bmatrix} \ddot{y}_1 \\ \ddot{y}_2 \\ \ddot{y}_3 \end{Bmatrix} + \begin{bmatrix} 1 & 0 & 0 \\ 0 & 1 & 0 \\ 0 & 0 & 1 \end{bmatrix}\begin{Bmatrix} y_1 \\ y_2 \\ y_3 \end{Bmatrix} = \begin{Bmatrix} 0 \\ 0 \\ 0 \end{Bmatrix}.} \tag{5.10j}$$

5.9 Matrix Equation of Motion from Lagrange's Equation

Lagrange's equation can be written in matrix form. Equation (5.115) can be written:

$$\frac{d}{dt}\left(\frac{\partial T}{\partial \dot{q}_1}\right)-\frac{\partial T}{\partial q_1}+\frac{\partial U}{\partial q_1}=Q_{nc(1)}$$

$$\frac{d}{dt}\left(\frac{\partial T}{\partial \dot{q}_2}\right)-\frac{\partial T}{\partial q_2}+\frac{\partial U}{\partial q_2}=Q_{nc(2)}$$

$$\frac{d}{dt}\left(\frac{\partial T}{\partial \dot{q}_3}\right)-\frac{\partial T}{\partial q_3}+\frac{\partial U}{\partial q_3}=Q_{nc(3)} \tag{5.141}$$

$$\vdots$$

$$\frac{d}{dt}\left(\frac{\partial T}{\partial \dot{q}_m}\right)-\frac{\partial T}{\partial q_m}+\frac{\partial U}{\partial q_m}=Q_{nc(m)}$$

or:

$$\frac{d}{dt}\begin{Bmatrix}\frac{\partial T}{\partial \dot{q}_1}\\ \vdots \\ \frac{\partial T}{\partial \dot{q}_m}\end{Bmatrix}-\begin{Bmatrix}\frac{\partial T}{\partial q_1}\\ \vdots \\ \frac{\partial T}{\partial q_m}\end{Bmatrix}+\begin{Bmatrix}\frac{\partial U}{\partial q_1}\\ \vdots \\ \frac{\partial U}{\partial q_m}\end{Bmatrix}=\begin{Bmatrix}Q_{nc(1)}\\ \vdots \\ Q_{nc(m)}\end{Bmatrix}. \tag{5.142}$$

Therefore, Lagrange's equation can be written:

$$\frac{d}{dt}\left\{\frac{\partial T}{\partial \dot{q}}\right\}-\left\{\frac{\partial T}{\partial q}\right\}+\left\{\frac{\partial U}{\partial q}\right\}=\{Q_{nc}\}. \tag{5.143}$$

The kinetic energy of the system in Figure 5.7, where u_i is replaced by q_i, can be written:

$$\begin{aligned} T &= \frac{1}{2}m_1\dot{q}_1^2+\frac{1}{2}m_2\dot{q}_2^2+\cdots+\frac{1}{2}m_n\dot{q}_n^2 \\ &= \frac{1}{2}\left(\dot{q}_1 m_1 \dot{q}_1+\dot{q}_2 m_2 \dot{q}_2+\cdots+\dot{q}_n m_n \dot{q}_n\right). \end{aligned} \tag{5.144}$$

0==

Equation (5.144) can be written:

$$T=\frac{1}{2}\left[\dot{q}_1 m_1+\dot{q}_2 m_2+\cdots+\dot{q}_n m_n\right]\begin{bmatrix}\dot{q}_1\\ \dot{q}_2\\ \vdots\\ \dot{q}_n\end{bmatrix} \tag{5.145}$$

or:

$$T=\frac{1}{2}\left[\dot{q}_1 \quad \dot{q}_2 \quad \cdots \quad \dot{q}_n\right]\begin{bmatrix}m_1 & 0 & \cdots & 0\\ 0 & m_2 & & \\ \vdots & & \ddots & \\ 0 & 0 & & m_n\end{bmatrix}\begin{bmatrix}\dot{q}_1\\ \dot{q}_2\\ \vdots\\ \dot{q}_n\end{bmatrix}. \tag{5.146}$$

Equation (5.146) can be written:

$$T=\frac{1}{2}\{\dot{q}\}^T[m]\{\dot{q}\}. \tag{5.147}$$

The mass matrix [m] in the above development is a diagonal matrix. However, the mass matrix can also assume the generalized form:

$$[m]=\begin{bmatrix}m_{11} & m_{12} & \cdots & m_{1n}\\ m_{21} & m_{22} & & \\ \vdots & & \ddots & \\ m_{n1} & & & m_{nn}\end{bmatrix}. \tag{5.148}$$

For linear systems, the mass matrix is a symmetric matrix where $m_{ij} = m_{ji}$.

The kinetic energy is a *quadratic function of the generalized velocities*. The kinetic energy by definition cannot be negative, and it vanishes only when the generalized velocities equal zero. Therefore, the kinetic energy is a *positive definite quadratic function of the generalized velocities*, and the form of the kinetic energy equation is a *positive definite quadratic* form. Consequently, the mass matrix [m] is always a *positive definite matrix*.

The potential energy in an elastic system that is associated with a force F_i that produces a displacement q_i can be expressed:

$$U_i = \frac{1}{2} F_i \, q_i . \tag{5.149}$$

The total potential energy of an elastic system is:

$$U = \frac{1}{2} \sum_{i=1}^{m} F_i \, q_i . \tag{5.150}$$

Substituting equation (5.132) into equation (5.150), where u_i is replaced with q_i, yields:

$$\begin{aligned} U &= \frac{1}{2} \sum_{i=1}^{m} q_i \sum_{j=1}^{m} k_{ij} \, q_j \\ &= \frac{1}{2} \sum_{i=1}^{m} \sum_{j=1}^{m} q_i \, k_{ij} \, q_j . \end{aligned} \tag{5.151}$$

Expanding the summations in equation (5.151) gives:

$$U = \frac{1}{2} \begin{Bmatrix} q_1 \, k_{11} \, q_1 + q_1 \, k_{12} \, q_2 + \cdots + q_1 \, k_{1n} \, q_n \\ q_2 \, k_{21} \, q_1 + q_2 \, k_{22} \, q_2 + \cdots + q_2 \, k_{2n} \, q_n \\ \vdots \\ q_n \, k_{n1} \, q_1 + q_n \, k_{n2} \, q_2 + \cdots + q_n \, k_{nn} \, q_n \end{Bmatrix} . \tag{5.152}$$

Equation (5.152) can be written:

$$U = \frac{1}{2} \begin{bmatrix} q_1 \, k_{11} + \cdots + q_1 \, k_{1n} + \cdots + \\ q_n \, k_{n1} + \cdots + q_n \, k_{nn} \end{bmatrix} \begin{bmatrix} q_1 \\ q_2 \\ \vdots \\ q_n \end{bmatrix} \tag{5.153}$$

or:

$$U = \frac{1}{2} \begin{bmatrix} q_1 & q_2 & \cdots & q_n \end{bmatrix} \begin{bmatrix} k_{11} & k_{12} & \cdots & k_{1n} \\ k_{21} & k_{22} & & \\ \vdots & & \ddots & \\ k_{n1} & & & k_{nn} \end{bmatrix} \begin{bmatrix} q_1 \\ q_2 \\ \vdots \\ q_n \end{bmatrix} . \tag{5.154}$$

Equation (5.154) can be written:

$$U = \frac{1}{2} \{q\}^T [k] \{q\} . \tag{5.155}$$

For linear systems, the stiffness matrix is a symmetric matrix where $k_{ij} = k_{ji}$.

The potential energy is a quadratic function of the generalized displacements. The potential energy of a system is specified in terms of a reference position. Therefore, it is usually specified to within an arbitrary constant. As has been discussed in previous chapters, the reference position of a vibration system is the equilibrium position of the system. Therefore, the generalized coordinates are defined such that they equal zero at the reference position, and the arbitrary constant is chosen such that the potential energy equals zero at the reference position. When the system is a stable system, the potential energy of the system is always zero at the equilibrium position and is greater than zero for any other position. Therefore, the potential energy for a stable system is a *positive definite quadratic function of the generalized coordinates*, and the form of the potential energy equation is a *positive definite quadratic form*. For this case, the stiffness matrix [k] is a *positive definite matrix*. A system where the mass and stiffness matrices are positive definite matrices is a *positive definite system*.

There are some cases when the potential energy of a system equals zero without the generalized coordinates equaling zero. For these cases, the potential energy is a *positive semi-definite quadratic function of the generalized coordinates*. Furthermore, the stiffness matrix [k] is a *positive semi-definite* matrix. A system where the mass matrix is a positive definite matrix and the stiffness matrix is a positive semi-definite matrix is a *semi-definite system*.

Lagrange's equation can be used to derive the equations of motion of a n-DOF vibration system in matrix form. This is accomplished by substituting equations (5.147) and (5.155) into equation (5.143) and performing the specified operations. The derivatives of a product of matrices is formed in the same manner that is used for scalar functions. However, the order of the matrix positions must be maintained. First, take the derivative of equation (5.147) with respect to $\dot{q}_r$. This yields:

$$\dot{q}_r \frac{\partial T}{\partial \dot{q}_r} = \frac{\partial}{\partial \dot{q}_r}\left[\frac{1}{2}\{\dot{q}\}^T[m]\{\dot{q}\}\right]$$
$$= \frac{1}{2}\left[\frac{\partial \{\dot{q}\}^T}{\partial \dot{q}_r}[m]\{\dot{q}\} + \{\dot{q}\}^T[m]\frac{\partial\{\dot{q}\}}{\partial \dot{q}_r}\right]$$
$$= \frac{1}{2}\left[\left\{\frac{\partial \dot{q}_i}{\partial \dot{q}_r}\right\}^T[m]\{\dot{q}\} + \{\dot{q}\}^T[m]\left\{\frac{\partial \dot{q}_i}{\partial \dot{q}_r}\right\}\right] \quad (5.156)$$

where i = 1, 2, 3, ..., n and r = 1, 2, 3, ..., n. Note that:

$$\left\{\frac{\partial \dot{q}_i}{\partial \dot{q}_r}\right\} = \{\delta_{ir}\} \quad \text{where} \quad \delta_{ir} = \begin{Bmatrix} 1 & \text{if} & i = r \\ 0 & \text{if} & i \neq r \end{Bmatrix}. \quad (5.157)$$

$\{\delta_{ir}\}$ is the Kronecker delta column vector whose elements equal zero when $i \neq r$ and equal one when $i = r$. Therefore, equation (5.156) becomes:

$$\frac{\partial T}{\partial \dot{q}_r} = \frac{1}{2}\left[\{\delta_{ir}\}^T[m]\{\dot{q}\} + \{\dot{q}\}^T[m]\{\delta_{ir}\}\right] \quad (5.158)$$

or:

$$\frac{\partial T}{\partial \dot{q}_r} = \{\delta_{ir}\}^T[m]\{\dot{q}\}. \quad (5.159)$$

Equation (5.159) can also be written:

$$\frac{\partial T}{\partial \dot{q}_r} = \lfloor m_r \rfloor\{\dot{q}\} \quad \text{where} \quad r = 1, 2, 3, \ldots, n \quad (5.160)$$

and where the matrix $\lfloor m_r \rfloor$ is a row matrix that is the same as the rth row of the mass matrix [m]. Equation (5.160) can be expanded to:

$$\begin{Bmatrix} \dfrac{\partial T}{\partial \dot{q}_1} \\ \dfrac{\partial T}{\partial \dot{q}_2} \\ \vdots \\ \dfrac{\partial T}{\partial \dot{q}_n} \end{Bmatrix} = \begin{Bmatrix} \lfloor m_1 \rfloor\{\dot{q}\} \\ \lfloor m_2 \rfloor\{\dot{q}\} \\ \vdots \\ \lfloor m_n \rfloor\{\dot{q}\} \end{Bmatrix} = \begin{Bmatrix} \lfloor m_1 \rfloor \\ \lfloor m_2 \rfloor \\ \vdots \\ \lfloor m_n \rfloor \end{Bmatrix}\{\dot{q}\}. \quad (5.161)$$

Noting that:

$$\begin{Bmatrix} \lfloor m_1 \rfloor \\ \lfloor m_2 \rfloor \\ \vdots \\ \lfloor m_n \rfloor \end{Bmatrix} = [m] \quad (5.162)$$

equation (5.161) can be written:

$$\left\{\frac{\partial T}{\partial \dot{q}}\right\} = [m]\{\dot{q}\} \quad (5.163)$$

and:

$$\frac{d}{dt}\left\{\frac{\partial T}{\partial \dot{q}}\right\} = [m]\{\ddot{q}\}. \quad (5.164)$$

When the kinetic energy is not a function of q:

$$\left\{\frac{\partial T}{\partial q}\right\} = 0. \quad (5.165)$$

Taking the derivative of equation (5.155) with respect to q_r yields:

$$\frac{\partial U}{\partial q_r} = \frac{\partial}{\partial q_r}\left[\frac{1}{2}\{q\}^T[m]\{q\}\right]$$
$$= \lfloor k_r \rfloor\{q\} \quad \text{where} \quad r = 1, 2, 3, \ldots, n. \quad (5.166)$$

Therefore:

$$\left\{\frac{\partial U}{\partial q}\right\} = [k]\{q\}. \tag{5.167}$$

Substituting equations (5.164), (5.165), and (5.167) into equation (5.143) yields:

$$\boxed{[m]\{\ddot{q}\} + [k]\{q\} = \{Q_{nc}\}.} \tag{5.168}$$

When the external forces equal zero, equation (5.168) becomes:

$$\boxed{[m]\{\ddot{q}\} + [k]\{q\} = \{0\}.} \tag{5.169}$$

5.10 Solution to the Matrix Equation of Motion for Unforced Vibration

To obtain the solution to equation (5.169), let:

$$\{q(t)\} = \{u\}\, T(t). \tag{5.170}$$

Therefore:

$$\{\ddot{q}(t)\} = \{u\}\, \ddot{T}(t). \tag{5.171}$$

Substituting equations (5.170) and (5.171) into equation (5.169) yields:

$$[m]\{u\}\,\ddot{T}(t) + [k]\{u\}\,T(t) = \{0\}. \tag{5.172}$$

Rearranging equation (5.172) gives:

$$\boxed{-\frac{\ddot{T}(t)}{T(t)} = \frac{[k]\{u\}}{[m]\{u\}} = \omega^2.} \tag{5.173}$$

The two sides of equation (5.173) are separated by letting both sides of the equation equal the separation constant ω^2. Therefore, equation (5.173) is separated into two equations:

$$\boxed{[m]\{u\} - \frac{1}{\omega^2}[k]\{u\} = \{0\}} \tag{5.174}$$

and:

$$\boxed{\ddot{T}(t) + \omega^2\, T(t) = 0.} \tag{5.175}$$

First, examine the solution to equation (5.174). Premultiply equation (5.174) by the inverse of the stiffness matrix. Therefore:

$$[k]^{-1}[m]\{u\} - \frac{1}{\omega^2}[k]^{-1}[k]\{u\} = [k]^{-1}\{0\}. \tag{5.176}$$

Relative to equation (5.176):

$$[k]^{-1}[m] = [D] \quad \text{and} \quad [k]^{-1}[k] = [I]. \tag{5.177}$$

[I] is the *identity matrix* [equation (5.130)] and [D] is called the *dynamical matrix*. Therefore, equation (5.176) can be written:

$$\boxed{\big[[D] - \lambda[I]\big]\{u\} = \{0\}} \tag{5.178}$$

where:

$$\boxed{\lambda = \frac{1}{\omega^2}} \tag{5.179}$$

Equation (5.178) can also be written:

$$\boxed{\big[\lambda[I] - [D]\big]\{u\} = \{0\}.} \tag{5.180}$$

For the nontrivial solution [$\{u\} \neq \{0\}$], equation (5.180) is satisfied when:

$$\boxed{\big|\lambda[I] - [D]\big| = 0.} \tag{5.181}$$

Equation (5.181) is referred to as the *characteristic determinant*. For a n-DOF vibration system, it results in an nth order polynomial equation in terms of λ. The polynomial equation is referred to as the *characteristic* or *frequency equation* of the system. The values of λ that are solutions to the polynomial equation are referred to as *characteristic values* or *eigenvalues* of the equation. If the number of degrees of freedom n is large, a numerical solution method must be used to solve for the roots to the polynomial equation.

Once the roots of the characteristic equation have been obtained, the corresponding resonance frequencies are obtained from equation (5.179), and the mode vector that corresponds to the ith resonance frequency is obtained by solving the matrix equation:

$$\boxed{\left[\lambda_i [I] - [D]\right]\left\{u^{(i)}\right\} = \{0\}.} \tag{5.182}$$

The mode vector is sometimes referred to as the *eigenvector* or the *modal vector*. Since the value of λ_i represents a singularity point in the matrix of equation (5.182), equation (5.182) only results in n - 1 independent equations. Therefore, the mode vector $\left\{u^{(i)}\right\}$ can only be determined to within an arbitrary constant.

EXAMPLE 5.11

Determine the resonance frequencies and corresponding modal vectors for the 3-DOF vibration system in Example 5.9 where:

$$m_1 = m_3 = m,\ m_2 = 4m, \text{ and } k_1 = k_2 = k_3 = k.$$

SOLUTION

The mass and inverse stiffness matrices are:

$$[m] = m\begin{bmatrix} 1 & 0 & 0 \\ 0 & 4 & 0 \\ 0 & 0 & 1 \end{bmatrix} \tag{5.11a}$$

and:

$$[k]^{-1} = \frac{1}{k}\begin{bmatrix} 1 & 1 & 1 \\ 1 & 2 & 2 \\ 1 & 2 & 3 \end{bmatrix}. \tag{5.11b}$$

The dynamic matrix is:

$$[D] = [k]^{-1}[m] = \frac{m}{k}\begin{bmatrix} 1 & 4 & 1 \\ 1 & 8 & 2 \\ 1 & 8 & 3 \end{bmatrix}. \tag{5.11c}$$

Substituting equation (5.11c) into equation (5.181) yields:

$$\left|\lambda\begin{bmatrix} 1 & 0 & 0 \\ 0 & 1 & 0 \\ 0 & 0 & 1 \end{bmatrix} - \frac{m}{k}\begin{bmatrix} 1 & 4 & 1 \\ 1 & 8 & 2 \\ 1 & 8 & 3 \end{bmatrix}\right| = 0 \tag{5.11d}$$

or:

$$\left|\lambda'\begin{bmatrix} 1 & 0 & 0 \\ 0 & 1 & 0 \\ 0 & 0 & 1 \end{bmatrix} - \begin{bmatrix} 1 & 4 & 1 \\ 1 & 8 & 2 \\ 1 & 8 & 3 \end{bmatrix}\right| = 0 \tag{5.11e}$$

where from equation (5.179):

$$\lambda' = \frac{\lambda k}{m} = \frac{k}{m\,\omega^2}. \tag{5.11f}$$

Equation (5.11e) can be written:

$$\begin{vmatrix} \lambda'-1 & -4 & -1 \\ -1 & \lambda'-8 & -2 \\ -1 & -8 & \lambda'-3 \end{vmatrix} = 0. \tag{5.11g}$$

Expanding equation (5.11g) yields:

$$\lambda'^3 - 12\,\lambda'^2 + 14\,\lambda' - 4 = 0. \tag{5.11h}$$

The three roots of equation (5.11h) are:

$$\lambda'_1 = 0.4605 \quad \lambda'_2 = 0.8095 \quad \lambda'_3 = 10.7300. \tag{5.11i}$$

Substituting the values for λ'_1 through λ'_3 into equation (5.11f) yields for the values of ω_i:

$$\omega_1 = 1.4736\sqrt{\frac{k}{m}} \tag{5.11j}$$

$$\omega_2 = 1.1115\sqrt{\frac{k}{m}} \tag{5.11k}$$

$$\omega_3 = 0.3053\sqrt{\frac{k}{m}}. \tag{5.11l}$$

The mode vectors are obtained from the equation:

$$\begin{bmatrix} \lambda'_i - 1 & -4 & -1 \\ -1 & \lambda'_i - 8 & -2 \\ -1 & -8 & \lambda'_i - 3 \end{bmatrix} \begin{Bmatrix} u_1^{(i)} \\ u_2^{(i)} \\ u_3^{(i)} \end{Bmatrix} = 0. \tag{5.11m}$$

Since the values of λ'_1 through λ'_3 represent *singularity points* in the determinant of the matrix in equation (5.11m), equation (5.11m) only results in two independent equations. Any two of the three equations represented by equation (5.11m) can be used to determine $u_2^{(i)}$ and $u_3^{(i)}$ in terms of $u_1^{(i)}$. The three equations are:

$$(\lambda'_i - 1)\, u_1^{(i)} - 4u_2^{(i)} - u_3^{(i)} = 0 \tag{5.11n}$$

$$-u_1^{(i)} - (\lambda'_i - 8)\, u_2^{(i)} - 2\, u_3^{(i)} = 0 \tag{5.11o}$$

$$-u_1^{(i)} - 8\, u_2^{(i)} + (\lambda'_i - 3)\, u_3^{(i)} = 0. \tag{5.11p}$$

Equations (5.11n) and (5.11o) can be used to obtain:

$$\frac{u_2^{(i)}}{u_1^{(i)}} = \frac{2\lambda'_i - 1}{\lambda'_i}. \tag{5.11q}$$

Equations (5.11n) and (5.11p) can be used to obtain:

$$\frac{u_3^{(i)}}{u_1^{(i)}} = \frac{2\lambda'_i - 1}{\lambda'_i - 1}. \tag{5.11r}$$

For $\lambda'_1 = 0.30779$:

$$\left\{u^{(1)}\right\} = u_1^{(1)} \begin{Bmatrix} 1.0000 \\ -0.1716 \\ 0.1464 \end{Bmatrix}. \tag{5.11s}$$

For $\lambda'_2 = 0.64298$:

$$\left\{u^{(2)}\right\} = u_1^{(2)} \begin{Bmatrix} 1.0000 \\ 0.7647 \\ -3.2493 \end{Bmatrix}. \tag{5.11t}$$

For $\lambda'_3 = 5.0468$:

$$\left\{u^{(3)}\right\} = u_1^{(3)} \begin{Bmatrix} 1.0000 \\ 1.9068 \\ 2.1028 \end{Bmatrix}. \tag{5.11u}$$

Figure 5.10 shows a graphical representation of the modal vectors.

The solution to equation (5.175) for the resonance frequency ω_i is:

$$T_i(t) = C_i \cos(\omega_i t + \phi_i). \tag{5.183}$$

The total solution to equation (5.178) associated with the system initial conditions is obtained by first writing equation (5.171):

$$\{q(t)\} = \sum_{i=1}^{n} \frac{\left\{u^{(i)}\right\}}{u_1^{(i)}}\, u_1^{(i)}\, T_i(t). \tag{5.184}$$

Next, substitute equation (5.183) into equation (5.184) to get:

$$\{q(t)\} = \sum_{i=1}^{n} \frac{\left\{u^{(i)}\right\}}{u_1^{(i)}}\, u_1^{(i)}\, C_i \cos(\omega_i t + \phi_i). \tag{5.185}$$

$\left\{u^{(i)}\right\} / u_1^{(i)}$ are the mode vectors normalized with respect to $u_1^{(i)}$, and ω_i are the corresponding resonance

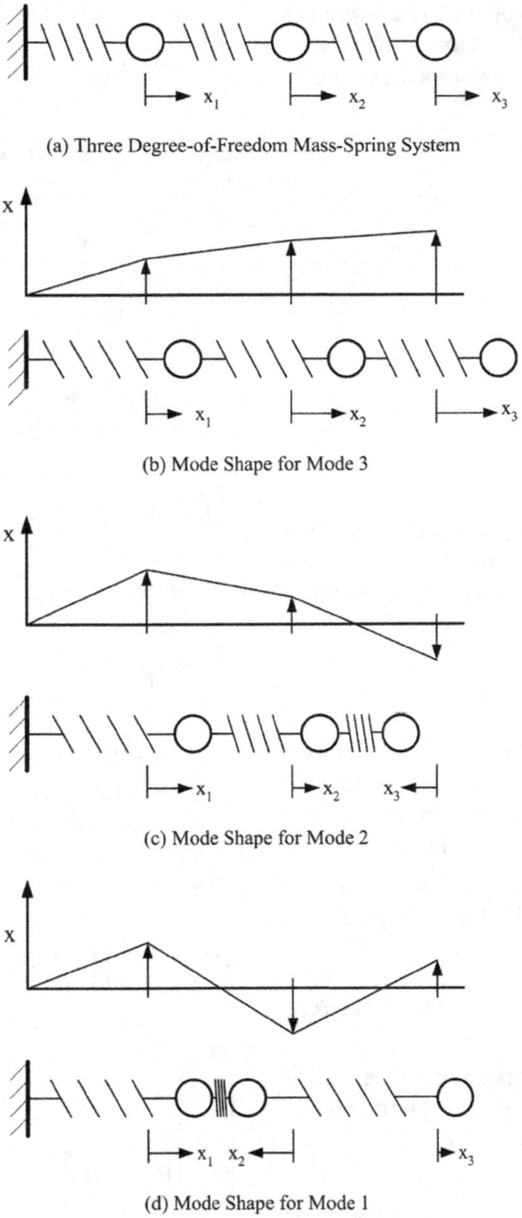

(a) Three Degree-of-Freedom Mass-Spring System

(b) Mode Shape for Mode 3

(c) Mode Shape for Mode 2

(d) Mode Shape for Mode 1

Figure 5.10 Mode Shapes for 3-DOF vibration System in Example 5.11

frequencies. The product $u_1^{(i)}\, C_i$ is the product of two arbitrary constants. It can be replaced with another arbitrary constant. Let:

$$u_1^{(i)}\, C_i = A_i. \tag{5.186}$$

Therefore, equation (5.185) can be written:

$$\{q(t)\} = \sum_{i=1}^{n} \frac{\{u^{(i)}\}}{u_1^{(i)}} A_i \cos(\omega_i t + \phi_i). \tag{5.187}$$

Equation (5.187) indicates that, for the case of unforced

vibration, the response vector $\{q(t)\}$ can be written as the sum of the normalized mode vectors, each with its associated resonance frequency, proportional amplitude, and relative phase.

The initial condition equations are:

$$\{q(0)\} = \sum_{i=1}^{n} \frac{\{u^{(i)}\}}{u_1^{(i)}} A_i \cos\phi_i \tag{5.188}$$

and:

$$\{\dot{q}(0)\} = \sum_{i=1}^{n} \frac{\{u^{(i)}\}}{u_1^{(i)}} A_i \, \omega_i \, \sin\phi_i . \tag{5.189}$$

These equations result in 2n independent equations that can be used to obtain the values for A_i and ϕ_i. n is the number of degrees-of-freedom of the system. Refer to Example 4.2 in Section 4.2 for an example.

5.11 Orthogonality of Mode Vectors

Normalization is a systematic way to scale the mode vectors obtained from matrix equations similar to equation (5.11m). Amplitudes of the elements of these mode vectors can only be determine to within an arbitrary constant. To normalize the mode vector $\{u^{(i)}\}$ where i = 1, 2, 3, ..., n, a scalar α is applied to the vector such that:

$$\alpha\{u^{(i)}\}^T\left[\alpha\{u^{(i)}\}\right] = 1. \tag{5.190}$$

Carrying out the prescribed operations in equation (5.190) and then rearranging the resulting terms yields:

$$\alpha = \frac{1}{\sqrt{\{u^{(i)}\}^T\{u^{(i)}\}}}. \tag{5.191}$$

The denominator of equation (5.191) is obtained from:

$$\sqrt{\{u^{(i)}\}^T\{u^{(i)}\}} = \sqrt{\sum_{j=1}^{n}\left[u_j^{(i)}\right]^2}. \tag{5.192}$$

Equation (5.192) can be written:

$$\left\|\{u^{(i)}\}\right\| = \sqrt{\sum_{j=1}^{n}\left[u_j^{(i)}\right]^2} \tag{5.193}$$

where $\left\|\{u^{(i)}\}\right\|$ is referred to as the *norm* of $\{u^{(i)}\}$. The value of the *normalized vector* $\{u_o^{(i)}\}$ is given by:

$$\{u_o^{(i)}\} = \frac{\{u^{(i)}\}}{\left\|\{u^{(i)}\}\right\|}. \tag{5.194}$$

Equation (5.194) is referred to as the *unit normal vector* that is characterized by the equation:

$$\{u_o^{(i)}\}^T\{u_o^{(i)}\} = 1. \tag{5.195}$$

If in addition:

$$\{u_o^{(i)}\}^T\{u_o^{(j)}\} = 0 \tag{5.196}$$

for the case when i $\neq$ j, the vectors $\{u_o^{(i)}\}$ are *orthogonal*. If the vectors $\{u_o^{(i)}\}$ are both *unit normal* and *orthogonal*, they are then *orthonormal vectors*. The *orthonormal matrix* [P] made up of the *orthonormal vectors* $\{u_o^{(i)}\}$ can be written:

$$[P] = \left[\{u_o^{(1)}\}\{u_o^{(2)}\}\{u_o^{(3)}\}\cdots\{u_o^{(n)}\}\right] \tag{5.197}$$

where:

$$\boxed{[P]^T[P] = [I] = \begin{bmatrix} 1 & 0 & \cdots & 0 \\ 0 & 1 & \cdots & 0 \\ \vdots & \vdots & \ddots & \vdots \\ 0 & 0 & \cdots & 1 \end{bmatrix}}. \tag{5.198}$$

EXAMPLE 5.12

Determine the orthonormal vectors of the vibration modes of the system in Examples 5.9 and 5.11.

SOLUTION

The normalized mode vectors from Example 5.11 are:

$$\left\{u^{(1)}\right\} = u_1^{(1)} \begin{Bmatrix} 1.0000 \\ -0.1716 \\ 0.1464 \end{Bmatrix} \quad (5.12a)$$

$$\left\{u^{(2)}\right\} = u_1^{(2)} \begin{Bmatrix} 1.0000 \\ 0.7647 \\ -3.2493 \end{Bmatrix} \quad (5.12b)$$

$$\left\{u^{(3)}\right\} = u_1^{(3)} \begin{Bmatrix} 1.0000 \\ 1.9068 \\ 2.1028 \end{Bmatrix}. \quad (5.12c)$$

The values for $\left\|\left\{u^{(1)}\right\}\right\|$, $\left\|\left\{u^{(2)}\right\}\right\|$, and $\left\|\left\{u^{(3)}\right\}\right\|$ are:

$$\left\|\left\{u^{(1)}\right\}\right\| = \sqrt{1^2 + (-0.1716)^2 + 0.1464^2}\, u_1^{(1)} = 1.0251\, u_1^{(1)} \quad (5.12d)$$

$$\left\|\left\{u^{(2)}\right\}\right\| = \sqrt{1^2 + 0.7647^2 + (-3.2493)^2}\, u_1^{(2)} = 3.4846\, u_1^{(2)} \quad (5.12e)$$

$$\left\|\left\{u^{(3)}\right\}\right\| = \sqrt{1^2 + 1.9068^2 + 2.1028^2}\, u_1^{(3)} = 3.0096\, u_1^{(3)}. \quad (5.12f)$$

The corresponding unit normal vectors are:

$$\boxed{\left\{u_o^{(1)}\right\} = \frac{u_1^{(1)}}{1.0251\, u_1^{(1)}} \begin{Bmatrix} 1.0000 \\ -0.1716 \\ 0.1464 \end{Bmatrix} = \begin{Bmatrix} 0.9755 \\ -0.1674 \\ 0.1428 \end{Bmatrix}} \quad (5.12g)$$

$$\boxed{\left\{u_o^{(2)}\right\} = \frac{u_1^{(2)}}{3.4846 u_1^{(2)}} \begin{Bmatrix} 1.0000 \\ 0.7647 \\ -3.2493 \end{Bmatrix} = \begin{Bmatrix} 0.2870 \\ 0.2194 \\ -0.9325 \end{Bmatrix}} \quad (5.12h)$$

$$\boxed{\left\{u_o^{(3)}\right\} = \frac{u_1^{(3)}}{3.0096 u_1^{(3)}} \begin{Bmatrix} 1.0000 \\ 1.9068 \\ 2.1028 \end{Bmatrix} = \begin{Bmatrix} 0.3323 \\ 0.6336 \\ 0.6987 \end{Bmatrix}.} \quad (5.12i)$$

The unit normal mode matrix [S] made up of the unit normal mode vectors obtained from the dynamical matrix is:

$$[S] = \begin{bmatrix} 0.9755 & 0.2870 & 0.3323 \\ -0.1674 & 0.2194 & 0.6336 \\ 0.1428 & -0.9325 & 0.6987 \end{bmatrix}. \quad (5.12j)$$

$[S]^T$ is:

$$[S]^T = \begin{bmatrix} 0.9755 & -0.1674 & 0.1428 \\ 0.2870 & 0.2194 & -0.9325 \\ 0.3323 & 0.6336 & 0.6987 \end{bmatrix}. \quad (5.12k)$$

and:

$$\boxed{\begin{aligned} [S]^T[S] &= \begin{bmatrix} 0.9755 & -0.1674 & 0.1428 \\ 0.2870 & 0.2194 & -0.9325 \\ 0.3323 & 0.6336 & 0.6987 \end{bmatrix} \times \begin{bmatrix} 0.9755 & 0.2870 & 0.3323 \\ -0.1674 & 0.2194 & 0.6336 \\ 0.1428 & -0.93256 & 0.6987 \end{bmatrix} \\ &= \begin{bmatrix} 1.0000 & 0.31790 & 0.4171 \\ 0.3179 & 1.0000 & -0.1101 \\ 0.4171 & -0.1101 & 1.0000 \end{bmatrix}. \end{aligned}} \quad (5.12l)$$

The diagonal terms in equation (5.12l) have values of 1.000, but the off-diagonal terms are not zero. Therefore, the mode vectors in the [S] matrix are unit normal vectors, but they are not orthonormal vectors. More will be discussed about this in Chapter 11. Basic matrix operations are described in Appendix 5A at the end of this chapter.

5.12 Semi-definite Systems

From equations (5.147) and (5.155), the kinetic and potential energies of a n-DOF vibration system can be expressed:

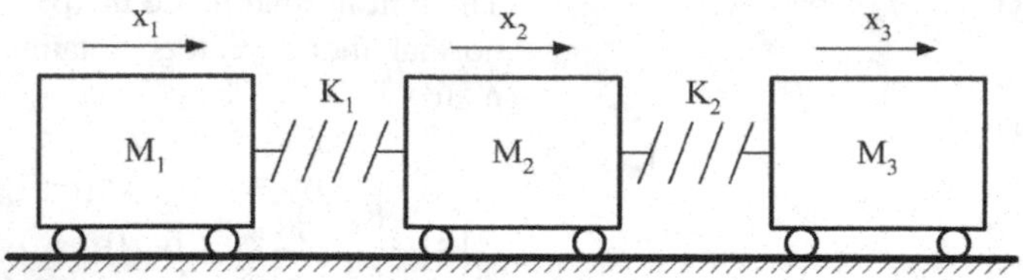

Figure 5.11 3-DOF Semi-definite Vibration System with Mass and Spring Elements

$$T = \frac{1}{2}\{\dot{q}\}^T [m] \{\dot{q}\} \tag{5.199}$$

and:

$$U = \frac{1}{2}\{q\}^T [k] \{q\} \tag{5.200}$$

where [m] and [k] are symmetric matrices. The kinetic energy is always positive; therefore, [m] is always a positive definite matrix. However, systems exist where the potential energy can equal zero without the corresponding displacements being equal to zero. The stiffness matrix [k] for such a system is a positive semi-definite matrix. A system where the mass matrix is a positive definite matrix and the stiffness matrix is a positive semi-definite matrix is a *semi-definite system*. The potential energy U = 0 does not correspond to the equilibrium position of a semi-definite system. A system that is not directly coupled to an inertial reference (or ground) is an example of a semi-definite system. Such systems are referred to as *unrestrained systems*. Figure 5.11 shows an example of an unrestrained semi-definite system. It consists of three mass elements (m_1, m_2 and m_3) that are coupled together by two spring elements (k_1 and k_2). None of the masses are coupled to "ground".

Pre-multiplying equation (5.174) by $\{u\}^T$ and rearranging the terms yields:

$$\omega^2 \{u\}^T [m]\{u\} = \{u\}^T [k]\{u\}. \tag{5.201}$$

For the ith mode:

$$\omega_i^2 \left\{u^{(i)}\right\}^T [m]\left\{u^{(i)}\right\} = \left\{u^{(i)}\right\}^T [k]\left\{u^{(i)}\right\}. \tag{5.202}$$

Let $u^{(o)}$ be a modal displacement vector that results in a potential energy that equals zero. Therefore, equation (5.202) becomes:

$$\omega_o^2 \left\{u^{(o)}\right\}^T [m]\left\{u^{(o)}\right\} = 0. \tag{5.203}$$

Since:

$$\left\{u^{(o)}\right\}^T [m]\left\{u^{(o)}\right\} \neq 0 \tag{5.204}$$

then $\omega_o = 0$. The mode that corresponds to ω_o is referred to as the *zero* or *rigid-body mode*.

EXAMPLE 5.13

Determine the resonance frequencies and corresponding modal vectors of the semi-definite system shown in Figure 5.11 when $M_1 = M_2 = M_3 = M$ and $K_1 = K_2 = K$.

SOLUTION

The kinetic and potential energy expressions for the system are:

$$T = \frac{1}{2} M_1 \dot{x}_1^2 + \frac{1}{2} M_2 \dot{x}_2^2 + \frac{1}{2} M_2 \dot{x}_3^2 \tag{5.13a}$$

and:

$$U = \frac{1}{2} K_1 (x_1 - x_2)^2 + \frac{1}{2} K_2 (x_2 - x_3)^2 . \tag{5.13b}$$

Lagrange's equation can be used to write these equations in matrix form. Therefore:

$$\begin{bmatrix} M_1 & 0 & 0 \\ 0 & M_2 & 0 \\ 0 & 0 & M_3 \end{bmatrix} \begin{Bmatrix} \ddot{x}_1 \\ \ddot{x}_2 \\ \ddot{x}_3 \end{Bmatrix} + \begin{bmatrix} K_1 & -K_1 & 0 \\ -K_1 & K_1+K_2 & -K_2 \\ 0 & -K_2 & K_2 \end{bmatrix} \begin{Bmatrix} x_1 \\ x_2 \\ x_3 \end{Bmatrix} = \begin{Bmatrix} 0 \\ 0 \\ 0 \end{Bmatrix}. \tag{5.13c}$$

Substituting the values for M_1, M_2, M_3, K_1, and K_2 yields:

$$M \begin{bmatrix} 1 & 0 & 0 \\ 0 & 1 & 0 \\ 0 & 0 & 1 \end{bmatrix} \begin{Bmatrix} \ddot{x}_1 \\ \ddot{x}_2 \\ \ddot{x}_3 \end{Bmatrix} + K \begin{bmatrix} 1 & -1 & 0 \\ -1 & 2 & -1 \\ 0 & -1 & 1 \end{bmatrix} \begin{Bmatrix} x_1 \\ x_2 \\ x_3 \end{Bmatrix} = \begin{Bmatrix} 0 \\ 0 \\ 0 \end{Bmatrix}. \tag{5.13d}$$

Letting:

$$x_1(t) = u_1\, e^{j\omega t};\ x_2(t) = u_2\, e^{j\omega t};\ x_3(t) = u_3\, e^{j\omega t} \tag{5.13e}$$

equation (5.13d) can be expressed:

$$\left[-\omega^2 M \begin{bmatrix} 1 & 0 & 0 \\ 0 & 1 & 0 \\ 0 & 0 & 1 \end{bmatrix} + K \begin{bmatrix} 1 & -1 & 0 \\ -1 & 2 & -1 \\ 0 & -1 & 1 \end{bmatrix} \right] \begin{Bmatrix} u_1 \\ u_2 \\ u_3 \end{Bmatrix} = \begin{Bmatrix} 0 \\ 0 \\ 0 \end{Bmatrix} \tag{5.13f}$$

or:

$$\begin{bmatrix} \lambda-1 & 1 & 0 \\ 1 & \lambda-2 & 1 \\ 0 & 1 & \lambda-1 \end{bmatrix} \begin{Bmatrix} u_1 \\ u_2 \\ u_3 \end{Bmatrix} = \begin{Bmatrix} 0 \\ 0 \\ 0 \end{Bmatrix} \tag{5.13g}$$

where:

$$\lambda = \frac{\omega^2 M}{K}. \tag{5.13h}$$

Setting the determinant of the matrix in equation (5.13g) equal to zero and simplifying yields:

$$\lambda(\lambda-1)(\lambda-3) = 0. \tag{5.13i}$$

Therefore:

$$\lambda_1 = 0; \quad \lambda_2 = 1; \quad \text{and} \quad \lambda_3 = 3 \tag{5.13j}$$

and:

$$\omega_1 = 0; \quad \omega_2 = \sqrt{\frac{M}{K}}; \quad \text{and} \quad \omega_3 = \sqrt{\frac{3M}{K}}. \tag{5.13k}$$

The modal vectors are obtained from the equations:

$$(\lambda-1)\,u_1 + u_2 = 0 \tag{5.13l}$$

$$u_1 + (\lambda-2)\,u_2 + u_3 = 0 \tag{5.13m}$$

$$u_2 + (\lambda-1)\,u_3 = 0. \tag{5.13n}$$

Using equations (5.13l) and (5.13m):

$$\frac{u_2}{u_1} = 1-\lambda \quad \text{and} \quad \frac{u_3}{u_1} = \left[(\lambda-2)(\lambda-1)-1\right]. \tag{5.13o}$$

The modal vectors that correspond to the three resonance frequencies are:

$$\boxed{\omega_1 = 0:} \quad \boxed{\left\{u^{(1)}\right\} = u_1^{(1)} \begin{Bmatrix} 1 \\ 1 \\ 1 \end{Bmatrix}} \tag{5.13p}$$

$$\boxed{\omega_2 = \sqrt{\frac{K}{M}}:} \quad \boxed{\left\{u^{(2)}\right\} = u_1^{(2)} \begin{Bmatrix} 1 \\ 0 \\ -1 \end{Bmatrix}} \tag{5.13q}$$

$$\boxed{\omega_3 = \sqrt{\frac{3\,K}{M}}:} \quad \boxed{\left\{u^{(3)}\right\} = u_1^{(3)}\begin{Bmatrix}1\\-2\\1\end{Bmatrix}.} \tag{5.13r}$$

For the first mode when $\omega_1 = 0$ and $u_1^{(1)} = u_2^{(1)} = u_2^{(1)}$, equation (5.13b) indicates the potential energy equals zero.

APPENDIX 5A - MATRIX ALGEBRA

A5.1 Introduction

This chapter discusses the development of systems of coupled 2nd order differential equations that describe the motions of multiple-DOF, lumped-parameter vibration systems. These systems of equations are presented in matrix format that must be solved by using operations associated with matrix algebra. Definitions and operations associated with matrices are presented in this appendix. These definitions and operations are used in Chapter 11 where principles associated with modal analysis are discussed.

A5.2 Definitions

Vector: a displacement vector associated with a n-DOF vibration system is expressed:

$$\{u\} = \begin{Bmatrix} u_1 \\ u_2 \\ \vdots \\ u_n \end{Bmatrix}. \qquad \text{(A1.1}$$

Matrix: a $n \times n$ matrix [A] is expressed:

$$[A] = \begin{bmatrix} a_{11} & a_{12} & \cdots & a_{1n} \\ a_{21} & a_{22} & \cdots & a_{2n} \\ \vdots & \vdots & \ddots & \\ a_{n1} & a_{n2} & & a_{nn} \end{bmatrix}. \qquad \text{(A5.2)}$$

For the elements a_{ij} of the matrix, i specifies the row number of the matrix and j specifies the column number. The numbers start at the upper left corner of the matrix and progressively increase to the lower right corner.

Determinant: the nth-order determinant of matrix [A] is expressed:

$$|A| = \begin{vmatrix} a_{11} & a_{12} & \cdots & a_{1n} \\ a_{21} & a_{22} & \cdots & a_{2n} \\ \vdots & \vdots & \ddots & \\ a_{n1} & a_{n2} & & a_{nn} \end{vmatrix}. \qquad \text{(A5.3)}$$

If the determinant |A| is a 2nd order determinant given by:

$$|A| = \begin{vmatrix} a_{11} & a_{12} \\ a_{21} & a_{22} \end{vmatrix}, \qquad \text{(A5.4)}$$

the amplitude of |A| is given by:

$$|A| = a_{11} \times a_{22} - a_{12} \times a_{21}. \qquad \text{(A5.5)}$$

Minor: The minor M_{ij} associated with an element a_{ij} in a determinant is a determinant where the ith row and the jth column from the original determinant are deleted. For example, if [A] is:

$$|A| = \begin{vmatrix} a_{11} & a_{12} & a_{13} \\ a_{21} & a_{22} & a_{23} \\ a_{31} & a_{32} & a_{33} \end{vmatrix}, \qquad \text{(A5.6)}$$

then M_{12} is given by:

$$M_{12} = \begin{vmatrix} a_{21} & a_{23} \\ a_{31} & a_{33} \end{vmatrix}. \qquad \text{(A5.7)}$$

Cofactor: The cofactor C_{ij} associated with an element, a_{ij}, of a determinant is expressed:

$$C_{ij} = (-1)^{i+j} M_{ij}. \qquad \text{(A5.8)}$$

With respect to equation (A5.7),

$$C_{12} = (-1)^{1+2} M_{12} = -M_{12}. \qquad \text{(A5.9)}$$

The numerical value of a determinant can be obtained by summing the cofactors associated with any row or column of the determinant. With regard to equation (A5.6), for row i:

$$|A| = a_{i1}C_{i1} + a_{i2}C_{i2} + a_{i3}C_{i3} = \sum_{j=1}^{3} a_{ij}C_{ij} \qquad \text{(A5.10)}$$

and for column j:

$$|A| = a_{1j}C_{1j} + a_{2j}C_{2j} + a_{3j}C_{3j} = \sum_{j=1}^{3} a_{ij}C_{ij} . \tag{A5.11}$$

Symmetric matrix: the matrix [A] is a symmetric matrix when the elements a_{ij} of the matrix have the characteristic:

$$a_{ij} = a_{ji} . \tag{A5.12}$$

Trace: the trace of a matrix [A] is the sum of the diagonal elements of the matrix:

$$\text{trace}[A] = a_{11} + a_{22} + \cdots + a_{nn} . \tag{A5.13}$$

Singular matrix: a singular matrix [A] is a matrix where the determinant |A| of the matrix equals zero.

Diagonal matrix: a diagonal matrix [A] is a matrix where all of the off-diagonal elements in the matrix equal zero:

$$[A] = \begin{bmatrix} a_{11} & 0 & \cdots & 0 \\ 0 & a_{22} & \cdots & 0 \\ \vdots & \vdots & \ddots & \\ 0 & 0 & & a_{nn} \end{bmatrix} . \tag{A5.14}$$

Unit or identity matrix: a unit or identity matrix [I] is a matrix where the diagonal elements equal 1 and the off-diagonal elements equal zero:

$$[I] = \begin{bmatrix} 1 & 0 & \cdots & 0 \\ 0 & 1 & \cdots & 0 \\ \vdots & \vdots & \ddots & \\ 0 & 0 & & 1 \end{bmatrix} . \tag{A5.15}$$

Transpose: The transpose matrix $[A]^T$ of matrix [A] is a matrix where the row elements are interchanged with the column elements. If matrix [A] is given by:

$$[A] = \begin{bmatrix} a_{11} & a_{12} & a_{13} \\ a_{21} & a_{22} & a_{23} \\ a_{31} & a_{32} & a_{33} \end{bmatrix} \tag{A5.16}$$

then:

$$[A]^T = \begin{bmatrix} a_{11} & a_{21} & a_{31} \\ a_{12} & a_{22} & a_{32} \\ a_{13} & a_{23} & a_{33} \end{bmatrix} . \tag{A5.17}$$

If [A] is a symmetric matrix:

$$[A]^T = [A] . \tag{A5.18}$$

If the vector {u} is given by:

$$\{u\} = \begin{Bmatrix} u_1 \\ u_2 \\ u_3 \end{Bmatrix} \tag{A5.19}$$

then:

$$\{u\}^T = \{u_1 \quad u_2 \quad u_3\} . \tag{A5.20}$$

Adjoint matrix: An adjoint of a square matrix [A] is the transpose of the matrix of cofactors of [A]. If the cofactor matrix [C] is given by:

$$[C] = \begin{bmatrix} C_{11} & C_{12} & C_{13} \\ C_{21} & C_{22} & C_{23} \\ C_{31} & C_{32} & C_{33} \end{bmatrix} \tag{A5.21}$$

then,

$$\text{adjoint}[A] = [C]^T \tag{A5.22}$$

or,

$$\text{adjoint}[A]=\begin{bmatrix} C_{11} & C_{21} & C_{31} \\ C_{12} & C_{22} & C_{32} \\ C_{13} & C_{23} & C_{33} \end{bmatrix}. \tag{A5.23}$$

Inverse matrix: the inverse of matrix [A] is given by:

$$[A]^{-1}=\frac{\text{adjoint}[A]}{|A|}. \tag{A5.24}$$

Skew symmetric matrix: the matrix [A] is skew symmetric if:

$$[A]=-[A]^T. \tag{A5.25}$$

Positive definite matrix: a matrix [A] is a positive definite matrix if:

$$\{u\}^T[A]\{u\}>0 \tag{A5.26}$$

for all vectors $[\{u\} \neq 0]$.

Semi-definite positive matrix: a matrix [A] is a semi-definite positive matrix if:

$$\{u\}^T[A]\{u\}\geq 0 \tag{A5.27}$$

for all vectors {u}. The matrix [A] is also referred to as a *positive matrix*.

Indefinite matrix: the matrix [A] is an indefinite matrix if:

$$\left(\{u\}^T[A]\{u\}\right)\left(\{v\}^T[A]\{v\}\right)<0 \tag{A5.28}$$

for all vectors {u} and {v}.

Orthogonal matrix: the matrix [A] is an orthogonal matrix if:

$$[A]^T[A]=[A][A]^T=[I]. \tag{A5.29}$$

A5.3 Vector and Matrix Operations

Addition of matrices:

$$[A]=\begin{bmatrix} a_{11} & a_{12} \\ a_{21} & a_{22} \end{bmatrix};\quad [B]=\begin{bmatrix} b_{11} & b_{12} \\ b_{21} & b_{22} \end{bmatrix}. \tag{A5.30}$$

Then:

$$[C]=\begin{bmatrix} c_{11} & c_{12} \\ c_{21} & c_{22} \end{bmatrix}=\begin{bmatrix} a_{11}+b_{11} & a_{12}+b_{12} \\ a_{21}+b_{21} & a_{22}+b_{22} \end{bmatrix}. \tag{A5.31}$$

Multiplication of a matrix by a constant:

$$\alpha[A]=\begin{bmatrix} \alpha a_{11} & \alpha a_{12} \\ \alpha a_{21} & \alpha a_{22} \end{bmatrix}. \tag{A5.32}$$

Multiplication of vectors:

$$\{u\}=\begin{Bmatrix} u_1 \\ u_2 \\ u_3 \end{Bmatrix};\quad \{v\}=\begin{Bmatrix} v_1 \\ v_2 \\ v_3 \end{Bmatrix}. \tag{A5.33}$$

Then:

$$\{u\}^T\{v\}=u_1v_1+u_2v_2+u_3v_3 \tag{A5.34}$$

and:

$$\{u\}\{v\}^T=\begin{bmatrix} u_1v_1 & u_1v_2 & u_1v_3 \\ u_2v_1 & u_2v_2 & u_2v_3 \\ u_2v_1 & u_2v_2 & u_3v_3 \end{bmatrix}. \tag{A5.35}$$

Multiplication of matrices:

$$[A]=\begin{bmatrix} a_{11} & a_{12} & a_{12} \\ a_{21} & a_{22} & a_{23} \\ a_{31} & a_{32} & a_{33} \end{bmatrix};\quad [B]=\begin{bmatrix} b_{11} & b_{12} & b_{12} \\ b_{21} & b_{22} & b_{23} \\ b_{31} & b_{32} & b_{33} \end{bmatrix}. \tag{A5.36}$$

Therefore

$$[D]=[A][B]=\begin{bmatrix} d_{11} & d_{12} & d_{12} \\ d_{21} & d_{22} & d_{23} \\ d_{31} & d_{32} & d_{33} \end{bmatrix} \tag{A5.37}$$

where:

$$d_{ij}=\sum_{k=1}^{3} a_{ik} b_{kj}. \tag{A5.38}$$

Multiplication of a matrix and vector:

$$[A]=\begin{bmatrix} a_{11} & a_{12} & a_{12} \\ a_{21} & a_{22} & a_{23} \\ a_{31} & a_{32} & a_{33} \end{bmatrix}; \quad \{u\}=\begin{Bmatrix} u_1 \\ u_2 \\ u_3 \end{Bmatrix}. \tag{A5.39}$$

Then:

$$\{v\}=\begin{bmatrix} a_{11} & a_{12} & a_{12} \\ a_{21} & a_{22} & a_{23} \\ a_{31} & a_{32} & a_{33} \end{bmatrix}\begin{Bmatrix} u_1 \\ u_2 \\ u_3 \end{Bmatrix}=\begin{Bmatrix} v_1 \\ v_2 \\ v_3 \end{Bmatrix} \tag{A5.40}$$

where:

$$v_i=\sum_{k=1}^{3} a_{ik}\, v_k \tag{A5.41}$$

and:

$$\{v\}^T=\begin{Bmatrix} u_1 \\ u_2 \\ u_3 \end{Bmatrix}^T\begin{bmatrix} a_{11} & a_{12} & a_{12} \\ a_{21} & a_{22} & a_{23} \\ a_{31} & a_{32} & a_{33} \end{bmatrix}=\begin{Bmatrix} v_1 \\ v_2 \\ v_3 \end{Bmatrix}^T \tag{A5.42}$$

where:

$$v_j=\sum_{k=1}^{3} u_k\, a_{kj}. \tag{A5.43}$$

Other matrix operations:

$$[[A]+[B]]^T=[A]^T+[B]^T \tag{A5.44}$$

$$[[A][B]]^T=[B]^T[A]^T \tag{A5.45}$$

$$\left[[A]^{-1}\right]^T=\left[[A]^T\right]^{-1} \tag{A5.46}$$

$$[[A][B]]^{-1}=[B]^{-1}[A]^{-1} \tag{A5.47}$$

A5.4 Eigenvalue Problem

The eigenvalue problem can be formulated with a matrix equation of the form:

$$[A]\{u\}=[\lambda]\{u\}. \tag{A5.48}$$

[A] is a $n \times n$ square matrix:

$$[A]=\begin{bmatrix} a_{11} & a_{12} & \cdots & a_{1n} \\ a_{21} & a_{22} & \cdots & a_{2n} \\ \vdots & \vdots & \ddots & \\ a_{n1} & a_{n2} & & a_{nn} \end{bmatrix} \tag{A5.49}$$

[λ] is a diagonal matrix:

$$[\lambda]=\begin{bmatrix} \lambda_1 & 0 & \cdots & 0 \\ 0 & \lambda_2 & \cdots & 0 \\ \vdots & \vdots & \ddots & \\ 0 & 0 & & \lambda_n \end{bmatrix}; \tag{A5.50}$$

and $\{u\}$ is a vector:

$$\{u\}=\begin{Bmatrix} u_1 \\ u_2 \\ \vdots \\ u_n \end{Bmatrix}. \tag{A5.51}$$

When equation (A5.48) is satisfied, the values for λ_i are the *eigenvalues*, and the vectors $\{u^{(i)}\}$ are the *eigenvectors* associated with each value for λ_i. With regard to a solution to equation (A5.48), one of the values for λ_i can equal zero. However, the corresponding eigenvector $\{u^{(i)}\}$ cannot be $\{0\}$.

The matrix equation for an undamped n-DOF, lumped-parameter vibration system can be written:

$$[m]\{\ddot{x}(t)\}+[k]\{x(t)\}=\{0\} \tag{A5.52}$$

where $\{x(t)\}=\{u\}e^{j\omega t}$, [m] is a diagonal mass matrix, [k] is the stiffness matrix, and {u} is the displacement vector associated with the spatial coordinates of the system.

Dynamical matrix solution: When [A] in equation (A5.48) is the *dynamical* matrix given by:

$$[A]=[k]^{-1}[m] \tag{A5.53}$$

the eigenvalues associated with equation (A5.48) are given in the form:

$$\lambda_i=\frac{1}{\omega_i^2} \tag{A5.54}$$

and the corresponding eigenvectors $\{u^{(i)}\}$ are given by:

$$\{u^{(i)}\}=\begin{Bmatrix} u_1^{(i)} \\ u_2^{(i)} \\ \vdots \\ u_n^{(i)} \end{Bmatrix} \tag{A5.55}$$

where there are n eigenvalues (and corresponding resonance frequencies) and n eigenvectors (mode vectors specified in terms of the system spatial coordinates). For the dynamical matrix solution $\{u^{(i)}\}$:

$$\begin{aligned}\{u^{(i)}\}^T\{u^{(j)}\}&=1 \quad \text{for} \quad i=j\\ &\neq 0 \quad \text{for} \quad i\neq j.\end{aligned} \tag{A5.56}$$

Therefore, the mode vectors $\{u^{(i)}\}$ are *unit normal vectors*. However, they are not orthogonal vectors. When $\{u^{(i)}\}$ are not unit normal vectors, they can be converted to unit normal vectors by:

$$\{u_o^{(i)}\}=\frac{\{u^{(i)}\}}{\left\|\{u^{(i)}\}\right\|} \tag{A5.57}$$

where:

$$\left\|\{u^{(i)}\}\right\|=\sqrt{\{u^{(i)}\}^T\{u^{(i)}\}}. \tag{A5.58}$$

$\left\|\{u^{(i)}\}\right\|$ is referred to as the *norm* of $\{u^{(i)}\}$. The mode matrix [S] in terms of the unit normal vectors is given by:

$$[S]=\left[\{u_o^{(1)}\}\ \{u_o^{(2)}\}\ \cdots\ \{u_o^{(n)}\}\right]. \tag{A5.59}$$

[S] is referred to as the *matrix of mode vectors*. When the eigenvectors $\{u^{(i)}\}$ associated with [S] are unit normal vectors, the diagonal elements of the multiplication of $[S]^T[S]$ will be 1; however,

$$[S]^T[S]\neq[I]. \tag{A5.60}$$

Symmetric eigenvalues solution: Let the system spatial coordinates associated with {x(t)} in equation (A5.52) be transferred into another set of coordinates associated with {q(t)} where $\{q(t)\}=\{v\}e^{j\omega t}$ by the equation:

$$\{x(t)\}=[m]^{-1/2}\{q(t)\}. \tag{A5,61}$$

Also, let the matrix [A] in equation (A5.48) be the *mass-normalized stiffness matrix* given by:

$$[A]=[m]^{-1/2}[k][m]^{-1/2}. \tag{A5.62}$$

For this case, the eigenvalues associated with equation (A5.48) are given in the form:

$$\lambda_i=\omega_i^2 \tag{A5.63}$$

and the corresponding eigenvectors $\{v^{(i)}\}$ are given by:

$$\left\{v^{(i)}\right\} = \begin{Bmatrix} v_1^{(i)} \\ v_2^{(i)} \\ \vdots \\ v_n^{(i)} \end{Bmatrix} \tag{A5.64}$$

where there are n eigenvalues (and corresponding resonance frequencies) and n eigenvectors (modal vectors in the transformed coordinate system). For this case:

$$\begin{aligned} \left\{v^{(i)}\right\}^T \left\{v^{(j)}\right\} &= 1 \quad \text{for} \quad i = j \\ &= 0 \quad \text{for} \quad i \neq j. \end{aligned} \tag{A5.65}$$

The modal vectors are unit normal vectors, and they are orthogonal. Therefore, they are called *orthonormal modal vectors*. The *orthonormal modal matrix* [P] is given by:

$$[P] = \left[\left\{v^{(1)}\right\} \; \left\{v^{(2)}\right\} \; \cdots \; \left\{v^{(n)}\right\}\right] \tag{A5.66}$$

where:

$$[P]^T [P] = [I]. \tag{A5.67}$$

The relation between [S] and [P] is:

$$[S] = [m]^{-1/2} [P]. \tag{A5.68}$$

A5.4 Other Definitions

Quadratic function: a quadratic function is a polynomial function of the form:

$$f(x) = a x^2 + b x + c \tag{A.5.69}$$

where $a \neq 0$.

Positive definite quadratic function: a quadratic function described by equation (A.5.69) where $f(x) > 0$.

Positive semi-definite quadratic function: a quadratic function described by equation (A5.69) where $f(x) \geq 0$.

PROBLEMS - CHAPTER 5

1. Use d'Alembert's principle to develop the equations of motion for the system shown in Figure P5.1. Write the equations in matrix format and in terms of x_1, x_2 and x_3.

2. Use d'Alembert's principle to develop the equations of motion for the system shown in Figure P5.2. Write the equations in matrix format and in terms of x_1, x_2 and x_3.

3. Use d'Alembert's principle to develop the equations of motion for the system shown in Figure P5.3. Write the equations in matrix format and in terms of θ_1, θ_2 and θ_3. Assume no slipping between between the rollers and surface.

4. Use d'Alembert's principle to develop the equations of motion of the system shown in Figure P5.4. Write the equation in matrix format and in terms of θ_1, θ_2 and θ_3. Assume no slipping between the rollers and surface.

5. Use d'Alembert's principle to develop the equations of motion of the system shown in Figure P5.5. Write the equations in matrix format and in terms of x_1, x_2, x_3 and x_4.

6. Use d'Alembert's principle to develop the equations of motion of the system shown in Fiugre P5.6. Write the equation in matrix format and in terms of θ_1, θ_2 and θ_3.

7. Use d'Alembert's principle to develop the equations of motion of the system shown in Figure P5.7.

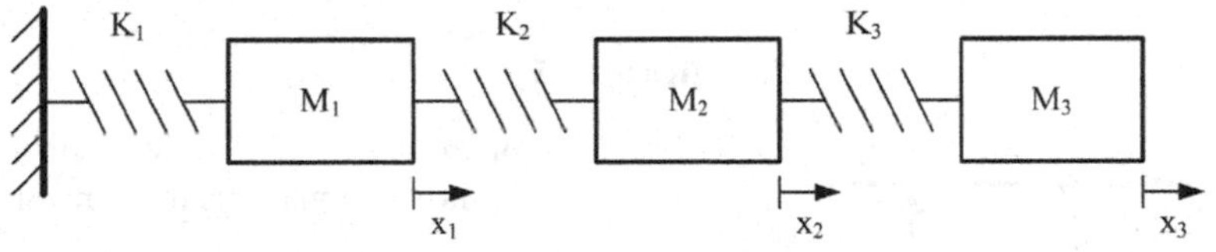

Figure P5.1

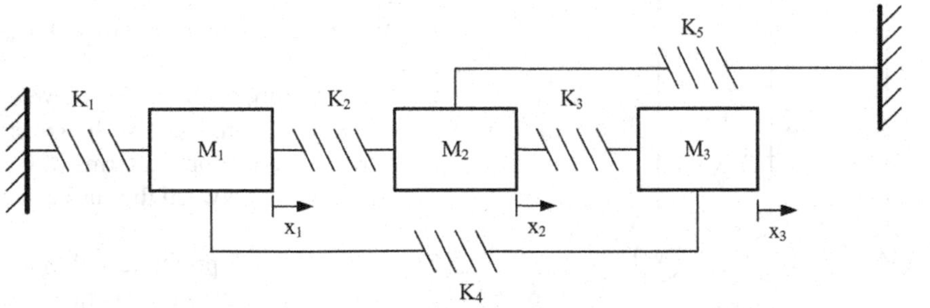

Figure P5.2

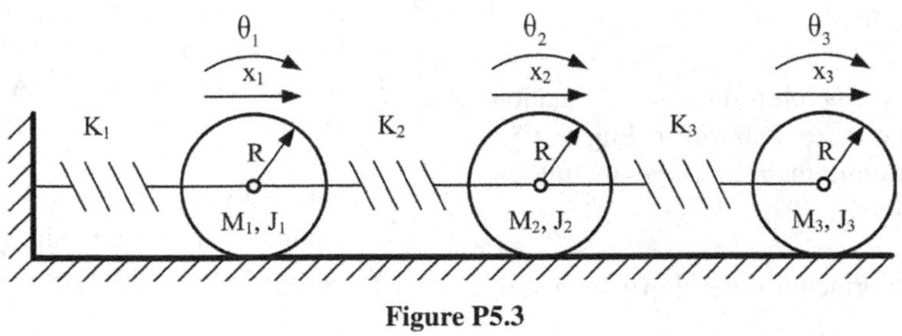

Figure P5.3

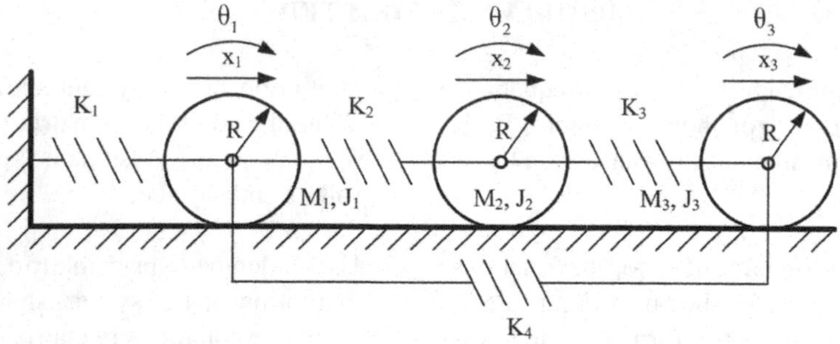

Figure P5.4

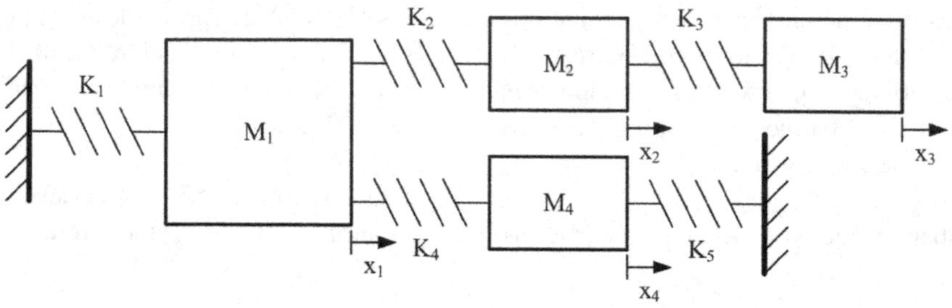

Figure P5.5

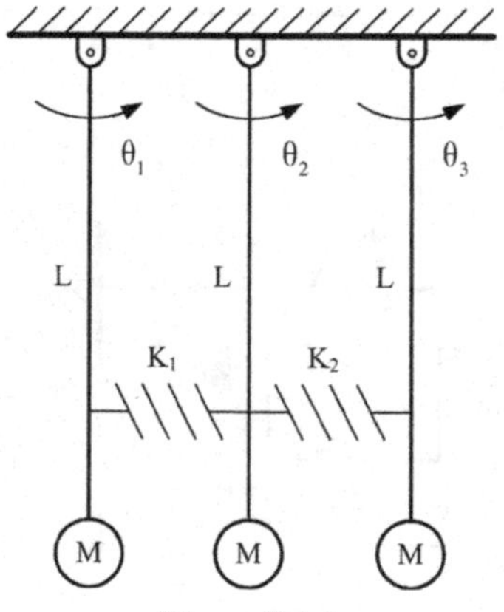

Figure P5.6

Write the equations in matrix format and in terms of x_1, x_2, x_3 and θ.

8. Use d'Alembert's principle to develop the equations of motion of the system shown in Fiugre P5.8. Write the equation in matrix format and in terms of θ_1, θ_2, θ_3 and y.

9. Use d'Alembert's principle to develop the equations of motion of the system shown in Figure P5.9. Write the equations in terms of x, θ and η.

10. Use d'Alembert's principle to develop the equations of motion of the system shown in Figure P5.10. Write the equations in terms of θ_1, θ_2 and η.

11. Use d'Alembert's principle to develop the equations of motion of the system shown in Figure P5.11. Write the equations in terms of θ and η. Assume no sllipping between the surface and roller.

12. Use Hamilton's principle to develop the equations of motion of the system shown in Figure P5.9. Write the equations in terms of x, θ and η.

13. Use Hamilton's principle to develop the equations of motion of the system shown in Figure P5.10. Write the equations θ_1, θ_2 and η. Assume no sllipping between the surface and roller.

14. Use Hamilton's principle to develop the equations of motion of the system shown in Figure P5.11. Write the equations in terms of θ and η.

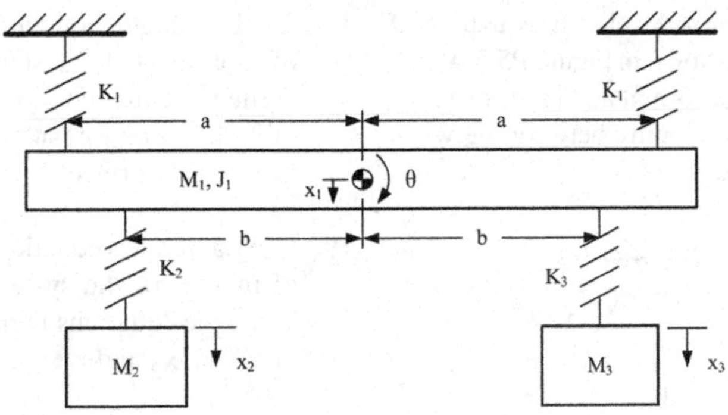

Figure P5.7

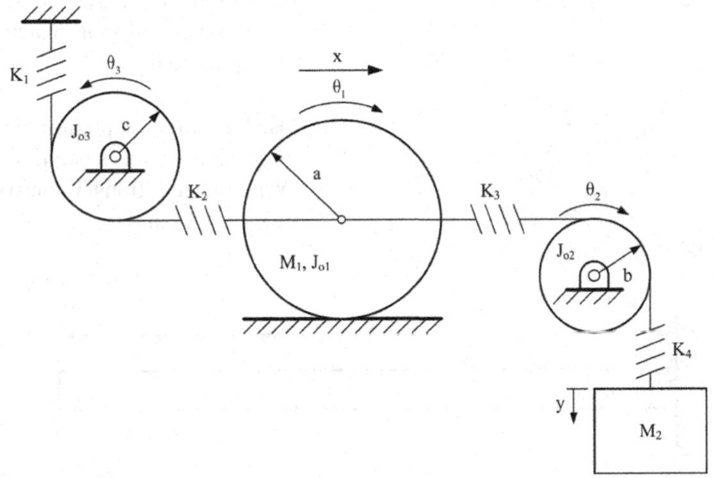

Figure P5.8

15. Use Lagrange's equation to develop the equations of motion for the system shown in Figure P5.1. Write the equations in matrix format and in terms of x_1, x_2 and x_3.

16. Use Lagrange's equation to develop the equations of motion for the system shown in Figure P5.2. Write the equations in matrix format and in terms of x_1, x_2 and x_3.

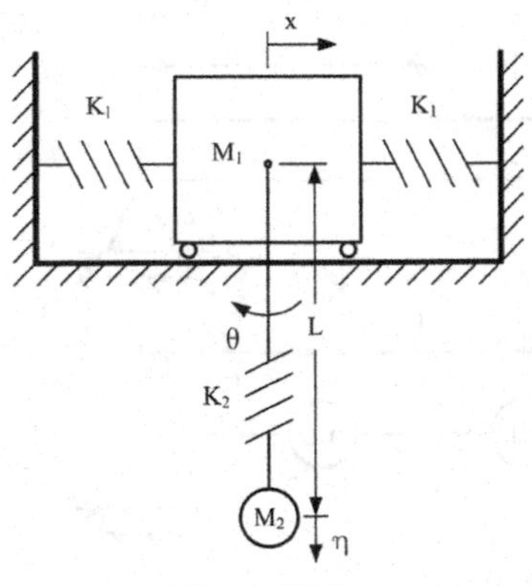

Figure P5.9

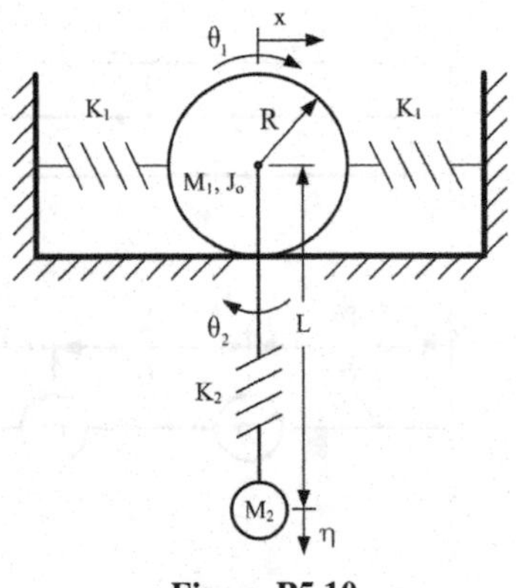

Figure P5.10

17. Use Lagrange's equation to develop the equations of motion for the system shown in Figure P5.3. Write the equations in matrix format and in terms of θ_1, θ_2 and θ_3. Assume on slipping between between the rollers and surface.

18. Use Lagrange's equation to develop the equations of motion of the system shown in Figure P5.4. Write the equation in matrix format and in terms of θ_1, θ_2 and θ_3. Assume no slipping between the rollers and surface.

19. Use Lagrange's equation to develop the equations of motion of the system shown in Figure P5.5. Write the equations in matrix format and in terms of x_1, x_2, x_3 and x_4.

20. Use Lagrange's equation to develop the equations of motion of the system shown in Fiugre P5.6. Write the equation in matrix format and in terms of θ_1, θ_2 and θ_3.

21. Use Lagrange's equation to develop the equations of motion of the system shown in Figure P5.7. Write the equations in matrix format and in terms of x_1, x_2, x_3 and θ.

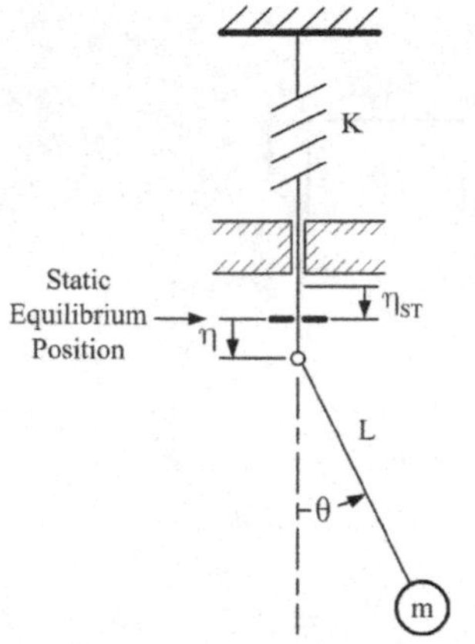

Figure P5.11

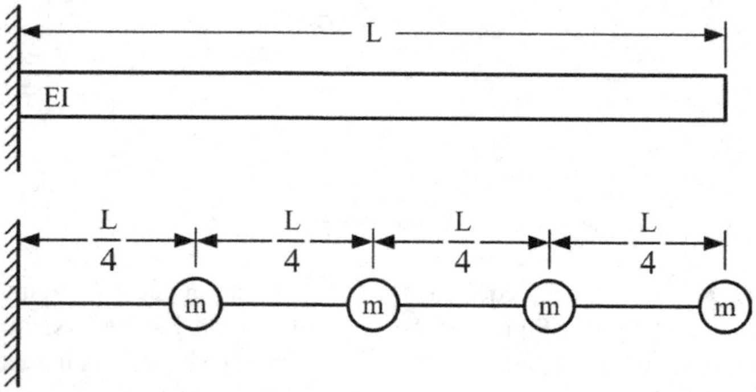

Figure P5.12

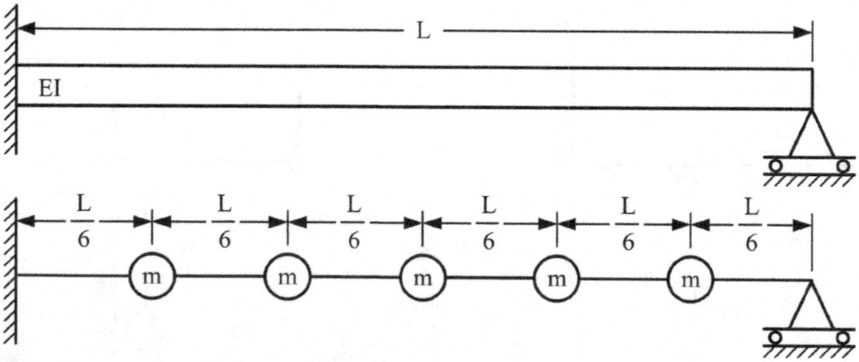

Figure P5.13

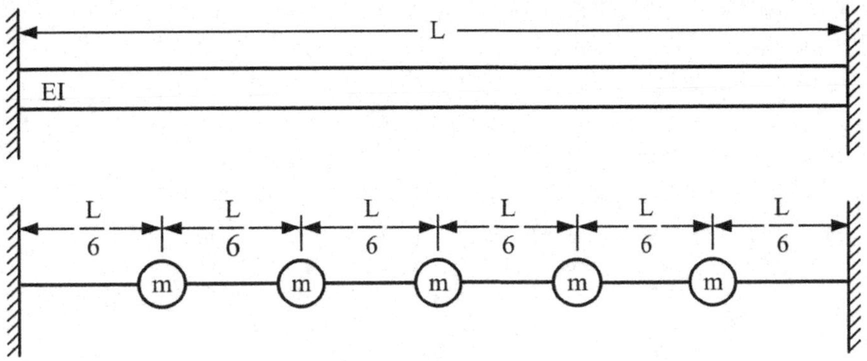

Figure P5.14

22. Use Lagrange's equation to develop the equations of motion of the system shown in Fiugre P5.8. Write the equation in matrix format and in terms of θ_1, θ_2 θ_3 and y.

23. Use Lagrange's equation to develop the equations of motion of the system shown in Figure P5.9. Write the equations in terms of x, θ and η.

24. Use Lagrange's equation to develop the equations of motion of the system shown in Figure P5.10. Write the equations in terms of θ_1, θ_2 and η. Assume no slipping between the surface and roller.

25. Use Lagrange's equation to develop the equations of motion of the system shown in Figure P5.11. Write the equations in terms of θ and η.

26. Figure P5.12 shows a 4-DOF approximation for a slender beam that is clamped at the left end and free at the right end. Use influence coefficients to develop the equations of motion for the beam.

27. Figure P5.13 shows a 5-DOF approximation for a slender beam that is clamped at the left end and simply supported at a position L from the left end. Use influence coefficients to develop the equations of motion for the beam.

28. Figure P5.14 shows a 5-DOF approximation for a slender beam that is clamped at the left and right ends. Use influence coefficients to develop the equations of motion for the beam.

CHAPTER 6
VIBRATION SYSTEMS
RESPONSE TO NON-PERIODIC EXCITATION

6.1 Introduction

Deterministic signals are one of two major classifications for signals that can excite vibration systems (See Figure 1.15.). They can be classified as *periodic* and *non-periodic* signals. *Periodic signal* repeats themselves at specified time intervals of nT, such that:

$$f(t) = f(t + nT) \quad \text{where} \quad n = 1, 2, 3, \ldots. \tag{6.1}$$

Discussions in the preceding five chapters addressed the response of vibration systems to periodic (harmonic) signals.

Non-periodic signal do not repeat themselves at specified time intervals of nT. Therefore:

$$f(t) \neq f(t + nT) \quad \text{where} \quad n = 1, 2, 3, \ldots. \tag{6.2}$$

Non-periodic signal can be classified as *almost-periodic* and *transient*. *Almost periodic signals* are comprised of harmonic components, and the response of vibration systems to these signals can be addressed as discussed in previous chapters.

Step and ramp functions shown in Figure 6.1 can be examined in a manner similar to transient signals. However, transient signals occur for a brief period of time and then disappear. For example, these signals can be square, triangular or sine pulses (Figure 6.2) and they can be signals that appear and then quickly decay with time to zero (Figure 6.3). Transient signals that excite vibration systems are often referred to as shock excitation.

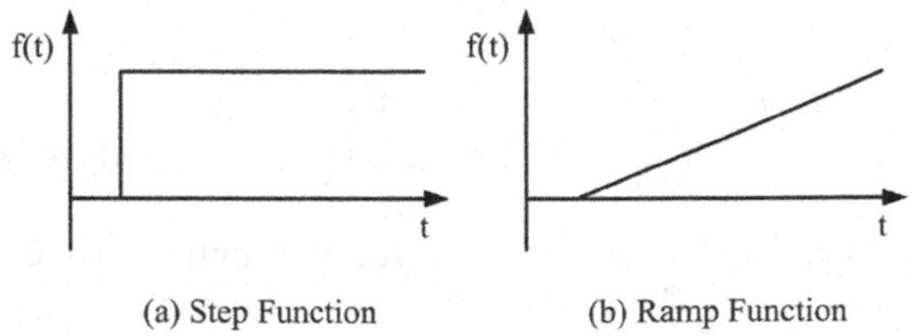

(a) Step Function (b) Ramp Function

Figure 6.1 Step and Ramp Functions

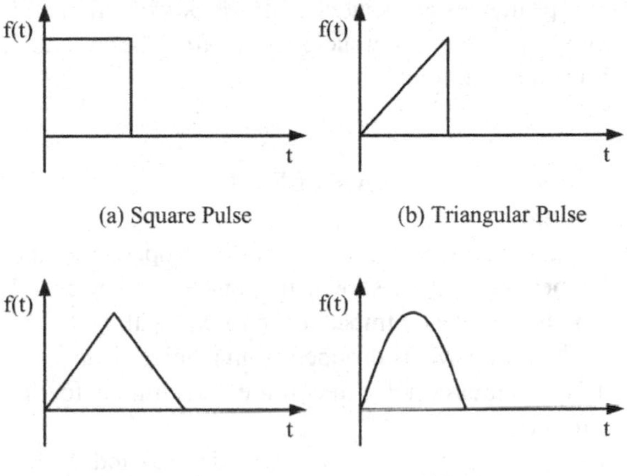

(a) Square Pulse (b) Triangular Pulse (c) Triangular Pulse (d) Sine Pulse

Figure 6.2 Transient Functions

The responses of linear vibration systems to transient excitation will be examined in this chapter. First, classical differential equation solution techniques will be used to examine the response of vibration systems to transient excitation. Then, Fourier and Laplace transform methods will be used to describe the response of vibration systems to transient signals.

6.2 Shock Excitation - Classical Approach

Vibration systems are often exposed to shock excitation.

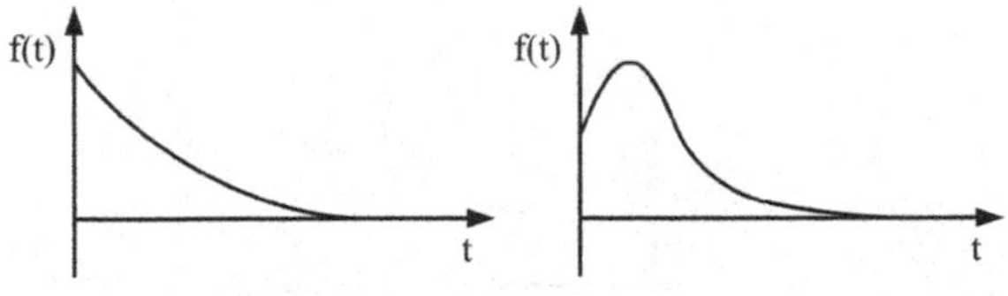

(a) Exponential Decay (b) Exponential Sine Decay

Figure 6.3 Transient Functions

Examples of shock excitation include, but are not limited to:

- The dropping of a delicate package onto a floor;
- A sudden jolt introduced to the base of an instrument support;
- The impact of one mass with another mass; and
- Pulse-type excitation associated with a punch-press.

Basic principles associated with shock excitation and related vibration system shock isolation can be presented with some examples.

EXAMPLE 6.1

Determine:

(a) The expression for the vibration response of the vibration system shown in Figure 6.4 associated with an inelastic impact of mass M_1 with mass M_2 when mass M_1 is dropped from a height h and

(b) The expression for the force transmitted to the foundation.

(c) Reduce the expressions obtained in (a) and (b) for the case when $\xi = 0$.

(d) What statements can be made with regard to the vibration isolation of the vibration system?

SOLUTION

(a) The vibration system in Figure 6.4 can be examined as if it were given an initial velocity v_o at time $t = 0$. It is first necessary to determine the velocity v_1 of mass M_1 at the time it impacts mass M_2. This is accomplished by noting the kinetic energy of mass M_1 upon impact with

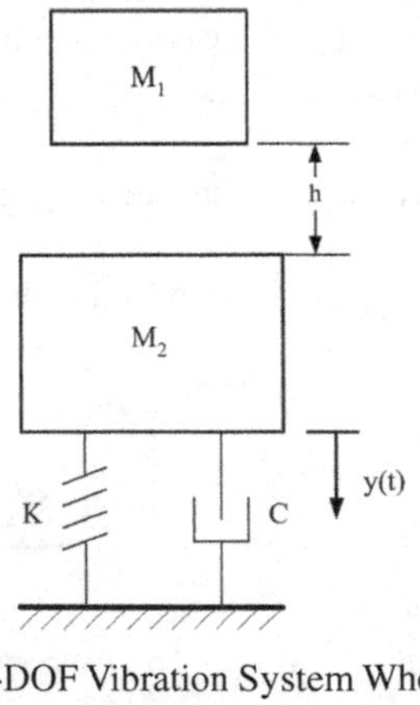

Figure 6.4 1-DOF Vibration System Where Mass M_1 Impacts Mass M_2 from a Height h

mass M_2 equals the potential energy of M_1 relative to M_2 before it is released and allowed to fall. Therefore:

$$M_1 g\,h = \frac{1}{2} M_1\, v_1^2 \tag{6.1a}$$

or:

$$v_1 = \sqrt{2gh}\,. \tag{6.1b}$$

The velocity v_o of masses M_1 and M_2 after impact is given by:

$$M_1\, v_1 = \left(M_1 + M_2\right) v_o \tag{6.1c}$$

or:

$$v_o = \frac{M_1}{M_1 + M_2}\, v_1\,. \tag{6.1d}$$

To determine the vibration response of M_1 and M_2 after impact, let $M = M_1 + M_2$ and assume $\xi < 1$. Therefore, the vibration system can be treated as an under damped system that is given an initial velocity v_o. The response of this system is given by equation (2.47) where:

$$Y = \frac{v_o}{\omega_d} \quad \text{and} \quad \phi = -\frac{\pi}{2} \tag{6.1e}$$

Therefore:

$$y(t) = \frac{v_o}{\omega_d}\, e^{-\xi\omega_n t} \sin\omega_d t \tag{6.1f}$$

where $\omega_d = \sqrt{1-\xi^2}\,\omega_n$ and $\omega_n = \sqrt{K/M}$ .Substituting equations (6.1b) and (6.1d)) into equation (6.1f) yields:

$$\boxed{y(t) = \frac{M_1\sqrt{2gh}}{M\,\omega_d}\, e^{-\xi\omega_n t} \sin\omega_d t\,.} \tag{6.1g}$$

(b) The expression for the force transmitted to the foundation is obtained from:

$$F_T = K y + C\dot{y}. \tag{6.1h}$$

Substituting equation (6.1f) into equation (6.1h) yields:

$$F_T = \frac{v_o}{\omega_d} e^{-\xi\omega_n t}\left[K \sin\omega_d t + C\begin{pmatrix}\omega_d \cos\omega_d t - \\ \xi\omega_n \sin\omega_d t\end{pmatrix}\right]. \tag{6.1i}$$

Factoring the term K from the bracketed term and noting that $C/K=2\xi/\omega_n$ gives:

$$F_T = \frac{v_o K}{\omega_d} e^{-\xi\omega_n t}\begin{bmatrix}(1-2\xi^2)\sin\omega_d t + \\ 2\xi\sqrt{1-\xi^2}\cos\omega_d t\end{bmatrix}. \tag{6.1j}$$

The above equation can be simplified and written:

$$F_T = \frac{v_o K}{\omega_d} e^{-\xi\omega_n t}\cos(\omega_d t + \phi) \tag{6.1k}$$

where:

$$\sqrt{(1-2\xi^2)^2 + \left(2\xi\sqrt{(1-\xi^2)}\right)^2} \approx 1 \tag{6.1l}$$

and:

$$\phi = \tan^{-1}\left[-\frac{1-2\xi^2}{2\xi\sqrt{1-\xi^2}}\right]. \tag{6.1m}$$

Substituting equations (6.1b) and (6.1d) into equation (6.1k) and noting $\omega_n = \sqrt{K/M}$ yields:

$$\boxed{F_T = \frac{M_1\sqrt{2gh}}{\sqrt{1-\xi^2}}\,\omega_n\, e^{-\xi\omega_n t}\cos(\omega_d t + \phi).} \tag{6.1n}$$

(c) When the damping ratio $\xi = 0$, equations (6.1g) and (6.1n) can be written:

$$\boxed{y(t) = \frac{M_1\sqrt{2gh}}{M\,\omega_n}\sin\omega_n t} \tag{6.1o}$$

and:

$$\boxed{F_T = M_1\sqrt{2gh}\,\omega_n \sin\omega_n t.} \tag{6.1p}$$

(d) The force transmitted to the foundation is normally of interest for the system shown in Figure 6.4. The transmitted force is reduced by lowering the value of the resonance frequency ω_n of the system and making the damping ratio ξ to be as close to zero as possible. Reducing the value of F_T by reducing ω_n results in a corresponding increase in the value of y(t).

EXAMPLE 6.2

Determine the expression for the displacement and for the acceleration of mass M in the box shown in Figure 6.5 after the box impacts the floor after falling from a height h. What statements can be made with regard to protecting mass M from this type of shock?

SOLUTION

This problem can be treated as if mass M in Figure 6.5 has an initial velocity v_o at the instant the box impacts the floor. The velocity v_o is $\sqrt{2gh}$. Assuming $\xi < 1$, the displacement of mass M is given by:

$$\boxed{y(t) = \frac{\sqrt{2gh}}{\omega_d} e^{-\xi\omega_n t}\sin\omega_d t.} \tag{6.2a}$$

The acceleration of mass M is obtained by taking the second derivative of equation (6.2a)) with respect to time to obtain:

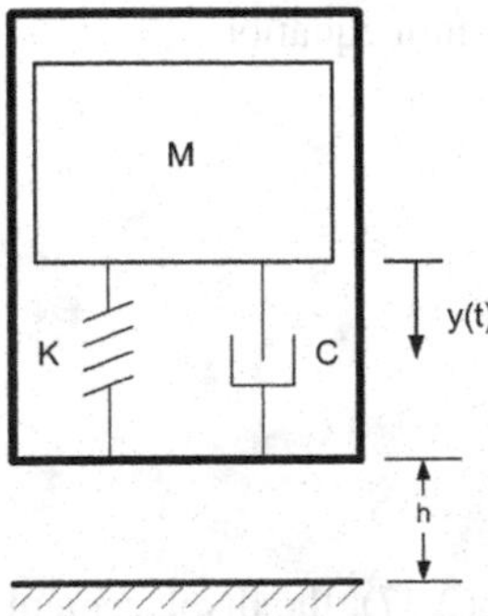

Figure 6.5 1-DOF Vibration System Where Box with Mass M Impacts Floor from Height h

$$\ddot{y}(t) = \frac{\sqrt{2gh}}{\sqrt{1-\xi^2}}\,\omega_n\, e^{-\xi\omega_n t}\begin{bmatrix}(2\xi^2-1)\sin\omega_d t - \\ 2\xi\sqrt{1-\xi^2}\cos\omega_d t\end{bmatrix}. \tag{6.2b}$$

Equation (6.2b) reduces to:

$$\boxed{\ddot{y}(t) = -\frac{\sqrt{2gh}}{\sqrt{1-\xi^2}}\,\omega_n\, e^{-\xi\omega_n t}\cos(\omega_d t + \phi)} \tag{6.2c}$$

where the amplitude associated with the coefficients of the sine and cosine terms and the amplitude of the phase angle ϕ are the same as those specified by equations (6.1l) and (6.1m).

For the shock excitation associated with this example, the acceleration (or deceleration) mass M experiences when the box impacts the floor must be reduced. This is accomplished by lowering the resonance frequency ω_n of the system inside the box and making the damping ratio ξ to be as close to zero as possible. This will result in a corresponding increase in the displacement y(t).

EXAMPLE 6.3

Determine the displacemented and the transmitted force to the foundation for a 1-DOF vibration system with mass, spring and damping elements (Figure 3.1) that is excited by an impulse. Assume that $\xi < 1$. What statements can be made with regard to vibration isolation of this system?

SOLUTION

This problem can be treated as an initial velocity problem. The initial velocity v_o of mass M is obtained from the impulse-momentum equation:

$$\text{Im pulse} = M\, v_o \tag{6.3a}$$

or:

$$v_o = \frac{\text{Im pulse}}{M}. \tag{6.3b}$$

Using equation (2.47), the displacement response can be expressed (Note that $y_o = 0$.):

$$\boxed{y(t) = \frac{\text{Im pulse}}{M\,\omega_d}\, e^{-\xi\omega_n t}\sin\omega_d t.} \tag{6.3c}$$

The expression for the transmitted force to the foundation for this problem has the same form as equation (6.1n). Therefore, the transmitted force F_T is given by:

$$\boxed{F_T = \frac{\text{Im pulse}}{\sqrt{1-\xi^2}}\,\omega_n\, e^{-\xi\omega_n t}\cos(\omega_d t + \phi)} \tag{6.3d}$$

where ϕ is given by equation (6.1m).

The comments associated with vibration isolation are the same as those in EXAMPLE 6.1(c).

EXAMPLE 6.4

Determine the expressions for the displacement and the transmitted force to the foundation for the 1-DOF vibration system in Figure 3.4 that is excited by a square pulse of duration T (Figure 6.6). What statements can be made with regard to vibration isolation of this system?

SOLUTION

The equation of motion for this system is given by:

$$M\,\ddot{y} + K\,y = f(t) \tag{6.4a}$$

where:

$$\begin{aligned} f(t) &= F_o \quad &&\text{for} \quad 0 \le t \le T \\ &= 0 \quad &&\text{for} \quad \text{all other values of } t. \end{aligned} \tag{6.4b}$$

The solution to equation (6.4a) is obtained by first considering both the complementary and particular solutions for time $t \le T$, The complementary solution

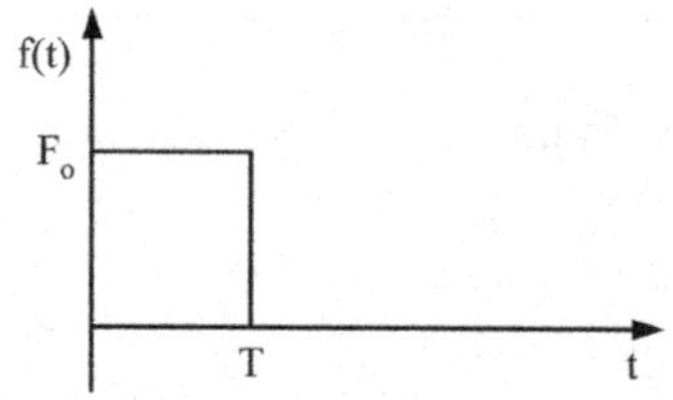

Figure 6.6 Square Pulse

is given by:

$$y_c(t) = A\cos\omega_n t + B\sin\omega_n t \tag{6.4c}$$

and the particular solution for the forcing function when $t \le T$ is given by:

$$y_p(t) = N. \tag{6.4d}$$

Substituting equation (6.4d) into equation (6.4a) and solving for N yields:

$$N = \frac{F_o}{K}. \tag{6.4e}$$

The total solution is $y(t) = y_c(t) + y_p(t)$, or:

$$y(t) = \frac{F_o}{K} + A\cos\omega_n t + B\sin\omega_n t. \tag{6.4f}$$

The values for the constants A and B are obtained by applying the initial conditions $y(0) = 0$ and $\dot{y}(0) = 0$. Therefore:

$$y(0) = 0 = \frac{F_0}{K} + A \quad \text{or} \quad A = -\frac{F_o}{K} \tag{6.4g}$$

and:

$$\dot{y}(0) = 0 = B\,\omega_n \quad \text{or} \quad B = 0. \tag{6.4h}$$

Therefore, the solution for y(t) for $t \le T$ is:

$$\boxed{y(t) = \frac{F_o}{K}(1 - \cos\omega_n t).} \tag{6.4i}$$

For time $t > T$ when $f(t) = 0$, only the complementary solution remains. Equation (2.21) gives the general form of the complementary solution to equation (6.4a) for the case where the initial displacement and velocity are given at time $t = 0$. For this problem, this equation can be extended for the case where the initial velocity and displacement are given for time $t = T$, or:

$$y_c(t) = \begin{bmatrix} y_T\cos\omega_n(t-T) + \\ \dfrac{v_T}{\omega_n}\sin\omega_n(t-T) \end{bmatrix} U(t-T) \tag{6.4j}$$

where $y_T = y(T)$ and $v_T = \dot{y}(T)$. $U(t - T)$ is the unit step function that has the following properties:

$$\begin{aligned} U(t-T) &= 1 \quad \text{for} \quad t \ge T \\ &= 0 \quad \text{for} \quad t < T. \end{aligned} \tag{6.4k}$$

Equation (6.4j) only has nonzero values for values of time $t \ge T$. Evaluating equation (6.4i) for both velocity and displacement at t=T yields:

$$y(T) = \frac{F_o}{K}(1 - \cos\omega_n T) \tag{6.4l}$$

$$\dot{y}(T) = \frac{F_o}{K}\omega_n \sin\omega_n T. \tag{6.4m}$$

Substituting equations (6.4l) and (6.4m) into equation (6.4j) and simplifying yields:

$$y(t) = Y\cos[\omega_n t + \phi]\,U(t-T) \tag{6.4n}$$

where:

$$Y = \frac{F_o}{K}\sqrt{(1-\cos\omega_n T)^2 + \sin^2\omega_n T} \tag{6.4o}$$

and:

$$\phi = \tan^{-1}\left[\frac{\sin\omega_n T}{1 - \cos\omega_n T}\right]. \tag{6.4p}$$

Multiplying and dividing equation (6.4o) by M and expanding and rearranging the terms under the radical gives:

$$Y = \frac{F_o}{M\omega_n^2}\sqrt{2(1-\cos\omega_n T)}. \tag{6.4q}$$

Noting that:

$$\omega_n = \frac{2\pi}{T_n}, \tag{6.4r}$$

where T_n is the period of vibration at resonance, equation (6.4q) can be written:

$$\boxed{\frac{Y}{\left(F_o / M\omega_n^2\right)} = \sqrt{2\left[1-\cos\left(2\pi\frac{T}{T_n}\right)\right]}.} \tag{6.4s}$$

The expression for the force transmitted to the foundation is given by:

$$f(t)_T = K\,y(t). \tag{6.4t}$$

The amplitude F_T of the transmitted force is obtained by substituting equation (6.4q) into equation (6.4t), or:

$$F_T = F_o\sqrt{2\left(1-\cos\omega_n T\right)}. \tag{6.4u}$$

The ratio of the transmitted force divided by the excitation force can be written:

$$\boxed{\frac{F_T}{F_o} = \sqrt{2\left[1-\cos\left(2\pi\frac{T}{T_n}\right)\right]}.} \tag{6.4v}$$

Equations (6.4s) and (6.4v) are plotted in Figure 6.7. The values for $Y/(F_o/M\omega_n^2)$ and F_T/F_o should be much less than one for acceptable vibration system shock isolation. Their values are less than one for values of $T/T_n < 0.17$. Therefore, T_n should be selected such that $T/T_n < 0.17$. Once T_n is specified, the corresponding value for thee system resonance frequency ω_n is given by equation (6.4r).

EXAMPLE 6.5

The base of a 1-DOF vibration system with a moving support similar to the one in Figure 3.28, but only with mass and spring elements0, is excited with the triangular pulse shown in Figure 6.8. Determine the expression for the displacement of mass M associated with the displacement pulse input to the base. What statements

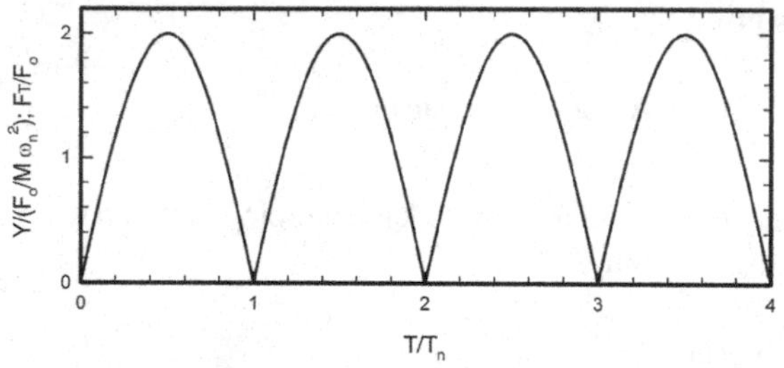

Figure 6.7 Ratios of $Y/(F_o/M\omega_n^2)$ and F_T/F_o as a Function of T/T_n for the 1-DOF Vibration System in EXAMPLE 6.4

can be made with regard to vibration isolation of this system?

SOLUTION

The equation of motion for the system described in this problem is:

$$M\ddot{y}_2 + K y_2 = K y_1 \tag{6.5a}$$

where:

$$y_1(t) = Y_1\frac{t}{T} \quad \text{for} \quad 0 \le t \le T$$
$$= 0 \quad \text{for} \quad \text{all other values of } t. \tag{6.5b}$$

The solution to equation (6.5a)) is obtained by first determining the complementary and particular solutions for time $t \le T$. The complementary solution is given by equation (6.4c). For the forcing function described by equation (6.5b), the particular solution is assumed to be:

$$y_p(t) = N\frac{t}{T}. \tag{6.5c}$$

Substituting equation (6.5c) into equation (6.5a) and

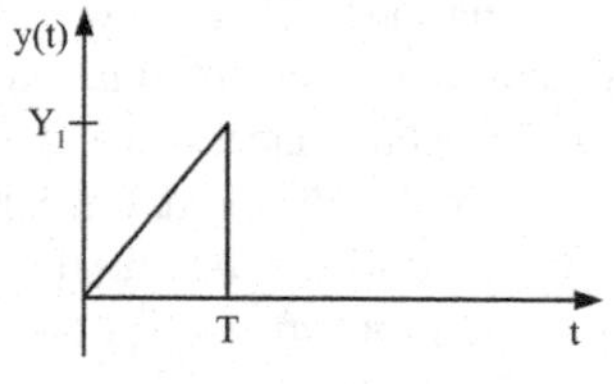

Figure 6.8 Triangular Pulse

solving for N yields:

$$N = Y_1 \,. \tag{6.5d}$$

The sum of the complementary solution plus the particular solution is:

$$y_2(t) = Y_1\,\frac{t}{T} + A\cos\omega_n t + B\sin\omega_n t \,. \tag{6.5e}$$

Applying the initial conditions $y_2(0) = 0$ and $\dot{y}_2(0) = 0$ yields:

$$A = 0 \qquad \text{and} \qquad B = -\frac{Y_1}{T\,\omega_n} \,. \tag{6.5f}$$

Therefore, the solution for $y_2(t)$ for $t \le T$ becomes:

$$\boxed{y_2(t) = \frac{Y_1}{T}\left(t - \frac{1}{\omega_n}\sin\omega_n t\right).} \tag{6.5g}$$

Evaluating $y_2(t)$ and $\dot{y}_2(t)$ at time $t = T$ and applying equation (6.4r) yields:

$$y_2(T) = \frac{Y_1}{T}\left[T - \frac{T_n}{2\pi}\sin\left(2\pi\frac{T}{T_n}\right)\right] \tag{6.5h}$$

and:

$$\dot{y}_2(T) = \frac{Y_1}{T}\left[1 - \cos\left(2\pi\frac{T}{T_n}\right)\right]. \tag{6.5i}$$

The complementary solution for $t > T$ is given by equation (6.4j). Substituting equations (6.5h) and (6.5i) into equation (6.4j) and simplifying gives:

$$y_2(t) = Y_2\cos\left[\omega_n(t - T) + \phi\right]U(t - T) \tag{6.5j}$$

where:

$$Y_2 = Y_1\sqrt{\left[1 - \frac{T_n}{2\pi T}\sin\left(2\pi\frac{T}{T_n}\right)\right]^2 + \left[\frac{T_n}{2\pi T}\left(1 - \cos\left(2\pi\frac{T}{T_n}\right)\right)\right]^2} \,. \tag{6.5k}$$

Expanding the expressions under the radical in equation (6.5k) and rearranging the terms yields:

$$\boxed{\frac{Y_2}{Y_1} = \sqrt{1 - \frac{T_n}{pT}\sin\left(2p\frac{T}{T_n}\right) + \left(\frac{T_n}{pT}\right)^2\sin^2\left(p\frac{T}{T_n}\right)} \,.} \tag{6.5l}$$

The equation for the phase angle ϕ is:

$$\boxed{\phi = \tan^{-1}\left[\frac{\frac{T_n}{2\pi T}\left[\cos\left(2\pi\frac{T}{T_n}\right) - 1\right]}{1 - \frac{T_n}{2\pi T}\sin\left(2\pi\frac{T}{T_n}\right)}\right].} \tag{6.5m}$$

The ratio Y_2/Y_1 is plotted in Figure 6.9. The value for Y_2/Y_1 should be much less than one for acceptable vibration system shock isolation. The value of Y_2/Y_1 equals one when $T/T_n = 0.35$. Therefore, T_n should be selected such that $T/T_n < 0.35$. Once T_n is specified, the corresponding value for the system resonance frequency ω_n is given by equation (6.4r).

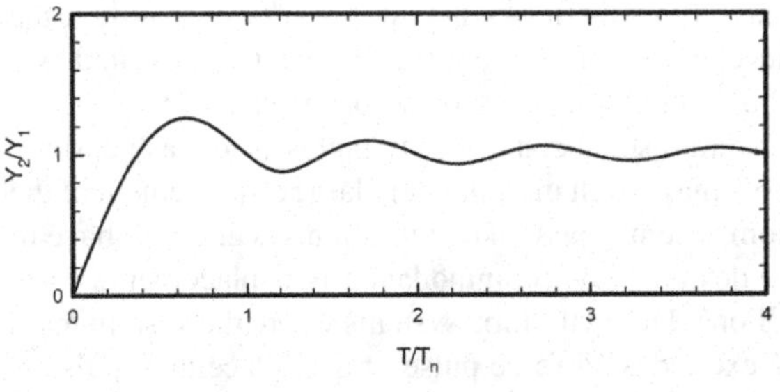

Figure 6.9 Ratio of Y_2/Y_1 as a Function of T/T_n for the 1-DOF Vibration System in EXAMPLE 6.5

6.3 Design Considerations for 1-DOF Vibration Systems that Are Subjected to Shock Excitation

Equations for the responses of several 1-DOF vibrating systems were developed in the preceding section for the cases when they were exposed to different types of shock excitation. The problems were divided into two categories: systems where the mass was subjected to shock excitation and systems where the foundation or base of the system was subjected to shock excitation.

For systems where the mass is excited by a shock pulse or force, it is usually necessary to reduce the force that is transmitted to the structure to which the system is attached. Equations (6.1p), (6.3d) and (6.4u) and Figure 6.7 indicate the value of the transmitted force can be reduced by making the system resonance frequency as low as possible. Once the resonance frequency of the vibration system has been selected, equations (6.1o), (6.3c), and (6.4q) can be used to select the value of the system mass that is necessary to achieve a desired value of displacement for the system mass. Situations where a small displacement is required will require a relatively large system mass. After the value of the system mass has been determined, the total stiffness of the resilient elements needed to support the system mass is obtained from $\omega_n = \sqrt{K/M}$.

For vibrating systems where a shock excitation is applied to the base or supporting structure of the system, it is usually necessary to reduce the resulting acceleration of the system mass. This is required to reduce the shock force to which the mass is exposed. Equation (6.2c) indicates the value of the acceleration is reduced by making the system resonance frequency as low as possible. The total stiffness of the resilient element(s) needed to support the mass is obtained from $\omega_n = \sqrt{K/M}$ where M is the value of the system mass. For situations where it is necessary to reduce the acceleration of a system component that is excited by a shock input to its base or supporting structure, select a system resonance frequency that is as low as possible. This may result in a relatively large displacement of the component (mass). The vibration system may have to be designed to accommodate this displacement.

For 1-DOF vibration systems where the system mass is excited by a force pulse or a displacement pulse is directed into the supporting structure of the system, the equation that describes the resulting motion of the system mass can become rather complicated and difficult to use when designing shock isolation into the system. Equation (6.51) is a good example of this. For most cases where mass or base excited shock excitation are encountered, it is usually desirable to select the resonance frequency of the vibration system such that $T/T_n \ll 1$. When a system with no damping is excited by shock excitation, the ratios $Y(F_o/M\omega_n^2)$, F_T/F_o, and Y_2/Y_1 are usually described by the same equation. Figures 6.7 and 6.9 indicate that for values of $T/T_n < 0.2$, the equation that describes the above ratios for a given type of pulse can be approximated by a straight line. For the square and triangular pulses discussed in EXAMPLES 6.4 and 6.5, this straight line approximations are given by:

$$F_T/F_o,\ Y/\left(F_o/M\omega_n^2\right),\ Y_2/Y_1 = 2\pi\frac{T}{T_n} \tag{6.3}$$

and:

$$F_T/F_o,\ Y/\left(F_o/M\omega_n^2\right),\ Y_2/Y_1 = \pi\frac{T}{T_n} \tag{6.4}$$

Table 6.1 gives the straight line approximations for the ratios $Y/(F_o/Mw)$, F_T/F_o, and Y_2/Y_1 as a function of T/T_n for several types of shock pulses when $T/T_n < 0.2$.

EXAMPLE 6.6

During a punching operation of a punch press machine, a square pulse in which $F_o = 1{,}000\ lb_f$ and $T = 10$ ms is generated. It is desired to limit the amplitude of the transmitted force to 10% of the amplitude of the applied force and the displacement of the punch press to 0.01 inch. Determine the weight of the mass needed to support the punch press and the value of the total stiffness of the springs necessary to support the weight of the entire system

SOLUTION

It is first necessary to determine the resonance frequency of the vibration isolation system. Since T/T_n will be less than 0.2 for this system, equation (6.3) can be used to obtain the resonance frequency.

$$\omega_n = \frac{1}{T}\frac{F_T}{F_o} \quad \text{or} \quad \omega_n = \frac{0.1}{0.01} = 10\,\frac{\text{rad}}{\text{s}}. \tag{6.6a}$$

Table 6.1 Approximate Functions for F_T/F_o, $Y/(F_o/M\omega_n^2)$ and Y_2/Y_1 as a Function of T/T_n for Different Types of Pulses for the Case when $T/T_n < 0.2$

Pulse Type	Equation of Pulse	Sketch of Pulse	F_T/F_o, $Y/(F_o/M w_n^2)$, Y_2/Y_1
Square Pulse	$f(t) = F_o \quad 0 \le t \le T$ $= 0 \quad t > T$	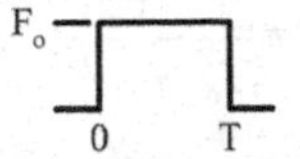	$2\pi \frac{T}{T_n}$
Sine Pulse	$f(t) = F_o \sin \pi \frac{t}{T} \quad 0 \le t \le T$ $= 0 \quad t > T$	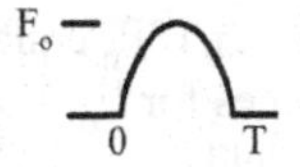	$1.27\pi \frac{T}{T_n}$
Versed-Sine Pulse	$f(t) = \frac{F_o}{2}\left(1 - \cos 2\pi \frac{t}{T}\right) \quad 0 \le t \le T$ $= 0 \quad t > T$	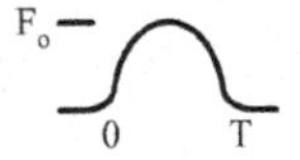	$\pi \frac{T}{T_n}$
Exponential Pulse	$f(t) = F_o \frac{1 - e^{2at/T}}{1 - e^a} \quad 0 \le t \le \frac{T}{2}$ $= F_o \frac{1 - e^{2a(1-t/T)}}{1 - e^a} \quad \frac{T}{2} < t \le T$ $= 0 \quad t > T$	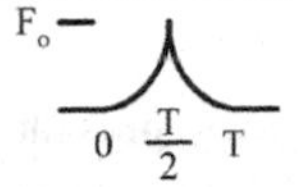 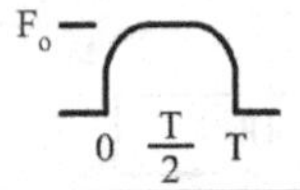	$0.314\pi \frac{T}{T_n} \quad (a = 2\pi)$ $1.67\pi \frac{T}{T_n} \quad (a = -2\pi)$
Quarter-Cycle Sine Pulse - Vertical Decay	$f(t) = F_0 \sin \frac{\pi}{2} \frac{t}{T} \quad 0 \le t \le T$ $= 0 \quad t > T$	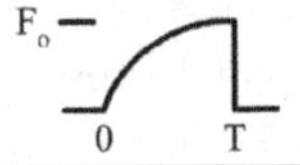	$1.27\pi \frac{T}{T_n}$
Half-Cycle Versed Sine Pulse - Vertical Decay	$f(t) = \frac{F_o}{2}\left(1 - \cos \pi \frac{t}{T}\right) \quad 0 \le t \le T$ $= 0 \quad t > T$	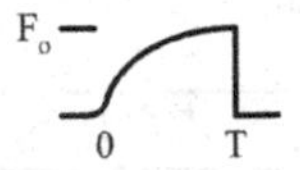	$\pi \frac{T}{T_n}$
Exponentual Pulse - Vertical Decay	$f(t) = F_o \frac{1 - e^{at/T}}{1 - e^a} \quad 0 \le t \le T$ $= 0 \quad t > T$	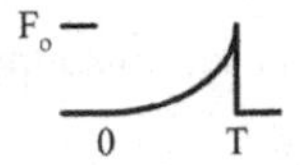	$0.314\pi \frac{T}{T_n} \quad (a = 2\pi)$
Quarter-Cycle Sine Pulse - Vertical Rise	$f(t) = F_o \cos \frac{\pi}{2} \frac{t}{T} \quad 0 \le t \le T$ $= 0 \quad t > T$	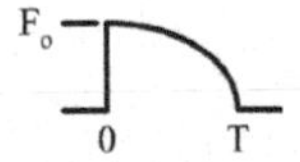	$1.27\pi \frac{T}{T_n}$
Half-Cycle Versed Sine Pulse - Vertical Rise	$f(t) = \frac{F_o}{2}\left(1 + \cos \pi \frac{t}{T}\right) \quad 0 \le t \le T$ $= 0 \quad t > T$	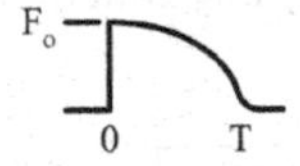	$\pi \frac{T}{T_n}$
Exponential Pulse - Vertical Rise	$f(t) = F_o \frac{1 - e^{a(1-t/T)}}{1 - e^a} \quad 0 \le t \le T$ $= 0 \quad t > T$	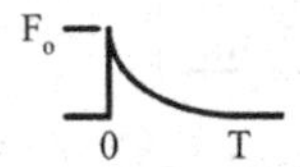	$0.314\pi \frac{T}{T_n} \quad (a = 2\pi)$
Triangular Pulse[1]	$f(t) = 2F_o \frac{t}{T} \quad 0 \le t \le \frac{T}{2}$ $= 2F_o\left(1 - \frac{t}{T}\right) \quad \frac{T}{2} < t \le T$ $= 0 \quad t > T$	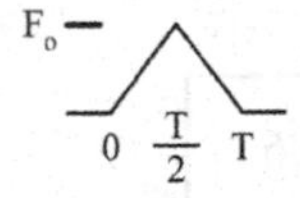	$\pi \frac{T}{T_n}$

[1] All triangular pulses, including those with vertical rise and vertical decay, have the same approximate equations for thev ratios F_T/F_o, $Y/(F_o/M\omega_n^2)$, Y_2/Y_1 .

The value of the supporting plus punch press mass is obtained from the equation:

$$\frac{Y}{F_o/M\omega_n^2} = 2\pi\frac{T}{T_n} \quad \text{or} \quad M = 2\pi\frac{T}{T_n}\frac{F_o}{Y\omega_n^2}. \tag{6.6b}$$

From the statement of the problem $2\pi T/T_n$ equals 0.1. Substituting this value and the values for F_o, Y and ω_n into equation (6.6b) and solving yields:

$$\boxed{M = 100\,\frac{lb_f - s^2}{in.}.} \tag{6.6c}$$

The total weight of the supporting mass plus the punch press mass is:

$$\boxed{We = 386 \times M \quad \text{or} \quad We = 38{,}600\ lb_f.} \tag{6.6d}$$

If, for example, the weight of the punch press is 3,000 lb_f, the weight of the supporting mass is:

$$\boxed{We_s = 38{,}600 - 3{,}000 \quad \text{or} \quad We_s = 35{,}600\ lb_f.} \tag{66e}$$

The total stiffness of the springs used to support the entire system is obtained from:

$$K = M\,\omega_n^2. \tag{6.6f}$$

Substituting the values of M and ω_n into equation (6.6f) yields:

$$\boxed{K = 100 \times 10^2 \quad \text{or} \quad K = 10{,}000\,\frac{lb_f}{in.}.} \tag{6.6g}$$

The resulting static deflection with the above mass and spring coefficient is:

$$\boxed{\delta = \frac{We}{K} \quad \text{or} \quad \delta = \frac{38{,}600}{10{,}000} = 3.86\ in.} \tag{6.6h}$$

6.4 Fourier Transforms

Fourier transforms can be used as a more generalized method for examining the response of linear vibration systems to non-periodic excitation. Figure 6.10 shows a complex periodic signal that consists of a series of square pulses of duration t' and with a period of repetition T. This signal can be represented by a complex Fourier series [equations (1.44) and (1.51) in Chapter 1], or:

$$f(t) = \sum_{n=-\infty}^{\infty} C_n\, e^{jn\omega_o t} \tag{6.5}$$

where:

$$C_n = \frac{1}{T}\int_{-T/2}^{T/2} f(t)\,e^{-jn\omega_o t}\,dt. \tag{6.6}$$

When a system is excited by a single square pulse, equations (6.5) and (6.6) can be used to describe the square pulse by letting $T \to \infty$. As $T \to \infty$, $\omega_o (= 2\pi/T) \to d\omega$ and $n\omega_o \to \omega$ in equations (6.5) and (6.6). Let $T \to \infty$ and note $T = 2\pi/\omega_o$. Then, equations (6.5) and (6.6) can be combined to yield:

$$f(t) = \int_{-\infty}^{\infty} \frac{1}{2\pi}\left[\int_{-\infty}^{\infty} f(t)\,e^{-j\omega t}\,dt\right] e^{j\omega t}\,d\omega. \tag{6.7}$$

The term within the brackets in equation (6.7) is defined as the *Fourier transform* $F(j\omega)$ of f(t) where:

$$F(j\omega) = \int_{-\infty}^{\infty} f(t)\,e^{-j\omega t}\,dt. \tag{6.8}$$

Therefore, equation (6.7) can be written:

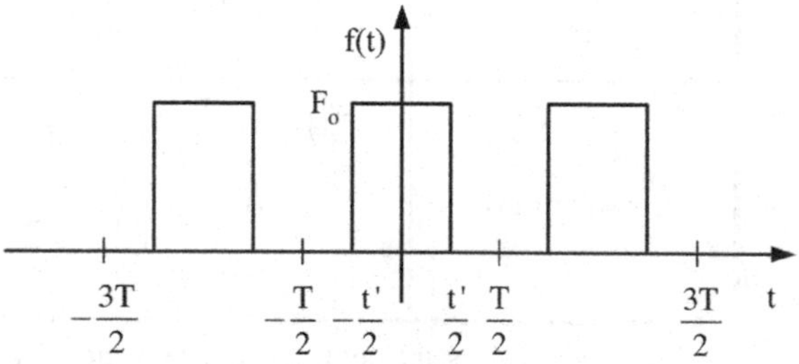

Figure 6.10 Series of Square Pulses

$$f(t) = \frac{1}{2\pi}\int_{-\infty}^{\infty} F(j\omega) e^{j\omega t}\, d\omega . \tag{6.9}$$

Equation (6.9) is defined as the *inverse Fourier transform* of F(jω). Equation (6.8) and (6.9) are defined as a Fourier transform pair. For the Fourier transform to exist, the integral of $|f(t)|$ from $-\infty$ to ∞ must be bounded, or:

$$\int_{-\infty}^{\infty} |f(t)|\, dt < \infty . \tag{6.10}$$

Equation (6.8) also requires:

$$f(t = \pm\infty) = 0 \qquad \text{and} \qquad \left.\frac{d^n f(t)}{dt^n}\right|_{t=\pm\infty} = 0 . \tag{6.11}$$

When analyzing a linear vibration system, the Fourier transform of the derivative of f(t) must be known where:

$$F\left[\frac{df(t)}{dt}\right] = \int_{-\infty}^{\infty} \frac{df(t)}{dt} e^{-jn\omega_o t}\, dt . \tag{6.12}$$

F[ ] denotes the Fourier transform of the function that is in the brackets. Equation (6.12) can be simplified by integrating by parts, where:

$$\int u\, dv = u v - \int v\, du . \tag{6.13}$$

To evaluate equation (6.12), let:

$$u = e^{-j\omega t} \qquad \text{and} \qquad v = f(t) . \tag{6.14}$$

Therefore:

$$du = -j\omega\, e^{-j\omega t}\, dt . \tag{6.15}$$

Substituting equations (6.14) and (6.15) into equation (6.13) yields:

$$F\left[\frac{df(t)}{dt}\right] = f(t) e^{-j\omega t}\Big|_{-\infty}^{\infty} + j\omega \int_{-\infty}^{\infty} f(t) e^{-j\omega t}\, dt . \tag{6.16}$$

Noting f(t = $\pm\infty$) = 0, equation (6.16) becomes:

$$F\left[\frac{df(t)}{dt}\right] = j\omega\, F(j\omega) . \tag{6.17}$$

In a similar manner, it can be shown that:

$$F\left[\frac{d^2 f(t)}{dt^2}\right] = (j\omega)^2\, F(j\omega) \tag{6.18}$$

and:

$$F\left[\frac{d^n f(t)}{dt^n}\right] = (j\omega)^n\, F(j\omega) . \tag{6.19}$$

EXAMPLE 6.7

A mass-spring-damper system similar to the one shown in Figure 3.1 in Chapter 3 is excited by a transient signal. Determine the Fourier transform of the response of the system.

SOLUTION

The equation of motion for the system is:

$$M\ddot{y} + C\dot{y} + K y = f(t) . \tag{6.7a}$$

The Fourier transform of this equation is:

$$(j\omega)^2 M Y[j\omega w] + (j\omega) C Y[j\omega] + K Y[j\omega] = F[j\omega] \tag{6.7b}$$

or:

$$\left\{(j\omega)^2 M + (j\omega) C + K\right\} Y[j\omega] = F[j\omega] . \tag{6.7c}$$

Rearranging equation (6.7c) yields:

$$Y[j\omega] = \left\{\frac{1}{(j\omega)^2 M + (j\omega) C + K}\right\} F[j\omega] \tag{6.7d}$$

or:

$$Y[j\omega]=\left\{\frac{1}{\left(K-M\omega^2\right)+j\omega C}\right\}F[j\omega]. \tag{6.7e}$$

The term in the brackets in equation (6.7e) can be written:

$$\left\{\frac{1}{\left(K-M\omega^2\right)+j\omega C}\right\}=\left\{\frac{(1/K)\,e^{-j\phi}}{\left(1-\frac{\omega^2}{\omega_n^2}\right)^2+\left(2\xi\frac{\omega}{\omega_n}\right)^2}\right\} \tag{6.7f}$$

where:

$$\omega_n=\sqrt{\frac{K}{M}} \quad \text{and} \quad \phi=\tan^{-1}\left[\frac{2\xi\frac{\omega}{\omega_n}}{1-\frac{\omega^2}{\omega_n^2}}\right]. \tag{6.7g}$$

Therefore, equation (6.7e) can be written:

$$Y[j\omega]=\left\{\frac{(1/K)\,e^{-j\phi}}{\left(1-\frac{\omega^2}{\omega_n^2}\right)^2+\left(2\xi\frac{\omega}{\omega_n}\right)^2}\right\}F[j\omega]. \tag{6.7h}$$

Equation (6.7h) in EXAMPLE 6.7 can be written:

$$Y[j\omega]=G[j\omega]\,F[j\omega] \tag{6.20}$$

where:

$$G[j\omega]=\frac{Y[j\omega]}{F[j\omega]}=\left\{\frac{(1/K)\,e^{-j\phi}}{\left(1-\frac{\omega^2}{\omega_n^2}\right)^2+\left(2\xi\frac{\omega}{\omega_n}\right)^2}\right\}. \tag{6.21}$$

$G(j\omega)$ is defined as the system *transfer function*. It is the complex ratio of the Fourier transform of the system response divided by the Fourier transform of the system input. Equation (6.21) implies the transfer function of a linear vibration system can be obtained by measuring the complex ratio of the Fourier transform of the system response divided by the Fourier transform of the system input. This can be accomplished experimentally with the use of a Fast Fourier Transform (FFT) analyzer.

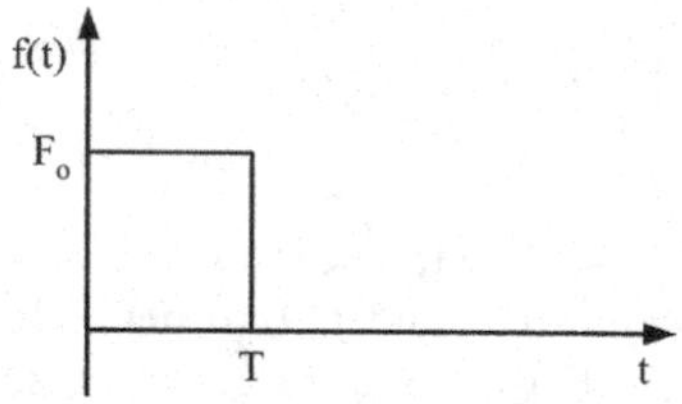

Figure 6.11 Square Pulse

EXAMPLE 6.8

Determine the Fourier transform of the square pulse shown in Figure 6.11.

SOLUTION

For this signal:

$$\begin{aligned} f(t)&=F_o && 0\le t\le T \\ &=0 && \text{for all other values of } t. \end{aligned} \tag{6.8a}$$

Therefore:

$$F[j\omega]=\int_0^T F_0\,e^{-j\omega t}\,dt \tag{6.8b}$$

or:

$$F[j\omega]=F_o\left.\frac{e^{-j\omega t}}{-j\omega}\right|_0^T. \tag{6.8c}$$

Reducing equation (6.8c) yields:

$$F[j\omega]=jF_o\left(\frac{e^{-j\omega T}-1}{\omega}\right). \tag{6.8d}$$

$F(j\omega)$ can be expressed:

$$F[j\omega]=\left|F[j\omega]\right|e^{j\phi(\omega)} \tag{6.8e}$$

where:

$$|F[j\omega]| = \sqrt{Re^2\{F[j\omega]\} + Im^2\{F[j\omega]\}} \tag{6.8f}$$

and:

$$\phi(\omega) = \tan^{-1}\left[\frac{Im\{F[j\omega]\}}{Re\{F[j\omega]\}}\right]. \tag{6.8g}$$

Expanding $e^{-j\omega T}$ in equation (6.8d) using Euler's equation and collecting terms yields:

$$F[j\omega] = \frac{F_o}{\omega}\left[\sin\omega T + j(\cos\omega T - 1)\right]. \tag{6.8h}$$

Substituting the real and imaginary parts of equation (6.8h) into equations (6.8f) and (6.8g) yields:

$$\boxed{|F[j\omega]| = \frac{F_o}{\omega}\sqrt{2 - 2\cos\omega T}} \tag{6.8i}$$

and:

$$\boxed{\phi(\omega) = \tan^{-1}\left[\frac{\cos\omega T - 1}{\sin\omega T}\right].} \tag{6.8j}$$

Noting that $2\sin^2 A = 1 - \cos 2A$, $|F(j\omega)|$ can be written:

$$\boxed{|F[j\omega]| = 2F_o\frac{\sin\left(\frac{\omega T}{2}\right)}{\omega}.} \tag{6.8k}$$

Equation (6.8k) indicates the amplitude of the Fourier transform of the square pulse shown in Figure 6.11 as a function of ω. Equation (6.8k) can also be written in terms of frequency f, or:

$$\boxed{|F[jf]| = \frac{F_o}{\pi}\frac{\sin(\pi f T)}{f}.} \tag{6.8l}$$

Equation (6.8l) is often referred to as the frequency spectrum of the square pulse.

Equation (6.8k) represents the *frequency spectrum* of a square pulse from $\omega = -\infty$ to $\omega = +\infty$. When the time function f(t) of a signal is real-valued, the $\pm$ frequency pairs of the Fourier transform of the signal add together to produce real-valued numbers. For this to occur:

$$F[j\omega] = F[-j\omega]^* \tag{6.22}$$

where $F[j\omega]^*$ is the *complex conjugate* of $F[j\omega]$. This requires the real part of the Fourier transform to be an even function:

$$Re\{F[j\omega]\} = Re\{F[-j\omega]\} \tag{6.23}$$

and the imaginary part of the Fourier transform to be an odd function:

$$Im\{F[j\omega]\} = -Im\{F[-j\omega]\}. \tag{6.24}$$

This insures the matched $\pm$ frequency components of the imaginary part of the Fourier transform cancel. Therefore, only the matched $\pm$ frequencies components of the real part of the Fourier transform have nonzero values when they are added together. Relative to the square pulse in EXAMPLE (6.8):

$$Re\{F[j\omega]\} = F_o\frac{\sin\omega T}{\omega} \tag{6.25}$$

which is an even function, and:

$$Im\{F[j\omega]\} = F_o\frac{\cos\omega T - 1}{\omega} \tag{6.26}$$

which is an odd function.

This discussion requires the amplitude $|F[j\omega]|$ of the Fourier transform of a real-valued signal be an even function, and the phase $\phi(\omega)$ be an odd function. Relative to the square pulse in EXAMPLE 6.8, equations (6.8j) and (6.8k) indicate this is true.

An FFT analyzer is often used to measure the frequency composition of a transient signal, such as a square pulse. Mathematically, the Fourier transform of the signal will give the frequency components of the signal from $-\omega$ to $+\omega$, where $-\omega$ is the lower frequency limit of the analysis and $+\omega$ is the upper frequency limit of the analysis. Negative frequencies have no physical meaning in the analysis of the frequency composition of a transient pulse or any other signal that is a function of time. Therefore, FFT analyzers are designed to measure only the positive frequency components of a signal. However, the negative frequency components of the signal represent energy that is present in the signal. To compensate for this, the negative frequencies of the Fourier transform are folded about the f = 0 axis and added to the positive frequency side of the spectrum. Therefore, the measured frequency spectrum of an FFT analyzer is $2|F[j\omega]|$, where ω is always greater than zero.

Fourier transforms can be used to obtain the frequency composition of non-periodic signals, and they can be used to obtain the frequency response of a linear vibration system that is excited by these signals. However, Fourier transforms have some very serious limitations. Equation (6.11) indicates Fourier transforms can only be obtained for non-periodic signals whose amplitudes and related first and higher order derivatives with respect to time go to zero as $t \to \infty$. There are many non-periodic signals, such as signals that are described by step functions, ramp functions, etc., for which this is not true. Therefore, it is not possible to obtain the Fourier transforms of these types of signals.

The time domain response of a linear vibration system that is excited by a non-periodic signal is often desired. When it is possible to obtain the Fourier transform of a non-periodic signal, it is generally possible to obtain the Fourier transform of the response of a linear vibration system that is excited by this signal. However, it often is extremely difficult, if not impossible, to use equation (6.7) to obtain the inverse Fourier transform of the frequency response of the system to determine its corresponding time-domain response. Therefore, it often is not possible to use Fourier transforms to obtain the time-domain response of a linear vibration system that is excited by a non-periodic signal.

6.5 Laplace Transforms

Relative to Fourier Transforms, f(t) can be multiplied by a convergence factor $e^{-\sigma t}$ to force the requirements of equation (6.8) to be met where:

$$\int_{-\infty}^{\infty} \left| f(t) e^{-\sigma t} \right| dt < \infty . \tag{6.27}$$

σ is arbitrary and can always be chosen such that f(t) $e^{-\sigma t}$ will always go to zero as t goes to infinity. There are some functions, such as where this is not true; however, for most signals of practical interest this is true. Substituting f(t) $e^{-\sigma t}$ into equation (6.8) instead of f(t) yields:

$$\hat{F}[j\omega] = \int_{-\infty}^{\infty} f(t) e^{-(\sigma + j\omega)t} \, dt . \tag{6.28}$$

The inverse transform becomes:

$$\hat{f}(t) = \frac{1}{2\pi} \int_{-\infty}^{\infty} \hat{F}[j\omega] e^{j\omega t} \, d\omega \tag{6.29}$$

where:

$$\hat{f}(t) = f(t) e^{-\sigma t} . \tag{6.30}$$

It is usually desirable to recover f(t) instead of the inverse transform $\hat{f}(t)$. Therefore, equation (6.29) can be written:

$$f(t) e^{-\sigma t} = \frac{1}{2\pi} \int_{-\infty}^{\infty} \hat{F}[j\omega] e^{j\omega t} \, d\omega \tag{6.31}$$

or:

$$f(t) = \frac{e^{\sigma t}}{2\pi} \int_{-\infty}^{\infty} \hat{F}[j\omega] e^{j\omega t} \, d\omega . \tag{6.32}$$

Since the integration is with respect to ω, $e^{\sigma t}$ can be brought inside the integral, or:

$$f(t) = \frac{1}{2\pi} \int_{-\infty}^{\infty} \hat{F}[j\omega] e^{(\sigma + j\omega)t} \, d\omega . \tag{6.33}$$

Now, define:

$$s = \sigma + j\omega \tag{6.34}$$

where:

$$ds = j\,d\omega \quad \text{or} \quad d\omega = \frac{1}{j}\,ds\,. \tag{6.35}$$

Substituting equations (6.34) and (6.35) into equations (6.28) and (6.33) and letting $\hat{F}[j\omega]$ become L[s], yields:

$$L[s] = \int_{-\infty}^{\infty} f(t)e^{-st}\,dt \tag{6.36}$$

and:

$$f(t) = \frac{1}{j2\pi}\int_{\sigma-j\infty}^{\sigma+j\infty} L[s]e^{st}\,ds\,. \tag{6.37}$$

Equation (6.36) is defined as the *two-sided Laplace transform* of f(t) and equation (6.37) is defined as the *inverse Laplace transform* of F(s). For the Laplace transform to exist, equation (6.27) must be satisfied.

Many situations exist where there is no information relative to the nature of the forcing function that excites a linear vibration system for some time $t \le t_o$. This situation can be handled by using a *one-sided Laplace transform.* For this case, the initial motion of the system at $t = t_o$ must be specified. Since t_o is often arbitrary, it can be specified as $t_o = 0$. Therefore, equation (6.36) becomes:

$$L[s] = \int_{0}^{\infty} f(t)e^{-st}\,dt\,. \tag{6.38}$$

As was the case with Fourier transforms, it is often necessary to know the Laplace transform of the derivative of f(t) with respect to time, or:

$$L\left[\frac{df(t)}{dt}\right] = \int_0^{\infty} \frac{df(t)}{dt}e^{-st}\,dt\,. \tag{6.39}$$

L [ ] denotes the *Laplace transform* of the function that is in the brackets. Equation (6.39) can be simplified by integrating by parts. Use equation (6.13) where:

$$u = e^{-st} \quad \text{and} \quad v = f(t)\,. \tag{6.40}$$

Therefore:

$$du = -s\,e^{-st}\,dt. \tag{6.41}$$

Substituting equations (6.40) and (6.41) into equation (6.13) yields:

$$L\left[\frac{df(t)}{dt}\right] = e^{-st} f(t)\Big|_0^{\infty} + s\int_0^{\infty} f(t)e^{-st}\,dt\,. \tag{6.42}$$

Noting that $s = \sigma + j\omega$ and that $f(t)\,e^{-st} = 0$ at $t = \infty$, equation (6.42) becomes:

$$L\left[\frac{df(t)}{dt}\right] = -f(0) + sF[s]\,. \tag{6.43}$$

In a like manner, it can be shown that:

$$L\left[\frac{d^2 f(t)}{dt^2}\right] = -\frac{df(t)}{dt}\bigg|_{t=0} - sf(0) + sF[s]\,. \tag{6.44}$$

and:

$$L\left[\frac{d^n f(t)}{dt^n}\right] = s^n F[s] - s^{n-1} f(0) - s^{n-2}\frac{df(t)}{dt}\bigg|_{t=0} - \dots - \frac{d^{n-1} f(t)}{dt^{n-1}}\bigg|_{t=0}\,. \tag{6.45}$$

Other operations associated with Laplace transforms that are important are:

$$L\left[f_1(t) \pm f_2(t)\right] = L\left[f_1(t)\right] \pm L\left[f_2(t)\right] \tag{6.46}$$

$$L\left[af(t)\right] = aF[s] \quad \text{where} \quad a = \text{const} \tag{6.47}$$

$$L\left[tf(t)\right] = -\frac{dF[s]}{ds} \tag{6.48}$$

$$L\left[\int f(t)dt\right] = \frac{1}{s}F[s] + \frac{1}{s}\int f(t)dt\Big|_{t=0^+} . \quad (6.49)$$

EXAMPLE 6.9

Determine the Laplace transform of a unit step function U(t) where U(t) is specified by:

$$\begin{aligned} U(t) &= 1 \quad \text{for} \quad t \geq 0 \\ &= 0 \quad \text{for} \quad t < 0 . \end{aligned} \quad (6.9a)$$

SOLUTION

All functions for one-sided Laplace transforms are defined as equal to zero for $t < 0$. Substituting the expression for U(t) into equation (6.38) yields:

$$F[s] = \int_0^\infty 1\, e^{-st}\, dt \quad \text{or} \quad F[s] = -\frac{e^{-st}}{s}\Big|_0^\infty . \quad (6.9b)$$

$e^{-\infty} = 0$; therefore, equation (6.9b) becomes:

$$\boxed{F[s] = \frac{1}{s}.} \quad (6.9c)$$

EXAMPLE 6.10

Determine the Laplace transform of a ramp function that is specified by:

$$f(t) = t\, U(t) . \quad (6.10a)$$

SOLUTION

Substituting the expression for f(t) into equation (6.38) yields:

$$F[s] = \int_0^\infty t\, e^{-st}\, dt \quad (6.10b)$$

or:

$$F[s] = \frac{e^{-st}}{s^2}(-st-1)\Big|_0^\infty . \quad (6.10c)$$

Therefore:

$$\boxed{F[s] = \frac{1}{s^2}.} \quad (6.10d)$$

EXAMPLE 6.11

Determine the Laplace transform of a sine function that is given by:

$$f(t) = \sin at\; U(t) . \quad (6.11a)$$

SOLUTION

Substituting the expression for f(t) into equation (6.38) yields:

$$F[s] = \int_0^\infty \sin at\; e^{-st}\, dt \quad (6.11b)$$

or:

$$F[s] = \frac{e^{-st}\left(-s\sin at - a\cos at\right)}{s^2 + a^2}\Big|_0^\infty . \quad (6.11c)$$

Therefore:

$$\boxed{F[s] = \frac{a}{s^2 + a^2}.} \quad (6.11d)$$

EXAMPLE 6.12

Determine the Laplace transform of an exponential function which is given by:

$$f(t) = e^{at}\; U(t) . \quad (6.12a)$$

SOLUTION

Substituting the expression for f(t) into equation (6.38) yields:

$$F[s] = \int_0^\infty e^{at}\; e^{-st}\, dt \quad (6.12b)$$

or:

$$F[s] = \frac{e^{-(s-a)t}}{-(s-a)}\Bigg|_0^{\infty}. \qquad (6.12c)$$

Therefore:

$$\boxed{F[s] = \frac{1}{s-a}.} \qquad (6.12d)$$

Listed below are some useful theorems associated with Laplace transforms:

Shifting Theorem:

When:

$$\boxed{L[f(t)] = L[s],} \qquad (6.50)$$

then:

$$L[f(t+T)]\,U(t-T) = e^{-sT}\,F[s]. \qquad (6.51)$$

Initial Value Theorem:

$$\boxed{\lim_{t\to 0} f(t) = \lim_{s\to\infty} sF[s].} \qquad (6.52)$$

Final Value Theorem:

$$\boxed{\lim_{t\to\infty} f(t) = \lim_{s\to 0} sF[s].} \qquad (6.53)$$

Figure 6.12 shows a sketch of a unit pulse. A *unit pulse* is defined:

$$f(t) = \frac{1}{\Delta T}[U(t-T) - U(t-T-\Delta T)] \qquad (6.54)$$

where T is the time at which the pulse occurs and ΔT is the duration of the pulse. The area within the unit pulse shown in Figure 6.12 is unity. The unit for a unit pulse is 1/s. If the pulse is applied at time t = 0, equation (6.54) becomes:

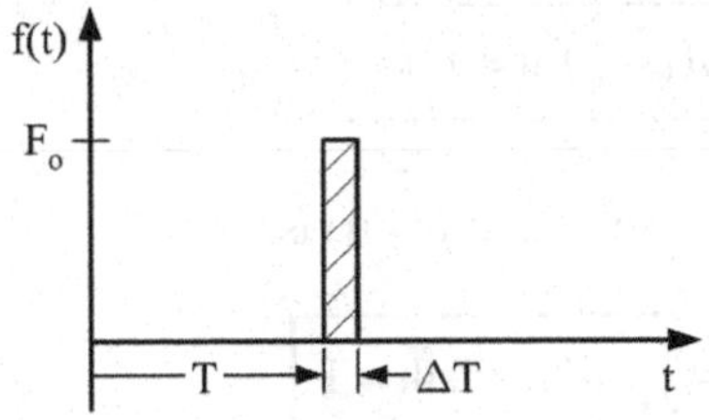

Figure 6.12 Unit Pulse

$$f(t) = \frac{1}{\Delta T}[U(t) - U(t-\Delta T)]. \qquad (6.55)$$

Using equation (6.46) and the shifting theorem, equation (6.51), the Laplace transform of a unit pulse that occurs at time T is:

$$L[\text{unit pulse}] = \frac{1}{\Delta T}\left[\frac{1}{s}e^{-sT} - \frac{1}{s}e^{-s(T+\Delta T)}\right] \qquad (6.56)$$

or:

$$L[\text{unit pulse}] = \frac{e^{-sT}}{s\Delta T}\left[1 - e^{-s\Delta T}\right]. \qquad (6.57)$$

In the limiting case where ΔT goes to zero and where the area within the pulse remains equal to unity:

$$\begin{aligned} f(t) &= \lim_{\Delta T\to 0} \frac{1}{\Delta T}[U(t-T) - U(t-T-\Delta T)] \\ &= \delta(t-T) \end{aligned} \qquad (6.58)$$

where the *dirac delta* function $\delta(\ )$ is defined as an *unit impulse*.

$$f(t) = \delta(t-T) \qquad (6.59)$$

is referred to as a unit impulse that occurs at time T.

The dirac delta function has the following properties:

$$\boxed{\begin{aligned} \delta(t-T) &= 1 \quad \text{for} \quad t = T \\ &= 0 \quad \text{for} \quad t \neq T \end{aligned}} \qquad (6.60)$$

$$\boxed{\int_0^{\infty} \delta(t-T)\,dt = 1} \qquad (6.61)$$

$$\int_0^\infty f(t)\delta(t-T)dt = f(t). \tag{6.62}$$

The Laplace transform of δ(t) is:

$$L[\delta(t)] = \int_0^\infty \delta(t)e^{-st}\,dt = 1. \tag{6.63}$$

The Laplace transform of δ(t - T) is:

$$L[\delta(t-T)] = \int_0^\infty \delta(t-T)e^{-st}\,dt = 1\,e^{-sT}. \tag{6.64}$$

EXAMPLE 6.13

A mass-spring-damper system similar to the one shown in Figure 3.1 in Chapter 3 is excited by an arbitrary force. Determine the Laplace transform of the response of the system.

SOLUTION

The equation of motion for this system is:

$$M\ddot{y} + C\dot{y} + K y = f(t). \tag{6.13a}$$

The Laplace transform of this equation is:

$$M\{s^2 Y[s] - s y(0) - \dot{y}(0)\} + C\{s Y[s] - y(0)\} + K Y[s] = F[s] \tag{6.13b}$$

or rearranging the terms:

$$[Ms^2 + Cs + K]Y[s] - M\dot{y}(0) - [Ms + C]y(0) = F[s]. \tag{6.13c}$$

Finally, the Laplace transform of the response is given by:

$$Y[s] = \frac{F(s)}{Ms^2 + Cs + K} + \frac{M\dot{y}(0) + [Ms + C]y(0)}{Ms^2 + Cs + K}. \tag{6.13d}$$

As was the case with Fourier transforms in EXAMPLE 6.7, an input- output transfer function relation also exists with Laplace transforms. If:

$$G[s] = \frac{1}{Ms^2 + Cs + K} \tag{6.65}$$

is defined as the *transfer function*. F(s) is defined as the *Laplace transform of the input*. Y(s) is defined as the *Laplace transform of the output*. $M\dot{y}(0) + [Ms + C]y(0)$ is defined as the *Laplace transform of the initial conditions*. Equation (6.13d) can be written in a general form that applies to all linear vibration systems:

$$Y[s] = G[s]\,F[s] + G[s]\begin{bmatrix}\text{Laplace Transform of} \\ \text{Initial Conditions}\end{bmatrix}. \tag{6.66}$$

Equation (6.66) is schematically shown in Figure 6.13.

EXAMPLE 6.14

The 2-DOF vibration system in Figure 6.14 is excited by an arbitrary force applied to mass M_1. Determine the Laplace transforms of $x_1(t)$ and $x_2(t)$ of the system that are associated with f(t). Assume all of the initial conditions are equal to zero.

SOLUTION

The equations of motion for this system are:

$$M_1\ddot{x}_1 + (K + K_1)x_1 - K x_2 = f(t) \tag{6.14a}$$

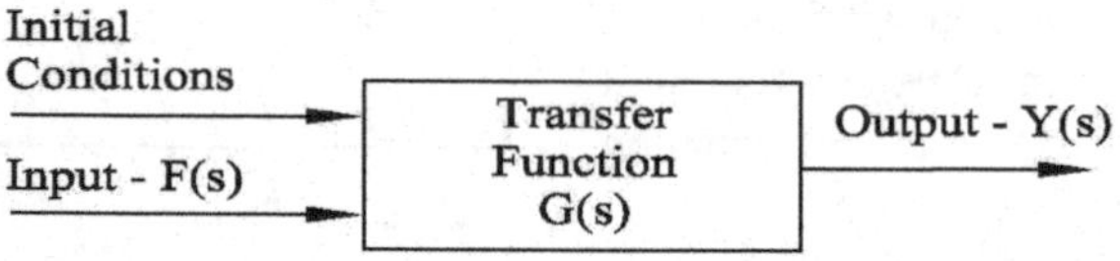

Figure 6.13 Vibration System Transfer Function

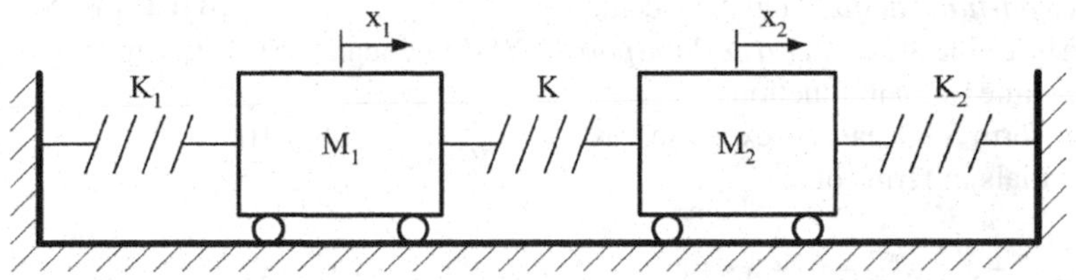

Figure 6.14 2-DOF Vibration System with Mass and Spring Elements

and:

$$M_2\,\ddot{x}_2 + (K + K_2)x_2 - K\,x_1 = 0 . \tag{6.14b}$$

The Laplace transforms of the above equations are:

$$M_1\,s^2\,X_1[s] + (K + K_1)X_1[s] - K\,X_2[s] = F[s] \tag{6.14c}$$

and:

$$M_2\,s^2\,X_2[s] + (K + K_2)X_2[s] - K\,X_1[s] = 0 . \tag{6.14d}$$

Rearranging the above equations yields:

$$\left(M_1\,s^2 + K + K_1\right)X_1[s] - K\,X_2[s] = F[s] \tag{6.14e}$$

and:1

$$\left(M_2\,s^2 + K + K_2\right)X_2[s] - K\,X_1[s] = 0 . \tag{6.14f}$$

Cramer's rule can be used to solve for $X_1[s]$ and $X_2[s]$. Therefore:

$$X_1[s] = \frac{F[s]\left(M_2\,s^2 + K + K_2\right)}{\left\{ M_1M_2\,s^4 + \begin{bmatrix} M_1\,(K + K_2) + \\ M_2\,(K + K_1) \end{bmatrix} s^2 + KK_1 + KK_2 + K_1\,K_2 \right\}} \tag{6.14g}$$

and:

$$X_1[s] = \frac{F[s]\,K}{\left\{ M_1M_2\,s^4 + \begin{bmatrix} M_1\,(K + K_2) + \\ M_2\,(K + K_1) \end{bmatrix} s^2 + KK_1 + KK_2 + K_1\,K_2 \right\}} \tag{6.14h}$$

When Laplace transforms are used to analyze a linear vibration system, the system of linear differential equations that describe the motion of the system masses can be converted into a corresponding system of linear algebraic equations. It is usually easier to manipulate the system of algebraic equations to solve for the system time domain response. However, to complete the process of determining the system response, the inverse Laplace transforms of the system response functions must be determined to obtain the time domain response of the system.

6.6 Inverse Laplace Transforms

The *inverse Laplace transform* is denoted by:

$$L^{-1}\{Y[s]\} = y(t) . \tag{6.67}$$

The inverse Laplace transform is obtained from equation (6.37). The use of equation (6.37) requires a substantial knowledge of complex variables. Fortunately, it is not necessary to evaluate the complex integral for many engineering problems. This has already been done, and the results for common functions are given in tables of Laplace transform pairs. Appendix 6A at the end of this chapter gives a short table of Laplace transform pairs that are important for the analysis of linear vibration systems. Therefore, to determine the inverse Laplace transform of the response of a linear vibration system, all that is necessary is to expand the Laplace transform of the system into its *partial fractions* and then identify each partial fraction in a table of *Laplace transform pairs* to obtain the proper time domain functions.

The Laplace transform Y[s] can be expressed as a ratio of two polynomials in terms of s:

$$Y[s]=\frac{A(s)}{B(s)}=\frac{a_m s^m + a_{m-1}s^{m-1}+\cdots+a_1 s+a_0}{s^n+b_{n-1}s^{n-1}+\cdots+b_1 s+b_0} \quad (6.68)$$

where m and n are positive integers. To reduce equation (6.68), it is generally assumed that Y[s] is a proper fraction where $n \geq m$. When Y[s] is not a proper fraction ($n < m$), the numerator is divided by the denominator to obtain Y[s] as a polynomial in terms of s plus a proper fraction. When the *method of partial fractions* is used, Y[s] can be expressed:

$$Y[s]=\frac{A(s)}{(s-s_1)(s-s_2)\cdots(s-s_n)} \quad (6.69)$$

where $s_{1,2,\ldots,n}$ are the roots of the equation B(s) = 0. If the equation is of degree n, it has n roots. These roots can be zero, real, complex, distinct, or repeating. Y[s] in equation (6.69) can be written:

$$Y[s]=\frac{A(s)}{B(s)}=\frac{A_1}{s-s_1}+\frac{A_2}{s-s_2}+\cdots+\frac{A_n}{s-s_n}. \quad (6.70)$$

When the method of partial fractions is used, equation (6.69) is rewritten into a sum of partial fraction expressions similar to those shown in equation (6.70). The inverse Laplace transforms of these partial fractions can then be obtained directly from a table of Laplace transform pairs.

Case 1: Roots of B(s) = 0 are real and distinct.

EXAMPLE 6.15

Determine the inverse Laplace transform of:

$$Y[s]=\frac{4s+10}{s^3+6s^2+5s}. \quad (6.15a)$$

SOLUTION

Y[s] in equation (6.15a) can be written:

$$Y[s]=\frac{4s+10}{s(s+1)(s+5)}. \quad (6.15b)$$

Using the partial fraction expansion, Y[s] becomes:

$$Y[s]=\frac{A_1}{s}+\frac{A_2}{s+1}+\frac{A_3}{s+5}. \quad (6.15c)$$

Writing equation (6.15c) over a common denominator yields:

$$Y[s]=\frac{A_1(s+1)(s+5)+A_2 s(s+5)+A_3 s(s+1)}{s(s+1)(s+5)}. \quad (6.15d)$$

Equations (6.15b) and (6.15d) are equal if their numerators are equal. Therefore, expanding the numerator of equation (6.15d), collecting terms and setting them equal to the numerator of equation (6.15b) gives:

$$(A_1+A_2+A_3)s^2+(6A_1+5A_2+A_3)s + 5A_1 = 4s+10. \quad (6.15e)$$

The above equation is valid only if the coefficients of similar powers of s on both sides of the equal sign are equal. Therefore:

$$A_1+A_2+A_3=0 \quad (6.15f)$$

$$6A_1+5A_2+A_3=4 \quad (6.15g)$$

$$5A_1=10. \quad (6.15h)$$

Solving equations (6.15f) through (6.15g) for A_1, A_2 and A_3 yields:

$$A_1=2 \qquad A_2=-\frac{3}{2} \qquad A_3=-\frac{1}{2}. \quad (6.15i)$$

Substituting these expressions into equation (6.15c)

gives:

$$Y[s] = \frac{2}{s} - \frac{3/2}{s+1} - \frac{1/2}{s+5}. \tag{6.15j}$$

The inverse Laplace transform of Y[s] is:

$$\boxed{y(t) = \left(2 - \frac{3}{2}e^{-t} - \frac{1}{2}e^{-5t}\right)U(t).} \tag{6.15k}$$

The unit step function U(t) is used at the end of equation (6.15k) to indicate that the solution only applies for the case where $t \geq 0$.

Case 2: Roots of B(s) are real and repeating.

EXAMPLE 6.16

Determine the inverse Laplace transform of:

$$Y[s] = \frac{s+5}{s^3(s+2)^2}. \tag{6.16a}$$

SOLUTION

Expanding equation (6.16a) using partial fractions yields:

$$Y[s] = \frac{A_{11}}{s^3} + \frac{A_{12}}{s^2} + \frac{A_{13}}{s} + \frac{A_{21}}{(s+2)^2} + \frac{A_{22}}{s+2}. \tag{6.16b}$$

Writing equation (6.16b) over a common denominator gives:

$$Y[s] = \frac{A_{11}(s+2)^2 + A_{12}s(s+2)^2 + A_{13}s^2(s+2)^2 + A_{21}s^3 + A_{22}s^2(s+2)}{s^3(s+2)^2}. \tag{6.16c}$$

Expanding the numerator of equation (6.16c), collecting terms and setting them equal to the numerator equation (6.16a) yields:

$$(A_{13} + A_{22})s^4 + (A_{12} + 4A_{13} + A_{21} + 2A_{22})s^3 + (A_{11} + 4A_{12} + 4A_{13})s^2 + (4A_{11} + 4A_{12})s + 4A_{11} = s+5. \tag{6.16d}$$

Equating coefficients of similar powers of s on both sides of the equal sign of equation (6.16d) and solving the resulting equations gives:

$$A_{11} = \frac{5}{4} \quad A_{12} = -1 \quad A_{13} = \frac{11}{16}$$
$$A_{21} = -\frac{3}{8} \quad A_{22} = -\frac{11}{16}. \tag{6.16e}$$

Therefore, equation (6.16b) becomes:

$$Y[s] = \frac{5/4}{s^3} - \frac{1}{s^2} + \frac{11/16}{s} - \frac{3/8}{(s+2)^2} - \frac{11/16}{s+2}. \tag{6.16f}$$

The inverse Laplace transform of equation (6.16f) is:

$$\boxed{y(t) = \left(\frac{5}{8}t^2 - t + \frac{11}{16} - \frac{3}{8}te^{-2t} - \frac{11}{16}e^{-2t}\right)U(t).} \tag{6.16g}$$

Case 3: Roots of B(s) are complex and distinct.

EXAMPLE 6.17

Determine the inverse Laplace transform of:

$$Y[s] = \frac{2s+5}{s(s^2+2s+5)}. \tag{6.17a}$$

SOLUTION

Using partial fractions, the above equation can be written:

$$Y[s] = \frac{A_1}{s} + \frac{A_2 s + A_3}{s^2 + 2s + 5}. \tag{6.17b}$$

Writing equation (6.17b) over a common denominator yields:

$$Y[s] = \frac{A_1\left(s^2+2s+5\right)+\left(A_2 s + A_3\right)s}{s\left(s^2+2s+5\right)}. \tag{6.17c}$$

Equating the numerators equations (6.17a) and (6.17c) gives:

$$A_1\left(s^2+2s+5\right)+\left(A_2 s + A_3\right)s = 2s+5. \tag{6.17d}$$

Therefore:

$$A_1 = 1 \qquad A_2 = -1 \qquad A_3 = 0. \tag{6.17e}$$

Equation (6.17b) now becomes:

$$Y[s] = \frac{1}{s} - \frac{s}{s^2+2s+5}. \tag{6.17f}$$

It is necessary to complete the squares in the numerator of the second term in equation (6.17f), or:

$$\begin{aligned} s^2+2s+5 &= s^2+2s+(1-1)+5 \\ &= (s+1)^2+2^2. \end{aligned} \tag{6.17g}$$

Therefore, equation (6.17f) can be written:

$$Y[s] = \frac{1}{s} - \frac{s}{(s+1)^2+2^2}. \tag{6.17h}$$

From the Laplace transform tables:

$$L^{-1}\left\{\frac{1}{(s+a)^2+b^2}\right\} = \frac{1}{b}e^{-at}\sin bt \tag{6.17i}$$

and:

$$L^{-1}\left\{\frac{s+a}{(s+a)^2+b^2}\right\} = e^{-at}\cos bt. \tag{6.17j}$$

Equation (6.17h) can be rewritten:

$$Y[s] = \frac{1}{s} - \frac{s+(1-1)}{(s+1)^2+2^2} \tag{6.17k}$$

or:

$$Y[s] = \frac{1}{s} - \frac{s+1}{(s+1)^2+2^2} + \frac{1}{(s+1)^2+2^2}. \tag{6.17l}$$

The inverse Laplace transform of Y(s) is:

$$\boxed{y(t) = \left(1 - e^{t}\cos 2t + \frac{1}{2}e^{-t}\sin 2t\right)U(t).} \tag{6.17m}$$

A slight variation of the method of partial fractions can be used to make the determination of inverse Laplace transforms easier. First, multiply both sides of equation (6.70) by ($s - s_n$) to obtain:

$$\begin{aligned}(s-s_n)Y[s] = {} & \frac{A_1(s-s_n)}{s-s_1} + \\ & \frac{A_2(s-s_n)}{s-s_2} + \cdots + A_n. \end{aligned} \tag{6.71}$$

When $s = s_n$, equation (6.71) becomes:

$$A_n = \left\{(s-s_n)Y[s]\right\}\Big|_{s=s_n}. \tag{6.72}$$

EXAMPLE 6.18

Evaluate the coefficients in equation (6.15b) in EXAMPLE 6.15, using equation (6.72).

SOLUTION

A_1 is obtained from:

$$A_1 = \left\{s\,\frac{4s+10}{s(s+1)(s+5)}\right\}\Bigg|_{s=0} \tag{6.18a}$$

or:

$$A_1 = \left\{ \frac{4s+10}{(s+1)(s+5)} \right\} \Bigg|_{s=0} = 2 . \tag{6.18b}$$

A_2 is obtained from:

$$A_2 = \left\{ (s+1) \frac{4s+10}{s(s+1)(s+5)} \right\} \Bigg|_{s=-1} \tag{6.18c}$$

or:

$$A_2 = \left\{ \frac{4s+10}{s(s+5)} \right\} \Bigg|_{s=-1} = -\frac{3}{2} . \tag{6.18d}$$

A_3 is obtained from:

$$A_3 = \left\{ (s+5) \frac{4s+10}{s(s+1)(s+5)} \right\} \Bigg|_{s=-5} \tag{6.18e}$$

or:

$$A_3 = \left\{ \frac{4s+10}{s(s+1)} \right\} \Bigg|_{s=-5} = -\frac{1}{2} . \tag{6.18f}$$

The values of A_1, A_2, and A_3 are the same as those obtained in EXAMPLE 6.15.

When a Laplace transform has repeated roots, equation (6.72) can be modified to:

$$A_{nk} = \frac{1}{(k-1)!} \left\{ \frac{d^{k-1} \left\{ (s-s_n)^p Y[s] \right\}}{ds^{k-1}} \right\} \Bigg|_{s=s_n} . \tag{6.73}$$

n is an integer representing the nth root of B(s) = 0; p is an integer representing the power to which the expression associated with the nth root is raised; and k is an integer that varies from k = 1 to k = p.

EXAMPLE 6.19

Determine the coefficients of equations (6.16a) in EXAMPLE 6.16, using equation (6.73):

SOLUTION

A_{11} is obtained from:

$$A_{11} = \left\{ s^3 \frac{s+5}{s^3 (s+2)^2} \right\} \Bigg|_{s=0} \tag{6.19a}$$

or:

$$A_{11} = \left\{ \frac{s+5}{(s+2)^2} \right\} \Bigg|_{s=0} = \frac{5}{4} . \tag{6.19b}$$

A_{12} is obtained from:

$$A_{12} = \left\{ \frac{d \left\{ s^3 \frac{s+5}{s^3 (s+2)^2} \right\}}{ds} \right\} \Bigg|_{s=0} \tag{6.19c}$$

or:

$$A_{12} = \left\{ \frac{1}{(s+2)^2} - \frac{2(s+5)}{(s+5)^3} \right\} \Bigg|_{s=0} = -1 . \tag{6.19d}$$

A_{13} is obtained from:

$$A_{13} = \frac{1}{2} \left\{ \frac{d^2 \left\{ s^3 \frac{s+5}{s^3 (s+2)^2} \right\}}{ds^2} \right\} \Bigg|_{s=0} \tag{6.19e}$$

or:

$$A_{13} = \frac{1}{2} \left\{ -\frac{4}{(s+2)^3} + \frac{6(s+5)}{(s+5)^4} \right\} \Bigg|_{s=0} = \frac{11}{16} . \tag{6.19f}$$

A_{21} is obtained from:

$$A_{21} = \left\{ (s+2)^2 \frac{s+5}{s^3 (s+2)^2} \right\} \Bigg|_{s=-2} \tag{6.19g}$$

or:

$$A_{21} = \left\{ \frac{s+5}{s^3} \right\}\Bigg|_{s=-2} = -\frac{3}{8}. \tag{6.19h}$$

Finally, A_{22} is obtained from:

$$A_{22} = \left\{ \frac{d\left\{ (s+2)^2 \dfrac{s+5}{s^3 (s+2)^2} \right\}}{ds} \right\}\Bigg|_{s=-2} \tag{6.19i}$$

or:

$$A_{22} = \left\{ \frac{1}{s^3} - \frac{3(s+5)}{s^4} \right\}\Bigg|_{s=-2} = -\frac{11}{16}.$$

These are the same results as those obtained in EXAMPLE 6.16.

EXAMPLE 6.20

The mass-spring-damper system in EXAMPLE 6.13 is excited by an impulse of f(t) = F δ(t - T) at time t = T. Determine the time domain response y(t) of this system to the impulse. Assume all initial conditions equal zero.

SOLUTION

The Laplace transform of f(t), using the shifting theorem, is:

$$F[s] = F e^{-sT}. \tag{6.20a}$$

Therefore, Y[s] in equation (6.13d) becomes:

$$Y[s] = \frac{F e^{-sT}}{M s^2 + C s + K}. \tag{6.20b}$$

The unit associated with δ(t - t) is 1/s. If the unit associated with f(t) is N, F must have the units of N•s . Dividing the numerator and denominator of the above equation by M and noting the C/M = $2\xi\omega_n$ and K/M = ω_n^2 yields:

$$Y[s] = \frac{(F/M) e^{-sT}}{s^2 + 2\xi\omega_n s + \omega_n^2}. \tag{6.20c}$$

Completing the square relative to the denominator gives:

$$s^2 + 2\xi\omega_n s + \left(\xi^2\omega_n^2 - \xi^2\omega_n^2\right) + \omega_n^2 = \left(s + \xi\omega_n\right)^2 + \omega_d^2. \tag{6.20d}$$

where:

$$\omega_d^2 = \left(1 - \xi^2\right)\omega_n^2. \tag{6.20e}$$

Therefore, equation (6.20c) becomes:

$$Y[s] = \frac{(F/M) e^{-sT}}{\left(s + \xi\omega_n\right)^2 + \omega_d^2}. \tag{6.20f}$$

Using the shifting theorem, the inverse Laplace transform of equation (6.20f) is:

$$y(t) = \frac{F}{M\omega_d}\left[e^{-\xi\omega_n (t-T)} \sin\omega_d (t-T)\right] U(t-T). \tag{6.20g}$$

If T = 0, y(t) becomes:

$$y(t) = \frac{F}{M\omega_d}\left[e^{-\xi\omega_n t} \sin\omega_d t\right] U(t). \tag{6.20h}$$

This is the same as equation (6.3c) in EXAMPLE 6.3.

EXAMPLE 6.21

Determine the response of the mass-spring-damper system in EXAMPLE 6.13 to the square pulse shown in Figure 6.11 for the case where C = 0. Assume all initial conditions equal zero.

SOLUTION

As is seen from Figure 6.11, a square pulse can be created from two step functions where:

$$f(t) = F_o \left[U(t) - U[t - T]\right]. \tag{6.21a}$$

The Laplace transform of f(t) is:

$$F[s]=F_o\left(\frac{1}{s}-\frac{e^{-sT}}{s}\right). \tag{6.21b}$$

Therefore, Y(s) in equation (6.13d) becomes:

$$Y[s]=\frac{F_o\left(1-e^{-sT}\right)}{s\left(Ms^2+K\right)} \tag{6.21c}$$

or:

$$Y[s]=\frac{F_o}{M}\left[\frac{1}{s\left(s^2+\omega_n^2\right)}-\frac{e^{-sT}}{s\left(s^2+\omega_n^2\right)}\right]. \tag{6.21d}$$

From the Laplace transform tables:

$$L^{-1}\left\{\frac{1}{s\left(s^2+a^2\right)}\right\}=\frac{1}{a^2}(1-\cos at). \tag{6.21e}$$

Therefore, y(t) is:

$$y(t)=\frac{F_o}{M\omega_n^2}\left\{\begin{matrix}(1-\cos\omega_n t)\,U(t)- \\ \left[1-\cos\omega_n(t-T)\right]U(t-T)\end{matrix}\right\}. \tag{6.21f}$$

For the case where $0\le t\le T$, y(t) is:

$$y(t)=\frac{F_o}{M\omega_n^2}(1-\cos\omega_n t)\,U(t). \tag{6.21g}$$

Noting that $M\omega_n^2=K$, this is the same as equation (6.4i) in EXAMPLE 6.4.

For the case where $t>T$, equation (6.21f) can be written:

$$y(t)=\frac{F_o}{M\omega_n^2}\left\{\cos\omega_n(t-T)-\cos\omega_n t\right\}U(t-T). \tag{6.21h}$$

Equation (6.21h) can be simplified to:

$$\boxed{y(t)=\frac{F_o}{M\omega_n^2}\sqrt{2(1-\cos\omega_n T)}\,\cos(\omega_n t+\phi)\,U(t-T)} \tag{6.21i}$$

where:

$$\boxed{\phi=\tan^{-1}\left[\frac{\sin\omega_n T}{1-\cos\omega_n T}\right].} \tag{6.21j}$$

Equations (6.21i) and (6.21j) are the same as equations (6.4n) through (6.4q) in EXAMPLE 6.4.

EXAMPLE 6.22

For the system in EXAMPLE 6.14, let:

$M_1=4$ kg, $M_2=1$ kg

$K_1=20$ N/m^2, $K=4$ N/m^2 and $K_2=2$ N/m^2.

Determine $x_1(t)$ and $x_2(t)$ for an input step force $f(t)=2U(t)$ applied to mass M_1.

SOLUTION

Substituting the M and K values into equation (6.14g) and (6.14h) yields:

$$X_1[s]=\frac{F[s](s^2+6)}{4(s^4+12s^2+32} \tag{6.22a}$$

or:

$$X_1[s]=\frac{F[s](s^2+6)}{4(s^2+4)(s^2+8)}. \tag{6.22b}$$

and:

$$X_2[s]=\frac{4\,F[s]}{4(s^4+12s^2+32} \tag{6.22c}$$

or:

$$X_2[s]=\frac{F[s]}{(s^2+4)(s^2+8)}. \tag{6.22d}$$

The Laplace transform of f(t) is:

$$F[s] = \frac{2}{s}. \tag{6.22e}$$

Therefore, $X_1(s)$ and $X_2(s)$ become:

$$X_1[s] = \frac{0.5(s^2+6)}{s(s^2+4)(s^2+8)} = \frac{A_1}{s} + \frac{A_2 s + A_3}{s^2+4} + \frac{A_4 s + A_4}{s^2+8}. \tag{6.22f}$$

and:

$$X_2[s] = \frac{2}{s(s^2+4)(s^2+8)} = \frac{B_1}{s} + \frac{B_2 s + B_3}{s^2+4} + \frac{B_4 s + B_4}{s^2+8} \tag{6.22g}$$

Using partial fraction expansions relative to the above equations:

$$A_1 = \frac{3}{32}; A_2 = -\frac{1}{16}; A_3 = 0; A_4 = -\frac{1}{32}; A_5 = 0 \tag{6.22h}$$

and:

$$B_1 = \frac{1}{16}; B_2 = -\frac{1}{8}; B_3 = 0; B_4 = \frac{1}{16}; B_5 = 0. \tag{6.22i}$$

Therefore, $X_1(s)$ and $X_2(s)$ can be written:

$$X_1[s] = \frac{3}{32s} - \frac{s}{16(s^2+4)} - \frac{s}{32(s^2+8)} \tag{6.22j}$$

and:

$$X_2[s] = \frac{1}{16s} - \frac{s}{8(s^2+4)} + \frac{s}{16(s^2+8)}. \tag{6.22k}$$

The inverse Laplace transforms of equations (6.22j) and (6.22k) are:

$$\boxed{x_1(t) = \left(\frac{3}{32} - \frac{1}{16}\cos 2t - \frac{1}{32}\cos 2\sqrt{2}\,t\right) U(t)} \tag{6.22l}$$

and:

$$\boxed{x_2(t) = \left(\frac{1}{16} - \frac{1}{8}\cos 2t + \frac{1}{16}\cos 2\sqrt{2}\,t\right) U(t).} \tag{6.22m}$$

In the above equations, $x_1(t) = 3/32$ m and $x_2(t) = 1/16$ m are the static displacements associated with a 2 N static force applied to mass M_1. $\omega_1 = 2$ rad/s and $\omega_2 = 2\sqrt{2}$ rad/s are the two resonance frequencies associated with the two-mass system in this example.

6.7 Convolution Integral

When the mass of a one-degree-of-freedom system is excited by a pulse of duration ΔT:

$$f(T)\Delta T = M v_o \tag{6.74}$$

where M is the mass and v_o is the initial velocity of the mass that is associated with the pulse excitation. Rearranging equation (6.74) yields:

$$v_o = \frac{f(T)\Delta T}{M}. \tag{6.75}$$

If the one-degree-of-freedom system is less than critically damped, the response of the system to the above initial velocity is:

$$y(t) = \frac{v_o}{\omega_d} e^{-\xi\omega_n t} \sin\omega_d t. \tag{6.76}$$

Substituting equation (6.75) into equation (6.76) yields:

$$y(t) = \frac{f(T)\Delta T}{M\omega_d} e^{-\xi\omega_n t} \sin\omega_d t. \tag{6.77}$$

If f(T) is a unit impulse:

$$f(t) = \frac{1}{\Delta T}. \tag{6.78}$$

Therefore, the response of the vibration system to a unit impulse applied at time t = 0 is:

$$y(t) = \frac{1}{M\omega_d} e^{-\xi\omega_n t} \sin \omega_d t \; U(t). \tag{6.79}$$

When the vibration system is excited by an unit impulse y(t) is designated g(t). Therefore, equation (6.79) is written:

$$g(t) = \frac{1}{M\omega_d} e^{-\xi\omega_n t} \sin \omega_d t \; U(t). \tag{6.80}$$

When the impulse is applied at a time τ, equation (6.80) becomes:

$$g(t+\tau) = \frac{1}{M\omega_d} e^{-\xi\omega_n (t-\tau)} \sin \omega_d (t-\tau) \; U(t-\tau). \tag{6.81}$$

Therefore, the response of any arbitrary system to a pulse f(t) of duration $\Delta\tau$ applied to the system at a time $t = \tau$ can be expressed:

$$y(t) = f(\tau)\Delta\tau \, g(t-\tau) \tag{6.82}$$

where $g(t - \tau)$ is the response of the system to an unit impulse applied to the system at $t = \tau$.

Referring to Figure 6.15, a non-periodic force that has a complicated waveform can be approximated by a series of consecutive pulses of pulse width $\Delta\tau$. The responses of a system to pulses applied at times τ_1, τ_2, τ_3, etc. are:

$$y_1(t) = f(\tau_1)\Delta\tau \, g(t-\tau_1) \tag{6.83}$$

$$y_2(t) = f(\tau_2)\Delta\tau \, g(t-\tau_2) \tag{6.84}$$

$$y_3(t) = f(\tau_3)\Delta\tau \, g(t-\tau_3). \tag{6.85}$$

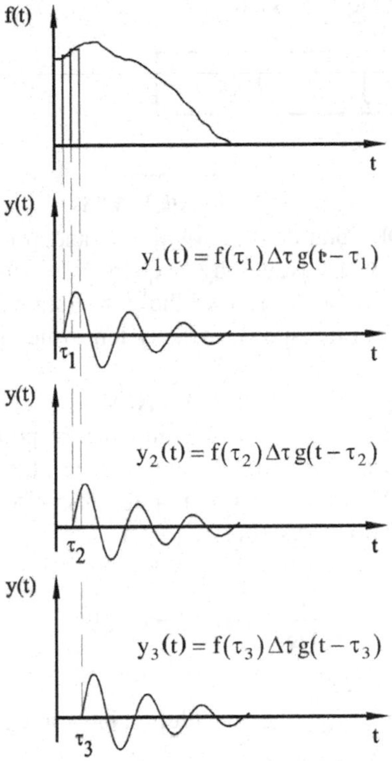

Figure 6.15 Response of a Vibration System to Successive Pulses

The principle of superposition can be used to obtain the total response of the system to all of the pulses. Therefore, the total response of the system is:

$$y(t) = y_1(t) + y_2(t) + y_3(t) + \cdots \tag{6.86}$$

Equation (6.86) can be written:

$$y(t) = \sum_{i=1}^{n} f(\tau_i) g(t-\tau_i) \, \Delta\tau. \tag{6.87}$$

When $\Delta\tau \rightarrow 0$, equation (6.87) becomes:

$$\boxed{y(t) = \int_0^t f(\tau) g(t-\tau) \, d\tau.} \tag{6.88}$$

Equation (6.88) is called the *convolution integral* where $f(\tau)$ is the expression associated with any arbitrary forcing function and $g(t - \tau)$ is the response of the system to a unit impulse applied at time $t = \tau$. Equation (6.88) can also be expressed:

$$\boxed{y(t)=\int_0^t f(t-\tau)g(\tau)\,d\tau .} \tag{6.89}$$

EXAMPLE 6.23

A 1-DOF vibration system with mass and spring elements (Figure 3.1)is excited by a square pulse similar to the one in Figure 6.11. Use the convolution integral to determine the response of the system to the square pulse.

SOLUTION

It is necessary to determine the response of the mass-spring system for the two cases where $0 \le t \le T$ and where $t > T$. The unit impulse response $g(t+\tau)$ of a mass-spring system is:

$$g(t-\tau)=\frac{1}{M\omega_n}\sin\omega_n(t-\tau)\,U(t-\tau). \tag{6.23a}$$

When $0 \le t \le T$, the convolution integral is:

$$y(t)=\frac{F_o}{M\omega_n}\int_0^t \sin\omega_n(t-\tau)\,d\tau \tag{6.23b}$$

or:

$$y(t)=\frac{F_o}{M\omega_n}\int_0^t \begin{pmatrix}\sin\omega_n t\cos\omega_n\tau - \\ \cos\omega_n t\sin\omega_n\tau\end{pmatrix} d\tau . \tag{6.23c}$$

Carrying out the above integration yields:

$$y(t)=\frac{F_o}{M\omega_n^2}\begin{bmatrix}\sin\omega_n t\sin\omega_n\tau\,\big|_o^t + \\ \cos\omega_n t\cos\omega_n\tau\,\big|_o^t\end{bmatrix}. \tag{6.23d}$$

Simplifying the above equation gives:

$$y(t)=\frac{F_o}{M\omega_n^2}\left[\sin^2\omega_n t+\cos^2\omega_n t-\cos\omega_n t\right] \tag{6.23e}$$

or:

$$y(t)=\frac{F_o}{M\omega_n^2}\left[1-\cos\omega_n t\right]U(t). \tag{6.23f}$$

Equation (6.23f) is the same as equation (6.21g) in EXAMPLE 6.21 for the case where $0 \le t \le T$.

For the case when $t > T$, the convolution integral becomes:

$$y(t)=\frac{1}{M\omega_n}\begin{bmatrix}\int_0^T F_o\sin\omega_n(t-\tau)\,d\tau + \\ \int_T^t (o)\sin\omega_n(t-\tau)\,d\tau\end{bmatrix} \tag{6.23g}$$

or:

$$y(t)=\frac{F_o}{M\omega_n}\int_0^T \sin\omega_n(t-\tau)\,d\tau . \tag{6.23h}$$

Carrying out the above integration and simplifying yields:

$$y(t)=\frac{F_o}{M\omega_n^2}\begin{bmatrix}\sin\omega_n t\sin\omega_n T + \\ \cos\omega_n t\cos_n T-\cos\omega_n t\end{bmatrix}. \tag{6.23i}$$

Noting that $\cos(A-B)=\cos A\cos B+\sin A\sin B$, the above equation becomes:

$$\boxed{y(t)=\frac{F_o}{M\omega_n^2}\begin{bmatrix}\cos\omega_n(t-T)- \\ \cos\omega_n t\end{bmatrix}U(t-T).} \tag{6.23j}$$

Equation (6.23j) is the same as equation (6.21h) in EXAMPLE 6.21.

EXAMPLE 6.24

The base of a 1-DOF vibration system similar to the one in Figure 3.28, but only with mass and sprint elements, is excited with the displacement input shown in Figure 6.16. Use the convolution integral to determine the response of the system to the displacement input for

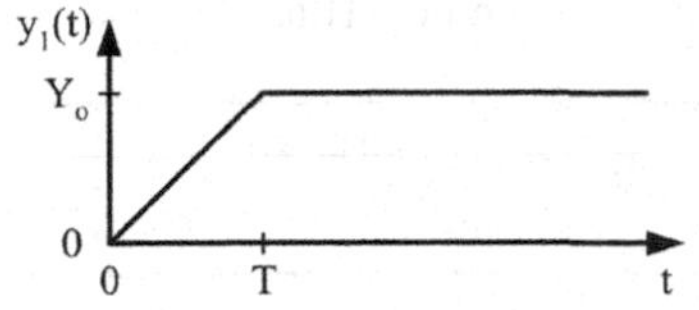

Figure 6.16 Excitation Function for EXAMPLE 6.24

the cases where $t \le T$ and $t > T$.

SOLUTION

First, it is necessary to determine the response of the system to a unit displacement impulse. The equation of motion for the system is:

$$M\ddot{y}_2 + K y_2 = K y_1 \tag{6.24a}$$

where y_2 is the response of the system mass and y_1 is the displacement input to the system base. The units on displacement impulse are in.-s (m•s). The Laplance transform of equation (6.24a) is:

$$Y_2(s) = \frac{K Y_2(s)}{M s^2 + K} \tag{6.24b}$$

or:

$$Y_2(s) = \frac{\omega_n^2 Y_1(s)}{s^2 + \omega_n^2}. \tag{6.24c}$$

The impulse response of equation (6.24c) is:

$$g(t-\tau) = \omega_n \sin[\omega_n (t-\tau)] U(t-\tau). \tag{6.24d}$$

It is necessary to obtain the response of the system for two time periods: $0 \le t \le T$ and $t > T$. The displacement input to the base is (Figure 6.16):

$$\begin{aligned} y_1(t) &= \frac{Y_o}{T} t \quad \text{for} \quad 0 \le t \le T \\ &= T \qquad \text{for} \quad t > T. \end{aligned} \tag{6.24e}$$

For the case when $0 \le t \le T$, the convolution integral is:

$$y_2(t) = \frac{Y_o \omega_n}{T} \int_0^T \tau \sin[\omega_n (t-\tau)] d\tau \tag{6.24f}$$

or:

$$y_2(t) = \frac{Y_o \omega_n}{T} \int_0^T \tau \begin{pmatrix} \sin\omega_n t \cos\omega_n \tau - \\ \cos\omega_n t \sin\omega_n \tau \end{pmatrix} d\tau. \tag{6.24g}$$

Carrying out the integrations in equation (6.24g) yields:

$$y_2(t) = \frac{Y_o}{\omega_n T} \left[\begin{matrix} \sin\omega_n t \begin{pmatrix} \cos\omega_n \tau + \\ \omega_n \tau \sin\omega_n \tau \end{pmatrix} \Bigg|_0^T - \\ \cos \sin\omega_n t \begin{pmatrix} \sin\omega_n \tau - \\ \omega_n \tau \cos\omega_n \tau \end{pmatrix} \Bigg|_0^T \end{matrix} \right]. \tag{6.24h}$$

Simplifying equation (6.24h) gives:

$$\boxed{y_2(t) = \frac{Y_o}{\omega_n T} (\omega_n t - \sin\omega_n t).} \tag{6.24i}$$

This is the same as equation (6.5g) in EXAMPLE 6.5.

For the case when $t > T$, the convolution equation becomes:

$$\begin{aligned} y_2(t) = \frac{Y_o \omega_n}{T} &\int_0^T \tau \sin[\omega_n (t-\tau)] d\tau + \\ &Y_o \omega_n \int_T^t \sin[\omega_n (t-\tau)] d\tau \end{aligned} \tag{6.24j}$$

or:

$$y_2(t) = \frac{Y_o \omega_n}{T} \left[\begin{matrix} \displaystyle\int_0^T \tau \begin{pmatrix} \sin\omega_n t \cos\omega_n \tau - \\ \cos\omega_n t \sin\omega_n \tau \end{pmatrix} d\tau + \\ \displaystyle T \int_T^t \begin{pmatrix} \sin\omega_n t \cos\omega_n \tau - \\ \cos\omega_n t \sin\omega_n \tau \end{pmatrix} d\tau \end{matrix} \right]. \tag{6.24k}$$

Carrying out the above integrations and simplifying yields:

$$y_2(t)=\frac{Y_o}{\omega_n T}\begin{pmatrix}\omega_n T-\sin\omega_n T+\sin\omega_n t\cos\omega_n T\\ -\cos\omega_n t\sin\omega_n T\end{pmatrix}. \qquad (6.24l)$$

Noting that $\sin(A-B)=\sin A\cos B-\cos A\sin B$, equation (6.14l) can be written:

$$\boxed{y_2(t)=\frac{Y_o}{\omega_n T}\begin{bmatrix}\omega_n T-\sin\omega_n T+\\ \sin\omega_n (t-T)\end{bmatrix}U(t-T).} \qquad (6.24m)$$

APPENDIX 6A - TABLES OF LAPLACE TRANSFORM PAIRS

Operation	f(t)	F[s] = F[f(t)]
Transform Integral	$f(t)$	$\int_0^{\infty} f(t)\, e^{-st}\, dt$
Constant multiplication	$a f(t)$	$a F[s]$
Addition and subtraction	$f_1(t) \pm f_2(t)$	$F_1[s] \pm F_2[s]$
First derivative	$\frac{df(t)}{dt} = \dot{f}(t)$	$sF[s] - f(0^+)$
Second derivative	$\frac{d^2 f(1)}{dt^2} = \ddot{f}(t)$	$s^2 F[s] - sf(0^+) - \dot{f}(0^+)$
n^{th} derivative	$\frac{d^n f(t)}{dt^n}$	$s^n F[s] - s^{n-1} f(0^+) - s^{n-2} \frac{df(0^+)}{dt} - \cdots - \frac{d^{n-1} f(0^+)}{d^{n-1}}$
Indefinite integral	$\int f(t) dt$	$\frac{1}{s}\{F[s] + f^{-1}(0^+)\}$
Definite integral	$\int_0^t f(t) dt$	$\frac{1}{s} F[s]$
Real time translation	$f(t-a)U(t-a)$	$e^{-as} F[s]$
Complex time translation	$e^{-at} f(t)$	$F[s+a]$
Periodic function	$f(t) = f(t+\tau)$	$\frac{1}{1-e^{-\tau s}} \int_0^{\tau} f(t) e^{-st} dt$
Convolution	$\int_0^t f_1(t) f_2(t) dt$	$F_1[s] F_2[s]$

Operation	f(t)	F[s] = F[f(t)]
Transform Integral	$f(t)$	$\int_0^\infty f(t)\, e^{-st}\, dt$
Constant multiplication	$a f(t)$	$a F[s]$
Addition and subtraction	$f_1(t) \pm f_2(t)$	$F_1[s] \pm F_2[s]$
First derivative	$\frac{df(t)}{dt} = \dot{f}(t)$	$sF[s] - f(0^+)$

F[s]	f(t), t $\geq$ 0
1	$\delta(t)$
$\dfrac{1}{s}$	$U(t)$
$\dfrac{1}{s^2}$	t
$\dfrac{1}{s^n}$ $n = 1, 2, 3, \cdots$	$\dfrac{t^{n-1}}{(n-1)!}$
$\dfrac{1}{s+a}$	e^{-at}
$\dfrac{1}{(s+a)^2}$	$t e^{-at}$
$\dfrac{1}{(s+a)^n}$ $n = 1, 2, 3, \cdots$	$\dfrac{1}{(n-1)!} t^{n-1} e^{-at}$
$\dfrac{s}{(s+a)^2}$	$(1-at)e^{-at}$
$\dfrac{1}{(s-a)(s-b)}$	$\dfrac{1}{(b-a)}\left(e^{-at} - e^{-bt}\right)$
$\dfrac{s}{(s-a)(s-b)}$	$\dfrac{1}{(a-b)}\left(a e^{-at} - b e^{-bt}\right)$
$\dfrac{1}{(s+a)(s+b)^2}$	$\dfrac{1}{(a-b)^2} e^{at} - \dfrac{(a-b)t - a}{(a-b)^2} e^{bt}$
$\dfrac{s}{(s+a)(s+b)^2}$	$-\dfrac{a}{(a-b)^2} e^{-at} - \dfrac{b(a-b)t - a}{(a-b)^2} e^{-bt}$

F[s]	f(t), t ≥ 0
$\dfrac{s^2}{(s+a)(s+b)^2}$	$\dfrac{a^2}{(a-b)^2}e^{-at}+\dfrac{b^2(a-b)t+b^2-2ab}{(a-b)^2}e^{-bt}$
$\dfrac{1}{s^2+a^2}$	$\dfrac{1}{a}\sin at$
$\dfrac{s}{s^2+a^2}$	$\cos at$
$\dfrac{1}{s^2-a^2}$	$\dfrac{1}{a}\sinh at$
$\dfrac{s}{s^2-a^2}$	$\cosh at$
$\dfrac{1}{s(s^2+a^2)}$	$\dfrac{1}{a^2}(1-\cos at)$
$\dfrac{1}{s^2(s^2+a^2)}$	$\dfrac{1}{a^3}(at-\sin at)$
$\dfrac{1}{(s+a)(s^2+b^2)}$	$\dfrac{1}{a^2+b^2}\left[e^{-at}+\dfrac{\sqrt{a^2+b^2}}{b}\sin(bt-\phi)\right]$ $\phi=\tan^{-1}\left(\dfrac{b}{a}\right)$
$\dfrac{s}{(s+a)(s^2+b^2)}$	$-\dfrac{1}{a^2+b^2}\left[a\,e^{-at}-\sqrt{a^2+b^2}\sin(bt-\phi)\right]$ $\phi=\tan^{-1}\left(\dfrac{a}{b}\right)$
$\dfrac{1}{(s+a)^2+b^2}$	$\dfrac{1}{b}e^{-at}\sin bt$

F[s]	f(t), t ≥ 0
$\frac{s+a}{(s+a)^2+b^2}$	$e^{-at}\cos bt$
$\frac{1}{s\left[(s+a)^2+b^2\right]}$	$\frac{1}{s\left[(s+a)^2+b^2\right]}$ $\phi=\tan^{-1}\left(\frac{b}{a}\right)$
$\frac{1}{s^2\left[(s+a)^2+b^2\right]}$	$\frac{1}{a^2+b^2}\left[t-\frac{2a}{a^2+b^2}+\frac{1}{b}e^{-at}\sin(bt+\phi)\right]$ $\phi=\tan^{-1}\left(\frac{2ab}{a^2-b^2}\right)$
$\frac{1}{(s^2+a^2)^2}$	$\frac{1}{2a^3}(\sin at-at\cos at)$
$\frac{s}{(s^2+a^2)^2}$	$\frac{t}{2a}\sin at$
$\frac{s^2}{(s^2+a^2)^2}$	$\frac{1}{2a}(\sin at+at\cos at)$
$\frac{s^2-a^2}{(s^2+a^2)^2}$	$t\cos at$
e^{-as}	$\delta(t-a)$
$\frac{e^{-as}}{s}$	$U(t-a)$

PROBLEMS - CHAPTER 6

1. Determine the expressions for (a) the displacement and (b) the transmitted force to the foundation for the 1-DOF vibration system in Figure 3.1, but only with mass and spring elements, that is excited by the triangular pulse of duration T shown in Figure P6.1 for the case when t > T. (c) What statements can be made with regard to thevibration isolation of this system?

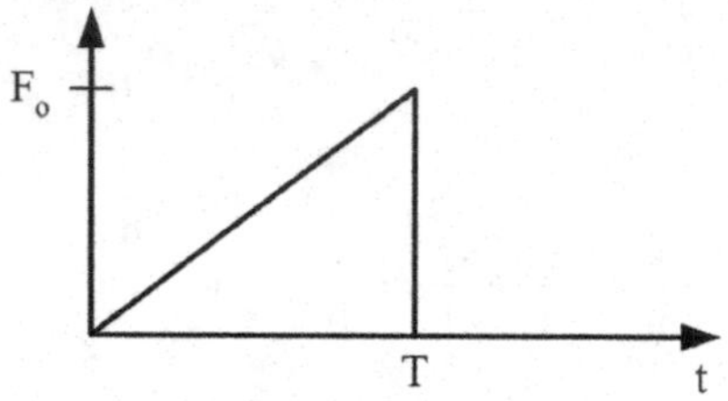

Figure P6.1

2. During a punching operation of a punch press machine, an exponential pulse with a vertical decay where F_o = 1,000 lb_f and T = 10 ms (Table 6.1) is generated. It is desired to limit the amplitude of the transmitted force to 5% of the amplitude of the applied force and the dynamic displacement of the punch press to 0.01 inch. Determine (a) the weight of the mass needed to support the punch press and (b) the value of the total stiffness coefficient of the spring elements needed to support the weight of the entire system.

3. The equation of the signal in Figure P6.1 is:

$$f(t) = F_o \frac{t}{T} \quad \text{for} \quad 0 \le t \le T$$
$$= 0 \quad \text{for} \quad \text{all other values of t.}$$

Determine the expression for the Fourier transform of the signal.

4. The equation of the signal in Figure P6.2 is:

$$f(t) = F_o \frac{1 - e^{t/T}}{1 - e} \quad \text{for} \quad 0 \le t \le T$$
$$= 0 \quad \text{for} \quad \text{all other values of t.}$$

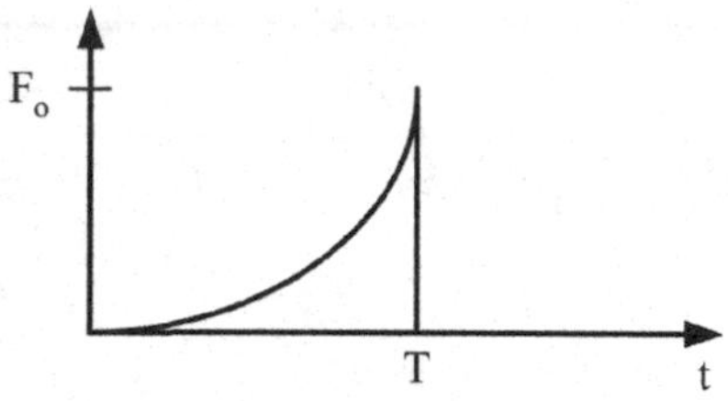

Figure P6.2

Determine the expression for the Fourier transform of the signal.

5. The equation of the signal in Figure P6.3 is:

$$f(t) = F_o \frac{2t}{T} \quad \text{for} \quad 0 \le t \le \frac{T}{2}$$
$$= 2F_o \left(1 - \frac{t}{T}\right) \quad \text{for} \quad \frac{T}{2} \le t \le T$$
$$= 0 \quad \text{for} \quad \text{all other values of t.}$$

Determine the expression for the Fourier transform of the signal.

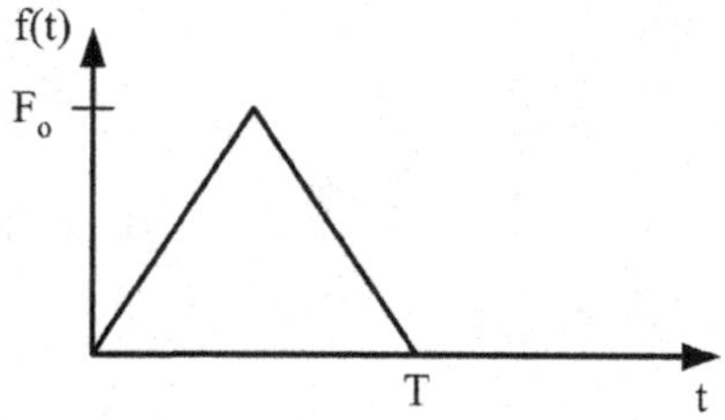

Figure P6.3

6. The equation of the signal in Figure P6.4 is:

$$f(t) = F_o \sin \pi \frac{t}{T} \quad \text{for} \quad 0 \le t \le T$$
$$= 0 \quad \text{for} \quad \text{all other values of t.}$$

Determine the expression for the Fourier transform of the signal.

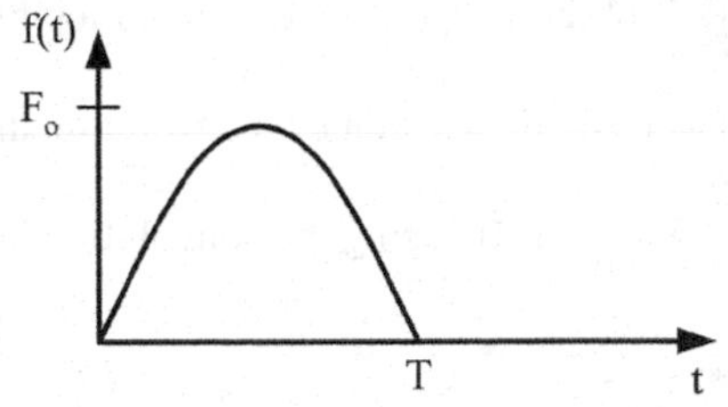

Figure P6.4

7. Determine the inverse Laplace transforms of the following function:

$$Y[s]=\frac{1}{(s+1)^2}$$

8. Determine the inverse Laplace transforms of the following function:

$$Y[s]=\frac{1}{(s+1)^2(s+2)}$$

9. Determine the inverse Laplace transforms of the following function:

$$Y[s]=\frac{1}{s^3+4s^2+3s}$$

10. Determine the inverse Laplace transforms of the following function:

$$Y[s]=\frac{5s+12}{s^2+4s+8}$$

11. Determine the inverse Laplace transforms of the following function:

$$Y[s]=\frac{10}{(s^2+4)(s^2+4s+8)}$$

12. Determine the inverse Laplace transforms of the following function:

$$Y[s]=\frac{s^2+4s+1}{(s+2)(s^2+9)}$$

13. Determine the inverse Laplace transforms of the following function:

$$Y[s]=\frac{s+4}{s(s^2 9s+20)}$$

14. Find the response x(t) for the 1-DOF vibration system with mass and spring elements in Figure P6.5 when it is excited by a forcing function given by Figure P6.1 where T = 0.01 s, using Laplace transforms, for the case when t > T. K = 36 lb_f/in. and M = 4 lb_f-s^2/in. The initial conditions are x(0) = 0 and $\dot{x}(0)=0$.

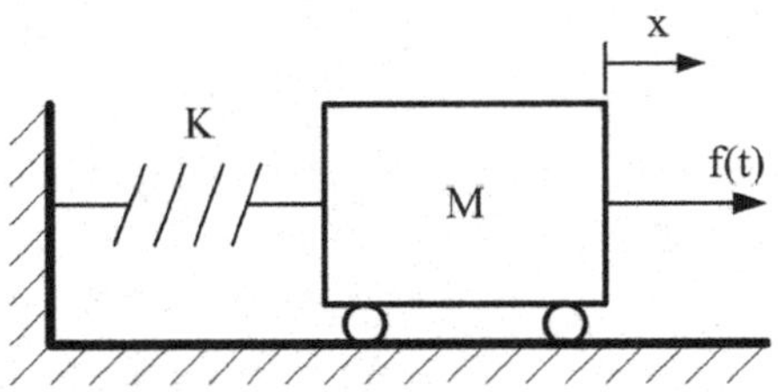

Figure P6.5

15. Find the response x(t) for 1-DOF vibration system with mass and spring elements in Figure P6.5 when it is excited by a forcing function given by Figure P6.3 where T = 0.01 s, using Laplace transforms for the case where t > T. K = 36 lb_f/in. and M = 4 lb_f-s^2/in. The initial conditions are x(0) = 0 and $\dot{x}(0)=0$.

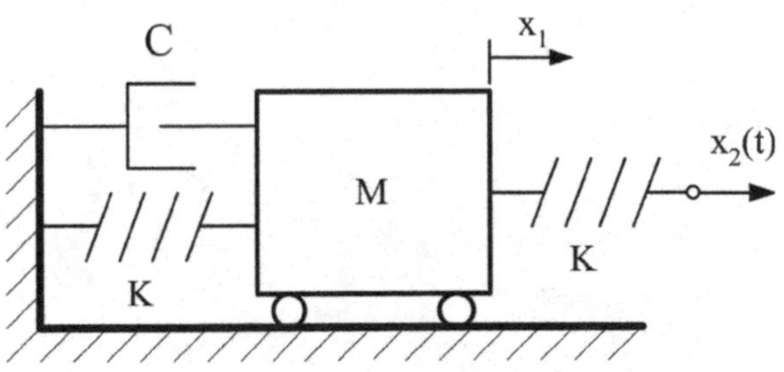

Figure P6.6

16. Find the response $x_1(t)$ for the 1-DOF vibration system with mass, spring and damping elements in Figure P6.6 when it is excited by a forcing function given by Figure P6.1 where T = 0.001 s, using Laplace transforms, for the case when $t > T$. M = 38.6 lb_f, C = 1.0 lb_f-s/in., and K = 20 lb_f/in. The initial conditions are $x(0) = 0$ and $\dot{x}(0) = 0$.

17. Work Problem 14, using the convolution integral.

18. Work Problem 15, using the convolution integral.

19. Work Problem 16, using the convolution integral.

CHAPTER 7
VIBRATION SYSTEMS RANDOM EXCITATION

7.1 Introduction

Random signals are one of two major signal classifications for signals that excite vibration systems. *Deterministic signals*, which are the other classification of signals, were discussed in preceding chapters. *Random signals* have no apparent pattern, and they cannot be described by explicit mathematical relations. They must be described in terms of probability statements and their statistical properties.

There are two classifications of random processes: *stationary* and *ergodic*. Several sample lengths or data records of a process can be grouped together to form an *ensemble* of data signals or records. When several ensembles of data from the same process have the same statistical properties, the process is *stationary*. When an ensemble of data records is stationary and when each data record in each ensemble has the same statistical properties, the process is *ergodic*. For a process to be ergodic, it must first be stationary. An ensemble of data records may be stationary, but the statistical properties of each data record in each ensemble may vary from those of the other data records. For this case, the process is stationary but not ergodic.

This chapter will primarily focus on statistical functions associated with signal processing, which will be covered in Chapter 10. Four basic types of statistical functions are used to describe the properties of random signals. They are *mean squared values* and *variances*, *probability functions*, *correlation functions*, and *spectral density functions*. The *mean squared values* and *variances* provide elementary information related to the amplitudes of signals. The *probability functions* yield information associated with the statistical properties of signals in the amplitude domain. The *correlation* and *power spectral density functions* provide information related to the statistical properties of signals in the time and frequency domains, respectively.

Discussions in this chapter introduce the analytical functions and relations that can be used to describe random signals. These include general functions that are used to describe the statistical properties of random signals in the amplitude, time and frequency domains. It is not the intent of this chapter to go into any of these areas in great detail. However, these areas are discussed in sufficient detail to provide a working knowledge of functional relations used to describe random signals.

7.2 Mean Squared Value, Mean Value, and Variance

The amplitude of a random signals can be described by its mean squared value. The *mean squared value* is the time average of the squared values of the signal over a specified time interval T:

$$\boxed{\left\langle x(t)^2 \right\rangle_t = \lim_{T\to\infty} \frac{1}{T}\int_0^T x(t)^2\, dt\,.} \tag{7.1}$$

The mean squared value can also be designated by Ψ_x^2.

The *root mean squared (rms) value* is another parameter used to specify the amplitude of a random signal. It is defined as the square root of the mean squared value:

$$\boxed{x_{rms} = \lim_{T\to\infty} \sqrt{\frac{1}{T}\int_0^T x(t)^2\, dt}\,.} \tag{7.2}$$

To measure the true rms value of a random signal, it is first necessary to square the signal, average the squared signal over time, and then take the square root of the result.

A random signal can be a combination of static (time invariant) and dynamic (time varying) components. The static component is described by the mean value of the signal. This is not to be confused with the rms value of

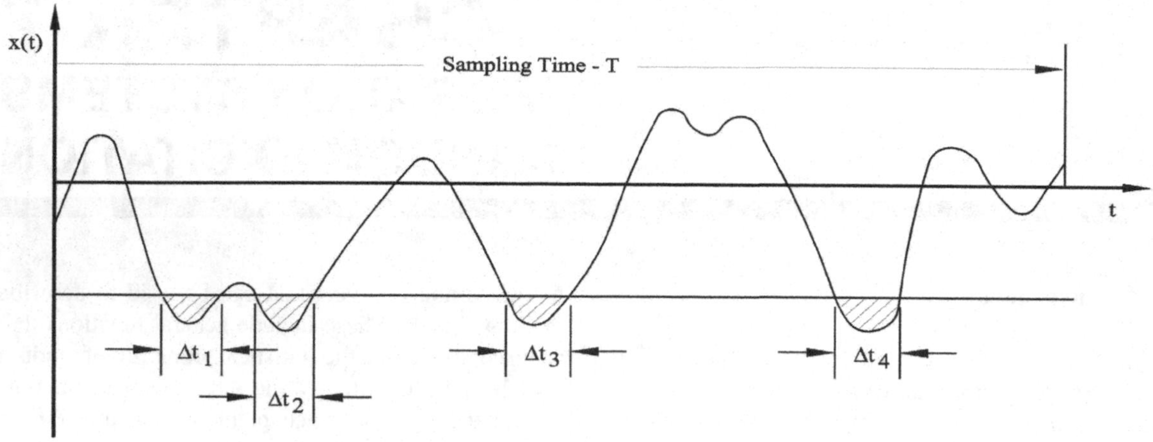

Figure 7.1 Time History of a Random Signal

the signal. The *mean value* is the time average of the signal:

$$\langle x(t)\rangle_t = \lim_{T\to\infty}\frac{1}{T}\int_0^T x(t)dt. \tag{7.3}$$

The mean value can also be denoted by μ_x. The dynamic component of the signal is described by the variance, which indicates the amount the signal varies about its mean value. The *variance* is the mean squared value about the mean:

$$\sigma_x^2 = \lim_{T\to\infty}\frac{1}{T}\int_0^T\left[x(t)-\mu_x\right]^2 dt. \tag{7.4}$$

Expanding the squared term in the integral and carrying out the prescribed operations yields:

$$\sigma_x^2 = \psi_x^2 - \mu_x^2. \tag{7.5}$$

The integrals related to the above functions (and many of those that will follow) are shown being integrated over the limit where T goes to infinity. This is not possible in real measurements associated with random processes. Therefore, T will always have a finite value. When measurements are made to obtain the values for statistical functions associated with a random process, care must be taken to ensure the integration time T is of a length that will result in measured values that are within acceptable amplitude limits.

7.3 Probability Functions

Referring to the random signal in Figure 7.1, it is often desired to know the probability that x(t) will be less than some specified value of x. This can be accomplished by counting the time intervals Δt_i for which $x(t) < x$ over the sample time T and then dividing by the sample time:

$$P\left[x(t) < x\right] = \lim_{T\to\infty}\frac{1}{T}\sum_{i=1}^{n}\Delta t_i. \tag{7.6}$$

The probability that x(t) will fall between the values of x and $x + \Delta x$ where $x + \Delta x > x$ is obtained from:

$$P\left[x < x(t) < x+\Delta x\right] = P\left[x(t) < x+\Delta x\right] - P\left[x(t) < x\right]. \tag{7.7}$$

If equation (7.7) represents a smooth curve, a first order *probability density function* p(x), which represents the slope of the probability function specified by equation (7.7) over a small interval Δx, can be defined such that:

$$p(x) = \lim_{\Delta x\to 0}\frac{P\left[x(t) < x+\Delta x\right] - P\left[x(t) < x\right]}{\Delta x} \tag{7.8}$$

or:

$$\boxed{p(x) = \frac{dP(x)}{dx}.} \tag{7.9}$$

Rearranging equation (7.9) yields:

$$\boxed{dP(x) = p(x)dx.} \tag{7.10}$$

Therefore, the probability that x(t) < x can be written:

$$\boxed{P[x(t) < x] = \int_{-\infty}^{x} p(\alpha)d\alpha.} \tag{7.11}$$

P(x) is called the *probability distribution function* or the *cumulative probability distribution function*. Equations (7.6) and (7.11) indicate that:

$$P(x) \to 0 \quad \text{as} \quad x \to -\infty \tag{7.12}$$

and:

$$P(x) \to 1 \quad \text{as} \quad x \to \infty. \tag{7.13}$$

Equation (7.7) indicates the probability a random signal x(t) will fall between x_1 and x_2 where $x_2 > x_1$ can be written:

$$P[x_1 < x(t) < x_2] = P[x(t) < x_2] - P[x(t) < x_1]. \tag{7.14}$$

Substituting equation (7.11) into equation (7.14) yields:

$$P[x_1 < x(t) < x_2] = \int_{-\infty}^{x_2} p(\alpha)d\alpha - \int_{-\infty}^{x_1} p(\alpha)d\alpha \tag{7.15}$$

or:

$$\boxed{P[x_1 < x(t) < x_2] = \int_{x_1}^{x_2} p(\alpha)d\alpha.} \tag{7.16}$$

The mean value, mean squared value, and variance can be written as a function of the probability density function. Consider a probability distribution curve that

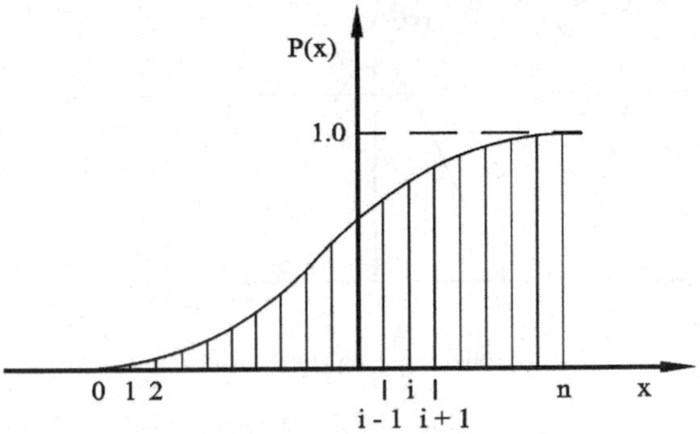

Figure 7.2 Segmented Probability Distribution Curve

is sliced into n segments as is indicated in Figure 7.2. The probability that x lies between) and x_{i+1} can be written:

$$\Delta P_i = P_{i+1} - P_{i-1}. \tag{7.17)}$$

Using this equation, the mean value μ_x can be expressed:

$$\mu_x = \frac{1}{P_n - P_0}\sum_{i=1}^{n-1} x_i (P_{i+1} - P_{i-1}). \tag{7.18}$$

Noting P_n-P_o = 1, letting ΔP_i go to zero, and noting x goes from -∞ to ∞, μ_x becomes:

$$\mu_x = \int_{-\infty}^{\infty} x\, dP(x). \tag{7.19}$$

Substituting equation (7.10) into equation (7.19) yields:

$$\boxed{\mu_x = \int_{-\infty}^{\infty} x\, p(x)dx.} \tag{7.20}$$

The mean squared value can be written in terms of the probability distribution function:

$$\psi_x^2 = \frac{1}{P_n - P_0}\sum_{i=1}^{n-1} x_i^2 (P_{i+1} - P_{i-1}). \tag{7.21}$$

Therefore:

$$\boxed{\psi_x^2 = \int_{-\infty}^{\infty} x^2\, dP(x)} \tag{7.22}$$

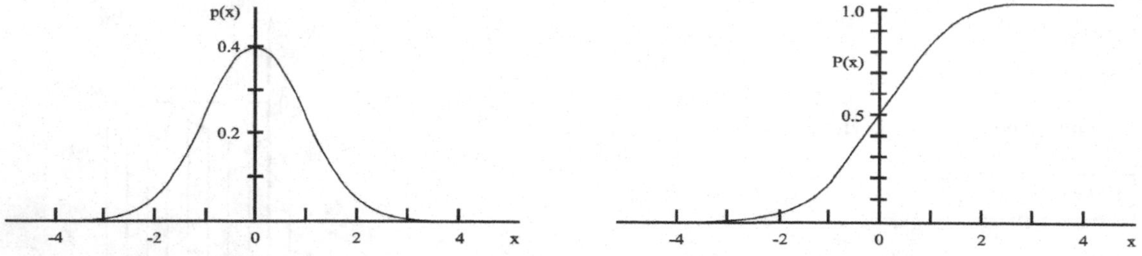

(a) Probability Density Function

(a) Probability Distribution Function

Figure 7.3 Gaussian Probability Density and Distribution Functions for $\sigma_x = 1.0$

or:

$$\psi_x^2 = \int_{-\infty}^{\infty} x^2\, p(x)\, dx. \tag{7.23}$$

Similarly, the variance can be written:

$$\sigma_x^2 = \int_{-\infty}^{\infty} (\mu_x - x)^2\, p(x)\, dx. \tag{7.24}$$

Random data that have values that can be either negative or positive often have a probability density function that is given by:

$$p(x) = \frac{1}{\sigma_x \sqrt{2\pi}} e^{-(x-\mu_x)/2\sigma_x^2} \tag{7.25}$$

where σ_x is called the *standard deviation*. The corresponding probability distribution function is:

$$P(x) = \int_{-\infty}^{x} \frac{1}{\sigma_x \sqrt{2\pi}} e^{-(\alpha-\mu_x)/2\sigma_x^2}\, d\alpha. \tag{7.26}$$

When $\mu_x = 0$, the expressions for the probability density distribution functions become:

$$p(x) = \frac{1}{\sigma_x \sqrt{2\pi}} e^{-x/2\sigma_x^2} \tag{7.27}$$

and:

$$P(x) = \int_{-\infty}^{x} \frac{1}{\sigma_x \sqrt{2\pi}} e^{-\alpha/2\sigma_x^2}\, d\alpha. \tag{7.28}$$

The probability functions defined by equations (7.25) - (7.28) are for a normal or Gaussian data distribution. Figure 7.3 shows the curves for the probability density and distribution functions for the case where $\mu_x = 0$ and $\sigma_x = 1$. When μ_x is a number other than zero, the bell-shaped probability density curve is centered over $x = \mu_x$ and the probability distribution function $P(x)$ equals 0.5 at $x = \mu_x$.

Random data that are restricted to positive values of x, such as the absolute value of x, tend to follow the Rayleigh distribution. The probability density function for the Rayleigh distribution is:

$$p(x) = \frac{x}{\sigma_x^2} e^{-x^2/2\sigma_x^2}. \tag{7.29}$$

Figure 7.4 shows the probability density functions for both *Gaussian* and *Rayleigh* distributions for different values of σ_x.

Figure 7.5 indicates, for a Gaussian distribution, the probability x lies between $-\lambda\sigma_x$ and $\lambda\sigma_x$ is given by equation (7.16) where:

$$P\left[-\lambda\sigma_x < x(t) < \lambda\sigma_x\right] = \int_{-\lambda\sigma_x}^{\lambda\sigma_x} \frac{\alpha}{\sigma_x^2} e^{-\alpha^2/2\sigma_x^2}\, d\alpha. \tag{7.30}$$

Similarly, for a Rayleigh distribution, the probability x exceeds a specified value of $\lambda\sigma_x$ is given by:

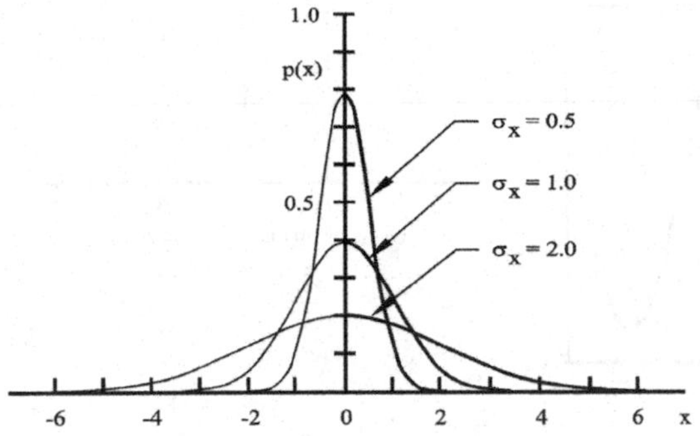

(a) Gaussian Probability Density Curves

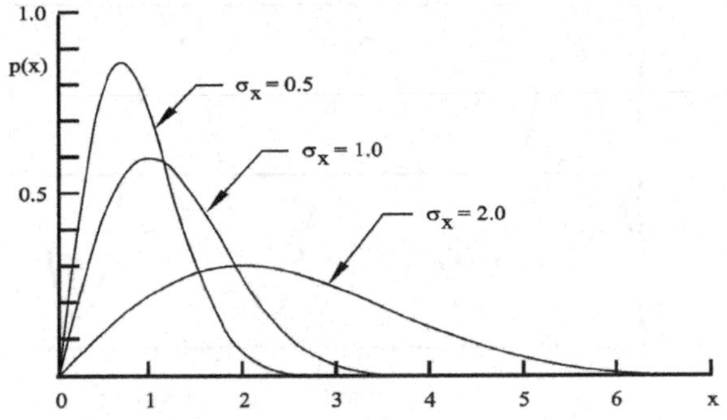

(b) Rayleigh Probability Density Curves

Figure 7.4 Gaussian and Rayleigh Probability Density Curves

$$P[x(t) > \lambda\sigma_x] = \int_{\lambda\sigma_x}^{\infty} \frac{\alpha}{\sigma_x^2} e^{-\alpha^2/2\sigma_x^2} d\alpha. \tag{7.31}$$

The results of equations (7.30) and (7.31) are tabulated in Table 7.1 for specified values of λ.

Normally, the standard deviation σ_x for Gaussian distributions is the positive square root of the variance σ_x^2. However, this is not the case for a Rayleigh distribution. From equation (7.5)

$$\boxed{\sigma_R^2 = \psi_x^2 - \mu_x^2} \tag{7.32}$$

where the subscript R denotes the Rayleigh variance. The mean value for the Rayleigh distribution is:

$$\mu_x = \int_{-\infty}^{\infty} \frac{\alpha^2}{\sigma_x\sqrt{2\pi}} e^{-(\alpha-\mu_x)/2\sigma_x^2} d\alpha \tag{7.33}$$

or:

Table 7.1 Tabulated Results of Equations (7.30) and (7.31)

λ	$P[-\lambda\sigma_x < x < \lambda\sigma_x]$	$P[x > \lambda\sigma_x]$
0	---	1.000
1	0.683	0.607
2	0.954	0.135
3	0.997	0.012

$$\mu_x = \sqrt{\frac{\pi}{2}}\sigma_x. \tag{7.34}$$

Similarly, the mean squared value the Rayleigh distribution is:

$$\psi_x^2 = \int_{-\infty}^{\infty} \frac{\alpha^3}{\sigma_x\sqrt{2\pi}} e^{-(\alpha-\mu_x)/2\sigma_x^2} d\alpha \tag{7.35}$$

or:

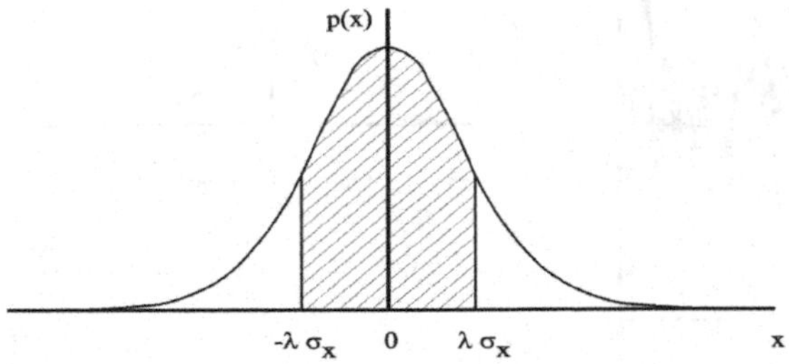

(a) Gaussian Distribution

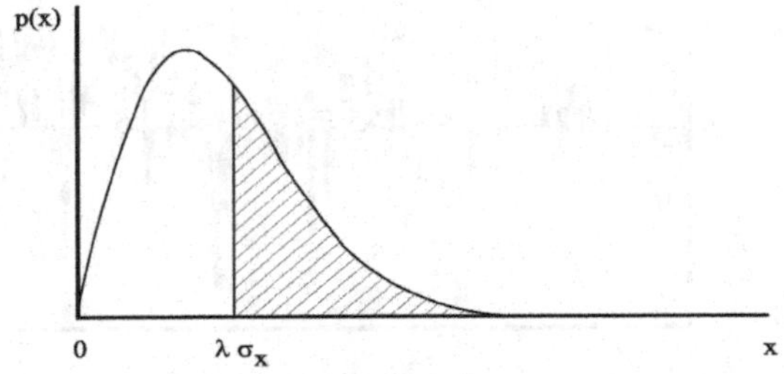

(b) Rayleigh Distribution

Figure 7.5 Gaussian and Rayleigh Distributions Associated with Table 7.1

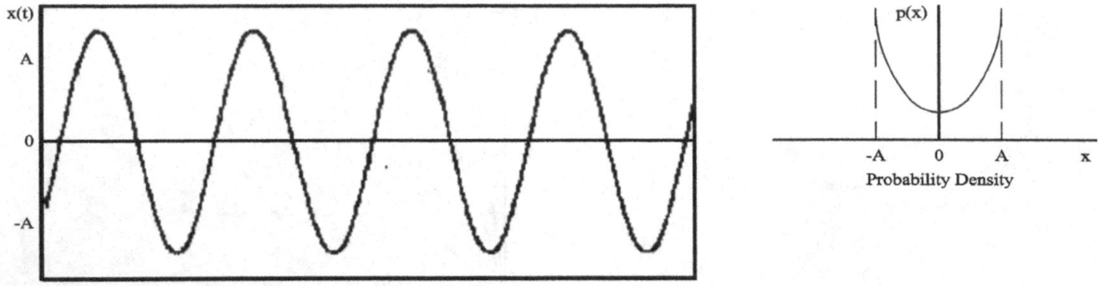

(a) Sinusoidal Signal

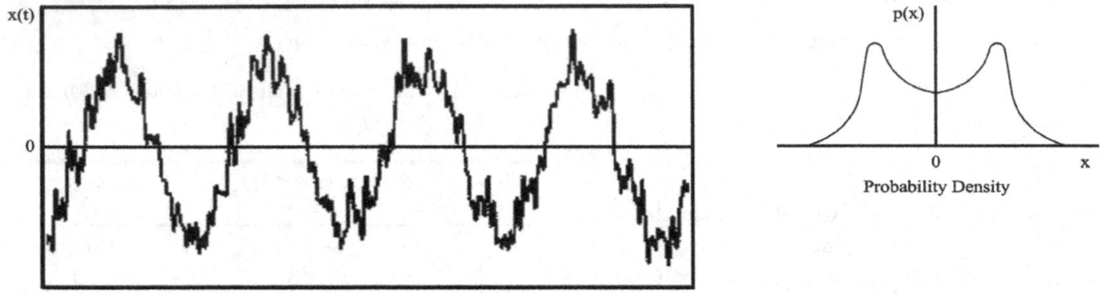

(b) Sinusoidal Signal Plus Wide-band White Noise

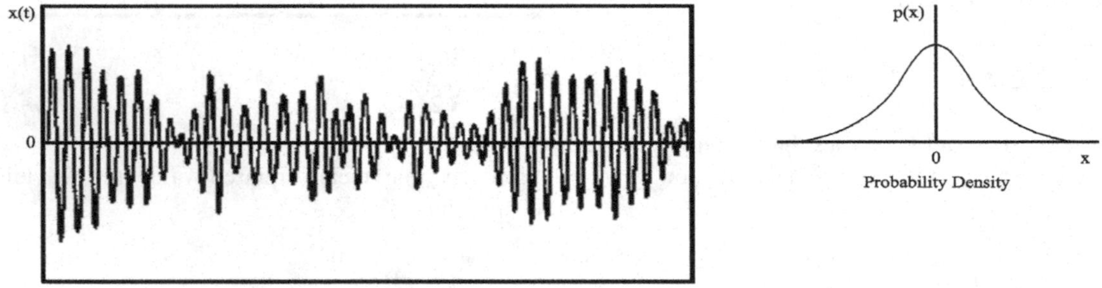

(c) Bandpass White Noise

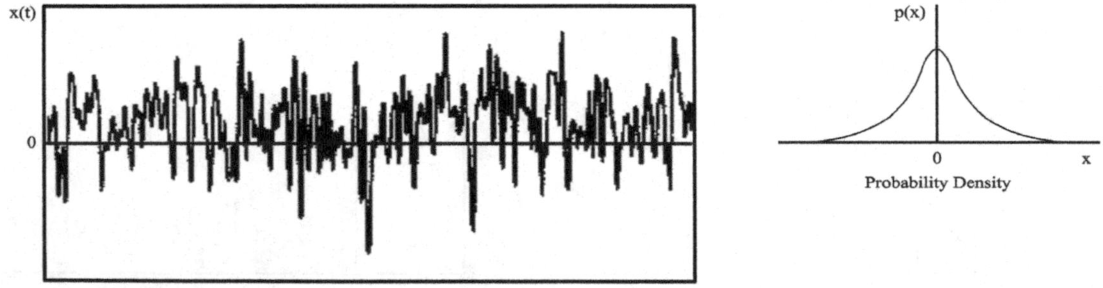

(d) Wide-band White Noise

Figure 7.6 Probability Density Plots of Common Signals

$$\boxed{\psi_x^2 = 2\sigma_x^2.} \tag{7.36}$$

Substituting equations (7.34) and (7.36) into equation (7.32) yields:

$$\sigma_R^2 = \left(2 - \frac{\pi}{2}\right)\sigma_x^2 \quad \text{or} \quad \sigma_R^2 = 0.43\,\sigma_x^2. \tag{7.37}$$

Therefore, the standard deviation for a Rayleigh distribution is:

$$\sigma_R = 0.66\,\sigma_x. \tag{7.38}$$

Probability density measurements can be used to evaluate the normality of data or the type of distribution it has, to predict extreme values in the data, to indicate nonlinear effects, and to estimate or predict damage or failure.

As an example of how probability density measurements can be used, consider the four signals shown in Figure 7.6. The probability distribution function associated with the sine signal is given by:

$$P(x) = \frac{1}{2} + \frac{1}{\pi}\sin^{-1}\frac{x}{A}. \tag{7.39}$$

Taking the derivative of equation (7.39) with respect to x yields:

$$\begin{aligned} p(x) &= \frac{1}{\pi\sqrt{A^2 - x^2}} && \text{for} \quad |x| < A \\ &= 0 && \text{for} \quad |x| > A. \end{aligned} \tag{7.40}$$

Figure 7.6(a) shows a plot of p(x) for a sine signal. The bell-shaped probability density plots for the bandpass and wide-band white noise signals indicate that these signals have a Gaussian distribution.

7.4 Measurement of Probability Density Values

It is possible to measure the probability density values associated with a random signal. From equation (7.8):

$$p(x) = \lim_{\Delta x \to 0} \frac{P[x(t) < x + \Delta x] - P[x(t) < x]}{\Delta x}. \tag{7.41}$$

The probability that x(t) is between x and x+Δx can be given by equations (7.6) and (7.7) where this time Δt_i represents the time intervals that x(t) is between x and x+Δx. Therefore, equation (7.41) becomes (refer to Figure 7.1):

$$p(x) = \lim_{T\to\infty} \lim_{\Delta x \to 0} \frac{1}{T\,\Delta x} \sum_{i=1}^{n} \Delta t_i. \tag{7.42}$$

Figure 7.7 shows a schematic diagram for measuring p(x). It is not possible to let T go to infinity or Δx to go to zero. Therefore, the measurement indicated by equation (7.42) represents an estimate $\hat{p}(x)$ for (t) where:

$$\boxed{\hat{p}(x) = \frac{1}{T\,\Delta x} \sum_{i=1}^{n} \Delta t_i.} \tag{7.43}$$

It is necessary to have an estimate of the error that is associated with $\hat{p}(x)$. The development of error expressions is rather complicated and beyond the scope of this chapter. Therefore, only the results are presented. For more complete information, refer to the references at the end of this chapter.

When sampling a random signal x(t), it is usually necessary to know how many discrete samples are required to describe x(t). When x(t) represents bandwidth limited Gaussian white noise, the required number n of statistically independent discrete samples necessary to described x(t) is given by:

$$\boxed{n = 2BT} \tag{7.44}$$

where B is the frequency bandwidth or bandpass of the signal and T is the sample time. n in equation (7.44) is often referred to as the *number of degrees of freedom* associated with x(t).

Relative to probability density measurements, the bias of $\hat{p}(x)$ in an amplitude window Δx is defined by:

$$\boxed{\text{bias}[\hat{p}(x)] = E[\hat{p}(x)] - p(x)} \tag{7.45}$$

and the variance of $\hat{p}(x)$ is defined by:

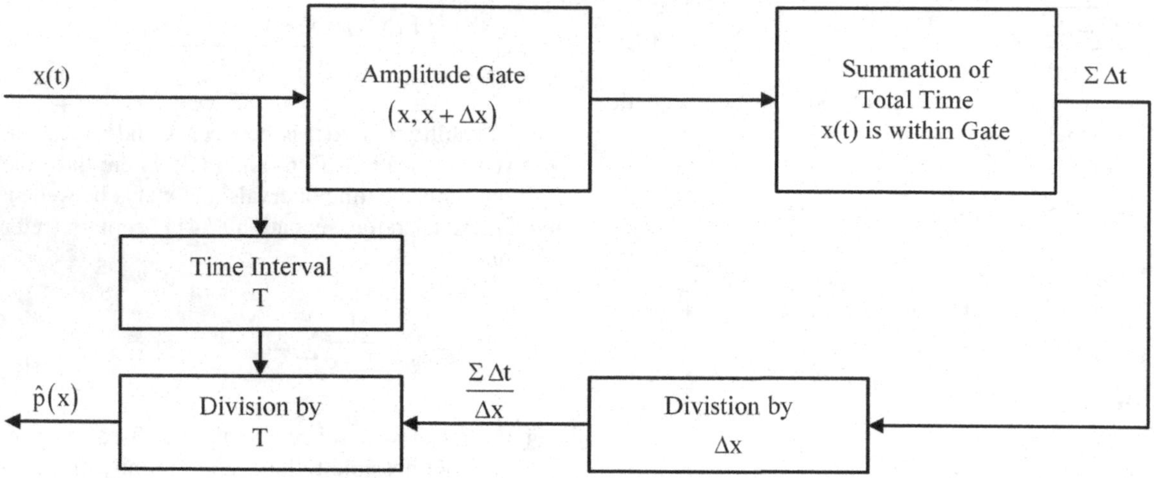

Figure 7.7 Probability Density Measurement

$$\boxed{\mathrm{VAR}\left[\hat{p}(x)\right]=E\left[\left(\hat{p}(x)-E\left[\hat{p}(x)\right]\right)^2\right].} \tag{7.46}$$

Note: E[ ] is defined in Section 7.4. The number of discrete samples that are necessary to reproduce x(t) contained in the amplitude window Δx is:

$$\boxed{n = 2BT\,\Delta x\,\hat{p}(x).} \tag{7.47}$$

For bandwidth limited Gaussian white noise, $\mathrm{bias}\left[\hat{p}(x)\right]$ and $\mathrm{VAR}\left[\hat{p}(x)\right]$ are given by:

$$\boxed{\mathrm{bias}\left[\hat{p}(x)\right]=\frac{\Delta x^2}{24}\frac{d^2\,p(x)}{dx^2}} \tag{7.48}$$

and:

$$\boxed{\mathrm{VAR}\left[\hat{p}(x)\right]=\frac{p(x)^2}{2BT\,\Delta x\,\hat{p}(x)}.} \tag{7.49}$$

The *mean squared error* $E\left[\left(\hat{p}(x)-E\left[\hat{p}(x)\right]\right)^2\right]$ is:

$$E\left[\left(\hat{p}(x)-E\left[\hat{p}(x)\right]\right)^2\right]=\mathrm{VAR}\left[\hat{p}(x)\right]+\mathrm{bias}\left[\hat{p}(x)\right]^2. \tag{7.50}$$

When the bias is small, the *normalized mean squared error* is:

$$\frac{E\left[\left(\hat{p}(x)-E\left[\hat{p}(x)\right]\right)^2\right]}{p(x)^2}=\frac{1}{2BT\,\Delta x\,\hat{p}(x)}. \tag{7.51}$$

Defining the *standard error* ε as the square root of the normalized mean squared error, for $\hat{p}(x)$:

$$\boxed{\varepsilon\left[\hat{p}(x)\right]=\frac{1}{\sqrt{2BT\,\Delta x\,\hat{p}(x)}}.} \tag{7.52}$$

Normally, the second derivative of p(x) is well behaved; Therefore, the bias can be kept small by making Δx as small as possible. Often the frequency bandpass associated with random data is fixed and Δx is specified to keep the bias small. Therefore, it is necessary to properly choose the sampling time T to insure that ε will be an acceptable value for measured values of $\hat{p}(x)$.

The standard error equations given by equation (7.52) and by the equations to follow can be used in two ways. They can be used to estimate the errors associated with measurements that have been made, and they can be used to determine the sampling times that are necessary to keep the standard error within acceptable limits.

7.5 Expected Values and Moments

Often when describing random signals the terms expected values and moments are used. The *expected value* E[ ] is defined as the sum of each of the real values of the function in the brackets multiplied by the probability of its occurrence and averaged over the probability of the occurrence of all real values of the function in the brackets. In terms of previous discussions, the *expected value* of x E[x] can be written:

$$\boxed{E[x]=\int_{-\infty}^{\infty} x\,p(x)\,dx} \tag{7.53}$$

or from equation (7.20):

$$\boxed{E[x]=\mu_x.} \tag{7.54}$$

The expected value of x is equal to the mean value of x. Similarly, the expected value of x^2 $E[x^2]$ is:

$$\boxed{E[x^2]=\int_{-\infty}^{\infty} x^2\,p(x)\,dx} \tag{7.55}$$

or from equation (7.23):

$$\boxed{E[x^2]=\psi_x^2.} \tag{7.56}$$

The expected value of x^2 is equal to the mean squared value of x. The variance can be written:

$$\boxed{E\left[(\mu_x-x)^2\right]=\int_{-\infty}^{\infty}(\mu_x-x)^2\,p(x)\,dx} \tag{7.57}$$

or from equation (7.24):

$$\boxed{E\left[(\mu_x-x)^2\right]=\sigma_x^2.} \tag{7.58}$$

The expected value of any arbitrary function g(x) of x is:

$$\boxed{E[g(x)]=\int_{-\infty}^{\infty} g(x)\,p(x)\,dx.} \tag{7.59}$$

The mean value and the mean squared value of x are referred to as the *first* and *second moments*, respectively, referenced to the probability density function p(x). Higher moments can be evaluated by multiplying the probability density function by higher powers of x. For most engineering problems, only the first and second moments, the mean value and the mean squared value, are used.

Comparing equations (7.54), (7.56) and (7.58) to equations (7.1), (7.3) and (7.4), it can be seen that if g(x) is a function of t, the expected value of g(t) can be expressed:

$$\boxed{E[g(t)]=\lim_{T\to\infty}\int_0^T g(t)\,dt.} \tag{7.60}$$

7.6 Correlation Functions

Sometimes it may be necessary to determine the relation of the values of a random signal at one instant in time to corresponding values of the signal at other instances in time. For example, one may want to identify the existence of periodic components that may be embedded in a signal that appears to be noise. The *autocorrelation function* can be used to do this. In signal processing, the *autocorrelation function* describes the correlation between values of a random signal at different times as a function of the time lag between values. Referring to Figure 7.8, the autocorrelation function $R_x(\tau)$ associated with a time delay τ between x(t) and x(t+τ) is obtained by multiplying x(t) by x(t+τ) and averaging over the sample time T where:

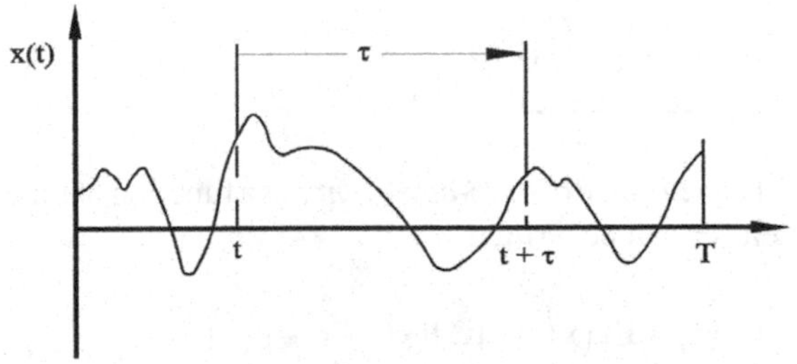

Figure 7.8 Autocorrelation of x(t)

$$\boxed{R_x(\tau) = \lim_{T\to\infty} \frac{1}{T} \int_0^T x(t)\,x(t+\tau)\,dt.} \tag{7.61}$$

In terms of expected values, $R_x(t)$ is expressed:

$$R_x(\tau) = E\left[x(t)\,x(t+\tau)\right]. \tag{7.62}$$

The autocorrelation function is always a real-valued even function that has a maximum at t = 0. In equation form:

$$R_x(\tau) = R_x(-\tau) \tag{7.63}$$

and:

$$R_x(0) \geq |R_x(\tau)| \quad \text{for all values of } \tau. \tag{7.64}$$

When $\tau = 0$, equation (7.61) is the same as equation (7.1). Therefore:

$$\boxed{R_x(0) = \psi_x^2.} \tag{7.65}$$

or the autocorrelation of x(t) at $\tau = 0$ is equal to the mean squared value of x(t).

When there are no periodicities in x(t), the square root of the autocorrelation function as t goes to infinity or gets very large is equal to the mean value of x(t), or:

$$\boxed{\mu_x = \sqrt{R_x(\infty)}.} \tag{7.66}$$

The autocorrelation function can be normalized with respect to $R_x(0)$. When this is done, $\rho_x(t)$ is called the *normalized autocorrelation coefficient* where:

$$\boxed{\rho_x(\tau) = \frac{R_x(\tau)}{R_x(0)}.} \tag{7.67}$$

The *variance* or *auto-covariance* as a function of time delay τ can be written:

$$C_x(\tau) = E\left[\left(x(t) - \mu_x\right)\left(x(t+\tau) - \mu_x\right)\right]. \tag{7.68}$$

Expanding the terms in the bracket, $C_x(\tau)$ becomes:

$$C_x(\tau) = E\begin{bmatrix} x(t)\,x(t+\tau) - \\ \mu_x\left\{x(t) + x(t+\tau)\right\} + \mu_x^2 \end{bmatrix}. \tag{7.69}$$

Noting that:

$$E\left[x(t) + y(t)\right] = E\left[x(t)\right] + E\left[y(t)\right] \tag{7.70}$$

$$E\left[a\,x(t)\right] = a\,E\left[x(t)\right] \quad \text{where} \quad a = \text{const.} \tag{7.71}$$

$$E[a] = a \quad \text{where} \quad a = \text{const.} \tag{7.72}$$

$C_x(t)$ becomes:

$$\boxed{C_x(\tau) = R_x(\tau) - \mu_x^2.} \tag{7.73}$$

When all periodicities are removed from a signal, equation (7.66) can be substituted for μ_x, or:

$$C_x(\tau) = R_x(\tau) - R_x(\infty). \tag{7.74}$$

When $\tau = 0$, $C_x(0)$ is the same as the variance where:

$$\boxed{\sigma_x^2 = R_x(0) - R_x(\infty).} \tag{7.75}$$

The *cross-correlation function* of two random signals describes the correlation between values of the two signal at different times as a function of the time lag between values. Figure 7.9 shows two random signals. The cross-correlation between two signals x(t) and y(t) can be determined by multiplying x(t) by y(t+τ) and averaging the results over the sample time T, or:

$$\boxed{R_{xy}(t) = \lim_{T\to\infty} \frac{1}{T} \int_0^T x(t)\,y(t+\tau)\,dt.} \tag{7.76)}$$

In terms of expected values, $R_{xy}(t)$ is expressed:

$$R_{xy}(t) = E\left[x(t)\,y(t+\tau)\right]. \tag{7.77}$$

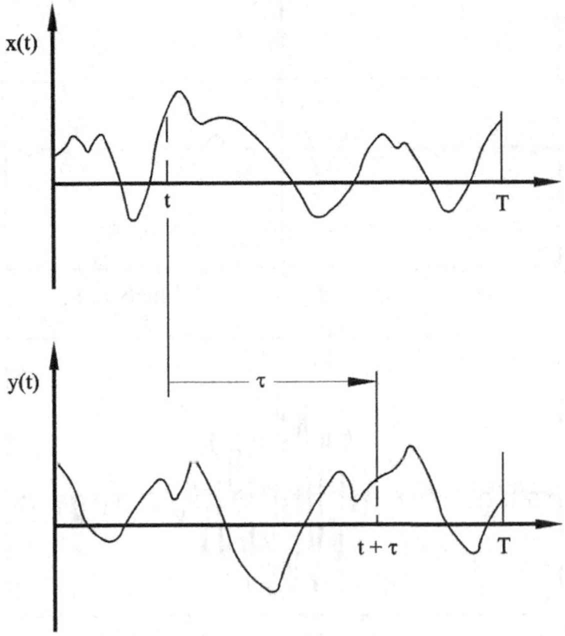

Figure 7.9 Cross-Correlations of x(t) and y(t+τ)

The cross correlation function is always real-valued. However, $R_{xy}(\tau)$ is not an even function and it may not always be maximum at $\tau = 0$. Some useful relationships are:

$$\boxed{R_{xy}(-\tau) = R_{yx}(\tau)} \tag{7.78}$$

$$\boxed{\left|R_{xy}(\tau)\right|^2 \leq R_x(0)R_y(0)} \tag{7.79}$$

$$\boxed{\left|R_{xy}(\tau)\right| \leq \frac{1}{2}\left[R_x(0) + R_y(0)\right].} \tag{7.80}$$

The *cross-covariance* as a function of time delay τ can be written:

$$\boxed{C_{xy}(\tau) = E\left[\left(x(t) - \mu_x\right)\left(y(t) - \mu_y\right)\right].} \tag{7.81}$$

In a manner similar to that used for the variance, $C_{xy}(\tau)$ reduces to:

$$\boxed{C_{xy}(\tau) = R_{xy}(\tau) - \mu_x\mu_y.} \tag{7.82}$$

The *normalized cross-covariance function* or the *correlation coefficient* is defined by:

$$\boxed{\rho_{xy}(\tau) = \frac{C_{xy}(\tau)}{\sqrt{C_x(0)C_y(0)}}} \tag{7.83}$$

where:

$$\boxed{C_x(0) = R_x(0) - \mu_x^2} \tag{7.84}$$

and:

$$\boxed{C_y(0) = R_x(0) - \mu_y^2.} \tag{7.85}$$

$\rho_{xy}(\tau)$ always satisfies the relation $-1 \leq \rho_{xy}(t) \leq 1$. When both μ_x and μ_y equal zero, $\rho_{xy}(t)$ becomes:

$$\boxed{\rho_{xy}(\tau) = \frac{R_{xy}(\tau)}{\sqrt{R_x(0)R_y(0)}}.} \tag{7.86}$$

When $\rho_{xy}(\tau)$ equals zero, the signals x(t) and y(t) are said to be uncorrelated. Even though it is not always true, for most engineering situations, when $\rho_{xy}(\tau)$ is zero for all values of τ, the two signals are normally independent.

Correlation measurements can be used to detect periodicities in random signals, to predict or detect deterministic signals that are buried in noise, to measure time delays associated with instruments or with vibration or sound transmission, and to locate disturbing sound or vibration sources. Figure 7.10 shows plots of the autocorrelation functions of some common signals. Figure 7.10a is an autocorrelation plot of a sinusoidal signal. Any noise signal that has a periodic signal buried in it will have an autocorrelation plot that will be somewhat similar to Figure 7.10a. One should note the changes in the autocorrelation plots when progressing from the sinusoidal signal in Figure 7.10a to the white noise signal in Figure 7.10f.

It is sometimes necessary to know the time delay or the time required for a signal to pass through a system. This can be determined by taking the cross-correlation of the output signal relative to the input signal of the system. The cross-correlation plot will be maximum or the input and output signals will have their greatest

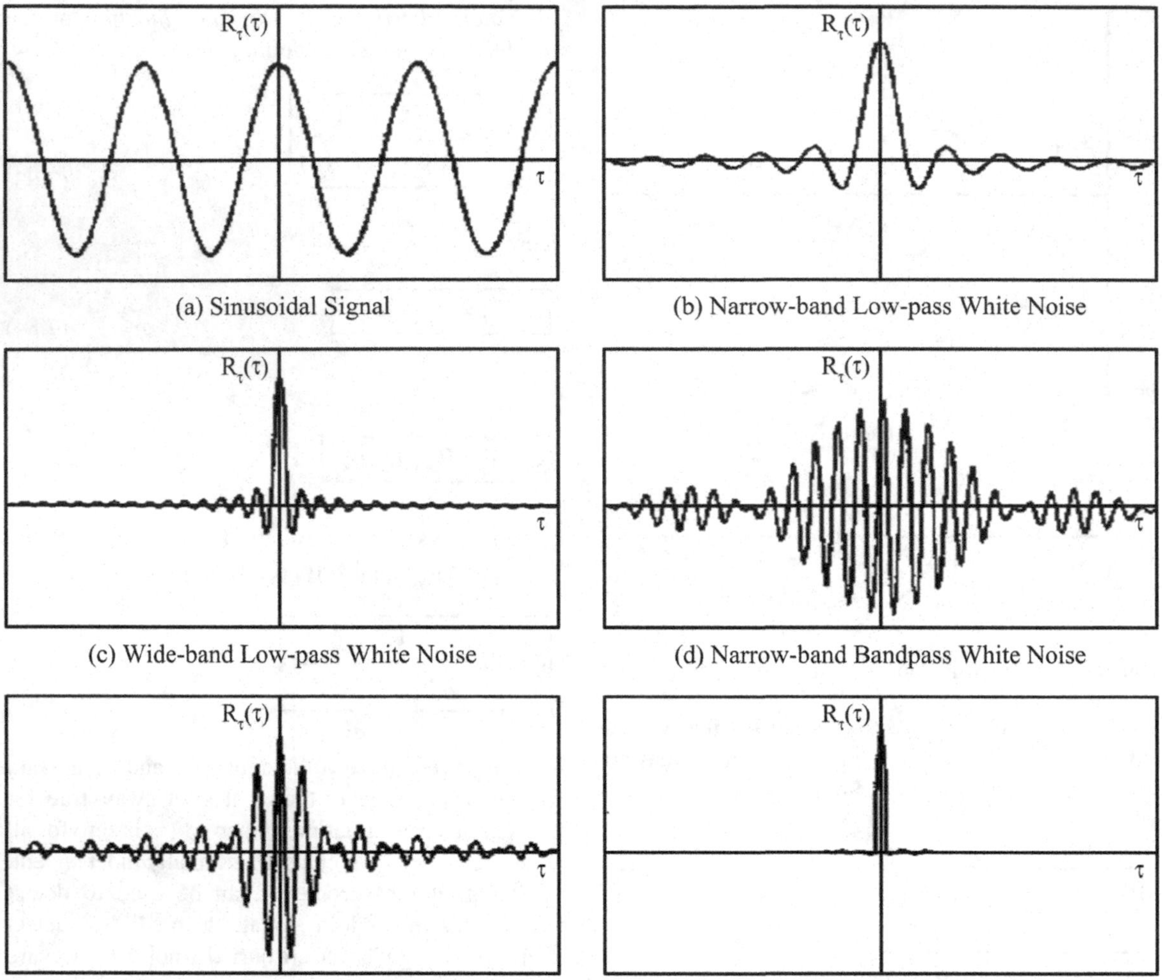

Figure 7.10 Autocorrelation Function Plots of Common Signals

correlation at a time delay equal to the time required for the signal to pass through the system.

Cross-correlation measurements can be used to determine transmission paths of vibration and sound signals. For example, examine the system shown in Figure 7.11. There are two apparent transmission paths for the vibration signal to get from the vibration sensor near the vibration source to the other sensor. These are labeled paths a and b. Obtaining the cross-correlation plot of the vibration signal farthest away from the vibration source relative to the closest signal indicates two peaks at τ_a and τ_b. The time delays associated with these peaks represent points of maximum correlation between the two signals or the times it takes the signals to travel paths a and b, reproductively.

Cross-correlation measurements can be used to detect signals buried in noise. When the signal to be detected can be recorded separately, cross-correlation measurements can be made between this recorded signal and the signal with this signal buried in noise. The net result is establishing a greater signal to noise ratio, making it easier to detect the signal buried in the noise.

EXAMPLE 7.1

Determine the autocorrelation function for a sinusoidal signal.

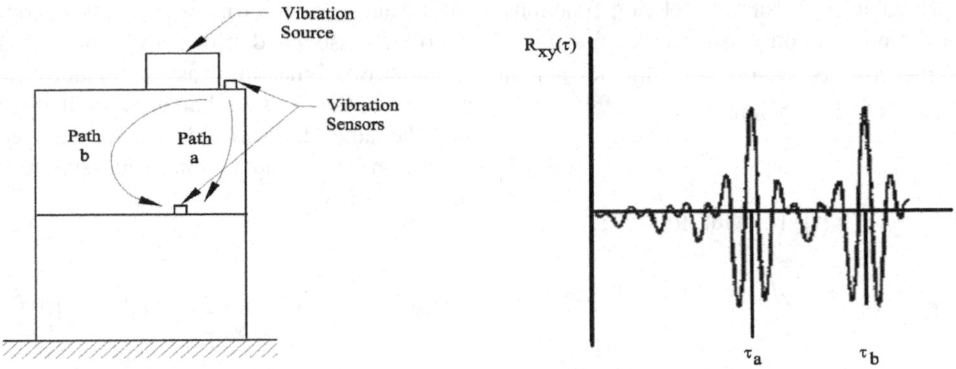

Figure 7.11 Possible Cross-correlation Plot Associated with Paths a and b between Two Vibration Sensors

SOLUTION

x(t) for a sinusoidal signal can be written:

$$x(t) = X \sin \omega t. \tag{7.1a}$$

Substituting this expression into equation (7.61) yields:

$$R_x(\tau) = \lim_{T\to\infty} \frac{1}{T} \int_0^T X^2 \sin \omega t \sin \omega (t + \tau) dt. \tag{7.1b}$$

Expanding the $\sin\omega(t + \tau)$ term and then collecting terms gives:

$$R_x(\tau) = \lim_{T\to\infty} \frac{X^2}{T} \int_0^T \begin{pmatrix} \sin^2 \omega t \cos \omega \tau + \\ \sin \omega t \cos \omega t \sin \omega \tau \end{pmatrix} dt. \tag{7.1c}$$

Reducing the above integral yields:

$$R_x(\tau) = \frac{X^2}{2} \cos \omega \tau \tag{7.1d}$$

or:

$$\boxed{R_x(\tau) = \frac{X^2}{2} \cos(2\pi f \tau).} \tag{7.1e}$$

Equation (7.1e) is plotted in Figure 7.10a.

EXAMPLE 7.2

Determine the autocorrelation function of the sum of two random signals given by:

$$z(t) = a\,x(t) + b\,y(t). \tag{7.2a}$$

SOLUTION

The autocorrelation function $R_z(\tau)$ is given by:

$$R_z(\tau) = E\left[\left(a\,x(t) + b\,y(t)\right)\begin{pmatrix} a\,x(t+\tau) + \\ b\,y(t+\tau) \end{pmatrix}\right]. \tag{7.2b}$$

Expanding the terms in the bracket yields:

$$R_z(\tau) = E\begin{bmatrix} a^2 x(t) x(t+\tau) + \\ ab \begin{pmatrix} x(t) y(t+\tau) + \\ y(t) x(t+\tau) \end{pmatrix} + \\ b^2 y(t) y(t+\tau) \end{bmatrix}. \tag{7.2c}$$

or:

$$\boxed{\begin{aligned} R_z(\tau) = {} & a^2 R_x(\tau) + \\ & ab\left[R_{xy}(\tau) + R_{yx}(\tau)\right] + b^2 R_y(\tau). \end{aligned}} \tag{7.2d}$$

7.7 Measurement of Correlation Functions

It is not possible to allow the integration time T to

go to infinity when measuring correlation functions. Therefore, it is possible only to measure an estimate $\hat{R}_{xy}(\tau)$ of the cross-correlation function $R_{xy}(\tau)$. In equation form, $\hat{R}_{xy}(\tau)$ is given by:

$$\begin{aligned}\hat{R}_{xy}(t) &= \frac{1}{T-\tau}\int_0^{T-\tau} x(t)y(t+\tau)dt \quad \text{for} \quad 0 \le t < T \\ &= \frac{1}{T-|\tau|}\int_{|\tau|}^{T-\tau} x(t)y(t+\tau)dt \quad \text{for} \quad -T \le t < 0.\end{aligned} \tag{7.87}$$

The autocorrelation function estimates $\hat{R}_x(\tau)$ and $\hat{R}_y(\tau)$ are just special forms of the cross-correlation function estimates. They are given by:

$$\begin{aligned}\hat{R}_{x}(t) &= \frac{1}{T-\tau}\int_0^{T-\tau} x(t)x(t+\tau)dt \quad \text{for} \quad 0 \le t < T \\ &= \frac{1}{T-|\tau|}\int_{|\tau|}^{T-\tau} x(t)x(t+\tau)dt \quad \text{for} \quad -T \le t < 0\end{aligned} \tag{7.88}$$

and:

$$\begin{aligned}\hat{R}_{y}(t) &= \frac{1}{T-\tau}\int_0^{T-\tau} y(t)y(t+\tau)dt \quad \text{for} \quad 0 \le t < T \\ &= \frac{1}{T-|\tau|}\int_{|\tau|}^{T-\tau} y(t)y(t+\tau)dt \quad \text{for} \quad -T \le t < 0.\end{aligned} \tag{7.89}$$

Figure 7.12 shows a schematic of the measurement setup for experimentally measuring the cross-correlation function. When measuring the autocorrelation functions either x(t) or y(t) is directed into both inputs.

Relative to correlation measurements, it can be shown that $\hat{R}_{xy}(\tau)$ is an unbiased estimate of $R_{xy}(\tau)$. The same is true for autocorrelation measurements. Therefore, there is no bias associated with correlation measurements. When the two signals in a cross-correlation measurement are limited to Gaussian, bandpass limited white noise with the same bandpass and when μ_x and μ_y equal zero, the variance associated with the measurement is given by:

$$VAR\left[\hat{R}_{xy}(\tau)\right] = \frac{1}{2BT}\left[\hat{R}_{xy}(\tau)^2 + \hat{R}_x(0)\hat{R}_y(0)\right]. \tag{7.90}$$

The normalized variance is:

$$\frac{VAR\left[\hat{R}_{xy}(\tau)\right]}{\hat{R}_{xy}(\tau)^2} = \frac{1}{2BT}\left[1 + \frac{\hat{R}_x(0)\hat{R}_y(0)}{\hat{R}_{xy}(\tau)^2}\right]. \tag{7.91}$$

Similarly, for the autocorrelation functions:

$$\frac{VAR\left[\hat{R}_{x}(\tau)\right]}{\hat{R}_{x}(\tau)^2} = \frac{1}{2BT}\left[1 + \frac{\hat{R}_x(0)^2}{\hat{R}_{x}(\tau)^2}\right] \tag{7.92}$$

and:

$$\frac{VAR\left[\hat{R}_{y}(\tau)\right]}{\hat{R}_{y}(\tau)^2} = \frac{1}{2BT}\left[1 + \frac{\hat{R}_y(0)^2}{\hat{R}_{y}(\tau)^2}\right]. \tag{7.93}$$

Therefore, the *normalized standard errors* for the cross-correlation and autocorrelation functions are:

$$\varepsilon\left[\hat{R}_{xy}(\tau)\right] = \sqrt{\frac{1}{2BT}\left[1 + \frac{\hat{R}_x(0)\hat{R}_y(0)}{\hat{R}_{xy}(\tau)^2}\right]} \tag{7.94}$$

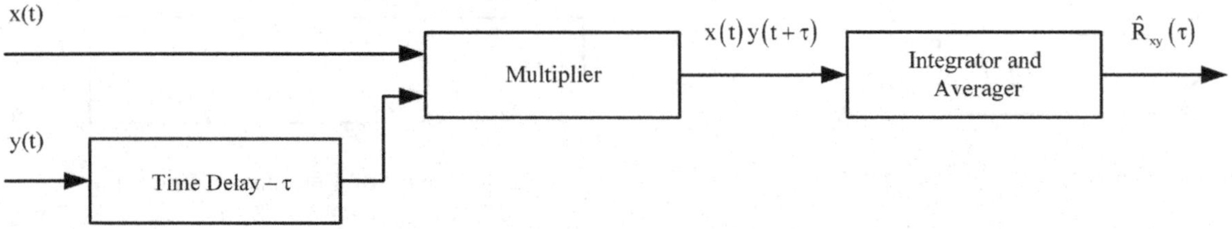

Figure 7.12 Correlations Measurements

$$\varepsilon\left[\hat{R}_x(\tau)\right]=\sqrt{\frac{1}{2BT}\left[1+\frac{\hat{R}_x(0)^2}{\hat{R}_x(\tau)^2}\right]} \quad (7.95)$$

$$\varepsilon\left[\hat{R}_y(\tau)\right]=\sqrt{\frac{1}{2BT}\left[1+\frac{\hat{R}_y(0)^2}{\hat{R}_y(\tau)^2}\right]}. \quad (7.96)$$

For the special case where $\tau = 0$, becomes the estimate for the mean squared value of x. Therefore, the variance of can be written:

$$\mathrm{VAR}\left[\hat{\psi}_x^2\right]=\frac{1}{BT}\hat{R}_x^2(0) \quad \text{or} \quad \mathrm{VAR}\left[\hat{\psi}_x^2\right]=\frac{\hat{\psi}_x^4}{BT}. \quad (7.97)$$

Therefore, the normalized standard error for the mean squared value is:

$$\varepsilon\left[\psi_x^2\right]=\frac{1}{\sqrt{BT}}. \quad (7.98)$$

7.8 Power Spectral Density Functions

The *power spectral density function* for a random signal describes the frequency composition of the signal in terms of the spectral density of its mean squared value. The mean squared value of a signal $\psi_x(f_o, \Delta f)^2$ contained in the frequency bandpass of $f_o - \Delta f/2$ to $f_o + \Delta f/2$, where f_o is the center frequency of the frequency bandpass and Δf is the width of the bandpass, is given by:

$$\psi_x(f_o,\Delta f)^2=\lim_{T\to\infty}\frac{1}{T}\int_0^T x(t,f_o,\Delta f)^2\,dt \quad (7.99)$$

$x(t, f_o, \Delta f)$ represents that portion of $x(t)$ that is contained in the frequency bandpass of $f_o - \Delta f/2$ and $f_o + \Delta f/2$. The *one-sided power spectral density function* $G_x(f)$ can be defined such that:

$$\psi_x(f_o,\Delta f)^2=\lim_{\Delta f\to 0} G_x(f)\Delta f \quad (7.100)$$

or:

$$G_x(f)=\lim_{\Delta f\to 0}\frac{\psi_x(f_o,\Delta f)^2}{\Delta f}. \quad (7.101)$$

Substituting equation (7.99) into equation (7.101) yields:

$$G_x(f)=\lim_{\Delta f\to 0}\lim_{T\to\infty}\frac{1}{T\Delta f}\int_0^T x(t,f_o,\Delta f)^2\,dt \quad (7.102)$$

for $0 \le f \le \infty$. Equation (7.10) is sometimes referred to as the *one-sided auto-power spectral density function*.

Just as the cross-correlation function can be used to determine the relation between two random signals in the time domain, the *cross-power spectral density function* can be used to determine the relation between two random signals in the frequency domain. Because the cross-correlation function is not an even function, the one-sided cross-power spectral density function is a complex quantity described by:

$$G_{xy}(f)=C_{xy}(f)-jQ_{xy}(f) \quad (7.103)$$

where the real part $C_{xy}(f)$ is called the *co-power spectral density function* and the imaginary part $Q_{xy}(f)$ is called the *quadrature spectral density function*. $C_{xy}(f)$ is given by:

$$C_{xy}(f)=\lim_{\Delta f\to 0}\lim_{T\to\infty}\frac{1}{T\Delta f}\int_0^T x(t,f_o,\Delta f)\,y(t,f_o,\Delta f)\,dt \quad (7.104)$$

and $Q_{xy}(f)$ is given by:

$$Q_{xy}(f)=\lim_{\Delta f\to 0}\lim_{T\to\infty}\frac{1}{T\Delta f}\int_0^T x(t,f_o,\Delta f)\,y^o(t,f_o,\Delta f)\,dt. \quad (7.105)$$

$x(t, f_o, \Delta f)$ and $y(t, f_o, \Delta f)$ are those portions of $x(t)$ and $y(t)$ that are contained in the frequency bandpass of Δf with a center frequency of f_o. $y^o(t, f_o, \Delta f)$ denotes a 90° phase shift relative to $y(t, f_o, \Delta f)$. $G_{xy}(f)$ can be written in the form:

$$G_{xy}(f)=\left|G_{xy}(f)\right|e^{-j\theta_{xy}(f)} \quad (7.106)$$

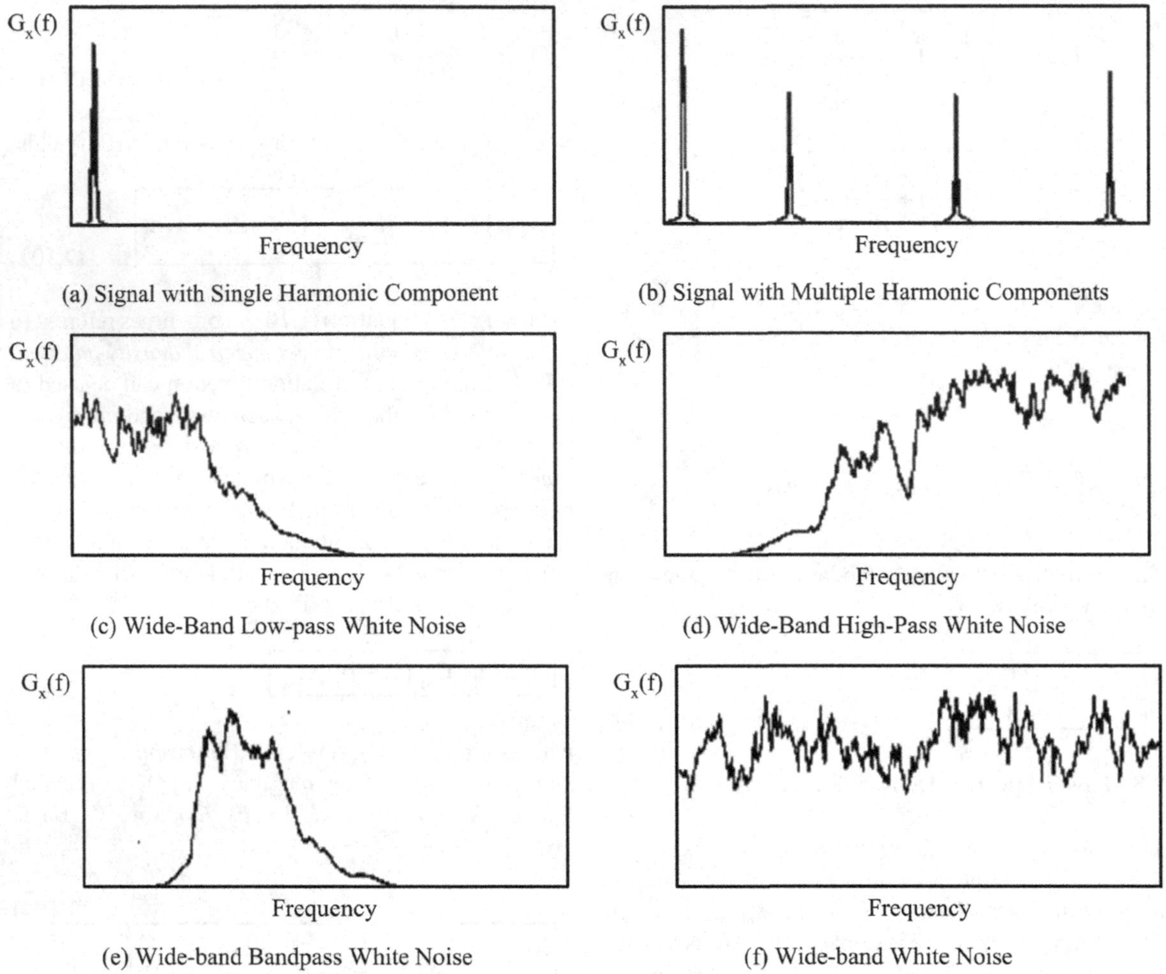

Figure 7.13 Auto-Power Spectral Density Plots of Common Signals

where:

$$\left|G_{xy}(f)\right| = \sqrt{C_{xy}(f)^2 + Q_{xy}(f)^2} \tag{7.107}$$

and:

$$\theta_{xy}(f) = \tan^{-1}\left[\frac{Q_{xy}(f)}{C_{xy}(f)}\right]. \tag{7.108}$$

Relative to the cross-power spectral density function:

$$C_{yx}(f) = C_{xy}(f) \tag{7.109}$$

and:

$$Q_{yx}(f) = -Q_{xy}(f). \tag{7.110}$$

$C_{yx}(f)$ and $Q_{yx}(f)$ are obtained by interchanging x(t) and y(t) in equations (7.104) and (7.105). Therefore, the following relations are true:

$$G_{yx}(f) = G_{xy}(f)^* \tag{7.111}$$

where $G_{xy}(f)^*$ is the complex conjugate of $G_{xy}(f)$ and:

$$G_{yx}(f) = G_{xy}(-f). \tag{7.112}$$

Another useful relation is:

$$\left|G_{xy}(f)\right|^2 \le G_x(f)G_y(f). \tag{7.113}$$

The *normalized cross-spectral density function* can be written:

$$\gamma_{xy}^2(f) = \frac{\left|G_{xy}(f)\right|^2}{G_x(f)G_y(f)} \quad \text{where} \quad \gamma_{xy}^2(f) \le 1. \tag{7.114}$$

$\gamma_{xy}(f)^2$ is called the *coherence function*, and it gives a measure of the correlation between two random signals in the frequency domain. When $\gamma_{xy}(f)^2$ equals zero at a particular frequency, signals x(t) and y(t) are said to be incoherent or uncorrelated at that frequency. When $\gamma_{xy}(f)^2$ equals zero at all frequencies, the two signals are totally incoherent or uncorrelated. When $\gamma_{xy}(f)^2$ equals one, the signals are said to be coherent or correlated.

Auto-spectral density measurements can be used to determine the frequency content of a random signal. This is particularly important when it is necessary to determine whether or not specific frequency components are present in the signal. Figure 7.13 shows the auto-power spectral density plots associated with some common signals. Cross-spectral density measurements can be used to determine the input-output relations and transfer function relations associated with complex systems. More is said about this in Section 7.11.

7.9 Measurement of Power Spectral Density Functions

Relative to equation (7.102) for auto-power spectral density measurements and to equations (7.103), (7.104) and (7.105.233) for cross-power spectral density measurements, it is not possible to allow the integration time T to go to infinity nor the filter bandpass Δf to go to zero. Therefore, it is only possible to measure estimates of the auto-power spectral density and the cross-power spectral density functions. In equation form:

$$\hat{G}_x(f) = \frac{1}{T\Delta f}\int_0^T x(t, f_o, \Delta f)^2 \, dt \tag{7.115}$$

and:

$$\hat{G}_{xy}(f) = \hat{C}_{xy}(f) - j\hat{Q}_{xy}(f) \tag{7.116}$$

where:

$$\hat{C}_{xy}(f) = \frac{1}{T\Delta f}\int_0^T x(t, f_o, \Delta f)\, y(t, f_o, \Delta f)\, dt. \tag{7.117}$$

and:

$$\hat{Q}_{xy}(f) = \frac{1}{T\Delta f}\int_0^T x(t, f_o, \Delta f)\, y^{o}(t, f_o, \Delta f)\, dt. \tag{7.118}$$

Figure 7.14 shows a schematic of the measurement setup for auto-power spectral density measurements and Figure 7.15 shows a schematic of the instrument setup for cross-spectral density measurements.

The measured auto-power spectral density values are not unbiased estimates of the auto-power spectral density function. The bias associated with auto-power spectral density measurements is given by:

$$\text{bias}\left[\hat{G}_x(f)\right] = E\left[\hat{G}_x(f)\right] - G_x(f). \tag{7.119}$$

For band-limited Gaussian white noise:

$$\text{bias}\left[\hat{G}_x(f)\right] = \frac{\Delta f^2}{24}\frac{d^2 G_x(f)}{df^2}. \tag{7.120}$$

The variance associated with auto-power spectral density measurements is:

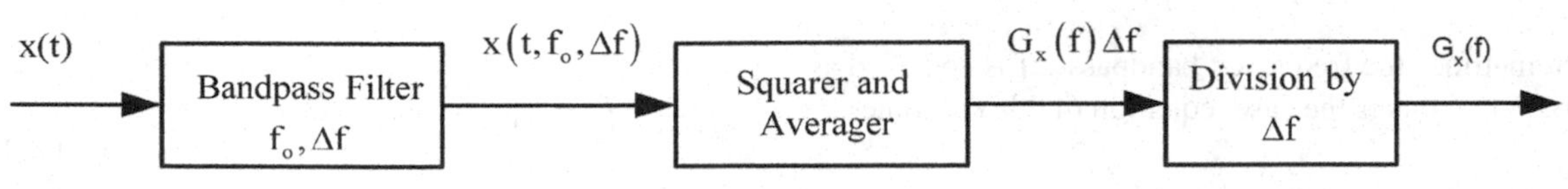

Figure 7.14 Auto-Power Spectral Density Measurements

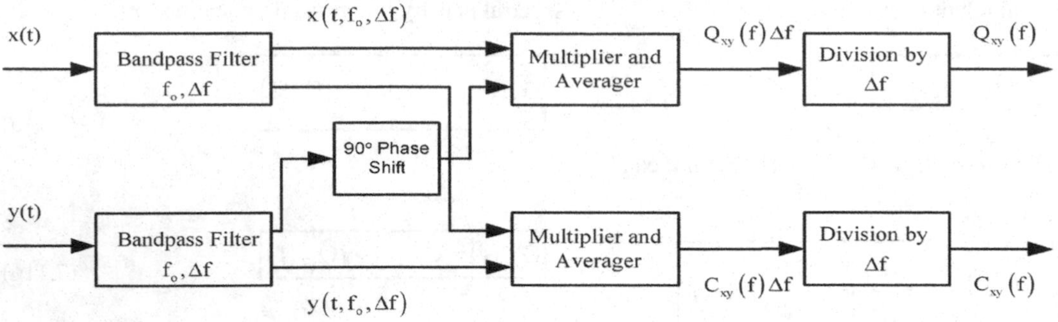

Figure 7.15 Cross-Power Spectral Density Measurements

$$\mathrm{VAR}\left[\hat{G}_x(f)\right]=E\left[\left\{\hat{G}_x(f)-E\left[\hat{G}_x(f)\right]\right\}^2\right] \tag{7.121}$$

or:

$$\mathrm{VAR}\left[\hat{G}_x(f)\right]=\frac{G_x(f)^2}{\Delta f\,T}. \tag{7.122}$$

Therefore, the mean squared error is:

$$E\left[\left(\hat{G}_x(f)-G_x(f)\right)^2\right]=\frac{G_x(f)^2}{\Delta f\,T}+\left[\frac{\Delta f^2}{24}\frac{d^2\,G_x(f)}{df^2}\right]^2. \tag{7.123}$$

Relative to auto-power spectral density measurements, the bias error can be minimized by making the frequency bandpass sufficiently small so that $G_X(f)$ is fairly flat over the bandpass. In some cases when Δf is not small, $E[\hat{G}_x(f)-G_x(f)]$ can be rather large, resulting in the bias error being a serious problem. When the bias error is small, the normalized standard error for the auto-power spectral density measurement is:

$$\boxed{\varepsilon\left[\hat{G}_x(f)\right]=\frac{1}{\sqrt{\Delta f\,T}}.} \tag{7.124}$$

Sometimes the frequency bandpass Δf is specified as B. When this is the case, equation (4.248) becomes:

$$\boxed{\varepsilon\left[\hat{G}_x(f)\right]=\frac{1}{\sqrt{BT}}.} \tag{7.125}$$

The errors associated with cross-power spectral density measurements are generally of the same relative magnitude as those associated with auto-power spectral density measurements.

7.10 Relation between Correlation and Power Spectral Density Functions

The correlation and power spectral density functions are Fourier transform pairs. For the auto-power spectral density and autocorrelation functions:

$$S_x(\omega)=\int_{-\infty}^{\infty}R_x(\tau)e^{-j\omega\tau}\,d\tau \tag{7.126}$$

or:

$$S_x(f)=\int_{-\infty}^{\infty}R_x(\tau)e^{-j2\pi f\tau}\,d\tau \tag{7.127}$$

and:

$$R_x(\tau)=\frac{1}{2\pi}\int_{-\infty}^{\infty}S_x(\omega)e^{j\omega\tau}\,d\omega \tag{7.128}$$

or:

$$R_x(\tau)=\int_{-\infty}^{\infty}S_x(f)e^{j2\pi f\tau}\,df. \tag{7.129}$$

where $S_x(f)$ is defined as the *two-sided auto-power spectral density function* for $-\infty \le f \le \infty$. Noting that $S_x(f) = S_x(-f)$ and $R_x(\tau) = R_x(-\tau)$ and using Euler's equation, equations (7.127) and (7.129) can be written:

$$S_x(f) = 2\int_0^\infty R_x(\tau)(\cos 2\pi f\tau - j\sin 2\pi f\tau)d\tau \tag{7.130}$$

$$R_x(\tau) = 2\int_0^\infty S_x(f)(\cos 2\pi f\tau + j\sin 2\pi f\tau)df \tag{7.131}$$

for $0 \le f \le \infty$. $S_x(f)$ and $R_x(\tau)$ are real-valued functions. Therefore, the imaginary parts of the above equations can be neglected, yielding:

$$\boxed{S_x(f) = 2\int_0^\infty R_x(\tau)\cos 2\pi f\tau\, d\tau} \tag{7.132}$$

and:

$$\boxed{R_x(\tau) = 2\int_0^\infty S_x(f)\cos 2\pi f\tau\, df} \tag{7.133}$$

for $-\infty \le f \le \infty$. The above equations are for the Fourier transform pairs of $S_x(f)$ and $R_x(\tau)$ for the case where f varies from $-\infty$ to ∞. It is not possible to measure the auto-power spectral density values at negative frequencies. Noting $S_x(f) = S_x(-f)$, all power spectrum density values at the corresponding positive frequencies result in measured power spectral density values equal to $2S_x(f)$. Therefore, the one-sided auto-power spectral density function $G_x(f)$ can be expressed:

$$G_x(f) = 2S_x(f). \tag{7.134}$$

Substituting equation (7.134) into equations (7.132) and (7.133) yields:

$$\boxed{G_x(f) = 4\int_0^\infty R_x(\tau)\cos 2\pi f\tau\, d\tau} \tag{7.135}$$

and:

$$\boxed{R_x(\tau) = \int_0^\infty G_x(f)\cos 2\pi f\tau\, df.} \tag{7.136}$$

EXAMPLE 7.3

Determine the autocorrelation function for a low-pass random signal where the one-sided auto-power spectral density function is given by:

$$\begin{aligned} G_x(f) &= X^2 \quad \text{for} \quad 0 \le f \le B \\ &= 0 \quad \text{for} \quad \text{all other values of f.} \end{aligned} \tag{7.3a}$$

SOLUTION

Substituting equation (7.3a) into equation (7.136) yields:

$$R_x = \int_0^B X^2 \cos 2\pi f\tau\, df \tag{7.3b}$$

or:

$$R_x = X^2 \left[\frac{1}{2\pi f}\sin 2\pi f\tau\right]\Bigg|_0^B. \tag{7.3c}$$

Therefore:

$$\boxed{R_x(\tau) = X^2 \frac{\sin 2\pi B\tau}{2\pi\tau}.} \tag{7.3d}$$

This is the function that is plotted in Figures 7.10b and 7.10c for different values of B.

EXAMPLE 7.4

Determine the autocorrelation function for a bandpass random signal where the one-sided auto-power spectral density function is given by:

$$\begin{aligned} G_x(f) &= X^2 \quad \text{for} \quad f_o - \frac{B}{2} \le f \le f_o + \frac{B}{2} \\ &= 0 \quad \text{for} \quad \text{all other values of f.} \end{aligned} \tag{7.4a}$$

SOLUTION

From the example 7.3:

$$R_x = \frac{X^2}{2\pi f}\left[\sin 2\pi\left(f_o + \frac{B}{2}\right)\tau - \sin 2\pi\left(f_o - \frac{B}{2}\right)\tau\right] \tag{7.4b}$$

or:

$$R_x = X^2 \frac{\sin \pi B \tau}{\pi \tau} \cos 2\pi f_o \tau. \tag{7.4c}$$

This is the function that is plotted in Figures 7.10d and 7.10e for different values of B.

Referring to equation (7.136), when t = 0:

$$R_x(0) = \int_0^\infty G_x(f)df \tag{7.137}$$

or:

$$\psi_x^2 = \int_0^\infty G_x(f)df. \tag{7.138}$$

Equation (7.138) forms the basis for an important check that can be made on the correctness of measured frequency data. Many times frequency data is obtained in the form of bandpass spectrum levels. Equation (7.138) can be written:

$$\psi_x^2 = \sum_{i=1}^{\infty} G_{x(i)}(f)\Delta f_i \tag{7.139}$$

or:

$$\psi_x^2 = \sum_{i=1}^{\infty} \psi_{x(i)}^2 \left(f_{o(i)}, \Delta f_i\right). \tag{7.140}$$

Therefore, the total mean squared value $\Psi_x{}^2$ of a random signal is equal to the sum of all of the individual mean squared values $\Psi_x(f_o, \Delta f)_i{}^2$ contained in the respective frequency bandpasses represented by $(f_o, \Delta f)_i$. The total mean squared value of a random signal can be measured with a true rms meter. This value should be nearly equal to the total mean squared value of the signal obtained from equation (7.140). This serves as a very good check on the correctness and accuracy of bandpass spectrum measurements.

EXAMPLE 7.5

A random vibration test specification requires: mean value of acceleration equals zero; one-sided, acceleration auto-power spectral density equals 0.3 $(m/s^2)^2/Hz$; and the frequency bandpass 30 Hz to 1,000 Hz. The auto-power density values are constant over the frequency bandpass. Determine the mean squared value and rms value of acceleration.

SOLUTION

From equation (7.140):

$$\psi_a^2 = 0.3 \times (1{,}000 - 30) \tag{7.5a}$$

or:

$$\psi_a^2 = 291\left(\frac{m}{s^2}\right)^2. \tag{7.5b}$$

The rms value of acceleration is:

$$a_{rms} = \sqrt{\psi_a^2} \quad \text{or} \quad a_{rms} = 17.1\frac{m}{s^2}. \tag{7.5c}$$

The cross-correlation and cross-power spectral density functions are also Fourier transform pairs. In equation form:

$$S_{xy}(f) = \int_{-\infty}^{\infty} R_{xy}(\tau)e^{-j2\pi f\tau}d\tau \tag{7.141}$$

and:

$$R_{xy}(\tau) = \int_{-\infty}^{\infty} S_{xy}(f)e^{j2\pi f\tau}df. \tag{7.142}$$

for $-\infty \le f \le \infty$. $S_{xy}(f)$ is defined as the *two-sided cross-power spectral density function* and is related to the one-sided cross-power spectral density function by:

$$G_{xy}(f) = 2S_{xy}(f). \tag{7.143}$$

Noting $G_{xy}(f) = C_{xy}(f) - jQ_{xy}(f)$, equation (7.142) can be written:

$$R_{xy}(\tau)=\frac{1}{2}\int_{-\infty}^{\infty}\left[C_{xy}(f)-jQ_{xy}(f)\right]e^{j2\pi f\tau}\,df. \tag{7.144}$$

Expanding $e^{j2\pi f\tau}$ using Euler's equation and retaining only the real part of the equation yields:

$$R_{xy}(\tau)=\int_{0}^{\infty}\left[C_{xy}(f)\cos 2\pi f\tau+Q_{xy}(f)\sin 2\pi f\tau\right]df \tag{7.145}$$

for $0\le f\le\infty$.

Equation (7.141) can be written:

$$G_{xy}(f)=2\left[\int_{-\infty}^{0}R_{xy}(\tau)e^{-j2\pi f\tau}\,d\tau+\int_{0}^{\infty}R_{xy}(\tau)e^{-j2\pi f\tau}\,d\tau\right] \tag{7.146}$$

for $0\le f\le\infty$. Reversing the limits of integration on the first integral and combining the two integrals yields:

$$G_{xy}(f)=2\int_{0}^{\infty}\left[R_{xy}(-\tau)e^{j2\pi f\tau}+R_{xy}(\tau)e^{-j2\pi f\tau}\right]d\tau \tag{7.147}$$

for $0\le f\le\infty$. Noting that $R_{xy}(-\tau)=R_{yx}(\tau)$ and using Euler's equation, $G_{xy}(f)$ can be written:

$$G_{xy}(f)=2\int_{0}^{\infty}\begin{bmatrix}\left[R_{xy}(\tau)+R_{yx}(\tau)\right]\cos 2\pi f\tau- \\ j\left[R_{xy}(\tau)-R_{yx}(\tau)\right]\sin 2\pi f\tau\end{bmatrix}d\tau. \tag{7.148}$$

Therefore:

$$\boxed{C_{xy}(f)=2\int_{0}^{\infty}\left[R_{xy}(\tau)+R_{yx}(\tau)\right]\cos 2\pi f\tau\,d\tau} \tag{7.149}$$

and:

$$\boxed{Q_{xy}(f)=2\int_{0}^{\infty}\left[R_{xy}(\tau)-R_{yx}(\tau)\right]\sin 2\pi f\tau\,d\tau.} \tag{7.150}$$

for $0\le f\le\infty$. It can be shown that:

$$\boxed{C_{xy}(f)=C_{xy}(-f)} \tag{7.151}$$

and:

$$\boxed{Q_{xy}(f)=-Q_{xy}(-f).} \tag{7.152}$$

EXAMPLE 7.6

Determine the auto-power spectral density function of the sum of two stationary random signals given by:

$$z(t)=a\,x(t)+b\,y(t). \tag{7.6a}$$

SOLUTION

The auto-power spectral density function of z(t) can be obtained by taking the Fourier transform of equation (7.2c). Substituting equation (7.2c) into equation (7.127) yields:

$$S_z(f)=\int_{-\infty}^{\infty}\begin{bmatrix}a^2R_x(\tau)+ \\ ab\left[R_{xy}(\tau)+R_{yx}(\tau)\right]+ \\ b^2R_y(\tau)\end{bmatrix}e^{-2\pi f\tau}\,d\tau. \tag{7.6b}$$

From equations (7.127) and (7.141), equation (7.6c) reduces to:

$$S_z(f)=a^2S_x(f)+ab\left[S_{xy}(f)+S_{yx}(f)\right]+ b^2S_y(f). \tag{7.6c}$$

Equation (7.6c) can also be written as a one-sided power spectral density function, or:

$$\boxed{G_z(f)=a^2G_x(f)+ab\left[G_{xy}(f)+G_{yx}(f)\right]+ b^2G_y(f).} \tag{7.6d}$$

7.11 Relation between Fourier Transforms and Power Spectral Density Functions

The expected value of $x(t)\bullet y(t)$ can be written:

$$E\left[x(t)y(t)\right] = \lim_{T\to\infty}\frac{1}{T}\int_{-T}^{T} x(t)y(t)dt. \quad (7.153)$$

x(t) can be expressed as the inverse Fourier transform of X(jω), or:

$$x(t) = \frac{1}{2\pi}\int_{-\infty}^{\infty} X(j\omega)e^{j\omega t}\, d\omega. \quad (7.154)$$

Substituting equation (7.154) into equation (7.153) yields:

$$E\left[x(t)y(t)\right] = \lim_{T\to\infty}\frac{1}{2\pi T}\int_{-T}^{T}\left[\int_{-\infty}^{\infty} X(j\omega)e^{j\omega t}\, d\omega\right]y(t)dt. \quad (7.155)$$

The above integral can be rearranged to give:

$$E\left[x(t)y(t)\right] = \lim_{T\to\infty}\frac{1}{2\pi T}\int_{-\infty}^{\infty} X(j\omega)\left[\int_{-T}^{T} y(t)e^{j\omega t}\, dt\right]d\omega. \quad (7.156)$$

Noting that:

$$Y(j\omega)^* = \lim_{T\to\infty}\int_{-T}^{T} y(t)e^{j\omega t}\, dt \quad (7.157)$$

where $Y(j\omega)^*$ is the complex conjugate of Y(jω), equation (7.156) becomes:

$$E\left[x(t)y(t)\right] = \lim_{T\to\infty}\frac{1}{2\pi T}\int_{-\infty}^{\infty} X(j\omega)Y(j\omega)^*d\omega \quad (7.158)$$

or:

$$E\left[x(t)y(t)\right] = \lim_{T\to\infty}\frac{1}{T}\int_{-\infty}^{\infty} X(jf)Y(jf)^*df \quad (7.159)$$

for $-\infty \le f \le \infty$. Noting:

$$E\left[x(t)y(t)\right] = R_{xy}(0) \quad (7.160)$$

and:

$$R_{xy}(0) = \int_{0}^{\infty} G_{xy}(f)df, \quad (7.161)$$

for $0 \le f \le \infty$, then:

$$G_{xy}(f) = \lim_{T\to\infty}\frac{1}{T}X(jf)Y(jf)^*. \quad (7.162)$$

In a similar manner, it can be shown that:

$$G_x(f) = \lim_{T\to\infty}\frac{1}{T}X(jf)X(jf)^* \quad (7.163)$$

and:

$$G_y(f) = \lim_{T\to\infty}\frac{1}{T}Y(jf)Y(jf)^*. \quad (7.164)$$

The input-output relation for a linear system can be expressed:

$$X_{out}(jf) = G(jf)Y_{in}(jf) \quad (7.165)$$

where $Y_{in}(jf)$ is the Fourier transform of the system input, $X_{out}(jf)$ is the Fourier transform of the system output, and G(jf) is the Fourier transform function or the frequency response function of the system in terms of frequency f. Multiplying both sides of equation (7.165) by:

$$\lim_{T\to\infty}\frac{1}{T}X_{out}(jf)^*$$

and multiplying and dividing the right side by $Y_{in}(jf)^*$ yields:

$$\lim_{T\to\infty} X_{out}(jf)X_{out}(jf)^* = \lim_{T\to\infty} G(jf)\frac{X_{out}(jf)^*}{Y_{in}(jf)^*}Y_{in}(jf)Y_{in}(jf)^*. \quad (7.166)$$

Noting:

$$G(jf)^* = \frac{X_{out}(jf)^*}{Y_{in}(jf)^*} \tag{7.167}$$

and:

$$|G(jf)|^2 = G(jf)G(jf)^*, \tag{7.168}$$

equation (7.166) becomes:

$$\lim_{T\to\infty} X_{out}(jf)X_{out}(jf)^* = \lim_{T\to\infty} |G(jf)|^2 Y_{in}(jf)Y_{in}(jf)^*. \tag{7.169}$$

Substituting equations (7.163) and (7.164) into equation (7.169) yields:

$$G_{x(out)} = |G(jf)|^2 G_{y(in)}. \tag{7.170}$$

Therefore, the auto-power spectral density function of the system output is equal to the auto-power spectral density function of the system input times the amplitude of the frequency response function squared.

Equation (7.170) assumes that the input and output signals are fully correlated or fully coherent. When this is the case and two of the three functions are known, the third function is specified. Often the frequency response function of a system is obtained by first measuring the power spectral density functions of the input and output and substituting the results into equation (7.170). Therefore:

$$\boxed{|G(jf)|^2 = \frac{G_{x(out)}}{G_{y(in)}}.} \tag{7.171}$$

The frequency response function G(jf) is a complex function given by:

$$\boxed{G(jf) = |G(jf)|e^{j\theta(f)}} \tag{7.172}$$

where $\Theta(f)$ is the phase between $G_{x(out)}$ and $G_{y(in)}$. Therefore, the instrument that is used to measure $G_{x(out)}$ and $G_{y(in)}$ must also be able to measure the phase between the two variables as a function of frequency.

Situations arise where the input and output signals are not fully correlated. This can occur when there are multiple inputs with a single output or when either the input or output signal (or both) is embedded in noise. For this situation multiply both sides of equation (7.165) by:

$$\lim_{T\to\infty} \frac{1}{T} Y_{in}(jf)^*$$

to obtain:

$$\lim_{T\to\infty} X_{out}(jf)Y_{in}(jf)^* = \lim_{T\to\infty} G(jf)Y_{in}(jf)Y_{in}(jf)^* \tag{7.173}$$

which reduces to:

$$G(jf) = \frac{G_{xy}(jf)}{G_{y(in)}(f)}. \tag{7.174}$$

Equation (7.174) can be written:

$$\boxed{G(jf) = \frac{|G_{xy}(jf)|e^{-j\theta_{xy}(f)}}{G_{y(in)}(f)}} \tag{7.175}$$

where:

$$|G_{xy}(f)| = \sqrt{C_{xy}(f)^2 + Q_{xy}(f)^2} \tag{7.176}$$

and:

$$\boxed{\theta_{xy}(f) = \tan^{-1}\left[\frac{Q_{xy}(f)}{C_{xy}(f)}\right].} \tag{7.177}$$

Therefore, both the amplitude and phase of the complex frequency response function can be obtained by dividing the cross-power spectral density function of the output relative to the input by the auto-power spectral density function of the input.

Two channel FFT analyzers are often used to compute the frequency response or transfer function of a system that is associated with the measured input and output signals of the system. To insure meaningful test results, the quality of the measurements must be determined. This is accomplished by the use of the coherence function. Equation (7.174) can be developed in terms of the coherence function. First, multiply both sides of equation (7.174) by:

$$\lim_{T\to\infty}\frac{1}{T}X_{out}(jf)^*Y_{in}(jf)=G_{xy}(jf)^* \tag{7.178}$$

to obtain:

$$\lim_{T\to\infty}\frac{1}{T}X_{out}(jf)^*Y_{in}(jf)G(jf)=\frac{\left|G_{xy}(jf)\right|^2}{G_{y(in)}(f)}. \tag{7.179}$$

Next, multiply and divide the left side of equation (7.179) by $Y_{in}(jf)^*$ and reduce the results to get:

$$\left|G(jf)\right|^2=\frac{\left|G_{xy}(jf)\right|^2}{G_{y(in)}(f)^2}. \tag{7.180}$$

Finally, multiply and divide the right side of equation (7.180) by $G_{x(out)}(f)$. The result is:

$$\left|G(jf)\right|^2=\frac{\left|G_{xy}(jf)\right|^2}{G_{x(out)}(f)G_{y(in)}(f)}\frac{G_{x(out)}(f)}{G_{y(in)}(f)} \tag{7.181}$$

Substituting equation (7.114) into equation (7.181) yields:

$$\boxed{\left|G(jf)\right|^2=\frac{G_{x(out)}(f)\gamma_{xy}^2(f)}{G_{y(in)}(f)}.} \tag{7.182}$$

When the input and output signals are fully correlated, $\gamma_{xy}^2(f)=1$ and equation (7.182) reduces to equation (7.171).

Situations exist where the output signal is embedded in noise. When this is the case, the part of the auto-power spectral density of the output signal S that is correlated to or coherent with the input signal is:

$$S=\gamma_{xy}^2(f)G_{x(out)}(f). \tag{7.183}$$

The remainder of the auto-power spectral density of the output signal N is associated with noise, or:

$$N=\left[1-\gamma_{xy}^2(f)\right]G_{x(out)}(f). \tag{7.184}$$

Therefore, the signal-to-noise ratio is:

$$\boxed{\frac{S}{N}=\frac{\gamma_{xy}^2(f)}{1-\gamma_{xy}^2(f)}.} \tag{7.185}$$

As a general rule, for a valid measurement using equation (7.175), $\gamma_{xy}^2(f)\geq 0.9$. This gives a signal-to-noise (S/N) ratio of 9 or a S/N ratio of 9.5 dB, which is acceptable for most measurements.

EXAMPLE 7.7

The frequency response function for a system in which a wave travels from one point to another point is given by:

$$G(jf)=A\,e^{-j\left(\frac{d}{c}\right)2\pi f} \tag{7.7a}$$

where A is a constant, d is the distance between the two points and c is the wave speed. The system input consists of low-pass white noise where:

$$\begin{aligned} G_y(f)&=Y^2 &&\text{for}\quad 0\leq f\leq B\\ &=0 &&\text{for}\quad \text{all other values of f.}\end{aligned} \tag{7.7b}$$

Determine the correlation coefficient of the output relative to the input and plot the results.

SOLUTION

From equation (7.174):

$$G_{xy}(jf)=A\,e^{-j2\pi f\tau_o}\quad\text{where}\quad \tau_o=\frac{d}{c}. \tag{7.7c}$$

Substituting equations (7.7b) and (7.7c) into equation (7.142) yields:

$$R_{xy}(\tau)=\frac{1}{2}\int_0^B AY^2\, e^{-j2\pi f\tau_o}\, e^{j2\pi f\tau}\, df \tag{7.7d}$$

or:

$$R_{xy}(\tau)=\frac{1}{2}\int_0^B AY^2\, e^{j2\pi f(\tau-\tau_o)}\, df. \tag{7.7e}$$

Evaluating the integral in equation (7.7e) gives:

$$R_{xy}(\tau)=AY^2\frac{1}{j2\pi(\tau-\tau_o)}e^{j2\pi f(\tau-\tau_o)}\Big|_0^B \tag{7.7f}$$

or:

$$R_{xy}(\tau)=\frac{AY^2\, e^{j2\pi B(\tau-\tau_o)}}{j2\pi\ (\tau-\tau_o)}. \tag{7.7g}$$

Expanding the complex exponential term using Euler's equation and neglecting the imaginary term, equation (7.7g) becomes:

$$R_{xy}(\tau)=AY^2\frac{\sin 2\pi B(\tau-\tau_o)}{2\pi(\tau-\tau_o)}. \tag{7.7h}$$

From equation (7.3d) in Example 7.3, $R_y(0)$ is given by:

$$R_y(0)=Y^2 B. \tag{7.7i}$$

Substituting equation (7.7a) into equation (7.168) yields:

$$|G(jf)|^2=A^2. \tag{7.7j}$$

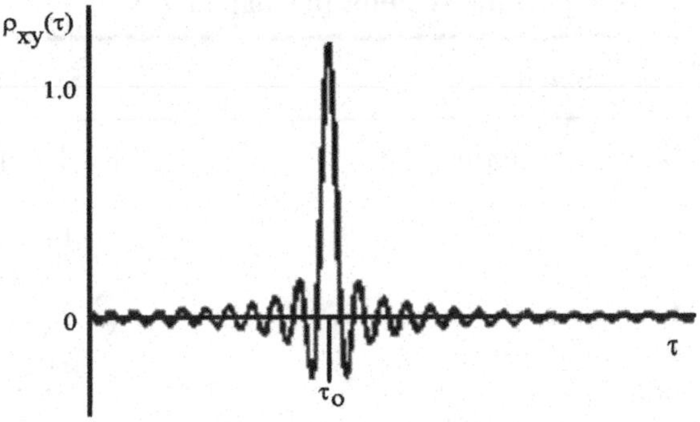

Figure 7.16 Correlation Coefficient for the System in EXAMPLE 7.6

Substituting equations (7.7b) and (7.7j) into equation (7.170) gives:

$$\begin{aligned} G_x(f) &= A^2Y^2 \quad \text{for} \quad 0\le f\le B \\ &= 0 \qquad\ \ \text{for} \quad \text{all other values of f}. \end{aligned} \tag{7.7k}$$

Substituting equation (7.7k) into equation (7.136) for $\tau = 0$ yields:

$$R_x(0)=\int_0^B A^2Y^2\, df \quad \text{or} \quad R_x(0)=A^2Y^2B. \tag{7.7l}$$

Finally, substituting equations (7.7h), (7.7i) and (7.7l) into equation (7.86) yields:

$$\boxed{\rho_{xy}(\tau)=\frac{\sin 2\pi B(\tau-\tau_o)}{2\pi B(\tau-\tau_o)}.} \tag{7.7m}$$

Equation (7.6m) is plotted in Figure 7.16.

7.12 Important Symbols in Chapter 7

Parameter	Symbol	English Units	SI Units
Fourier Transform*	F{ }, $F(j\omega)$, etc.	( ) $(rad/sec)^{-1}$	( ) $(rad/s)^{-1}$
		( ) sec	( ) s
	F(jf), etc	( )/Hz	( )/Hz
		( ) sec	( ) s
Mean squared value*	Ψ_x^2, $\Psi_x(f_o,\Delta f)^2$	$( )^2$	$( )^2$
Mean value*	μ_x	( )	( )
Variance*	σ_x^2	$( )^2$	$( )^2$
Standard deviation*	σ_x	( )	( )
Probability function	P[x]	dimensionless	dimensionless
Probability density function	p(x)	1/( )	1/( )
Correlation function*	$R_x(t)$, $R_y(t)$, $R_{xy}(t)$	$( )^2$	$( )^2$
Normalized correlation function	$\rho_x(t)$, $\rho_y(t)$, $\rho_{xy}(t)$	dimensionless	dimensionless
Power spectral density function*	$G_x(f)$, $G_y(f)$, $G_{xy}(f)$	$( )^2$/Hz	$( )^2$/Hz
Coherence function	$\gamma_{xy}(f)^2$	dimensionless	dimensionless

* The bracket ( ) implies that the units in the bracket are those of the signal being analyzed. For example, if the

PROBLEMS - CHAPTER 7

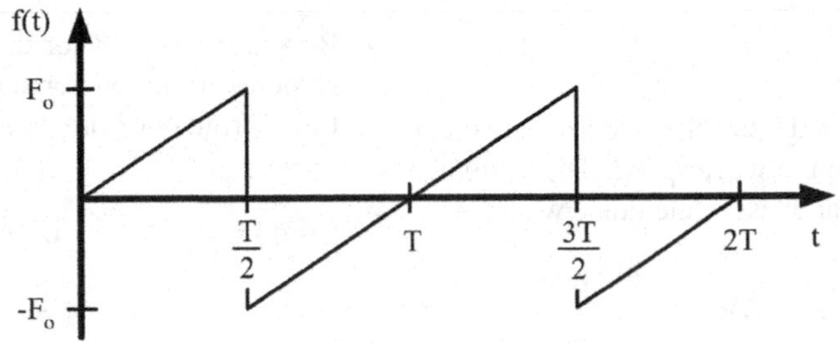

Figure P7.1

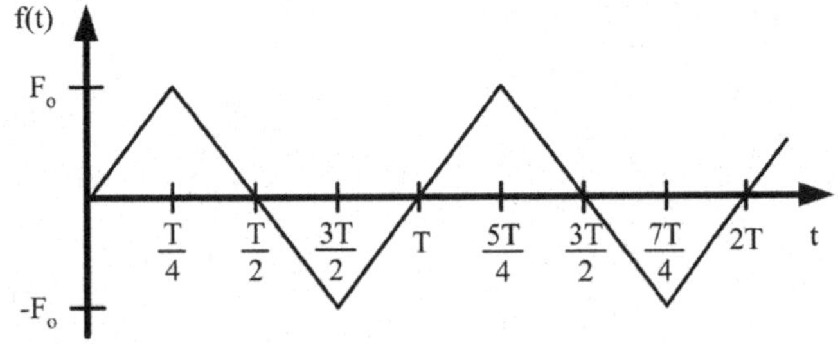

Figure P7.2

1. Determine the mean squared and rms values of the signal shown in Figure P7.1.

2. Determine the mean squared and rms values of the signal shown in Figure P7.2.

3. A 1-DOF, mass-spring-damper system with a resonance frequency ω_n and a damping ration $\xi = 0.1$ is excited by the force:

$$f(t) = F_o \cos(0.5\omega_n t) + 0.5\,F_o \cos(\omega_n t) + 2\,F_o \cos(3\omega_n t)$$

Determine the mean squared value of f(t) and the mean squared value of the amplitude of the response of the system.

4. Determine the autocorrelation function of the signal shown in Figure P7.1.

5. Determine the autocorrelation function of the signal shown in Figure P7.2.

6. A signal has a constant and time-varying component that is given by the equation:

$$f(t) = A + B\sin\omega t.$$

Determine (a) E[y(t)] and (b) E[y(t)^2].

7. An acceleration signal has the following rms values in the indicated frequency ranges:

Δf_i - Hz	44-88	88-177	177-355
a_{rms} - m/s^2	1.2	0.2	1.5

Δf_i - Hz	355-710	710-1,420
a_{rms} - m/s^2	7.3	8.2

Determine the total mean squared and rms values of the signal.

8. The frequency response function for a system where a wave traves from one point to another point is

given by:

$$G(jf) = A\,e^{-j\frac{d}{c}2\pi f}$$

where A is a constant, d is the distance between the two points, and c is the wave speed. The system imput consists of bandpass white noise where:

$$G(jf) = Y^2 \quad \text{for} \quad f_o - \frac{B}{2} \le f \le f_o + \frac{B}{2}$$
$$= 0 \quad \text{for} \quad \text{all other values of f.}$$

Determine the correlation coefficient of the output relative to the input and plot the results.

9. Repeat Problem 8 for the case where frequency response function for the system where a wave traves from one point to another point is given by:

$$G(jf) = A\,e^{-j\frac{d}{c}2\pi f} + 0.5A\,e^{-j\frac{2d}{c}2\pi f}.$$

CHAPTER 8
VIBRATION OF CONTINUOUS SYSTEMS

8.1 Introduction

Vibrating systems in preceding chapters were modeled as lumped masses connected together by spring and damping elements. When set in motion by initial conditions or applied forces and moments, these masses then oscillate about their static equilibrium positions as a function of time. The modeling of lumped parameter vibration systems results in a set of differential equations that describe the motions of the system masses.

Vibration systems exist that cannot be modeled as a system of lumped masses connected together by spring and damping elements. These systems can often be modeled as a system with a distributed mass that has elastic characteristics. These systems are referred to as continuous vibration systems. Vibration analyses of these systems result in partial differential equations that describe the motions of the distributed masses as functions of space and time.

This chapter will start with the analysis of the transverse vibration of a string and progress to the analysis of the transverse vibration of membranes and a plate. While the analyses of these systems are fairly basic, much can be learned from the developments of their respective partial differential equations and the related solutions.

8.2 Transverse Vibration of a String

Transverse vibration of a string is associated with a string that is stretched and fixed at both ends. An example is a stretched guitar string. When the string is plucked, its resulting vibration motion is perpendicular to the length of the string.

Some simplifying assumptions must be made before the equation of motion for a vibrating string can be developed. They include

1. Hooke's law applies where stress is linearly proportional to strain.
2. There are no shear forces in the string.
3. There are no bending moments acting on the string.
4. The weight of the string is very small and is distributed evenly along the length of the string.
5. The vibration amplitudes of the string are very small.
6. The tension applied to the ends of the string remains constant and acts evenly throughout the string.
7. There is no energy dissipation in the string when it vibrates.

Figure 8.1 shows an incremental segment of a vibrating string where x is the position coordinate, t is time, and y(x,t) is displacement in the transverse or vertical direction as a function of x and t. Summing the forces acting on the string in the y direction yields:

$$\sum F_y = (m\,dx)\frac{\partial^2 y(x,t)}{dt^2} \tag{8.1}$$

where m is the mass per unit length of the string. From Figure 8.1:

$$\sum F_y = T\sin\theta\big|_{x+dx} - T\sin\theta\big|_{x} \tag{8.2}$$

where T is the tension force acting on the string. A Taylor series expansion of a function of the form f(x + dx) can be stated:

$$f(x+dx) = f(x) + \frac{df(x)}{dx}dx + \frac{d^2 f(x)}{dx^2}dx^2 + \cdots \tag{8.3}$$

which for very small values of f(x) and dx reduces to:

$$f(x+dx) = f(x) + \frac{df(x)}{dx}dx. \tag{8.4}$$

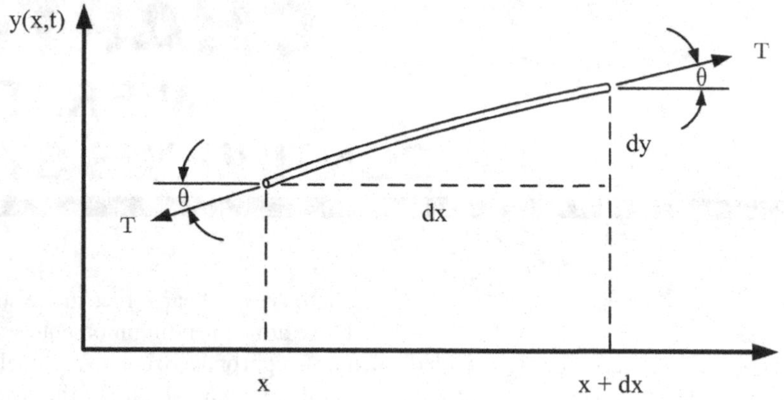

Figure 8.1 Free Body Diagram of an Incremental Segment of a Vibrating String in Tension

Equating equation (8.1) and equation (8.2) yields:

$$m\frac{\partial^2 y(x,t)}{\partial t^2}dx = T\sin\theta + \frac{\partial(T\sin\theta)}{\partial x}dx - T\sin\theta \tag{8.5}$$

or:

$$m\frac{\partial^2 y(x,t)}{\partial t^2} = \frac{\partial(T\sin\theta)}{\partial x}. \tag{8.6}$$

Since y(x,t) is assumed to be very small, θ can also be assumed to be very small. Therefore:

$$\sin\theta \approx \theta \approx \tan\theta \quad \text{where} \quad \tan\theta = \frac{\partial y(x,t)}{\partial x}. \tag{8.7}$$

Substituting equation (8.7) into equation (8.6) and noting T is constant yields:

$$\frac{\partial^2 y(x,t)}{\partial^2 t} = \frac{T}{m}\frac{\partial^2 y(x,t)}{\partial x^2} \tag{8.8}$$

The units on mass per unit length m are lb_f-s^2/in./in. (kq/m) and on tension T is lb_f (N). Therefore, the units for T/m are in.^{2}/s^2 (m^2/s^2), which are the units for velocity squared. T/m specifies the *speed of propagation* or *wave speed* c for transverse waves in a string that form the vibration modes in a string when it is excited by initial conditions. Therefore, equation (8.8) can be written:

$$\boxed{\frac{\partial^2 y(x,t)}{\partial^2 t} - c^2\frac{\partial^2 y(x,t)}{\partial x^2} = 0} \tag{8.9}$$

where:

$$\boxed{c = \sqrt{\frac{T}{m}}.} \tag{8.10}$$

Equation (8.9) is called the *equation for transverse vibration of a string*.

8.3 Solution to the Equation for Transverse Vibration of a String

When finite boundary conditions exist on a string, the *method of separation of variables* can be used to solve equation (8.9). To solve this equation, first assume a solution of the form:

$$\boxed{y(x,t) = X(x)T(t)} \tag{8.11}$$

Substituting equation (8.11) into equation (8.9) and carrying out the prescribed operations yields:

$$X(x)\frac{\partial^2 T(t)}{\partial t^2} - c^2\,T(t)\frac{\partial^2 X(x)}{\partial x^2} = 0. \tag{8.12}$$

Noting T(t) is only a function of t and X(x) is only a function of x and rearranging the terms in equation

(8.12) gives:

$$\frac{1}{T(t)}\frac{\partial^2 T(t)}{\partial t^2} = \frac{c^2}{X(x)}\frac{\partial^2 X(t)}{\partial x^2}. \quad (8.13)$$

Separating the variables and setting each side of the above equation equal to the *separation variable* ω^2 yields:

$$\frac{1}{T(t)}\frac{d^2 T(t)}{dt^2} = -\omega^2 \quad (8.14)$$

or:

$$\boxed{\frac{d^2 T(t)}{dt^2} + \omega^2 T(t) = 0} \quad (8.15)$$

and:

$$\frac{c^2}{X(x)}\frac{d^2 X(x)}{dx^2} = -\omega^2 \quad (8.16)$$

or:

$$\boxed{\frac{d^2 X(x)}{dx^2} + k^2 X(x) = 0.} \quad (8.17)$$

where ω is frequency and the *wave number* k is given by:

$$\boxed{k = \frac{\omega}{c}.} \quad (8.18)$$

Equations (8.15) and (8.17) are second order ordinary differential equations that have the solutions:

$$T(t) = A\cos\omega t + B\sin\omega t \quad (8.19)$$

and:

$$X(x) = C\cos kx + D\sin kx. \quad (8.20)$$

Substituting equations (8.19) and (8.20) into equation (8.11), the response y(x,t) becomes:

$$\boxed{y(x,t) = (A\cos\omega t + B\sin\omega t)\times (C\cos kx + D\sin kx).} \quad (8.21)$$

where the constants A and B are obtained from the initial conditions, and the constants C and D are obtained from the boundary conditions.

EXAMPLE 8.1

Determine the frequencies and mode shapes of the vibration modes that can exist on a string that has the following boundary conditions: $y(x,t)|_{x=0} = 0$ and $y(x,t)|_{x=L} = 0$. Develop the expressions for determining the constants associated with the initial conditions.

SOLUTION

To determine the constants C and D of equation (8.21), first write the equation (8.21) as:

$$y(x,t) = T(t)(C\cos kx + D\sin kx) \quad (8.1a)$$

Applying the boundary condition y(0,t) = 0 yields:

$$y(x,t)\|_{x=0} = T(t)C = 0 \quad (8.1b)$$

or:

$$C = 0. \quad (8/1c)$$

Applying the boundary condition y(L,t) = 0 gives:

$$\boxed{y(x,t)|_{x=L} = T(t)D\sin kL = 0} \quad (8.1d)$$

or:

$$D\sin kL = 0. \quad (8.1e)$$

The boundary condition at x = L is satisfied either when D = 0 or when $\sin kL = 0$ Since the condition D = 0 results in the trivial solution y(x,t) = 0 for all values of x, the term $\sin kL = 0$ is chosen to satisfy the boundary condition $y(x,t)|_{x=L} = 0$. The condition sin kL = 0 is satisfied when:

$$kL = \pi, 2\pi, 3\pi, \cdots \quad \text{or} \quad k_n L = n\pi. \tag{8.1f}$$

Therefore, the *wave number* k_n can be written:

$$\boxed{k_n = \frac{n\pi}{L}.} \tag{8.1g}$$

Noting $k = \omega/c$ yields:

$$\boxed{\omega_n = \frac{n\pi c}{L}.} \tag{8.1h}$$

which are the *resonance frequencies* associated with the normal modes of vibration that can exist on a string that is clamped at both ends. The function $\sin k_n x$ is the *mode function* of the nth vibration mode that can exist on the string. Figure 8.2 shows the shapes of the first five vibration modes that can exist on a string that is clamped at each end.

The solution for the nth term can be written:

$$\boxed{y_n(x,t) = (A_n \cos\omega_n t + B_n \sin\omega_n t)\sin k_n x} \tag{8.1i}$$

where k_n and ω_n are specified by equations (8.1g) and (8.1h), respectively. Multiplying the constants A and B in equation (8.21) by D results in the constants A_n and B_n.

Equation (8.1i) indicates there are n resonance frequencies and corresponding vibration modes that can exist on a string that is clamped at both ends. Therefore, the most general solution that obeys the stated boundary conditions for the string is the sum of all of the individual mode functions, or:

$$\boxed{y(x,t) = \sum_{n=1}^{\infty}(A_n \cos\omega_n t + B_n \sin\omega_n t)\sin k_n x.} \tag{8.1j}$$

The amplitudes of and the phase relations between each of the mode functions is obtained by applying the initial conditions to determine the expressions for A_n and B_n.

The expressions for A_n and B_n are obtained by applying the initial conditions and using the orthogonality conditions that exist between the normal vibration modes of vibration that can exist on the string. Let the initial displacement and velocity at time $t = 0$ be:

$$y(x,0) = y_o(x) \tag{8.1k}$$

and:

$$\frac{\partial y(x,0)}{\partial t} = v_o(x). \tag{8.1l}$$

A_n is obtained by multiplying both sides of equation (8.1j) by the mode function $\sin k_m x$ and then integrating the results with respect to x over the interval $x = 0$ to $x = L$. Because of the orthogonality relationship associated with sine and cosine functions [see equations (1.48) - (1.50)], the only non-zero term in the right hand side of equation (8.1j) when the integral is evaluated at time $t = 0$ is for the case where $k_m = k_n$. Therefore, equation (8.1j) reduces to:

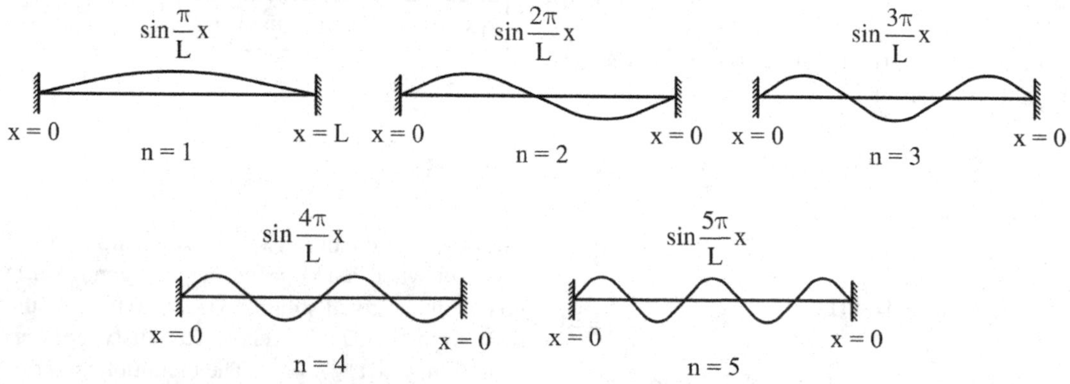

Figure 8.2 Graphis of First Five Modes that Can Exist on a String that Is Clamped at Both Ends

$$\int_0^L y_o(x)\sin k_n x\,dx = \frac{L}{2}A_n \tag{8.1m}$$

or:

$$\boxed{A_n = \frac{2}{L}\int_0^L y_o(x)\sin k_n x\,dx.} \tag{8.1n}$$

B_n is obtained by first taking the derivative of both sides of equation (8.1j) with respect to t and then repeating the same procedures used to determine the expression for A_n. This yields:

$$\int_0^L v_o(x)\sin k_n x\,dx = \frac{L}{2}\omega_n B_n \tag{8.1o}$$

or:

$$\boxed{B_n = \frac{2}{L\omega_n}\int_0^L v_o(x)\sin k_n x\,dx.} \tag{8.1p}$$

EXAMPLE 8.2

A stretched string of length L is plucked at the position L/4 by producing an initial displacement h and then releasing the string. Determine the expression for the amplitudes of A_n and B_n and plot the resulting wave form at representative values of time between t = 0 and t = L/c, using the expressions for the first ten terms $(n = 1 - 10)$ in the series solution.

SOLUTION

The initial velocity of the string is zero. Therefore, $B_n = 0$. Referring to Figure 8.3 for time t = 0, the expression y(x,0) for the initial displacement can be written:

$$\begin{aligned} y(x,0) &= \frac{4h}{L}x \quad \text{for} \quad 0 \le x \le \frac{L}{4} \\ &= \frac{4h}{3L}(L-x) \quad \text{for} \quad \frac{L}{4} < x \le L. \end{aligned} \tag{8.2a}$$

Substituting the expressions for the initial displacement into equation (8.1n) yields:

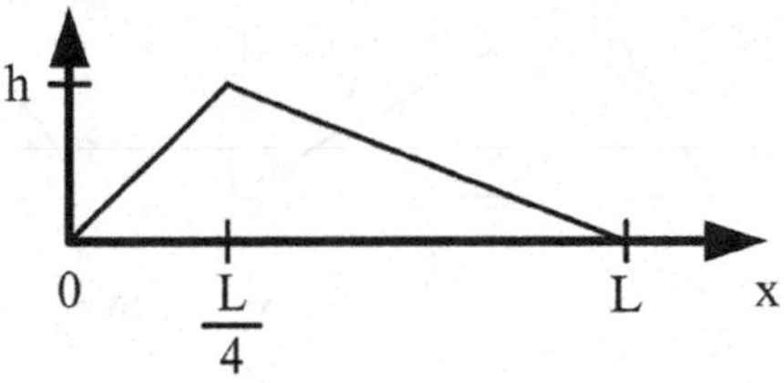

Figure 8.3 Initial Condition for EXAMPLE 8.22

$$A_n = \frac{2}{L}\left(\begin{aligned} &\int_0^{L/4}\frac{4h}{L}x\sin k_n x\,dx + \\ &\int_{L/4}^{L}\frac{4h}{3L}(L-x)\sin k_n x\,dx \end{aligned}\right). \tag{8.2b}$$

or rearranging terms:

$$A_n = \frac{8h}{L^2}\left(\begin{aligned} &\int_0^{L/4} x\sin k_n x\,dx + \frac{L}{3}\int_{1/4}^{L}\sin k_n x\,dx + \\ &\frac{1}{3}\int_{L/4}^{L} x\sin k_n x\,dx \end{aligned}\right). \tag{8.2c}$$

Evaluating the integrals in equation (8.2c) and combining the resulting expressions yields:

$$A_n = \frac{32h}{3L^2}\frac{\sin\left(\frac{k_n L}{4}\right)}{k_n}. \tag{8.2d}$$

For a string clamped at each end, k_n is given by equation (8.1g). Therefore, equation (8.2d) becomes:

$$\boxed{A_n = \frac{32h}{3(n\pi)^2}\sin\left(\frac{n\pi}{4}\right).} \tag{8.2e}$$

Substituting the expressions for A_n and B_n into equation (8.1j) yields:

$$\boxed{y(x,t) = \sum_{n=1}^{\infty}\frac{32h}{3(n\pi)^2}\sin\left(\frac{n\pi}{4}\right)\cos\omega_n t\,\sin k_n x.} \tag{8.2f}$$

where:

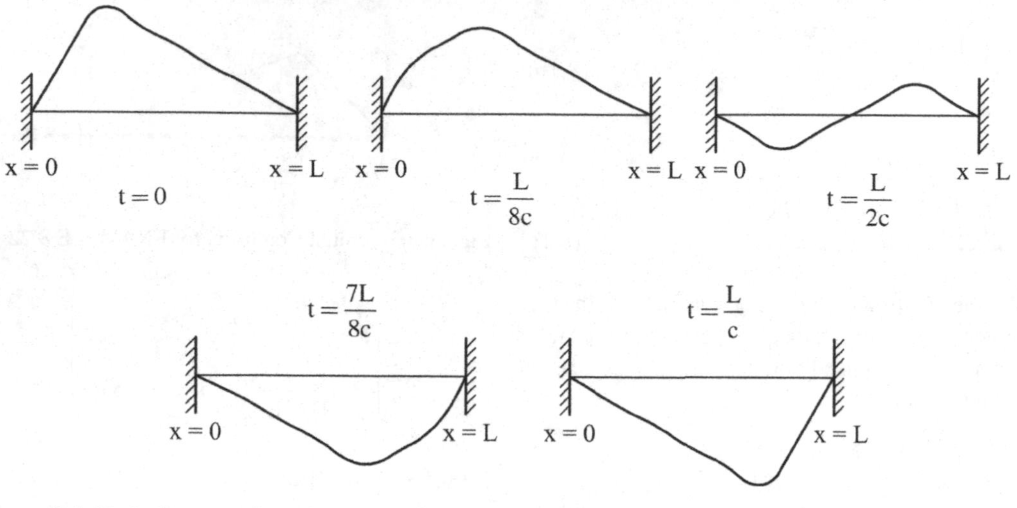

Figure 8.4 Mode Shapes at Specified Values of Time Associated with an Intial Triangular Wave Form at Time t = 0 Obtained from the First Ten Terms of Equation (8.2f)8u

$$\boxed{\omega_n = \frac{n\pi c}{L}} \tag{8.2g}$$

and:

$$\boxed{k_n = \frac{n\pi}{L}.} \tag{8.2h}$$

Figure 8.4 shows the resulting wave forms obtained by summing the first ten terms of the equation (8.2f) at the indicated values for time times.

EXAMPLE 8.3

A stretched string of length L is struck with an initial velocity v_O between x = 3L/8 and x = 5L/8. Determine the expression for the amplitudes of A_n and B_n and plot the resulting wave form at representative values of time between t = 0 and t = L/c using the first ten terms (n = 1 to n = 10) in the series solution.

SOLUTION

Since the initial displacement of the string equals zero, $e^{j(\omega t-kx)}$ Referring to Figure 8.5 for time t = 0, the expression $\partial y(x,0)/\partial t$ for the initial velocity can be written:

$$\begin{aligned} \frac{\partial y(x,0)}{\partial t} &= v_o \quad \text{for} \quad \frac{3L}{8} \le x \le \frac{5L}{8} \\ &= 0 \quad \text{for} \quad \text{all other values of x.} \end{aligned} \tag{8.3a}$$

Substituting the expressions for the initial velocity into equation (8.1p) yields:

$$B_n = \frac{2}{L\omega_n} \int_{3L/8}^{5L/8} v_o \sin k_n x \, dx. \tag{8.3b}$$

Evaluating the integral in equation (8.3b) and combining the terms yields:

$$B_n = \frac{2 v_o}{L\omega_n k_n}\left[\cos\left(k_n \frac{3L}{8}\right) - \cos\left(k_n \frac{5L}{8}\right)\right]. \tag{8.3c}$$

or noting $k_n = n\pi/L$ and $\omega_n = n\pi c/L$, equation (8.3c) reduces to:

$$\boxed{B_n = \frac{2 v_o L}{c(n\pi)^2}\left[\cos\left(\frac{3n\pi}{8}\right) - \cos\left(\frac{5n\pi}{8}\right)\right].} \tag{8.3d}$$

Therefore, equation (8.1j) for the stated initial conditions can be written:

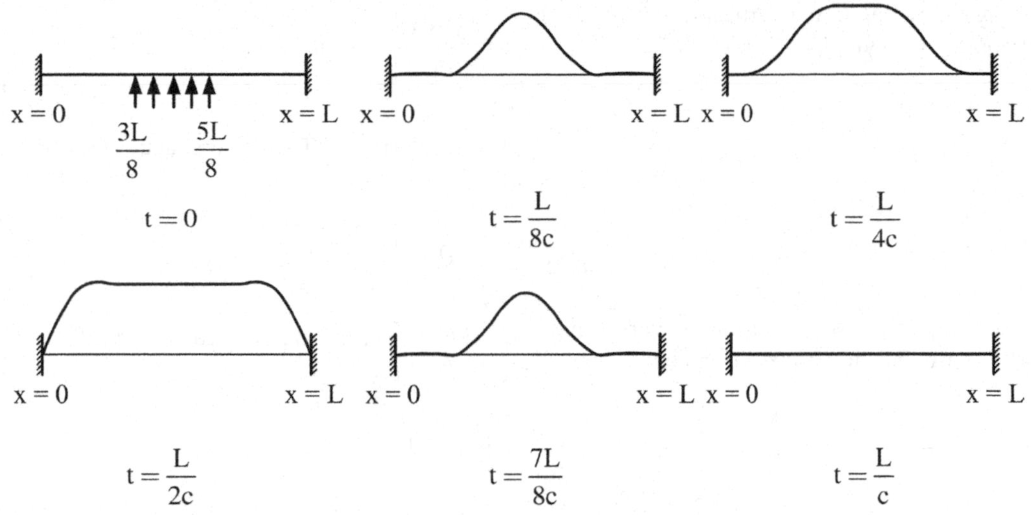

Figure 8.5 Mode Shapes at Specific Values of Time Associated with an Initial Velocity v_o between $x = 3L/8$ and $x = 5L/8$ Obtained from the First Ten Terms of Equation (8.3e)

$$y(x,t) = \sum_{n=1}^{\infty} \frac{2 v_o L}{c(n\pi)^2} \left[\cos\left(\frac{3n\pi}{8}\right) - \cos\left(\frac{5n\pi}{8}\right) \right] \sin \omega_n t \sin k_n x. \tag{8.3e}$$

Figure 8.5 shows the resulting mode shapes obtained by summing the first ten terms of equation (8.3e) at the indicated values for time.

8.4 Forced Vibration of a String

An external force can be applied to the end of a semi-infinite string. For this analysis, let the end of the string at $x = 0$ be constrained to move only in the vertical direction. Assume the end of the string at $x = \infty$ is attached to a rigid support, and assume the string is stretched to a constant tension T (Figure 8.6). The vertical driving force can be expressed:

$$f(t) = F e^{j\omega t}. \tag{8.22}$$

For this example, the driving force will create waves on the string that will travel in the positive x direction. Since the string is infinitely long, there will be no reflected waves that travel in the negative direction. Therefore, the response y(x,t) of the string can be written:

$$y(x,t) = \mathbf{A} e^{j(\omega t - kx)} \quad \text{where} \quad \mathbf{A} = A e^{j\theta}. \tag{8.23}$$

The complex amplitude **A** indicates there may be a phase relation between the driving force on the string and the response of the string.

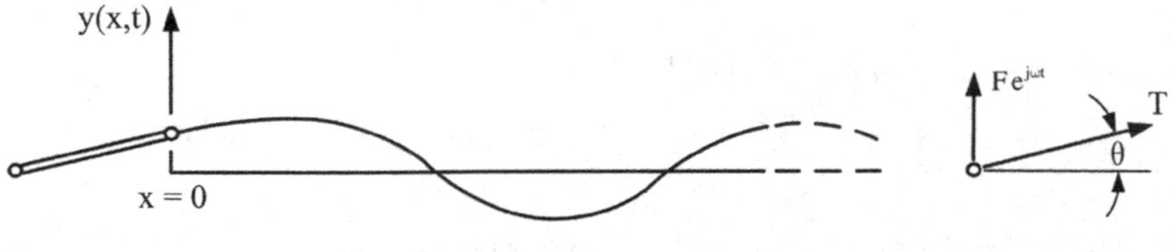

(a) Diagram of Semi-Infinite String Forced at x = 0

(b) Free Body Diagram of Forces Acting on Hinge

Figure 8.6 Semi-Infinite String Forced at x = 0

Summing the forces in the vertical direction at the driving point (x = 0) yields:

$$\sum F_y = 0 = F e^{j\omega t} + T\sin\theta \tag{8.24}$$

or:

$$F e^{j\omega t} = -T\sin\theta. \tag{8.25}$$

When the amplitude of vibration is very small, $\sin\theta \approx \theta \approx \tan\theta$. The tan θ at x = 0 can be written:

$$\tan\theta = \frac{\partial y(x,t)}{\partial x}. \tag{8.26}$$

Therefore, equation (8.25) becomes:

$$F e^{j\omega t} = -T\frac{\partial y(x,t)}{\partial x}. \tag{8.27}$$

Substituting equation (8.23) into the equation (8.27) and evaluating the derivative at x = 0 yields:

$$F e^{j\omega t} = jk\mathbf{A}T e^{j\omega t} \tag{8.28}$$

or:

$$\mathbf{A} = -j\frac{F}{Tk}. \tag{8.29}$$

Noting $-j = e^{-j\pi/2}$, equation (8.29) can be written:

$$\mathbf{A} = \frac{F}{Tk}e^{-j\frac{\pi}{2}}. \tag{8.30}$$

Substituting equation (8.30) into equation (8.23) yields:

$$y(x,t) = \frac{F}{Tk}e^{j\left(\omega t - kx - \frac{\pi}{2}\right)}. \tag{8.31}$$

The driving point mechanical impedance Z_m of the string is the ratio of the driving force divided by the string velocity at x = 0. The velocity at x = 0 is:

$$\left.\frac{\partial y(x,t)}{\partial t}\right|_{x=0} = j\omega\frac{F}{Tk}e^{j\left(\omega t - \frac{\pi}{2}\right)}. \tag{8.32}$$

$k = \omega/c$, and $j = e^{j\pi/2}$. Therefore, equation (8.32) becomes:

$$\left.\frac{\partial y(x,t)}{\partial t}\right|_{x=0} = \frac{Fc}{T}e^{j\omega t}. \tag{8.33}$$

Dividing equation (8.33) by equation (8.22) and noting $c = \sqrt{T/m}$ yields:

$$\boxed{Z_m = mc.} \tag{8.34}$$

Z_m is purely resistive, and it is a function of only the mass per unit length of the string and the wave speed c of the string. Since Z_m is a function of the physical properties of the string, it is referred to as the *characteristic mechanical impedance* for transverse vibration of a string and is designated Z_c instead of Z_m. Therefore, equation (8.34) is written:

$$\boxed{Z_c = mc.} \tag{8.35}$$

The driving point mechanical impedance of a semi-infinite string is purely resistive. This indicates all of the energy delivered by the driving force to the string is transmitted to the string.

The energy that is directed into the string over a specified period of time can be determined by finding the expression for the time average power input to the string. The average power input to the string is obtained from the equation:

$$\left\langle \Pi_{in} \right\rangle_t = \frac{1}{T}\int_0^T f(t)\left.\frac{\partial y(x,t)}{\partial t}\right|_{x=0} dt. \tag{8.36}$$

Using complex numbers and letting:

$$v(0,t) = \left.\frac{\partial y(x,t)}{\partial t}\right|_{x=0}, \tag{8.37}$$

the average power can be obtained from:

$$\langle \Pi_{in} \rangle_t = \frac{1}{2}\mathrm{Re}\left[\mathbf{f}(t) \times \mathbf{v}(0,t)^*\right] \tag{8.38}$$

where $\mathbf{v}(0,t)^*$ is the complex conjugate of v(0,t). Substituting equations (8.22) and (8.37) into equation (8.38) yields:

$$\langle \Pi_{in} \rangle_t = \frac{1}{2}\frac{F^2 c}{T}. \tag{8.39}$$

Noting:

$$V_m\big|_{x=0} = V_o = \frac{Fc}{T} \tag{8.40}$$

and:

$$Z_m\big|_{x=0} = \frac{F}{V_o}, \tag{8.41}$$

equation (8.39) can be expressed:

$$\boxed{\langle \Pi_{in} \rangle_t = \frac{1}{2} Z_m\big|_{x=0} V_o^2.} \tag{8.42}$$

Consider the case where the string is forced at x = 0 and clamped at x = L. The expression for the driving force is given by equation (8.22). The driving force causes a wave that travels in the positive direction that can be described by the expression $e^{j(\omega t - kx)}$.. The clamp at x = L results in a wave that travels in the negative direction that can be described by the expression $e^{j(\omega t + kx)}$. Therefore, the response y(x,t) can be expressed:

$$y(x,t) = \mathbf{A}e^{j(\omega t - kx)} + \mathbf{B}e^{j(\omega t + kx)}. \tag{8.43}$$

The partial derivative of y(x,t) with respect to x evaluated at x = 0 is given by:

$$\frac{\partial y(x,t)}{\partial x}\bigg|_{x=0} = -jk\mathbf{A}e^{j\omega t} + jk\mathbf{B}e^{j\omega t}. \tag{8.44}$$

The sum of the forces at x = 0 is given by equation (8.27). Therefore, substituting equations (8.22) and (8.44) into equation (8.27) yields:

$$F = jkT(\mathbf{A} - \mathbf{B}). \tag{8.45}$$

Applying the boundary condition $y(x,t)\big|_{x=L} = 0$ at $x = L$ gives:

$$\mathbf{A}e^{-jkL} + \mathbf{B}e^{jkL} = 0. \tag{8.46}$$

Eliminating **B** from equations (8.45) and (8.46) and noting:

$$\cos kL = \frac{e^{jkL} + e^{-jkL}}{2} \tag{8.47}$$

yields:

$$\mathbf{A} = \frac{F e^{jkL}}{j2Tk\cos kL}. \tag{8.48}$$

Substituting equation (8.48) into equation (8.46) gives:

$$\mathbf{B} = \frac{-F e^{-jkL}}{j2Tk\cos kL}. \tag{8.49}$$

The response y(x,t) is obtained by substituting the expressions for **A** and **B** into equation (8.43):

$$y(x,t) = \frac{F}{Tk\cos kL}\left[\frac{e^{jk(L-x)} - e^{-jk(L-x)}}{2j}\right]e^{j\omega t}. \tag{8.50}$$

Noting:

$$\sin k(L-x) = \frac{e^{jk(L-x)} - e^{-jk(L-x)}}{2j}, \tag{8.51}$$

$T = mc^2$, and $k = \omega/c$, equation (8.50) becomes:

$$\boxed{y(x,t) = \frac{F\sin k(L-x)}{Z_c\,\omega\cos kL}e^{j\omega t}.} \tag{8.52}$$

Equation (8.52) indicates the maximum values of y(x,t) occur when $\cos kL = 0$ or when:

$$\boxed{k_n = \left(\frac{2n-1}{2}\right)\frac{\pi}{L} \quad \text{for} \quad n = 1, 2, 3, \cdots.} \tag{8.53}$$

The wave number k can be written:

$$k = \frac{2\pi}{\lambda} \tag{8.54}$$

where λ is the wavelength. Substituting equation (8.54) into equation (8.53) and rearranging the terms yields:

$$L = \frac{2n-1}{4}\lambda \quad \text{for} \quad n = 1, 2, 3, \cdots. \tag{8.55}$$

The length of the string for given values of n can be expressed in terms of the corresponding wavelengths where $L = \lambda/4$ for $n = 1$, $L = 3\lambda/4$ for $n = 2$, $L = 5\lambda/4$ for $n = 3$, etc. Therefore, the frequencies that result in maximum values for y(x,t) for a string that is forced at $x = 0$ and clamped at $x = L$ are often referred to as quarter wavelength frequencies.

Equation (8.52) indicates the minimum values of y(x,t) occur when $\cos kL = \pm 1$ or when:

$$\boxed{k_n = \frac{n\pi}{L} \quad \text{for} \quad n = 1, 2, 3, \cdots.} \tag{8.56}$$

In a manner similar to that discussed above, these lengths of the string can be written in terms of the wavelengths given by equation (8.56):

$$L = \frac{n\lambda}{2} \quad \text{for} \quad n = 1, 2, 3, \cdots \tag{8,57}$$

where $L = \lambda/2$ for $n = 1$, $L = \lambda$ for $n = 2$, $L = 3\lambda/2$ for $n = 3$, etc. Therefore, the frequencies that result in minimum values for y(x,t) for a string that is forced at $x = 0$ and clamped at $x = L$ are often referred to as half wavelength frequencies.

The driving point mechanical impedance is obtained by first determining the velocity at $x = 0$ where:

$$\left.\frac{\partial y(x,t)}{\partial t}\right|_{x=0} = j\frac{F \sin kL}{Z_c \cos kL} e^{j\omega t} \tag{8.58}$$

or:

$$\left.\frac{\partial y(x,t)}{\partial t}\right|_{x=0} = j\frac{F}{Z_c} \tan kL \; e^{j\omega t}. \tag{8.59}$$

Dividing equation (8.22) by equation (8.59) yields:

$$\boxed{\mathbf{Z}_m = -jZ_c \cot kL.} \tag{8.60}$$

Figure 8.7 shows a plot of $|\mathbf{Z}_m/Z_c|$ as a function of kL.

Figure 8.7 indicates the amplitude of $\mathbf{Z}_m$ is zero when the velocity at $x = 0$ is a maximum or when the excitation frequency corresponds to one of the quarter wavelength frequencies [equation (8.53)]. This implies a driving force with a small amplitude at $x = 0$ results in a large motion of the string at that point.

Figure 8.7 indicates $\mathbf{Z}_m$ goes to infinity when the excitation frequency corresponds to one of the half wavelength frequencies [equation (8.56)]. This indicates, regardless of the amplitude of the driving force, the motion of the string at $x = 0$ is equal to zero even though the motion at other points along the string is non-zero.

The driving point mechanical impedance of a string clamped at $x = L$ is purely reactive. This indicates there is no net transfer of energy between the driving source and the string. The energy for this case is transferred back and forth between the source and string so that even though the instantaneous value of energy is not zero, the energy averaged over n cycles is equal to zero. This can be verified by showing the time average power transmitted to a string that is clamped at $x = L$ is equal to zero. Substituting equations (8.22) and (8.58) into equation (8.37) yields:

$$\langle \Pi_{in} \rangle_t = \frac{1}{2}\text{Re}\left[-j\frac{F^2}{Z_c}\tan kL\right] \tag{8.61}$$

or:

$$\boxed{\langle \Pi_{in} \rangle_t = 0.} \tag{8.62}$$

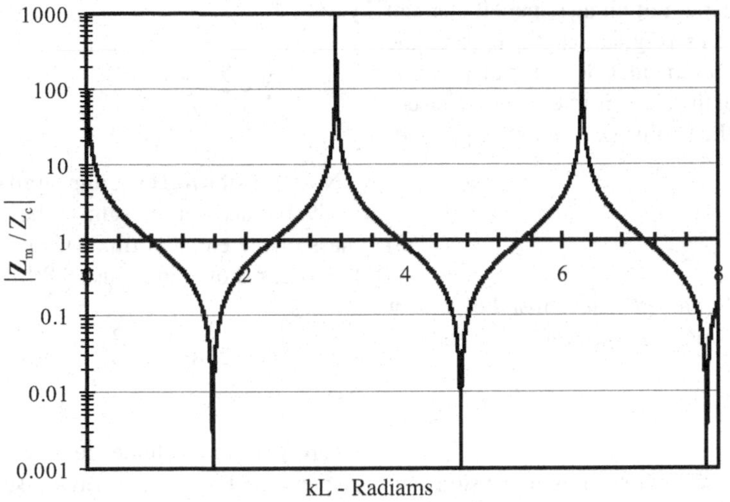

Figure 8.7 $|Z_m / Z_c|$ vs. kL for a String Forced at x = 0 and Clamped at x = L

8.5 Longitudinal Vibration of a Bar

Longitudinal vibration in a bar generates oscillating stress waves that travel in a direction that is parallel to the axis or length of the bar. Simplifying assumptions must be made before the equation of motion for longitudinal vibration in a bar can be developed. These assumptions include:

1. Hooke's law applies.
2. There are no shear forces acting across the bar.
3. There are no bending moments acting on the bar.
4. The amplitudes of vibration are very small.
5. The width of the bar is much less than its length
6. The material in the bar is homogeneous.
7. There is no energy dissipation in the bar.

Figure 8.8(a) shows an incremental element of a bar. x is the position coordinate along the length of the bar, $\zeta(x,t)$ is the longitudinal displacement of an element of the bar in the x direction, and t is time. The cross section dimensions of a bar are much less than its length. Therefore, the longitudinal motion described by $\zeta(x,t)$ is the same at any point in a cross section element of the bar.

To develop the equation of motion for longitudinal vibration of bars, the *strain* in the bar associated with

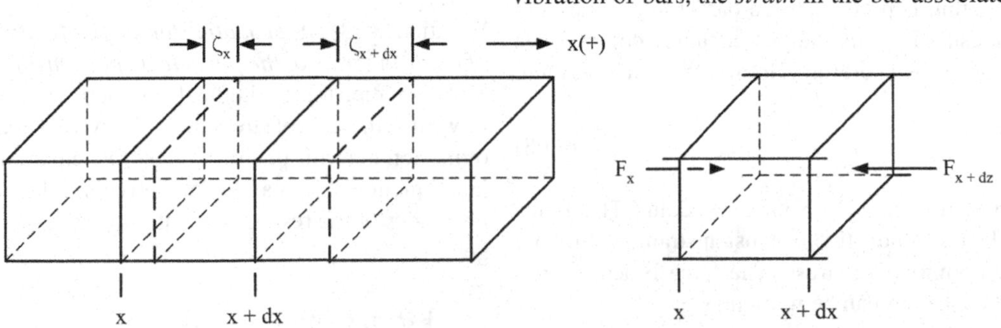

(a) Longitudinal Displacement of an Element in a Bar

(b) Positive Compressive Force Acting through an Element in a Bar

Figure 8.8 Longitudinal Displacement of and Compressive Force Acting through an Element in a Bar vibration must be specified. *Strain* is defined as the ratio of the change in the length of an incremental element of the bar divided by its original length. If dx is the original length of an incremental element and $d\zeta(x,t)$ is the change in length of the element after it moves a very small distance, then the strain $\varepsilon(x,t)$ can be expressed:

$$\varepsilon(x,t) = \frac{d\varsigma(x,t)}{dx}. \tag{8.63}$$

Figure 8.8 indicates the change in length of the element after it moves a very small distance can be given by:

$$d\varsigma(x,t) = \varsigma_{x+dx} - \varsigma_x. \tag{8.64}$$

Expanding ζ_{x+dx} in a Taylor series expansion and neglecting the second and higher order terms yields:

$$d\varsigma(x,t) = \frac{\partial \varsigma(x,t)}{\partial x} dx \tag{8.65}$$

or:

$$\frac{d\varsigma(x,t)}{dx} = \frac{\partial \varsigma(x,t)}{\partial x}. \tag{8.66}$$

Therefore, the strain can be written:

$$\varepsilon(x,t) = \frac{\partial \varsigma(x,t)}{\partial x}. \tag{8.67}$$

When strain is produced in a bar, elastic forces and stresses result. For this analysis, assume a compressive stress is positive. Therefore, Hooke's law can be written:

$$\sigma = -E\varepsilon \tag{8.68}$$

where σ is stress and E is Young's modulus. The minus sign indicates a negative or tension strain results in a positive compressive stress. The force F acting on a element of the bar can be written:

$$F = S\sigma \tag{8.69}$$

where S is the cross section area of the bar. Substituting equations (8.67) and (8.68) into the equation (8.69) yields:

$$F = -ES\frac{\partial \varsigma(x,t)}{\partial x}. \tag{8.70}$$

Next, sum the forces acting on an incremental element of the bar and set the sum of these forces equal to the mass of the element times its acceleration. Summing the forces shown in Figure 8.8(b) gives:

$$F_x - F_{x+dx} = \rho S\, dx \frac{\partial^2 \varsigma(x,t)}{\partial t^2} \tag{8.71}$$

where ρ is the volume density of the bar material. Expanding F_{x+dx} in a Taylor series expansion and neglecting the higher order terms yields:

$$-\frac{\partial F}{\partial x} = \rho S \frac{\partial^2 \varsigma(x,t)}{\partial t^2}. \tag{8.72}$$

Substituting equation (8.70) into equation (8.72) yields:

$$\boxed{\frac{\partial^2 \varsigma(x,t)}{\partial t^2} - c_l^2 \frac{\partial^2 \varsigma(x,t)}{\partial x^2} = 0} \tag{8.73}$$

where the *longitudinal wave speed* c_l in the bar is:

$$\boxed{c_l = \sqrt{\frac{E}{\rho}}.} \tag{8.74}$$

Equation (8.73) is the *equation for longitudinal vibration of a bar in terms of the particle displacement* $\zeta(x,t)$.

The equation for longitudinal vibration of a bar can also be written in terms of stress $\sigma(x,t)$. To write the equation of motion in terms of stress, substitute equation (8.69) into equation (8.70) and take the second derivative of both sides of the resulting equation with respect to t to get:

$$-\frac{1}{E}\frac{\partial^2 \sigma(x,t)}{\partial t^2} = \frac{\partial^3 \varsigma(x,t)}{\partial x\, \partial t^2}. \tag{8.75}$$

Next, substitute equation (8.69) into equation (8.72)

and take the derivative of the resulting equation with respect to x to obtain:

$$-\frac{1}{\rho}\frac{\partial^2 \sigma(x,t)}{\partial x^2} = \frac{\partial^3 \varsigma(x,t)}{\partial t^2 \partial x}. \tag{8.76}$$

Finally, equating equations (8.75) and (8.76) yields:

$$\boxed{\frac{\partial^2 \sigma(x,t)}{\partial t^2} - c_1^2 \frac{\partial^2 \sigma(x,t)}{\partial x^2} = 0.} \tag{8.77}$$

8.6 Solution to the Equation for the Longitudinal Vibration of a Bar

When finite boundary conditions exist at both ends of a bar, the method of separation of variables can be used to obtain the solution for equation (8.73). Therefore, $\zeta(x,t)$ can be written:

$$\boxed{\varsigma(x,t) = (A\cos\omega t + B\sin\omega t)(C\cos kx + D\sin kx)} \tag{8.78}$$

where A and B are determined from the initial conditions and C and D are obtained from the boundary conditions.

There are two boundary conditions that can exist at the end of a bar. The first applies to the case where the end of the bar is rigidly clamped. If the location of the boundary is designated x = b, this boundary condition is specified by:

$$\boxed{\varsigma(x,t)\big|_{x=b} = 0.} \tag{8.79}$$

The second is for the case when the end of the bar is free and is not attached to anything. For this case, the force at the boundary is equal to zero. Using equation (8.70), this boundary condition is specified by:

$$\boxed{\left.\frac{\partial \varsigma(x,t)}{\partial x}\right|_{x=b} = 0.} \tag{8.80}$$

EXAMPLE 8.4

Determine the resonance frequencies and corresponding mode functions of the longitudinal vibration modes that can exist on a bar of length L that is free at both ends.

SOLUTION

Since the bar is free at both ends, the boundary conditions at x = 0 and x = L are:

$$\left.\frac{\partial \varsigma(x,t)}{\partial x}\right|_{x=0} = 0 \tag{8.4a}$$

and:

$$\left.\frac{\partial \varsigma(x,t)}{\partial x}\right|_{x=L} = 0. \tag{8.4b}$$

Taking the derivative of equation (8.77) with respect to x yields:

$$\frac{\partial \varsigma(x,t)}{\partial x} = T(t)(-C\, k\sin kx + D\, k\cos kx). \tag{8.4c}$$

Applying the boundary condition at x = 0 results in D = 0. Applying the boundary condition at x = L yields:

$$\sin kx = 0. \tag{8.4d}$$

This condition is satisfied when:

$$kL = n\pi \quad \text{for} \quad n = 1, 2, 3, \cdots. \tag{8.4e}$$

Therefore, k_n and ω_n can be written:

$$\boxed{k_n = \frac{n\pi}{L} \quad \text{for} \quad n = 1, 2, 3, \cdots} \tag{8.4f}$$

and:

$$\boxed{\omega_n = \frac{n\pi c}{L} \quad \text{for} \quad n = 1, 2, 3, \cdots.} \tag{8.4g}$$

The solution for $\zeta(x,t)$ is:

$$\boxed{\varsigma(x,t) = \sum_{n=1}^{\infty}(A_n\cos\omega_n t + B_n\sin\omega_n t)\cos k_n x.} \tag{8.4h}$$

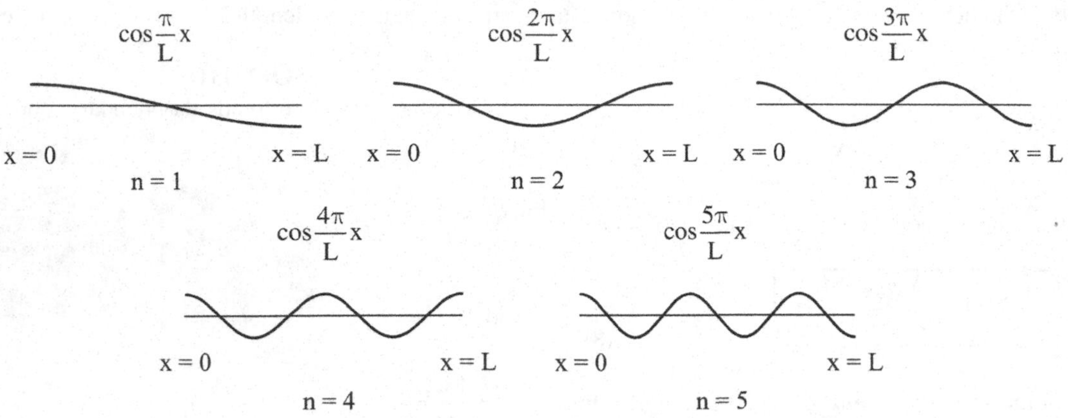

Figure 8.9 Graphs of the First Five Longitudinal Vibration Modes that Can Exist on a Bar that Is Free at Both Ends

A_n and B_n are determined from the initial conditions. Figure 8.9 shows the first five longitudinal vibration modes that can exist on a bar that is free at both ends.

EXAMPLE 8.5

A tension force equal to F is applied to both ends of a bar (Figure 8.10). At time t = 0 the tension force is released. Determine the expression for the amplitudes of A_n and B_n.

SOLUTION

The initial velocity for this problem is:

$$\left.\frac{\partial \varsigma(x,t)}{\partial t}\right|_{x=0} = 0. \tag{8.5a}$$

The initial displacement is determined by first integrating equation (8.63) with respect to x to obtain:

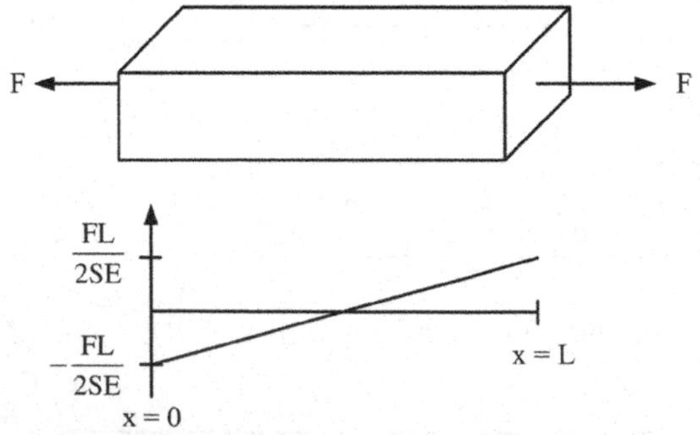

Figure 8.10 Initial Displacement of a Bar Associated with a Tension Force F Applied to Both Ends

$$\varsigma(x,t)\big|_{t=0} = \varepsilon(x,t)\big|_{t=0}\, x. \tag{8.5b}$$

Noting $\varepsilon = \sigma/E$ and $\sigma = F/S$, equation (8.5b) becomes:

$$\varsigma(x,t)\big|_{t=0} = -\frac{F}{SE}x. \tag{8.5c}$$

For this problem, because the bar is free at both ends and an equal force is applied to each end, the boundary at x = 0 will experience a displacement of:

$$\varsigma(x,t)\big|_{\substack{t=0\\x=0}} = -\frac{FL}{2SE} \tag{8.5d}$$

and the boundary at x = L will experience a displacement of:

$$\varsigma(x,t)\big|_{\substack{t=0\\x=L}} = \frac{FL}{2SE}. \tag{8.5e}$$

See Figure 8.10. As is indicated by equation (8.5c) and Figure 8.10, the initial longitudinal displacement along the bar is a linear function of x. Therefore, the initial displacement of the bar at time t = 0 can be expressed:

$$\varsigma(x,t)\big|_{t=0} = \frac{FL}{2ES}\left(\frac{2x}{L} - 1\right). \tag{8.5f}$$

Since the ends of the bar are free, the results of EXAMPLE 8.4 describe the vibration response of the bar subject to the boundary conditions, or:

$$\varsigma(x,t) = \sum_{n=1}^{\infty} (A_n \cos\omega_n t + B_n \sin\omega_n t)\cos k_n x. \tag{8.5g}$$

where $k_n = n\pi/L$ and $\omega_n = n\pi c/L$. Applying the orthogonality conditions that exist relative to sine and cosine functions to equation (8.5g) [see equations (1.48) - (1.50)] using the mode function $\cos k_n L$ yields $B_n = 0$ and:

$$A_n = \frac{2}{L}\int_0^L \varsigma(x,t)\big|_{t=0} \cos k_n x \, dx. \tag{8.5h}$$

Substituting equation (8.5f) into equation (8.5h) yields:

$$A_n = \frac{2}{L}\int_0^L \frac{FL}{2ES}\left(\frac{2x}{L} - 1\right)\cos k_n x \, dx. \tag{8.5i}$$

Rearranging the terms in equation (8.5f) gives:

$$A_n = \frac{F}{ES}\left[\frac{2}{L}\int_0^L x\cos k_n x \, dx - \int_0^L \cos k_n x \, dx\right]. \tag{8.5j}$$

Evaluating the integrals in equation (8.5j), combining the resulting terms, and noting $k_n = n\pi/L$ yields:

$$A_n = \frac{2FL}{ES}\frac{(\cos n\pi - 1)}{(n\pi)^2} \tag{8.5k}$$

or:

$$\boxed{A_n = -\frac{4FL}{ES(n\pi)^2}} \tag{8.5l}$$

for n = odd integers. Therefore, the response of the bar for this example can be written:

$$\boxed{\varsigma(x,t) = -\frac{4FL}{ES}\sum_{n=odd}^{\infty}\frac{1}{(n\pi)^2}\cos\omega_n t \cos k_n x.} \tag{8.5m}$$

k_n and ω_n are given by equations (8.4f) and 8.4g), respectively, in EXAMPLE 8.4 .

EXAMPLE 8.6

Figure 8.11 shows a thin bar that is clamped at x = 0 and that has an attached mass M at x = L. Determine the frequencies and mode functions of the longitudinal vibration modes that can exist on the bar.

SOLUTION

Applying the boundary condition $\zeta(0,t) = 0$ to equation (8.78) yields C = 0. Therefore, $\zeta(x,t)$ becomes:

$$\varsigma(x,t) = (A_n \cos\omega_n t + B_n \sin\omega_n t)\sin k_n x. \tag{8.6a}$$

Therefore, the mode function is specified by $\sin k_n x$.

Next, determine the boundary condition that exists at x = L. Figure 8.11 indicates the force F the bar exerts on mass M and, conversely, mass M exerts on the bar. Using equation (8.69) and noting the force F that acts on the bar is a negative tension force, the expression for the force acting on the bar at x = L can be written:

$$F = ES\frac{\partial\varsigma(x,t)}{\partial x}\bigg|_{x=L}. \tag{8.6b}$$

The force F acting on mass M is equal to the mass times its acceleration, or:

$$M\frac{\partial^2\varsigma(x,t)}{\partial t^2}\bigg|_{x=L} = -F. \tag{8.6c}$$

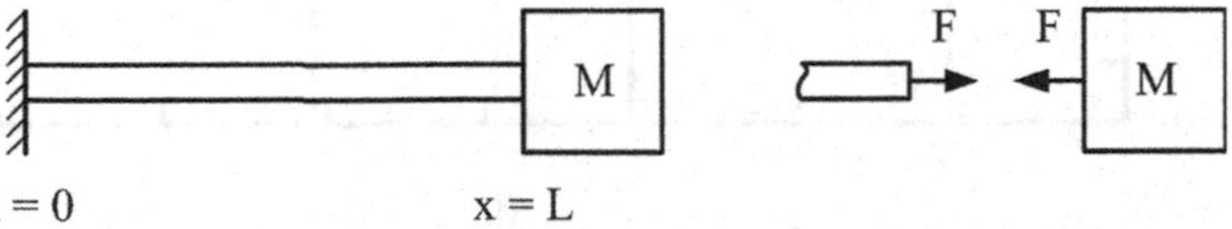

Figure 8.1 1 Boundart Condition at x = L for a Bar that Has a Lumped Mass M at x = L

Equating equations (8.6b) and (8.6c) yields:

$$M\frac{\partial^2 \varsigma(x,t)}{\partial t^2}\bigg|_{x=L} = -ES\frac{\partial \varsigma(x,t)}{\partial x}\bigg|_{x=L}. \tag{8.6d}$$

Substituting equation (8.6a) into equation (6.6d) and carrying out the prescribed operations yields:

$$\omega_n^2 M \sin k_n L = ES k_n \cos k_n L \tag{8.6e}$$

or:

$$\tan k_n L = \frac{ES}{M c_1 \omega_n}. \tag{8.6f}$$

Noting the mass density ρ of the bar can be written [equation (8.74)]:

$$\rho = E/c_1^2 \tag{8.6g}$$

and:

$$\rho = \frac{M_b}{SL} \tag{8.6h}$$

where M_b is the total mass of the bar and equating equations (8.6g) and (8.6h) yields:

$$S = \frac{M_b c_1^2}{EL}. \tag{8.6i}$$

Substituting equation (8.6i) into equation (8.6f) and noting $k_n = \omega_n/c_1$ gives:

$$\boxed{\tan k_n L = \frac{M_b}{M}\frac{1}{k_n L}.} \tag{8.6j}$$

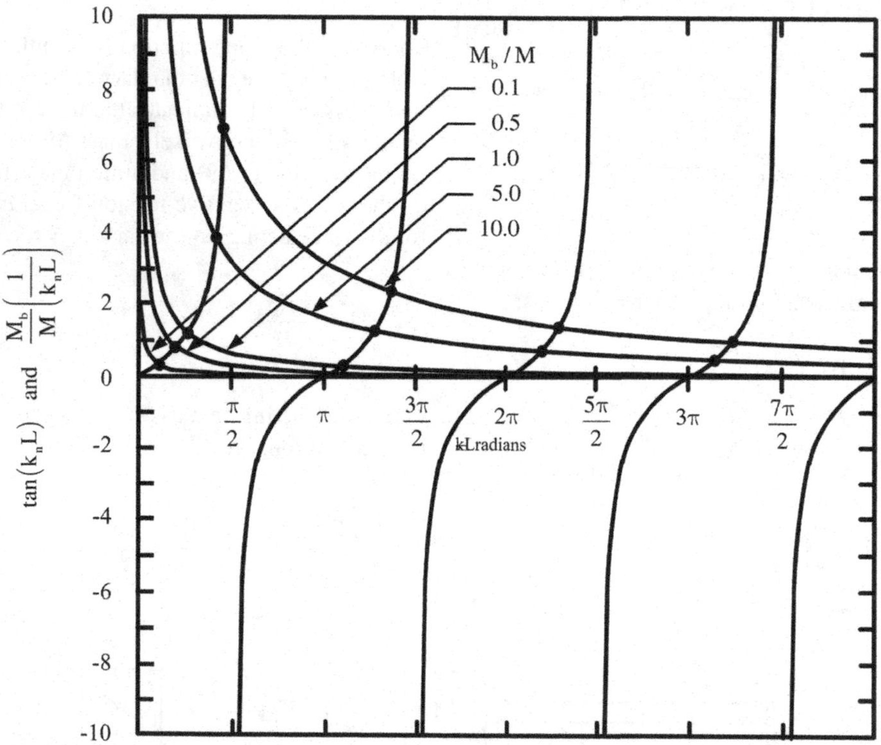

Figure 8.12 $\tan(k_nL)$ and $\frac{M_b}{M}\left(\frac{1}{k_nL}\right)$ vs. k_nL

Equation (8.6j) is a transcendental equation in terms of k_nL. It can be solved by plotting the functions tan k_nL and $(M_b/M)(1/k_nL)$ on the same graph. The values of k_n and subsequently ω_n, which satisfy equation (8.6j), are specified by the points of intersection of the two functions. Figure 8.12 shows the roots to equation (8.6j) for several values of M_b/M.

8.7 Forced Longitudinal Vibration of Bars

An external force can be applied to a semi-infinite bar at x = 0 as shown in Figure 8.13. The harmonic force can be expressed:

$$f(t) = F e^{j\omega t}. \tag{8.81}$$

This force will excite only positive stress waves in the bar. Therefore, the response $\zeta(x,t)$ can be expressed:

$$\varsigma(x,t) = \mathbf{A} e^{j(\omega t - kx)} \quad \text{where} \quad \mathbf{A} = A e^{j\theta}. \tag{8.82}$$

The driving force must equal the force specified by equation (8.70) at x = 0:

$$F e^{j\omega t} = -ES \frac{\partial \varsigma(x,t)}{\partial x}\bigg|_{x=0}. \tag{8.83}$$

Substituting equation (8.82) into equation (8.83) and evaluating the derivative at x = 0 yields:

$$F = -ES(-jk\mathbf{A}) \quad \text{or} \quad A = -j\frac{F}{ESk}. \tag{8.84}$$

Therefore, the response for $\zeta(x,t)$ can be written:

$$\varsigma(x,t) = \frac{F}{ESk} e^{j\left(\omega t - kx - \frac{\pi}{2}\right)}. \tag{8.85}$$

The driving point mechanical impedance is the ratio of the driving force divided by the longitudinal velocity of the bar at x = 0. The velocity for the bar at x = 0 is:

Figure 8.13 Semi-Infinite Bar Forced at x = 0

$$\frac{\partial \varsigma(x,t)}{\partial t}\bigg|_{x=0} = \frac{F\omega}{ESk} e^{j\omega t}. \tag{8.86}$$

Noting $k = \omega / c_1$, equation (8.86) becomes:

$$\frac{\partial \varsigma(x,t)}{\partial t}\bigg|_{x=0} = \frac{Fc_1}{ES} e^{j\omega t}. \tag{8.87}$$

Dividing equation (8.83) by equation (8.87) yields:

$$Z_m = \frac{ES}{c_1}. \tag{8.88}$$

Noting $c_1^2 = E/\rho$ and $\rho S = m$ where m is the mass per unit length of the bar, the driving point mechanical impedance can be written:

$$\boxed{Z_m = c_1 m.} \tag{8.89}$$

Equation (8.89) is a function of only the characteristic parameters of the bar. Therefore, for this case, Z_m is called the characteristic *mechanical impedance* Z_c for longitudinal vibration of a bar, or:

$$\boxed{Z_c = c_1 m.} \tag{8.90}$$

Z_m is purely resistive. As a result, all of the energy delivered to the bar by the driving force at x = 0 is transmitted to the bar. It can be shown that the time averaged input power at x = 0 is:

$$\boxed{\langle \Pi_{in} \rangle_t = \frac{1}{2} Z_m V_o^2} \tag{8.91}$$

where Z_m is the mechanical impedance specified by equation (8.89) and V_o is the amplitude of the longitudinal velocity at x = 0 (equation (8.87)).

Figure 8.14 shows the case where a bar is forced at x = 0 and clamped at x = L. The expression for the driving force is given by equation (8.81). However, since both positive and negative stress waves exist in the bar

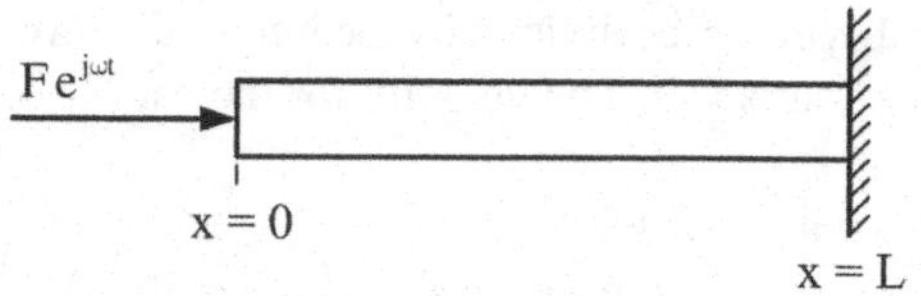

Figure 8.14 Bar of Length L Forced at x = 0

because of the finite boundary condition at x = L, the response of the bar must be expressed:

$$\varsigma(x,t) = \mathbf{A}e^{j(\omega t - kx)} + \mathbf{B}e^{j(\omega t + kx)} \tag{8.92}$$

where:

$$\mathbf{A} = Ae^{j\theta 1} \quad \text{and} \quad \mathbf{B} = Be^{j\theta 2}. \tag{8.93}$$

Since the bar is clamped at x = L, the boundary condition at this location is given by:

$$\varsigma(x,t)\big|_{x=L} = 0. \tag{8.94}$$

The sum of the forces at x = 0 is obtained from equation (8.83). Taking the derivative of equation (8.92) with respect to x and evaluating the results at x = 0 yields:

$$\left.\frac{\partial \varsigma(x,t)}{\partial x}\right|_{x=0} = -jk\mathbf{A}e^{j\omega t} + jk\mathbf{B}e^{j\omega t}. \tag{8.95}$$

Substituting equation (8.95) into equation (8.83) gives:

$$F = jkES(\mathbf{A} - \mathbf{B}). \tag{8.96}$$

Applying the boundary condition at x = L yields:

$$\mathbf{A}e^{-jkL} + \mathbf{B}e^{jkL} = 0. \tag{8.97}$$

Solving for **A** and **B** in equations (8.96) and (8.97) yields:

$$\mathbf{A} = \frac{Fe^{jkL}}{j2ESk\cos kL} \tag{8.98}$$

and:

$$\mathbf{B} = \frac{-Fe^{-jkL}}{j2ESk\cos kL}. \tag{8.99}$$

Substituting equations (8.98) and (8.99) into equation (8.92) and noting $E = \rho c_l^2$, $m = \rho S$, and $k = \omega / c_l$ yields:

$$\varsigma(x,t) = \frac{F\sin[k(L-x)]}{Z_c\,\omega\cos kL}e^{j\omega t}. \tag{8.100}$$

It can be shown that the driving point mechanical impedance of a bar excited by a longitudinal harmonic force at x = 0 is given by:

$$\boxed{\mathbf{Z}_m = -jZ_c\cot kL.} \tag{8.101}$$

Since the expressions for the response ζ(x,t) and the driving point mechanical impedance $\mathbf{Z}_m$ for forced longitudinal vibration of a bar that is clamped at x = L are the same as the corresponding expressions for forced vibration of a string that is clamped at x = L [equations (8.51) and (8.59)], the comments made with regard to quarter and half wavelength frequencies, the driving point mechanical impedance, and the time averaged power for forced vibration of a string also apply to forced longitudinal vibration of a bar.

8.8 Transverse Vibration of a Beam

A beam is capable of supporting both longitudinal and transverse vibration. In many cases, because of the coupling that exists between longitudinal and bending strains in a beam, it is difficult to produce one type of vibration without exciting the other. However, only transverse vibration of a beam will be examined in this section. The following assumptions must be made:

1. The beam has a uniform cross section.
2. The beam is symmetric about its neutral axis.
3. The lateral dimensions are small compared to the length of the beam.
4. EI is constant.
5. Plane surfaces in the beam remain plane.
6. There are no net longitudinal forces.
7. The angular or rotary inertial of the beam is very small and can be neglected.
8. The vibration amplitudes of the beam are very small.

For the following development, $\zeta(x,t)$ is the longitudinal displacement variable, $y(x,t)$ is the transverse displacement variable, x is the position along the beam, and t is time. If a beam, which is symmetric about its neutral axis, is bent and has a radius of curvature R as indicated in Figure 8.15, the length dx of the neutral axis will remain unchanged. However, the area of the beam above the neutral axis will be in tension while the area below the neutral axis will be in compression. Using the sign convention that relates stress and strain given by $\sigma = -E\varepsilon$), equation (8.69) indicates the incremental tensile force dF acting on an incremental area dx above the neutral axis can be written:

$$dF = -E\,dS\frac{\partial \varsigma(x,t)}{\partial x}. \tag{8.102}$$

Similarly, the incremental compressive force dF acting on an incremental area dS below the neutral axis can be expressed:

$$dF = E\,dS\frac{\partial \varsigma(x,t)}{\partial x}. \tag{8.103}$$

The length of the arc dx of the neutral axis is given by:

$$dx = R\,d\phi. \tag{8.104}$$

Similarly, the length of the arc of any longitudinal section of the beam at a location other than the neutral axis can be expressed:

$$dx + d\varsigma(x,t) = (R + r)\,d\phi. \tag{8.105}$$

Substituting equation (8.104) into equation (8.105) for $d\phi$ and reducing the resulting expression yields:

$$\frac{d\varsigma(x,t)}{dx} = \frac{r}{R}. \tag{8.106}$$

Equation (8.65) indicates:

$$\frac{d\,\varsigma(x,t)}{dx} = \frac{\partial \varsigma(x,t)}{\partial x}. \tag{8.107}$$

Substituting equation (8.106) into equation (8.102) gives:

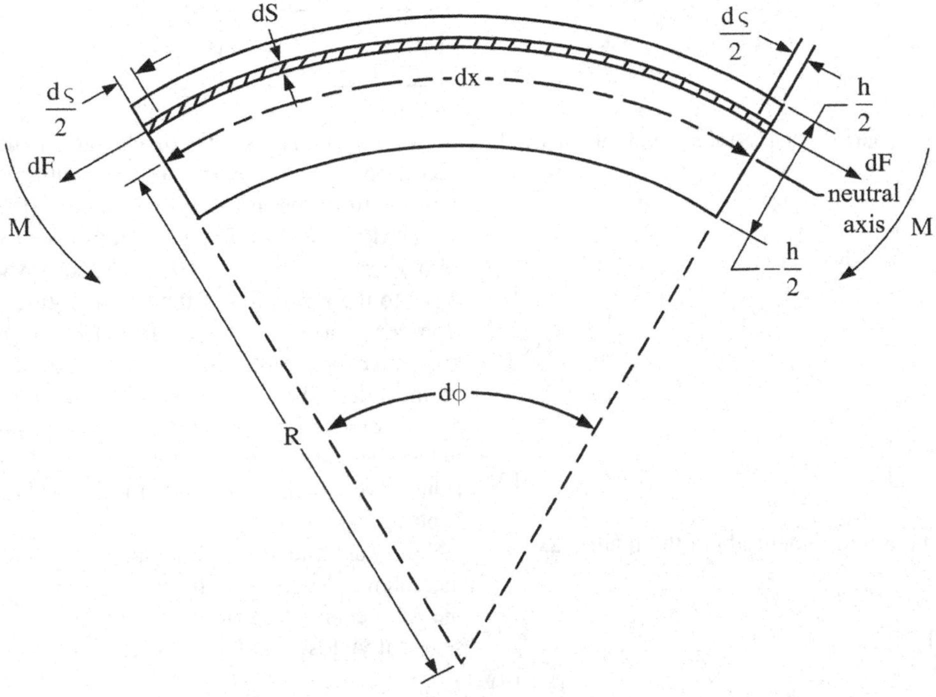

Figure 8.15 Bending Strains and Stresses in a Beam

$$dF = -E\,dS\frac{r}{R}. \tag{8.108}$$

Integrate equation (8.108) over the thickness h of the beam to obtain:

$$F = -E\int_{-\frac{h}{2}}^{\frac{h}{2}} \frac{r}{R}\,dS \tag{8.109}$$

where h is the height of the beam. The incremental area dS = b dr where b is the width of the beam. Therefore, equation (8.109) becomes:

$$F = -\frac{Eb}{R}\int_{-\frac{h}{2}}^{\frac{h}{2}} r\,dr \quad \text{or} \quad F = 0. \tag{8.110}$$

Equation (8.110) indicates there are no net longitudinal forces acting on the beam. The negative tensile forces above the neutral axis equal the positive compressive forces below the neutral axis.

The moment M acting about the neutral axis is obtained from:

$$M = \int_{-\frac{h}{2}}^{\frac{h}{2}} r\,dF. \tag{8.111}$$

Substituting equation (8.108) into equation (8.111) yields:

$$M = -\frac{E}{R}\int_{-\frac{h}{2}}^{\frac{h}{2}} r^2\,dr. \tag{8.112}$$

Noting:

$$\boxed{I = \int_{-\frac{h}{2}}^{\frac{h}{2}} r^2\,dr} \tag{8.113}$$

where I is the area moment about the neutral axis of the beam:

$$M = -\frac{EI}{R}. \tag{8.114}$$

In general, the radius of curvature R of a beam is not constant but is a function of the position along the neutral axis. It can be shown that:

$$R = \frac{\left[1 + \left(\frac{\partial y(x,t)}{\partial x}\right)^2\right]^{3/2}}{\frac{\partial^2 y(x,t)}{\partial x^2}}. \tag{8.115}$$

When the vibration amplitudes are very small,:

$$\frac{\partial y(x,t)}{\partial x} \ll 1. \tag{8.116}$$

Therefore, equation (8.115) reduces to:

$$R = \frac{1}{\frac{\partial^2 y(x,t)}{\partial x^2}}. \tag{8.117}$$

Substituting equation (8.117) into equation (8.114) yields:

$$\boxed{M = -EI\frac{\partial^2 y(x,t)}{\partial x^2}.} \tag{8.118}$$

$\partial^2 y(x,t)/\partial x^2$ is the rate of change of the slope of curvature $\partial y(x,t)/\partial x$ of the beam with respect to x. Moving from the left side of the beam (Figure 8.15) toward $d\phi/2$, $\partial y(x,t)/\partial x$ goes from a positive value to zero. Therefore, $\partial^2 y(x,t)/\partial x^2$ is negative. Moving from $d\phi/2$ to the right side of the beam, $\partial y(x,t)/\partial x$ goes from zero to a negative value. Therefore, $\partial^2 y(x,t)/\partial x^2$ is negative for the entire increment dx. Consequently, the moment described by equation (8.118) is positive. If a counter clockwise moment is defined as being positive, the moment at the right end of the beam (Figure 8.15) is negative and the moment at the left end of the beam is positive.

Next, sum the forces and moments acting on the incremental segment of a beam (Figure 8.16). Summing the moments acting about the left end of the beam segment yields:

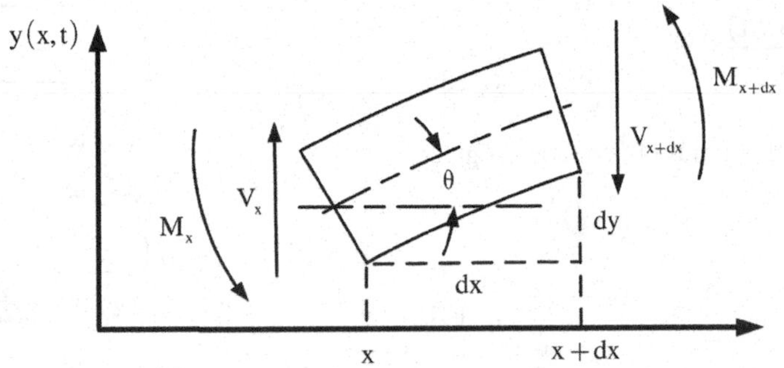

Figure 8.16 Bending Moments and Shear Forces Acting on a Beam Element

$$\sum M = M_x - M_{x+dx} - V_{x+dx}\, dx. \tag{8.119}$$

Expanding M_{x+dx} and V_{x+dx} in a Taylor series expansion and neglecting the higher order terms gives:

$$M_{x+dx} = M_x + \frac{\partial M}{\partial x} dx \tag{8.120}$$

and:

$$V_{x+dx} = V_x + \frac{\partial V}{\partial x} dx. \tag{8.121}$$

Substituting equations (8.120) and (8.121) into equation (8.119) yields:

$$\sum M = -\frac{\partial M}{\partial x} dx - V_x\, dx - \frac{\partial V}{\partial x} dx^2. \tag{8.122}$$

Assuming the rotary inertia term to be very small resulting in $\sum M = 0$ and neglecting the higher order term, equation (8.122) becomes:

$$\sum M = 0 = -\frac{\partial M}{\partial x} dx - V_x\, dx \tag{8.123}$$

or:

$$V_x = -\frac{\partial M}{\partial x}. \tag{8.124}$$

Substituting equation (8.118) into equation (8.124) yields:

$$\boxed{V_x = EI\frac{\partial^3 y(x,t)}{\partial x^3}.} \tag{8.125}$$

Summing the forces acting on the beam (Figure 8.16) in the y direction gives:

$$\sum F_y = V_x - V_{x+dx} \tag{8.126}$$

or:

$$\sum F_y = -\frac{\partial V}{\partial x} dx. \tag{8.127}$$

Since the beam has dynamic motion in the y direction:

$$\sum F_y = M\frac{\partial^2 y(x,t)}{\partial t^2} \quad \text{where} \quad M = \rho S\, dx. \tag{8.128}$$

M is the total mass of the beam segment, ρ is the mass density of the beam, and S is the cross sectional area of the beam. Equating equations (8.127) and equation (8.128) yields:

$$m\frac{\partial^2 y(x,t)}{\partial t^2} = -\frac{\partial V}{\partial x} \tag{8.129}$$

where m is the mass per unit length of the beam. Substituting equation (8.125) into equation (8.129) gives:

$$\frac{\partial^2 y(x,t)}{\partial t^2}+\frac{EI}{m}\frac{\partial^4 y(x,t)}{\partial x^4}=0. \tag{8.130}$$

Noting $c_1^2 = E/\rho$ and the radius of gyration κ is given by $\kappa^2 = I/S$:

$$\boxed{\frac{EI}{m}=c_1^2\,\kappa^2.} \tag{8.131}$$

Therefore, equation (8.130) becomes:

$$\boxed{\frac{\partial^2 y(x,t)}{\partial t^2}+c_1^2\,\kappa^2\,\frac{\partial^4 y(x,t)}{\partial x^4}=0.} \tag{8.132}$$

Equation (8.132) is the *equation for transverse vibration of a beam.* It is a second order partial differential equation with respect to t and a forth order partial differential equation with respect to x. Therefore, to obtain the complete solution for this equation, two initial conditions and four boundary conditions must be specified. The initial conditions are given by:

$$\boxed{y(x,t)\big|_{t=0}=y_o(x)} \tag{8.133}$$

and:

$$\boxed{\left.\frac{\partial y(x,t)}{\partial t}\right|_{t=0}=v_o(x).} \tag{8.134}$$

Three sets of end or boundary conditions can be specified for the transverse vibration of a beam. They are:

1. Free end [Figure 8.17(a)],
2. Simply supported end [Figure 8.17(b)], and
3. Clamped end [Figure 8.17(c)].

The equations that evolved during the development of equation (8.132) can be used to specify the equations for the different sets of boundary conditions.

For the free end, there are no restraining forces or moments acting on the end. Therefore, the moment M [equation (8.118)] and the shear force V [equation (8.125)] are zero. As a result, the boundary conditions

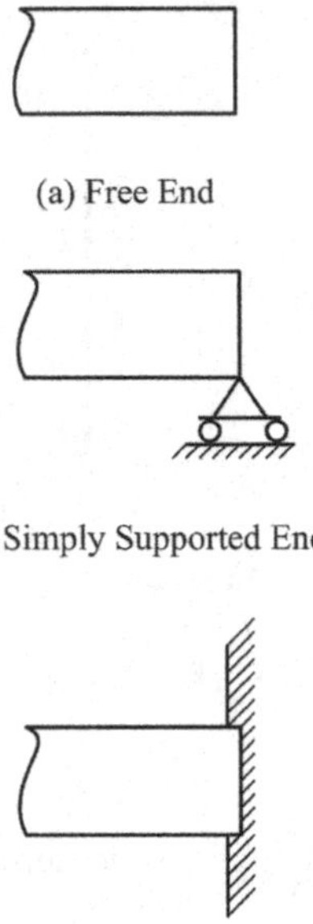

(a) Free End

(b) Simply Supported End

(c) Clamped End

Figure 8.17 Transverse Vibration Boundary Condition that Can Exist on a Beam

for a free end are from equation (8.118):

$$\boxed{\left.\frac{\partial^2 y(x,t)}{\partial x^2}\right|_{x=b}=0} \tag{8.135}$$

and from equation (8.125):

$$\boxed{\left.\frac{\partial^3 y(x,t)}{\partial x^3}\right|_{x=b}=0.} \tag{8.136}$$

For the simply supported end, the beam is free to rotate at the end. However, the beam is constrained from moving in the vertical direction. Therefore, the moment and displacement at the boundary are equal to zero. The boundary conditions for a simply supported end are:

$$\boxed{y(x,t)\big|_{x=b}=0} \tag{8.137}$$

and from equation (8.118):

$$\left.\frac{\partial^2 y(x,t)}{\partial x^2}\right|_{x=b} = 0. \tag{8.138}$$

For the clamped end, both a restraining moment and shear force are applied to the beam at the boundary. The beam is constrained from moving in the transverse direction and from rotating at the end (the slope of the beam at the boundary equals zero). Therefore, the boundary conditions for a clamped end are given by:

$$y(x,t)\big|_{x=b} = 0 \tag{8.139}$$

and:

$$\left.\frac{\partial y(x,t)}{\partial x}\right|_{x=b} = 0. \tag{8.140}$$

8.9 Solution to the Equation for Transverse Vibration of a Beam

The solution to equation (8.132) is obtained by using separation of variables. Let:

$$y(x,t) = \Psi(x)T(t). \tag{8.141}$$

Substituting equation (8.141) into equation (8.132) and carrying out the prescribed operations yields:

$$\Psi(x)\frac{\partial^2 T(t)}{\partial t^2} + c_1^2\kappa^2 T(t)\frac{\partial^4 \Psi(x)}{\partial x^4} = 0 \tag{8.142}$$

or:

$$\Psi(x)\frac{\partial^2 T(t)}{\partial t^2} = -c_1^2\kappa^2 T(t)\frac{\partial^4 \Psi(x)}{\partial x^4}. \tag{8.143}$$

Separating the variables gives:

$$\frac{d^2 T(t)}{dt^2} + \omega^2 T(t) = 0 \tag{8.144}$$

and:

$$\frac{d^4 \Psi(x)}{dx^4} - \frac{\omega^2}{c_1^2\kappa^2}\Psi(x) = 0. \tag{8.145}$$

The solution to equation (8.144) is:

$$T(t) = A\cos\omega t + B\sin\omega t. \tag{8.146}$$

Before solving equation (8.145), write:

$$\frac{\omega^2}{c_1^2\kappa^2} = k_b^4 \quad \text{where} \quad k_b = \frac{\omega}{c_b} \tag{8.147}$$

and where c_b is the *wave speed of a bending wave traveling on the beam.* Solving equation (8.147) for c_b yields:

$$c_b = \sqrt{c_1\kappa\omega}. \tag{8.148}$$

The wave speed c_b of a bending wave traveling on a beam is a function of frequency. As a result, c_b is referred to as the *phase velocity relative to the frequency* ω. Substituting equation (8.147) into equation (8.145) yields:

$$\frac{d^4 \Psi(x)}{dx^4} - k_b^4\Psi(x) = 0. \tag{8.149}$$

Now let:

$$\Psi(x) = Ce^{sx}. \tag{8.150}$$

Substituting equation (8.150) into equation (8.149) yields:

$$\left(s^4 - k_b^4\right)Ce^{sx} \quad \text{or} \quad s^4 = k_b^4. \tag{8.151}$$

Therefore:

$$s^2 = \pm k_b^2 \tag{8.152}$$

or:

$$s = k_b;\ s = -k_b;\ s = jk_b;\ \text{and}\ s = -jk_b. \tag{8.153}$$

Since s has four roots, equation (8.150) becomes:

$$\Psi(x) = C_1 e^{k_b x} + C_2 e^{-k_b x} + C_3 e^{jk_b x} + C_4 e^{-jk_b x}. \tag{8.154}$$

Substituting the relations:

$$e^{\pm k_b x} = \cosh k_b x \pm \sinh k_b x \tag{8.155}$$

and:

$$e^{\pm jk_b x} = \cos k_b x \pm \sin k_b x \tag{8.156}$$

into the equation (8.154) and rearranging the terms yields:

$$\boxed{\Psi(x) = D_1 \cosh k_b x + D_2 \sinh k_b x + D_3 \cos k_b x + D_4 \sin k_b x.} \tag{8.157}$$

The total solution for y(x,t) is obtained by substituting equations (8.146) and (8.157) into equation (8.141).

$$\boxed{y(x,t) = (A\cos\omega t + B\sin\omega t)\times \begin{pmatrix} D_1 \cosh k_b x + D_2 \sinh k_b x + \\ D_3 \cos k_b x + D_4 \sin k_b x \end{pmatrix}.} \tag{8.158}$$

A and B in equation (8.158) are determined by applying the initial conditions, and D_1 through D_4 are obtained from the four boundary conditions.

As was previously mentioned, the wave speed c_b of a bending wave is a function of frequency and it increases with increasing frequency. As a result, high frequency waves travel at greater speeds than do low frequency waves. Therefore, an initial wave form which is composed of many frequency components at time t = 0 is altered as the wave travels along a beam at times t greater than zero. This is because the higher frequency waves outrun the lower frequency waves. This phenomenon is similar to the transmission of light waves through glass. The different frequency components of the light waves (colors) travel at different speeds. Hence, dispersion results. Therefore, bending or transverse waves on a beam are called *dispersive waves*.

EXAMPLE 8.7

Determine the frequencies and mode shapes of the free transverse vibration modes that can exist on a beam that is simply supported at x = 0 and x = L (Figure 8.18).

SOLUTION

The boundary conditions for this problem are:

$$y(x,t)\big|_{x=0} = 0 \quad \text{and} \quad \left.\frac{\partial^2 y(x,t)}{\partial x^2}\right|_{x=0} = 0 \tag{8.7a}$$

and:

$$y(x,t)\big|_{x=L} = 0 \quad \text{and} \quad \left.\frac{\partial^2 y(x,t)}{\partial x^2}\right|_{x=L} = 0. \tag{8.7b}$$

Since only the frequencies and mode shapes for the free bending modes on the beam are desired, equation (8.157) will be used to solve this problem. The second derivative of equation (8.157) with respect to x is:

$$\frac{\partial^2 \Psi(x)}{\partial x^2} = k_b^2 \begin{pmatrix} D_1 \cosh k_b x + D_2 \sinh k_b x - \\ D_3 \cos k_b x - D_4 \sin k_b x \end{pmatrix}. \tag{8.7c}$$

Applying the boundary condition $y(x,t)\big|_{x=0}$ yields:

$$\begin{aligned}\Psi(0) = D_1 \cosh(0) + D_2 \sinh(0) - \\ D_3 \cos(0) - D_4 \sin(0) \\ = 0\end{aligned} \tag{8.7d}$$

or:

$$D_1 + D_3 = 0. \tag{8.7e}$$

Similarly, applying the boundary condition $\partial^2 y(x,t)/\partial x^2\big|_{x=0} = 0$ gives:

$$D_1 - D_3 = 0. \tag{8.7f}$$

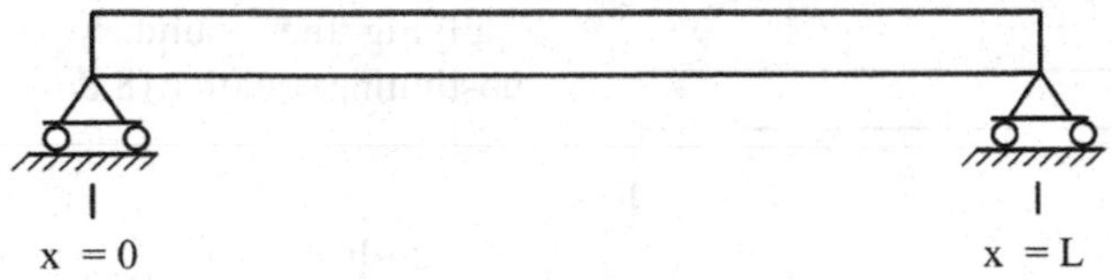

Figure 8.18 Beam that Is Simply Supported at Both Ends

The only way that equations (8.7e) and (8.7f) can be satisfied is for:

$$D_1 = 0 \quad \text{and} \quad D_3 = 0. \tag{8.7g}$$

Therefore, equations (8.157) and (8.7c) reduce to:

$$\Psi(x) = D_2 \sinh k_b x + D_4 \sin k_b x \tag{8.7h}$$

and:

$$\frac{\partial^2 \Psi(x)}{\partial x^2} = k_b^2 (D_2 \sinh k_b x - D_4 \sin k_b x). \tag{8.7i}$$

Applying the two boundary conditions at x = L yields:

$$D_2 \sinh(k_b L) + D_4 \sin(k_b L) = 0 \tag{8.7j}$$

and:

$$D_2 \sinh(k_b L) - D_4 \sin(k_b L) = 0. \tag{8.7k}$$

The only way the above two equations can be satisfied is for both of the terms in each equation to be equal to zero. Since:

$$\sinh(k_b L) \neq 0 \quad \text{then} \quad D_2 = 0. \tag{8.7l}$$

D_4 cannot equal zero. Therefore:

$$\sin(k_b L) = 0 \quad \text{or} \quad k_b L = n\pi, 2n\pi, 3n\pi, \cdots \tag{8.7m}$$

Consequently:

$$\boxed{k_{b(n)} = \frac{n\pi}{L} \quad \text{for} \quad n = 1, 2, 3, \cdots} \tag{8.7n}$$

and:

$$\omega_n = \frac{n\pi c_{b(n)}}{L} \quad \text{for} \quad n = 1, 2, 3, \cdots. \tag{8.7o}$$

However, noting $c_{b(n)}^2 = c_1 \kappa \omega_n$, ω_n in equation (8.7o) becomes:

$$\boxed{\omega_n = \left(\frac{n\pi}{L}\right)^2 c_1 \kappa \quad \text{for} \quad n = 1, 2, 3, \cdots.} \tag{8.7p}$$

Therefore, the frequencies of the free transverse vibration modes that can exist on a beam that is simply supported at both ends are given by equation (8.7p) and the mode shapes are given by:

$$\boxed{\Psi_n(x) = D_{4(n)} \sin(k_{b(n)} x)} \tag{8.7q}$$

where $k_{b(n)}$ is given by equation (8.7n). The total solution is:

$$\boxed{y(x,t) = \sum_{n=1}^{\infty} (A_n \cos \omega_n t + B_n \sin \omega_n t) \sin k_{b(n)} x.} \tag{8.7r}$$

where A_n and B_n are determined by the initial conditions.

EXAMPLE 8.8

Determine the frequencies and mode shapes of the free transverse vibration modes that can exist on a beam that is clamped at x = 0 and is free at x = L (Figure 8.19).

SOLUTION

The boundary conditions for this problem are:

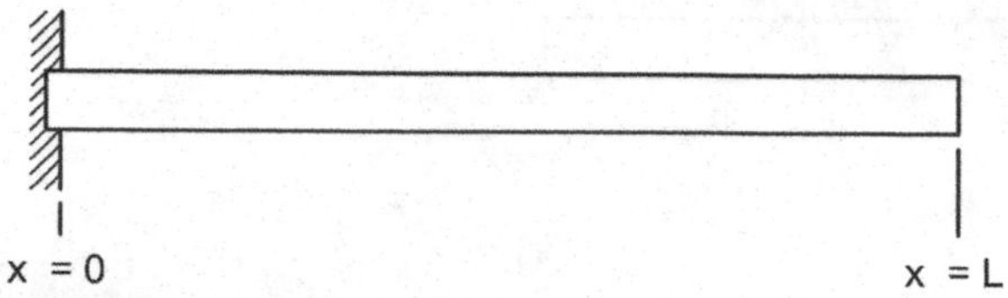

Figure 8.19 Beam that Is Clamped at x = 0 and Free at x = L

$$y(x,t)\big|_{x=0} = 0 \quad \text{and} \quad \left.\frac{\partial y(x,t)}{\partial x}\right|_{x=0} = 0 \tag{8.8a}$$

and:

$$\left.\frac{\partial^2 y(x,t)}{\partial x^2}\right|_{x=L} = 0 \quad \text{and} \quad \left.\frac{\partial^3 y(x,t)}{\partial x^3}\right|_{x=L} = 0. \tag{8.8b}$$

Equation (8.157) will be used to solve this problem. The first derivative of equation (8.157) with respect to x is:

$$\frac{\partial \Psi(x)}{\partial x} = k_b \begin{pmatrix} D_1 \sinh k_b x + D_2 \cosh k_b x - \\ D_3 \sin k_b x + D_4 \cos k_b x \end{pmatrix}. \tag{8.8c}$$

The second derivative of equation (8.157) with respect to x is given by equation (8.7c). The third derivative of equation (8.157) with respect to x is:

$$\frac{\partial^3 \Psi(x)}{\partial x^3} = k_b^3 \begin{pmatrix} D_1 \sinh k_b x + D_2 \cosh k_b x + \\ D_3 \sin k_b x - D_4 \cos k_b x \end{pmatrix}. \tag{8.8d}$$

Applying the boundary conditions at x = 0 yield:

$$\Psi(x)\big|_{x=0} = 0 = D_1 + D_3 \tag{8.8e}$$

and:

$$\left.\frac{\partial \Psi(x)}{\partial x}\right|_{x=0} = 0 = D_2 + D_4. \tag{8.8f}$$

Therefore:

$$D_3 = -D_1 \quad \text{and} \quad D_4 = -D_2. \tag{8.8g}$$

Applying the boundary conditions at x = L and substituting equation (8.8g) into the resulting equations yield:

$$D_1 \begin{pmatrix} \cosh k_b L + \\ \cos k_b L \end{pmatrix} + D_2 \begin{pmatrix} \sinh k_b L + \\ \sin k_b L \end{pmatrix} = 0 \tag{8.8h}$$

and:

$$D_1 \begin{pmatrix} \sinh k_b L - \\ \sin k_b L \end{pmatrix} + D_2 \begin{pmatrix} \cos k_b L + \\ \cos k_b L \end{pmatrix} = 0 \tag{8.8i}$$

Write the above two equations in matrix form:

$$\begin{bmatrix} \begin{pmatrix} \cosh k_b L + \\ \cos k_b L \end{pmatrix} & \begin{pmatrix} \sinh k_b L + \\ \sin k_b L \end{pmatrix} \\ \begin{pmatrix} \sinh k_b L - \\ \sin k_b L \end{pmatrix} & \begin{pmatrix} \cos k_b L + \\ \cos k_b L \end{pmatrix} \end{bmatrix} \begin{Bmatrix} D_1 \\ D_2 \end{Bmatrix} = 0. \tag{8.8j}$$

To obtain the non-trivial solution to the above equation, the determinant of the matrix must equal zero, or:

$$\begin{vmatrix} \begin{pmatrix} \cosh k_b L + \\ \cos k_b L \end{pmatrix} & \begin{pmatrix} \sinh k_b L + \\ \sin k_b L \end{pmatrix} \\ \begin{pmatrix} \sinh k_b L - \\ \sin k_b L \end{pmatrix} & \begin{pmatrix} \cos k_b L + \\ \cos k_b L \end{pmatrix} \end{vmatrix} = 0. \tag{8.8k}$$

Expanding the above determinant and noting:

$$\cos^2 k_b L + \sin^2 k_b L = 1 \tag{8.8l}$$

and:

$$\cosh^2 k_b L - \sinh^2 k_b L = 1 \tag{8.8m}$$

yields:

$$2 + 2\cos k_b L \cosh k_b L = 0 \tag{8.8n}$$

or:

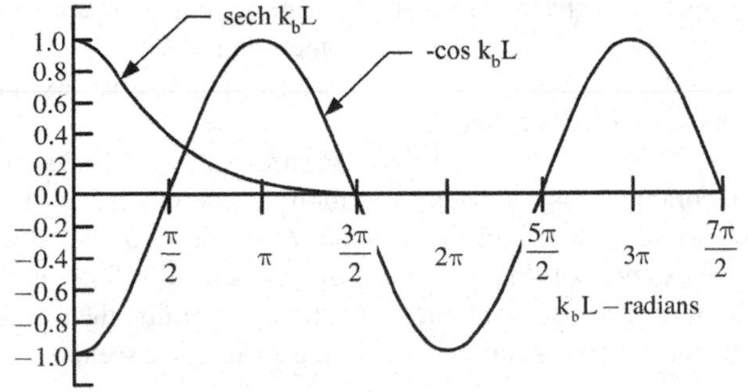

Figure 8.20 sech k_bL and -cos k_bL vs. k_bL

$$\boxed{\operatorname{sech} k_bL = -\cos k_bL}. \tag{8.8o}$$

Equation (8.8o)is a transcendental equation in terms of k_bL. It can be solved by plotting the functions sech k_bL and cos k_bL on the same graph. The values of $k_{b(n)}$ which satisfy equation (8.8o) are the points of intersection of the two functions. Figure 8.20 shows a plot of the two functions in equation (8.8o) vs. k_bL. Once $k_{b(n)}$ has been determined, ω_n can be obtained from the equation (8.147):

$$\boxed{\omega_n = k_{b(n)}^2\, c_1\, \kappa.} \tag{8.8p}$$

Once the values for $k_{b(n)}$ have been determined, the expression for the corresponding mode shapes can be obtained from equation (8.8j). Since the matrix in equation (8.8j) is singular (its determinant equals zero), it results in only one independent equation. Therefore, either equation (8.8h) or (8.8i) can be used to solve for D_2 in terms of D_1. Using equation (8.8h) yields:

$$D_2 = -\left(\frac{\cosh k_bL + \cos k_bL}{\sinh k_bL + \sin k_bL}\right) D_1. \tag{8.8q}$$

Therefore, the equation for the mode shapes can be written:

$$\boxed{\Psi_n(x) = D_{1(n)}\left[\left(\cosh k_{b(n)}x - \cos k_{b(n)}x\right) - \left(\frac{\cosh k_{b(n)}L + \cos k_{b(n)}L}{\sinh k_{b(n)}L + \sin k_{b(n)}L}\right)\left(\begin{matrix}\sinh k_{b(n)}x - \\ \sin k_{b(n)}x\end{matrix}\right)\right].} \tag{8.8r}$$

Figure 8.21 shows a plot of the first five transverse

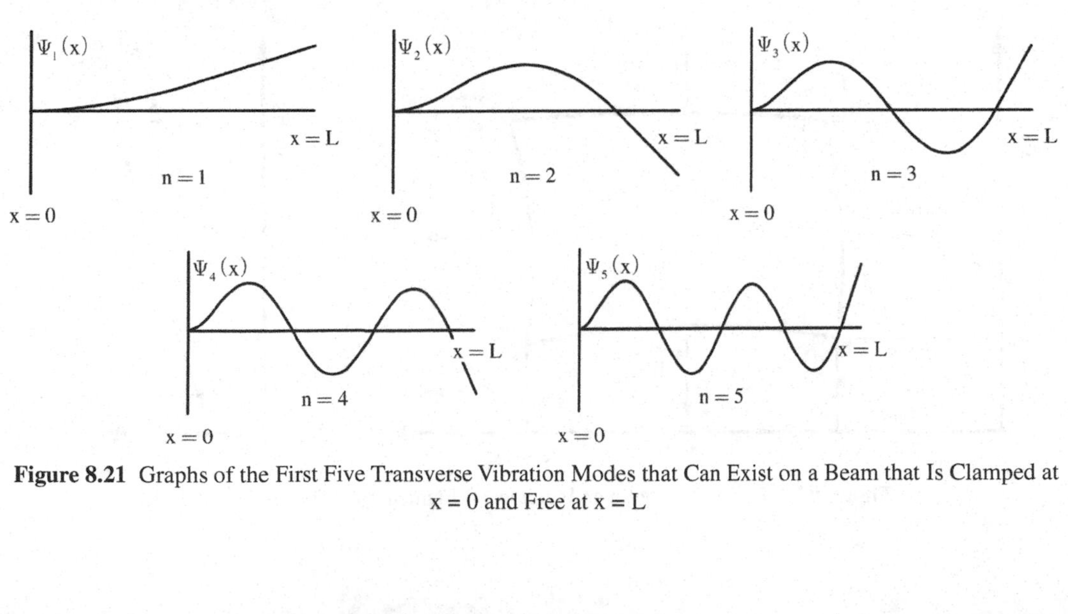

Figure 8.21 Graphs of the First Five Transverse Vibration Modes that Can Exist on a Beam that Is Clamped at x = 0 and Free at x = L

vibration modes that can exist on a beam that is clamped at x = 0 and is free at x = L.

8.10 Transverse Vibration of a Thin Membrane

There are occasions when continuous systems that are described with more than one spatial coordinate can experience vibration. One example of this type of vibration is the transverse vibration of a thin membrane. As before, it is necessary to make some simplifying assumptions. They are:

1. Hooke's law applies.
2. There are no shear forces acting on the membrane.
3. There are no bending moments acting on the membrane.
4. The mass of the membrane is very small and is evenly distributed throughout the membrane.
5. The vibration amplitudes are very small.
6. The tension applied to the sides of the membrane remains constant and acts evenly throughout the membrane.
7. There is no energy dissipated in the membrane.

Figure 8.22 shows an incremental segment of a vibrating membrane. With regard to this figure, x and y are the two spatial coordinates, z(x,y,t) is the displacement coordinate in the transverse direction and t is time. Summing the forces in the z direction relative to the x direction yields:

$$\sum F_{z(x)} = T'dy\sin\theta\big|_{x+dx} - T'dy\sin\theta\big|_{x} \tag{8.159}$$

where T' is the force per unit length acting on the sides of the membrane. Expanding equation (8.159) using a Taylor series expansion and noting for θ very small:

$$\sin\theta \approx \frac{\partial z(x,y,t)}{\partial x} \tag{8.160}$$

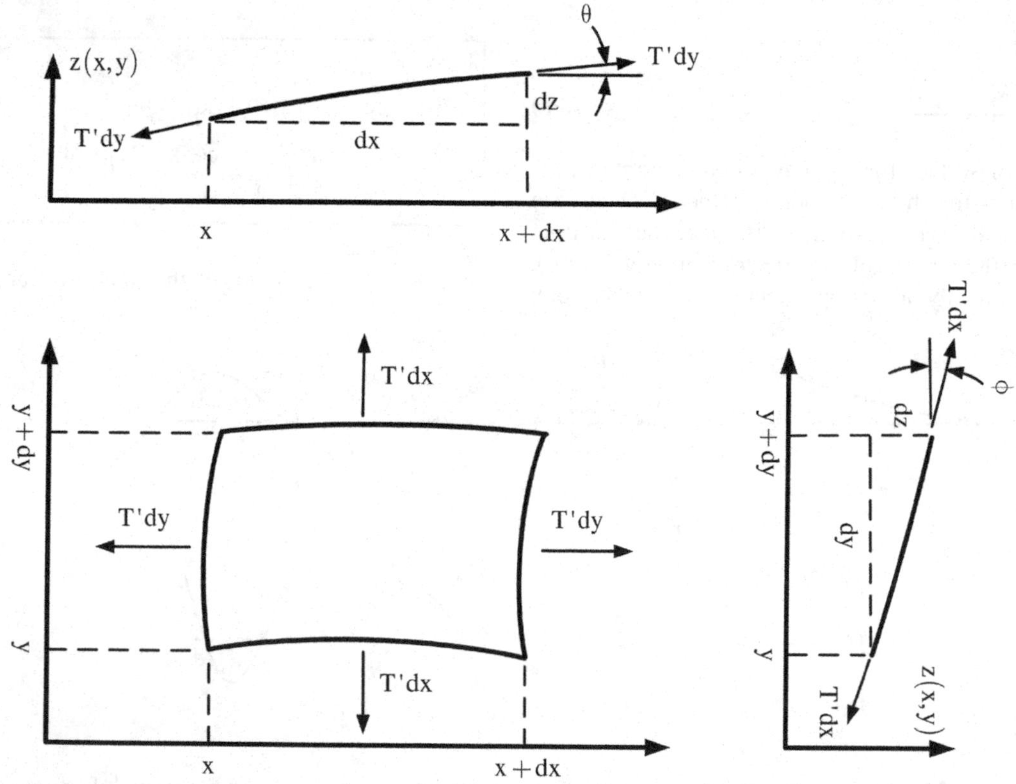

Figure 8.22 Forces Acting on an Incremental Element of a Thin Membrane

gives:

$$\sum F_{z(x)} = T'\frac{\partial^2 z(x,y,t)}{\partial x^2} dx\, dy. \tag{8.161}$$

Summing the forces in the z direction relative to the y direction yields:

$$\sum F_{z(y)} = T' dx \sin\phi\big|_{y+dy} - T' dx \sin\phi\big|_{y}. \tag{8.162}$$

Expanding equation (8.162) using a Taylor series expansion and noting for ϕ very small:

$$\sin\phi \approx \frac{\partial z(x,y,t)}{\partial y} \tag{8.163}$$

gives:

$$\sum F_{z(y)} = T'\frac{\partial^2 z(x,y,t)}{\partial y^2} dx\, dy. \tag{8.164}$$

The sum of the vertical forces relative to both the x and y directions equals the mass of the elemental segment of the membrane times its acceleration. Therefore:

$$m' dx\, dy \frac{\partial^2 z(x,y,t)}{\partial t^2} = T'\frac{\partial^2 z(x,y,t)}{\partial x^2} dx\, dy + T'\frac{\partial^2 z(x,y,t)}{\partial y^2} dx\, dy \tag{8.165}$$

or:

$$\boxed{\frac{\partial^2 z(x,y,t)}{\partial t^2} - c^2\left[\frac{\partial^2 z(x,y,t)}{\partial x^2} + \frac{\partial^2 z(x,y,t)}{\partial y^2}\right] = 0} \tag{8.166}$$

where *wave speed in the thin membrane* is:

$$\boxed{c = \sqrt{\frac{T'}{m'}}} \tag{8.167}$$

and m' is the mass per unit area of the membrane. Equation (8.166) can be written:

$$\frac{\partial^2 z(x,y,t)}{\partial t^2} - c^2 \nabla^2 z(x,y,t) = 0. \tag{8.168}$$

$\nabla^2(\ )$ is called the *Laplacian operator*. In Cartesian coordinates, it is written:

$$\boxed{\nabla^2 z(x,y,t) = \frac{\partial^2 z(x,y,t)}{\partial x^2} + \frac{\partial^2 z(x,y,t)}{\partial y^2}.} \tag{8.169}$$

Equation (8.168) is the *two-dimensional equation for transverse vibration of a thin membrane.*

8.11 Solution to the Equation for Transverse Vibration of a Thin Membrane

When finite boundary conditions exist for a thin membrane, the method of separation of variables can be used to solve equation (8.168) where:

$$z(x,y,t) = X(x)Y(y)T(t). \tag{8.170}$$

Substituting equation (8.170) into equation (8.168) and carrying out the prescribed operations yields:

$$c^2\left[\frac{1}{X(x)}\frac{d^2 X(x)}{dx^2} + \frac{1}{Y(y)}\frac{d^2 Y(y)}{dy^2}\right] = \frac{1}{T(t)}\frac{d^2 T(t)}{dt^2}. \tag{8.171}$$

Separating the time and spatial variables gives:

$$\frac{1}{T(t)}\frac{d^2 T(t)}{dt^2} = -\omega^2 \tag{8.172}$$

or:

$$\boxed{\frac{d^2 T(t)}{dt^2} + \omega^2 T(t) = 0} \tag{8.173}$$

and:

$$\boxed{\frac{1}{X(x)}\frac{d^2X(x)}{dx^2}+\frac{1}{Y(y)}\frac{d^2Y(y)}{dy^2}=-k^2.} \quad (8.174)$$

where:

$$\boxed{k=\frac{\omega}{c}.} \quad (8.175)$$

k is called the *wave number*. Separating equation (8.174) yields:

$$\frac{1}{X(x)}\frac{d^2X(x)}{dx^2}=-k_x^2 \quad (8.176)$$

or:

$$\boxed{\frac{d^2X(x)}{dx^2}+k_x^2X(x)=0} \quad (8.177)$$

and:

$$\frac{1}{Y(y)}\frac{d^2Y(y)}{dy^2}=-k_y^2. \quad (8.178)$$

or:

$$\boxed{\frac{d^2Y(y)}{dy^2}+k_y^2Y(y)=0} \quad (8.179)$$

where:

$$\boxed{k=\sqrt{k_x^2+k_y^2}.} \quad (8.180)$$

The solutions for equations (8.173), (8.177) and (8.179) can be written:

$$\boxed{T(t)=A\cos\omega t+B\sin\omega t} \quad (8.181)$$

$$\boxed{X(x)=C\cos k_xx+D\sin k_xx} \quad (8.182)$$

$$\boxed{Y(y)=E\cos k_yy+\sin F\sin k_yy.} \quad (8.183)$$

Noting z(x,y,t) = X(x) Y(y) T(t) yields:

$$\boxed{z(x,y,t)=\begin{pmatrix}A\cos\omega t+\\ B\sin\omega t\end{pmatrix}\begin{pmatrix}C\cos k_xx+\\ D\sin k_xx\end{pmatrix}\begin{pmatrix}E\cos k_yy+\\ F\sin k_yy\end{pmatrix}.} \quad (8.184)$$

A and B are determined from the two initial conditions; C and D are obtained from the two boundary conditions for the x direction; and E and F are specified by the two boundary conditions for the y direction. Equation (8.184) can be written:

$$\boxed{z(x,y,t)=\Psi(x,t)(A\cos\omega t+B\sin\omega t)} \quad (8.185)$$

where:

$$\boxed{\Psi(x,y)=\begin{pmatrix}C\cos k_xx+\\ D\sin k_xx\end{pmatrix}\begin{pmatrix}E\cos k_yy+\\ F\sin k_yy\end{pmatrix}.} \quad (8.186)$$

Ψ(x,y) is the *mode function* for standing vibration modes that can exist on a thin membrane.

EXAMPLE 8.9

Determine the frequencies and corresponding mode functions for the free transverse vibration modes that can exist on a membrane that is clamped on all four sides.

SOLUTION

The two boundary conditions in the x direction are:

$$z(x,y,t)\big|_{x=0}=0 \quad \text{and} \quad z(x,y,t)\big|_{x=a}=0. \quad (8.9a)$$

where a is the length of the membrane in the x direction. In terms of equation (8.182), the boundary conditions can be written X(0) = 0 and X(a) = 0. Applying the first boundary condition yields:

$$X(0)=0=C\cos(0)+D\sin(0) \quad \text{or} \quad C=0. \quad (8.9b)$$

Applying the second boundary condition gives:

$$X(a) = 0 = D\sin(k_x a) \quad \text{or} \quad \sin(k_x a) = 0. \tag{8.9c}$$

Therefore:

$$\boxed{k_{x(m)} = \frac{m\pi}{a} \quad \text{for} \quad m = 1, 2, 3, \cdots.} \tag{8.9d}$$

The two boundary conditions in the y direction are:

$$z(x,y,t)\big|_{y=0} = 0 \quad \text{and} \quad z(x,y,t)\big|_{y=b} = 0. \tag{8.9e}$$

where b is the length of the membrane in the y direction. In terms of equation (8.178), the boundary conditions can be written Y(0) = 0 and Y(b) = 0. Applying the first boundary condition gives:

$$Y(0) = 0 = E\cos(0) + F\sin(0) \quad \text{or} \quad E = 0. \tag{8.9f}$$

Applying the second boundary condition yields:

$$Y(b) = 0 = F\sin(k_y b) \quad \text{or} \quad \sin(k_y b) = 0. \tag{8.9g}$$

Therefore:

$$\boxed{k_{y(n)} = \frac{n\pi}{b} \quad \text{for} \quad n = 1, 2, 3, \cdots.} \tag{8.9h}$$

Substituting equations (8.9d) and (8.9h) into equations (8.180) and noting $\omega = k c$ yields:

$$\boxed{k_{mn} = \sqrt{\left(\frac{m\pi}{a}\right)^2 + \left(\frac{n\pi}{b}\right)^2}} \tag{8.9i}$$

and:

$$\boxed{\omega_{mn} = c\sqrt{\left(\frac{m\pi}{a}\right)^2 + \left(\frac{n\pi}{b}\right)^2}.} \tag{8.9j}$$

The mode function for this problem is:

$$\Psi_{mn}(x,y) = D_{mn}\sin(k_{x(m)}x)\sin(k_{y(n)}y) \tag{8.9k}$$

or:

$$\boxed{\Psi_{mn}(x,y) = D_{mn}\sin\left(\frac{m\pi}{a}x\right)\sin\left(\frac{n\pi}{b}y\right).} \tag{8.9l}$$

Figure 8.23 shows plots of the mode shapes for the (m = 1 n = 1), (m = 2 n = 1, (m = 1 n = 2) and (m = 2 n = 2) transverse vibration modes that can exist on the membrane. The corresponding frequencies at which these mn vibration modes exist are given by equation (8.9j). The complete response for transverse vibration of the membrane is given by:

$$\boxed{y(x,y,t) = \sum_{m=1}^{\infty}\sum_{n=1}^{\infty}\begin{pmatrix} A_{mn}\cos\omega_{mn}t + \\ B_{mn}\sin\omega_{mn}t \end{pmatrix}\sin(k_{x(m)}x)\sin(k_{y(n)}y).} \tag{8.9m}$$

The constants A_{mn} and B_{mn} are obtained by applying the initial conditions.

Equation (8.168) also describes the vibration characteristics of a thin circular membrane where the Laplacian operator for cylindrical coordinates (Figure 8.24) is given by:

$$\boxed{\nabla^2 z(r,\theta,t) = \frac{1}{r}\frac{\partial}{\partial r}\left[r\frac{\partial z(r,\theta,t)}{\partial r}\right] + \frac{1}{r^2}\frac{\partial^2 z(r,\theta,t)}{\partial\theta^2}.} \tag{8.187}$$

z(r,θ,t) can be written:

$$\boxed{z(r,\theta,t) = F(r)G(\theta)T(t).} \tag{8.188}$$

Substituting equation (8.188) into equation (8.168) and carrying out the prescribed operations yield:

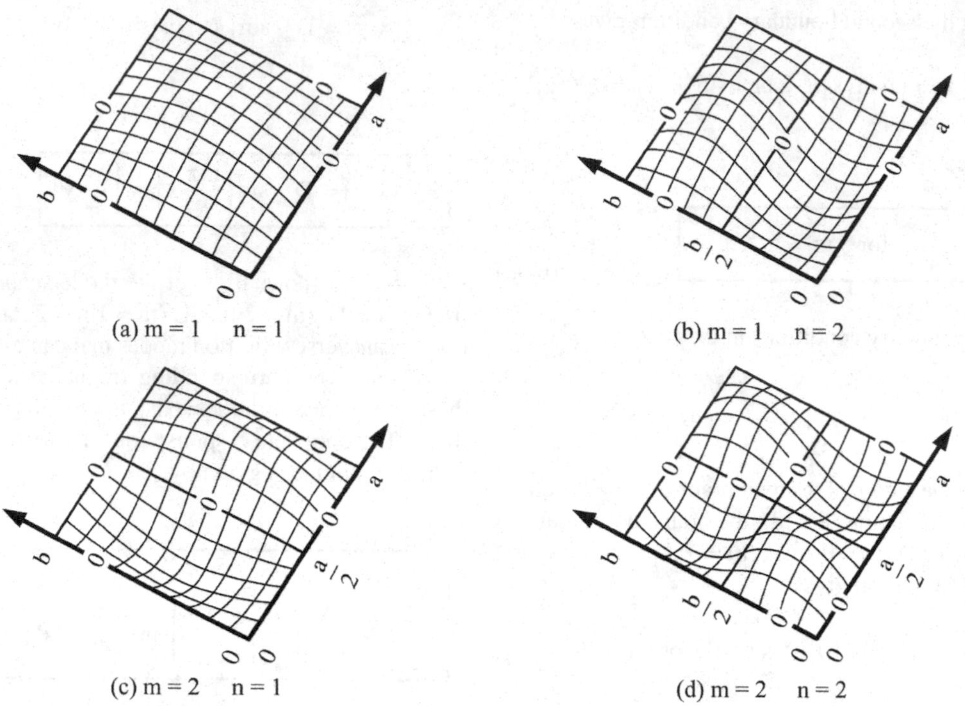

Figure 8.23 Mode Shapes of the Indicated Transverse Vibration Modes that Can Exist on a Thin Membrane that Is Clamped on All Four Sides

$$c^2 \left\{ \frac{1}{r\,F(r)} \frac{d}{dr}\left[r \frac{d\,F(r)}{\partial r} \right] + \frac{1}{r^2\,G(\theta)} \frac{d^2\,G(\theta)}{d\theta^2} \right\} = \frac{1}{T(t)} \frac{d^2\,T[t]}{dt^2}. \tag{8.189}$$

Separating the time and spatial variables gives:

$$\boxed{\frac{d^2\,T[t]}{dt^2} + \omega^2\,T(t) = 0} \tag{8.190}$$

and:

$$\frac{1}{r\,F(r)} \frac{d}{\partial r}\left[r \frac{d\,F(r)}{dr} \right] + \frac{1}{r^2\,G(\theta)} \frac{d^2\,G(\theta)}{d\theta^2} + k^2 = 0. \tag{8.191}$$

Equation (8.191) can be separated into two separate equations:

$$\frac{r}{F(r)} \frac{d}{dr}\left[r \frac{d\,F(r)}{dr} \right] + k^2 r^2 = -\frac{1}{G(\theta)} \frac{d^2\,G(\theta)}{d\theta^2} = m^2. \tag{8.192}$$

The two ordinary differential equations that result from equation (8.192) are:

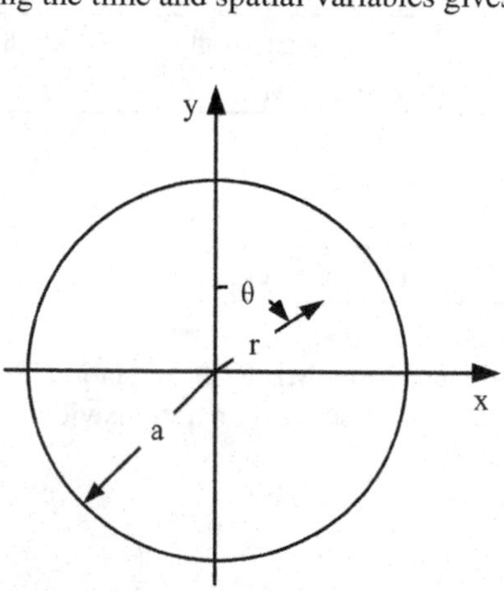

Figure 8.24 Cylindrical Coordinates

$$\frac{d^2 G(\theta)}{d\theta^2} + m^2 G(\theta) = 0 \quad (8.193)$$

and:

$$r\frac{d}{dr}\left[r\frac{dF(r)}{dr}\right] + (k^2r^2 - m^2)F(r) = 0. \quad (8.194)$$

The solution to equation (8.190) is equation (8.181). No definite boundary conditions are specified for the θ direction. However, there is a periodicity requirement such that:

$$G(\theta = 0) = G(\theta = 2\pi). \quad (8.195)$$

This results in a solution to equation (8.193) of the form:

$$G(\theta) = A_\theta \cos(m\theta) + B_\theta \sin(m\theta). \quad (8.196)$$

Rearranging equation (8.194) yields:

$$r^2 \frac{d^2 F(r)}{dr^2} + r\frac{dF(r)}{dr} + (k^2r^2 - m^2)F(r) = 0. \quad (8.197)$$

Equation (8.197) is *Bessel's equation of order m.* Its solution is given by:

$$F(r) = A_r J_m(kr) + B_r Y_m(kr). \quad (8.198)$$

$J_m(kr)$ is called the *Bessel function of the first kind of order m*, and $Y_m(kr)$ is called the *Bessel function of the second kind of order m*. Substituting equations (8.181), (8.196) and (8.198) into equation (8.188) yields:

$$z(r,\theta,t) = \begin{pmatrix} A\cos\omega t + \\ B\sin\omega t \end{pmatrix} \times \begin{bmatrix} A_\theta \cos(m\theta) + \\ B_\theta \sin(m\theta) \end{bmatrix} \times \left[A_r J_m(kr) + B_r Y_m(kr)\right]. \quad (2.199)$$

EXAMPLE 8.10

Determine the frequencies and corresponding mode functions of the free transverse vibration modes that can exist on a circular membrane of radius a that is clamped along its outer edge.

SOLUTION

The condition specified by equation (8.196) can be met by either a sine or cosine function for a circular membrane. For this example let:

$$G(\theta) = A_\theta \cos(m\theta). \quad (8.10a)$$

The motion of the membrane must be bounded at its center. Since $Y_m(kr)$ is always unbounded at r = 0, B_r = 0. Therefore, equation (8.198) becomes:

$$F(r) = A_r J_m(kr). \quad (8.10b)$$

The motion of the membrane at r = a is zero. This boundary condition is satisfied when:

$$J_m(\beta_{mn}) = 0 \quad \text{where} \quad \beta_{mn} = k_{mn}a. \quad (8.10c)$$

Table 8.1 gives specified values of β_{mn} for which the

Table 8.1 Values of β_{mn} for Which Equation 10c is Satisfied

	β_{mn}					
m \ n	1	2	3	4	5	6
0	2.40482	5.52007	8.65327	11.79153	14.93091	18.07106
1	3.83171	7.01559	10.17347	13.32369	16.47063	19.61596
2	5.13562	8.41724	11.61984	14.79595	17.95982	21.11700
3	6.38016	9.76102	13.01520	16.22347	19.40942	22.58273
4	7.58834	11.06471	14.37254	17.61597	20.82693	24.01902
5	8.77145	12.23860	15.70017	18.98013	22.21780	25.43034
6	9.93611	13.58929	17.00382	20.32079	23.58608	26.82015

equation (8.10c) is satisfied. The frequency for the mn mode is:

$$\boxed{\omega_{mn} = \frac{\beta_{mn}\, c}{a}.} \tag{8.10d}$$

The mode functions for the transverse vibration of a circular membrane of radius a that is clamped along its outer edge can be expressed:

$$\boxed{\Psi_{mn}(r,\theta) = A_{mn}\cos(m\theta)\, J_m\left(\beta_{mn}\frac{r}{a}\right).} \tag{8.10e}$$

Figure 8.25 shows a plot of some typical free transverse vibration modes that can exist on circular membrane of radius a that is clamped along its outer edge. The complete response for transverse vibration of the circular membrane is:

$$\boxed{z(r,\theta,t) = \sum_{m=1}^{\infty}\sum_{n=1}^{\infty}\begin{pmatrix} A_{mn}\cos\omega_{mn} t + \\ B_{mn}\sin\omega_{mn} t \end{pmatrix}\cos(m\theta)\, J_m\left(\beta_{mn}\frac{r}{a}\right).} \tag{8.10f}$$

The constants A_{mn} and B_{mn} are obtained by applying the initial conditions.

8.12 Transverse Vibration of Thin Plates

The equation for the transverse vibration of a thin plate can be developed in a manner similar that used in Section 8.8 for the transverse vibration of a bar. However, this development would be much more complicated. With some effort, the equation for transverse vibration of a plate can be shown to be:

$$\boxed{\nabla^4 w(x,y,t) + h\left(\frac{1-\nu^2}{EI}\right)\rho\frac{\partial^2 w(x,y,t)}{\partial t^2} = 0} \tag{8.200}$$

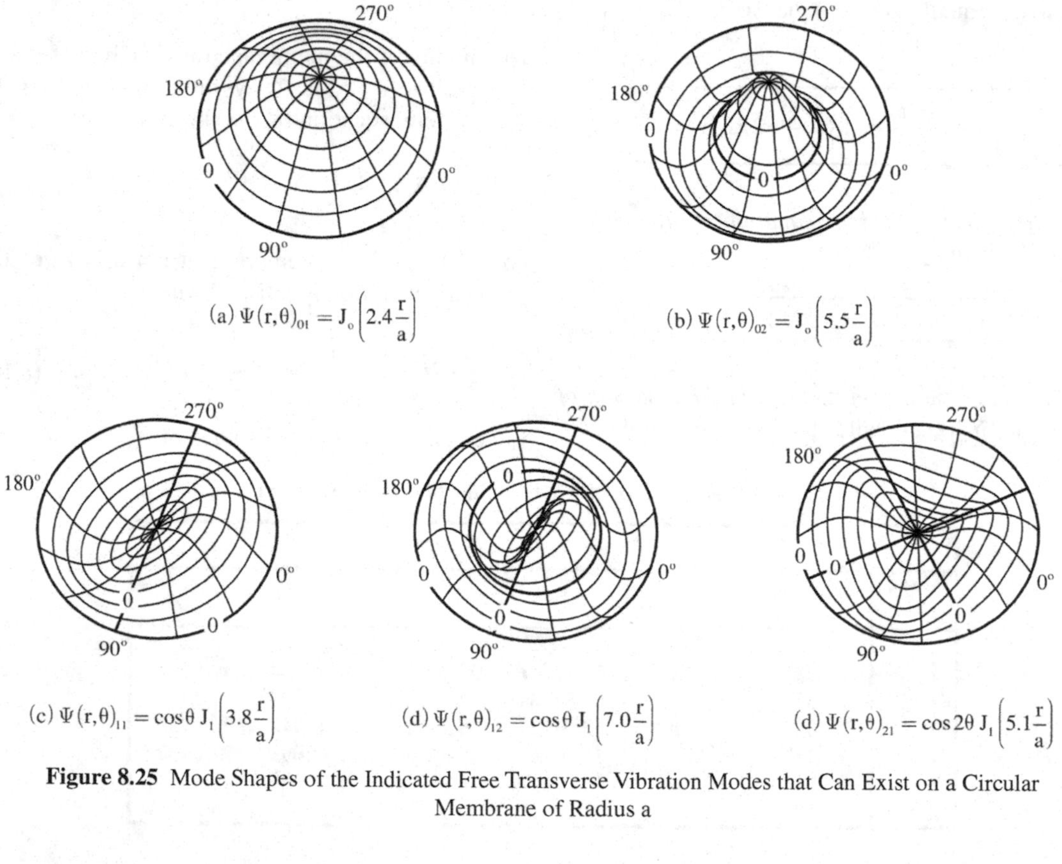

(a) $\Psi(r,\theta)_{01} = J_0\left(2.4\frac{r}{a}\right)$

(b) $\Psi(r,\theta)_{02} = J_0\left(5.5\frac{r}{a}\right)$

(c) $\Psi(r,\theta)_{11} = \cos\theta\, J_1\left(3.8\frac{r}{a}\right)$

(d) $\Psi(r,\theta)_{12} = \cos\theta\, J_1\left(7.0\frac{r}{a}\right)$

(d) $\Psi(r,\theta)_{21} = \cos 2\theta\, J_1\left(5.1\frac{r}{a}\right)$

Figure 8.25 Mode Shapes of the Indicated Free Transverse Vibration Modes that Can Exist on a Circular Membrane of Radius a where for Cartesian coordinates:

$$\nabla^4 w(x,y,t) = \frac{\partial^4 w(x,y,t)}{\partial x^4} + 2\frac{\partial^4 w(x,y,t)}{\partial x^2\,\partial y^2} + \frac{\partial^4 w(x,y,t)}{\partial y^4}. \tag{8.201}$$

h is the plate thickness, ν is Poisson's ratio, E is Young's modulus, I is the area moment of the plate and ρ is the mass density of the plate. Noting the longitudinal wave speed c_l for a plate is given by:

$$c_l = \sqrt{\frac{E}{\rho(1-\nu^2)}} \tag{8.202}$$

and the radius of gyration κ is given by:

$$\kappa^2 = \frac{I}{h} \tag{8.203}$$

equation (8.200) becomes:

$$c_l^2\,\kappa^2\,\nabla^4 w(x,y,t) + \frac{\partial^2 w(x,y,t)}{\partial t^2} = 0. \tag{8.204}$$

Equation (8.204) contains fourth order derivatives with respect to both x and y. Therefore, eight boundary conditions are needed, four for the x direction and four for the y direction, to solve the equation. Boundary conditions similar to those described in Section 8.8 for the transverse vibration of a beam also apply for the transverse vibration of thin plates.

If a plate is simply supported along its four edges, it can be shown that for the mnth mode:

$$w_{mn}(x,y,t) = A_{mn}\sin(k_m x)\sin(k_m y)\cos(\omega_{mn} t + \phi_{mn}) \tag{8.205}$$

where:

$$k_m = \frac{m\pi}{l_x} \quad \text{and} \quad k_n = \frac{n\pi}{l_y} \tag{8.206}$$

for $m = 1, 2, 3, \cdots$ and $n = 1, 2, 3, \cdots$. l_x and l_y are the lengths of the plate in the x and y directions, respectively. Substituting equation (8.205) into equation (8.203) and carrying out the prescribed operations yields:

$$\left(k_m^4 + 2k_m^2\,k_n^2 + k_n^2\right) - \frac{\omega_{mn}^2}{c_l^2\,\kappa^2} = 0 \tag{8.207}$$

or:

$$\omega_{mn} = c_l\,\kappa\left(k_m^2 + k_n^2\right). \tag{8.208}$$

Noting:

$$k_{b(mn)}^2 = k_m^2 + k_n^2 \tag{8.209}$$

where $k_{b(mn)}$ is the bending wave number for the mnth mode, equation (8.208) can be written:

$$\omega_{mn} = c_l\,\kappa\,k_{mn}^2. \tag{8.210}$$

Noting:

$$k_{mn} = \frac{\omega_{mn}}{c_{b(mn)}} \tag{8.211}$$

$c_{b(mn)}$ can be written:

$$c_{b(mn)} = \sqrt{c_l\,\kappa\,\omega_{mn}}. \tag{8.212}$$

Like the bending wave speed for a beam, the bending wave speed of a plate associated with the mnth mode is a function of the frequency for the mnth mode. Therefore, the free bending waves that exist on a plate are dispersive waves.

The total response for the transverse vibration of a thin plate that is simply supported along its four edges is:

$$w(x,y,t) = \sum_{m=1}^{\infty}\sum_{n}^{\infty} A_{mn}\sin(k_m x)\sin(k_n y)\cos(\omega_{mn}t+\phi_{mn}). \quad (8.213)$$

The constants A_{mn} and ϕ_{mn} are obtained by applying the initial conditions.

8.13 Important Symbols in Chapter 8

Parameter	Symbol	English Units	SI Units
Mass	M	lb_f-sec^2/in.	kg
Mass/unit length	m	lb_f-sec^2/$in.^2$	kg/m
Mass densigy	ρ	lb_f-sec^2/$in.^4$	kg/m^3
Young's modulus	E	lb_f/$in.^2$	N/m^2
Poisson's ratio	ν	dimensionless	dimensionless
Area moment	I	$in.^4$	m^4
Radius of gyration	κ	in.	m
Wave speed	c	in./sec	m/s
Longitudinal wave speed of bar or plate	c_l	in./sec	m/s
Bending wave speed	c_b	in./sec	m/s
Wave number	k	1/in.	1/m
Bending wave number	k_b	1/in.	1/m
Frequency	ω, f	rad/sec, Hz	rad/s, Hz
Wavelength	λ	in.	m
Transverse displacement of string or bar	y(x,t)	in.	m
Longitudinal displacement of a bar	ζ(x,t)	in.	m
Transverse displacement of a thin membrane	z(x,y,t)	in.	m
Transverse displacement of a plate	w(x,y,t)	in.	m
Force	F	lb_f	N
Moment	M	lb_f-in.	N•m
Stress	σ	lb_f/$in.^2$	N/m^2
Strain	ε	in./in.	m/m
Mechanical impedance	Z_m	lb_f-sec/in.	N•s / m
Characteristic mechanical impedance	Z_c	lb_f-sec/in.	N•s / m
Power	Π	lb_f-in./sec	watts

PROBLEMS - CHAPTER 8

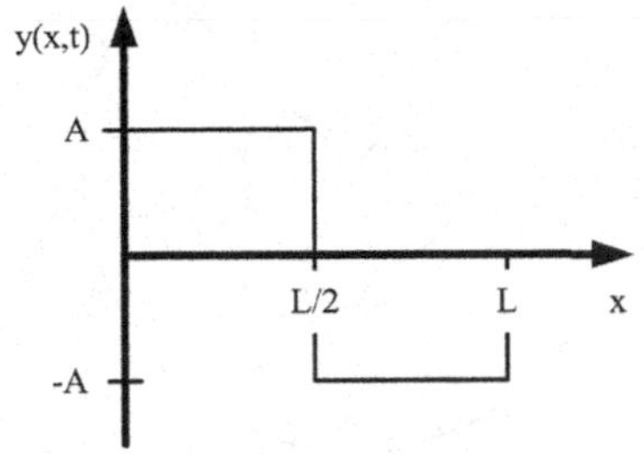

Figure P8.1

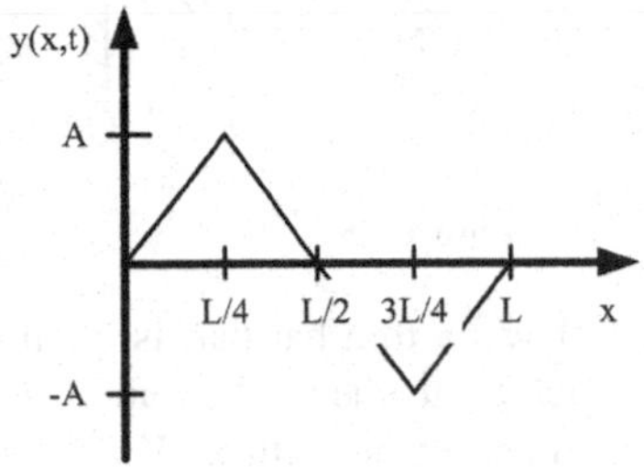

Figure P8.2

1. A string with clamped ends at x = 0 and x = L is initally displaced as is shown in Figure P8.1. The expression for the initial displacement is:

$$y(x,0) = A \quad \text{for} \quad 0 \le x < \frac{L}{2}$$
$$= -A \quad \text{for} \quad \frac{L}{2} \le x \le L.$$

At time t = 0, the string is released.
(a) Determine the expressions for the mode functions and the resonant frequencies.
(b) Which of the normal modes are missing from the series solution for the problem? That is, which modes do not participate in the motion of the string? Can you explain why?
(c) Write the complete series solution for the problem.

2. A string with clamped ends at x = 0 and x = L is initially displaced as is shown in figure P8.2. At time t = 0, the string is released.
(a) Determine the expressions for the mode functions and the resonant frequencies.
(b) Which of the normal modes are missing from the series solution for the problem? That is, which modes do not participate in the motion of the string? Can you explain why?
(c) Write the complete series solution for the problem.

3. A stretched string that is clamped at x = 0 and x = L is struck such that the initial velocity of the string from x = 0 to x = L/4 and from x = 3L/4 to x = L is zero. From x = L/4 to x = 3L/4 the velocity is given by:

$$\dot{y}(x,0) = 4v_o\left(x - \frac{L}{4}\right) \quad \text{for} \quad \frac{L}{4} \le x < \frac{L}{2}$$
$$= 4v_o\left(\frac{3L}{4} - x\right) \quad \text{for} \quad \frac{L}{2} \le x \le \frac{3L}{4}.$$

(a) Determine the expressions for the mode functions and the resonant frequencies.(b) Write the complete series solution for the problem.

4. A steel wire that has a mass of 0.015 kg and a length of 1.5 m is stretched to a tension of 15 N. What is the fundamental resonance frequency of vibration for the wire?

5. A steel bar that is clamped at x = 0 and free at x = L has the stress state that is given by:

$$\sigma(x,0) = A \quad \text{for} \quad 0 \le x < \frac{L}{2}$$
$$= 0 \quad \text{for} \quad \frac{L}{2} \le x \le L.$$

(a) Determine the expressions for the frequencies and mode functions for the longitudinal vibration modes that can exist on the bar.
(b) Write the complete series solution for the problem.

6. A steel bar is clamped at x = 0 and x = L. Determine the expressions for the frequencies and mode functions for the longitudinal vibration modes that can exist on the bar.

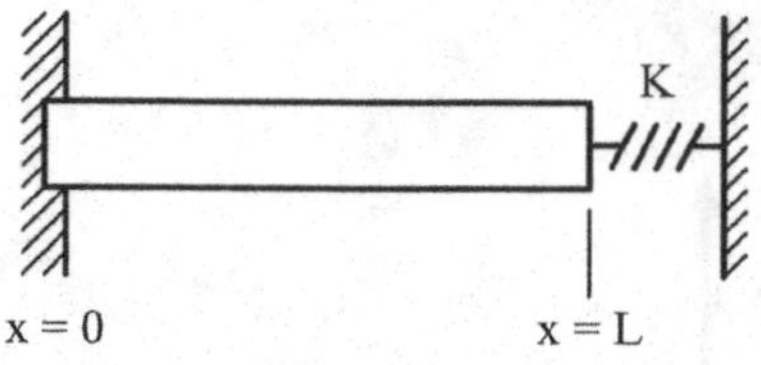

Figure P8.3

7. Figure P8.3 shows a thin bar that is clamped at x = 0 and which is attached to "ground" at x = L by means of a spring of stiffness K. Determine the frequencies and mode functions of the free longitudinal vibration modes that can exist on the bar.

8. A thin bar of length L is driven by a longitudinal force f(t) = F $e^{j\omega t}$ at x = 0 and is free at x = L.
(a) Determine the expression that describes the longitudinal displacement of the bar.
(b) Write the expression for the input mechancial impedance of the bar.

9. The beam shown in Figure 8.11 is subjected to transverse vibration. Determine the frequencies and mode functions of the free bending vibration modes that can exist on the bar.

10. A steel beam is clamped at x = 0 and x = L. Determine the expressions for the mode functions and the resonance frequencies for free transverse vibration that can exist on the bar.

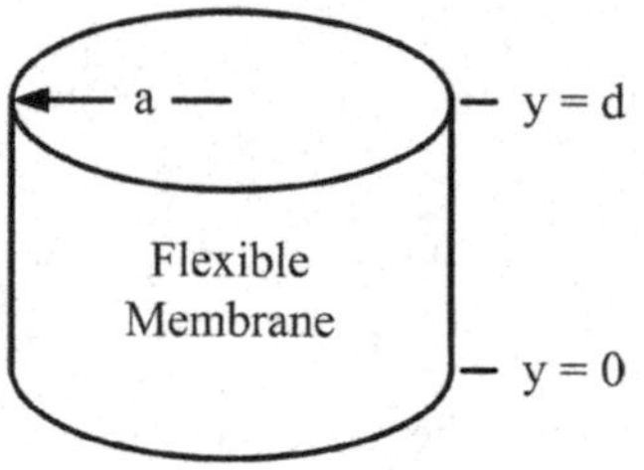

Figure P8.4

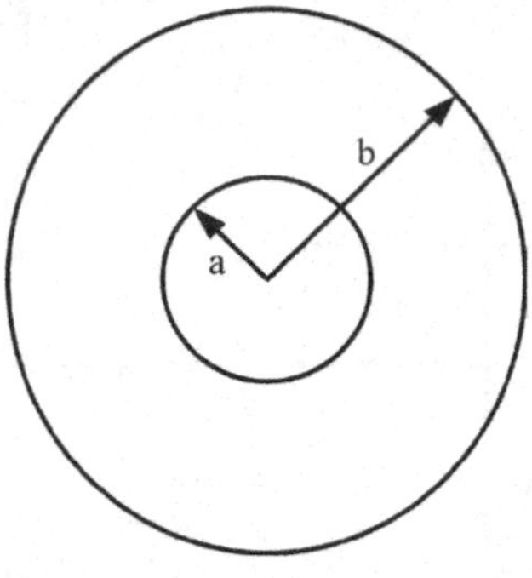

Figure P8.5

11. A flexible membrane is stretched between two circular hoops as is indicated in Figure P8.4. Determine the resonant frequencies of this membrane. Hint: Uncurl the membrane and lay it out flat. It is then obvious that the boundary conditions are:

$$z(x,y,t)=0 \quad \text{at} \quad y=0 \quad \text{and} \quad y=d$$

$$z(0,y,t)=z(2\pi a,y,t).$$

12. Given a square membrane and a circular membrane of the same surface area and material and under the same tension, which membrane has the lower resonance frequency? What is the percent difference between the lowest resonance frequency associated with each membrane?

14. Determine the equation that specifies the frequencies of an annular membrane of inner radius a and outer radius b (Figure P8.5).

CHAPTER 9
VIBRATION MEASUREMENTS

9.1 Introduction

Reasons why vibration measurements are made include:

- *Machinery vibration monitoring/analysis* - rotor dynamics and balancing and machine vibration monitor related to preventative maintenance;
- *Environmental monitoring and control* - parts quality and defects testing, mechanical system vibration control, and duplicating environmental shock and vibration environments in laboratory testing;
- *Structural testing* - structural resonance identification, experimental modal analyses, and structural vibration animation;
- *Seismic vibration* - full-scale building, bridge, etc. shock and vibration testing;
- *Shock, ride quality and motion control* - active vibration control, crash testing of vehicles, and mechanical system vibration reduction and noise control;
- *Flight shock and vibration testing* - shock and vibration testing of aircraft and rocket components and systems.

Figure 9.1 shows a schematic of a basic vibration measurement setup. First, is the structure, machine or system component on which vibration measurements are made. The vibration transducer, which most often is an accelerometer, is attached to the structure, machine or system component at an appropriate point. The signal from the accelerometer is passed through a signal conditioning amplifier. If the accelerometer is a charge-type accelerometer, the signal conditioning amplifier is a charge amplifier that converts the charge signal from the accelerometer to a corresponding voltage signal proportional to the acceleration amplitude. If the accelerometer is an IEPE (Integrated Electronics Piezo Electric) accelerometer, the signal conditioning amplifier is an IEPE power supply. The power supply is a constant-current regulated DC voltage source that supplies power to the accelerometer and receive the voltage signal proportional to acceleration amplitude from the accelerometer. The accelerometer voltage signal is transferred to a frequency analyzer. The frequency analyzer processes the acceleration signal and presents the related acceleration amplitude from the structure, machine or system component as a function of frequency.

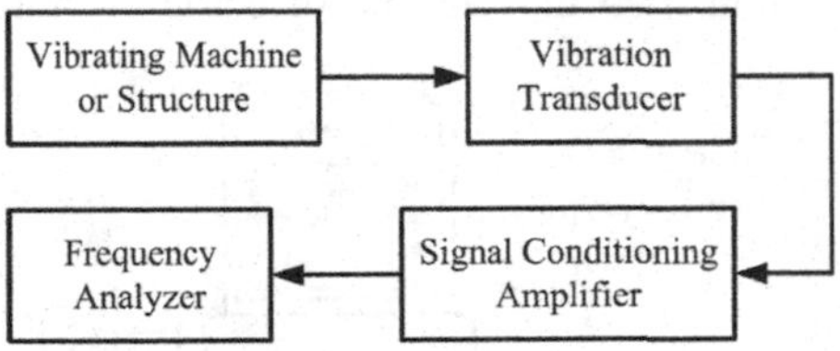

Figure 9.1 Basic Vibration Measurement Setup

9.2 Vibration Transducers

The primary device for measuring acceleration is the accelerometer. There are three basic types of accelerometers: stain gauge, piezo-resistive and piezoelectric. In *strain gauge accelerometers*, the seismic mass force associated with acceleration causes a deflection of the beams that support the mass (Figure 9.2). This deflection results in a change in resistance in the silicon or foil strain gauge elements attached to the beams. This change in resistance is typically sensed by a wheatstone bridge (Figure 9.3). The beam deflection is proportional to the seismic mass acceleration, and the wheatstone bridge outputs a voltage proportional to the acceleration.

A *piezo-resistive accelerometer* has a seismic mass placed on top of a piezo-resistive substrate (Figure 9.4). The force generated by the acceleration of the seismic mass changes the resistance of the piezo-resistive substrate. This change is sensed by a wheatstone

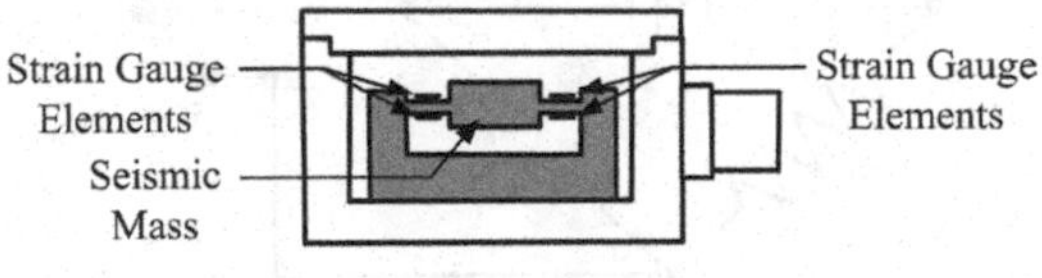

Figure 9.2 Strain Gauge Accelerometer

This chapter was written in collaboration with Thomas Lagö, Ph.D.

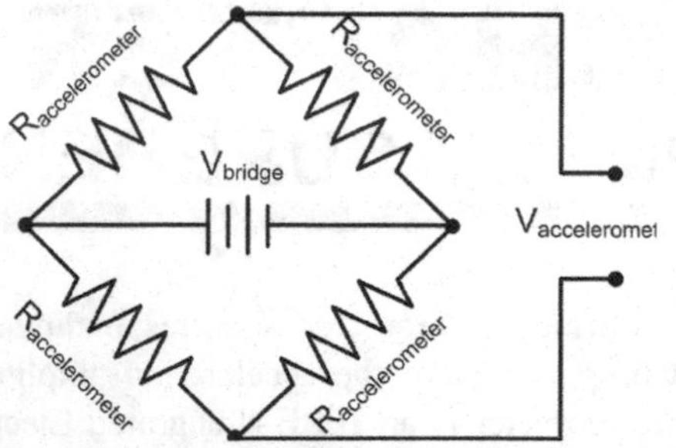

Figure 9.3 Wheatsone Bridge for Strain Guage Acclerometer

bridge that outputs a voltage signal proportional to the acceleration of the seismic mass (Figure 9.5). Both strain gauge and piezo-resistive accelerometers can measure acceleration frequencies down to zero Hz.

Piezoelectric accelerometers are generally the accelerometer of choice for most vibration measurements. The basic elements of a piezoelectric accelerometer are a piezoelectric crystal that supports a seismic mass (Figure 9.6). Piezoelectric crystals are naturally occurring quartz crystals or man-made polycrystalline ceramic crystals that produce a charge output when they are compressed, flexed or subjected to shear forces. When the seismic mass experiences an acceleration, it generates an inertia force that compresses and flexes the piezoelectric crystal. This compressing and flexing

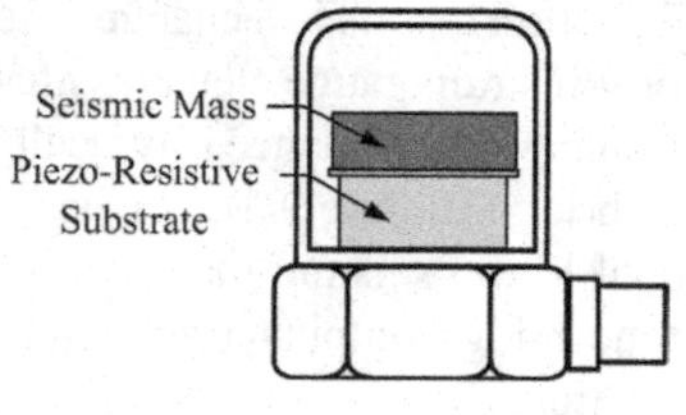

Figure 9.4 Piezo-resistive Acclerometer

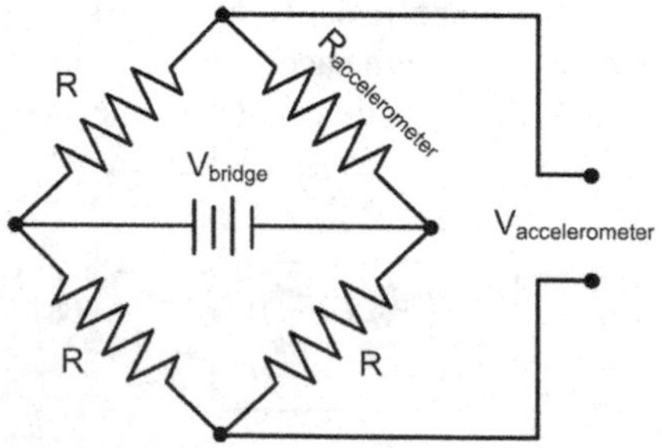

Figure 9.5 Wheatsone Bridge for Piezo-resistive Acclerometer

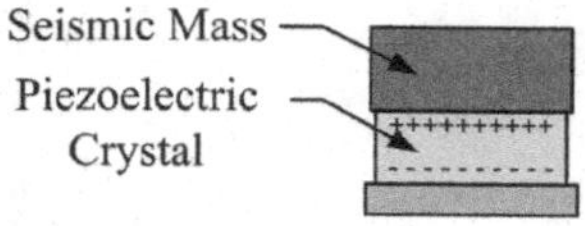

Figure 9.6 BasicElementsofaPiezo-electricaccelerometner

generates a charge in the crystal that is proportional to the inertia force of the mass. The weight of the seismic mass is constant. Therefore, the charge generated in the piezoelectric crystal is also proportional to the acceleration of the mass. The charge output of the crystal (measured in terms of Pico-coulombs/g) can be directed to a charge amplifier that converts the charge signal to a corresponding voltage signal proportional to acceleration. It can also be directly converted by internal electronics in the accelerometer to a voltage signal that is proportional to acceleration. There are four basic types of piezoelectric accelerometers. They are:

1. Single-ended compression accelerometer,
2. Isolated compression accelerometer,
3. Shear accelerometer, and
4. Flexural accelerometer.

Single-ended compression accelerometer: The piezoelectric crystal in a single-ended compression accelerometer is mounted directly to the base of the accelerometer (Figure 9.7). The seismic mass is attached to the crystal by a setscrew, bolt or fastener. These accelerometers contain few parts, are easy to fabricate and have high mass-crystal resonance frequencies. However, they have a very high thermal transient sensitivity and they have a high base strain sensitivity. Base strain occurs when the accelerometer base is subjected to bending or torque or to distortion caused either by mechanical movement or thermal stresses in the structure to which the accelerometer is attached (Figure 9.8). Resulting strain waves then travel unimpeded into the piezoelectric crystal. This

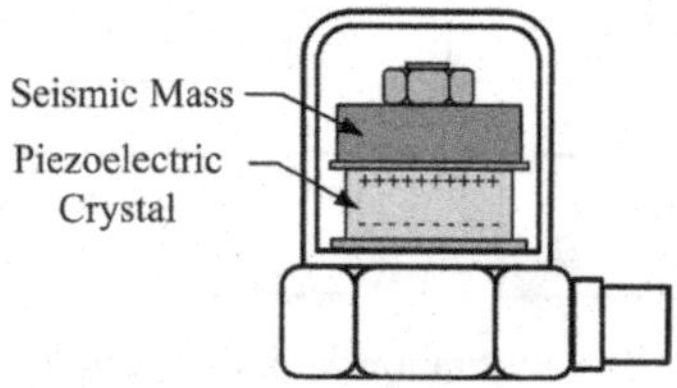

Figure 9.7 Single-ended Compression Accelerometer

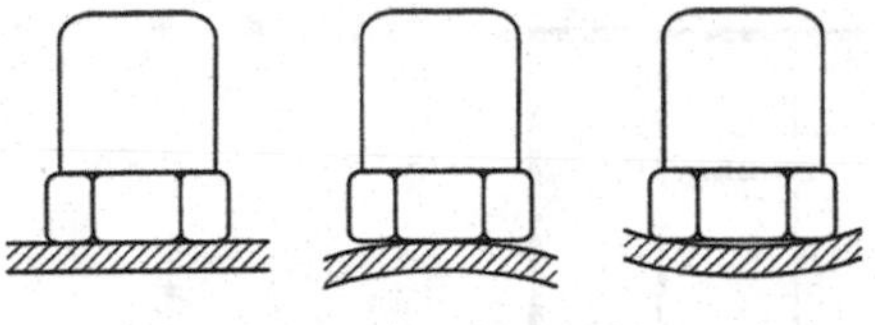

Figure 9.8 Base Strain Caused by Bending of the Structure to which an Accelerometer is Attached

signal is superimposed onto the charge signal created by the seismic mass acceleration. These two charge signals cannot be separated, resulting in a contaminated acceleration signal from the accelerometer.

Isolated compression accelerometer: Two methods can be used to minimize the effects of base strain. First, the piezoelectric crystal can be mounted on an isolation washer between the crystal and the accelerometer base as shown in Figure 9.9 (a). Second, the piezoelectric crystal can be mounted on a tapered isolation base, as shown in Figure 9.9 (b). Both methods isolate the crystal from strain waves associated with base strain.

Shear accelerometer: The seismic mass in a shear accelerometer is oriented so that the piezoelectric crystal experiences a shear force, rather than a compression-tensile force (Figure 9.10). Shear accelerometers come in small sizes. They have very low base strain sensitivity, which make them excellent for use on flexible structures. Strain waves associated with base strain are shunted at the post that supports the piezoelectric crystal and seismic mass. They also have low thermal transient sensitivity,

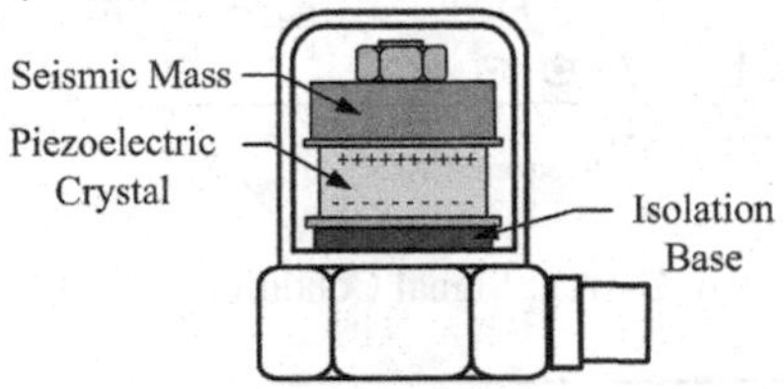

(a) Washer Isolation Base

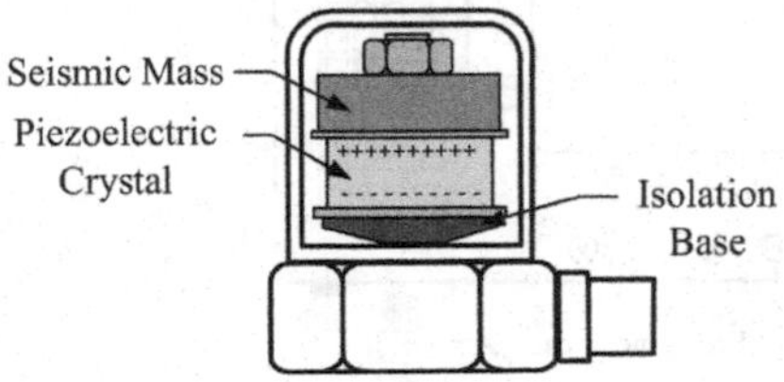

(b) Tappered Isolation Base

Figure 9.9 Isolation Compression Accelerometers

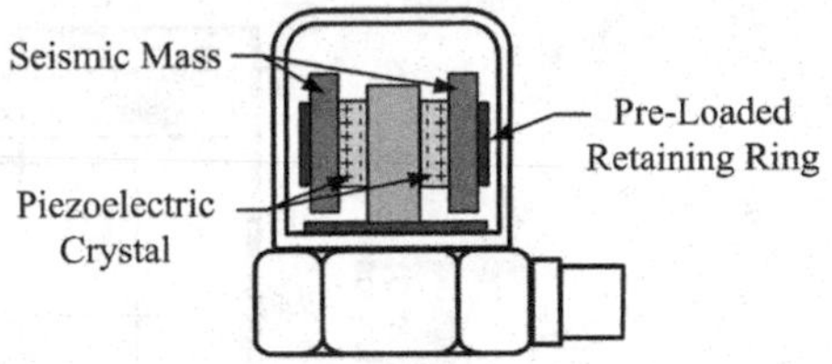

Figure 9.10 Shear Accelerometer

which allows for low frequency operation in thermally unstable environments.

Flexural accelerometer: The seismic mass in a flexural accelerometer is centered over a fulcrum as shown in Figure 9.11. The piezoelectric crystal is attached to the top of the seismic mass. When the mass experiences an acceleration, it flexes, causing the piezoelectic crystal to flex, which creates a charge that is proportional to acceleration. Flexural accelerometers have few parts and are inexpensive to make. They come in small sizes and have low profiles. They have low thermal transient sensitivities and low base strain sensitivities. Strain waves associated with base strain are shunted by the fulcrum that supports the piezoelectric crystal and seismic mass. The mass/crystal in the accelerometer has a low resonance frequency, which limits the high frequency range of the accelerometer.

Figure 9.12 shows the amplitude and frequency ranges associated with strain gauge, piezo-resistive and piezoelectric accelerometers.

9.3 Piezoelectric Accelerometer Characteristics

Piezoelectric accelerometers generate a charge signal when the piezoelectric crystal in the accelerometer is loaded in compression/tension or shear or is flexurally loaded by a seismic mass attached to the crystal that experiences acceleration. Electronically, there are two basic types of piezoelectric accelerometers: (1) charge accelerometer and (2) Integrated Electronics Piezo Electric (IEPE) accelerometer. Integrated

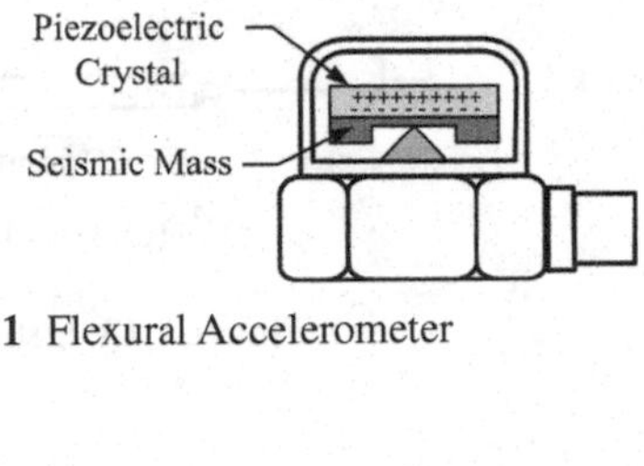

Figure 9.11 Flexural Accelerometer

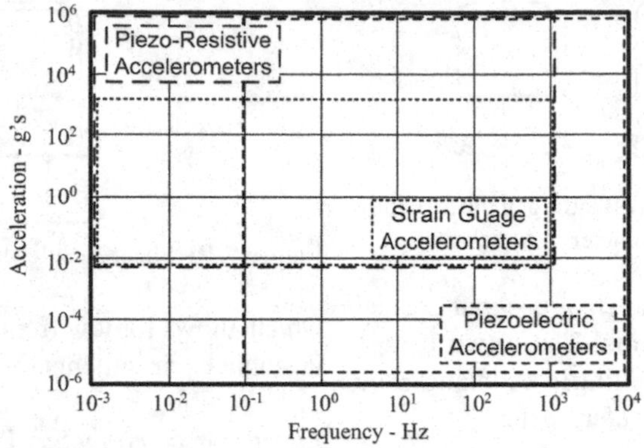

Figure 9.12 Frequency and Amplitude Ranges Associated with Strain Gauge, Piezo-resistive and Piezoelectric Accelerometers

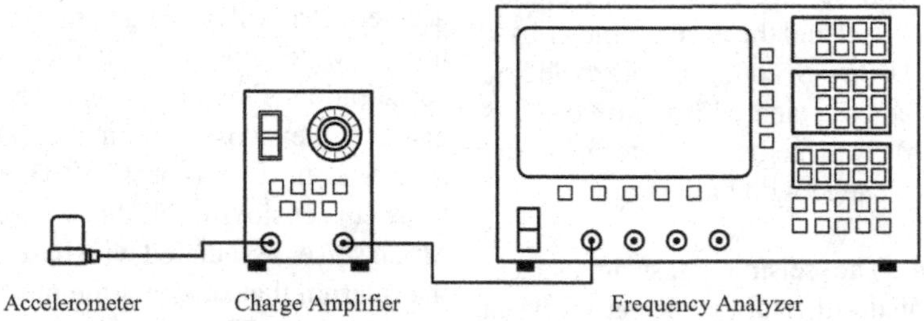

(a) Accelerometer with a Charge Amplifier

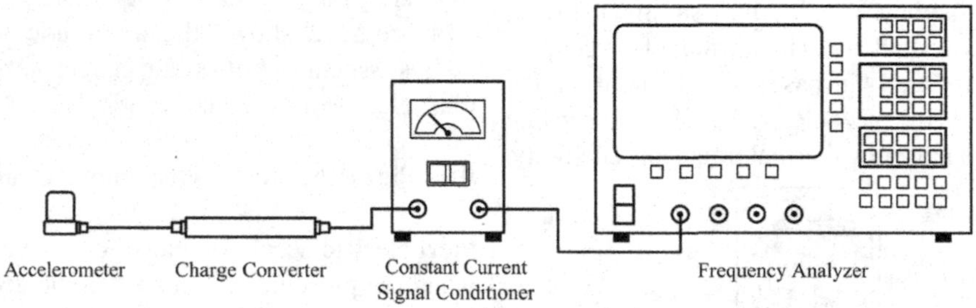

(b) Accelerometer with a Charge Converter and Constant Current Signal Conditioner

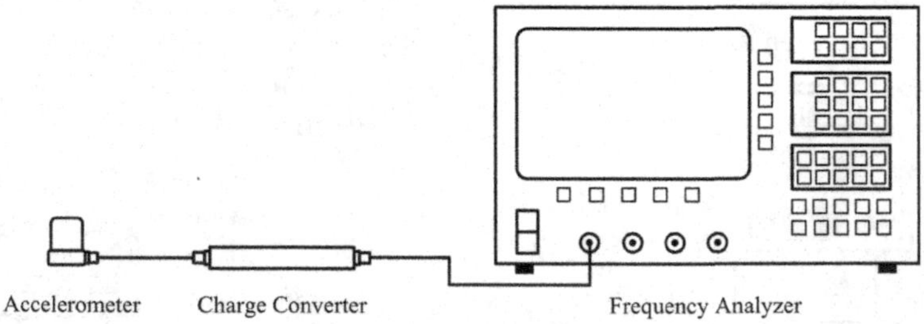

(c) Accelerometer with a Charge Converter

Figure 9.13 Instrument Setups for a Charge Accelerometer

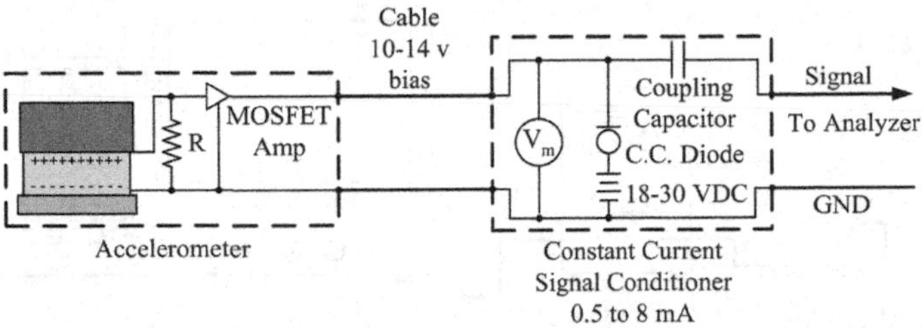

Figure 9.14 IEPE Accelerometer with Constant Current Signal Conditioner

Circuit Piezoelectric (ICP) and Constant Current Piezoelectric (CCP) are other terms that are used for IEPE accelerometers.

Charge accelerometers are self-generating accelerometers that do not need internal electronics to generate an output signal. In response to acceleration, they generate a high-impedance output charge signal measured in Pico-coulombs (pC) per g (m/s^2). This signal is directed to a charge amplifier, often located several feet away, that converts the charge signal to a low-impedance voltage signal in mV per g (m/s^2) that can be directed to a recording instrument or frequency analyzer. [Figure 9.13 (a)] Charge amplifiers have adjustable conversion factors in converting from pC/g (pC/m/s^2) to mV/g (mV/m/s^2). A charge converter can also be used to convert the accelerometer charge signal to a voltage signal [Figures 9.13 (b) and (c)]. The charge converter is powered either by a constant current signal conditioner used to power an IEPE accelerometer [Figures 9.13 (b) and 9.14) or from a DC constant current signal from the input terminal of a readout device or frequency analyzer [Figure 9.13 (c)]. Other general characteristics of charge accelerometers include:

- They have high impedance output signals that require low noise cables that are easily corrupted by EMI and RFI signals.
- They often require expensive charge amplifiers that must be powered by 110/220 V AC power.
- All high impedance accelerometer components must be kept dry.
- Their signal/noise ratio is limited by cable length.
- The capacitance of the accelerometer cable, which varies with cable length, affects the discharge time constant of the accelerometer.
- They are difficult to use in contaminated environments.
- They have a high temperature range (up to 1,000 °F (540 °C).
- The accelerometer signal normalization, dynamic (amplitude) range, lower frequency limit and frequency range can be adjusted at the charge amplifier.

Charge accelerometers are most often used for high temperature and high radiation environments.

IEPE accelerometers have built-in electronics within the accelerometer that directly convert the high impedance piezoelectric crystal charge signal to a low impedance accelerometer output voltage signal in mV/g (mV/m/s^2). The accelerometer is not self-generating. It must receive a minimum 2 mA constant current DC supply voltage from either a constant current signal conditioner or directly from an input channel of a readout device or frequency analyzer that can supply a constant current supply voltage to the accelerometer (Figures 9.14 and 9.15). The accelerometer acceleration signal is superimposed on a 10-14 V DC bias voltage (Figure 9.16). If the IEPE accelerometer receives its constant current DC supply voltage from a constant current signal conditioner, the acceleration signal from the signal conditioner is directed to an input channel of a readout device or frequency analyzer [Figure 9.15 (a)]. If the IEPE accelerometer is powered by a DC constant current signal from the input terminal of a readout device or frequency analyzer, the acceleration signal from the accelerometer is directed into the input terminal of the readout device or frequency analyzer. [Figure 9.15 (b)]. General characteristics of IEPE accelerometers include:

- They have integral built-in electronics that give the accelerometer a low output impedance, which allows the accelerometer output signal to be transmitted over low-cost standard two-wire coaxial cables with long cable lengths and to be transmitted through harsh environments without loss in signal quality.
- They have fixed sensitivities and discharge time constants that are not affected by cable length.

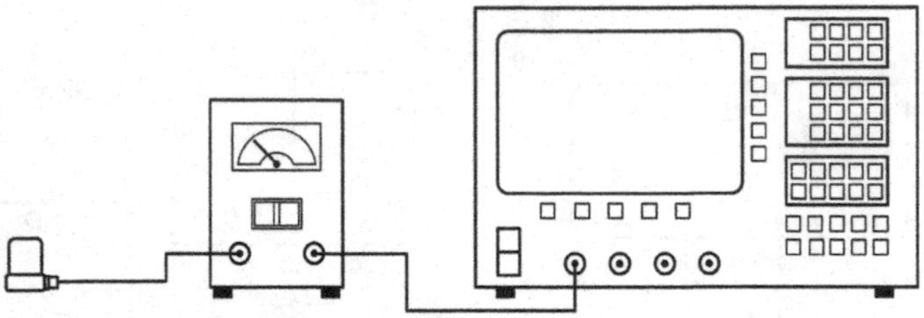

(a) Accelerometer with a Constant Current Signal Conditioner

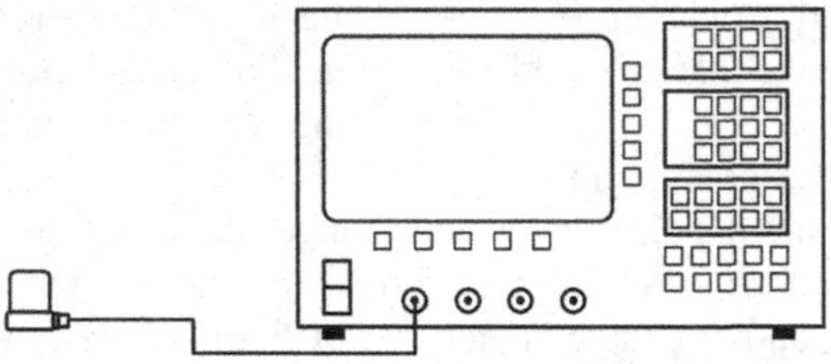

(b) Accelerometer with Acceleration Signal Directed into Frequency Analyzer

Figure 9.15 Instrument Setups for a IEPE Accelerometer

- They are simple to operate.
- They can have a high frequency response range.
- External constant current signal conditioners are powered by batteries. They are also built into frequency analyzers and other instruments. This makes them more convenient to use in field vibration measurements.

IEPE accelerometers can be used for almost all types of vibration testing. These include modal analysis and machinery monitoring and cryogenic, shock, seismic, high frequency, structural, flight, underwater and human vibration testing.

WARNING: *The internal resistance (or impedance) of an IEPE accelerometer should never be measured by directly placing ohm meter probes across the input terminals of the accelerometer. The constant voltage supplied by the ohm meter will destroy the internal electronics in the accelerometer.*

The dynamic range of an IEPE accelerometer is

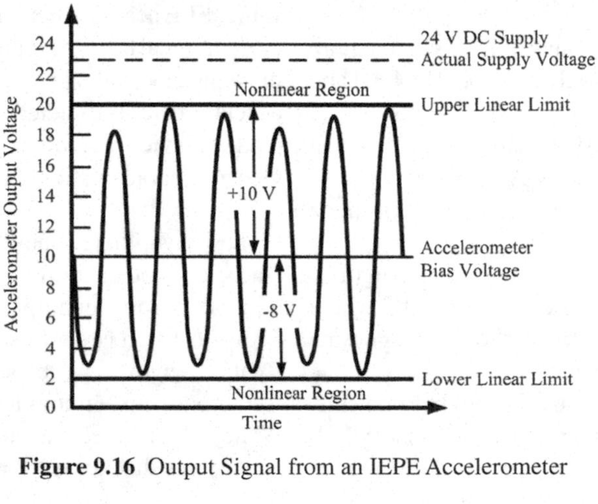

Figure 9.16 Output Signal from an IEPE Accelerometer

determined by the supply voltage minus 1 V drop across the C.C. diode, the bias voltage from the constant current signal conditioner and the accelerometer sensitivity [mV/g (mV/m/s^2)]. The specified maximum output voltage of IEPE accelerometers is $\pm$5 V. The maximum over-range of IEPE accelerometers is $\pm$10 V. The accelerometer amplitude linearity within the $\pm$5 V output range is better than 1 percent of accelerometer full-scale output. Typical accelerometer sensitivities are:

- Seismic accelerometers: 1-100 V/g (.1 to 10 V/m/s^2) - 0.05-5 g dynamic range.
- General purpose accelerometers: 10-100 mV/g (1-10 mV/m/s^2) - 50-500 g dynamic range.
- Shock accelerometers: .05-5 mV/g (.005-.5 mV/m/s^2) - 1,000-100,000 g dynamic range.

Figure 9.16 shows the accelerometer output voltage for the case where the constant current signal conditioner has a 24 V DC supply voltage and a 10 V DC bias voltage. The linear accelerometer output range for this case is -8 to +10 V. If the accelerometer output voltage stays within this range, it will be linearly related to the piezoelectric crystal output and there will be no clipping of the accelerometer output signal. If the supply voltage from the signal conditioner is reduced to 18 V DC, the actual supply voltage will be 17 V DC, and the accelerometer output range will be -8 to +7 V (Figure 9.17). For this case, accelerometer output voltages above 17 V will be clipped. This will result in erroneous analyzer results being generated by a frequency analyzer.

The above illustration demonstrates the importance of ensuring the accelerometer internal electronics are not overloaded. When an IEPE accelerometers is used, the DC supply voltage and current and the DC bias voltage of the constant current signal conditioner or from the input channel of the readout device or frequency analyzer should be known. The time-domain output from the accelerometer should be monitored to ensure its output voltage stays within the range of $\pm$ 5 V.

Figure 9.18 shows the amplitude ratio (output divided by input acceleration) and phase responses associated with a piezoelectric accelerometer. The frequency response ranges of these accelerometers are normally specified by the $\pm$ 5 percent and $\pm$ 3 dB response ranges of the accelerometer. These are ranges over which the accelerometer amplitude ratio and phase are within the amplitude tolerance ranges that are specified. The upper limit of the frequency range is significantly less than the resonance frequency of the seismic mass/piezoelectric crystal system. Because of the coupling capacitor in the constant current signal conditioner and the capacitance of the cable between the accelerometer and signal conditioner, the amplitude ratio of a piezoelectric accelerometer starts to decrease and the phase response starts to increase at frequencies around 2-3 Hz. A strain gauge and piezo-resistive accelerometer will have similar responses, except their amplitude ratio will be 0 dB and their phase will be 0 degrees down to zero Hz.

The crystal in a piezoelectric accelerometer will produce a charge output in both the on-axis and cross-axis (transverse) directions of the accelerometer (Figure 9.19). The transverse direction is oriented 90 degrees relative to the on-axis direction. The on-axis crystal charge output is representative of the acceleration the accelerometer is designed to measure. If at the same time the accelerometer is being excited in its on-axis direction it is also excited in a transverse direction, the

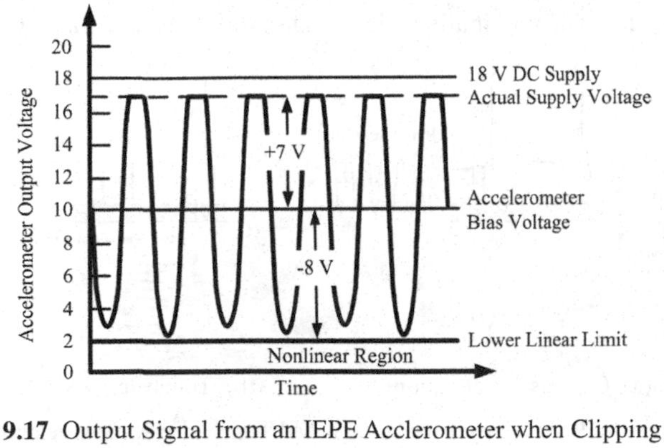

Figure 9.17 Output Signal from an IEPE Acclerometer when Clipping Occurs

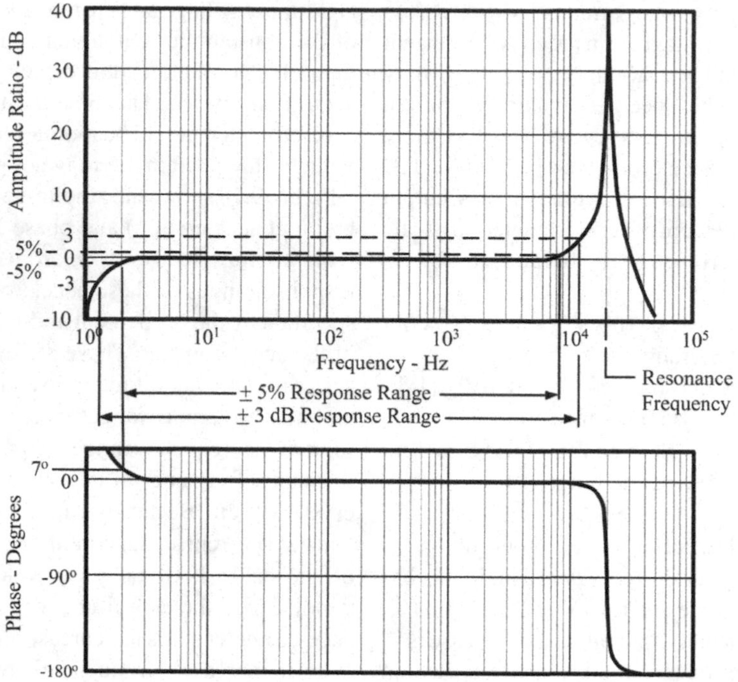

Figure 9.18 Typical Piezoelectric Accelerometer Frequency Range

accelerometer crystal will produce charge outputs for both directions. The internal accelerometer electronics cannot distinguish between these two charge signals. Therefore, the transverse charge signal can contaminate the on-axis signal. The transverse sensitivities of piezoelectric accelerometers are normally less than 5% of their on-axis sensitivities. Consequently, the transverse charge signal does not contaminate the on-axis signal for these accelerometers.

General purpose piezoelectric accelerometers nominally weigh less than 1 gm (< 0.034 oz) to over 80 gm (>2.8 oz). Accelerometers that nominally weigh 25-35 gm (0.88-1.23 oz) can be used for most structural and machine vibration measurements. However, these accelerometers should not be used for vibration measurements on lightweight structural systems, thin panels, or very small pieces of equipment. The mass of the accelerometer in these situations can be of the same order of magnitude as the effective mass of the structure to which it is attached. This can cause a condition known as *mass loading*. The accelerometer mass, when coupled to the structure mass, changes the frequency response of the structure. To prevent mass loading, the accelerometer mass should be less than one-tenth of the effective mass

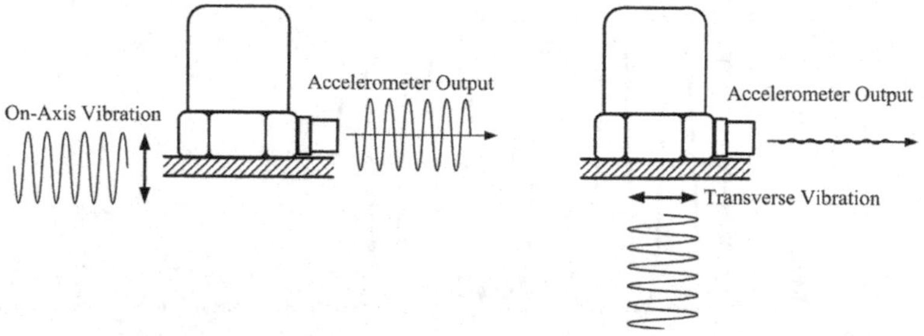

(a) On-axis Acceleration (b) Transverse Axis Acceleration

Figure 9.19 Accelerometer On-axis and Transverse Axis Acceleration of the structure or piece of equipment to which it is attached. Accelerometers that weigh less than 1 gm (< 0.034 oz) have nominal acceleration sensitivities in the range of 10 mV/g (1 mV/m/s²). These accelerometers should be used on lightweight structures. High sensitivity accelerometers [100-1,000 mV/g (10-100 mV/m/s²)] that nominally weigh 10-80 gms (0.34-2.8 oz) should be used in situations where extremely low acceleration values must be measured. These usually include cases where floor or machinery vibration measurements must be made to check for a manufacturer's compliance for vibration exposure to sensitive optical, photo masking, electronic or other similar equipment.

Accelerometers can be used to simultaneously measure acceleration in two or three mutually orthogonal directions. Accelerometers that simultaneously measure acceleration in two orthogonal directions are referred to as *bi-axial accelerometers*. They have two seismic mass/piezoelectric crystal configurations that are oriented perpendicular to each other and that are encapsulated in the same structural enclosure. Accelerometers that simultaneously measure acceleration in three mutually orthogonal directions are referred to as *tri-axial accelerometers*. These accelerometers have three seismic mass/piezoelectric crystal configurations that are oriented orthogonal to each other and that are encapsulated in the same structural enclosure.

Figure 9.20 shows the amplitude and frequency ranges for typical accelerometer applications.

9.4 Shock and Overload Protection of Accelerometers

Seismic, general purpose and shock accelerometers can be damaged when they are exposed to accelerations or mechanical shocks that exceed their maximum rated values. Piezoelectric crystals and internal accelerometer circuits can be damage when accelerometers are exposed to accelerations that significantly exceed their rated design values. For these cases, the voltage output of the accelerometer will be significantly greater than the maximum specified value of 5 V for an IEPE accelerometer, and they can be greater than the 10 V maximum allowable value. Therefore, care must be exercised to ensure that an accelerometer is used within the dynamic amplitude range for which it is designed.

Accelerometer piezoelectric crystals and internal electronics can also be damaged when the accelerometer is exposed to a high sudden shock. This can occur when an accelerometer experiences a high-g impact input for which it is not designed. This can also occur when the accelerometer:

- Is dropped onto a hard surface,
- Is impacted by a hard object,
- Receives a high-amplitude shock impact from a nearby event,
- Is dropped or slammed onto a surface when a magnetic mount {Figure 9.24) is used, or
- Is carelessly removed from a cemented base mounting (Figure 9.22).

Care must be exercised to ensure these events do not occur.

9.5 Accelerometer Mountings

The method of mounting an accelerometer to a vibrating

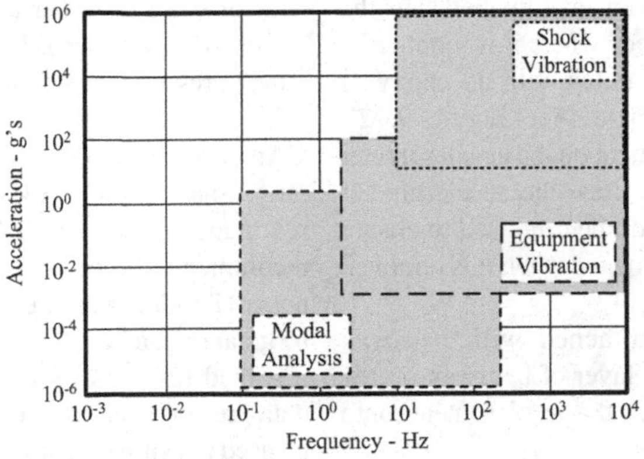

Figure 9.20 Some Typical Amplitude and Frequency Ranges of Accelerometers

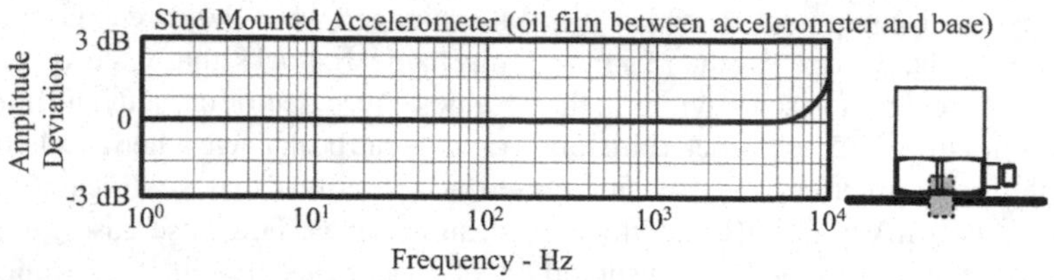

Figure 9.21 Stud Mounted Accelerometer (Oil Film between Accelerometer and Base)

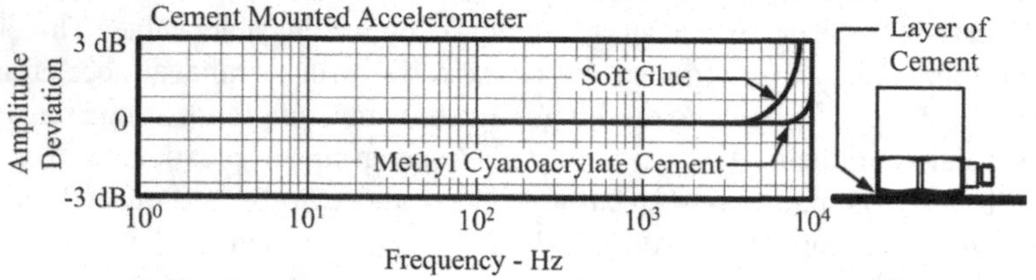

Figure 9.22 Cement Mounted Accelerometer

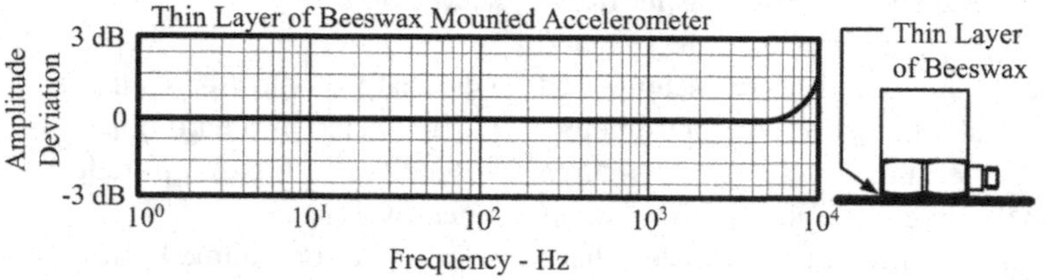

Figure 9.23 Thin Layer of Beexwax Mounted Accelerometer

surface can significantly influence its performance. Accelerometers should be attached to vibrating surfaces according to manufacturer's instructions. The following methods can normally be used to mount accelerometers:

- Clean the surface where an accelerometer is to be mounted. The surface must be flat.
- Accelerometers can be mounted with a threaded metal stud (Figure 9.21). Apply an oil film between the accelerometer and the surface where it is mounted.
- Accelerometers can be mounted with an epoxy or methyl cyanoacrylate cement (See Figure 9.22). Place a thin layer of the cement on the accelerometer mounting surface and then press the accelerometer onto the surface. Care should be exercised to ensure the accelerometer is not damaged when it is removed from the surface.
- Accelerometers can be attached with beeswax (Figure 9.23). Place a thin layer of beeswax on the accelerometer and then press the accelerometer onto the surface.
- Accelerometers can be mounted with the use of a double sided adhesive (See Figure 9.24). Place the adhesive on the accelerometer mounting surface and then press the accelerometer onto the surface.
- A magnet can be used to attach large accelerometers to metal vibrating surfaces (Figure 9.25). Attach the accelerometer to a magnetic base and then place the magnetic base on the metal surface.
- An accelerometer can be attached (usually with a metal stud) to either a short or long pointed metal probe (Figure 9.26). The pointed end of the probe is then pressed against vibration surfaces.

An accelerometer should be used over a frequency range that is well below its fundamental mounted resonance frequency. When the manufacturer's recommended mounting guidelines are available, the upper vibration measurement frequency limit should not be greater than 20 percent of the manufacturer's quoted mounted resonance frequency. When manufacturers' data are not available, Figures 9.21 through 9.26 can be used to estimate anticipated accelerometer mounted frequency ranges. When using these figures for a

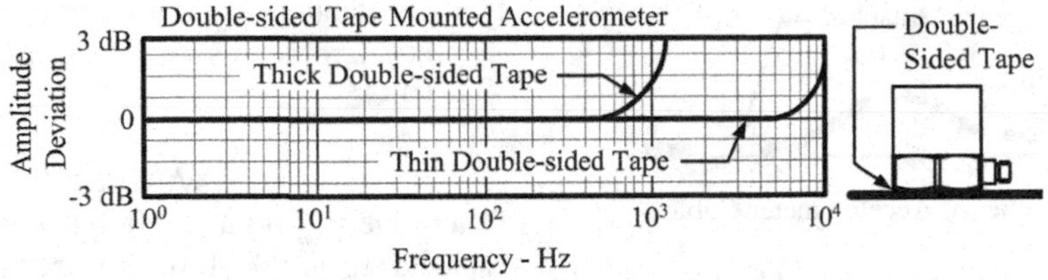

Figure 9.24 Double-Sided Tape Mounted Accelerometer

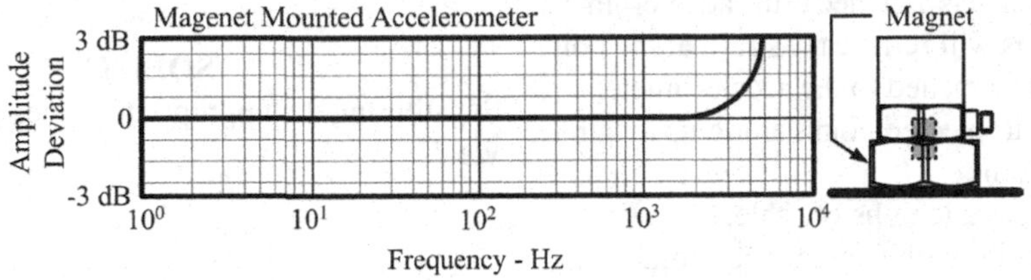

Figure 9.25 Magnetic Mounted Accelerometer

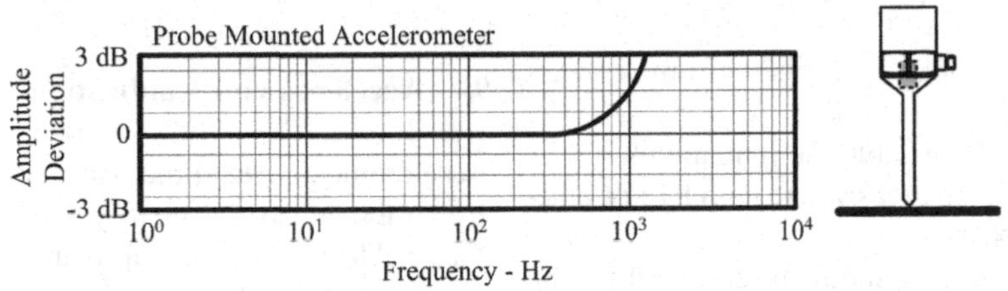

Figure 9.26 Probe Mounted Accelerometer

specified mounting arrangement, the accelerometer should be used over a mounting frequency range where the amplitude deviation is zero.

The following guidelines apply for the use of different accelerometer mounting methods:

- Stud- and cement-mounted accelerometers and accelerometers mounted with beeswax can be used for vibration measurements where the upper vibration measurement frequency limit does not exceed 5,000 Hz. An accelerometer mounted with beeswax is often preferred for most vibration measurements. Cement-mounted accelerometers are used when very high vibration amplitudes and moderately high shock levels are present. Stud-mounted accelerometer are used when very high shock levels are present.
- Magnetically-mounted accelerometers can be used for vibration measurements where the upper vibration measurement frequency limit does not exceed 1,000 Hz. A magnetically-mounted accelerometer is normally used when a large accelerometer (> 50 gm) is mounted to large machine structural bases and large pieces of equipment.
- Hand-held, probe-mounted accelerometers can be used for vibration measurements where the upper vibration measurement frequency limit does not exceed 300 Hz. Hand-held, probe-mounted accelerometers are recommended only for exploratory purposes. They should not be used for vibration measurements where accurate results are required.
- Using double-sided adhesive tape to mount an accelerometer is normally not recommended.

9.6 Accelerometer Cables

The accuracy and reliability of a vibration measurement system can be significantly affected by the cable between the accelerometer and charge amplifier, constant current signal conditioner or frequency analyzer input terminal. Some general rules for accelerometer cables include:

- When an accelerometer is exposed to high acceleration or shock inputs, make sure that the cable connector is tightly threaded onto the accelerometer. In some cases,

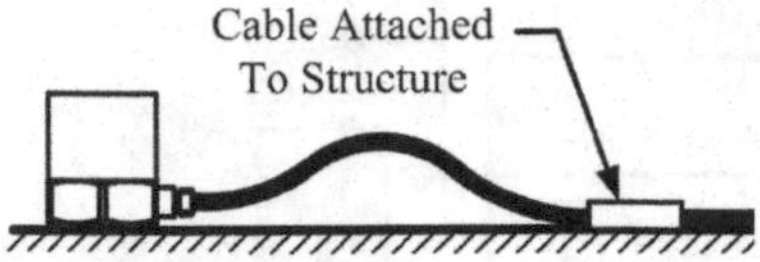

Figure 9.27 Strain Relieved Accelerometer Cable

this may require the use of a "thread locking" liquid.
- Strain relieve the accelerometer cable as close as possible to the point it is attached to the accelerometer (Figure 9.27). This will relieve the strain at the point where the cable is attached to the accelerometer.
- Secure the cable at selected points along its length to minimize its motion.
- "Figure 8" wrap extra lengths of cable.
- Use low-noise cables with charge accelerometers.
- Do not "kink" low-noise cables.
- Use twisted pairs of wires in cables in high magnetic areas.
- Do not run cables parallel to or in the same conduit with power lines.
- Cross power lines with cables at right angles.
- Use shielded cables in areas where high RFI or other electrical noise exists.
- The outer cable covering should be compatible with its environment (heat, chemicals, etc.)
- Keep cable connectors clean, especially in high impedance charge systems.
- Properly seal cable connectors for underwater testing or in high humidity areas.
- Use hard-lined cables in areas with very high temperatures.
- Check cable capacitance when testing at very high frequencies.

Capacitive loading of long cable runs can distort higher frequency components in a signal from an accelerometer. The following equation can be used to calculate the upper frequency limit F_{max} (Hz) of a long cable:

$$F_{max} = \frac{10^9 (I-1)}{2\pi C V} \tag{9.1}$$

where:

C = Cable capacitance (pF)
V = Peak voltage output from the accelerometer (V)
I = Constant current output from the accelerometer signal conditioner (mA)
π = 3.14159.

EXAMPLE 9.1

Determine F_{max} for a 150 ft long cable where the cable capacitance is 30 pF/ft, the accelerometer output is ± 5 V, and the constant current supply of the signal conditioner is 2 mA.

SOLUTION

Substituting the appropriate values into equation (9.1) yields:

$$F_{max} = \frac{10^9 (2-1)}{2 \times 3.14159 \times 30 \times 150 \times 5} = 7{,}074 \text{ Hz}. \tag{9.1a}$$

9.7 Accelerometer Calibration

Accelerometers, real-time analyzers, and FFT analyzers are calibrated at the factory with traceability to the National Institute of Standards and Technology (NIST). The calibration of a measurement system consisting of an accelerometer and real-time analyzer or FFT analyzer should be checked before each series of measurements. A vibration calibrator that generates a known acceleration signal can be used to accomplish this. Vibration calibrators generate a calibration acceleration signal with a value of 386 in./s^2 (1 g) or 10 m/s^2 at a frequency of 159.2 Hz.

Real-time analyzers normally can be calibrated so the instrument reads out directly in engineering units. The exact calibration procedure may vary from manufacturer to manufacturer. However, it generally goes as follows:

- Enter the accelerometer sensitivity [mv/g, mv/(m/s^2), mv/(in./s^2), etc.] in the analyzer setup menu as instructed by the instrument manufacturer.
- Place the accelerometer on the vibration calibrator, and turn the calibrator on.
- If the calibration signal is 1 g, the instrument should read 1 g, 9.81 m/s^2, or 386 in./s^2.
- If the instrument does not read out one of these exact values, then adjust the accelerometer sensitivity in the analyzer setup menu until the correct calibration acceleration value appears.

Some real-time analyzers have an internal automatic vibration calibration capability. Read the manufacturer's instrument manual to correctly use this capability.

Some hand-held real-time analyzers have both real-time analyzer and FFT analyzer capabilities. For the real-time analyzer to be used for making vibration measurements, it must go down to the 1 Hz 1/3 octave frequency band. The FFT analyzer can always be used for vibration analyses. Caution should be exercised with AC filters in a hand-held real-time analyzer. They may affect the FFT response at very low frequencies (< 1 Hz).

Hand-held real-time analyzers sometimes only read out dB values. Therefore, a special calibration procedure must be followed to convert acceleration levels in dB to their corresponding values in engineering units [mv/g, mv/(m/s²), mv/(in./s²), etc.].

- Set the hand-held analyzer to the real-time analyzer mode and activate the 1/3 octave band filters.
- The hand-held analyzer will normally have an accelerometer preamplifier that can be plugged into the meter. Attach the accelerometer cable to the preamplifier, place the accelerometer on the vibration calibrator, and turn the calibrator on.

Read the calibration level in dB in the 160 Hz 1/3 octave band. The equation for the acceleration level, L_a, is:

$$\boxed{L_{a(cal)} = 20 \log_{10}\left[\frac{a_{rms(cal)}}{a_{ref}}\right]} \tag{9.2}$$

$a_{rms(cal)}$ is the rms value of the acceleration calibration signal (g, m/s², in./s²., etc.) and a_{ref} is the reference acceleration value (g, m/s², in./s². etc.). $a_{rms(cal)}$ and a_{ref} must have the same units. a_{ref} is a function of internal analyzer settings and the accelerometer sensitivity. a_{ref} is obtained by rearranging equation (9.2) to obtain:

$$a_{ref} = a_{rms(cal)} \times 10^{-L_{a(cal)}/20} \tag{10-3}$$

If $a_{rms(cal)}$ has a value of 386 in./s², equation (9.3) becomes:

$$a_{ref} = 386 \times 10^{-L_{a(cal)}/20} \tag{9.4 U.S.}$$

and a_{ref} will have the units of in./s². If $a_{rms(cal)}$ has a value of 9.81 m/s², equation (9.3) becomes:

$$\boxed{a_{ref} = 9.81 \times 10^{-L_{a(cal)}/20}} \tag{9.4 SI}$$

and a_{ref} will have the units of m/s².

- If the FFT mode of the hand-held analyzer is used, change to the FFT mode.
- The equation for the acceleration levels that are measured is:

$$\boxed{L_a = 20 \log_{10}\left[\frac{a_{rms}}{a_{ref}}\right].} \tag{9.5}$$

- The acceleration values in engineering units are obtained by rearranging Equation 9.5 to:

$$\boxed{a_{rms} = a_{ref} \times 10^{L_a/20}.} \tag{9.6}$$

a_{rms} and a_{ref} must have the same units.

9.8 Accelerometer Selection

The characteristics of several types of accelerometers and transducer signal conditioning equipment have been presented in this chapter. These include strain gauge, piezo-resistive, and charge type and IEPE piezoelectric accelerometers. When selecting any accelerometers, the manufacturer's specifications, guidelines for and cautions related to use, and transducer limitations must be carefully reviewed. When selecting any accelerometer, the following should be considered:

- *Accelerometer casing*: The outer case of an accelerometer must be robust and rugged enough to adequately protect the inner components of the accelerometer in the measurement environment where it will be used.
- *Accelerometer sensitivity*: The accelerometer sensitivity must be appropriate for the anticipated vibration amplitudes to be measured. The selected sensitivity must maximize the usable measurement dynamic range and signal-to-noise ratio of the accelerometer and ensure the accelerometer and signal conditioning electronics are not overloaded during a vibration measurement.

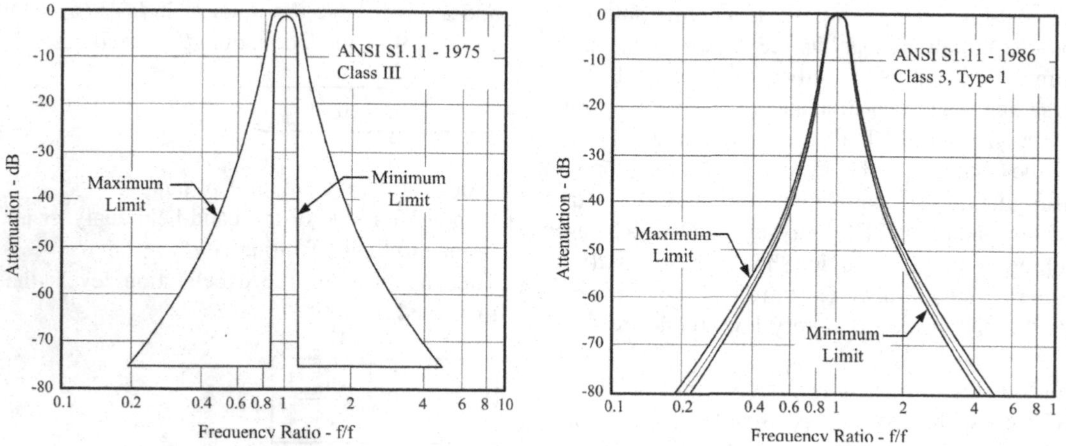

(a) ANSI S1.11 - 1975 Requirements (b) ANSI S1.11 - 1986 Requirements

Figure 9.28 1/1 and 1/3 Octave Filter Characteristics

- *Accelerometerfrequencybandwidth*:Theaccelerometer and accelerometer mounting configuration must be appropriate for the desired lower and upper frequency measurement limits.
- *Acclamatory mounting*: The accelerometer mounting must be appropriate for the measurement environment and the anticipated vibration amplitudes to be measured.
- *Accelerometer mass*: The mass of an accelerometer must not significantly add to the effective mass of the area of a structure or a machine to which it is attached. As a general rule, the mass of the accelerometer should be less than 10 percent of the effective mass of the area of a structure or a machine to which it is attached. Accelerometers that are too heavy will cause *mass loading* of the structure or machine. This will change its vibration characteristics.

9.9 Frequency Analyzers

Two types of frequency analyzers can be used to investigate vibration: (1) real-time and (2) FFT analyzers.

Real-time analyzers used for vibration measurements contain parallel 1/1, 1/3, 1/12 and 1/24 octave band filters.These filters simultaneously process those parts of the acceleration signal contained within each filter passband. There are three 1/3 octave band filters in an octave band, twelve 1/12 octave band filters in an octave band, and twenty-four 1/24 octave band filters in an octave band. Figures 9.28(a) and 9.28(b) show plots of the required 1/3 octave band filter attenuation characteristics required by ANSI S1.11.

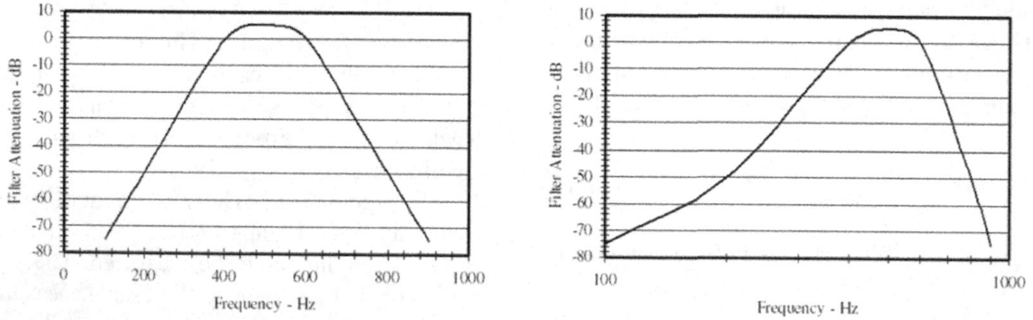

(a) Digital Filter Characteristics Plotted on Linear Scale (b Digital Filter Characteristics Plotted on Logrithmic Scale

Figure 9.29 Sample Digital Filter Characterics

Table 9.1 Octave and Third Octave Filter Center Frequencies and Passbands

Filter Band Number	Third Octave Filter Center Frequency	Third Octave Filter Passband	Octave Filter Center Frequency	Octave Filter Passband
	Hz	Hz	Hz	Hz
-	0.8	0.708 - 0.891		
-	1	0.891 - 1.12	1	0.708 - 1.41
1	1.25	1.12 - 1.41		
2	1.6	1.41 - 1.78		
3	2	1.78 - 2.24	2	1.41 - 2.82
4	2.5	2.24 - 2.82		
5	3.15	2.82 - 3.55		
6	4	3.55 - 4.47	4	2.82 - 5.62
7	5	4.47 - 5.62		
8	6.3	5.62 - 7.08		
9	8	7.08 - 8.91	8	5.62 - 11.2
10	10	8.91 - 11.2		
11	12.5	11.2 - 14.1		
12	16	14.1 - 17.8	16	11.2 - 22.4
13	20	17.8 - 22.4		
14	25	22.4 -28.2		
15	31.5	28.2 - 35.5	31.5	22.4 - 44.7
16	40	35.5 - 44.7		
17	50	44.7 - 56.2		
18	63	56.2 - 70.8	63	44.7 - 89.1
19	80	70.8 - 89.1		
20	100	89.1 - 112		
21	125	112 - 141	125	89.1 - 178
22	160	141 - 178		
23	200	178 - 224		
24	250	224 - 282	250	178 - 355
25	315	282 - 355		
26	400	355 - 447		
27	500	447 - 562	500	355 - 708
28	630	562 - 708		
29	800	708 - 891		
30	1000	891 - 1120	1000	708 - 1410
31	1250	1120 - 1410		
32	1600	1410 - 1780		
33	2000	1780 - 2240	2000	1410 - 2820
34	2500	2240 - 2820		

Octave and fractional octave band filters are proportional bandwidth filters that require symmetry with regard to logarithmic frequency. The pre-1975 ANSI S1.11 requirements shown in Figure 9.28(a) were initially specified for analog filters. Lower order digital filters used in some computer software and digital-based instruments are symmetric with regard to linear frequency as shown in Figure 9.29(a). Figure 9.29(b) indicates, however, that the filter attenuation characteristics of these filters are not symmetric with regard to logarithmic frequency. Therefore, digital octave and fractional octave band filters created with these lower order algorithms will yield results that differ from those obtained from corresponding analog filters. Consequently, ANSI S1.11 was revised in 1986 [Figure 9.28(b)] to require digital octave and fractional octave band filters be created with higher order algorithms that will yield digital filters that are symmetric with regard

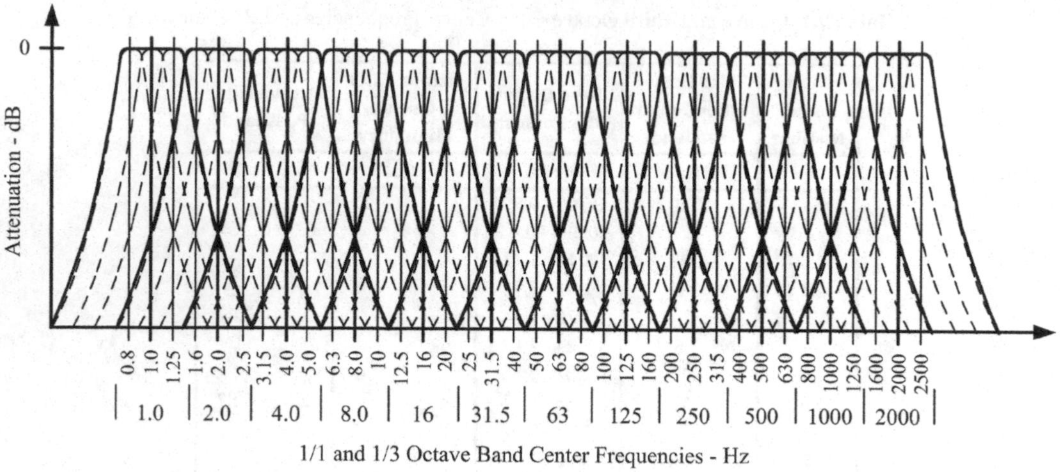

Figure 9.30 1/1 and 1/3 Octave Band Filter Sets

to logarithmic frequency.

Real-time analyzers that are used to analyze vibration should have, at a minimum, 1/3 octave band filters that go down to at least 0.8 Hz. Figure 9.30 and Table 9.1 show 1/1 and 1/3 octave band filter sets with 1/1 octave band filters from 1 to 2,000 Hz and 1/3 octave band filters from 0.8 to 2,500 Hz. Note that there are three 1/3 octave band filters associated with each 1/1 octave band filter. Normally, 1/1 octave band filters are not used for vibration measurements.

FFT analyzers repetitively capture consecutive acceleration signal segments and mathematically process these segments to obtain the frequency and amplitude content of the signal. Whereas the filter passbands of 1/3, 1/12 and 1/24 octave band filters increase with increasing center frequency, the frequency bandpass associated with FFT analyzers remains constant with frequency. The frequency resolution of FFT analyzers is specified in terms of the number of spectral lines contained in the overall frequency bandpass of the analysis. For example, if the overall frequency bandpass of interest for an analysis is from 0 to 200 Hz and the analyzer is set to 400 spectral lines, the frequency bamdpass associated with each spectral line will be 200/400 or 0.5 Hz.

Several companies manufacture FFT frequency analyzers that are self-contained units or PC based systems. Many of these analyzers have up to 32,000 spectral lines of resolution. A few manufactures of hand-held real-time analyzers have options that include a FFT analyzer capability. For reasonable frequency resolution, these analyzers should have a minimum of 1,600 spectral lines of resolution.

QUESTIONS - CHAPTER 9

1. List the three basic types of accelerometers and briefly describe they work.

2. List the four basic types of piezoelectric accelerometers and describe their characteristics.

3. Describe how charge piezoelectric accelerometers work.

4. Describe how Integrated Electronics Piezo Electric (IEPE) accelerometers work.

5. What are the general characteristics of charge accelerometers.

6. What are the general characteristics of IEPE accelerometers.

7. Describe the conditions in which clipping can occur with IEPE accelerometers.

8. Figure 9.18 shows that amplitude ratio and phase characteristics of a piezoelectric accelerometer with a resonance frequency of 20,000 Hz. Using Figure 3.28 and starting with equations (3.205) through (3.207), develop the expressions for the acceleration amplitude ratio and phase shown in Figure 9.18. Plot the results given by the resulting equations for the case where the resonance frequency of the accelerometer is 20,000 Hz.

9. What are bi-axial and tri-axial accelerometers?

10. List the general rules associated with accelerometer cables.

11. Determine F_{max} for a 200 ft long accelerometer cable where the cable capacitance is 28 pF/ft, the accelerometer output is ± 5 V, and the constant current supply of the signal conditioner is 4 mA.

12. What factors should be considered when selecting an accelerometer?

CHAPTER 10
DIGITAL PROCESSING OF VIBRATION SIGNALS

10.1 Introduction

Signal processing is the analysis, manipulation, and interpretation of signals. Unknown signals can be deterministic, random, or both (Figure 10.1). Sinusoidal and complex periodic excitations have been discussed throughout this book. Non-harmonic excitation was discussed in Chapter 6, and random signals were described in Chapter 7.

Unknown signals can be a combination of different signal types (Figure 10.1) that can be analyzed with a Fast Fourier Transform (FFT) analyzer. The following signal types can be easily processed by a FFT analyzer:

Sinusoidal, complex periodic and almost periodic signals. These signals are made up of a single or a combination of multiple sinusoidal signal elements. These signals are the easiest to process and interpret.

Transient signals. These signals exist for only a short period of time and then disappear. To conduct FFT analyses of these signals, care must be exercised to properly capture these signals in digital format in the time domain.

Ergodic random signals. The frequency domain properties of an ergodic random signal can be characterized by conducting a FFT analysis of a single data record that contains multiple random signal time segments.

If a random signal is non-ergodic but stationary, the number of signal time segments must be increased to the point where the FFT analysis will produce frequency domain properties of the ensemble of the time segments that become constant and that can be repeated for other similar ensembles of signal time segments. If a random signal is non-stationary, multiple ensembles of signal time segments must be processed to the point where the variation of the frequency domain properties between the ensembles of time segments are within acceptable limits.

A primary objective of signal processing is to identify, quantify and interpret the individual elements contained within an unknown signal by using appropriate mathematical tools. Principles associated with the processing of sinusoidal, complex and almost periodic, transient, and ergodic random signals will be discussed in this chapter.

The science of digital signal processing is very broad and encompasses many areas other than mechanical vibration. Detailed math concepts associated with many digital signal processing (DSP) operations are beyond the

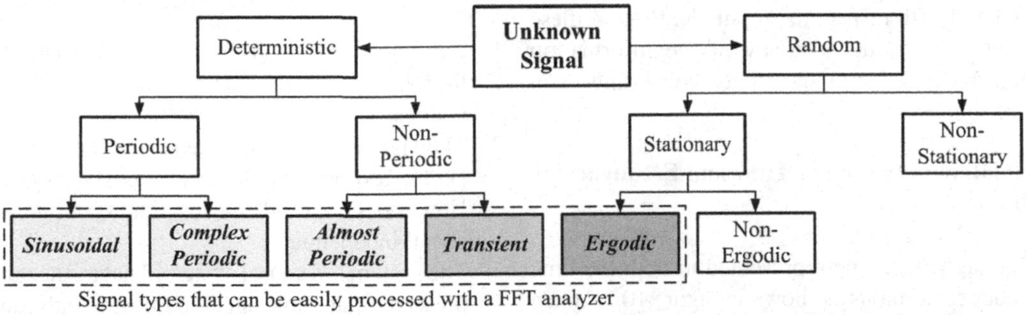

Figure 10.1 Unknown Signal Divided into Categories that Can Be Processed with Mathematical Tools

This chapter was written in collaboration with Thomas Lagö, Ph.D.

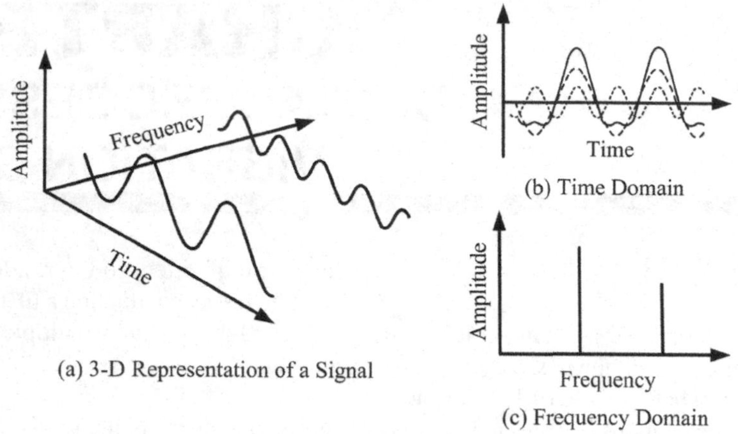

Figure 10.2 Time and Frequency Domain representation of a Vibration Signal

scope of this chapter, and therefore, are not discussed.

Discussions in this chapter are restricted to the analysis of vibration signals. Vibration of physical systems can be measured with accelerometers that were discussed in Chapter 9. When coupled to appropriate signal conditioning equipment, they output continuous analog voltage signals proportional to vibration amplitudes that vary with time. Discussions in this chapter address principles associated with digital signal processing (DSP) concepts that are commonly used in fast Fourier transform (FFT) analyzers to process and analyze these analog acceleration signals.

With the advent of digital computers and related digital signal processing, analog voltage signals from accelerometers and other related vibration transducers are converted to digital signals before they are digitally processed and analyzed. Digital signals are different from analog signals in two important aspects. They are discretely sampled with respect to time, and they are quantized with respect to amplitude. Both of these operations impact the accuracy of the information extracted from analog signals when converting them to digital signals.

10.2 Relation between the Time and Frequency Domains

Vibration signals can be represented in both the time and frequency domains as shown in Figure 10.2. When conducting Fourier series or transform operations, there are no real differences in the time and frequency data that describe a particular signal. The perceived differences arise in how the data are viewed - in either the time domain or the frequency domain. Specific information may be easier to perceive in a particular domain. This is why the two domains are used to examine different types of information that may be contained within a signal. When something is changed in or removed from one domain, there is a corresponding change in the other domain. By viewing the data in the time domain [Figure 10.2(b)], the sum of the two sinusoids in Figure 10.2(a) will be seen. When viewing in the frequency domain [Figure 10.2(c)], the two frequency components will be identified. Each viewpoint (time or frequency) helps to identify information contained in the data that comprise the signal. When viewing the signal in either the time or frequency domain, the data are not changed. Only the viewpoint is changed.

Signal processing operations associated with mechanical vibration typically assume:

1. The frequency response function (FRF) that describes the relation between the input and output of a system is linear.
2. The signal that excites the system and the related system response is either deterministic or stationary. However, linear and time invariant systems are input signal independent.
3. The system is a causal system. That is, the output of the system at some specified time t_o only depends on the input to the system at values of $t \leq t_o$. This assumption is only imposed for real physical systems.

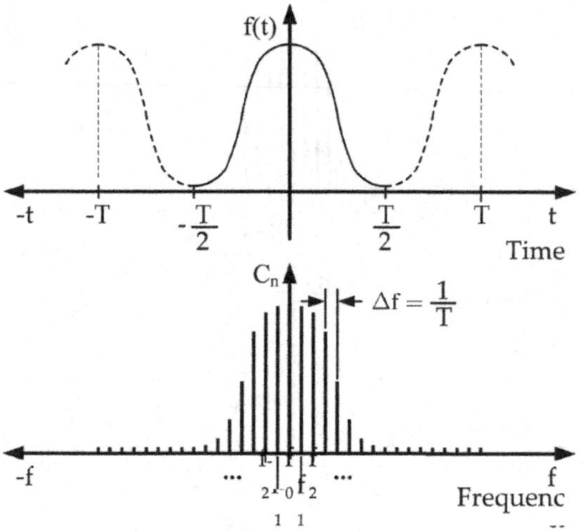

Figure 10.3 Fourier Series Representation of a Signal

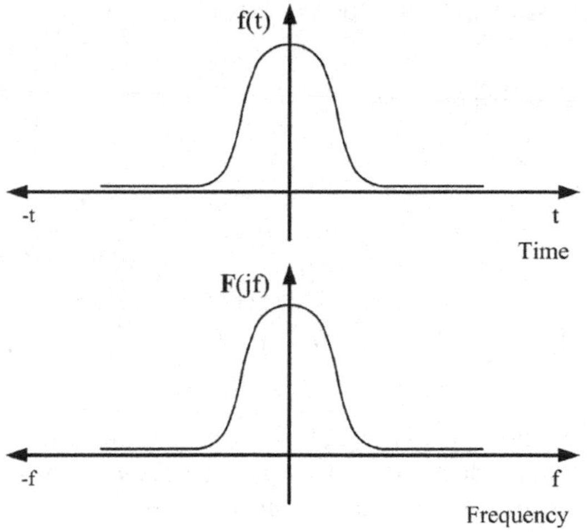

Figure 10.4 Fourier Transform Representation of a

4. The system is strictly stable.
5. The system is time invariant.

Normally, only positive values of time and frequency are considered when investigating the vibration response of a system. Sometimes complex data values will exists when transforming data from one domain (time or frequency) to the other. An understanding of the existence of these values and why they exist is important even though they do not exist in real life.

Fourier Series

Vibration signals that are *complex periodic* repeat themselves over a time period T (Figure 10.3). This signal class has an easy-to-use solid mathematical foundation. Consequently, vibrations signals are often, and sometimes incorrectly, assumed to fit into this category. Therefore, its properties must be clearly understood. Complex periodic signals are composed of a series of sinusoidal components (periodic) whose higher frequencies are integral multiples of the lowest frequency component of the signal. Using complex rotating phasors, the signal can be expressed as a *complex Fourier series* where:

$$f(t) = \sum_{n=-\infty}^{\infty} C_n\, e^{j2\pi f_n t} \tag{10.1}$$

where n is an integer that varies from $-\infty$ to $+\infty$ and:

$$\Delta f = \frac{1}{T} \quad \text{and} \quad f_n = n\,\Delta f. \tag{10.2}$$

C_n is given by:

$$C_n = \frac{1}{T}\int_0^T f(t)\, e^{-j2\pi f_n t}\, dt. \tag{10.3}$$

C_1, C_2, etc. in equation (10.3) are the amplitudes of the frequency components contained within f(t) that correspond to f_1, f_2, etc. in Figure 10.3. The units of C_n are the engineering units (EU) associated with f(t). Figure 10.3 indicates the Fourier series representation of a signal f(t) is continuous with time in the time domain and has discrete values at f_1, f_2, etc. in the frequency domain. The signal is periodic in the time domain and non-periodic in the frequency domain.

Fourier Transforms (Continuous in Both Time and Frequency Domains)

The *Fourier transform* can be used to transform a signal from the time to frequency domain when the signal is a transient signal that goes to zero over a finite period of time (Figure 10.4). The *Fourier transform* of a signal is obtained from:

$$\mathbf{F}(jf) = \int_{-\infty}^{\infty} f(t)\, e^{-j2\pi f t}\, dt. \tag{10.4}$$

For the Fourier transform to exist:

$$\int_{-\infty}^{\infty} |f(t)|\, dt < \infty \tag{10.5}$$

and:

$$f(t=\pm\infty)=0 \quad \text{and} \quad \left.\frac{d^n f(t)}{dt^n}\right|_{t=\pm\infty}=0. \tag{10.6}$$

The time domain signal can be obtained from the frequency domain signal by taking the *inverse Fourier transform* of the frequency signal:

$$f(t)=\int_{-\infty}^{\infty} \mathbf{F}(j2\pi f)e^{j2\pi f t}\, df. \tag{10.7}$$

The units of **F**(jf) are either $EU \cdot s$ (transients), EU (harmonic components) or EU/Hz (broadband) where EU is the engineering units associated with f(t). Figure 10.4 indicates that the Fourier transform representation of f(t) given by equations (10.4) and (10.7) are continuous and go to zero in both the time and frequency domains.

Fourier Transforms - Sampled Time Functions

Figure 10.5 represents a case where f(t) is digitally sampled in the time domain. For this case, time functions are represented as a time series that is given by a sequence of values at discrete, equally spaced points in time. This is the reverse of the Fourier Series. Because of the symmetry of the Fourier transform pairs, the frequency spectrum becomes periodic with a frequency that equals the *sampling frequency* f_s, which is given by:

$$f_s=\frac{1}{\Delta t} \tag{10.8}$$

where Δt is the time interval between sampled points. The *Fourier transform pair* for this case is:

$$\mathbf{F}(jf)=\sum_{n=-\infty}^{\infty} f(n\Delta t)e^{-j2\pi f n\Delta t} \tag{10.9}$$

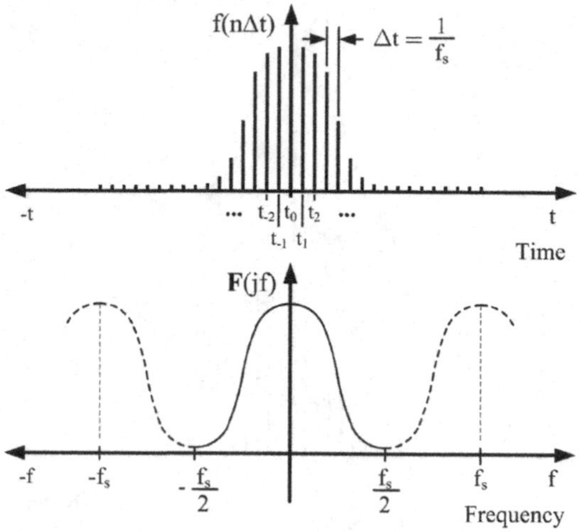

Figure 10.5 Fourier Transform Representation of a Signal - Sampled Time Functions

$$f(n\Delta t)=\frac{1}{f_s}\int_{-\frac{f_s}{2}}^{\frac{f_s}{2}} \mathbf{F}(jf)e^{j2\pi f n\Delta t}\, df. \tag{10.10}$$

The units of **F**(jf) are either $EU \cdot s$ (transient), EU (harmonic components) or EU/Hz (broadband) where EU are the engineering units associated with f(t).

Discrete Fourier Transforms (DFT)

Both time and frequency functions can be represented as time and frequency series that are given by a sequence of values at discrete equally spaced points in time or as a function of frequency (Figure 10.6). Because of the digital sampling, both the time signal and frequency spectrum will be periodic. The *Discrete Fourier Transform (DFT)* pair for a digitally sampled signal is:

$$\mathbf{F}_k=\frac{1}{N}\sum_{n=0}^{N-1} f_n e^{-j2\pi k\frac{n}{N}} \tag{10.11}$$

$$f_n=\sum_{k=0}^{N-1} \mathbf{F}_k e^{j2\pi n\frac{k}{N}}. \tag{10.12}$$

N is the number of digitized points of f(t). The units of $\mathbf{F}_k$ are either $EU \cdot s$ (transient), EU (harmonic components) or EU/Hz (broadband) where EU are the engineering

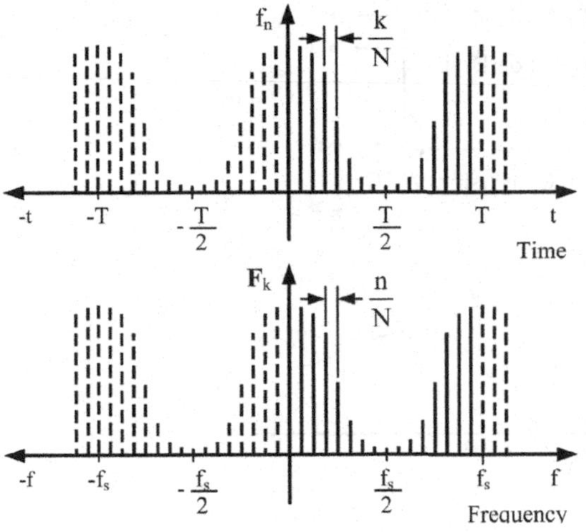

Figure 10.6 Fourier Transform Representation of a Signal - Discrete Fourier Transform

units associated with f(t).

Using Euler's formula, equation (10.11) can be written:

$$\mathbf{F}_k = \frac{1}{N}\left\{ \begin{array}{l} \sum_{n=0}^{N-1} f_k \cos\left[2\pi k\left(\frac{n}{N}\right)\right] - \\ j\sum_{n=0}^{\infty} f_k \sin\left[2\pi k\left(\frac{n}{N}\right)\right] \end{array} \right\}. \tag{10.13}$$

$\mathbf{F}_k$ can be written as:

$$\mathbf{F}_k = |\mathbf{F}_k| e^{j\phi}. \tag{10.14}$$

$|\mathbf{F}_k|$ is given by:

$$|\mathbf{F}_k| = \sqrt{\mathrm{Re}[\mathbf{F}_k]^2 + \mathrm{Im}[\mathbf{F}_k]^2} \tag{10.15}$$

where:

$$\mathrm{Re}[\mathbf{F}_k] = \frac{1}{N}\sum_{n=0}^{N-1} f_k \cos\left[2\pi k\left(\frac{n}{N}\right)\right] \tag{10.16}$$

$$\mathrm{Im}[\mathbf{F}_k] = -\frac{1}{N}\sum_{n=0}^{N-1} f_k \sin\left[2\pi k\left(\frac{n}{N}\right)\right]. \tag{10.17}$$

ϕ is given by:

$$\phi = \tan^{-1}\left[\frac{\mathrm{Im}[\mathbf{F}_k]}{\mathrm{Re}[\mathbf{F}_k]}\right]. \tag{10.18}$$

N^2 complex multiplications are required to obtain the DFT. Therefore, if N = 1024, then 1,048,576 complex multiplications are required to obtain the frequency components. Utilizing symmetry effects in the FFT matrix, the *Fast Fourier Transform (FFT)* can be used to calculate the Fourier transform for equally spaced discrete points in the time domain. Therefore, the FFT only requires N ln N multiplications [the upper or lower half (includes points on the diagonal) of the matrix]. For N = 1024, only 7,098 multiplications are thus required to obtain the frequency information. This is a reduction factor of 148. The FFT algorithm is used by modern FFT analyzers for this reason.

Normally, the FFT does not produce the same exact frequency data as the DFT. Selective frequency spacing can be used for the DFT, and it is not necessary to calculate all frequency lines. The DFT and FFT produce the same results when the time block data are multiples of 2^n and the discrete time points are equally spaced. If n < 16, the DFT is often faster than the FFT. However, for n > 16 and larger data record lengths, the FFT is typically faster. The number for n when a DFT will be faster than the FFT varies with implementation hardware and software. The FFT is always faster when FFT analyzers use at least 1024 data points when processing a signal.

10.3 Digital Quantization of Analog Signals

Figure 10.7 shows a block diagram of the *analog-to-digital (A/D) conversion* process. An analog signal (in the case of vibration an acceleration voltage signal) is directed into a *sample/hold (S/H)* circuit and then into an *analog-to-digital (A/D) converter*. The S/H circuit captures a specified time block of the analog signal and holds it constant while the A/D converter converts the analog signal into a series of digitized points. Figure 10.8(a) shows a time block of an analog signal. Figure 10.8(b) shows the corresponding digitally sampled analog signal that is held by the S/H circuit. The analog signal is digitally sampled at specific time segments.

Figure 10.8(c) shows an expanded segment of the

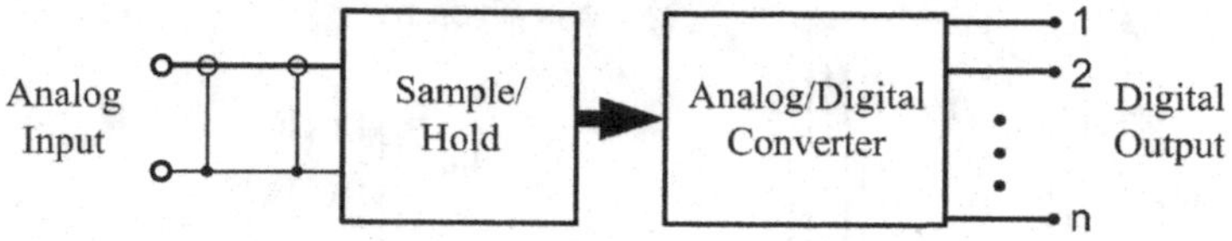

Figure 10.7 Analog-to-Digital Conversion (ADC)

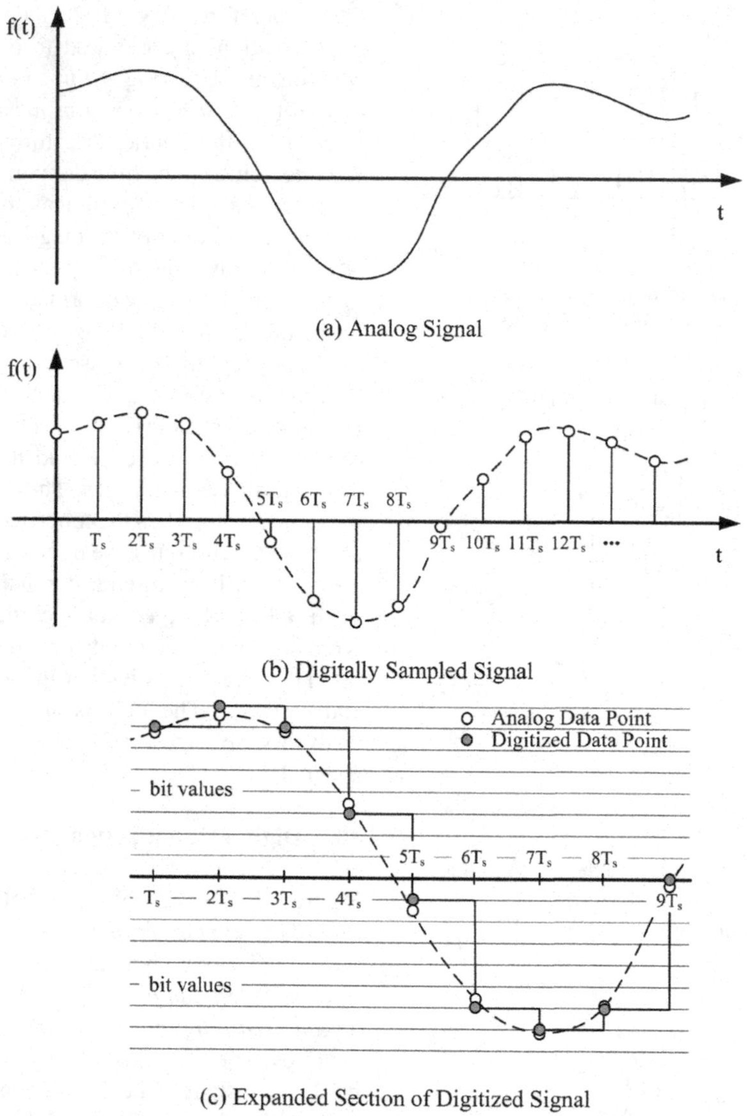

Figure 10.8 Conversion of an Analog Signal to a Digital Signal

digitized signal. The actual digitized points are not necessarily the same as the corresponding digital points that fall on the analog curve. The bit lines in the figure represent the bit values associated with the resolution of the A/D converter. When a digital point falls in between two bit lines, the amplitude of the digitized point is the value of the closest bit line. The error associated with the digitized point is the difference in amplitude between the digital and corresponding digitized point. The errors between the digital and corresponding digitized points create the background noise associated with the A/D converter that is called *quantization noise*.

The *resolution* of an A/D converter indicates the number of discrete values it can produce over a specified voltage range of an analog signal. The digital values of an A/D converter are electronically stored in binary form specified as *bits*. The number of discrete values or levels that are available in the A/D converter is specified as 2 raised to the power n (2^n) where n is the number of bits associated with the A/D converter. An 8-bit A/D converter, for example, can digitize an analog voltage to a resolution of 1 in 2^8 or 256 different digital levels. A 12-bit A/D converter can digitize an analog voltage to a resolution of 1 in 2^{12} or 4,096 different digital levels.

The resolution, Q, of an A/D converter can thus be expressed by:

$$Q = \frac{e_{FSR}}{2^n} \tag{10.19}$$

where n is the number of bits associated with the A/D converter and e_{FSR} is the full-scale voltage range of the A/D converter.

EXAMPLE 10.1

Determine the voltage resolution in an 8-bit and a 12-bit A/D converter if the voltage range of the A/D converters is ± 5 V.

SOLUTION

The value of e_{FSR} for the two A/D converters is:

$$e_{FSR} = 5 - (-5) = 10\,\text{V}. \tag{10.1a}$$

The resolution of the 8-bit A/D converter is:

$$Q = \frac{10}{2^8} = \frac{10}{256} = 0.039\,\text{V}. \tag{10.1b}$$

Therefore, the resolution of the 8-bit A/D converter is 39 mV/level.

The resolution of the 12-bit A/D converter is:

$$Q = \frac{10}{e^{12}} = \frac{10}{4{,}096} = 0.00244\,\text{V} \tag{10.1c}$$

Therefore, the resolution of the 12-bit A/D converter is 2.44 mV/level.

The *dynamic range (DR)* and *signal-to-noise ratio (SNR)* of an A/D system indicate the amplitude range over which a signal can be properly analyzed. The theoretical dynamic range DR of a fixed-point A/D converter in decibels (dB) is given by:

$$DR(dB) = 20\log_{10}\left(2^n\right) \approx 6.02\,n \tag{10.20}$$

where n is the number of bits in the A/D converter. A 12-bit A/D converter will have a dynamic range of around 72 dB. The signal-to-noise ratio of a fixed-point A/D converter is approximately given by:

$$SNR(dB) \approx 20\log_{10}\left(2^n\sqrt{1.5}\right) \approx 6.02\,n + 1.76. \tag{10.21}$$

A 12-bit A/D converter will thus have a theoretical SNR value of around 74 dB. The actual DR and SNR values of a measurement system can be limited by distortion, analog noise, missing codes, spurious and other error components that will change the DR and SNR values for the vibration transducers and related signal conditioning devices that comprise the system. A good rule of thumb is, therefore, to subtract at least two bits from the A/D converter. This will give a good estimate on the expected DR and SNR values for the A/D converter. For a 12 bit A/D converter, the actual DR value would thus be about 60 dB instead of the theoretically stipulated value of 72 dB. The SNR value would be around 62 dB instead of 74 dB. There are A/D converter systems for vibration measurements that produce a lot less than this. Therefore, A/D converter systems should be checked by using a sinusoid signal at full range amplitude for the system and then check the noise floor for the system.

FFT analyzers will normally have frequency passbands from DC-51 kHz down to DC-100 Hz. This filtering introduces averaging that can increase the dynamic range of the analyzer to values that will be larger than those discussed above. Theoretically, for every factor

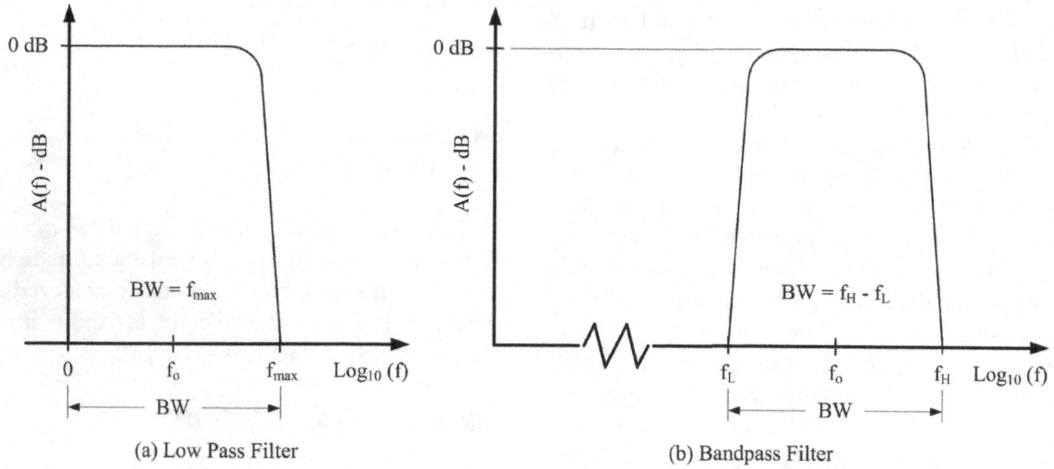

Figure 10.9 Schematic Representations of a Low Pass and Bandpass Filters

of two decrease in frequency range, the dynamic range should improve 3 dB. This does not always occur, but the dynamic range will be improved when the analyzer frequency bandpass setting is decreased.

10.4 Digital Sampling of Analog Signals

Let f(t) be a continuous signal with respect to time t [Figure 10.8(a)]. When the signal is digitally sampled every T_s seconds, the digitized signal f[n] is given by [Figure 10.8(c)]:

$$f[n] = f(nT_s) \quad \text{for} \quad n = 0,1,2,3,\cdots. \tag{10.22}$$

The corresponding *sampling frequency* f_s is given by:

$$f_s = \frac{1}{T_s}. \tag{10.23}$$

f[n] only has discrete values at the specific time intervals nT.

The sampling frequency must be high enough to accurately capture all of the frequencies contained in an analog signal. The required sampling frequency is specified by the *Nyquist-Shannon sampling theorem*, which states[1, 2]:

> If a function f(t) contains no frequencies higher than f_{max} Hz, then f(t) is completely determined by giving its ordinates at a series of points spaced $1/(2f_{max})$ seconds apart.

This theorem assumes a baseband frequency passband of $0 \le f \le f_{max}$ where the filter passband BW is f_{max} [Figure 10.9(a)]. The general sampling theorem is based on bandwidth and not frequency.

From the standpoint of mechanical vibration, let f(t) represent a continuous analog signal and **F**(jf) represents a continuous Fourier transform of f(t). Furthermore, let f(t) be band limited to a one-sided baseband bandwidth BW of $0 \le f \le f_{max}$ where:

$$\mathbf{F}(jf) = 0 \quad \text{for all} \quad f > f_{max}. \tag{10.24}$$

When this is the case, the sampling frequency f_s must meet the requirement:

$$f_s > 2f_{max} \quad \text{or} \quad f_{max} < \frac{f_s}{2}. \tag{10.25}$$

$2f_{max}$ is the *Nyquist rate* and $f_s/2$ or f_{max} is the *Nyquist frequency*. The corresponding *sampling interval* must meet the requirement:

1. C. E. Shannon, "Communication in the presence of noise", Proc. Institute of Radio Engineers, vol. 37, no. 1, pp. 10-21, Jan. 1949. Reprint as classic paper in: Proc. IEEE, Vol. 86, No. 2, (Feb 1998).
2. T. L. Lagö (Feb 2002), "Digital Sampling According to Nyquist and Shannon," Sound and Vibration.

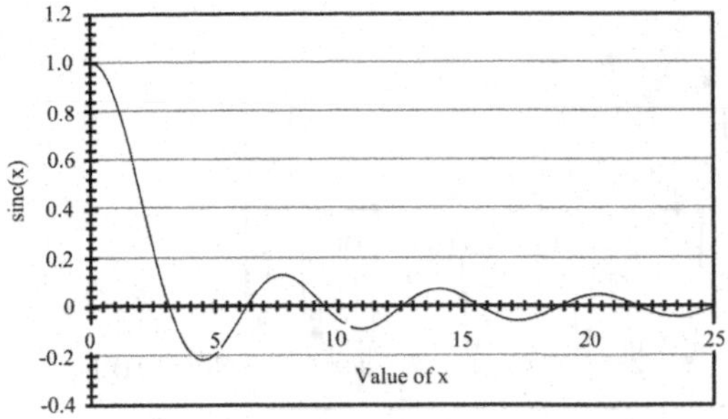

(a) Normalized Sinc Function in the Time Domain

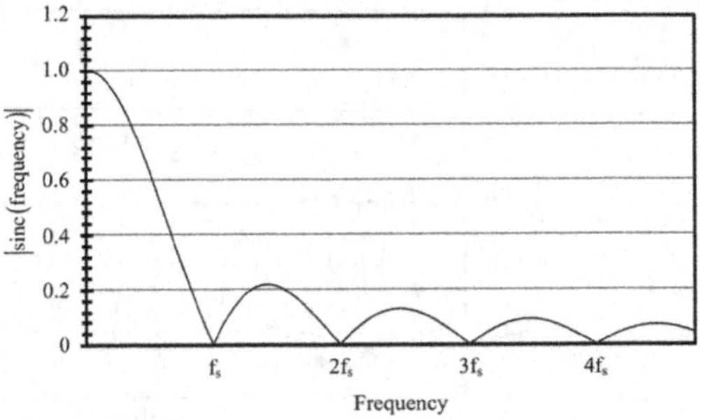

(b) Sinc Filter in the Frequency Domain

Figure 10.10 Sinc Function

$$T_s < \frac{1}{2f_{max}}. \tag{10.26}$$

Sometimes it may be necessary to digitize an analog signal that contains frequencies over a specified frequency passband BW given by $f_L \le f \le f_U$ where f_L and f_U are the lower and upper frequency band limits (where $f_L \neq 0$) [Figure 10.9(b)]. This correspond to the general sampling theorem. For this case the filter bandwidth BW is:

$$BW = f_H - f_L \tag{10.27}$$

and:

$$\mathbf{F}(jf) = 0 \quad \text{for all} \quad |f| < f_L \text{ and } |f| > f_U. \tag{10.28}$$

For passband signals, only frequency information within the filter passband BW is of interest. It is, therefore, possible to fulfill the requirements of the Nyquist-Shannon sampling theorem when the sampling frequency is greater than twice the passband filter bandwidth[2]. This represents a significant savings in unusable frequency lines in a FFT analysis compared to an analysis with a baseband frequency bandwidth of $0 \le f \le f_H$, and then using only the frequency lines within the filter frequency passband of $f_L \le f \le f_H$. A sampling rate of 2.56 x BW is commonly used in FFT analyzers that can do real zoom measurements[2]. This sampling rate is independent of the 2.56 x f_{max} sampling rate used for the frequency bandwidth of $0 \le f \le f_{max}$. This process is often referred to as a super heterodyne analysis and will create a complex time signal. This is only associated with "zooming" into the specified filter passband BW and has nothing to do with the real world signal, which is always real. Some modern FFT analyzers can achieve a sampling rate of 2.1 x BW by using anti-aliasing filters with very steep filter slopes associated with advanced digital signal processing technology.

When the conditions specified by equations (10.22) and (10.24) are met, the continuous analog signal f(t) can be completely reconstructed from the digitized signal $f[n] = f(nT_s)$ by applying the *Shannon interpolation formula*[1]:

$$f(t) = \sum_{n=0}^{\infty} f(nT_s)\,\text{sinc}\left[\frac{t - nT_s}{T_s}\right] \tag{10.29}$$

where sinc(x) is the *sinc function* given by [Figure 10.10(a)]:

$$\text{sinc}(x) = \frac{\sin(x)}{x}. \tag{10.30}$$

10.5 Aliasing

Ideally, one would like an analog signal that has been reconstructed from digitized points in time [equation 10.29)] to exactly overlay the original analog signal. With respect to this potential reconstruction, Figure 10.11 addresses two conditions in which a 10 Hz sinusoidal wave form is digitized. The continuous solid curve represents the analog signal before it enters an A/D converter. The solid circles represent the digitized values of the signal leaving the A/D converter.

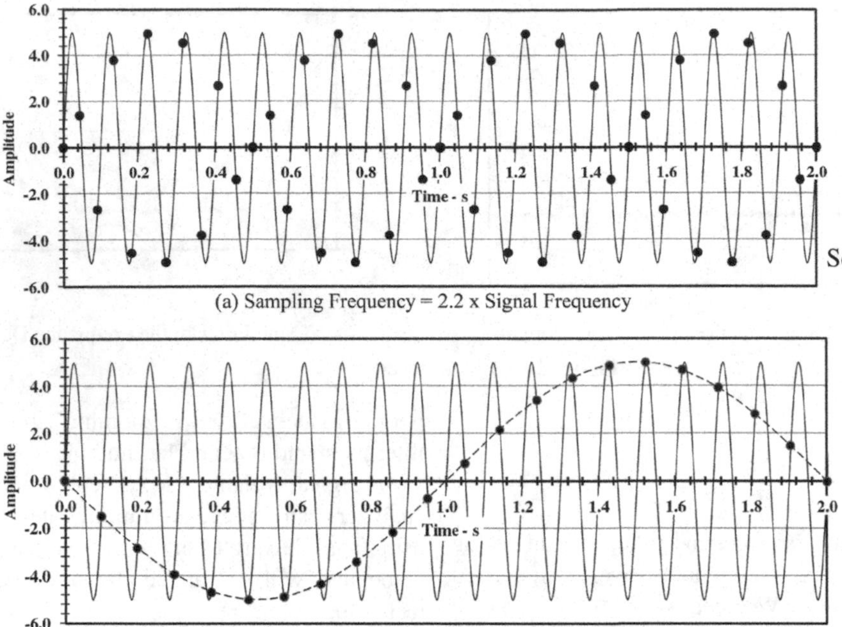

(a) Sampling Frequency = 2.2 x Signal Frequency

(b) Sampling Frequency = 1.05 x Signal Frequency

Figure 10.11 Examples of Different Sampling Frequencies

The digitized values in Figure 10.11(a) are for a sampling frequency of 2.2 times the frequency of the original analog signal. This meets the requirements of the Nyquist-Shannon sampling theorem. There are 2.2 digitized points per signal cycle. The digitized points represent a unique set of points from which the signal can be completely reconstructed using equation (10.29). Also, the amplitude and frequency of the original signal can be obtained by taking the FFT of the digitized points.

The Nyquist-Shannon sampling theorem assumes an infinite record length. In reality, however, record lengths are finite. Their lengths are the sampling interval T_{sb}. Within the sampling interval T_{sb}, a FFT analyzer will convert analog signal segments into blocks of digitized points in increments of 2 raised to the n power (2^n). If the number of digitized points N in the time interval T_{sb} relative to the sampling frequency f_s (or f_s/f_{sb}) is too small, the peak values of the lower frequency elements in the analog signal may not be correctly identified. For digitized data blocks where N is greater than or equal to 2^{10} (1,024) digitized points, this error will be small. When digitized data block sizes (N) are less than 2^7 (128) samples, this error can be significant.

As will be discussed in Section 10.8, the data records in the time domain are multiplied by a *uniform rectangular time window*. The inverse of this time window in the frequency domain is the *inverse sinc function* [Figure 10.10(b)]. Digitized data points are multiplied by an inverse sinc function in the frequency domain. Noting that f_s in Figure 10.10(b) equals ($1/T_{sb}$), if T_{sb} is too small, the primary lobe of the inverse sinc function ($0 \le f \le f_s$) in the frequency domain can be rather wide. This can make it difficult to accurately identify the peak values of the lower frequency elements in an analog signal.

The digitized values in Figure 10.11(b) are for a sampling frequency of 1.05 times the frequency of the original analog signal. This sampling rate does not meet the requirements of the Nyquist-Shannon sampling theorem and errors will occur. Instead of representing a sine wave with a frequency of 10 Hz, the sine wave represented by the digitized points has a frequency of 2 Hz. This phenomenon of the original 10 Hz frequency sine wave being digitally transformed into a 2 Hz sine wave is called *aliasing or folding*. Since the digitized points only trace out a 2 Hz sine waves, the original signal cannot be reconstructed from these points. A FFT analysis of the these points will yield a frequency of only 2 Hz. The 2 Hz signal is referred to as an *alias*

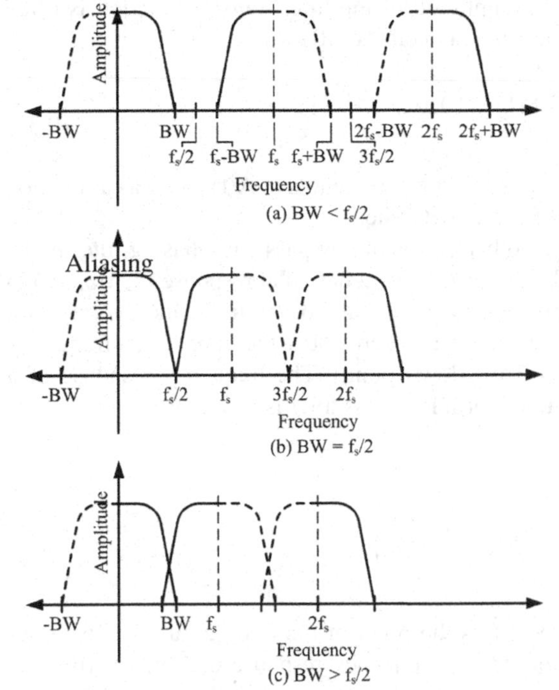

Figure 10.12 Aliasing in the Frequency Domain

of the original 10 Hz signal.

Because of the periodic nature of digital Fourier transform pairs (refer to Figure 10.6), left-for-right flipped images of the original frequency spectrum are produced during the digitizing process that are centered at multiples of the sampling frequency (f_s, $2f_s$, $3f_s$, etc.) as shown in Figure 10.12. During the computation of the fast Fourier transform of a vibration signal, these multiple image spectra, along with their respective components, are folded about points on the frequency axis equal to multiples of $f_s/2$ when nothing is done to prevent this from occurring. When this does occur, the original and all image frequency components in the frequency range from zero to infinity will end up in the frequency range of zero to $f_s/2$. This phenomenon in the frequency domain is also referred to as *aliasing*,

An anti-aliasing filter with a filter passband of $0 \leq f \leq$ BW can be used to prevent aliasing from occurring in the frequency domain. So long as BW is less than $f_s/2$, the image spectra centered about f_s (and consequently higher frequency image spectra) and the original spectrum do not overlap each other as shown in Figure 10.12(a), and aliasing does not occur. When BW equals $f_s/2$, the original and image spectra just begin to overlap each other as shown Figure 10.12(b). Theoretically, aliasing begins to occur at this point. When BW is less than $f_s/2$, the original and image spectra overlap each other as shown in Figure 10.12(c), and aliasing occurs.

When the image spectra overlaps the original spectrum as shown in Figure 10.12(c), the frequency contents of the two signals add together, forming a single combined and aliased digital signal. It is not possible to separate the overlapping frequencies of the original and folded spectra. Thus, amplitude and frequency information contained in the original analog signal become distorted and are lost. Steps must be taken to ensure this does not occur.

In general, BW should be less then $f_s/2$. When BW comes too close to $f_s/2$, because of errors in the digitizing process and leakage that can occur with anti-aliasing filters, distortions can also occur in the digitized signal at frequencies in the vicinity of BW.

10.6 Anti-Aliasing Filters

Quite often an analog signal will contain signal elements with frequencies above the frequency range of interest in a FFT analysis, or the analog signal may contain high frequency noise that must be filtered out. To address these issues, a low-pass *anti-aliasing filter* is used to limit the frequency passband of the analog signal (Figure 10.13), which normally ranges from zero to f_{max}, that will be analyzed. This is done to ensure that the requirements of the Nyquist-Shannon sampling theorem can be approximately satisfied when digitizing the analog signal. The design and mathematical models used for anti-aliasing filters are typically based on white noise as the input signal to the filter.

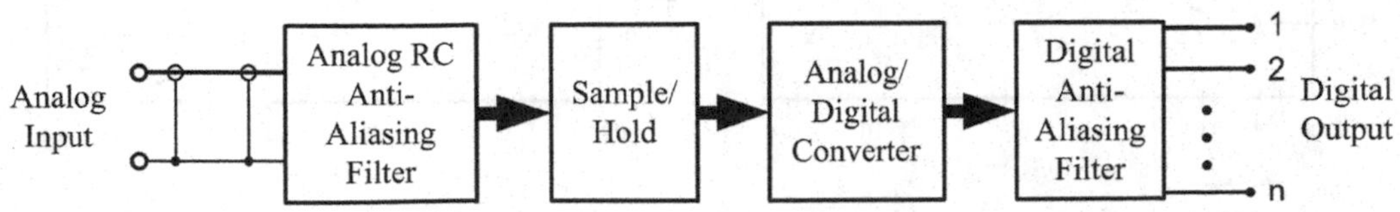

Figure 10.13 Analog-to-Digital Conversion with a combination of analog and digital Anti-Aliasing Filters

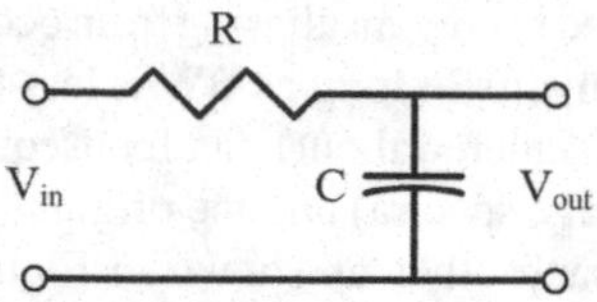

Figure 10.14 RC Filter

The simplest low-pass anti-aliasing filter is a RC filter (Figure 10.14). The filter response is given by:

$$G^2(\omega) = \frac{1}{1+(\omega RC)^2} \quad \text{where} \quad G(\omega) = \frac{V_{out}}{V_{in}} \tag{10.31}$$

or:

$$G^2(f) = \frac{1}{1+\left(\frac{f}{f_c}\right)^2} \tag{10.32}$$

and:

$$\phi(\omega) = -\tan^{-1}(\omega RC) \quad \text{or} \quad \phi(f) = -\tan^{-1}\left(\frac{f}{f_c}\right) \tag{10.33}$$

where G(f) is the filter transfer function, $\phi(f)$ is the phase lag between the filter output and input, and f_c is the *filter corner frequency* given by:

$$\omega_c = \frac{1}{RC} \quad \text{or} \quad f_c = \frac{1}{2\pi RC}. \tag{10.34}$$

The amplitude of the filter transfer function is often specified in decibels (dB) where:

$$A(f) = 10\log_{10}\left[G^2(f)\right]. \tag{10.35}$$

Figure 10.15 shows a plots of A(f) and ϕ as a function of f/f_c for a RC filter.

Another common low-pass anti-aliasing filter is the *Butterworth filter*. It has a flat response (no ripples) in the frequency passband of the filter, and the response rolls off towards zero above the filter corner frequency f_c (-3 dB down point). The frequency response of a Butterworth low-pass filter is given by:

$$G^2(f) = \frac{1}{1+\left(\frac{f}{f_c}\right)^{2n}} \tag{10.36}$$

where n is the order of the filter. Figure 10.16 shows plots of the frequency response of Butterworth low-pass filters of orders 1 through 5. The slope of the filter response beyond the filter corner frequency is 20n dB/decade or 6n dB/octave.

Aliasing is always possible in analyses of analog and digital vibration signals. Two issues are important when addressing this potential. The first is the minimization of the number of unusable FFT frequency lines in the filter transition band between f_c and f_{max} (Figure 10.17). The number of unusable lines is a function of the steepness of the slopes of the FFT analyzer anti-aliasing filters and the analyzer sampling rate. The steeper the slopes of the anti-aliasing filters in relation to the dynamic range of the analyzer A/D converters and analyzer sampling rate, the lower the number of unusable frequency lines.

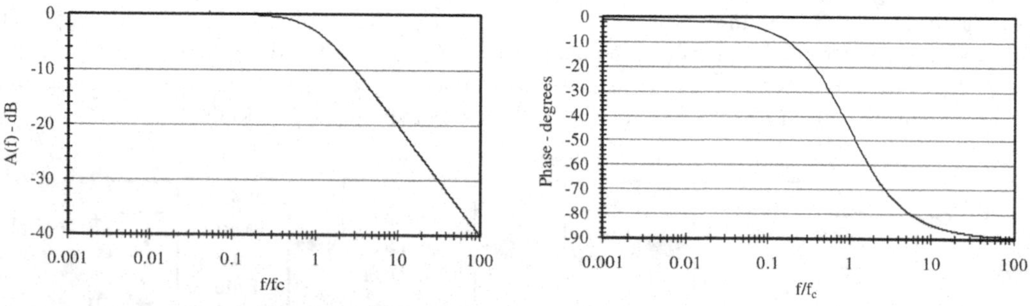

Figure 10.15 Gain and Phase Response of an RC Filter

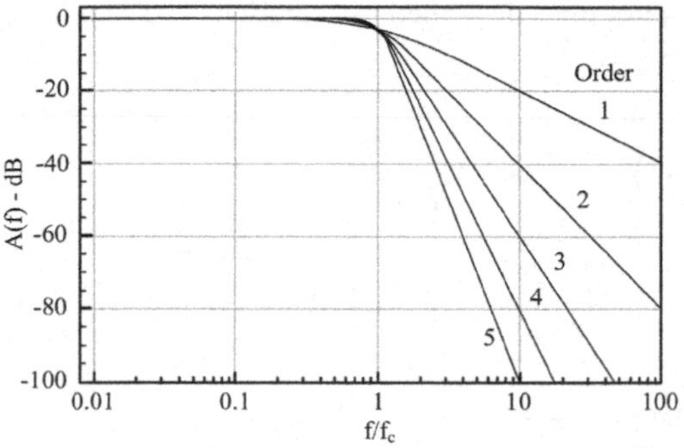

Figure 10.16 Gain Response of a Butterworth Filter

The second is the dynamic range of the FFT analyzer. The greater the dynamic range, the greater the potential SNR of FFT analyses.

A relationship exists between the upper frequency limit f_{max} of an anti-aliasing filter baseband and the filter slope, dynamic range, and FFT analyzer sampling frequency. The quantization noise floor, which limits the dynamic range of a FFT analyzer, is set by the number of bits associated with the analyzer A/D converters. If the A/D converter dynamic range is tied to the anti-aliasing filter baseband bandwidth and slope as schematically shown in Figure 10.17, the following "rule of thumb" for the sampling rate f_s can be applied[2]:

$$f_s = f_c + f_{max} \tag{10.37}$$

where f_c is the filter corner frequency and f_{max} is the upper frequency limit of the filter baseband or f_L for a passband filter [Figure 10.9(b)]. The filter attenuation at f_{max} is -X (the desired analyzer dynamic range). The following must be met for equation (10.37) to apply:

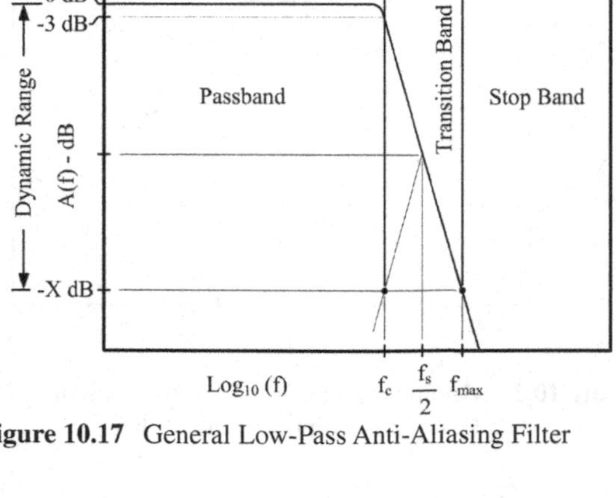

Figure 10.17 General Low-Pass Anti-Aliasing Filter

1. The analog signal noise floor is below the quantization noise floor of the A/D converter.
2. The anti-aliasing filter continues to attenuate the analog signal above its cutoff frequency.
3. The logarithmic slope of the anti-aliasing filter (in dB/Hz) is linear.
4. The upper frequency limit associated with the analog signal is assumed to be associated with *white noise*, and therefore, is theoretically infinite. This implies that the analog signal amplitudes associated with *signal noise* for frequencies above f_{max} either must stay constant or decrease with increasing frequency. This often is not the case when using accelerometers. The amplitudes of acceleration signals often increase with increasing frequency.
5. Their are no filtered signal elements in the analog signal at frequencies above f_{max} that have amplitudes above -X.

When the above conditions are met, the sampling rate f_s specified by equation (10.38) will be conservative, and aliasing will begin to occur when f_s is less than $(f_c + f_{max})/2$. When any of the above conditions are not met (particularly conditions 4 and 5), equation (10.38) should not be used to specify f_s. When this is the case, f_s should be given by:

$$f_s = 2f_{max}\,. \tag{10.38}$$

The quantization dynamic range associated with A/D converters can exceed 100 dB when elliptical-Cauer or FIR digital filters are used. To maximize the number of usable frequency lines in a FFT analysis using these A/D converters, anti-aliasing filters with very steep slopes in the filter transition band must be used. Slopes of 100 to 200 dB per octave are often required. These steep slopes are normally achieved by placing an elliptical-Cauer anti-aliasing filter before an A/D converter or a digital FIR filter after the A/D converter.

Figure 10.18 shows a comparison of Butterworth, Chebyshev, and elliptical-Cauer filter frequency responses. Very steep filter slopes can be achieved in the transition band of an elliptical or Cauer filter by allowing "ringing" to occur in the filter passband and stop band. The amplitude of the ringing level in the filter passband of an elliptical-Cauer filter can be controlled to within small value, but 0.1 dB is common for FFT analyzers.

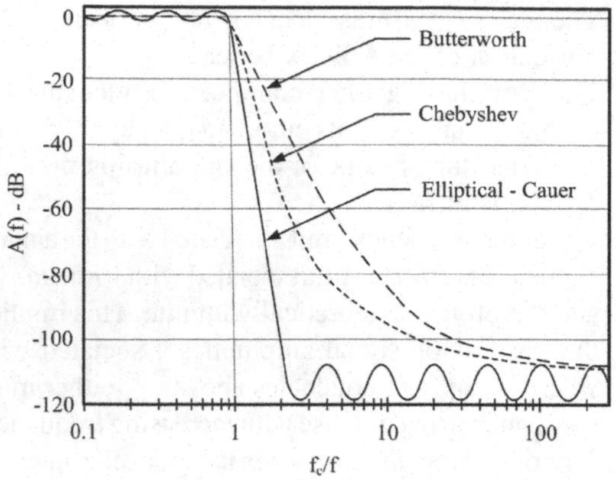

Figure 10.18 Anti-aliasing Low Pass Filters

The ringing in the stop band can be set to occur below the quantization noise floor of the A/D converter.

Anti-alias filters introduce a phase delay (often referred to as a time delay versus frequency) between the filter input and output. Refer to equation (10.33) and Figure 10.15 for a RC filter. The value of the phase delay as a function of frequency is a function of the filter transfer function. The value of the phase delay for most types of anti-aliasing filters varies with frequency.

Group delay is commonly used to describe the phase delay as a function of frequency for anti-aliasing filters. The group delay D(f) is the derivative of the phase function associated with the filter transfer function, or[3]:

$$D(f) = -\frac{d\phi(f)}{df} \tag{10.39}$$

where D(f) has the unit of s. The group delay for the RC low-pass filter described by equations (10.31) through (10.33) is:

$$D(f) = \frac{d}{df}\left[\tan^{-1}\left(\frac{f}{f_c}\right)\right] \tag{10.40}$$

or:

$$D(f) = \frac{1/f_c}{1+\left(\frac{f}{f_c}\right)^2}. \tag{10.41}$$

Figure 10.19 shows a plot of the group delay for the RC filter shown in Figure 10.15. Figure 10.20 shows plots of the group delays for 8 pole constant delay, Butterworth, and elliptical-Cauer filters[3]. The actual time delays AD(f) associated with anti-aliasing filters is given by:

$$AD(f) = \frac{D(f)}{f_c} \tag{10.42}$$

where the units on AD(f) are s/Hz.

The group delay for the ADC part of a FFT analyzer (Figure 10.21) will affect the time domain signal directed into the input terminal of the analyzer. The group delay of the DAC part of the analyzer (see Figure 10.45) will affect the reconstruction of the time domain signal from its digitized elements at the output terminal of the analyzer. The effects of the input group delay can be corrected in the DAC part of the

3. *Analog and Digital Products Design/Selection Guide*, Frequency Devices, Inc., 1784 Chessie Lane, Ottawa, IL 61350.

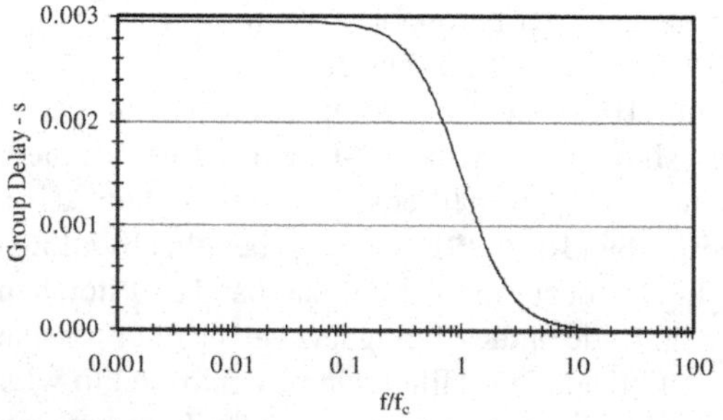

Figure 10.19 Group Delay for RC Filter

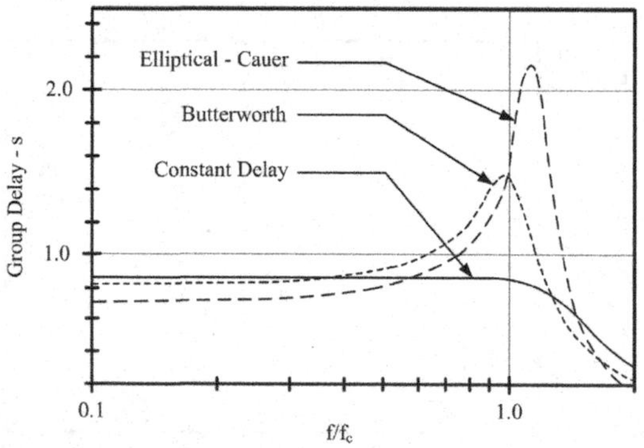

Figure 10.20 Group Delay for Butterworth, Constant Delay, and Elliptical-Cauer Filters

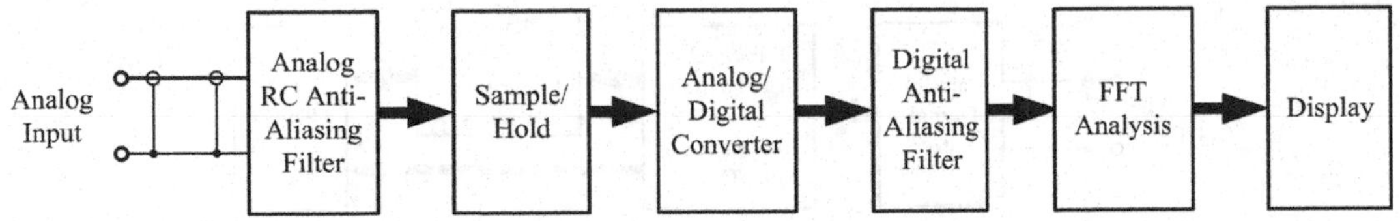

Figure 10.21 Simple Schematic of a FFT Analyzer

analyzer if its characteristics are known. This is not so important with FFT analyses that are conducted in the frequency domain. However, it is especially important for transient vibration measurements, particularly when it is desired to reconstruct transient signals in the time domain from digitized components of the signals. Also, in multi-channel FFT analyzers, the anti-aliasing filters for all of the channels must have matched group delay characteristics. The above requirements can be achieved by using digital FIR filters. These filters can be designed so that all filters have excellent and equal phase characteristics.

A Delta-Sigma A/D converter arrangement can be used to economically achieve a constant group delay through the anti-aliasing filters of the arrangement (Figures 10.13 and 10.21). An initial simple RC low-pass filter is placed before the A/D converter. The corner frequency for this filter is set substantially higher than the desired sampling frequency f_s for a FFT analysis so that the group delay of the output signal from the filter at frequencies between zero and $f_s/2$ will be constant. The sampling interval T_s [equation (10.26)] in the A/D converter for digitizing the analog signal is set very small [$T_s << 1/(2f_{max})$], and consequently, the sampling frequency will be significantly greater than the desired sampling frequency f_s of the analysis. The significantly over sampled digitized signal is then passed through a digital FIR anti-aliasing filter with a very steep slope in the filter transition band. The digital FIR filter is designed to reduce the upper frequency limit of the signal to $f_s/2$. This removes the signal noise, frequency content above $f_s/2$, and the periodic replications of the signal that arise from the digitizing process. The dynamic range of the signal is also improved.

10.7 Dynamic Signal Analysis

Two types of frequency analyzers are common when conducting dynamic signal analyses: (1) *real-time analyzers* and (2) *FFT analyzers. Real-time analyzers* normally use constant percentage bandwidth filters. Examples of constant percentage bandwidth filters are 1/1 octave, 1/3 octave, 1/12 octave, and 1/24 octave filters. There are three 1/3 octave band filters in each 1/1 octave frequency band, twelve 1/12 octave band filters in each 1/1 octave frequency band, and twenty-four 1/24 octave band filters in each 1/1 octave frequency band. The frequency passbands of these filters are a constant percentage of the filter center frequency (logarithmic in frequency). For example, the center frequencies of 1/1 octave band filters double for each successively higher filter center frequency (1.0. 2.0, 4.0, 8.0, 16, 31.5, 63,

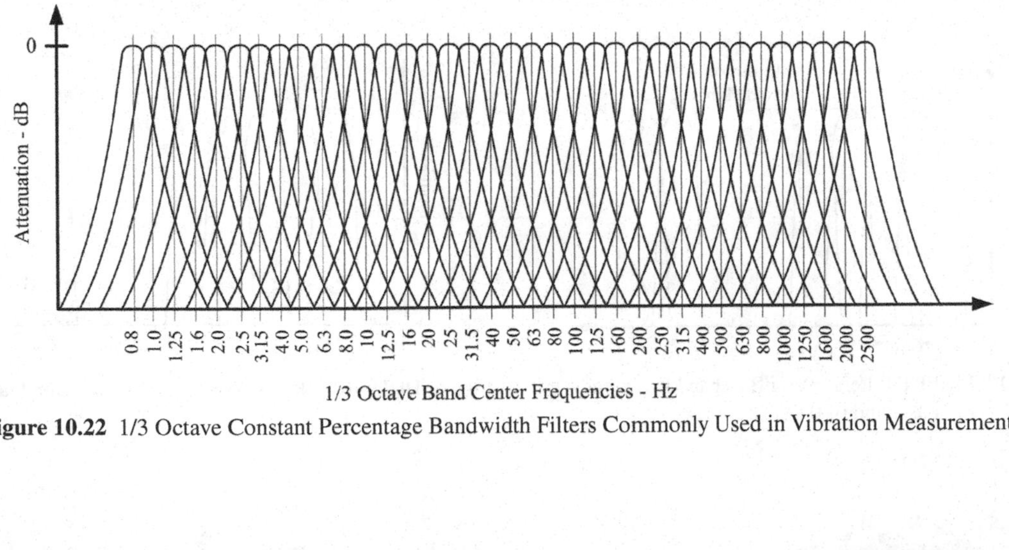

Figure 10.22 1/3 Octave Constant Percentage Bandwidth Filters Commonly Used in Vibration Measurements

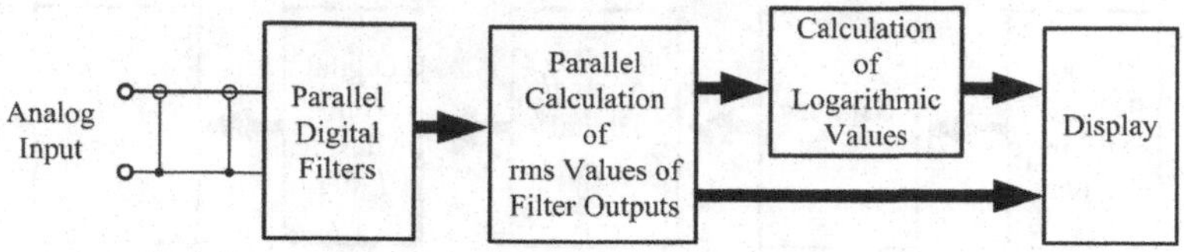

Figure 10.23 Simple Schematic of a Real-Time Analyzer

125, 250, 500, 1000, 2000, etc. Hz). The frequency passbands of 1/1 octave band filters also double for each successively higher filter center frequency. Figure 10.22 shows a 1/3 octave band filter set that is typically used for vibration measurements. The filter center frequencies are presented on a logarithmic scale in Figure 10.22. The frequency passbands of the 1/3 octave band filters appear to be the same on a logarithmic frequency scale.

Figure 10.23 shows a simple schematic of a *real-time analyzer*. The potential negative effects of aliasing that must be considered when analog signals are digitized before they are processed are not a factor in real-time analyzers. Anti-aliasing filters are not needed for real-time analyzers because analog signals directed into these analyzer are not digitized. They are simultaneously filtered by a set of parallel digital filters. The analog filtered signals within the respective filter passbands are directly processed to determine their rms values and their corresponding log values, when desired, and then displayed on the analyzer display screen.

Figure 10.21 shows a typical schematic of a *FFT analyzer* used for vibration analyses. The analog signal is digitized using a Delta-Sigma D/A converter arrangement (Figure 10.21). The digitized signal is then processed to perform the FFT analysis.

Whereas the frequency passbands of octave and fractional octave band filters increase with increasing filter center frequency, a FFT "filter bank" uses linear frequency. All filters have the same bandwidth, are symmetric relative to linear frequency, and are equally spaced. Figure 10.24 shows a representation of the filter passbands of a FFT analyzer for a baseband frequency passband of 0 to $f_s/2$. Figure 10.25 shows a similar representation for a passband filter with a frequency passband of f_L to f_U. The frequency axes in Figures 10.24 and 10.25 are linear axes.

FFT analyzers convert analog signals into blocks of digitized points in increments of 2 raised to the n power (2^n). For example, analog signals can be digitized into data blocks of:

2^8	256 points
2^9	512 points
2^{10}	1,024 points
2^{11}	2,048 points
2^{12}	4,096 points
⋮	⋮

The relationship between the number of digitized points N in the time period T_{sb} of the signal block and the upper frequency limit of the baseband bandwidth (BW) of 0 to $f_s/2$ is given by:

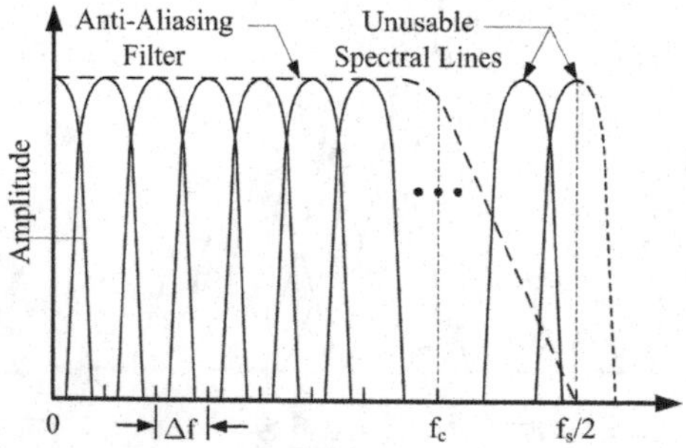

Figure 10.24 FFT Filters over a Baseband Frequency Passband of 0 to $f_s/2$

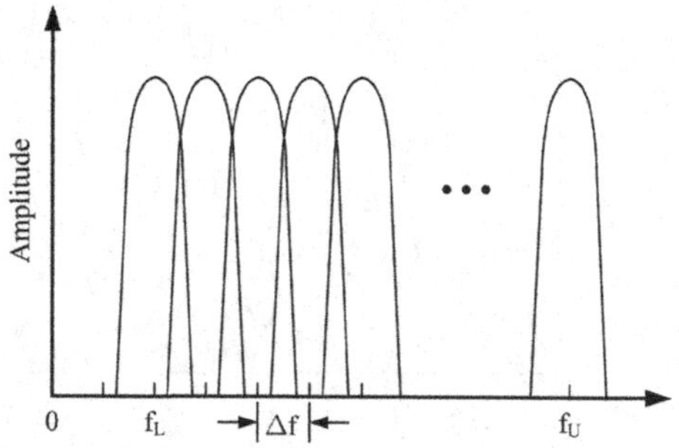

Figure 10.25 FFT Filters over a Frequency Bandpass of f_L to f_U

$$\left[\frac{f_s}{2}\right] = \frac{N}{2}\frac{1}{T_{sb}}. \quad (10.43)$$

$f_s/2$ is the upper frequency limit of the baseband [Figures 10.9(a), 10.17, and 10.24]. The number of useful spectral lines in the signal block is less. How much less is a function of the steepness of the slope of the anti-aliasing filter. The steeper the slope, the greater the number of useful spectral lines.

The total number of spectral lines in a FFT analysis equals N/2, assuming that a real FFT is obtained. That is, there are N/2 spectral lines from DC to $f_s/2$. The spectral lines between the corner frequency f_c of the anti-aliasing filter and $f_s/2$ are in the filter transition band (Figure 10.17 and 10.24). They are contained in the FFT analysis but should not be used. Most FFT analyzers block the viewing of these spectral lines.

The number of unusable (or unviewable) spectral lines in the filter transition band is a function of the steepness of the slope of the anti-aliasing filter. Anti-aliasing filters for some FFT analyzers are designed so that 78.125 percent of the total number of spectral lines (for example, 800 lines out of 1024) within the filter baseband of zero to $f_s/2$ are contained within the filter passband of zero to f_c and are viewable in a FFT analysis (2.56:1 in terms of sampling rate). This percent value can vary between different FFT analyzers if their sampling rates are different. For some modern FFT analyzers, the number of viewable spectral lines within the filter passband can be as high as 95 percent (for example, 980 lines out of 1024) of the frequency lines within the filter baseband.

When performing a zoom analysis, complex FFT values are obtained, and the time block size is equal to N digitized points. Frequency and time information are never lost. N real digitized points result in N/2 complex FFT values.

From Fourier transform theory, the number of spectral lines $N_{(\text{spectral lines})}$ in a FFT analysis is given by (when 78.125 percent of the total number of frequency lines are usable):

$$N_{(\text{spectral lines})} = \frac{0.78125\,N}{2}. \quad (10.44)$$

Substituting equation (10.44) into equation (10.43) yields:

$$f_c = \frac{N_{(\text{spectral lines})}}{T_{sb}}. \quad (10.45)$$

The values for f_c (upper frequency limit of FFT analysis) and $N_{(\text{spectral lines})}$ (number of spectral lines in FFT analysis) are typically selected as settings in FFT analyzers. When this is done, the time period T_{sb} of the digitized signal blocks is:

$$T_{sb} = \frac{N_{(\text{spectral lines})}}{f_c}. \quad (10.46)$$

The frequency bandwidth Δf of the FFT spectral lines is:

$$\Delta f = \frac{f_c}{N_{(\text{spectral lines})}}. \quad (10.47)$$

The sampling rate T_s or the time span Δt between digitized points is given by:

$$T_s = \frac{T_{sb}}{N}. \quad (10.48)$$

The lowest frequency f_{min} that a FFT analysis can process is limited by the time span T_s of the sampled data blocks. f_{min} is given by:

$$f_{min} = \frac{1}{T_{sb}}. \quad (10.49)$$

EXAMPLE 10.2

The desired passband frequency limit of an analog signal is 1,000 Hz. The desired number of spectral lines in the FFT analysis is 800. Determine:
(a) the time period T_{sb} of the data blocks,
(b) the number of digitized points N in the data blocks,
(c) the sampling rate T_s,
(d) the frequency bandwidth Δf associated with each spectral line, and
(e) the lowest frequency the FFT analysis can process.

SOLUTION

(a) The time period of the data blocks is obtained from equation (10.46):

$$T_{sb} = \frac{800}{1,000} = 0.8 \text{ s}. \tag{10.2a}$$

(b) The number of digitized points in the data block is obtained by rearranging equation (10.44):

$$N = \frac{2\, N_{(\text{spectral lines})}}{0.78125} = 2,048. \tag{10.2b}$$

(c) The data sampling rate is obtained from equation (10.48):

$$T_s = \frac{0.8}{2,048} = 390.6\ \mu\text{s}. \tag{10.2c}$$

(d) The frequency bandwidth associated with each spectral line is obtained from equation (10.47):

$$\Delta f = \frac{1,000}{800} = 1.25 \text{ Hz}. \tag{10.2d}$$

(e) The lowest frequency the FFT analysis can process is given by equation (10.49):

$$f_{min} = \frac{1}{0.8} = 1.25 \text{ Hz}. \tag{10.2e}$$

10.8 FFT Time Domain Windows

When a FFT analyzer performs a FFT analysis, it digitizes a signal segment of an analog signal within the *signal block* time span T_{sb} as is shown in Figure 10.26(b). This signal segment is then replicated, and the original and replicated segments are linked together (repeated in time from $-\infty$ to $+\infty$) as is shown in Figure 10.26(c) to perform the FFT analysis. Most often the beginning and end points of these signal segments do not match up. As a result, signal discontinuities represented by the dashed lines in Figure 10.26(c) occur at the points where these signal segments are linked together. This results in a phenomenon called *leakage*. Leakage is a bleeding of signal frequency content from one FFT spectral line to adjacent lines. The FFT peak frequency

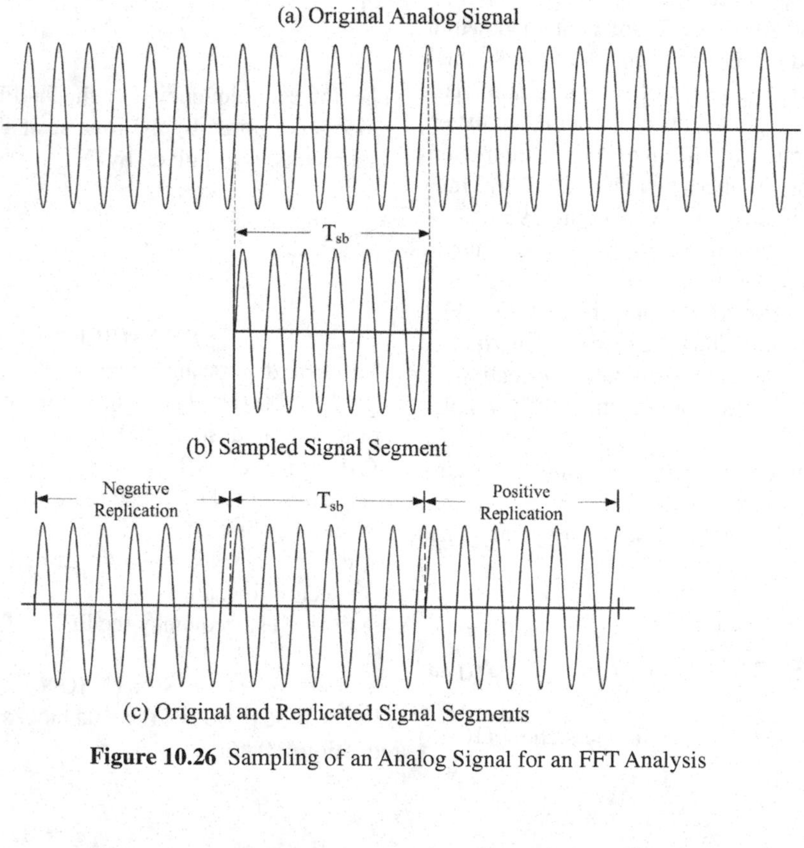

Figure 10.26 Sampling of an Analog Signal for an FFT Analysis

amplitudes will decrease and the resulting frequency spreading can result in a very distorted representation of the FFT spectrum.

Rectangular, Hanning, and Flat-top Time Domain Windows

FFT analyzers use time domain windows that are designed to decrease leakage. Three common time domain windows are shown in Figure 10.27. Figure 10.27(a) shows a *rectangular time domain window*. This window passes the analog signal within the sampling time span without any modifications and thus this windows is often referred to as "none" or "rectangular". When the end points of the signal time segments do not perfectly match up when they are linked together to perform the FFT analysis, leakage will occur. Figure 10.27(b) shows a *Hanning time domain window*. This window shapes the amplitudes of the analog signal within the sampling time span so that the portions of the signal on either end of the sampling time span are forced to progressively go to zero while the portion of the signal within the center of the sampling time

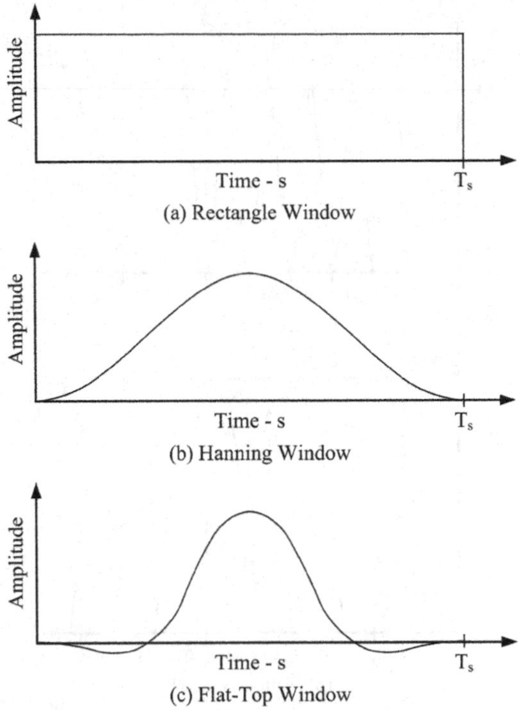

Figure 10.27 Typical FFT Time Domain Windows

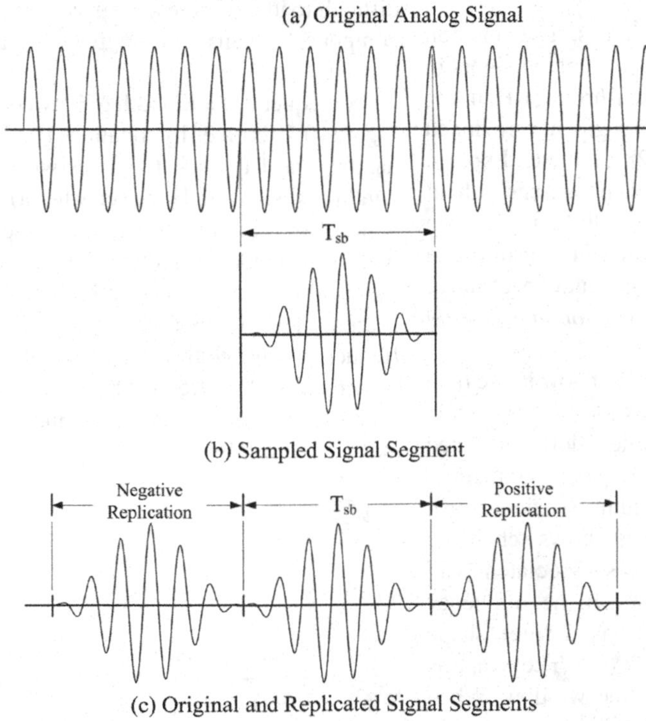

Figure 10.28 Sampling of the Analog Signal in Figure 10.15 Using a Hanning Window

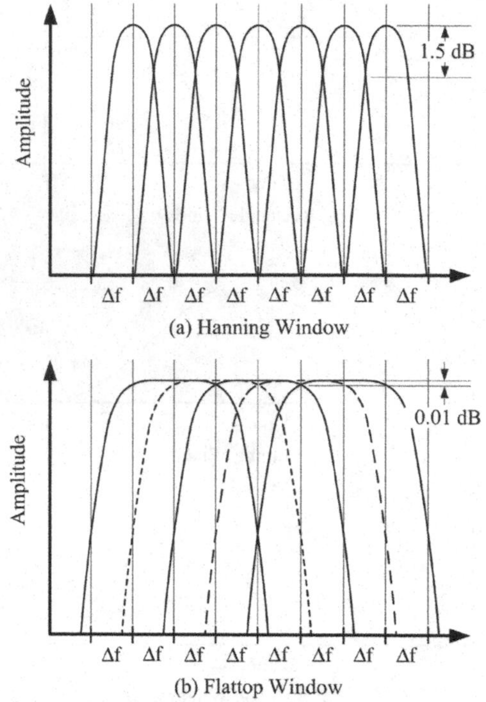

Figure 10.29 FFT Frequency Domain Filters

span maintains its original amplitudes. The function is a half sine. Figure 10.28 shows the analog signal in Figure 10.26 that has been processed through a Hanning window. When the sampled signal segments are linked together as shown in Figure 10.28(c), the end points, which have zero amplitude, are linked together without discontinuities. This minimizes the leakage that can occur with the rectangular window while resulting in an accurate representation of the frequency spectrum of the signal. Figure 10.27(c) shows a *flattop time domain window*.

The effective width of the flattop window in the time domain is smaller than the effective width of the Hanning window. Therefore, the effective filter width associated with the flattop window in the frequency domain is wider than the effective filter width associated with the Hanning window. Figure 10.29 shows schematic representations of the filter shapes associated with the Hanning and flattop windows in the frequency domain. The Hanning window has a filter shape with a rounded tops [Figure 10.29(a)]. The shape is the frequency transform of the time window. When the signal component contained within a frequency passband Δf is broadband, it will be more accurately measured when using the Hanning window. As a result, the Hanning window is used to conduct FFT analyses of broadband analog signals. However, when a signal component within a frequency passband Δf is periodic (or harmonic) with a frequency that is near to either the upper or lower boundaries of the frequency passband associated with the Hanning window, the signal can be attenuated by up to 1.5 dB (16 percent) when the Hanning window is used. This reduction is generally unacceptable if accurate measurements associated with periodic signal components are desired.

Challenges associated with the Hanning window and periodic signals can be resolved with the use of the flattop window. The flattop window has a broader flat filter passband in the frequency domain as shown in Figure 10.29(b). The maximum attenuation that can occur with a periodic signal within the filter passband Δf is 0.01 dB (1 percent). This increase in accuracy with periodic signals comes with a price. The top of the filter passband has been flattened at the expense of widening the filter passband. As a result, a loss in frequency resolution occurs with the flattop window, as compared to the Hanning window. It can be more difficult to identify closely spaced harmonic frequency components in an analog signal when a flattop window is used.

The frequency bandwidth Δf between spectral lines in Figure 10.29 is given by equation (10.47). The frequency bandwidth of the spectral lines when the rectangular window is used is Δf. However, when a Hanning or flattop window is used, the effective time span of the signal within the window is reduced because the signal within the window is attenuated at both ends of the window [Figures 10.27(b) and (c)] . As a result, the effective frequency bandwidth associated with each spectral line is increased. The effective frequency bandwidth Δf_e of each spectral line when a Hanning or flattop window is used is given by:

$$\Delta f_e = B_e \, \Delta f \tag{10.50}$$

where:

$$B_e = \frac{\text{mean}(\mathbf{w} \cdot \mathbf{w}^*)}{\text{mean}(\mathbf{w}) \cdot \text{mean}(\mathbf{w})} \tag{10.51}$$

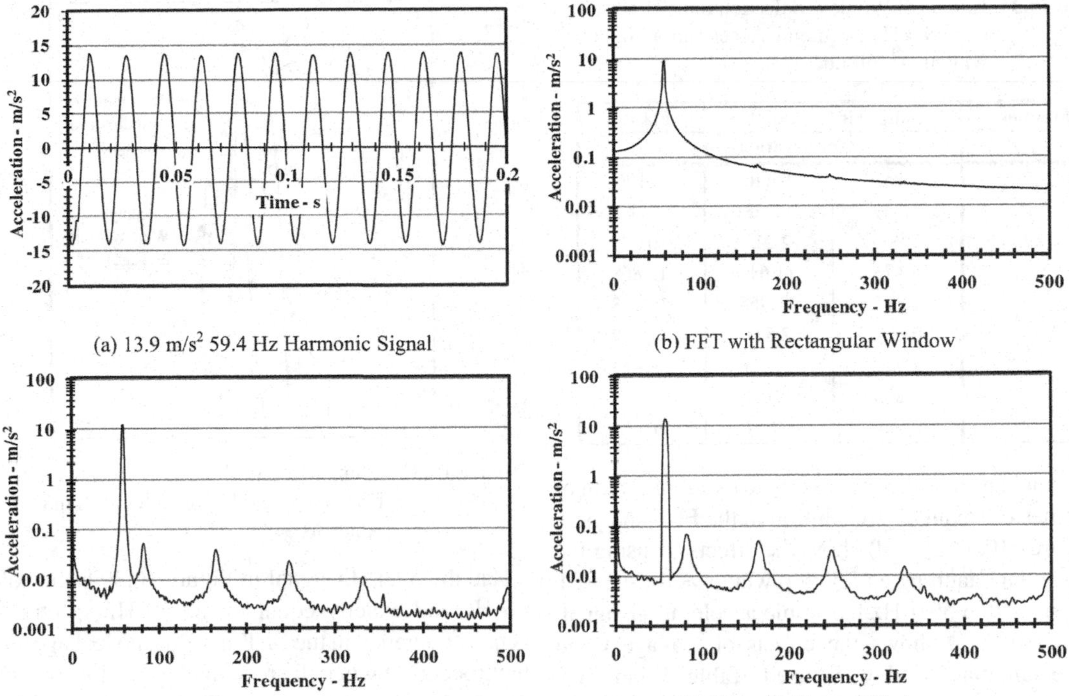

(a) 13.9 m/s² 59.4 Hz Harmonic Signal

(b) FFT with Rectangular Window

(c) FFT with Hanning Window

(d) FFT with Flattop Window

Figure 10.30 FFT Analysis of a Harmonic Analog Acceleration Wave Form

and **w** is the Fourier transform of the function that describes the window shape in the time domain. Values for B_e for rectangular, Hanning, and flattop windows are given in Table 10.1. Figure 10.29 shows that when a Hanning or flattop window is used, the frequency spacing Δf between FFT spectral lines remains unchanged. However, the effective bandwidth associated with each spectrum line when a Hanning or Flattop window is used is spread over adjacent spectral lines in the FFT spectrum. For the Hanning window, the effective bandwidth of each spectral line is 1.5 times the frequency spacing Δf between spectral lines. For the flattop window, the effective bandwidth of each spectral line is 3.43

Table 10.1 Values of B_e

Window Type	Value of B_e
Rectangular	1.00
Hanning	1.50
Flattop[1]	3.43

[1] Value may vary, depending on flattop window used.

times the frequency spacing Δf between spectral lines. Therefore, spectral lines associated with the effective bandwidth will overlap when the flattop window is used. This value varies with different flattop windows. The ISO standard has specified a ISO flattop window, making sure that its parameters are well understood.

Figure 10.30 demonstrates the effects of the rectangular, Hanning, and flattop windows on a FFT analysis of a 59.4 Hz harmonic acceleration signal that has a peak amplitude of 13.9 m/s². The results are for a sampling time window T_{sb} of 0.8 s with a delta frequency Δf of 1.25 Hz. The selected overall frequency passband is 0 to 500 Hz, and the resulting number of spectral lines is 400. With this FFT configuration, 59.4 Hz falls close to the lower boundary of the 1.25 Hz frequency passband centered at 60 Hz.

Table 10.2 shows the numerical FFT analyses results for the 55 to 65 Hz spectral lines. Notice that, since 59.4 Hz is near the lower boundary of the 1.25 Hz frequency passband centered at 60 Hz, it is also detected and has a significant presence in the frequency passband centered at 58.75 Hz. To more clearly identify the 59.4

Table 10.2 Effects of Window Selection on FFT Results of a 59.4 Hz Harmonic Acceleration Signal with an Amplitude of 13.9 m/s²

Frequency	Rectangualr	Hanning	Flattop
Hz	Acceleration - m/s²		
55	1.253	0.110	0.196
56.25	1.749	0.328	3.644
57.5	2.900	2.215	10.971
58.75	8.473	11.643	13.868
60	9.195	11.965	13.878
61.25	2.981	2.511	11.219
62.5	1.779	0.346	3.909
63.75	1.268	0.114	0.236
65	0.986	0.052	0.009

Hz frequency, it will be necessary to use a minimum of 800 lines of frequency resolution in the FFT analysis.

Figures 10.30(a) - (d) show the effects of using the rectangular, Hanning and flattop windows for the FFT analysis of the 59.4 Hz harmonic acceleration signal. Figure 10.30(b) shows the effects of leakage when the rectangular window is used. Table 1 indicates the measured amplitude of the signal is significantly attenuated because of leakage. An examination of Figures 10.30(c) and (d) also indicates the low-level frequency peaks in the signal above 60 Hz are totally obscured by the effects of leakage. Figure 10.30(c) shows the effects of using the Hanning window. The effects of leakage are minimized. However, because the signal frequency is close to the lower boundary of the 60 Hz spectral line, the percent (%) attenuation of the signal for this line is:

$$\% \text{ attenuation} = 100\,\frac{13.9-11.965}{13.9} = 13.9\,\%. \tag{10.52}$$

This is consistent with what would be expected with the use of the Hanning window. Figure 10.30(d) shows the effects of using the flattop window. The figure shows the effects of the wider effective filter bandwidth associated with the flattop window. The percent attenuation of the signal amplitude for the 60 Hz spectral line is:

$$\% \text{ attenuation} = 100\,\frac{13.9-13.878}{13.9} = 0.16\,\%. \tag{10.53}$$

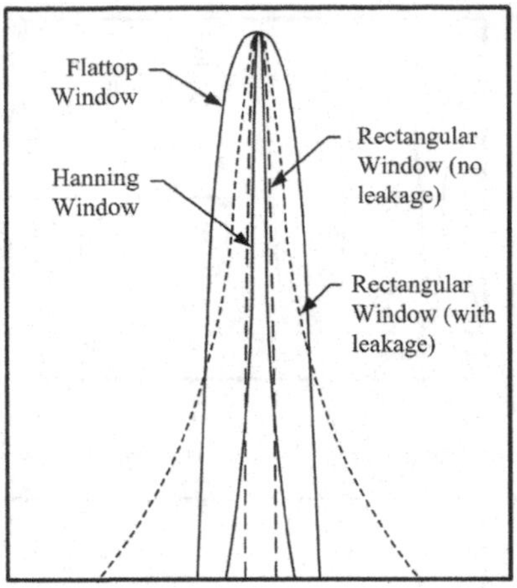

Figure 10.31 Comparison of the Use of Rectangular, Hannning, and Flattop Windows in an FTT Analysis

Had the 59.4 Hz signal in Figure 10.30 been exactly 60 Hz and exactly equal to the 60 Hz spectral line center frequency in the FFT analysis, no leakage would have occurred when the rectangular window was used. The 60 Hz component of the FFT analysis would have appeared as a narrow straight band on the frequency axis in Figure 10.30(a) and the rest of the spectrum would have look similar to those in Figures 10.30(c) and (d).. Figure 10.31 shows a comparison of the FFT spectrum shapes associated with a single spectral line when a rectangular window with and with leakage, a Hanning window, and a flattop window are used.

Transient Signals

It is sometimes necessary to conduct FFT analyses of transient signals. Figure 10.32(a) shows a representative transient signal, which decays to zero over the desired sampling time span. Using a Hanning window to process this signal will result in a spectrum similar to the one shown in Figure 10.30(c) (only without the low-level higher frequency peaks). This would indicate that the energy contained in the transient signal is primarily confined to a narrow frequency passband centered about the frequency of the signal. In fact, this is not characteristic of transient signals. The energy contained in transient signals is generally spread over a wider passband of frequencies, and the actual spectrum of the signal in Figure 10.32(a) should look more like the spectrum in Figure 10.30(b) (only with a less

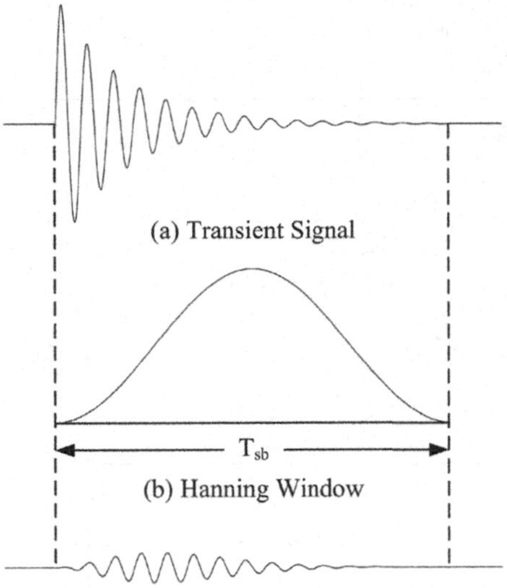

(c) Windowed Transient Signal

Figure 10.32 Windowed Transient Signal with a Hanning Window

predominant peak). The shorter the time span of the transient signal in Figure 10.32(a), the broader will be the spread of spectral energy about the frequency of the signal.

The beginning and end points of transient signals normally have zero (or near zero) amplitudes. Therefore, leakage will be low and the rectangular time window will normally yield accurate results when conducting FFT analyses of these signals. Transient signals are called *self-windowing functions* and time windows, such as the Hanning or flattop window, should not be used when conducting FFT analyses of these signals.

Force and Response Windows

FFT analyzers are often used to determine the vibration response characteristics of structural systems to shock and vibration inputs. This is accomplished by determining the input-output relationships or transfer functions associated with these systems. This type of experimental analysis with a FFT analyzer requires a FFT analyzer with a minimum of two input channels. The analytical bases for determining system frequency response functions are discussed in Chapter 7.

To experimentally determine the structural response characteristics or transfer function of a structural system with a FFT analyzer, an input is applied to some point on the system by means of an impact hammer. This hammer will have an embedded force transducer that can measure the force of the hammer strike. The corresponding response of the system to this input is measured either at the input location or another location on the system with an accelerometer. The input transient and related response signals are then processed by the FFT analyzer to obtain the transfer or system response function of the system.

The impact hammer is used to strike the structural system at a specified location. The force transducer in the hammer measures the amplitude of the input force pulse to the system. Figure 10.33(b) shows the normalized frequency content of a 1 ms rectangular force pulse [Figure 10.33(a)]. As the time span T of the force pulse approaches zero (becomes an impulse), the normalized amplitude of the frequency spectrum will approach 1 at all frequencies. The input associated with the rectangular pulse will excite all system resonance frequencies that are contained within the pulse frequency bandwidth.

In reality, a force hammer strike usually does not

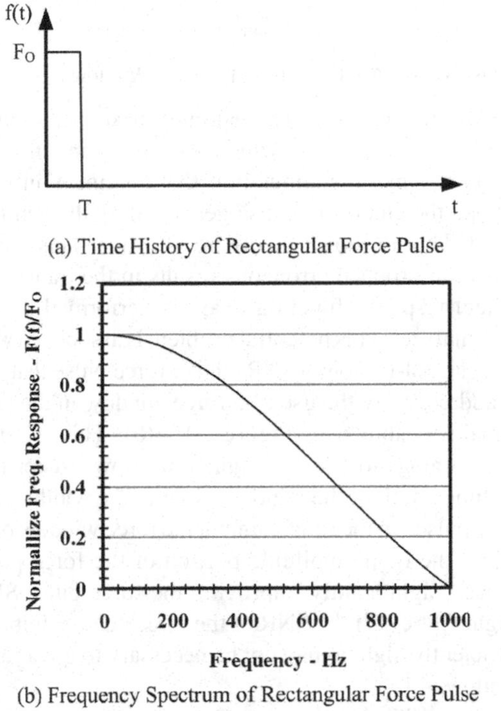

Figure 10.33 Time and Frequency Characteristics of a 1 ms Rectangular Force Pulse

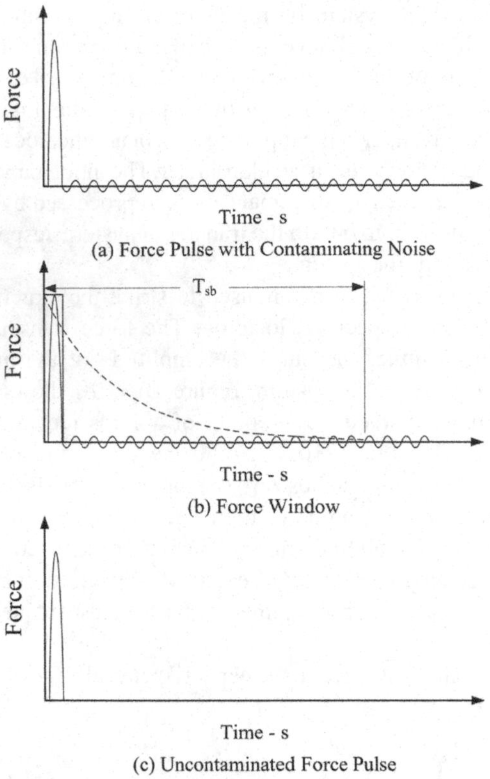

Figure 10.34 Application of the Force Window

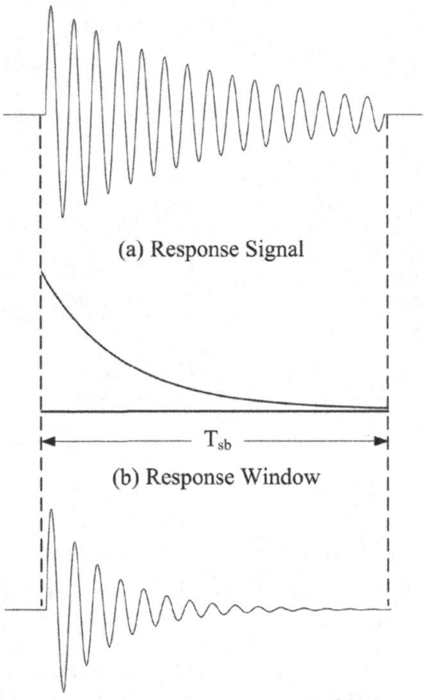

Figure 10.35 Application of the Response Window

produce a pure pulse. The hammer strike can produce a pulse of finite width, which also may contain noise or other signal contamination that continues in time beyond the end of the pulse generated by the hammer strike. This is shown in Figure 10.34(a). This signal noise can produce erroneous results in the transfer or system response function analysis performed by the FFT analyzer. This potential problem is associated with a low signal-to-noise (SNR) in the force pulse that can be addressed by the use of a force window in the time domain as shown in Figure 10.34(b). This window looks similar to the rectangular window except that the time width of the window is set to the width of the force pulse. In the time domain, the force window only allows the high-amplitude portion of the force pulse to pass, significantly improving the force pulse SNR [Figure 34(c)]. If the SNR of the force pulse is initially sufficiently high, it may not be necessary to use a force window.

Figure 10.35(a) shows a typical structural system response signal that does not go to zero amplitude by the end of the sampling time span T_{sb}. To prevent the effects of leakage from producing erroneous results in the transfer function or system response analysis performed by the FFT analyzer, the amplitude of the response function must be forced to go to zero by the end of the sampling time span. This can be accomplished in the time domain with the use of the response window. The response window is an exponential function e^{-at} that decays with time as shown in Figure 10.35(b). The value of a in the exponential function is chosen so that, when the structural system response values are multiplied by e^{-at}, the system response will go to zero by the end of the sampling time span as shown in Figure 10.35(c). The use of the exponential decay function in the response window introduces additional damping to the response signal. If the identification of damping in a structural system is important, care must be taken in the proper use the response window.

The outputs of the force and response windows must be matched to properly obtain the transfer or system response function. To accomplish this, the force pulse signal within the force window time span is exponentially attenuated with a similar exponential decay function

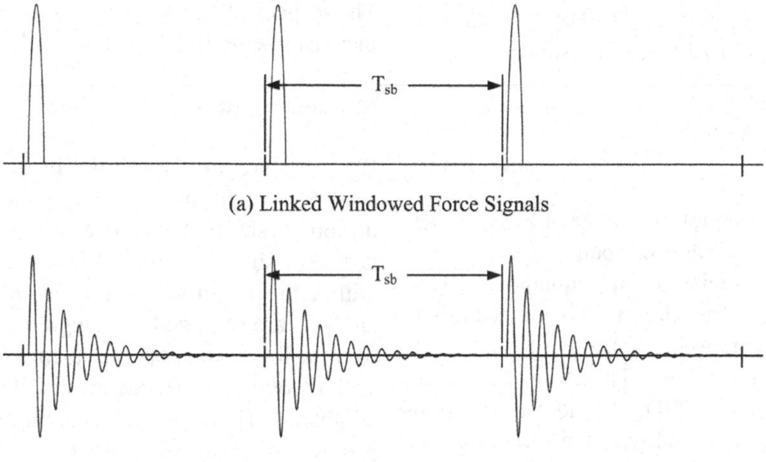

Figure 10.36 Linked Windowed Force and Response Signals

to the one that is used to attenuate the response signal within the response window [Figure 10.34(b)] . The force signal is then forced to zero outside the force window.

Normally, hammer strikes must be used to acquire several input force and system response signals that are necessary for the FFT analyzer to perform a structural system transfer or system response function analysis. These signals must then be linked together as shown in Figure 10.36 to perform the FFT analysis. The proper use of the force and response windows in the data collection will ensure that the effects of leakage will be minimized in the analysis.

A pre-trigger setting must be used for the hammer strikes to make sure that the first couple of digitized time data points at the beginning of the force pulse within the force time window are zero. This is necessary to ensure the correct calculation of the transfer or system response function during the FFT analysis.

10.9 Spectral Functions and Units

The function and resulting units that should be used to quantify the spectral amplitudes of a FFT analysis depend on the type of signal being analyzed. There are three basic types of signals: complex or almost periodic, transient, and random.

Spectral Functions

The primary spectral function in a frequency analysis is the *mean squared value* $\Psi_f(f_o,\Delta f_c)^2$ within a specified effective filter bandwidth Δf_e:

$$\Psi_f(f_o,\Delta f_e)^2 = \frac{1}{T}\int_0^T f(t,f_o,\Delta f_e)^2\,dt \tag{10.54}$$

where T is the time span of the signal and f_o is the center frequency of the effective filter bandwidth. Because power is normally proportional to engineering units (EU) squared, equation (10.54) is often referred to as the *power spectrum (PS)* of f(t) and has units of EU^2. When equation (10.54) is divided by the effective filter bandwidth Δf_e, the resulting expression is the *power spectral density (PSD)* $G_f(f)$ of f(t) given by:

$$G_f(f) = \frac{1}{\Delta f_e}\left[\frac{1}{T}\int_0^T f(t,f_o,\Delta f_e)^2\,dt\right]. \tag{10.55}$$

The PSD units are EU^2/Hz. Often it is desired to present the results of a frequency analysis in terms of the root-mean-squared (rms) value of the engineering units. These results are obtained by taking the square roots of equations (10.54) and (10.54). When this is done, the PS becomes the spectrum (S) of f(t) with units of EU_{rms}, and the power spectral density becomes the spectral density (SD) with units of $EU_{rms}/\sqrt{Hz}$. The PSD can be written in terms of the Fourier transform of f(t):

$$G_f(f)=\frac{1}{T}\mathbf{F}(jf)\cdot\mathbf{F}(jf)^* \tag{10.56}$$

where **F**(jf)* is the complex conjugate of **F**(jf) and T is the analog signal sampling time span.

The discussions associated with equations (10.54) through (10.54) assume that f(t) is deterministic (complex or almost periodic or transient) or ergodic and is continuous over time. Therefore, the values associated with the PS, PSD, S, and SD functions represent averages integrated over the sampling time period T. When a Hanning or flattop window is used for the FFT analysis, the effective filter bandwidth Δf_e in equations (10.56) and (10.57) and implied in equation (10.58) is given by equation (10.52).

Periodic Signals

Periodic signals contain only harmonic components. They can be signals with only a single sinusoidal component or with multiple sinusoidal components. The sampling filter bandwidth Δf of the FFT analyzer should be selected small enough so that only a single periodic frequency contained within a signal will be present within an individual filter bandwidth and represented on a corresponding single spectral line in a FFT analysis. The flattop window should be used to ensure that the peak values of the sinusoidal elements within a signal can be accurately measured. The use of the flattop window, as opposed to the Hanning window, may have to be balanced with the requirement of being able to identify closely-spaced sinusoidal elements within a signal. All the power in any single filter bandwidth will be concentrated at a single frequency, will be independent of the filter bandwidth, and will be presented at a single spectral line. For this case, the spectrum (S) in units of EU_{rms} or power spectrum (PS) in units of EU^2 should be used to quantify the FFT values. If, for whatever reason, the measurement filter bandwidth Δf and the corresponding effective filter bandwidth Δf_e are changed, the corresponding S and PS values within the respective filter bandwidths will not change. This will not be the case if spectral density (SD) or power spectral density (PSD) were to be used. The respective SD and PSD values will decrease with increasing effective filter bandwidth Δf_e.

Random Signals

Random signals normally have spectra that are reasonably continuous in the frequency domain. They do contain sharp spectral peaks associated with specific harmonic signal elements. The amplitudes of the power within the effective filter bandwidths Δf_e of a FFT analysis are obtained by integrating the signal power over the lower and upper frequency limits of Δf_e. There will be continuous frequency distributions within these passbands. Therefore, the Hanning window should be used when the signal is an ergodic random signal. The power within the effective frequency passbands and presented at the corresponding spectral lines must be normalized with respect to Δf_e. The power, therefore, must be specified in terms of the spectral density (SD) in units of $EU_{rms}/\sqrt{Hz}$ or power spectral density (PSD) in units of EU^2/Hz. The sampling frequency passband Δf should be selected small enough so that SD and PSD will have small variations within the respective effective frequency passbands of the analysis. When this is done, the SD and PSD values will not significantly change should the sampling frequency passband be changed. On the other hand, if the spectra (S) or power spectra (PS) is used, the measured S or PS values will increase with increasing bandwidths.

Transient Signals

Transient signals start and end with zero (or near zero) amplitude. They contain finite energy that is dependent on the signal length. The longer the signal length, the greater the energy contained in the signal. Transient signals, like continuous signals, will contain energy that is continuously distributed over frequency. Therefore, the energy in the signal must be normalized with respect to filter bandwidth Δf, and it must also be rescaled with respect to signal length. This is accomplished by rewriting equations (10.55) and (10.56):

$$E_f(f)=\frac{1}{\Delta f}\left[\int_0^{T_s} f(t,f_o,\Delta f)^2\,dt\right] \tag{10.57}$$

and:

$$E_f(f) = \mathbf{F}(jf) \cdot \mathbf{F}(jf)^* \tag{10.58}$$

where the integration time in equation (10.57) and used to calculate the Fourier transform in equation (10.58) is the time span T_{sb} of the transient signal. Note that the rectangular time window is used when obtaining the FFT of transient signals. Equations (10.57) and (10.58) give the energy spectral density (ESD) with units of $EU^2 \cdot s / Hz$.

Combined Signals

Some signals can contain both significant periodic and random components, as well as, possible transient components. An example of such a signal is the acceleration signal of an angle grinder interacting with a workpiece. The signal will contain a strong harmonic component that will correspond to the rotational speed (rpm) of the grinder. This component will be associated with the combined unbalance of the grinder rotational elements and abrasive wheel. The signal will also contain a significant broadband random component that will be associated with vibration caused by the interaction of the grinder wheel with the workpiece. Possible transient components associated with impact forces acting on the grinder due to uneven material properties, etc. may be present.

For this case, it may be necessary to conduct three separate FFT analyses of the acceleration signal. The first will be an analysis designed to measure the spectrum (S) or power spectrum (PS) related to the harmonic components of the signal that are associated with the unbalance of the grinder rotating elements and abrasive wheel. The second will be an analysis designed to measure the spectral density (SD) or power spectral density (SPD) associated with the random (broad band) portion of the signal caused by the interaction of the grinder wheel with the workpiece. The third will focus on attempting to analyze the transients in the data by using energy spectral density (ESD). The results of these FFT analyses will then need to be assessed to totally characterize the vibration signal of the angle grinder.

Basic assumptions must often be made relative to identifying potential signal types (periodic, random, and transient) that may be present in a vibration signal to perform a correct assessment of the resulting FFT spectral values. However, sometimes there may be insufficient knowledge of a vibration signal to make appropriate pre-assessments as to the signal types that may be contained within the signal. When this is the case, different FFT analyses of the vibration signal must be conducted. These analyses will use different time domain windows (rectangular, Hanning, or flattop) and data block sizes with corresponding different values of Δf and measure different spectral values [spectrum (S), spectral density (SD), power spectrum (PS), power spectral density (PSD), and energy spectral density (EPD) values].

Using different values of Δf by changing the number of spectral lines in the overall frequency passband of different FFT analyses, the results of the analyses can be overlaid to identify the signal types present in a vibration signal. When conducting these comparisons, the number of spectral lines in the different FFT analyses should be varied by a minimum factor of four. For example, if 1024 spectral lines are used for the first analysis, a minimum of 4096 spectral lines should be used for the second analysis.

When periodic components exist in the vibration signal and the flattop window is used for the FFT analysis, the spectral (S) or power spectral (PS) values of the spectral lines that correspond to the frequencies of sinusoidal components within the signal will overlay (or nearly overlay) each other when Δf is varied between different FFT analyses of the signal. If random (broadband) components also exist within the same vibration signal, the spectral lines associated with the S or PS values that correspond to these components will not overlay each other. They will increase with increasing Δf.

If the spectral units in the above analyses are changed to spectral density (SD) or power spectral density (PSD) and a Hanning window is used for the analyses, the SD or PSD values that correspond the random (broadband) spectral lines will not change when Δf is change; they will overlay (or nearly overlay) each other. However, the S or PS values that correspond to the frequencies of the sinusoidal components in the signal will decrease when Δf is increased.

The above procedure can be used to identify transient signal components within a vibration signal that may contain other signal types. For this case, energy spectral density (EPD) should be used as the units for the measurements. However, because of the requirement of using the rectangular window to quantify the spectral contents associated with the transient signal components

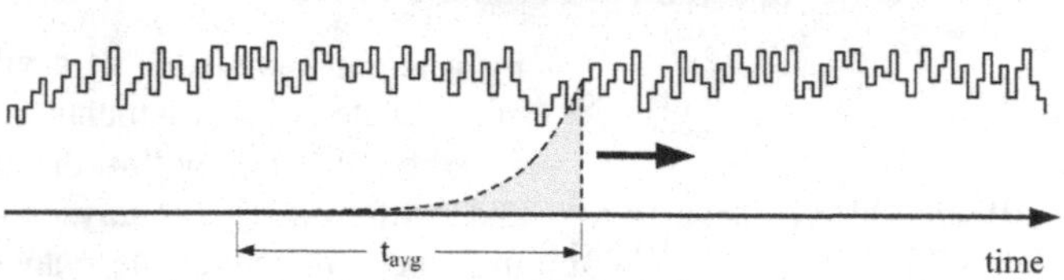

(a) Exponential Averaging

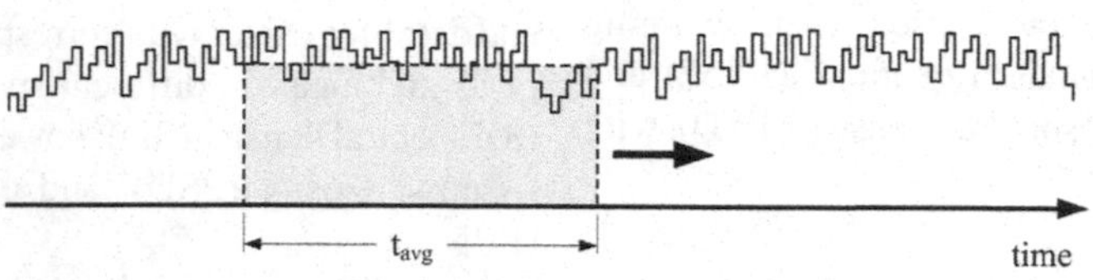

(a) Linear Averaging

Figure 10.37 Exponential and Linear Averaging

and the related potential problems associated with leakage that may be present because of other signal types that may be present, the process of quantifying the transient components within the vibration signal can be difficult.

10.10 Averaging

Exponential and Linier Averaging

When a FFT analyzer conducts a FFT analysis, it digitizes an analog signal and collects data blocks in increments of 256, 512, 1,024, 2,048, etc. points, depending on the analysis frequency passband and the number of spectral lines selected on the FFT analyzer. A FFT analysis is

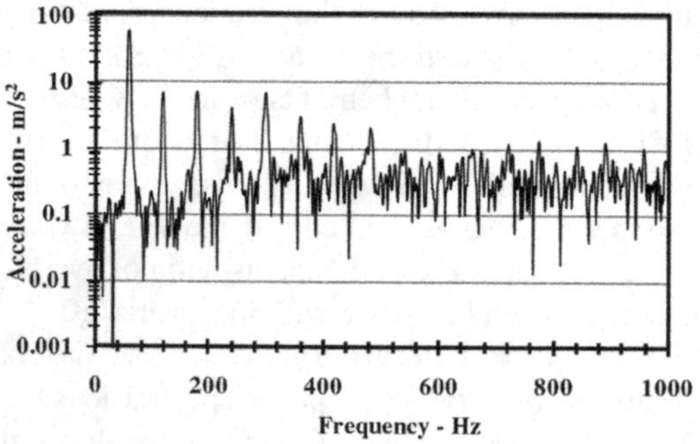

(a) 1 Average

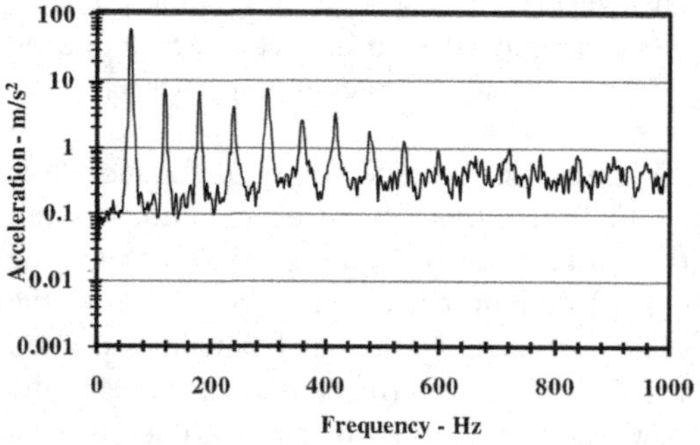

(b) 5 Averages

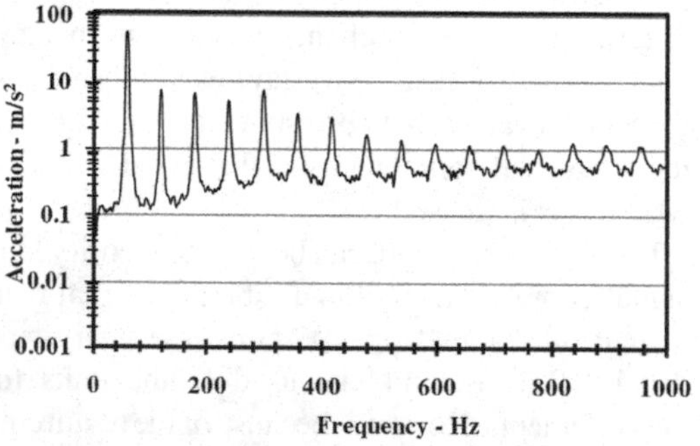

(c) 25 Averages

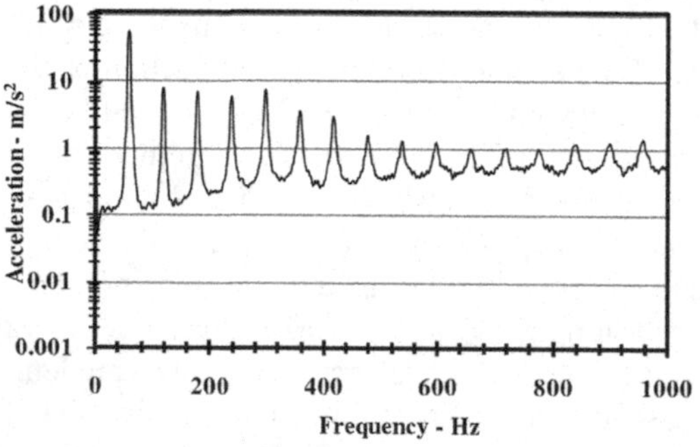

(d) 50 Averages

Figure 10.38 Linear Averaging

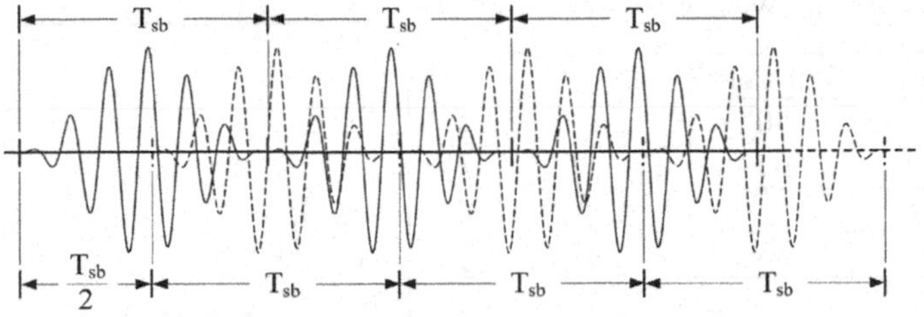

Figure 10.39 50 Per Cent Overlap with Hanning Window

conducted on each data block after it is collected, after which another data block is collected and processed. This process is repeated until the number of data blocks selected for the FFT analysis have been processed.

FFT analyses presume ergodic signals, which rarely occur in real life. Therefore, averaging techniques must be used to properly combine averaging with signal time variance. Normally, only exponential and linear averaging can be selected on FFT analyzers. *Exponential averaging* is very useful when signal amplitudes slowly vary over time. For *exponential averaging*, the FFT analysis results of data blocks are averaged in a manner that weighs the most recent results more heavily and the more distant result less heavily with the most distance result eventually not being averaged into the results [(a)]. Exponential averaging is often used when it is desired to observe smoothed-out time-varying trends of a long analog data record.

The sampling time span T_{sb} of digitized data blocks is often very short. These data blocks contain information to be extracted by the FFT analysis, as well as, signal noise. *Linear averaging* is used to significantly improve the signal-to-noise ratio (SNR) of a FFT analysis of a set of data records. For *linear averaging*, the results of each data block FFT analysis is added to the previous FFT analysis results and the resulting spectral amplitudes are divided by the number of data records being averaged. The results of the FFT analysis of each data block are weighted evenly in the averaging process [Figure 10.37(b)]. Figure 10.38 shows the linear averaged results of a FFT analysis of a vibration signal in which 1, 5, 25, and 50 data blocks were averaged. The figure clearly shows the improvement in SNR that results as more data blocks are linearly averaged.

Averaging and Windows (Overlapping)

When using a flattop or Hanning window, energy is removed from the data and time variance will normally increase. *Overlapping* of consecutive data blocks can be used to recover the lost energy. Figure 10.39 shows an example of the overlapping process for the case where the Hanning window is used. The overlap is normally specified as a percent of the time span T_{sb} of the data records. The overlap of the data records in Figure 10.39 is $T_{sb}/2$ for an overlap of 50 percent.

There is an optimal overlap for each time window (Table 10.3). The time variance will decrease up to some percentage of overlap because new data is added to the averaging process of the data records. No overlap is necessary for the rectangular window because none of the data values in the sampled time blocks are attenuated by windowing.

10.11 Digital-to-Analog (D/A) Conversion

There are particular operations where it is desired to convert a digital vibration signal to a corresponding analog signal. One operation is where the digital output of an electromechanical shaker controller must be converted to a corresponding analog signal that is then directed into the power amplifier that drives the electromechanical shaker.

The digital-to-analog (D/A) conversion process is

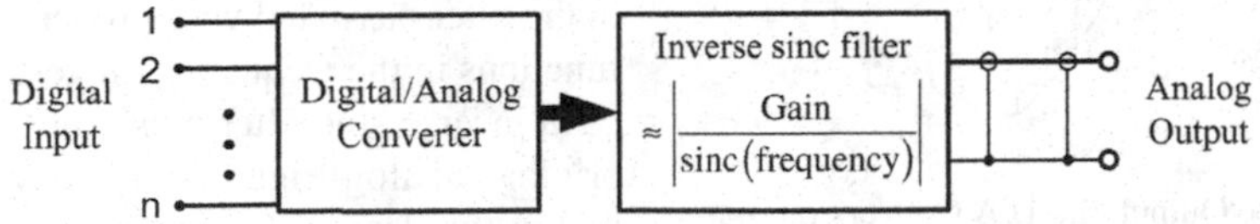

Figure 10.40 Digital-to-Analog Conversion Process

Table 10.3 Optimum Overlaps for Time Windows

Hanning Window	**Per Cent**
Minimum Overlap	50
Optimum Overlap	63
Flattop Window	
Optimum Overlap	83

straight forward and is schematically shown in Figure 10.40. The digital signal sequence is converted to a corresponding analog signal by using an inverse sinc algorithm in the D/A converter that applies equation (10.30) to the points in the digital signal sequence. The output of this process for a harmonic signal element is shown in Figure 10.41. The resulting staircase representation of the harmonic analog signal will add high frequency components to the converted digital signal that were not present in the original analog signal before it was digitized or that are not present in a digital signal created by a digital shaker controller. Since the transfer function associated with the algorithm that converts the digital sequence of points to a corresponding analog signal is a sinc function, the staircase representation of the analog signal can be removed by passing the signal through an inverse sinc filter (Figure 10.42). The outputs of the transfer functions associated with the sinc and inverse sinc functions will cancel each other, eliminating the staircase response and resulting in a smooth sine function.

In theory, the use of an inverse sinc filter is a simple process. However, there are application hurdles that must be overcome. Figure 10.42 shows a plot of the amplitude of the sinc function where frequency is the argument. Figure 10.43 shows the corresponding amplitude of the inverse sinc function. The "zeros" in the sinc function go to infinity in the inverse sinc function. This filter characteristic is highly undesirable in an inverse sinc filter and impossible to realize in an inverse sinc filter. Therefore, it is necessary to develop an approximate inverse sinc filter that will have a bounded response in the filter passband with proper amplitude and phase functions in the frequency range of interest.

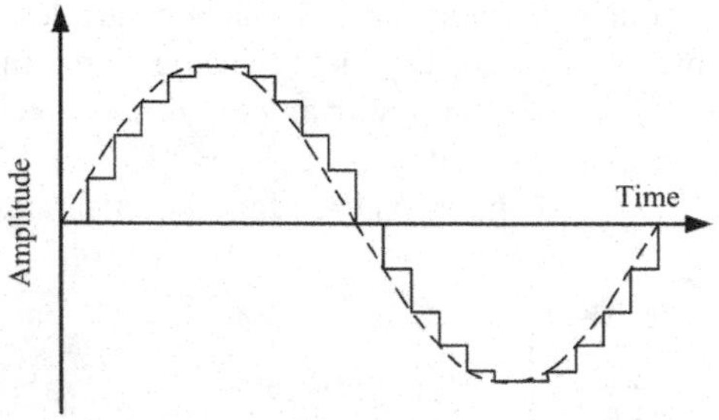

Figure 10.41 Staircase Output of a D/A Conversion of a Harmonic Signal Element

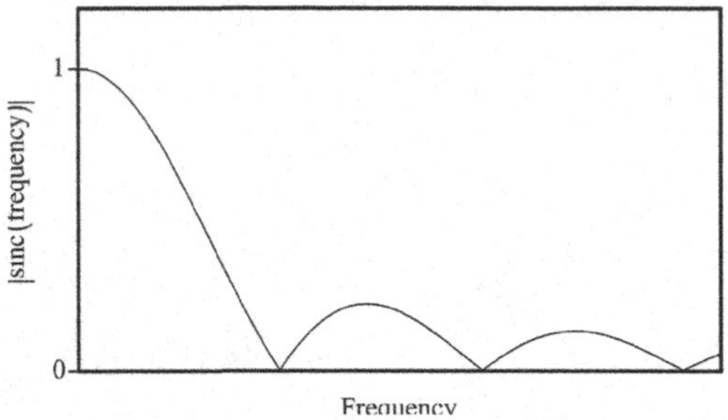

Figure 10.42 Representation of the Amplitude of the Inverse Sinc Function with Frequency as an Argument

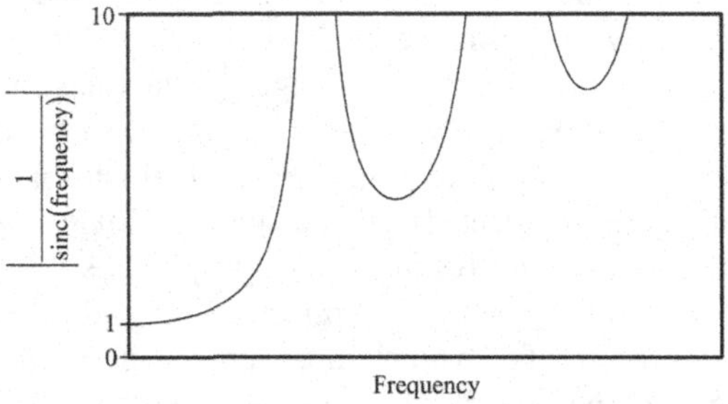

Figure 10.43 Representation of the Amplitude of the Inverse Sinc Function with Frequency as an Argument

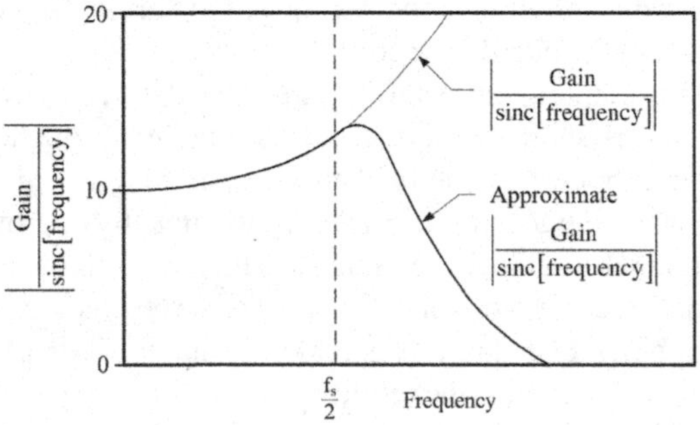

Figure 10.44 Approximate Inverse Sinc Filter

An inverse sinc filter must only accommodate those original analog signal frequency elements that are within the desired frequency passband of the analog

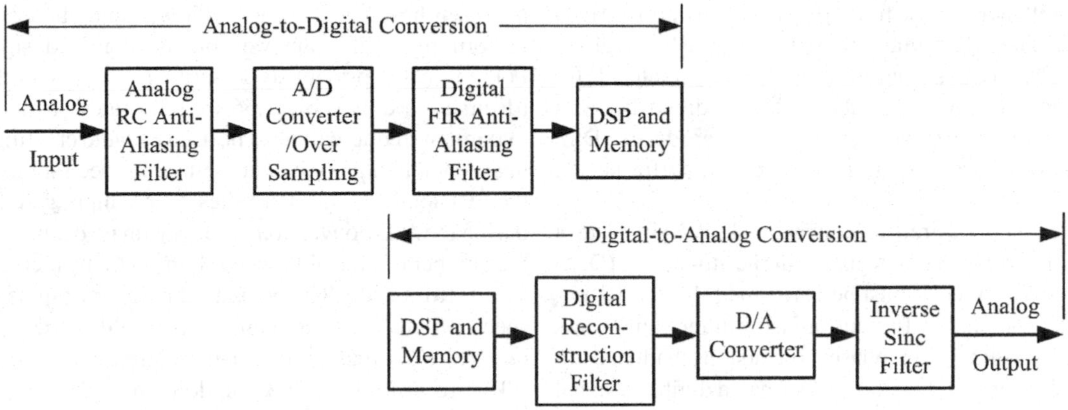

Figure 10.45 A/D and D/A Conversion Processes

signal anti-aliasing filter, which is specified by $f_s/2$. Figure 10.44 shows the response curve of a possible inverse sinc filter. This approximate inverse sinc filter can be used so long as its response corresponds to the theoretical response of the actual inverse sinc function that falls within the filter passband of 0 to $f_s/2$. The gain level of the inverse sinc function shown in Figure 10.45 allows the amplitude of the filter output to be adjusted so it will correspond to the amplitude of the original analog signal.

In the process of converting an analog signal to a corresponding sequence of digitized points in time, the amplitude and phase relationships between all of the frequency elements within the signal must be accurately accounted for when the signal is passed through anti-aliasing filters during the A/D conversion process. This is required so that the analog signal can be accurately reconstructed from the resulting digitized signal during the D/A conversion process. This can be accomplished by knowing the group phase characteristics of the anti-aliasing filters used during the A/D conversion process. This information can then be used to compensate for any phase distortions in the digitized time signal that were introduced during the A/D conversion procession during the D/A conversion process.

The above can be accomplished by placing an initial analog RC anti-aliasing filter that is used in the frequency range where a linear phase relationship exists between the filter input and output before the signal is passed through an A/D converter as shown in Figures 10:13 and 10.21. The signal is then over sampled in the A/D converter by using a sampling frequency that is much higher than that required for the desired frequency passband of the signal. The signal is then passed through a digital FIR anti-aliasing filter where the desired frequency passband of the signal and the sampling interval between the digitized points are accurately controlled and where the amplitude and phase relationships between the filter input and output can be more easily handled. The resulting digitized signal can then be played back through a D/A converter as shown in Figure 10.40 to reconstruct the signal in the time domain.

10.12 Sparse Sampling

FFT analyzers use *sparse sampling* in both the time and frequency domains. By using an interpolation technique, information can be revealed that is initially "hidden" in the data.

The analog-to-digital conversion process is often performed with a fast sampling Delta-Sigma A/D conversion process. The cut-off frequency of the analog RC anti-aliasing filter for this process is typically set much higher than the desired frequency range for a digital signal processing (DSP) operation (Figure 10.45). When this is the case, the desired sampling rate of the conversion process will be significantly lower than that allowed by the RC anti-aliasing filter. The corresponding sampling frequency will also be substantially less than the cut-off frequency of the RC filter. For this case, the A/D converter will sample at the higher rate allowed by the analog RC anti-aliasing filter. For lower desired analyses frequency ranges, the digital FIR anti-aliasing filter will *decimate* (reduce the digital sampling rate and corresponding cut-off frequency for the digital anti-aliasing filter) and digitally filter the digitized signal.

This will increase the time span between digitized data points. The FIR digital filter will also have linear group delays between its input and output, which is valuable for time domain transient analyses. The modified digitized signal will then be placed in memory where desired DSP operations, such as FFT analyses, can be performed on it.

When it is desired to convert a digital signal to a corresponding analog signal, a digital-to-analog (D/A) conversion process must be performed (Figure 10.45). So that the analog RC anti-aliasing filter can be less complex with fewer phase conversion challenges, a digital reconstruction of the decimated digitized signal is carried out before the D/A conversion. Additional data points are inserted in between the actual digitized data points during this reconstruction, using a sinc interpolation process. As a result, the conversion rate of the D/A conversion process will be higher than the sampling rate associated with the corresponding A/D conversion/decimation process. The reconstructed analog signal is then passed through a sinc filter to remove the stair steps shown in Figure 10.41.

Another approach to digital reconstruction can be carried out by using a ten-times *over sampling* during the A/D process with no digital decimation. This implies that the A/D portion of Figure 10.45 will deliver ten times more digitized data points than the sampling theorem requires. As a result, it will not be necessary to perform a ten-point sinc interpolation in between data points during reconstruction of the digitized signal. This approach, however, is not desirable when simultaneously performing a FFT analysis on the digitized signal. For this case, previously described sparse sampling (decimation down to 2:2.56 or 2:2.1 rates versus the maximum frequency) must be used to avoid creating an unacceptably high number of unusable spectral lines in the FFT analysis. If a ten-times over sampling is used during the A/D conversion process with no decimation, the number of unusable spectral lines will be increased by a factor of ten. This increased number of unusable spectral lines without decimation would increase the required FFT analysis computation time.

The resolution of a time domain signal created by converting a frequency domain signal to the corresponding time domain signal can be improved by using *zero-padding*. As an example, a time domain data block with only 1,024 data points can be *up sampled* by a factor of seven by adding another 7,168 zeroes to the end of the data block. When the up sampled time domain data block is Fourier transformed into the frequency domain, there will be 8,192 data points in the frequency domain data block. The Fourier transform, by means of interpolation, will increase the number of data points in the frequency domain by a factor of 7. However, this will not change the frequency resolution of the data. When the frequency data are inverse transformed back into the time domain, the inverse Fourier transform will also interpolate the data points, resulting in a plot of the time domain data that will look "smoother". This is a reconstruction method that is very useful at times.

QUESTIONS - CHAPTER 10

1. Describe the relationship that exists between the time and frequency data associated with an acceleration signal.

2. List the assumptions that are associated with the signal processing of vibration signals.

3. Discuss the relationships that exist between the time and frequency domains for the following situations: (a) Fourier series, (b) Fourier transform (continuous in both time and frequency domain, (c) Fourier transform (sampled time domain), and (d) discrete Fourier transform.

4. What is the source of the quantization noise in the in the analog-to-digital conversion process?

5. What determines the resolution of an A/D converter?

6. What is the resolution of a 16-bit A/D converter that has a voltage range of $\pm$ 5 V?

7. How do you determine the dynamic range of an A/D converter?

8. What is the signal-to-noise ratio of an A/D converter?

9. What is the sampling frequency of the analog-to-digital process?

10. State the Niquist-Shannon sampling theorem.

11. What is the significance of the Shannon interpolation formula?

12. What is the sinc function?

13. What is aliasing?

14. Why does aliasing occur?

15. What is an anti-aliasing filter?

16. Why are anti-aliasing filters used?

17. What is the passband of a filter?

18. What is the transition band of a filter?

19. What is the stop band of a filter?

20. What is the group delay of a filter?

21. What are the two types of dynamic signal analyzers?

22. Describe how a real-time analyzer works.

23. Describe how an FFT analyzer works.

24. With respect the use of an anti-aliasing filter with an FFT analyzer, what is meant by usable and unusable spectral lines?

25. The desired passband frequency limit of an analog signal is 2,000 Hz. The desired number of spectral lines in the FFT analysis is 1,600. Determine:
    (a) the time period T_{sb} of the data blocks,
    (b) the number of digitized points N in the data blocks,
    (c) the sampling rate T_s,
    (d) the frequency bandwidth Δf associated with each spectral line, and
    (e) the lowest frequency the FFT analysis can process.

26. What is a rectangular FFT time window, and when can it be used?

27. What is a Hanning FFT time window, and when is it normally used?

29. What is a flattop FFT time window, and when is it normally used?

30. What is a force FFT time window, and when is it normally used?

31. What is a response FFT time window, and when is it normally used?

32. What is exponential averaging?

33. What is linear averaging?

34. What is overlapping?

35. What percent overlap should be used with a Hanning window?

36. What percent overlap should be used with a flattop window?

37. What process is used to convert a digital time signal to its corresponding analog time signal?

38. Why is it necessary to use a sinc filter when converting a digital time signal to its corresponding analog time signal?

39. What is sparse sampling?

40. Why is sparse sampling used?

CHAPTER 11
ANALYTICAL AND EXPERIMENTAL MODAL ANALYSES

11.1 Introduction

Vibration systems are often analytically or experimentally analyzed to identify their vibration characteristics. These are usually structural and mechanical systems that have many degrees of freedom. The ability to determine the vibration characteristics of these systems is important because:

- The systems' resonance frequencies, mode vectors, and damping ratios must be identified.
- Analytical or numerical models of the systems must be developed and validated.
- Whether or not important vibration design criteria associated with the systems have been met must be determined.
- It may be necessary to diagnose vibration problems associated with the systems that are related to one or more system resonance frequencies or to prevent vibration problems from occurring.

Analytical and experimental modal analyses are used to identify the vibration characteristics of structural and mechanical systems. Topics related to these analysis methods are covered in earlier chapters. Chapter 5 presents introductory concepts associated with analytical modal analysis. Chapter 7 covers concepts associated with random vibrations. Chapter 9 addresses concepts associated with vibration measurements. Chapter 10 discusses concepts associated with digital processing of vibration signals. This chapter will expand upon many of these concepts. It will:

- Further develop math concepts associated with analytical modal analysis and present modeling and solution methods for equations associated with n-DOF vibration systems using principles related to the *dynamical matrix* and the *symmetric eigenvalue solution* methods.
- Provide practical guidelines for experimental modal analysis that include descriptions of test equipment, experimental measurement protocols, and vibration system parameter extraction methods that can be used to conduct evaluative modal tests. These guidelines will include methods for extracting vibration system parameters associated with resonance frequencies, mode vectors, and damping ratios from experimental modal test data.
- Present numerical examples that use eigenvalue/eigenvector solution and modeling techniques associated with commercially available math software.

11.2 Introduction to Modal Analysis

Modal analysis is a process whereby the response of a n-DOF vibration system can be described by:

- The system's resonance frequencies and corresponding mode vectors in terms of the system's *spatial coordinates* and
- The superposition of the responses of a corresponding set of uncoupled 1-DOF vibration systems in terms of the system's *modal coordinates* (Figure 11.1).

The *spatial coordinates* of the n-DOF vibration system describe the system's motion in terms of its physical coordinates. The *modal coordinates* of the corresponding uncoupled 1-DOF vibration systems are a transformed set of coordinates. The 1-DOF equations associated with the modal coordinates can be more easily solved to determine the responses of the 1-DOF vibration systems. The system response in terms of the spatial coordinates

This chapter was written in collaboration with George Ladkany.

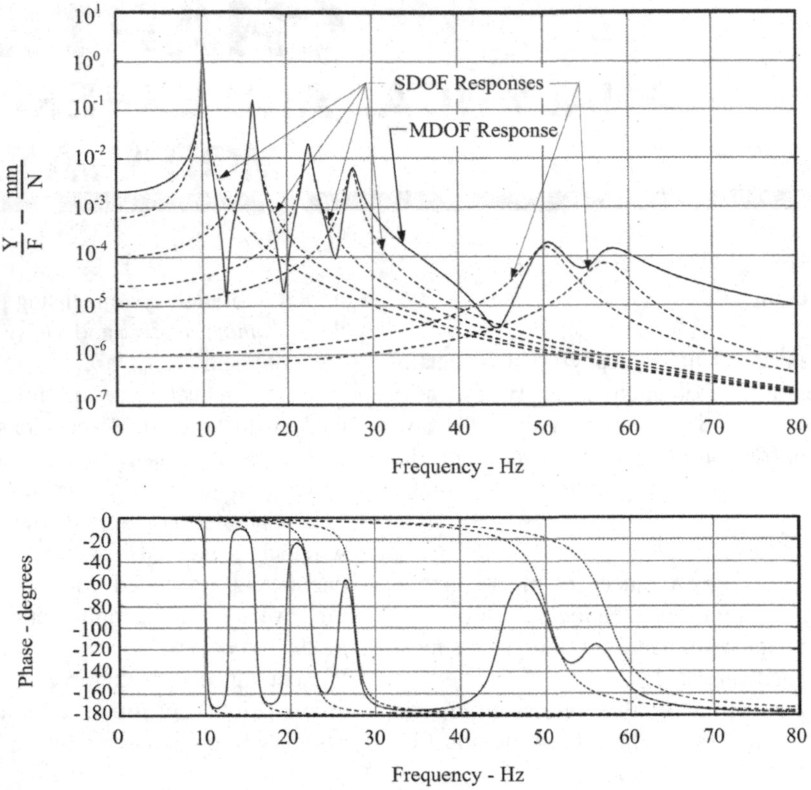

Figure 11.1 Decoupled 1-DOF Vibration System Responses Associated with a n-DOF Vibration System

can then be obtained by transforming the individual 1-DOF responses back into the n-DOF vibration system spatial coordinates.

The coupled set of system equations for a n-DOF vibration system can be written in matrix form:

$$[m]\{\ddot{x}(t)\}+[k]\{x(t)\}=\{0\} \tag{11.1}$$

where [m] is the mass matrix; [k] is the stiffness matrix; and {x(t)} is the vector set of spatial coordinates that describe the motions of the system mass elements. As will be discussed in this chapter, equation (11.1) can be transformed into another set of coordinates q(t) where the equation becomes;

$$[I]\{\ddot{q}(t)\}+[\tilde{K}]\{q(t)\}=\{0\} \tag{11.2}$$

where $[I]$ is the *identity matrix* and $[\tilde{K}]$ is the *mass-normalized stiffness matrix*. Equation (11.2) can be easily solved to obtain the system resonance frequencies and corresponding system orthonormal mode vectors.

11.3 The Eigenvalue Problem

Consider the 3-DOF vibration system in Figure 11.2. The equations of motion for the system can be written in matrix form as:

$$\begin{bmatrix} m_1 & 0 & 0 \\ 0 & m_2 & 0 \\ 0 & 0 & m_3 \end{bmatrix}\begin{Bmatrix} \ddot{x}_1(t) \\ \ddot{x}_2(t) \\ \ddot{x}_3(t) \end{Bmatrix}+ \begin{bmatrix} k_1+k_2 & -k_2 & 0 \\ -k_2 & k_2+k_3 & -k_3 \\ 0 & -k_3 & k_3 \end{bmatrix}\begin{Bmatrix} x_1(t) \\ x_2(t) \\ x_3(t) \end{Bmatrix}=\begin{Bmatrix} 0 \\ 0 \\ 0 \end{Bmatrix}. \tag{11.3}$$

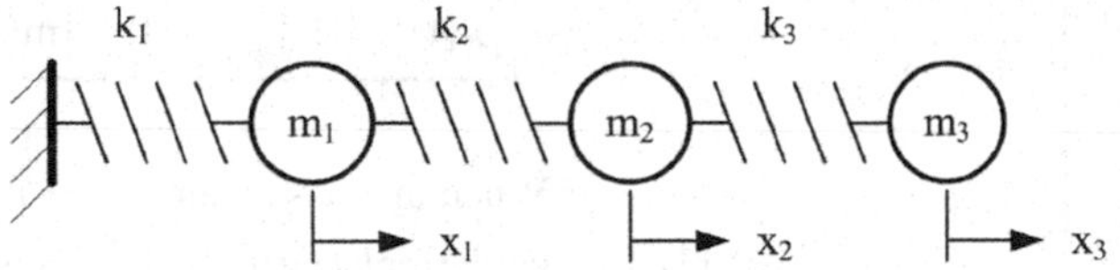

Figure 11.2 3-DOF Vibration System witn Mass and Spring Elements

An eigenvalue solution technique can be used to solve for the system resonance frequencies and mode vectors in terms of the systems spatial coordinates. The solution form for equation (11.3) when $\{x(t)\} = \{u\}e^{j\omega t}$ can be written:

$$\left[-\omega^2[D]-[I]\right]\{u\}=0 \quad \text{or} \quad [D]\{u\}=\lambda[I]\{u\}. \tag{11.4}$$

For an n-DOF vibration system, $[D]$ is the $n \times n$ *dynamical matrix* given by:

$$[D]=[k]^{-1}[m]. \tag{11.5}$$

λ_i for $i = 1, 2, 3, \ldots, n$ are the *eigenvalues* associated with equation (11.4). n is the number of degrees of freedom for the system specified by equation (11.4). The corresponding resonance frequencies ω_i are obtained from:

$$\lambda_i = \frac{1}{\omega_i^2} \tag{11.6}$$

The *eigenvectors* associated with the ith eigenvalue λ_i are expressed as:

$$\left\{u^{(i)}\right\} = \begin{Bmatrix} u_1^{(i)} \\ u_2^{(i)} \\ \vdots \\ u_n^{(i)} \end{Bmatrix}. \tag{11.7}$$

The eigenvectors $\left\{u^{(i)}\right\}$ for $i = 1, 2, 3, \ldots, n$ correspond to the spatial coordinates of the system. The matrix [S] comprised of the eigenvectors $\left\{u^{(i)}\right\}$ can be written:

$$[S]=\left[\left\{u^{(1)}\right\} \; \left\{u^{(2)}\right\} \; \cdots \; \left\{u^{(n)}\right\}\right]. \tag{11.8}$$

$[S]$ is referred to as the *matrix of mode vectors*. The vectors that make up the matrix [S] are *unit normal mode vectors*. However, they are not orthonormal mode vectors.

11.4 The Symmetric Eigenvalue Solution

The *symmetric eigenvalue solution technique* can be used to solve for the system eigenvalues and eigenvectors associated with equation (11.3). This solution technique can be used to more conveniently convert from the system spatial coordinates into the corresponding system mode coordinates. It is the least computational intensive. For a vibration system with n-degrees-of-freedom, it requires a computer program to perform only n^3 floating point operations (flops) while other solution methods can require $7n^3$ flops or more.

Orthogonal eigenvectors can be transformed from one coordinate system to another. This can be done in both numerical and experimental modal analyses. A coupled system of equations sprcified in terms of the spatial coordinates of a n-DOF vibration system can be transformed into a corresponding set of uncoupled 1-DOF equations in terms of the modal coordinates of the system.

The symmetric eigenvalue solution technique is formulated by first normalizing the stiffness matrix $[k]$ with respect to the mass matrix $[m]$ If the mass matrix is symmetric and positive definite, it can be decomposed into:

$$[m]=[L][L]^T \tag{11.9}$$

where [L] is the decomposition matrix. If $[m]$ is a diagonal matrix specified by:

$$[m]=\begin{bmatrix} m_1 & 0 & \cdots & 0 \\ 0 & m_2 & \cdots & 0 \\ \vdots & \vdots & \ddots & \\ 0 & 0 & & m_n \end{bmatrix}, \tag{11.10}$$

then $[L]$ becomes:

$$\boxed{[L]=[m]^{1/2}=\begin{bmatrix} \sqrt{m_1} & 0 & \cdots & 0 \\ 0 & \sqrt{m_2} & \cdots & 0 \\ \vdots & \vdots & \ddots & \\ 0 & 0 & & \sqrt{m_n} \end{bmatrix}} \tag{11.11}$$

and $[L]^{-1}$ is:

$$\boxed{[L]^{-1}=[m]^{-1/2}=\begin{bmatrix} \frac{1}{\sqrt{m_1}} & 0 & \cdots & 0 \\ 0 & \frac{1}{\sqrt{m_2}} & \cdots & 0 \\ \vdots & \vdots & \ddots & \\ 0 & 0 & & \frac{1}{\sqrt{m_n}} \end{bmatrix}}. \tag{11.12}$$

When $[m]$ is not a diagonal matrix, $[L]$ can be calculated using a *Cholesky decomposition*, which is easily calculated by using any mathematical software package.

Modal analysis can be used to uncouple a system of coupled system equations associated with the spatial coordinates of a n-DOF vibration system into a corresponding system of uncoupled 1-DOF vibration system equations in terms of the system modal coordinates. Consider an undamped n-DOF vibration system described by:

$$[m]\{\ddot{x}(t)\}+[k]\{x(t)\}=\{0\}. \tag{11.13}$$

Equation (11.13) can be transformed into a new set of coordinates $\{q(t)\}$ by the equation:

$$\boxed{\{x(t)\}=[L]^{-1}\{q(t)\}=[m]^{-1/2}\{q(t)\}.} \tag{11.14}$$

When the mass matrix $[m]$ is a diagonal matrix, $[L]^{-1}$ can be replaced by $[m]^{-1/2}$. Substituting equation (11.14) into equation (11.13) and pre-multiplying the resulting equation by $[m]^{-1/2}$ yields:

$$[m]^{-1/2}[m][m]^{-1/2}\{\ddot{q}(t)\}+ [m]^{-1/2}[k][m]^{-1/2}\{q(t)\}=\{0\}. \tag{11.15}$$

Noting that:

$$[m]^{-1/2}[m][m]^{-1/2}=[I] \tag{11.16}$$

where $[I]$ is the identity matrix, equation (11.15) becomes:

$$[I]\{\ddot{q}(t)\}+[m]^{-1/2}[k][m]^{-1/2}\{q(t)\}=\{0\}. \tag{11.17}$$

The *mass-normalized stiffness matrix* $[\tilde{K}]$ can be defined as:

$$\boxed{[\tilde{K}]=[m]^{-1/2}[k][m]^{-1/2}.} \tag{11.18}$$

Substituting equation (11.18) into equation (11.17) yields:

$$\boxed{[I]\{\ddot{q}(t)\}+[\tilde{K}]\{q(t)\}=\{0\}.} \tag{11.19}$$

$[m]$ is a symmetric matrix. Therefore, $[\tilde{K}]$ is also a symmetric matrix. The solution form for equation (11.19) when $\{q(t)\}=\{v\}e^{j\omega t}$ can be written:

$$\boxed{[\tilde{K}]\{v\}=\lambda[I]\{v\}} \tag{11.20}$$

where the eigenvalues are:

$$\boxed{\lambda_i=\omega_i^2} \tag{11.21}$$

and corresponding eigenvectors are:

$$\left\{v^{(i)}\right\} = \begin{Bmatrix} v_1^{(i)} \\ v_2^{(i)} \\ \vdots \\ v_n^{(i)} \end{Bmatrix}. \tag{11.22}$$

$i = 1, 2, 3, \ldots, n$ in equations (11.21) and (11.22) where n is the number of system degrees of freedom. The eigenvectors $\left\{v^{(i)}\right\}$ are referred to as the *orthonormal mode vectors* associated with the transformed coordinates $\{q(t)\}$. A matrix $[P]$ can be written:

$$[P] = \left[\left\{v^{(1)}\right\} \; \left\{v^{(2)}\right\} \; \cdots \; \left\{v^{(n)}\right\}\right]. \tag{11.23}$$

$[P]$ is referred to as the *orthonormal modal matrix* and satisfies the relation:

$$[P]^T [P] = [I] \quad \text{where} \quad [P]^T = [P]^{-1}. \tag{11.24}$$

Noting that $\left\{u^{(i)}\right\}$ are the eigenvectors associated with $\{x(t)\}$ and $\left\{v^{(i)}\right\}$ are the eigenvectors associated with $\{q(t)\}$, from equation (11.14):

$$\left\{u^{(i)}\right\} = [m]^{-1/2} \left\{v^{(i)}\right\}. \tag{11.25}$$

Similarly, $[S]$ is related to $[P]$ by:

$$[S] = [m]^{-1/2} [P]. \tag{11.26}$$

A second coordinate transformation can be made such that:

$$\{q(t)\} = [P]\{r(t)\} \tag{11.27}$$

where r(t) is the system's *modal coordinates*. Substituting equation (11.27) into equation (11.19) and pre-multiplying the resulting equation by $[P]^T$ yields:

$$[P]^T [I][P]\{\ddot{r}(t)\} + [P]^T [\tilde{K}][P]\{r(t)\} = \{0\}. \tag{11.28}$$

Noting that:

$$[P]^T [I][P] = [I] \tag{11.29}$$

and:

$$[P]^T [\tilde{K}][P] = [\Lambda_K] = \begin{bmatrix} \omega_1^2 & 0 & \cdots & 0 \\ 0 & \omega_2^2 & \cdots & 0 \\ \vdots & \vdots & \ddots & \\ 0 & 0 & & \omega_n^2 \end{bmatrix}, \tag{11.30}$$

equation (11.27) becomes:

$$[I]\{\ddot{r}(t)\} + [\Lambda_K]\{r(t)\} = \{0\}. \tag{11.31}$$

For a n-DOF vibration system, equation (11.31) represents a set of n uncoupled second order differential equation of the form:

$$\ddot{r}_i(t) + \omega_i^2 \, r_i(t) = 0 \tag{11.32}$$

where $i = 1, 2, 3, \ldots, n$. The coupled second order differential equations represented in equation (11.13) in terms of the system spatial coordinates $\{x(t)\}$ have been transformed into a corresponding set of uncoupled second order differential equations in terms of the system modal coordinates $\{r(t)\}$. For example, the 3-DOF vibration system in Figure 11.2 can be transformed into three uncoupled 1-DOF vibration systems as shown in Figure 11.3.

The responses associated with the modal coordinates $\{r(t)\}$ can be transformed back into the spatial coordinates $\{x(t)\}$ by substituting equation (11.27) into equation (11.14). Therefore:

$$\{x(t)\} = [m]^{-1/2} [P]\{r(t)\}. \tag{11.33}$$

Substituting equation (11.26) into equation (11.33) yields:

$$\{x(t)\} = [S]\{r(t)\}. \tag{11.34}$$

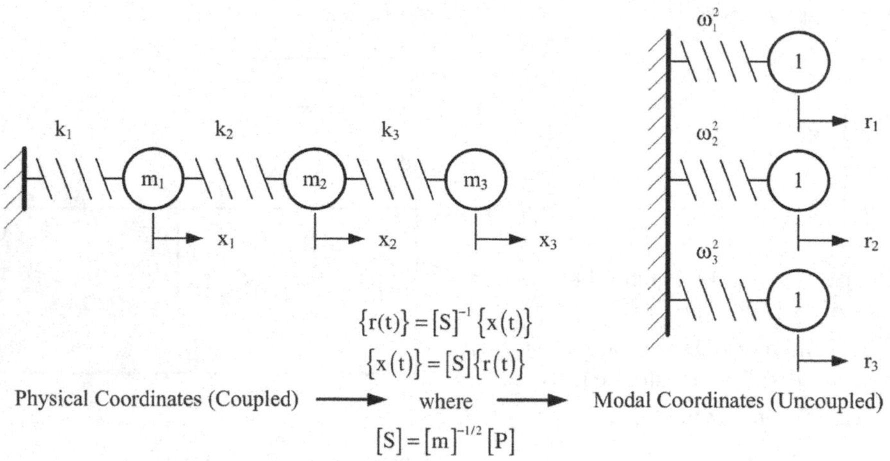

Figure 11.3 Three Uncoupled 1-DOF Vibration Systems Associated with the 3-DOF Vibration System in Figure 11.2

Also,

$$\boxed{\{\dot{x}(t)\} = [S]\{\dot{r}(t)\}.} \tag{11.35}$$

Equation (11.34) can be rearranged to:

$$\boxed{\{r(t)\} = [S]^{-1}\{x(t)\}.} \tag{11.36}$$

The initial values $\{r(0)\}$ and $\{\dot{r}(0)\}$ for an initial value problem associated with $\{r(t)\}$ can be obtained from:

$$\boxed{\{r(0)\} = [S]^{-1}\{x(0)\}; \quad \{\dot{r}(0)\} = [S]^{-1}\{\dot{x}(0)\}.} \tag{11.37}$$

The results of this section can be used to conduct an analytical modal analysis of a vibration system. Computer software similar to MATLAB® can be used to perform the required matrix operations. The recommended steps for the modal analysis are:

1. Develop the system equations of motion, and put them in matrix form as indicated by equations (11.3) and (11.13).
2. Obtain $[m]^{1/2}$ or $[L]$. Then determine $[m]^{-1/2}$ or $[L]^{-1}$.
3. Obtain $[\tilde{K}]$ using equation (11.18).
4. Obtain the symmetric eigenvalue solution for $\lambda_i = \omega_i^2$ and $[P]$ associated with equation (11.20).
5. Verify that the matrix $[P]$ is an orthonormal matrix using equation (11.24).
6. Determine the responses of the n 1-DOF equations specified by equations (11.31) and (11.32) in terms of the modal coordinates $\{r(t)\}$.
7. Determine the response of the n-DOF vibration system by transforming to the system spatial coordinates $\{x(t)\}$ using equation (11.34).

EXAMPLE 11.1

The values of the masses in equation (11.3) are:

$m_1 = 2$ kg, $m_2 = 3$ kg, and $m_3 = 2.5$ kg

and the values of the stiffness coefficients are:

$k_1 = 8{,}000$ N/m, $k_2 = 6{,}000$ N/m, and $k_3 = 10{,}000$ N/m.

(a) Use the dynamical matrix to obtain the system resonance frequencies and corresponding mode vectors. (b) Show the mode vectors are unit normal vectors, but not orthonormal vectors. (c) Determine the [P] matrix from the [S] matrix and show that [P] is an orthonormal matrix.

SOLUTION

Shown below is a MATLAB® script used to obtain the results asked for in this example:

```
m=[2 0 0; 0 3 0; 0 0 2.5];
k=[14000 -6000 0;-6000 16000 -10000;0 -10000...
10000];

% THE DYNAMICAL MATRIX METHOD

m_dynamical = k^(-1)*m

[s,lamda]=eig(m_dynamical)

% It is the Mode Vector Matrix in Physical
% Coordinates

omega=sqrt(diag(1./lamda))

s_transpose=transpose(s)

% The s matrix is a unit normal matrix.
% It is not an orthonormal matrix

s_unit_normal=s_transpose*s

% CALCULATE the p matrix from the s matrix

p_from_s=m^(1/2)*s

% At this point, p is not a unit normal matrix
% Obtain p as a unit normal matrix

p(:,1)=p_from_s(:,1)/norm(p_from_s(:,1));
p(:,2)=p_from_s(:,2)/norm(p_from_s(:,2));
p(:,3)=p_from_s(:,3)/norm(p_from_s(:,3));

p

% Check to see if p is an orthonormal matrix

p_transpose=transpose(p)

p_orthonormal=p_transpose*p
```

The script results are shown below:

```
m_dynamical =

 1.0e-003 *

 0.2500 0.3750 0.3125
 0.2500 0.8750 0.7292
 0.2500 0.8750 0.9792

s =

 -0.2907 -0.8018 0.6628
 -0.6255 -0.2673 -0.6160
 -0.7241 0.5345 0.4257

lamda =

 0.0018 0 0
 0 0.0002 0
 0 0 0.0001

omega =

 23.3422
 77.4597
 98.9367

s_transpose =

 -0.2907 -0.6255 -0.7241
 -0.8018 -0.2673 0.5345
 0.6628 -0.6160 0.4257

s_unit_normal =

 1.0000 0.0132 -0.1156
 0.0132 1.0000 -0.1392
 -0.1156 -0.1392 1.0000

p_from_s =

 -0.4111 -1.1339 0.9373
 -1.0833 -0.4629 -1.0670
 -1.1449 0.8452 0.6731

p =

 -0.2524 -0.7620 0.5964
 -0.6651 -0.3111 -0.6789
 -0.7029 0.5680 0.4283

p_transpose =

 -0.2524 -0.6651 -0.7029
 -0.7620 -0.3111 0.5680
 0.5964 -0.6789 0.4283

p_orthonormal =

 1.0000 0.0000 -0.0000
 0.0000 1.0000 -0.0000
 -0.0000 -0.0000 1.0000
```

The off-diagonal terms in the s_unit_normal matrix are not equal to zero. Therefore, it is a unit normal modal matrix but not an orthonormal modal matrix. The off-diagonal terms in the p_orthonormal matrix equal zero. Therefore, p is an orthonormal modal matrix.

EXAMPLE 11.2

Repeat EXAMPLE 11.1, using the symmetric eigenvalue

solution method. Determine the [S] matrix from the [P] matrix.

SOLUTION

Shown below is a MATLAB® script used to obtain the results asked for in this example:

```
m=[2 0 0; 0 3 0; 0 0 2.5];
k=[14000 -6000 0;-6000 16000 -10000;0 -10000...
10000];

% USE THE SYMETRIC MATRIX METHOD

k_mass_normalized = m^(-1/2)*k*m^(-1/2)

[p,lamda]=eig(k_mass_normalized)

%The p matrix is an orthononormal matrix

%Calculate the resonance frequencies (rad/s)

omega=sqrt(diag(lamda))

%Check to see that p is an orthonormal matrix

p_transpose=transpose(p)

p_orthonormal=p_transpose*p

%CALCULATE the p matrix from the s matrix

s_from_p=m^(-1/2)*p

%At this point, s is not a unit normal matrix
%Convert s to a unit normal matrix

s(:,1)=s_from_p(:,1)/norm(s_from_p(:,1));
s(:,2)=s_from_p(:,2)/norm(s_from_p(:,2));
s(:,3)=s_from_p(:,3)/norm(s_from_p(:,3));

s

%Check to see if s is a unit normal matrix
%but not an orthonormal matrix

s_transpose=transpose(s)

s_unit_normal=s_transpose*s
```

The script results are shown below:

```
k_mass_normalized =

1.0e+003 *

7.0000 -2.4495 0
-2.4495 5.3333 -3.6515
0 -3.6515 4.0000

p =

-0.2524 0.7620 -0.5964
-0.6651 0.3111 0.6789
-0.7029 -0.5680 -0.4283

lamda =

1.0e+003 *

0.5449 0 0
0 6.0000 0
0 0 9.7885

omega =

23.3422
77.4597
98.9367

p_transpose =

-0.2524 -0.6651 -0.7029
0.7620 0.3111 -0.5680
-0.5964 0.6789 -0.4283

p_orthonormal =

1.0000 -0.0000 -0.0000
-0.0000 1.0000 -0.0000
-0.0000 -0.0000 1.0000

s_from_p =

-0.1785 0.5388 -0.4217
-0.3840 0.1796 0.3920
-0.4445 -0.3592 -0.2709

s =

-0.2907 0.8018 -0.6628
-0.6255 0.2673 0.6160
-0.7241 -0.5345 -0.4257

s_transpose =

-0.2907 -0.6255 -0.7241
0.8018 0.2673 -0.5345
-0.6628 0.6160 -0.4257
```

```
s_unit_normal =

 1.0000 -0.0132 0.1156
-0.0132 1.0000 -0.1392
 0.1156 -0.1392 1.0000
```

The three calculated resonance frequencies in EXAMPLES 11.1 and 11.2 are presented in reverse order, but they have the same values for both examples. The mode vectors in the [S] and [P] matrices are presented in reverse order in the respective matrices, but the corresponding vectors are the same in both examples.

Converting from [S] to [P] or from [P] to [S] is a two step process. First, equation (11.26) is used to do the initial transformation. Then, the vectors in the transformed matrix must be converted to their respective unit normal representations.

Example 11.3

The 3-DOF vibration system in EXAMPLES 11.1 and 11.2 is given the following initial conditions:

$$x_1(0) = 10 \text{ mm};\ x_2(0) = 0 \text{ and } x_3(0) = 0$$

$$\dot{x}_1(0) = 0;\ \dot{x}_2(0) = 0 \text{ and } \dot{x}_3(0) = 20 \text{ mm/s}.$$

Determine the responses for $x_1(t)$, $x_2(t)$, $x_3(t)$, $\dot{x}_1(t)$, $\dot{x}_2(t)$, and $\dot{x}_3(t)$. Plot the responses for a time period of 0 to 2 s.

SOLUTION

The MATLAB® script for obtaining the responses for $x_1(t), x_2(t), x_3(t), v_1(t), v_2(t)$, and $v_3(t)$ is given in Appendix 11A. The output of this script is given below:

```
M =

 2.0000 0 0
 0 3.0000 0
 0 0 2.5000

K =

 14000 -6000 0
 -6000 16000 -10000
 0 -10000 10000

r =

 3

L =

 1.4142 0 0
 0 1.7321 0
 0 0 1.5811

K_sym =

 1.0e+003 *

 7.0000 -2.4495 0
 -2.4495 5.3333 -3.6515
 0 -3.6515 4.0000

P =

 -0.5964 -0.7620 0.2524
 0.6789 0.3111 0.6651
 -0.4283 0.5680 0.7029

lambda =

 1.0e+003 *

 9.7885 0 0
 0 6.0000 0
 0 0 0.5449

P_reverse =

 0.2524 -0.7620 -0.5964
 0.6651 -0.3111 0.6789
 0.7029 0.5680 -0.4283

P_ortho =

 1.0000 -0.0000 -0.0000
 -0.0000 1.0000 0.0000
 -0.0000 0.0000 1.0000

omega =

 98.9367
 77.4597
 23.3422

omega_reverse =

 23.3422
 77.4597
 98.9367

S_from_P =
```

```
 0.1785 -0.5388 -0.4217
 0.3840 -0.1796 0.3920
 0.4445 0.3592 -0.2709

S_norm =

 0.2907 -0.8018 -0.6628
 0.6255 -0.2673 0.6160
 0.7241 0.5345 -0.4257

S_transpose =

 0.2907 0.6255 0.7241
-0.8018 -0.2673 0.5345
-0.6628 0.6160 -0.4257

S_unit_normal =

 1.0000 -0.0132 -0.1156
-0.0132 1.0000 0.1392
-0.1156 0.1392 1.0000

x_0 =

 0.0100
 0
 0

v_0 =

 0
 0
 0.2000

r_0 =

 0.0022
-0.0072
-0.0054

rdot_0 =

 0.1364
 0.1207
-0.0862
```

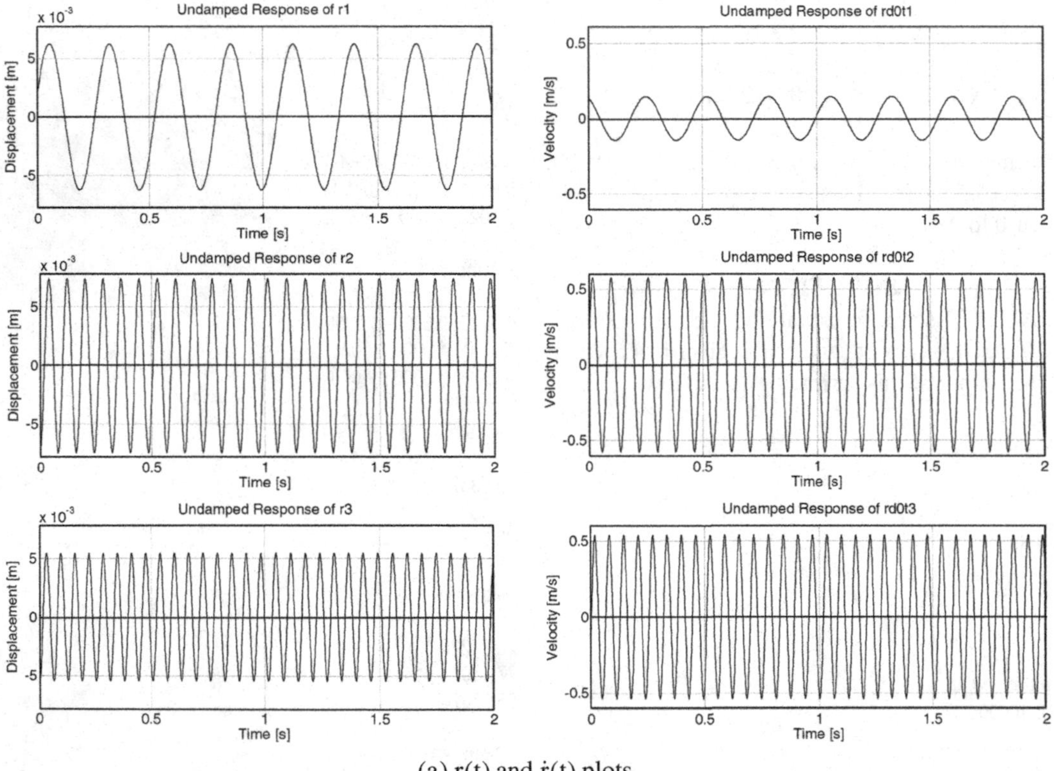

(a) r(t) and ṙ(t) plots

Figue 11.4 Displacement and velocity plots for EXAMPLE 11.4

T =

2

The displacement and velocity MATLAB® plots for the modal and spatial coordinate systems are shown in Figure 11.4.

11.5 Introduction to Experimental Modal Analysis

Experimental modal analysis (often referred to as *modal testing*) is the process of identifying by means of appropriated experimental procedures the *modal parameters* of a n-DOF vibration system. The modal parameters include system resonance frequencies, mode vectors, damping ratios, and modal scaling. Experimental modal analysis employs modal test protocols and vibration system parameter extraction algorithms. This and following sections will not present a comprehensive discussion of these protocols and algorithms. However, they will provide basic test protocols, experimental techniques, and technical insights that can be used to perform informative modal tests. The information that will be presented will be practical. It will give guidance for performing modal tests that require a minimal amount of commercially available and relatively inexpensive equipment and related transducers.

Four basic assumptions must be made with regard to any physical structure on which an experimental modal analysis is to be performed. They are:

- *The structure can be modeled as a linear system.* With this assumption, the motion of the structure is characterized by a set of controlled experiments in which the motion of the structure is linearly related to the input forces used to excite the structure. This requires the use of the principle of superposition, often referred to as the Prony method. This assumption is not valid for structures that have non-linear behavioral

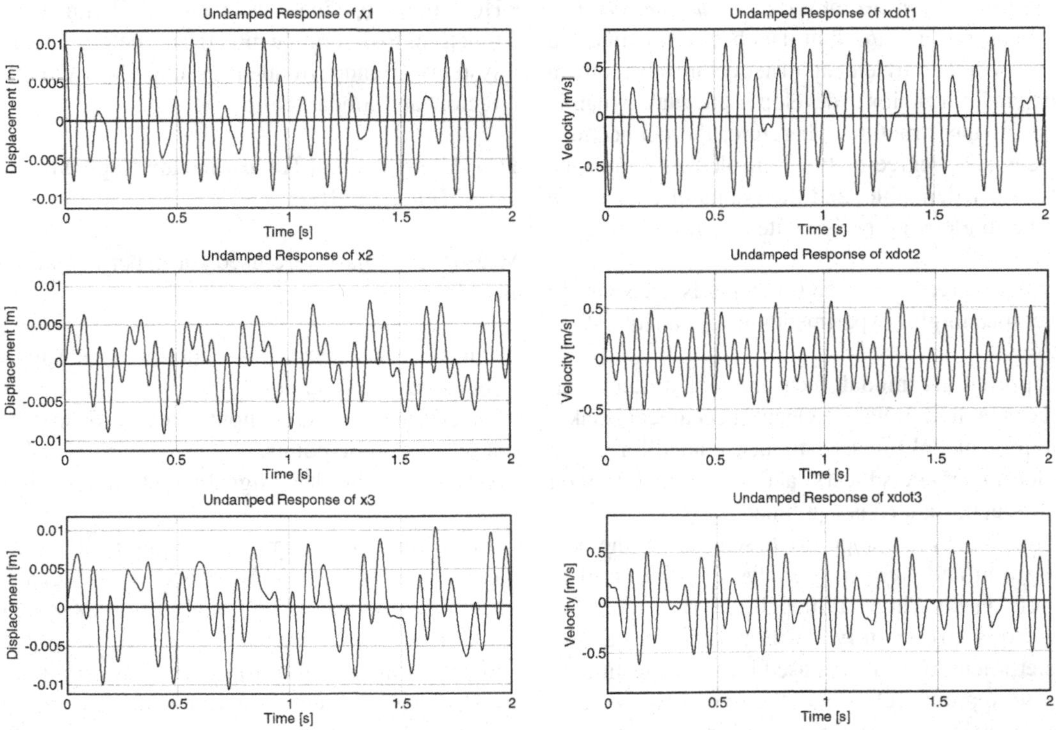

(b) x(t) and $\dot{x}(t)$ plots

Figue 11.4 Displacement and velocity plots for EXAMPLE 11.4

characteristics. However, when an experimental modal analysis is conducted on a non-linear structure, the associated linear model that is developed is assumed to provide a reasonable approximation of the actual system response.

- *The physical characteristics of the structure are time invariant.* The values of physical parameters that describe the dynamic response of the structure, such as mass, damping and stiffness, do not vary with time.
- *The forces applied to the structure and the resulting displacements of the structure obey Maxwell's reciprocity theorem.* A force applied at node point a will result in a corresponding response at node point b. When the same force is applied at node point b, it will result in the same response at node point a. Therefore, the frequency response function H_{ab} of node point a relative to node point b will equal the frequency response function H_{ba} of node point b relative to node point a.
- *The movements of assigned degree-of-freedom elements of a structure selected to describe the structure's motion are observable and measurable.* The measured input/output data between connected degree-of-freedom elements of a structure must contain sufficient information to develop an adequate dynamic model of the structure. Therefore, the connections between the degree-of-freedom elements must be structurally definable, and the points of connection must be able to be properly excited by an applied force.

Information relative to the following is necessary to conduct meaningful experimental modal analyses:

- An appropriate modal measurement system that includes a hydraulic or electromechanical shaker system or modal impact hammer, related force and accelerometer transducers, and signal conditioning and frequency analysis equipment.
- The features of and appropriate settings for the data acquisition and frequency analysis system used to perform an experimental modal analysis.
- How to appropriately calibrate the force and accelerometer transducers used for modal testing.
- How to appropriately select the frequency range of interest, estimate the number of degrees-of-freedom associated with the experimental modal analysis, and determine the cutoff frequency for a specified force-input device.
- How to appropriately select the signal processing FFT window for the input and output signals associated with the modal measurements to eliminate leakage. After selecting the type of excitation system to be used for the modal testing, the proper FFT window for that excitation can be selected.
- How to appropriately select the appropriate signal processing overlap associated with the selected FFT window and the required signal averaging to maximize the achievable dynamic range and signal-to-noise ratio associated with the modal measurements. This depends on the type of excitation system used for the modal testing.
- How to determine if the coherence between the input and output signals are high enough to result in a valid measurement.
- Concepts and principles associated with system resonances and the system properties that exist at the resonance frequencies.
- Concepts and principles associated with spectral density and transfer functions.
- How to best use the experimentally obtained transfer functions to determine modal parameters that include system resonance frequencies, damping ratios, modal vectors and modal scaling.

11.6 Measurement Hardware for Experimental Modal Analysis

Measurement hardware used for experimental modal analyses includes:

- Transducers to measure the motion at selected node points on a structure,
- Systems and devices to input a force into the structure at selected node points,
- Appropriate transducer signal conditioning equipment; and
- DSP (digital signal process) equipment to capture and process related motion and force signals and then perform desired FFT analyses on these signals.

Accelerometers and force transducers that are used for experimental modal analyses are either charge or integrated electronic piezoelectric (IEPE) transducers. A discussion associated with charge and IEPE accelerometers is presented in Chapter 9. The discussions associated with accelerometers also apply

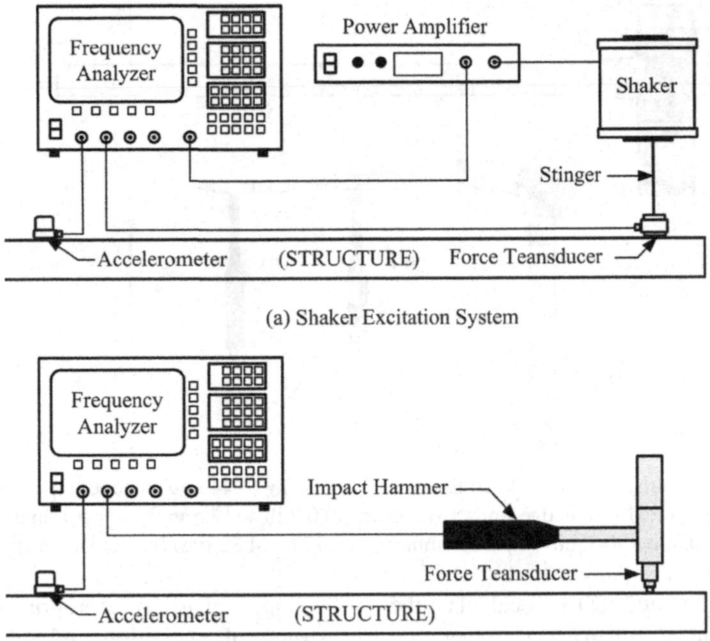

(a) Shaker Excitation System

(b) Impact Hammer Excitation System

Figure 11 .5 Modal Excitation Systems

to force transducers.

Two methods are normally used to direct a force input into a structure at selected node points. One method employs the use of a shaker system, as shown in Figure 11.5(a). The shaker is attached to a node point on the structure by means of a *stinger*. The stinger is a thin metal rod positioned between the node point and the shaker head. A force transducer is positioned between the stinger and the point of attachment to the structure to measure the force that is directed into the structure at the node point. Hydraulic shakers are normally used for frequencies from 0.1 to 50 Hz. Electromechanical shakers have frequency ranges that vary from approximately 5 to 4,000 Hz, depending on the size and stroke displacement of the shaker. The force range of electromechanical shakers varies from around 10 lb_f (44.5 N) to several thousand pounds force (Newtons), depending on the size of the shaker. Smaller shakers have lower force and higher frequency ranges, while larger shakers have higher force and lower frequency ranges.

The second method for directing a force input into a structure at a selected node point on the structure employs the use of an impulse (or impact) hammer [Figure 11.5(b)]. Selected impulse hammers are shown in Figure 11.6. These hammers have specially designed impact heads (or tips) that control the widths of the force pulse and related usable frequency bandwidths associated with a specific hammer. A force transducer is positioned between the hammer mass and impact tip to measure the input force signal associated the hammer. The 12 lb_f (5.443 kg) sledge hammer is used on large or heavy structures. The 3 lb_f (1.361 kg) sledge hammer is used on medium size or weight structures. The 0.3 lb_f (90.7 gm) and 0.7 lb_f (317.5 gm) hammers are used on light-weight structures. The 0.17 oz (4.82 gm) hammer is used on ultra-lightweight structures, such as computer hard-drive discs. ONLY THE USE OF IMPULSE HAMMERS AS A MEANS OF DIRECTING A FORCE INTO A STRUCTURE AT SELECTED NODE POINTS WILL BE DISCUSSED IN THE FOLLOWING SECTIONS.

The signals from the accelerometer(s) and force transducer are directed into the input channels of a FFT analyzer (Figure 11.5). The FFT analyzer should have DSP firmware and software that will appropriately filter the signal in the time and frequency domains to prevent aliasing and time window the signals in the time

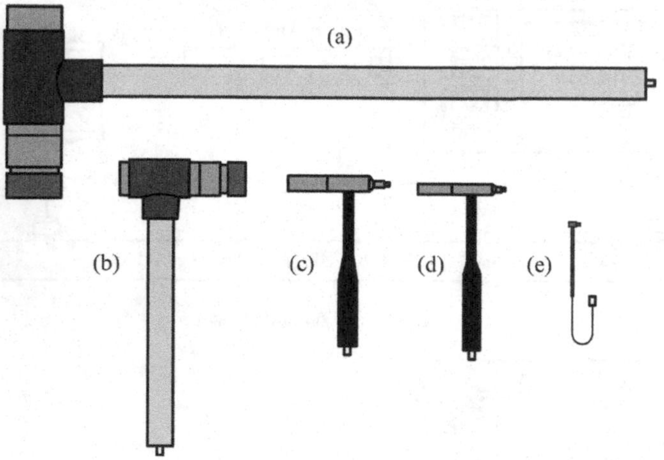

Figure 11.6 Impulse Hammer Modal Excitation - (a) 12 lb_f (5.443 kg) Sledge Impact Hammer; (b) 3 lb_f (1.361 kg) Sledge Impact Hammer; (c) 0.7 lb_f (317.5 gm) Impact Hammer; (d) 0.3 lb_f (90.7 gm) Impact Hammer; (e) 0.17 oz (4.82 gm) Impact Hammer

domain to prevent leakage and that will calculate the transfer functions associated with matched acceleration and force signals. Chapter 10 presents discussions associated with digital signal processing of vibration signals. Analog-to-digital conversion, digital sampling, anti-aliasing filters, dynamic signal analysis, FFT time domain windows, and averaging are discussed in the chapter.

11.7 Measurement Protocols Associated with an Experimental Modal Analysis

Many force-input/acceleration-response configurations can be used for experimental modal analyses that require multiple force inputs and response outputs to a FFT analyzer. However, the most basic configuration is one that requires only one impulse hammer force input and one accelerometer response output. This configuration requires only a two-channel FFT analyzer. THIS IS THE ONLY CONFIGURATION THAT WILL BE DISCUSSED IN THIS AND THE FOLLOWING SECTIONS.

Transducer Selections: The following guidelines apply to the selection of an accelerometer and impulse hammer to be used for an experimental modal analysis:

1. Use an accelerometer that is appropriately sized for the structure that is to be examined and that has a voltage sensitivity appropriate for the anticipated structural excitation and related response (see Chapter 9). The size and weight of an accelerometer that is used on a lightweight structure should be selected so that *mass loading* of the structure by the accelerometer will not occur. This is a condition where the accelerometer, if it is too heavy, will materially add to the mass of the structure, altering its frequency response characteristics. The accelerometer should be selected that will yield the largest possible dynamic amplitude range and the greatest possible signal-to-noise ratio. Make sure the voltage sensitivity of the accelerometer is such that voltage overloads and clipping in the accelerometer signal conditioning amplifier (if used) and in the related FFT analyzer input amplifier do not occur during a test.
2. Select an impulse hammer that is sized appropriately for the size and type of structure that is to be tested. Most impulse hammers have several impact heads (tips) that range from very soft to very hard. These tips will yield different pulse widths and related usable frequency bandwidths for the hammer. The softer the tip, the wider the resulting hammer pulse width and the narrower the corresponding usable frequency bandwidth for the hammer. Conversely, the harder the tip, the narrower the resulting hammer pulse width and the wider the corresponding usable frequency bandwidth for the hammer. Select an

impulse hammer tip that is appropriated for the desired frequency bandwidth of the experimental modal analysis to be conducted.

Digital Sampling Requirements Associated with Impulse Hammer Impacts: The requirements for digital sampling of analog signals is discussed in Section 10.4. The Niquist-Shannon sampling theorem indicates the digital sampling interval T_s must meet the requirement:

$$\boxed{T_s < \frac{1}{2f_{max}}} \tag{11.38}$$

where f_{max} is the upper frequency limit of the frequency bandwidth of the desired FFT analysis. Because of the short time duration of the force pulse associated with an impulse hammer impact, to obtain a reasonable digitized representation of the force pulse, T_s should be:

$$\boxed{T_s < \frac{T_p}{5}} \tag{11.39}$$

where T_p is the time duration of the force pulse. Substituting equation (11.39) into equation (11.38), the suggested value for f_{max} becomes:

$$\boxed{f_{max(p)} > \frac{5}{2T_p}} \tag{11.40}$$

where $f_{max(p)}$ is the suggested value for f_{max} that will result in a digital sampling interval that meets the requirement of equation (11.38).

FFT Analyzer Set-Up: The following are suggested guidelines for setting up a FFT analyzer for an experimental modal analysis:

1. Enter the accelerometer and force transducer voltage sensitivities as required into the FFT analyzer being used for the experimental modal analysis. Set the FFT analyzer input gain settings for the accelerometer and force channels so voltage overloading does not occur.
2. Within the FFT analyzer setting menu, set the desired upper frequency limit f_{max} for the experimental modal analysis and specify the desired number of spectral lines for the analysis. This will set the FFT analyzer digital sampling frequency f_s digital sampling interval T_s and measurement time period T_{sb} for each captured data block (see Section 10.7). If $f_{max(p)}$ is greater than the initial desired value for f_{max}, then f_{max} should be changed to $f_{max(p)}$. This may require increasing the number of spectral lines to achieve the desired sampling frequency.
3. For impact tests using the measurement system in Figure 11.5(b), within the FFT analyzer setting menu, select an appropriate input-trigger so that a complete impact signal data block can be correctly captured. Depending on the FFT analyzer, this often requires the use of an appropriate "pre-trigger" or a "negative trigger delay" setting. If the shaker measurement system in Figure 11.5(a) is used, set the input trigger setting to *Free-Run*.
4. Within the FFT analyzer setting menu, appropriately set the analyzer force and response time windows (see Section 10.8). The force time window should be set so that only the force pulse from the impulse hammer is captured. Beyond the end of the force pulse, the impulse hammer force time signal must be set to zero for the remainder of the measurement time period. The response time window should be set so that the measured acceleration response of the structure goes to zero by the end of the analyzer measurement time period. This is necessary so that leakage does not occur (see Section 10.8). Depending on the response time window setting, artificial damping (damping not present in the structure to be tested) can be introduced into the experimental modal analysis results.
5. Within the FFT analyzer setting menu, select the following measurement outputs to be displayed: un-windowed and windowed force responses as a function of time; un-windowed and windowed acceleration responses as a function of time; power spectral densities of the force input and acceleration response as a function of frequency; the amplitude and phase of the measured accelerance (acceleration/force) as a function of frequency; the real and imaginary values of the accelerance as a function of frequency; and the coherence associated with the measured accelerance as a function of frequency.
6. Within the FFT analyzer setting menu, select the number of data records that are to be "linearly" averaged for each test series in the experimental modal analysis.

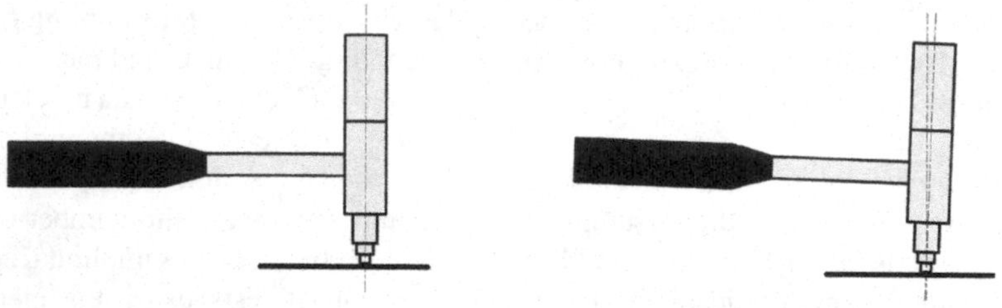

Figure 11.7 Proper Use of an Impact Hammer

7. Within the FFT analyzer setting menu, select the data record "accept/reject" feature, so the experimental results associated with each individual test (hammer strike) within a test series can be reviewed before it is accepted to be averaged with the results associated with the other individual data records in the test series. This allows for bad results to be rejected and not averaged with the other results in a test series.

Proper Use of Impulse Hammers: An impulse hammer must be appropriately selected to match the flexibility and size of the structure to be impacted with the hammer. An impulse hammer with the appropriate tip should be selected to yield the desired or largest possible excitation frequency bandwidth and signal-to-noise ratio. The sledge hammers [Figure 11.6(a)] should be used on large more massive structures. The smaller hammers should be used on smaller, lighter weight flexible structures.

Art and finesse are often associated with the proper use of an impulse hammer. This is particularly true for small hammers. Therefore, the use of an impulse hammer may require practice before it can be used to generate a proper impulse force into a structure. Care must be exercised to ensure the hammer strikes the surface of a structure squarely, as shown in Figure 11.7(a). Striking the surface at an angle, as shown in Figure 11.7(b), can result in a distorted hammer force pulse. Care must also be exercised on lightweight flexible structures to avoid double or multiple hammer impacts that can occur with a single hammer strike. These occur when the structure's surface springs back after an impact and strikes the hammer tip before it can be properly moved away from the surface. Even though, the FFT analyzer force window will only allow the first pulse associated with the impact to pass through for the FFT analysis of the force and resulting structure acceleration signals, the second and possible additional impacts will result in a significant distortion in the acceleration signal. Multiple impulse hammer impacts can be detected and avoided by monitoring the force versus time measured output of an FFT analyzer.

Pre-Test Protocols: The following protocols are suggested for pre-tests associated with an experimental modal analysis:

1. Check the calibration of the accelerometer and the force transducer in the impulse hammer. More will be said about this letter in this chapter.
2. Label the measurement node points on the structure to be tested. Choose which of the following two modal tests protocols will be used. *Method 1*: A single node point is selected on which the accelerometer is placed for all individual modal tests in a test series. A different node point is impacted by the impulse hammer for each successive test in the test series. *Method 2*: A single node point is selected to be impacted by the impulse hammer for all individual modal tests in a test series. The accelerometer is moved to a different node point for each successive test in the test series. Maxwell's reciprocity theorem indicates that both methods will produce the same experimental modal analysis results. Convenience and the physical and dynamic characteristics of the structure to be tested often determine which of the two methods is appropriate for a given series of modal tests.
3. Conduct preliminary impact tests on the structure to be examined to ensure that the transducer signal conditioning and FFT input amplifier gain settings for each channel are appropriately set for the tests to be conducted. Make sure the resulting signal levels for each input channel are above the noise floor of the FFT analyzer and signal overload clipping does

not occur in any channel.

4. Conduct preliminary impact tests on the structure to be examined to ensure the force and response time windows are appropriately set. Make adjustments to the force and response time window settings as necessary to ensure the requirements associated with each window are met before proceeding to the full experimental modal analysis.
5. During the tests in (4), check for anomalies in the measurement outputs associated with the force and response time responses and the structure accelerance. The anomalies may be an indication that FFT settings or measurement protocols must be addressed and modified.
6. With regard to the measured accelerance, check the related coherence output (see Sections 7.8 and 7.11). The coherence should be greater than 0.75 at frequencies near a structural resonance and greater than 0.9 at all other frequencies in the measurement frequency passband.

11.8 Accelerometer and Force Transducer Calibration

The calibration of the accelerometer and the force transducer in the impulse hammer used for an experimental modal analysis should be checked before the measurements associated with an experimental modal analysis begin. Section 9.7 in Chapter 9 presents a discussion on the calibration procedure for the accelerometer.

Figure 11.8 shows the measurement setup for calibrating the force transducer in the impulse hammer. The mass supported with two thin flexible cords as shown in Figure 11.8 will be constrained to move parallel to the ground when it is struck with the impulse hammer. The response of the mass in the frequency domain when struck by the impulse hammer is given by:

$$\boxed{m(f) = \frac{F}{a}(f)} \tag{11.41}$$

where m is the value of the mass (lb_f-s^2/in., kg), a is the value of the acceleration (in./s^2, m/s^2) associated with the force pulse, F (lb_f-s^2/in, N). F/a is defined as the effective mass (see Section 3.4). Use the following protocols to check the calibration of the force transducer:

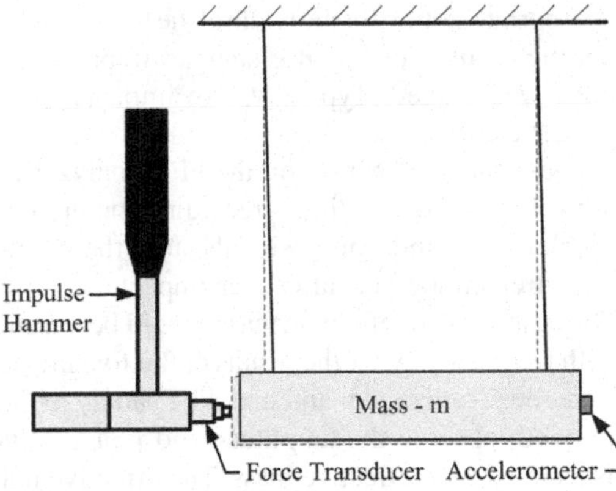

Figure 11.8 Accelerometer and Force Transducer Calibration

1. Within the FFT analyzer input menu, set the manufacture's calibrated transducer voltage sensitivities for both the accelerometer and the impulse hammer force transducer. If the accelerometer has been calibrated per the procedures presented in Section 9.7, use the resulting calibration voltage sensitivity for the accelerometer.
2. Place the impulse hammer head (tip) that will be used for the experimental modal analysis on the hammer.
3. Within the FFT analyzer setting menu, select the measurement output of effective mass amplitude and phase as functions of frequency and the measurement outputs of force and acceleration responses as functions of time.
4. Both the force and acceleration analog signals will be pulses. Therefore, within the FFT analyzer setting menu, select the force window for both the force and acceleration channels.
5. Within the FFT analyzer setting menu, select an appropriate input-trigger so that a complete signal data block can be correctly captured. Depending on the FFT analyzer, this often requires the use of an appropriate "pre-trigger" or a "negative trigger" delay" setting.
6. Within the FFT analyzer setting menu, set the value for $f_{max(p)}$ per the requirements of equation (11.40). Set the number of spectral lines to give a digital sampling frequency f_s of less than or equal to 1 Hz. It may be necessary to strike the mass with the impulse hammer a couple of times and observe the force measurement output as a function of time to determine the time width of the force pulse and an

appropriate setting for the width of the force window.

7. Set the number of impulse hammer impacts to be linearly averaged. Typically, five impacts should be sufficient.
8. Depress the start button on the FFT analyzer and impact the mass five times, recording the digitized acceleration and force signals and the related measurement outputs after each impact.
9. The amplitude of effective mass should be constant with frequency. After the results of the five impacts have been averaged and the FFT analyzer has stopped, observe the amplitude and phase values associated with effective mass. The effective mass value should equal the weighed value of the mass, and the phase value should equal zero as functions of frequency.
10. If the measured value of effective mass does not equal the weighed value of the mass, adjust the voltage sensitivity of the force transducer in the FFT analyzer input menu and repeat the test. It usually is necessary to repeat the test only once before the measured value of effective mass equals the weighed value of the mass. When it does, the force sensitivity setting in the FFT analyzer has been calibrated.

EXAMPLE 11.4

Figure 11.9 shows the initial measured values of effective mass and phase associated with impacts of an 18 kg mass per the test procedures described above. The measured value of effective mass in the flat portion of the effective mass amplitude curve is 16.84 kg, which is less than the weighed value of 18 kg for the mass. The listed calibration sensitivity value for the force transducer in the impulse hammer is 0.2382 mV/N. Determine the correction factor for the force transducer calibration sensitivity value and the corrected force transducer calibration sensitivity value. Repeat the calibration tests and plot the results.

SOLUTION

The force transducer calibration correction factor is obtained by dividing the measured value of the effective mass by the corresponding weighed value, or:

$$\text{corr. factor} = \frac{16.8412}{18} = 0.9356. \tag{11.4a}$$

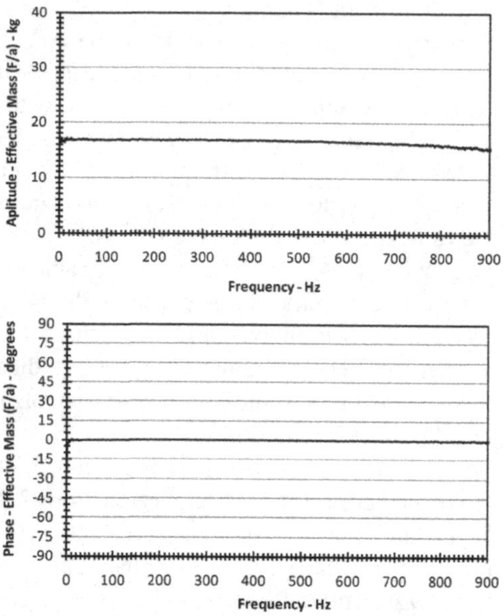

Figure 11.9 Effective Mass Amplitude and Phase vs. Frequency Plots - First Calibration Test EXAMPLE 11.4

The corrected force transducer calibration sensitivity value is:

$$\boxed{\begin{aligned}\text{Force Trans. Cal.} &= 0.9356 \times 0.2382 \,\frac{\text{mV}}{\text{N}}\\ &= 0.2230 \,\frac{\text{mV}}{\text{N}}.\end{aligned}} \tag{11.4b}$$

Figure 11.10 shows the measured values of effective mass and phase for the repeated calibration test. The measured value of effective mass in the flat portion of the effective mass amplitude curve is 18.1535 kg, which is nearly equation to the weighed value of 18 kg for the mass.

Note that the phase values associated with the corresponding effective mass values plotted in both Figures 11.9 and 11.10 are nearly equal to zero.

11.9 Development of the Transfer Function Matrix

Experimentally measured transfer functions can be used to determine the resonance frequencies and mode vectors associated with a structure. Assume harmonic

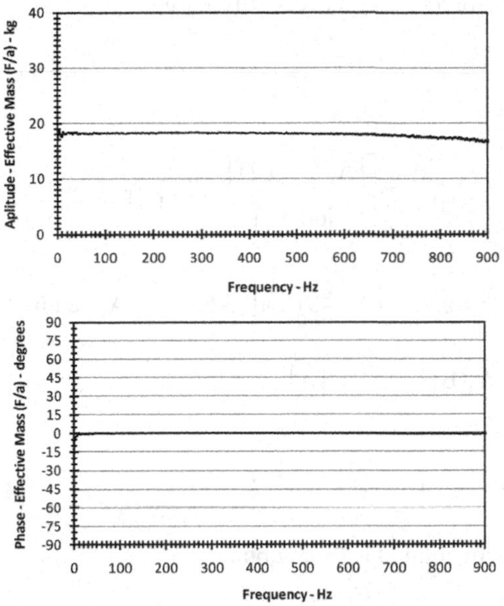

Figure 11.10 Effective Mass Amplitude and Phase vs. Frequency Plots - Second Calibration Test EXAMPLE 11.4

excitation and start with the equation:

$$[m]\{\ddot{x}(t)\}+[c]\{\dot{x}(t)\}+[k]\{x(t)\}=\{f(t)\}. \tag{11.42}$$

Next, let:

$$\{x(t)\}=\{u\}e^{j\omega t} \quad \text{and} \quad \{f(t)\}=\{f\}e^{j\omega t}. \tag{11.43}$$

Substituting equation (11.43) into equation (11.42) and carrying out the prescribed operations yields:

$$\left[[k]-\omega^2[m]+j\omega[c]\right]\{u\}=\{f\}. \tag{11.44}$$

Equation (11.44) can be written:

$$\boxed{\{u\}=\left[H_{ab}(\omega)\right]_{Disp}\{f\}} \tag{11.45}$$

where $\left[H_{ab}^{(i)}(\omega)\right]_{Disp}$ is the *dynamic compliance transfer function matrix*. The rows denoted by the subscript a in $H_{ab}(\omega)$ correspond to the structure response locations. The columns denoted by the subscript b correspond to the structure force input locations. The transfer function matrix is given by:

$$\boxed{\left[H_{ab}(\omega)\right]_{Disp}=\left[[k]-\omega^2[m]+j\omega[c]\right]^{-1}.} \tag{11.46}$$

The transformations used in Section 11.4 and the eigenvalue/eigenvector solution obtained from equation (11.20) can be applied to equation (11.42). For the first transformation let:

$$\boxed{\{x(t)\}=[m]^{-1/2}\{q(t)\}.} \tag{11.47}$$

Substituting equation (11.47) into equation (11.42) and pre-multiplying by $[m]^{-1/2}$ yields:

$$\boxed{[I]\{\ddot{q}(t)\}+[\tilde{C}]\{\dot{q}(t)\}+[\tilde{K}]\{q(t)\}=[m]^{-1/2}\{f(t)\}} \tag{11.48}$$

where:

$$\boxed{\{q(t)\}=\{v\}e^{j\omega t}} \tag{11.49}$$

$$\boxed{\{u\}=[m]^{-1/2}\{v\}} \tag{11.50}$$

$$\boxed{[m]^{-1/2}[m][m]^{-1/2}=[I]} \tag{11.51}$$

$$\boxed{[m]^{-1/2}[c][m]^{-1/2}=[\tilde{C}]} \tag{11.52}$$

$$\boxed{[m]^{-1/2}[k][m]^{-1/2}=[\tilde{K}]} \tag{11.53}$$

$$\boxed{\lambda_i=\omega_i^2 \qquad \text{where} \qquad i=1,2,3,\ldots,n.} \tag{11.54}$$

ω_i in equation (11.54) are the undamped resonance frequencies associated with the eigenvalues λ_i that are derived from the symmetric eigenvalue solution to equation (11.42) for the case where there is no damping. $[P]$ is the orthonormal modal matrix associated with the eigenvectors $\{v^{(i)}\}$.

For the second transformation let:

$$\boxed{\{q(t)\}=[P]\{r(t)\}.} \tag{11.55}$$

Substituting equation (11.55) into equation (11.48) and pre-multiplying the results by $[P]^T$ yields:

$$\boxed{[I]\{\ddot{r}(t)\}+[\Lambda_C]\{\dot{r}(t)\}+[\Lambda_K]\{r(t)\}=[P]^T[m]^{-1/2}\{f(t)\}} \tag{11.56}$$

where:

$$\boxed{[P]^T[I][P]=[I]} \tag{11.57}$$

$$\boxed{[P]^T[m]^{-1/2}[c][m]^{-1/2}[P]=[\Lambda_C]} \tag{11.58}$$

$$\boxed{[\Lambda_C]=\begin{Bmatrix} 2\xi_1\omega_1 & 0 & \cdots & 0 \\ 0 & 2\xi_2\omega_2 & \cdots & 0 \\ \vdots & \vdots & \ddots & \\ 0 & 0 & & 2\xi_n\omega_n \end{Bmatrix}} \tag{11.59}$$

$$\boxed{[P]^T[m]^{-1/2}[k][m]^{-1/2}[P]=[\Lambda_K]} \tag{11.60}$$

$$\boxed{[\Lambda_K]=\begin{bmatrix} \omega_1^2 & 0 & \cdots & 0 \\ 0 & \omega_2^2 & \cdots & 0 \\ \vdots & \vdots & \ddots & \\ 0 & 0 & & \omega_n^2 \end{bmatrix}.} \tag{11.61}$$

Multiply the left side of equations (11.58) and (11.60) by $[m]^{1/2}[P]$ and the right side by $[P]^T[m]^{1/2}$ to get:

$$[c]=[m]^{1/2}[P][\Lambda_C][P]^T[m]^{1/2} \tag{11.62}$$

$$[k]=[m]^{1/2}[P][\Lambda_K][P]^T[m]^{1/2}. \tag{11.63}$$

Noting $[m]$ can be written:

$$[m]=[m]^{1/2}[P][P]^T[m]^{1/2} \tag{11.64}$$

and substituting equations (11.62) and (11.63) into equation (11.46) and simplifying yields:

$$[H_{ab}(\omega)]_{Disp}= \left[[m]^{1/2}[P]\begin{bmatrix}[\Lambda_K]-\omega^2[I]+\\ j\omega[\Lambda_C]\end{bmatrix}[P]^T[m]^{1/2}\right]^{-1}. \tag{11.65}$$

Use equations (A5.45) and (A5.47) in Appendix 5A:

$$[[A][B]]^T=[B]^T[A]^T \tag{11.66}$$

$$[[A][B]]^{-1}=[B]^{-1}[A]^{-1}. \tag{11.67}$$

and equation (11.26) to obtain:

$$[S]^{-1}=[P]^T[m]^{1/2} \quad \text{where} \quad [P]^T=[P]^{-1} \tag{11.68}$$

$$[S]^T=[P]^T[m]^{-1/2} \tag{11.69}$$

$$[S]^{-T}=[m]^{1/2}[P]. \tag{11.70}$$

Substituting equations (11.68) and (11.70) into equation (11.65) yields:

$$[H_{ab}(\omega)]_{Disp}= \left[[S]^{-T}[[\Lambda_K]-\omega^2[I]+j\omega[\Lambda_C]][S]^{-1}\right]^{-1}. \tag{11.71}$$

Applying equation (11.67), equation (11.71) becomes:

$$\boxed{[H_{ab}(\omega)]_{Disp}= [S][[\Lambda_K]-\omega^2[I]+j\omega[\Lambda_C]]^{-1}[S]^T.} \tag{11.72}$$

From equation (11.8), $[S]$ is the system matrix of mode vectors given by:

$$\boxed{[S]=[\{u^{(1)}\} \;\; \{u^{(2)}\} \;\; \cdots \;\; \{u^{(n)}\}].} \tag{11.73}$$

$[\Lambda_C], [\Lambda_K]$, and $[I]$ are diagonal matrices, and the columns of $[S]$ are the mode vectors associated with the undamped version of equation (11.42). Therefore, equation (11.73) can be written as a sum of n matrices rather than the product of three matrices:

$$\left[H_{ab}(\omega)\right]_{Disp} = \sum_{i=1}^{n}\left[\frac{\{u^{(i)}\}\{u^{(i)}\}^T}{(\omega_i^2-\omega^2)+j2\xi_i\omega_i\omega}\right]. \tag{11.74}$$

$\{u^{(i)}\}\{u^{(i)}\}^T$ is a $n \times n$ matrix [refer to equation A5.35 in Appendix 5A]. When velocity is used instead of displacement, equation (11.74) becomes:

$$\left[H_{ab}(\omega)\right]_{vel} = \sum_{i=1}^{n}\left[\frac{\omega^2\{u^{(i)}\}\{u^{(i)}\}^T}{(\omega_i^2-\omega^2)+j2\xi_i\omega_i\omega}\right]. \tag{11.75}$$

Equation (11.75) is the *mobility transfer function matrix*. When acceleration is used instead of displacement, equation (11.74) becomes:

$$\left[H_{ab}(\omega)\right]_{Acc} = \sum_{i=1}^{n}\left[\frac{\omega^4\{u^{(i)}\}\{u^{(i)}\}^T}{(\omega_i^2-\omega^2)+j2\xi_i\omega_i\omega}\right]. \tag{11.76}$$

Equation (11.76) is the *accelerance transfer function matrix*.

11.10 Determination of Modal Parameters

The dynamic compliance transfer function matrix for the ith mode specified in equations (11.74) for all the possible input/output combinations in a n-DOF vibration system has the form:

$$\left[H_{ab}^{(i)}(\omega)\right]_{Disp} = \begin{bmatrix} H_{11}^{(i)} & H_{12}^{(i)} & \cdots & H_{1n}^{(i)} \\ H_{21}^{(i)} & H_{22}^{(i)} & \cdots & H_{2n}^{(i)} \\ \vdots & \vdots & \ddots & \\ H_{n1}^{(i)} & H_{n2}^{(i)} & & H_{nn}^{(i)} \end{bmatrix}_{Disp} \times$$
$$= \frac{\begin{bmatrix} u_1^{(i)}u_1^{(i)} & u_1^{(i)}u_2^{(i)} & \cdots & u_1^{(i)}u_n^{(i)} \\ u_2^{(i)}u_1^{(i)} & u_2^{(i)}u_2^{(i)} & \cdots & u_2^{(i)}u_n^{(i)} \\ \vdots & \vdots & \ddots & \\ u_n^{(i)}u_1^{(i)} & u_n^{(i)}u_2^{(i)} & & u_n^{(i)}u_n^{(i)} \end{bmatrix}}{(\omega_i^2-\omega^2)+j2\xi_i\omega_i\omega}. \tag{11.77}$$

where the matrix rows designate the structure response locations, the matrix columns designate the structure excitation locations, and n is the number of degrees-of-freedom associated with the system. The dynamic compliance transfer function associated with the ith mode for an individual structure excitation location b and an individual structure response location a is:

$$H_{ab}^{(i)}(\omega) = \frac{\bar{u}_a}{f_b} = \frac{u_a^{(i)}u_b^{(i)}}{(\omega_i^2-\omega^2)+j2\xi_i\omega_i\omega}. \tag{11.78}$$

When reasonable spacing exists between a system's resonance frequencies, the summation in equation (11.74) evaluated in the region of the ith resonance frequency ω_i will be dominated by the system's response at that frequency. Therefore, the amplitude of the dynamic compliance transfer function, equation (11.78), evaluated at ω_i is:

$$\left|H_{ab}^{(i)}(\omega_i)\right|_{Disp} = \frac{u_a^{(i)}u_b^{(i)}}{2\xi_i\omega_i^2}. \tag{11.79}$$

When velocity is measured, the amplitude of the mobility function evaluated at ω_i is:

$$\left|H_{ab}^{(i)}(\omega_i)\right|_{Vel} = \frac{u_a^{(i)}u_b^{(i)}}{2\xi_i}. \tag{11.80}$$

When acceleration is measured, the amplitude of the accelerance function evaluated at ω_i is:

$$\left|H_{ab}^{(i)}(\omega_i)\right|_{Acc} = \frac{\omega_i^2 u_a^{(i)} u_b^{(i)}}{2\xi_i}. \tag{11.81}$$

There are n^2 elements in the transfer function matrix in equation (11.77), but only n of the elements are unique. Therefore, only n measurements of $\left[H_{ab}^{(i)}(\omega)\right]_{Disp}$ are necessary for determining the mode vectors, $\{u^{(i)}\}$. This can be accomplished by measuring the values of $H_{ab}^{(i)}(\omega)$ in a row in equation (11.77) where, for example:

$$\left[H_{2b}^{(i)}(\omega)\right]_{Disp} = \begin{bmatrix} \cdots & \cdots & \cdots & \cdots \\ H_{21}^{(i)}(\omega) & H_{22}^{(i)}(\omega) & \cdots & H_{2n}^{(i)}(\omega) \\ \cdots & \cdots & \cdots & \cdots \\ \cdots & \cdots & \cdots & \cdots \end{bmatrix}_{Disp} \tag{11.82}$$

or:

$$\left[\left|H_{2b}^{(i)}(\omega_i)\right|\right]_{Disp} = \frac{\begin{bmatrix} \cdots & \cdots & \cdots & \cdots \\ u_2^{(i)}u_1^{(i)} & u_2^{(i)}u_2^{(i)} & \cdots & u_2^{(i)}u_n^{(i)} \\ \cdots & \cdots & \cdots & \cdots \\ \cdots & \cdots & \cdots & \cdots \end{bmatrix}}{2\xi_i\omega_i^2}. \tag{11.83}$$

This corresponds to placing an accelerometer at structure location 2 and sequentially applying a force at locations 1 to n. Alternately, this can be accomplished by measuring the values of $\left[H_{ab}^{(i)}(\omega)\right]_{Disp}$ in a column in equation (11.77) where, for example:

$$\left[H_{a2}^{(i)}(\omega)\right]_{Disp} = \begin{bmatrix} \cdots & H_{12}^{(i)}(\omega) & \cdots & \cdots \\ \cdots & H_{22}^{(i)}(\omega) & \cdots & \cdots \\ \cdots & \vdots & \cdots & \cdots \\ \cdots & H_{n2}^{(i)}(\omega) & \cdots & \cdots \end{bmatrix}_{Disp} \tag{11.84}$$

or:

$$\left[\left|H_{a2}^{(i)}(\omega_i)\right|\right]_{Disp}^{(i)} = \frac{\begin{bmatrix} \cdots & u_1^{(i)}u_2^{(i)} & \cdots & \cdots \\ \cdots & u_2^{(i)}u_2^{(i)} & \cdots & \cdots \\ \cdots & \vdots & \cdots & \cdots \\ \cdots & u_n^{(i)}u_2^{(i)} & \cdots & \cdots \end{bmatrix}}{2\xi_i\omega_i^2}. \tag{11.85}$$

This corresponds to applying a force to structure at location 2 and sequentially placing an accelerometer at locations 1 to n.

The experimentally determined unit normal mode vectors can be obtained from the displacement, velocity or acceleration values. They all will result in the same unit normal mode vectors. Normally, accelerance is measured when conducting modal tests. Therefore, equation (11.81) is used to obtain the mode vectors. When velocity is used to obtain the mode vectors, the ω_i^2 term is eliminated from equation (11.81). Noting that:

$$\dot{u}_a^{(i)} = \omega_i\, u_a^{(i)} \quad \text{and} \quad \dot{u}_b^{(i)} = \omega_i\, u_b^{(i)}, \tag{11.86}$$

equation (11.81) can be rearrange to yield:

$$\dot{u}_a^{(i)}\dot{u}_b^{(i)} = 2\xi_i \left|H_{ab}^{(i)}(\omega_i)\right|_{Acc}. \tag{11.87}$$

Assume for a 3-DOF vibration system that a force is applied at structure location 2 and acceleration is measured at structure locations 1 through 3. For this case, use equations (11.85) and (11.87). For the first resonance frequency ω_1:

$$\dot{u}_1^{(1)}\dot{u}_2^{(1)} = 2\xi_1 \left|H_{12}^{(1)}(\omega_1)\right|_{Acc} \tag{11.88}$$

$$\dot{u}_2^{(1)}\dot{u}_2^{(1)} = 2\xi_1 \left|H_{22}^{(1)}(\omega_1)\right|_{Acc} \tag{11.89}$$

$$\dot{u}_3^{(1)}\dot{u}_2^{(1)} = 2\xi_1 \left|H_{32}^{(1)}(\omega_1)\right|_{Acc}. \tag{11.90}$$

For the second resonance frequency ω_2:

$$\boxed{\dot{u}_1^{(2)}\dot{u}_2^{(2)} = 2\xi_2 \left|H_{12}^{(2)}(\omega_2)\right|_{Acc}} \tag{11.91}$$

$$\boxed{\dot{u}_2^{(2)}\dot{u}_2^{(2)} = 2\xi_2 \left|H_{22}^{(2)}(\omega_2)\right|_{Acc}} \tag{11.92}$$

$$\boxed{\dot{u}_3^{(2)}\dot{u}_2^{(2)} = 2\xi_2 \left|H_{32}^{(2)}(\omega_2)\right|_{Acc}.} \tag{11.93}$$

For the third resonance frequency ω_3:

$$\boxed{\dot{u}_1^{(3)}\dot{u}_2^{(3)} = 2\xi_3 \left|H_{12}^{(3)}(\omega_3)\right|_{Acc}} \tag{11.94}$$

$$\boxed{\dot{u}_2^{(3)}\dot{u}_2^{(3)} = 2\xi_3 \left|H_{22}^{(3)}(\omega_3)\right|_{Acc}} \tag{11.95}$$

$$\boxed{\dot{u}_3^{(3)}\dot{u}_2^{(3)} = 2\xi_3 \left|H_{32}^{(3)}(\omega_3)\right|_{Acc}.} \tag{11.96}$$

Each set of three equations has three unknowns. Therefore, the values of $\dot{u}_1$, $\dot{u}_2$, and $\dot{u}_3$ can be determined for the mode vectors associated with each of the three resonance frequencies. The corresponding unit normal values associated with u_1, u_2, and u_3 are obtained when the corresponding unit normal mode vectors are obtained.

For this example, the sources of the measured parameters are:

1. ω_1, ω_2, and ω_3 are the three resonance frequencies obtained from the accelerance measurements.
2. $\left|H_{12}^{(1)}(\omega_1)\right|_{Acc}$ is the accelerance amplitude at the frequency ω_1 at location 1 for the force applied to location 2. $\left|H_{22}^{(1)}(\omega_1)\right|_{Acc}$ is the accelerance amplitude at the frequency ω_1 at location 2 for the force applied to location 2. $\left|H_{32}^{(1)}(\omega_1)\right|_{Acc}$ is the amplitude of the accelerance amplitude at the frequency ω_1 at location 3 for the force applied to location 2.
3. $\left|H_{12}^{(2)}(\omega_2)\right|_{Acc}$ is the accelerance amplitude at the frequency ω_2 at location 1 for the force applied to location 2. $\left|H_{22}^{(2)}(\omega_2)\right|_{Acc}$ is the accelerance amplitude at the frequency ω_2 at location 2 for the force applied to location 2. $\left|H_{32}^{(2)}(\omega_2)\right|_{Acc}$ is the amplitude of the accelerance amplitude at the frequency ω_2 at location 3 for the force applied to location 2.
4. $\left|H_{12}^{(3)}(\omega_3)\right|_{Acc}$ is the accelerance amplitude at the frequency ω_3 at location 1 for the force applied to location 2. $\left|H_{22}^{(3)}(\omega_3)\right|_{Acc}$ is the accelerance amplitude at the frequency ω_3 at location 2 for the force applied to location 2. $\left|H_{32}^{(3)}(\omega_3)\right|_{Acc}$ is the amplitude of the accelerance amplitude at the frequency ω_3 at location 3 for the force applied to location 2.
5. ξ_1 is the damping ratio associated with $\left|H_{12}^{(1)}(\omega_1)\right|_{Acc}$, $\left|H_{22}^{(1)}(\omega_1)\right|_{Acc}$, and $\left|H_{32}^{(1)}(\omega_1)\right|_{Acc}$. ξ_1 is obtained from the equation:

$$\boxed{\xi_1 = \frac{1}{2}\frac{\left(\dfrac{\omega_r}{\omega_q}\right)-1}{\left(\dfrac{\omega_r}{\omega_q}\right)+1}.} \tag{11.97}$$

ω_q and ω_r are the lower (ω_q) and upper (ω_r) frequency values associated with the half-power points of the real part of the transfer function in the region of ω_1 for each curve [see Figure 3.10 and equation (3.94) in Section 3.5].

6. ξ_2 is the damping ratio associated with $\left|H_{12}^{(2)}(\omega_2)\right|_{Acc}$, $\left|H_{22}^{(2)}(\omega_2)\right|_{Acc}$, and $\left|H_{32}^{(2)}(\omega_2)\right|_{Acc}$. ξ_2 is obtained from equation (11.97) where ω_q and ω_r are the lower (ω_q) and upper (ω_r) frequency limits associated with the half-power points of the real part of the transfer function in the region of ω_2 for each curve.
7. ξ_3 is the damping ratio associated with $\left|H_{12}^{(3)}(\omega_3)\right|_{Acc}$, $\left|H_{22}^{(3)}(\omega_3)\right|_{Acc}$, and $\left|H_{32}^{(3)}(\omega_3)\right|_{Acc}$. ξ_3 is obtained from equation (11.97) where ω_q and ω_r are the lower (ω_q) and upper (ω_r) frequency limits associated with the half-power points of the real part of the transfer function in the region of ω_3 for each curve.
8. The phase associated with the three elements in each of the three mode vectors is obtained from the signs of each of the imaginary parts of the three accelerance functions in the regions of ω_1, ω_2, and ω_3. If the sign is positive, the phase is positive. If the sign is negative, the phase is negative.

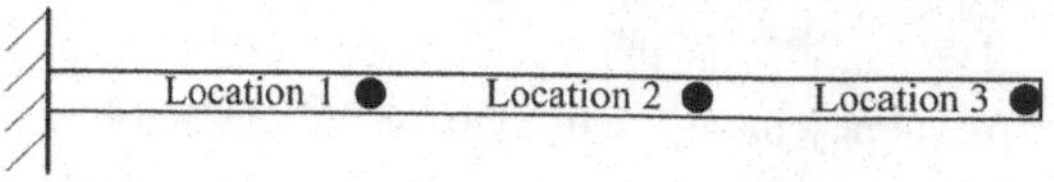

Figure 11.11 Impact Test Locations on a Cantilevered

EXAMPLE 11.5

Modal measurements were made on the cantilevered beam in Figure 11.11. The beam was divided into three sections with node points at Locations 1, 2 and 3. For each individual test, the impulse hammer was used to impact the beam at Location 1. For each test, the accelerometer was sequentially moved from Location 1 to Location 3, as shown in Figures 11.12, 11.13 and 11.14. The parameters that were recorded as a function of frequency were $|H_{ab}(f)|$, $Re[H_{ab}(f)]$, and $Im[H_{ab}(f)]$. The resulting curves are shown in Figures 11.12, 11.13 and 11.14. The following parameter values were extracted from the curves:

Resonance Frequencies and corresponding peak accelerance values:

Peak Accelerance Values			
Resonance freq. - Hz	i = 1 43.1	i = 2 267.8	i = 3 748.1
$\lvert H_{12}\rvert_{Acc(i)}$	13.44	175.18	419.47
$\lvert H_{22}\rvert_{Acc(i)}$	44.50	129.47	385.09
$\lvert H_{32}\rvert_{Acc(i)}$	79.90	281.47	543.95

Phase sign:

Phase Sign			
Resonance Freq. - Hz	i= 1 43.1	i = 2 267.8	i = 3 748.1
$Im[H_{12}(f_i)]$sign	1	1	-1
$Im[H_{22}(f_i)]$sign	1	1	1
$Im[H_{32}(f_i)]$sign	1	-1	-1

Lower and upper frequency values for calculation of damping ratios:

Lower and Upper Frequency Values for ξ Calculations			
Resonance Freq. - Hz	i = 1 43.1	i = 2 267.8	i = 3 748.1
$f_{a(12)(i)}$ - Hz	42.5	266.9	746.3
$f_{b(12)(i)}$ - Hz	43.7	268.1	749.4
$f_{a(22)(i)}$ - Hz	42.5	266.8	746.3
$f_{b(22)(i)}$ - Hz	43.8	268.8	749.4
$f_{a(32)(i)}$ - Hz	41.8	266.3	745.7
$f_{b(321)(i)}$ - Hz	43.8	268.1	748.8

Calculate the damping ratios and mode vectors associated with the three resonance frequencies.

SOLUTION

The damping ratios are obtained by substituting the appropriate values for f_a and f_b into equation (11.97). f_a and f_b are the lower and higher frequency values associated with the minima and maxima of the real values of the accelerance curves in the regions of f_i shown in Figures 11.12, 11.13 and 11.14.

Equations (11.88) through (11.96) are used to obtain the mode vectors associated with the three resonance frequencies. Solving these equations) for $\dot{u}_1^{(i)}$, $\dot{u}_2^{(i)}$, and $\dot{u}_3^{(i)}$ yields:

$$\dot{u}_2^{(i)} = \sqrt{2\xi_{i(avg)}\left|H_{22}^{(i)}(f_i)\right|_{Acc}} \tag{11.5a}$$

$$\dot{u}_1^{(i)} = \frac{2\xi_{i(avg)}\left|H_{21}^{(i)}(f_i)\right|_{Acc}}{\dot{u}_2^{(i)}} \tag{11.5b}$$

$$u_3^{(i)} = \frac{2\xi_{i(avg)}(2\pi f_i)^2\left|H_{31}^{(i)}(f_i)\right|_{Acc}}{u_2^{(i)}}. \tag{11.5c}$$

A MATLAB® script for calculating the unit normal mode vectors from the above data is presented in Appendix B. The output of this script is presented below.

```
Res_Freq =

43.1000 267.8000 748.1000
```

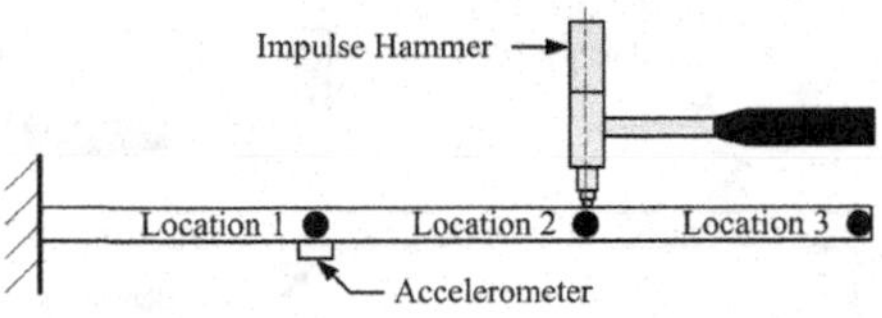

(a) Impulse Hammer Location 2; Accelerometer Location 1

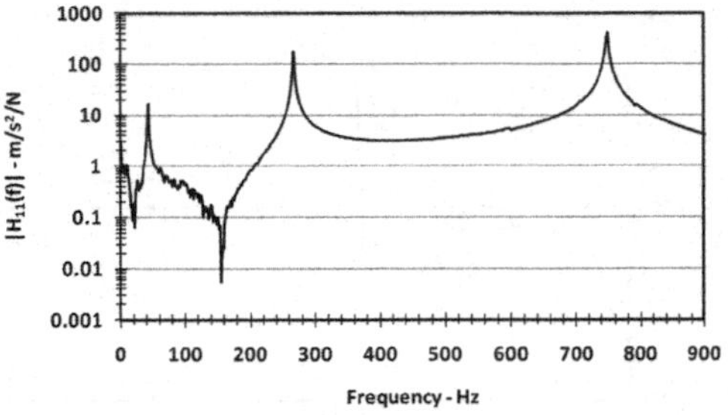

(b) $|H_{12}(f)|$ versus Frequency

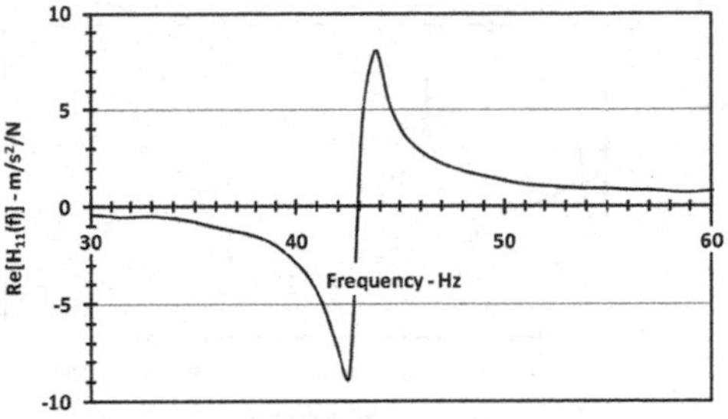

(e) $Re[H_{12}(f)]$ versus Frequency in Region of f_1

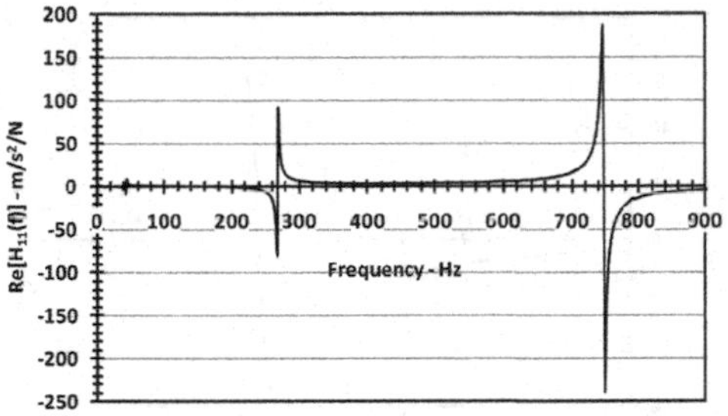

(c) $Re[H_{12}(f)]$ versus Frequency

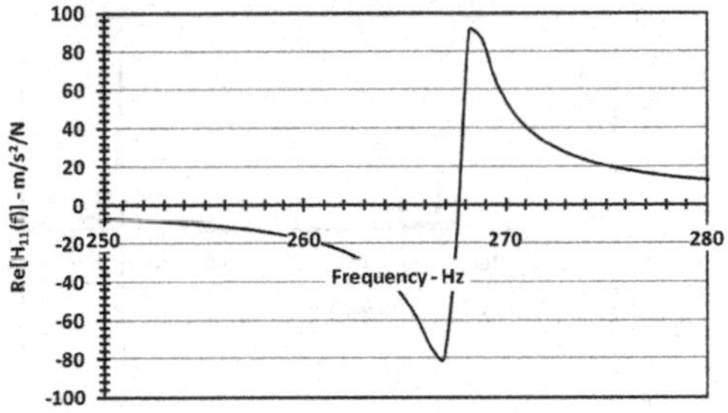

(f) $Re[H_{12}(f)]$ versus Frequency in Region of f_2

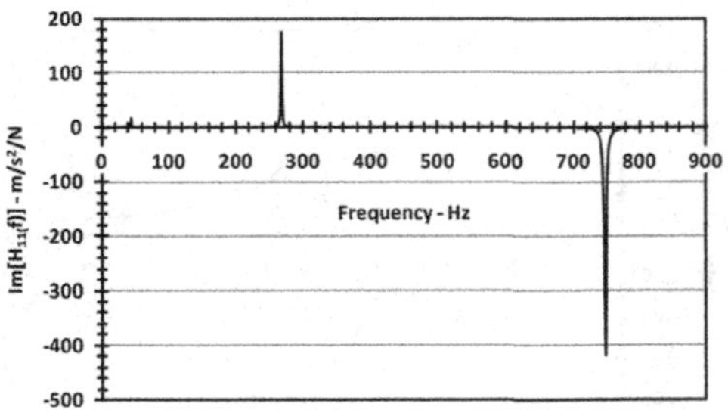

(d) $Im[H_{12}(f)]$ versus Frequency

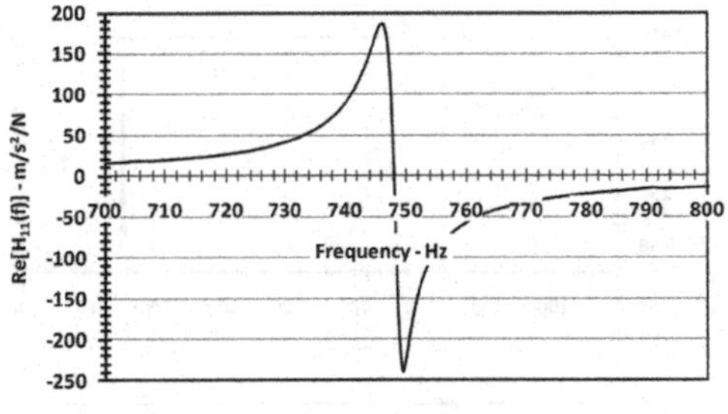

(g) $Re[H_{12}(f)]$ versus Frequency in Region of f_3

Figure 11.12 Measured Beam Modal Values for Hammer Location 2 and Response Location 1

```
H_Accel_Max =

 13.4400 175.1800 419.4500
 44.5000 129.4700 385.0900
 79.9000 281.4700 543.9500
```

```
Phase_Sign =

 1 1 -1
 1 1 1
 1 -1 -1
```

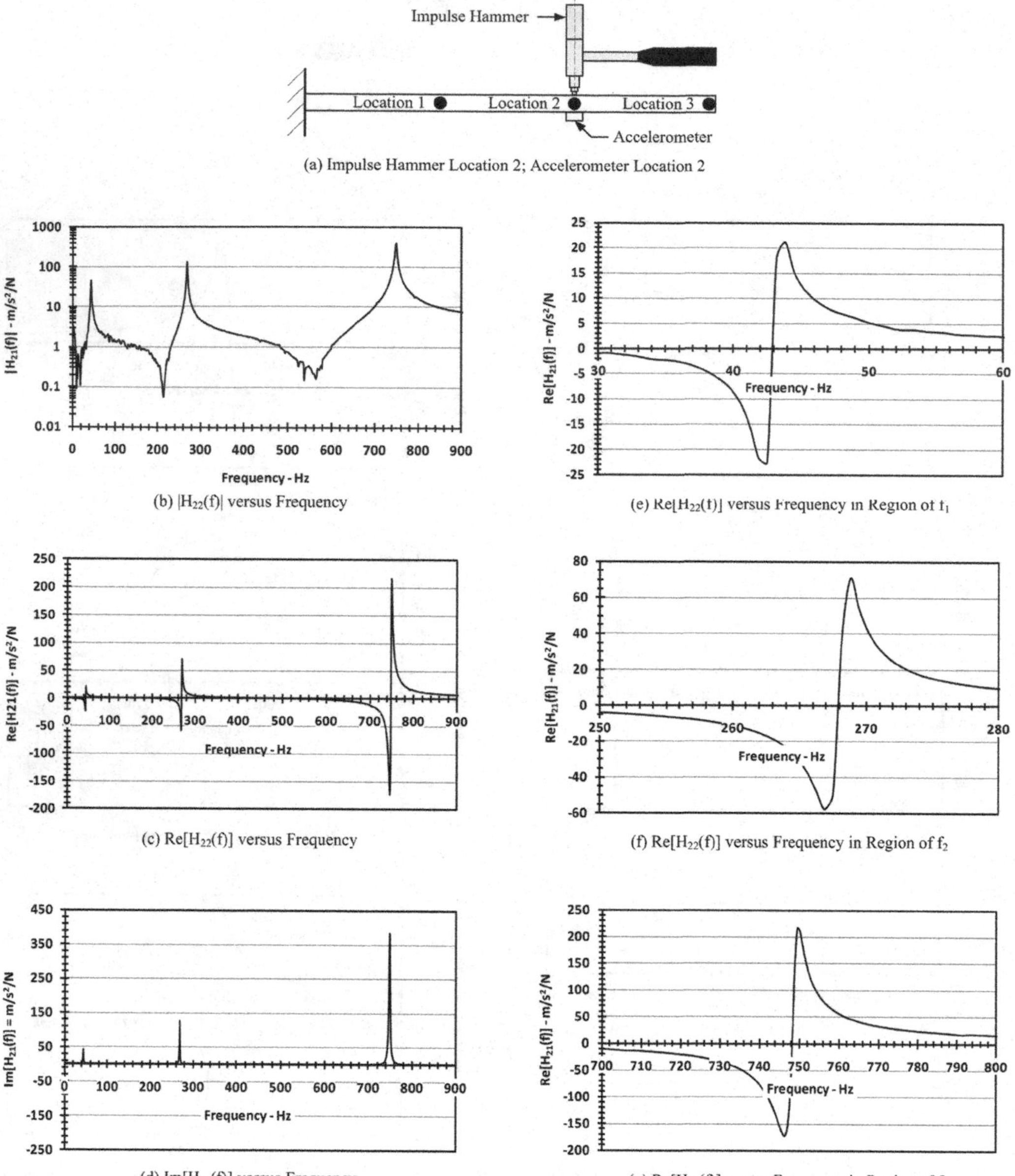

(a) Impulse Hammer Location 2; Accelerometer Location 2

(b) $|H_{22}(f)|$ versus Frequency

(e) $Re[H_{22}(f)]$ versus Frequency in Region of f_1

(c) $Re[H_{22}(f)]$ versus Frequency

(f) $Re[H_{22}(f)]$ versus Frequency in Region of f_2

(d) $Im[H_{22}(f)]$ versus Frequency

(g) $Re[H_{22}(f)]$ versus Frequency in Region of f_3

Figure 11.13 Measured Beam Modal Values for Hammer Location 2 and Response Location 2

```
H_Real_Freq_Low =

 42.5000 266.9000 746.3000
 42.5000 266.8000 746.3000
 41.8000 266.3000 745.7000
```

```
H_Real_Freq_High =

 43.7000 268.1000 749.4000
 43.8000 268.8000 749.4000
 43.8000 268.1000 748.8000
```

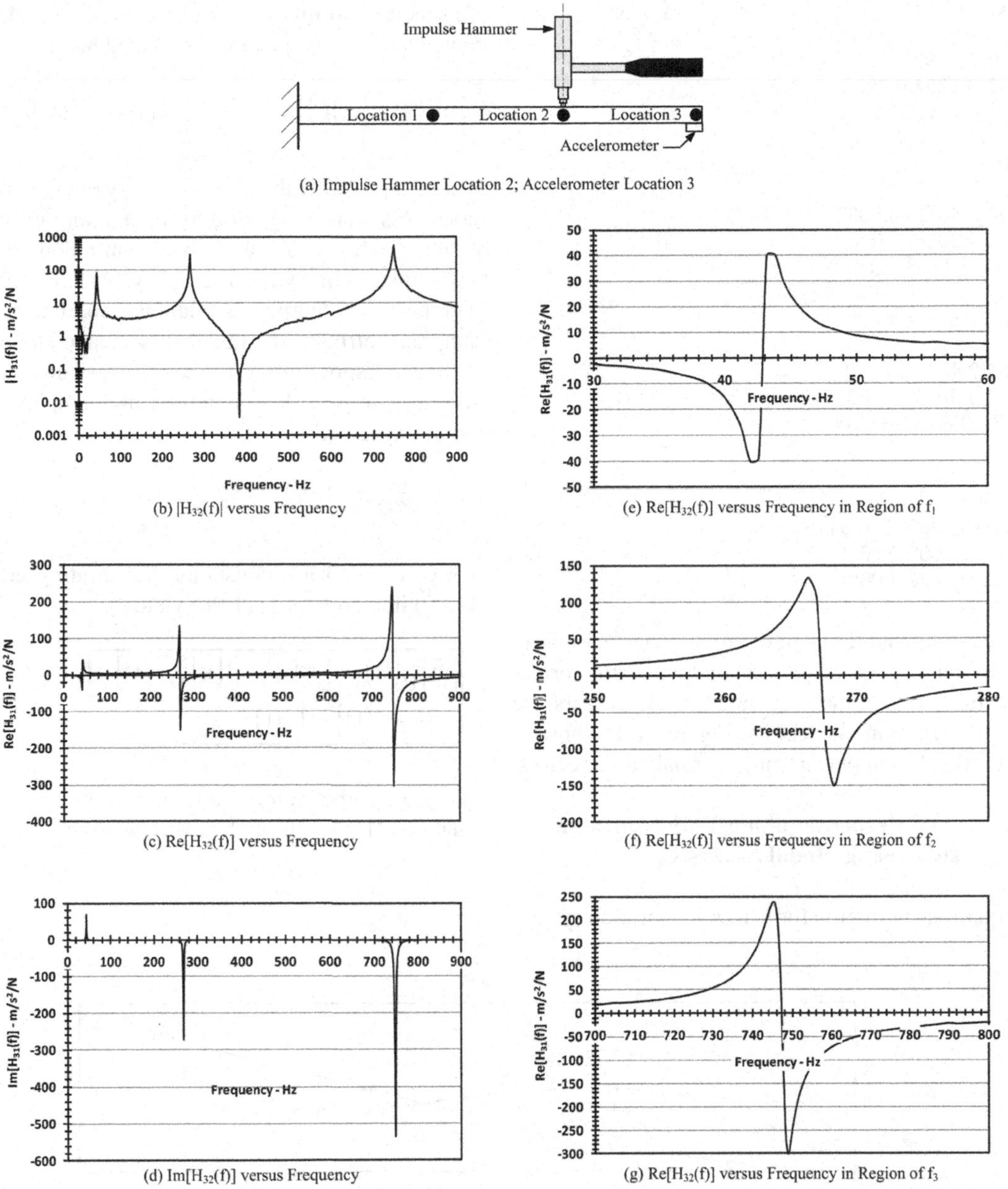

Figure 11.14 Measured Beam Modal Values for Hammer Location 2 and Response Location 3

```
damp =

 0.0070 0.0011 0.0010
 0.0075 0.0019 0.0010
 0.0117 0.0017 0.0010
```

```
damp_avg =

 0.0087 0.0016 0.0010
```

```
udot =

 0.2661 0.8593 -0.9732
 0.8812 0.6351 0.8935
 1.5822 -1.3807 -1.2621

u_norm =

 0.1454 0.4922 -0.5327
 0.4814 0.3638 0.4890
 0.8644 -0.7908 -0.6908

u_norm_transpose =

 0.1454 0.4814 0.8644
 0.4922 0.3638 -0.7908
 -0.5327 0.4890 -0.6908

u_unit_normal =

 1.0000 -0.4369 -0.4391
 -0.4369 1.0000 0.4620
 -0.4391 0.4620 1.0000
```

The off-diagonal terms in the u_unit_normal matrix are non-zero. Therefore, it is not an orthonormal modal matrix, and the associated mode vectors are not orthonormal mode vectors. Figure 11.15 shows a MATLAB® plots of the three unit normal mode vectors.

11.11 Forced Responses of a n-DOF Vibration System using Modal Analysis

The equations of motion for a n-DOF vibration system are specified in matrix form by equation (11.42). For convenience, the equation is repeated here:

$$[m]\{\ddot{x}(t)\}+[c]\{\dot{x}(t)\}+[k]\{x(t)\}=\{f(t)\}. \tag{11.98}$$

When damping in the system is very small (damping ratios less than 0.1), *proportional damping* can be assumed. This is a reasonable assumption for must structural and many mechanical systems. For the case of proportional damping, the damping coefficients in the damping matrix [c] are modeled as linear combinations of the corresponding terms in the mass and stiffness matrices [m] and [k]. Therefore, the damping matrix can be expressed:

$$[c]=\alpha[m]+\beta[k] \tag{11.99}$$

where α and β are constants. Substituting equation (11.99) into equation (11.98) yields:

$$\boxed{\begin{aligned}&[m]\{\ddot{x}(t)\}+[\alpha[m]+\beta[k]]\{\dot{x}(t)\}+\\&[k]\{x(t)\}=\{f(t)\}.\end{aligned}} \tag{11.100}$$

Performing the same transformations as used in equations (11.47) through (11.61) gives:

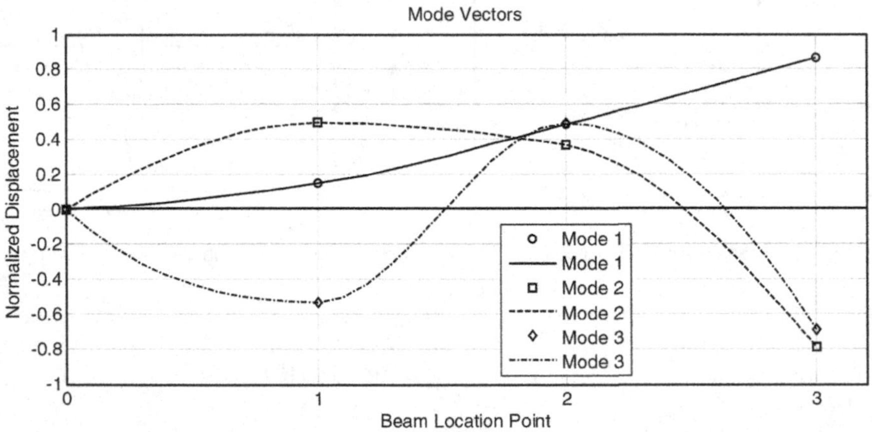

Figure 11.15 Plots of the Unit Normal Mode Vectors for the 3-DOF Vibration System in EXAMPLE 11.5

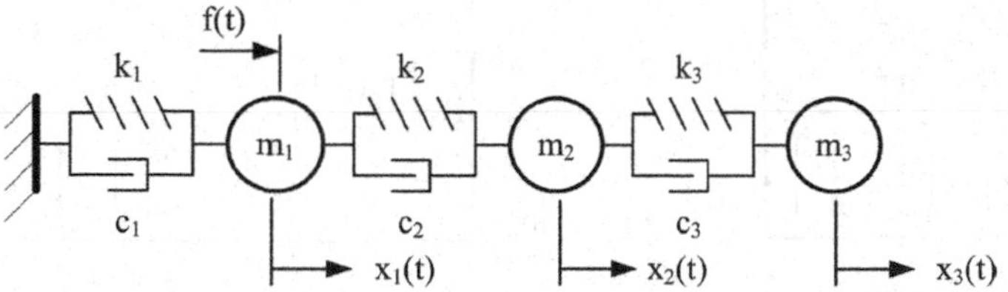

Figure 11.16 3-DOF Vibration System with Mass, Spring and Damping Elements and with a Forcing Function Acting on Mass 1

$$[I]\{\ddot{r}(t)\}+\left[\alpha[I]+\beta[\Lambda_K]\right]\{\dot{r}(t)\}+[\Lambda_K]\{r(t)\}=[S]^T\{f(t)\}. \tag{11.101}$$

The damping matrix in equation (11.101) can be written:

$$[\Lambda_C]=\alpha[I]+\beta[\Lambda_K] \tag{11.102}$$

where:

$$2\xi_i\omega_i=\alpha+\beta\omega_i^2 \tag{11.103}$$

for $i = 1, 2, 3, \ldots, n$. Substituting equation (11.102) into equation (11.101) yields:

$$[I]\{\ddot{r}(t)\}+[\Lambda_C]\{\dot{r}(t)\}+[\Lambda_K]\{r(t)\}=[S]^T\{f(t)\}. \tag{11.104}$$

The individual equations of motion associated with the modal coordinates $r_i(t)$ can be written:

$$\ddot{r}_i(t)+2\xi_i\omega_i\dot{r}_i(t)+\omega_i^2 r_i(t)=F_i(t) \tag{11.105}$$

where:

$$F_i(t)=\sum_{j=1}^{n} s_{ij}^T f_j(t). \tag{11.106}$$

s_{ij}^T are the row and column elements of the $[S]^T$ matrix. $i = 1, 2, 3, \ldots, n$ are the row numbers of the $[S]^T$ matrix. $j = 1, 2, 3, \ldots, n$ are the column numbers of the $[S]^T$ matrix and the element numbers of the $\{f(t)\}$ vector.

After the displacements in terms of the modal coordinates $r_i(t)$ have been determined, the corresponding displacements in terms of the system spatial coordinates $x_i(t)$ can be obtained from:

$$\{x(t)\}=[S]\{r(t)\}. \tag{11.107}$$

As an example, examine the response of the3-DOF, vibration system with mass, spring and damping elements in Figure 11.16 where a force is applied to mass 1. For this case:

$$[S]=\begin{bmatrix} u_1^{(1)} & u_1^{(2)} & u_1^{(3)} \\ u_2^{(1)} & u_2^{(2)} & u_2^{(3)} \\ u_3^{(1)} & u_3^{(2)} & u_3^{(3)} \end{bmatrix} \tag{11.108}$$

$$[S]^T=\begin{bmatrix} u_1^{(1)} & u_2^{(1)} & u_3^{(1)} \\ u_1^{(2)} & u_2^{(2)} & u_3^{(2)} \\ u_1^{(3)} & u_2^{(3)} & u_3^{(3)} \end{bmatrix}. \tag{11.109}$$

The three uncoupled equations of motion are:

$$\ddot{r}_1(t)+2\xi_1\omega_1\dot{r}_1(t)+\omega_1^2 r_1(t)=u_1^{(1)}f(t) \tag{11.110}$$

$$\ddot{r}_2(t)+2\xi_2\omega_2\dot{r}_2(t)+\omega_2^2 r_2(t)=u_1^{(2)}f(t) \tag{11.111}$$

$$\ddot{r}_3(t)+2\xi_3\omega_3\dot{r}_3(t)+\omega_3^2 r_3(t)=u_1^{(3)}f(t). \tag{11.112}$$

Assume the forcing function f(t) is harmonic and given by $f(t) = Fe^{j\omega t}$. The responses associated with each of the modal coordinates are:

$$\mathrm{Re}\left[\frac{r_1(\omega)}{F}\right]=\frac{\frac{u_1^{(1)}}{\omega_1^2}\left(1-\frac{\omega^2}{\omega_1^2}\right)}{\left(1-\frac{\omega^2}{\omega_1^2}\right)^2+\left(2\xi_1\frac{\omega}{\omega_1}\right)^2} \tag{11.113}$$

$$\mathrm{Im}\left[\frac{r_1(\omega)}{F}\right]=\frac{-j\frac{u_1^{(1)}}{\omega_1^2}\left(2\xi_1\frac{\omega}{\omega_1}\right)}{\left(1-\frac{\omega^2}{\omega_1^2}\right)^2+\left(2\xi_1\frac{\omega}{\omega_1}\right)^2} \tag{11.114}$$

$$\left|\frac{r_1(\omega)}{F}\right|=\sqrt{\left|\mathrm{Re}\left[\frac{r_1(\omega)}{F}\right]\right|^2+\left|\mathrm{Im}\left[\frac{r_1(\omega)}{F}\right]\right|^2} \tag{11.115}$$

$$\theta_1(\omega)=\tan^{-1}\left[\frac{-\left|\mathrm{Im}\left[r_1(\omega)/F\right]\right|}{\left|\mathrm{Re}\left[r_1(\omega)/F\right]\right|}\right] \tag{11.116}$$

$$\mathrm{Re}\left[\frac{r_2(\omega)}{F}\right]=\frac{\frac{u_1^{(2)}}{\omega_2^2}\left(1-\frac{\omega^2}{\omega_2^2}\right)}{\left(1-\frac{\omega^2}{\omega_2^2}\right)^2+\left(2\xi_2\frac{\omega}{\omega_2}\right)^2} \tag{11.117}$$

$$\mathrm{Im}\left[\frac{r_2(\omega)}{F}\right]=\frac{-j\frac{u_1^{(2)}}{\omega_2^2}\left(2\xi_2\frac{\omega}{\omega_2}\right)}{\left(1-\frac{\omega^2}{\omega_2^2}\right)^2+\left(2\xi_2\frac{\omega}{\omega_2}\right)^2} \tag{11.118}$$

$$\left|\frac{r_2(\omega)}{F}\right|=\sqrt{\left|\mathrm{Re}\left[\frac{r_2(\omega)}{F}\right]\right|^2+\left|\mathrm{Im}\left[\frac{r_2(\omega)}{F}\right]\right|^2} \tag{11.119}$$

$$\theta_2(\omega)=\tan^{-1}\left[\frac{-\left|\mathrm{Im}\left[r_2(\omega)/F\right]\right|}{\left|\mathrm{Re}\left[r_2(\omega)/F\right]\right|}\right] \tag{11.120}$$

$$\mathrm{Re}\left[\frac{r_3(\omega)}{F}\right]=\frac{\frac{u_1^{(3)}}{\omega_3^2}\left(1-\frac{\omega^2}{\omega_3^2}\right)}{\left(1-\frac{\omega^2}{\omega_3^2}\right)^2+\left(2\xi_3\frac{\omega}{\omega_3}\right)^2} \tag{11.121}$$

$$\mathrm{Im}\left[\frac{r_3(\omega)}{F}\right]=\frac{-j\frac{u_1^{(3)}}{\omega_3^2}\left(2\xi_3\frac{\omega}{\omega_3}\right)}{\left(1-\frac{\omega^2}{\omega_3^2}\right)^2+\left(2\xi_3\frac{\omega}{\omega_3}\right)^2} \tag{11.122}$$

$$\left|\frac{r_3(\omega)}{F}\right|=\sqrt{\left|\mathrm{Re}\left[\frac{r_3(\omega)}{F}\right]\right|^2+\left|\mathrm{Im}\left[\frac{r_3(\omega)}{F}\right]\right|^2} \tag{11.123}$$

$$\theta_3(\omega)=\tan^{-1}\left[\frac{-\left|\mathrm{Im}\left[r_3(\omega)/F\right]\right|}{\left|\mathrm{Re}\left[r_3(\omega)/F\right]\right|}\right]. \tag{11.124}$$

The corresponding system responses with respect to the spatial coordinates are:

$$\mathrm{Re}\left[\frac{x_1(\omega)}{F}\right]=u_1^{(1)}\,\mathrm{Re}\left[\frac{r_1(\omega)}{F}\right]+u_1^{(2)}\,\mathrm{Re}\left[\frac{r_2(\omega)}{F}\right]+u_1^{(3)}\,\mathrm{Re}\left[\frac{r_3(\omega)}{F}\right] \tag{11.125}$$

$$\mathrm{Im}\left[\frac{x_1(\omega)}{F}\right]=u_1^{(1)}\,\mathrm{Im}\left[\frac{r_1(\omega)}{F}\right]+u_1^{(2)}\,\mathrm{Im}\left[\frac{r_2(\omega)}{F}\right]+u_1^{(3)}\,\mathrm{Im}\left[\frac{r_3(\omega)}{F}\right] \tag{11.126}$$

$$\left|\frac{x_1(\omega)}{F}\right|=\sqrt{\mathrm{Re}\left[\frac{x_1(\omega)}{F}\right]^2+\mathrm{Im}\left[\frac{x_1(\omega)}{F}\right]^2} \tag{11.127}$$

$$\phi_1(\omega) = \tan^{-1}\left[\frac{\mathrm{Im}[x_1(\omega)/F]}{\mathrm{Re}[x_1(\omega)/F]}\right] \tag{11.128}$$

$$\mathrm{Re}\left[\frac{x_2(\omega)}{F}\right] = u_2^{(1)}\,\mathrm{Re}\left[\frac{r_1(\omega)}{F}\right] + u_2^{(2)}\,\mathrm{Re}\left[\frac{r_2(\omega)}{F}\right] + u_2^{(3)}\,\mathrm{Re}\left[\frac{r_3(\omega)}{F}\right] \tag{11.129}$$

$$\mathrm{Im}\left[\frac{x_2(\omega)}{F}\right] = u_2^{(1)}\,\mathrm{Im}\left[\frac{r_1(\omega)}{F}\right] + u_2^{(2)}\,\mathrm{Im}\left[\frac{r_2(\omega)}{F}\right] + u_2^{(3)}\,\mathrm{Im}\left[\frac{r_3(\omega)}{F}\right] \tag{11.130}$$

$$\left|\frac{x_2(\omega)}{F}\right| = \sqrt{\mathrm{Re}\left[\frac{x_2(\omega)}{F}\right]^2 + \mathrm{Im}\left[\frac{x_2(\omega)}{F}\right]^2} \tag{11.131}$$

$$\phi_2(\omega) = \tan^{-1}\left[\frac{\mathrm{Im}[x_2(\omega)/F]}{\mathrm{Re}[x_2(\omega)/F]}\right] \tag{11.132}$$

$$\mathrm{Re}\left[\frac{x_3(\omega)}{F}\right] = u_3^{(1)}\,\mathrm{Re}\left[\frac{r_1(\omega)}{F}\right] + u_3^{(2)}\,\mathrm{Re}\left[\frac{r_2(\omega)}{F}\right] + u_3^{(3)}\,\mathrm{Re}\left[\frac{r_3(\omega)}{F}\right] \tag{11.133}$$

$$\mathrm{Im}\left[\frac{x_3(\omega)}{F}\right] = u_3^{(1)}\,\mathrm{Im}\left[\frac{r_1(\omega)}{F}\right] + u_3^{(2)}\,\mathrm{Im}\left[\frac{r_2(\omega)}{F}\right] + u_3^{(3)}\,\mathrm{Im}\left[\frac{r_3(\omega)}{F}\right] \tag{11.134}$$

$$\left|\frac{x_3(\omega)}{F}\right| = \sqrt{\mathrm{Re}\left[\frac{x_3(\omega)}{F}\right]^2 + \mathrm{Im}\left[\frac{x_3(\omega)}{F}\right]^2} \tag{11.135}$$

$$\phi_3(\omega) = \tan^{-1}\left[\frac{\mathrm{Im}[x_3(\omega)/F]}{\mathrm{Re}[x_3(\omega)/F]}\right]. \tag{11.136}$$

EXAMPLE 11.6

Assume the values of the masses and the stiffness coefficients of the mass and spring elements in Figure 11.16 are the same as those given in EXAMPLE 11.3. Proportional damping is assumed where $\alpha = 0.03$ and $\beta = 0.0001$. Determine and plot the results for $x_1(\omega)/F$, $x_2(\omega)/F$ and $x_3(\omega)/F$.

SOLUTION

MATLAB® script for calculating and plotting the amplitude and phase associated with $x_1(f)/F$, $x_2(f)/F$ and $x_3(f)/F$ as a function of frequency is presented in Appendix 11C. The output of this script is given below:

```
M =

 2.0000 0 0
 0 3.0000 0
 0 0 2.5000

K =

 14000 -6000 0
 -6000 16000 -10000
 0 -10000 10000

alpha =

 0.0300

beta =

 1.0000e-004

r =

 3

L =

 1.4142 0 0
 0 1.7321 0
 0 0 1.5811
```

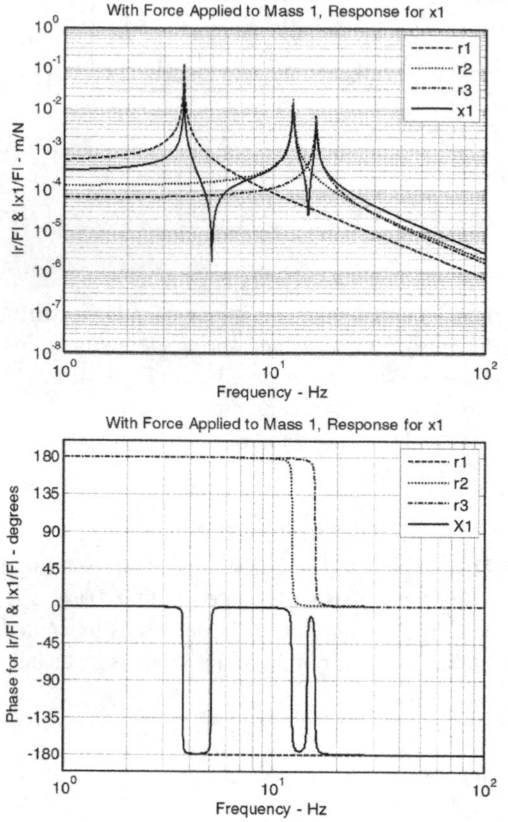

Figure 11.17 Plots associated with $x_1(\omega)$ in EXAMPLE 11.6

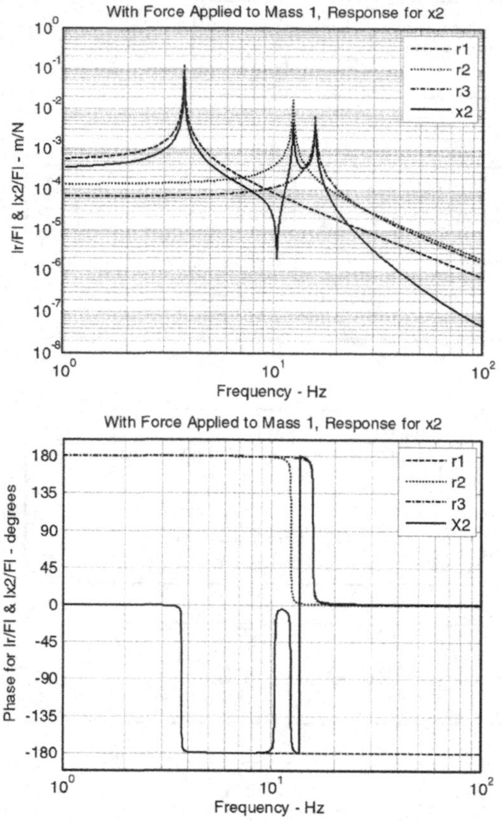

Figure 11.18 Plots associated with $x_2(\omega)$ in EXAMPLE 11.6

```
K_sym =

 1.0e+003 *

 7.0000 -2.4495 0
 -2.4495 5.3333 -3.6515
 0 -3.6515 4.0000

P =

 -0.5964 -0.7620 0.2524
 0.6789 -0.3111 0.6651
 -0.4283 0.5680 0.7029

lambda =

 1.0e+003 *

 9.7885 0 0
 0 6.0000 0
 0 0 0.5449

P_reverse =

 0.2524 -0.7620 -0.5964
 0.6651 -0.3111 0.6789
 0.7029 0.5680 -0.4283

P_ortho =

 1.0000 -0.0000 -0.0000
 -0.0000 1.0000 0.0000
 -0.0000 0.0000 1.0000

omega =

 98.9367
 77.4597
 23.3422
```

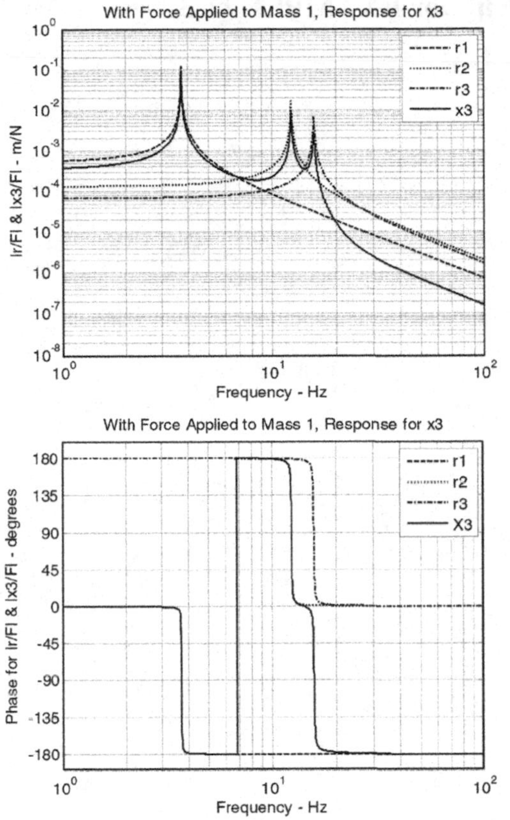

Figure 11.19 Plots associated with $x_3(\omega)$ in EXAMPLE 11.6

omega_reverse =

23.3422
77.4597
98.9367

freq =

3.7150
12.3281
15.7463

damp =

0.0018
0.0041
0.0051

S_from_P =

0.1785 -0.5388 -0.4217
0.3840 -0.1796 0.3920
0.4445 0.3592 -0.2709

S_norm =

0.2907 -0.8018 -0.6628
0.6255 -0.2673 0.6160
0.7241 0.5345 -0.4257

S_transpose =

0.2907 0.6255 0.7241
-0.8018 -0.2673 0.5345
-0.6628 0.6160 -0.4257

S_unit_normal =

1.0000 -0.0132 -0.1156
-0.0132 1.0000 0.1392
-0.1156 0.1392 1.0000

force =

1
0
0

F =

0.2907
-0.8018
-0.6628

Figures 11.17 through 11.19 show MATLAB® plots of the results.

Appendix 11A - MATLAB® Script for EXAMPLE 11.3

```
%% MATLAB Script for EXAMPLE 11.3
%% Calculates the Modal Parameters and Response of a n-DOF,
%% Mass-Spring System to Initial Conditions

clear all
close all
clc

%% Function Summary
% Caluculates:
% omega = Undamped Natural Frequencies
% P = Orthonormal Eigenvectors
% S = Unit normal mode vectors in Physical Coordinates
% X = Displacement time response in spatial coordinatese
% Xdot = velocity time response in spatial coordinates
% R = Displacement time response in modal coordinates
% Rdot = velocity time response in modal coordinates
% if Intial conditions are included System Response will also be plotted
% In vector format:
% x_0 = Inital displacement [x1(0); x2(0); ...xn(0)] in spatidal
% coordinates (m)
% v_0 = Initial velocities [v1(0); v2(0);...vn(0)] in spatiaol
% coordinates (m/s)
% r_0 = Inital displacement [r1(0); r2(0); ...rn(0)] in modal
% coordiantes (m)
% rdot_0 = Initial velocities [rdo1(0); rdot2(0);...rdotn(0)] in modal
% coordinates (m/s)
% T = Desired simulation time in seconds

%% Symetric Eigenvalue Solution
%Input mass and stiffmess matrices
M=[2 0 0; 0 3 0; 0 0 2.5];
M
K=[14000 -6000 0;-6000 16000 -10000;0 -10000 10000];
K

[r c]=size(M);
r

L=chol(M); %Cholesky Decomposition
L
K_sym=inv(L)*K*inv(L); %Mass-Normalized Stiffness
K_sym
[P,lambda]=eig(K_sym);%Eigenvalue Solution
P
lambda

%Reverse the order of the column vectors in P
for i=1:r
 n=r-i+1;
 P_reverse(:,i)=P(:,n);
```

```
end
P_reverse
P=P_reverse;

P_Transpose=transpose(P);
P_ortho=P_Transpose*P;
P_ortho

omega=sqrt(diag(lambda)); %Resonance Frequencies
omega

%Reverse the order of the resonance frequencies
for i=1:r
 n=r-i+1;
 omega_reverse(i,:)=omega(n,:);
end
omega_reverse
omega=omega_reverse;

S_from_P=inv(L)*P; %Mode Shape Matrix
S_from_P

%Convert S_from_P to a unit normal matrix
for i=1:r
 S_norm(:,i)=1/norm(S_from_P(:,i))*S_from_P(:,i);
end
S_norm
S=S_norm;
S_inv=inv(S);

S_transpose=transpose(S);
S_transpose
S_unit_normal=S_transpose*S; %Check that S is a unit normal matrix
S_unit_normal

%% Free Response (if I.C. are provided)
nargin=3;

if nargin > 2

%Specify initial conditions
%Units are m and m/s
 x_0=[0.01;0;0];
 x_0
 v_0=[0;0;0.2];
 v_0
 r_0 = S_inv * x_0;
 r_0
 rdot_0 = S_inv*v_0;
 rdot_0

 %% Time Solution In Modal Coordinates
 T=2; %Time span of response analysis
```

```
T = 2
dt=1/1000; %Plot time resolution
t=0:dt:T;

%Convert units from m and m/s to mm and mm/s
for i=1:r
R(i,:)=r_0(i,:)*cos(omega(i,:)*t) +(rdot_0(i,:)/omega(i,:))*sin(omega(i,:)*t);
Rdot(i,:)=-omega(i,:)*r_0(i,:)*sin(omega(i,:)*t)+rdot_0(i,:)*cos(omega(i,:)*t);
end

%% Time Solution In PHYSICAL Coordinates
X=S*R;
Xdot=S*Rdot;

%% Plotting
low1=min(min(R, [] , 2)); low1a=1.05*low1;
high1=max(max(R, [] , 2)); high1a=1.05*high1;
for i=1:r
figure
h=gca;
plot(t,R(i,:), 'LineWidth', 1.5, 'Color', 'black');
ylim([low1a, high1a]);
grid on
set(h,'FontSize',18);
xlabel('Time [s]','FontSize',18);
ylabel('Displacement [m]','FontSize',18);
title(sprintf('Undamped Response of r%d',i),'FontSize',18);
end
low2=min(min(Rdot, [] , 2)); low2a=1.05*low2
high2=max(max(Rdot, [] , 2)); high2a=1.05*high2
for i=1:r
figure
h=gca;
plot(t,Rdot(i,:), 'LineWidth', 1.5, 'Color', 'black');
ylim([low2a, high2a]);
grid on
set(h,'FontSize',18)
xlabel('Time [s]','FontSize',18);
ylabel('Velocity [m/s]','FontSize',18);
title(sprintf('Undamped Response of rd0t%d',i),'FontSize',18);
end
low3=min(min(X, [] , 2)); low3a=1.05*low3
high3=max(max(X, [] , 2)); high3a=1.05*high3
for i=1:r
figure
h=gca;
plot(t,X(i,:), 'LineWidth', 1.5, 'Color', 'black');
ylim([low3a, high3a]);
grid on
set(h,'FontSize',18)
xlabel('Time [s]','FontSize',18);
ylabel('Displacement [m]','FontSize',18);
title(sprintf('Undamped Response of x%d',i),'FontSize',18);
```

```
 end
 low4=min(min(Xdot, [] , 2)); low4a=1.05*low4
 high4=max(max(Xdot, [] , 2)); high4a=1.05*high4
 for i=1:r
 figure
 h=gca;
 plot(t,Xdot(i,:), 'LineWidth', 1.5, 'Color', 'black');
 ylim([low4a, high4a]);
 grid on
 set(h,'FontSize',18);
 xlabel('Time [s]','FontSize',18);
 ylabel('Velocity [m/s]','FontSize',18);
 title(sprintf('Undamped Response of xdot%d',i),'FontSize',18);
 end
end
```

Appendix 11B - MATLAB® Script for EXAMPLE 11.5

```
%% MATLAB Script for EXAMPLE 11.5
%% Calculaates the unit-normal mode vectors of a 3-DOF
%% approximation of a clamped-free beam from measured data

clear all
close all
clc
format short

%% Input measuired data
% Input measured resonace frequencies
Res_Freq=[43.1 267.8 748.1];
Res_Freq

% Input the peak acceleration transfer function values
% at the three resonance frequencies
H_Accel_Max=[13.44 175.18 419.45; 44.50 129.47 385.09; 79.90 281.47 543.95];
H_Accel_Max

% Input the phase sign at the three resonance frequencies
Phase_Sign=[ 1 1 -1; 1 1 1; 1 -1 -1];
Phase_Sign

% Input lower frequency value forthe real H_Accel max/min at the
% three resonance frequencies
H_Real_Freq_Low=[42.5 266.9 746.3; 42.5 266.8 746.3; 41.8 266.3 745.7];
H_Real_Freq_Low

% Input higher frequency value for the real H_Accel max/min at the
% three resonance frequencies
H_Real_Freq_High = [43.7 268.1 749.4; 43.8 268.8 749.4; 43.8 268.1 748.8];
H_Real_Freq_High

%% Caluation damping ratios associated with the three resonace frequeices
%% for the three sets of test data
for i=1:3
 damp(i,1)=0.5*((H_Real_Freq_High(i,1)/H_Real_Freq_Low(i,1)-1)/...
 (H_Real_Freq_High(i,1)/H_Real_Freq_Low(i,1)+1));
 damp(i,2)=0.5*((H_Real_Freq_High(i,2)/H_Real_Freq_Low(i,2)-1)/...
 (H_Real_Freq_High(i,2)/H_Real_Freq_Low(i,2)+1));
 damp(i,3)=0.5*((H_Real_Freq_High(i,3)/H_Real_Freq_Low(i,3)-1)/...
 (H_Real_Freq_High(i,3)/H_Real_Freq_Low(i,3)+1));
end
damp

% Calculate the average damping ratio for each of the three resonance frequencies
for i=1:3
 damp_avg(:,i)=(damp(1,i)+ damp(2,i)+damp(3,i))/3;
end
damp_avg
```

```
%% Calculate the non-normalized mode vectors
for i=1:3
 udot(2,i)=sqrt(2*damp_avg(:,i)*H_Accel_Max(2,i))*...
 Phase_Sign(2,i);
end
for i=1:3
 udot(1,i)=(2*damp_avg(:,i)*H_Accel_Max(1,i)/udot(2,i))*...
 Phase_Sign(1,i);
end
for i=1:3
 udot(3,i)=(2*damp_avg(:,i)*H_Accel_Max(3,i)/udot(2,i))*...
 Phase_Sign(3,i);
end
udot

%% Calculate the unit normald mode vectors
u_norm(:,1)=udot(:,1)/norm(udot(:,1));
u_norm(:,2)=udot(:,2)/norm(udot(:,2));
u_norm(:,3)=udot(:,3)/norm(udot(:,3));
u_norm

u_norm_transpose=transpose(u_norm);
u_norm_transpose

u_unit_normal=u_norm_transpose*u_norm;
u_unit_normal

%% Plot the Mode Vectors
x=0:1:3;
y1(:,1)=0;
y2(:,1)=0;
y3(:,1)=0;
for i=1:3
 n=i+1;
 y1(:,n)=u_norm(i,1);
 y2(:,n)=u_norm(i,2);
 y3(:,n)=u_norm(i,3);
end

xi=0:.1:3;
y1isp=interp1(x,y1,xi,'pchip');
y2isp=interp1(x,y2,xi,'pchip');
y3isp=interp1(x,y3,xi,'pchip');

figure
plot(x,y1,'o',xi,y1isp,x,y2,'s',xi,y2isp,'--',x,y3,'d',xi,y3isp,'-.',...
 'LineWidth',1.5,'color','black');
legend('Mode 1','Mode 1','Mode 2','Mode 2','Mode 3','Mode 3',0)
axis([0,3.2,-1,1])
grid on
h=gca;
set(h,'xtick',[0:1:3]);
set(h,'ytick',[-1:0.2:1]);
```

```
set(h,'FontSize',12);
xlabel('Beam Location Point','FontSize',12);
ylabel('Normalized Displacement','FontSize',12);
title('Mode Vectors');
```

Appendix 11C - MATLAB® Script for EXAMPLE 11.6

```
%% MATLAB Script for EXAMPLE 11.6
%% Calulates the Dynamic Compliance Functions of a 3-DOF
%% Mass-Spring-Damper System from the Systems Modal Parameters

clear all
close all
clc

%% Function Summary
% Caluculates:
% omega = Undamped resonance rrequencies (rad/s)
% Freq = Undamped resonance frequency (Hz)
% P = Orthonormal eigenvectors
% S = Unit normal mode vectors in physical coordinates
% X = Dynamic complinace in spatial coordinatese
% R = Dynaamic compliance in modal coordinates

%% Symetric Eigenvalue Solution

%Input mass and stiffmess matrices
M=[2 0 0; 0 3 0; 0 0 2.5];
M
K=[14000 -6000 0;-6000 16000 -10000;0 -10000 10000];
K

%Specify alpa and beta
alpha=0.03;
alpha
beta=0.0001;
beta

[r c]=size(M);
r

L=chol(M); %Cholesky Decomposition
L
K_sym=inv(L)*K*inv(L); %Mass-Normalized Stiffness
K_sym
[P,lambda]=eig(K_sym);%Eigenvalue Solution
P
lambda

%Reverse the order of the column vectors in P_from_P
for i=1:r
 n=r-i+1;
 P_reverse(:,i)=P(:,n);
end
P_reverse
P=P_reverse;

P_Transpose=transpose(P);
```

```
P_ortho=P_Transpose*P;
P_ortho

omega=sqrt(diag(lambda)); %Natural Frequencies
omega

%Reverse the order of the resonance frequencies
for i=1:r
 n=r-i+1;
 omega_reverse(i,:)=omega(n,:);
end
omega_reverse
omega=omega_reverse;
freq=omega/(2*pi);
freq

for i=1:r
 damp1(i,:)=alpha+beta*omega(i,:)^2;
 damp(i,:)=damp1(i,:)/(2*omega(i,:));
end
damp

S_from_P=inv(L)*P; %Mode Shape Matrix
S_from_P

%Convert S1 to a unit normal matrix
for i=1:r
 S_norm(:,i)=1/norm(S_from_P(:,i))*S_from_P(:,i);
end
S_norm
S=S_norm;
S_inv=inv(S);

S_transpose=transpose(S);
S_transpose
S_unit_normal=S_transpose*S; %Check that S is a unit normal matrix
S_unit_normal

%% Calculate r( ) and x( ) Amplitude and Phase Values

%Specify F (The mass to which the force is applied)
force=[1; 0; 0];
force
F=S_transpose*force;
F

frequencies=1:0.01:100;

r1_real=(F(1,:)./(omega(1,:).^2)).*(1-(frequencies./freq(1,:)).^2)./...
 ((1-(frequencies/freq(1,:)).^2).^2+(2.*damp(1,:).*frequencies./freq(1,:)).^2);
r1_imag=(F(1,:)./(omega(1,:).^2)).*(-(2.*damp(1,:)*frequencies./...
 freq(1,:)))./((1-(frequencies/freq(1,:)).^2).^2+(2.*damp(1,:).*frequencies/freq(1,:)).^2);
r1_amp=sqrt(r1_real.^2+r1_imag.^2);
```

```
r1_phase=atan2(r1_imag,r1_real).*180./pi;

r2_real=(F(2,:)./(omega(2,:).^2)).*(1-(frequencies./freq(2,:)).^2)./...
 ((1-(frequencies/freq(2,:)).^2).^2+(2.*damp(2,:).*frequencies./freq(2,:)).^2);
r2_imag=(F(2,:)./(omega(2,:).^2)).*(-(2.*damp(2,:)*frequencies./...
 freq(2,:)))./((1-(frequencies/freq(2,:)).^2).^2+(2.*damp(2,:).*frequencies/freq(2,:)).^2);
r2_amp=sqrt(r2_real.^2+r2_imag.^2);
r2_phase=atan2(r2_imag,r2_real).*180./pi;

r3_real=(F(3,:)./(omega(3,:).^2)).*(1-(frequencies./freq(3,:)).^2)./...
 ((1-(frequencies/freq(3,:)).^2).^2+(2.*damp(3,:).*frequencies./freq(3,:)).^2);
r3_imag=(F(3,:)./(omega(3,:).^2)).*(-(2.*damp(3,:)*frequencies./...
 freq(3,:)))./((1-(frequencies/freq(3,:)).^2).^2+(2.*damp(3,:).*frequencies/freq(3,:)).^2);
r3_amp=sqrt(r3_real.^2+r3_imag.^2);
r3_phase=atan2(r3_imag,r3_real).*180./pi;

Amp_Total(1,:)=sqrt((S(1,1).*r1_real+S(1,2).*r2_real+S(1,3).*r3_real).^2+...
 (S(1,1).*r1_imag+S(1,2).*r2_imag+S(1,3).*r3_imag).^2);
Amp_Total(2,:)=sqrt((S(2,1).*r1_real+S(2,2).*r2_real+S(2,3).*r3_real).^2+...
 (S(2,1).*r1_imag+S(2,2).*r2_imag+S(2,3).*r3_imag).^2);
Amp_Total(3,:)=sqrt((S(3,1).*r1_real+S(3,2).*r2_real+S(3,3).*r3_real).^2+...
 (S(3,1).*r1_imag+S(3,2).*r2_imag+S(3,3).*r3_imag).^2);

Phase_Total(1,:)=atan2((S(1,1).*r1_imag+S(1,2).*r2_imag+S(1,3)*r3_imag),...
 (S(1,1).*r1_real+S(1,2).*r2_real+S(1,3).*r3_real)).*180./pi;
Phase_Total(2,:)=atan2((S(2,1).*r1_imag+S(2,2).*r2_imag+S(2,3)*r3_imag),...
 (S(2,1).*r1_real+S(2,2).*r2_real+S(2,3).*r3_real)).*180./pi;
Phase_Total(3,:)=atan2((S(3,1).*r1_imag+S(3,2).*r2_imag+S(3,3)*r3_imag),...
 (S(3,1).*r1_real+S(3,2).*r2_real+S(3,3).*r3_real)).*180./pi;

%% Plot Amplitude and Phase Results
for i=1:r
 figure(i)
 h=gca;
 loglog(frequencies,r1_amp,'--',frequencies,r2_amp,':',frequencies,r3_amp,'-.',...
 frequencies,Amp_Total(i,:),'LineWidth',1.5,'color','black')
 grid on
 set(h,'ytick', [1E-8 1E-7 1E-6 1E-5 1E-4 1E-3 1E-2 1E-1 1E0])
 set(h,'FontSize',12)
 if i==1;legend('r1','r2','r3','x1');end
 if i==2;legend('r1','r2','r3','x2');end
 if i==3;legend('r1','r2','r3','x3');end
 title(sprintf('With Force Applied to Mass 1, Response for x%d',i),'FontSize',12);
 xlabel('Frequency - Hz','FontSize',12)
 if i ==1; ylabel('|r/F| & |x1/F| - m/N','FontSize',12);end
 if i ==2; ylabel('|r/F| & |x2/F| - m/N','FontSize',12);end
 if i ==3; ylabel('|r/F| & |x3/F| - m/N','FontSize',12);end
end

for i=1:r
 figure(i+3)
 h=gca;
 semilogx(frequencies,r1_phase,'--',frequencies,r2_phase,':',frequencies,r3_phase,'-.',...
```

```
frequencies,Phase_Total(i,:),'LineWidth',1.5,'color','black')
grid on
set(h,'ytick', [-180 -135 -90 -45 0 45 90 135 180])
set(h,'FontSize',12)
if i==1;legend('r1','r2','r3','X1');end
if i==2;legend('r1','r2','r3','X2');end
if i==3;legend('r1','r2','r3','X3');end
title(sprintf('With Force Applied to Mass 1, Response for x%d',i),'FontSize',12);
xlabel('Frequency - Hz','FontSize',12)
if i==1;ylabel('Phase for |r/F| & |x1/F| - degrees','FontSize',12);end
if i==2;ylabel('Phase for |r/F| & |x2/F| - degrees','FontSize',12);end
if i==3;ylabel('Phase for |r/F| & |x3/F| - degrees','FontSize',12);end
```

PROBLEMS and LABORATORY EXERCISES - CHAPTER 11

PROBLEMS

1. The influence coefficient matrix for the transverse vibration of the approximation for the cantilevered beam shown in Figure P11.1 is:

$$[a]=\frac{L^3}{EI}\begin{bmatrix}0.333 & 0.211 & 0.104 & 0.029\\0.211 & 0.141 & 0.073 & 0.021\\0.104 & 0.073 & 0.042 & 0.013\\0.029 & 0.021 & 0.013 & 0.005\end{bmatrix}.$$

Not that $[k]=[a]^{-1}$. The mass matrix is:

$$[m]=m\begin{bmatrix}1 & 0 & 0 & 0\\0 & 1 & 0 & 0\\0 & 0 & 1 & 0\\0 & 0 & 0 & 1\end{bmatrix}.$$

(a) Use the symmetrical eigenvalue solution method to obtain the values for the resonance frequencies and the corresponding mode vectors in terms of the modal coordinates.
(b) Show that the mode vectors are orthonormal mode vectors.

2. The mass and spring stiffness coefficient values for the vibration system with mass and spring elements in Figure P11.2 are:

M_1 = 6 kg; M_2 = 15 kg; and M_3 = 10 kg
K_1 = 15,000 N/m; K_2 = 20,000 N/m;
K_3 = 10,000 N/m; and K_4 = 15,000 N/m.

(a) Develop the equations of motion for the vibration system.
(b) Use the symmetrical eigenvalue solution method to obtain the values for the resonance frequencies and the corresponding mode vectors in terms of the modal coordinates.
(c) Show that the mode vectors are orthonormal mode vectors.

3. (a) Repeat Problem 2 using the dynamical matrix method. Obtain the resonance frequencies and the mode vectors in terms of the spacial coordinates.
(b) Show that the mode vectors are unit normal vectors, but not orthonormal vectors.
(c) Obtain the orthonoral modal matrix from the unit normal mode matrix.

4. The mass and spring stiffness coefficient values for the vibration system with mass and spring elements in Figure P11.3 are:

M_1 = 16 kg; M_2 = 6 kg; M_3 = 8 kg; and M_4 = 5 kg
K_1 = 25,000 N/m; K_2 = 15,000 N/m;
K_3 = 9,000 N/m; K_4 = 12,000 N/m. and
K_5 = 10,000 N.m.

(a) Develop the equations of motion for the vibration system.
(b) Use the symmetrical eigenvalue solution

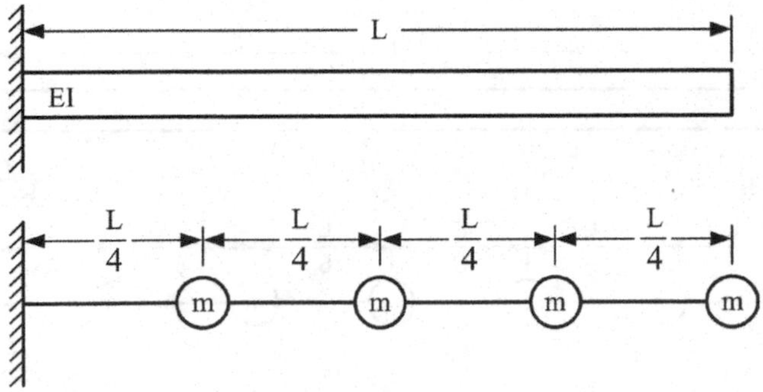

Figure P11.1

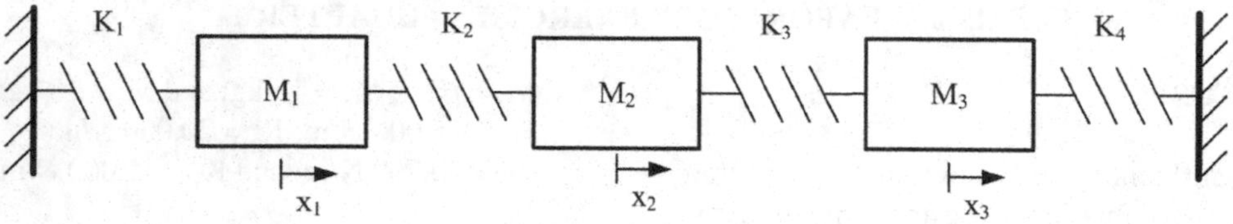

Figure P11.2

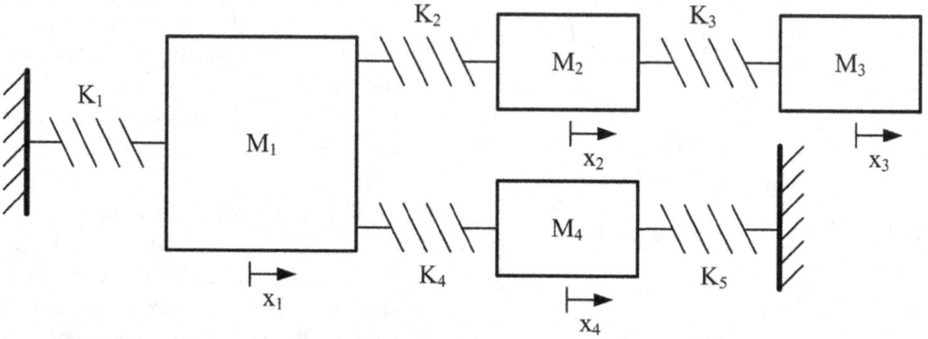

Figure P11.3

method to obtain the values for the resonance frequencies and the corresponding mode vectors in terms of the modal coordinates.
(c) Show that the mode vectors are orthonormal mode vectors.

5. (a) Repeat Problem 4 using the dynamical matrix method. Obtain the resonance frequencies and the mode vectors in terms of the spacial coordinates.
(b) Show that the mode vectors are unit normal vectors, but not orthonormal vectors.
(c) Obtain the orthonoral modal matrix from the unit normal mode matrix.

6. Figure P11.4 shows a 5-DOF approximation for a slender beam that is clamped at the left end and simply supported at a position L from the left end.
(a) Use influence coefficients to develop the equations of motion. Not that $[k]=[a]^{-1}$.
(b) se the symmetrical eigenvalue solution method to obtain the values for the resonance frequencies and the corresponding mode vectors in terms of the modal coordinates.
(c) Show that the mode vectors are orthonormal mode vectors.

7. Figure P11.5 shows a 5-DOF approximation for a slender beam that is clamped at the left end and

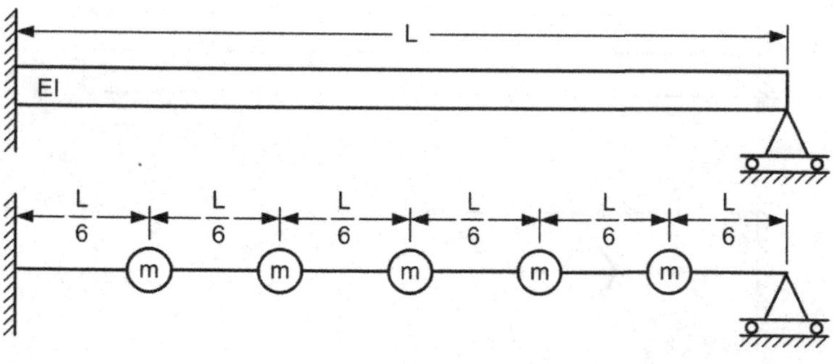

Figure P11.4

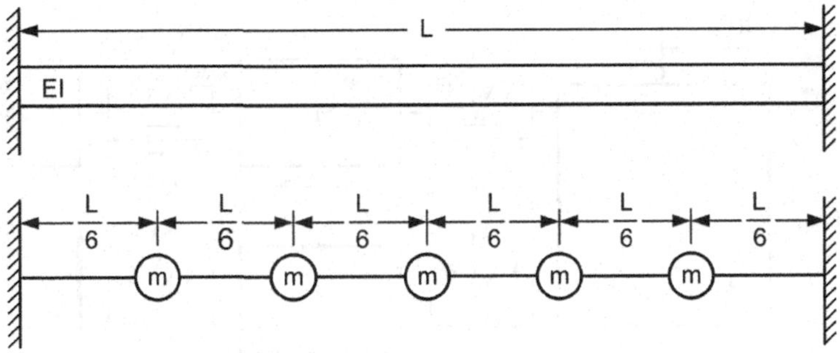

Figure P11.5

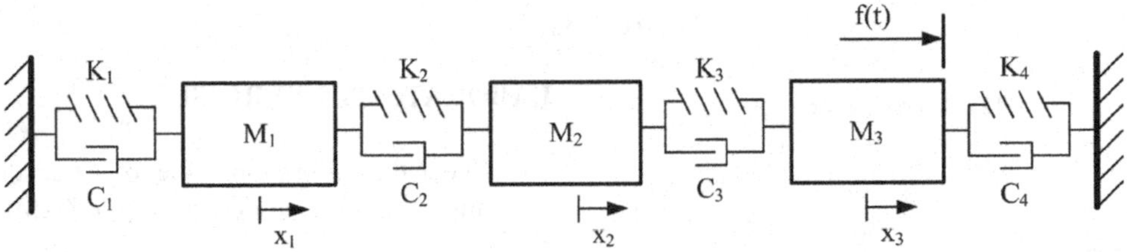

Figure P11.6

clamped at a position L from the left end.

(a) se influence coefficients to develop the equations of motion. Not that $[k]=[a]^{-1}$.

(b) Use the symmetrical eigenvalue solution method to obtain the values for the resonance frequencies and the corresponding mode vectors in terms of the modal coordinates.

(c) Show that the mode vectors are orthonormal mode vectors.

8. The vibration system with mass and spring elements in Figure 11.P6 has a harmonic force $f(t)=F e^{j\omega t}$ applied to mass M_3. Proportional damping is assumed where $\alpha = 0.002$ and $\beta = 0.0005$.

(a) Use the modal analysis procedure presented in Section 11.11 and the results of Problems 2 and 3 to obtained the forced responses for masses M_1, M_2, M_3 and M_4.

(b) On individual graphs, plots the responses $x_1(f)$, $x_2(f)$ and $x_3(f)$ as functions of frequency, f.

9. The vibration system with mass and spring elements in Figure 11.P7 has a harmonic force $f(t)=F e^{j\omega t}$ applied to mass, M_1. Proportional damping is assumed where $\alpha = 0.002$ and $\beta = 0.0005$.

(a) Use the modal analysis procedure presented in Section 11.11 and the results of Problems 4 and 5 to obtained the forced responses for masses M_1, M_2 and M_3.

(b) On individual graphs, plots the responses $x_1(f)$, $x_2(f)$, $x_3(f)$ and $x_4(f)$ as functions of frequency, f.

10. Repeat EXAMPLE 11.5 for the case where the harmonic force is acting on mass m_2.

11. Repeat EXAMPLE 11.5 for the case whee the harmonic force is acting on mass m_3.

12. Figures P11.8 - P11.10 show plots of the accelerance amplitudes at locations 1, 2 and 3, respectively, for the case where the cantilevered beam shown in Figure 11.11 is impacted at location 1 . Table P11.1 shows the peak accelerance values at the indicated resonance frequencies. Table P11.2 shows the phase signs at the indicated resonance frequencies. Table P11.3 shows the lower and upper frequency values for calculation of damping ratios at the indicated resonance frequencies. Modify the MATLAB® script in Annex B to obtain the resulting unit normal modal vectors and plot the unit normal mode vectors as shown in Figure 11.15

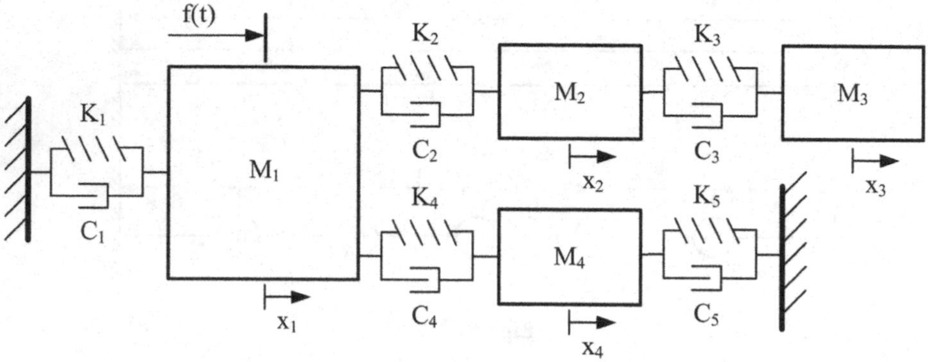

Figure P11.7

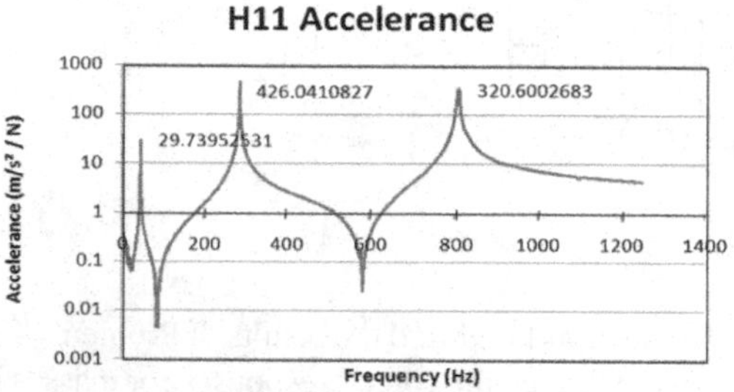

Figure P11.8

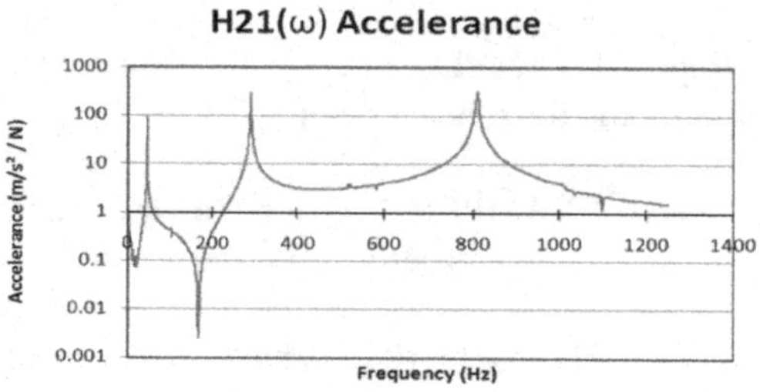

Figure P11.9

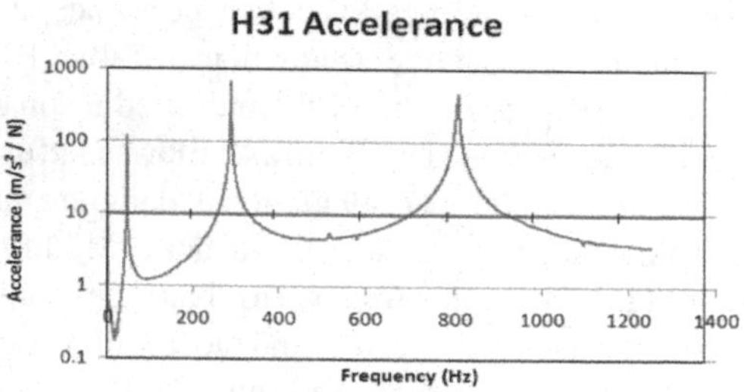

Figure P11.10

LABORATORY EXERCISES

1. Check the calibration of the force transducer in an impact hammer as described in EXAMPLE 11.3.

2. Repeat the modal measurements and the related data reduction, analysis and plots described in EXAMPLE 11.4. Impact the beam at location 2 in this laboratory exercise.

3. Repeat the modal measurements and the related data reduction, analysis and plots described in EXAMPLE 11.4. Impact the beam at location 1 in this laboratory exercise.

Table P11.1

Peak Acceleration Amplitudes - m/s^2			
Resonance	i = 1	i = 2	i = 3
Freq - Hz	46.70	290.80	808.70
$\lvert H_{11}(f) \rvert$	29.74	439.29	350.45
$\lvert H_{21}(f) \rvert$	98.35	294.98	324.42
$\lvert H_{31}(f) \rvert$	116.03	662.50	452.33

Table P11.2

Phase			
Resonance	**i = 1**	**i = 2**	**i = 3**
Freq - Hz	**46.70**	**290.80**	**808.70**
$Im[H_{11}(f)]$	1	1	1
$Im[H_{21}(f)]$	1	1	-1
$Im[H_{31}(f)]$	1	-1	1

Table P11.3

Lower and Upper Frequency Values for ξ Callculations			
Resonance	**i = 1**	**i = 2**	**i = 3**
Freq - Hz	**46.70**	**290.80**	**808.70**
$f_{a(11)}$	46.48	290.04	805.66
$f_{b(11)}$	47.07	291.21	810.55
$f_{a(21)}$	46.48	290.23	805.27
$f_{b(21)}$	47.07	291.60	810.35
$f_{a(311)}$	46.29	289.45	804.10
$f_{b(31)}$	46.88	290.82	809.18

CHAPTER 12
BUILDING MECHANICAL EQUIPMENT VIBRATION ISOLATION

12.1 Vibration in Buildings

Mechanical vibration and vibration induced sound are often major problems in modern buildings. Mechanical equipment associated with building heating, ventilating and air conditioning (HVAC) systems generate and transmit vibration to building structures. Large roof mounted air conditioning systems in large buildings, roof mounted packaged air handling units located on light weight roof structures, and mechanical equipment rooms located on intermediate floors in buildings result in vibration generating mechanical equipment being located adjacent to and above occupied areas in buildings.

Figure 12.1 schematically presents an example of the adverse effects of structure borne vibration generated by mechanical equipment. There are two vibration sources in the figure: the cooling tower on the roof above the room and the pump to the left of the room. Vibration sources in the cooling tower are a large propeller fan, a water circulation pump, and falling water. These vibration sources are not properly isolated from the building structure. As a result, cooling tower vibration is transmitted to the roof deck, which provides a vibration transmission path to the walls and floor of the room below. The pump to the left of the room is mounted directly to the floor, and the pump inlet and discharge pipes are directly attached to the building structure. As a result, pump vibration is transmitted to the building structure through the floor and pipes. The structure borne vibration from the cooling tower, pump and pipes generates structure borne sound that radiates into the room from the roof deck and room walls and transmits unacceptable vibration to the room floor. This chapter will address how structure borne vibration can be reduced to acceptable amplitudes.

Building occupant complaints associated with building vibration normally take three forms:

- The amplitude of perceived vibration is of sufficient magnitude to cause concern or alarm.

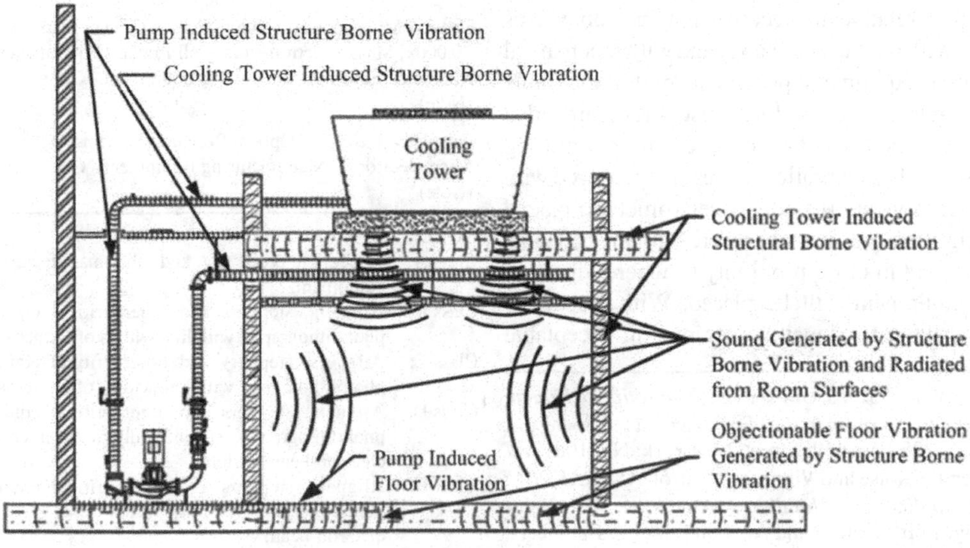

Figure 12.1 The Presence of Structure-Borne Vibration in a Building

- Vibration energy from mechanical equipment is transmitted throughout a building's structure and radiated from building surfaces as objectionable structure borne sound.
- Vibration present in a building interferes with the proper operation of sensitive equipment and instrumentation.

Information presented in this chapter is an extension of and builds upon the analytical and related information covered in Chapter 3.

12.2 Vibration Isolation System Design Criteria

Chapter 3 presented basic 1-DOF analytical concepts that can be used to design vibration isolation systems used to reduce:

- The vibration amplitudes of mechanical and related equipment that vibrate and the vibration energy this equipment transmit to building structures to which they are attached; and
- The vibration energy transmitted from vibrating building structures to vibration sensitive mechanical, optical, electronic, instrument and computer systems attached to these structures.

Acceptable vibration criteria must be specified when minimizing negative effects of building vibration is an important design objective. Building vibration criteria are specified relative to acceptable vibration values associated with human occupancy and with mechanical and sensitive equipment present in buildings. Table 12.1 and Figures 12.2 and 12.3 present recommended vibration values for vibration present in building structures.[1,2] These vibration values are measured with vibration transducers (usually accelerometers) placed on building structures in areas where human occupants are present and in close proximity to where vibration sensitive equipment will be placed. With respect to vibration criteria for sensitive equipment, acceptable

Table12.1 Vibration Criteria Curves for Acceptable Vibration in Buildings for Continuous Vibration (Curves Refer to Values Specified in Figures 12.3 and 12.4[1, 2]

Human Occupant and Equipment Requirements		Curve
Human Occupancy:	**Time of Day**	
Workshops	All	J
Office Areas	All	I
Residential (Good Environmental Standards)	0700-2200	H-I
	2200-0700	G
Hospital Operating Rooms and Critical Work Areas	All	F
Equipment Requirements:		
Computer Areas		H
Bench Microscopes up to 100x Magnification; Laboratory Robots		F
Bench Microscopes up to 400x Magnification; Optical and Other Precision Balances; Coordinate Measuring Machines; Metrology Laboratories; Optical Comparators;Microelectronics Manufacturing Equipment - Class A (Note)		E
Micro-Surgery, Eye Surgery, Neuro-Surgery; Bench Microscope at Magnification Greater than 400x; Optical Equipment on Isolation Tables; Microelectronic Manufacturing Equipment - Class B (Note)		D
Electron Microscopes up to 30,000x Magnification; Microtomes; Magnetic Resonance Imagers; Microelectronics Manufacturing Equipment - Class C (Note)		C
Electron Microscopes at Magnification Greater than 30,000x; Mass Spectrometers; Cell Implant Equipment; Microelectronics Manufacturing Equipment - Class D (Note)		B
Unisolated Laser and Optical Research Systems; Microelectronics Manufacturing Equipment - Class E (Note)		A

NOTE:

Class A: Inspection, probe test, and other manufacturing support equipment.

Class B: Aligners, steppers and other critical equipment for photolithography with line widths of 3 microns or more.

Class C: Aligners, steppers and other critical equipment for photolithography with line widths of 1 micron.

Class D: Aligners, steppers and other critical equipment for photolithography with line widths of 1/2 micron; includes electron-beam systems.

Class E: Aligners, steppers and other critical equipment for photolithography with line widths of 1/4 micron; includes electron-beam systems.

1. ANSI Standard S2.70, *Guide to the Evaluation of Human Exposure to Vibration in Buildings,* Acoustical Society of America, 335 East 45th Street, New York, NY 10017.
2. Chapter 48, Noise and Vibration Control, *ASHRAE 2015 HVAC Applications Handbood*, American Society for Heating, Refrigerating and Air Conditioning Engineers, 1791 Tullie Circle, N.E.n Atlanta, GA 30329.

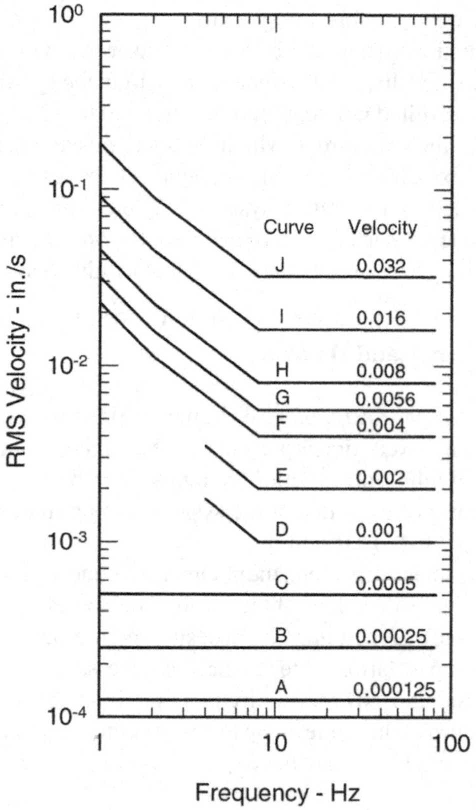

Figure 12.2 (U.S.) Building Vibration Criteria Measured on the Building Structure[1, 2]

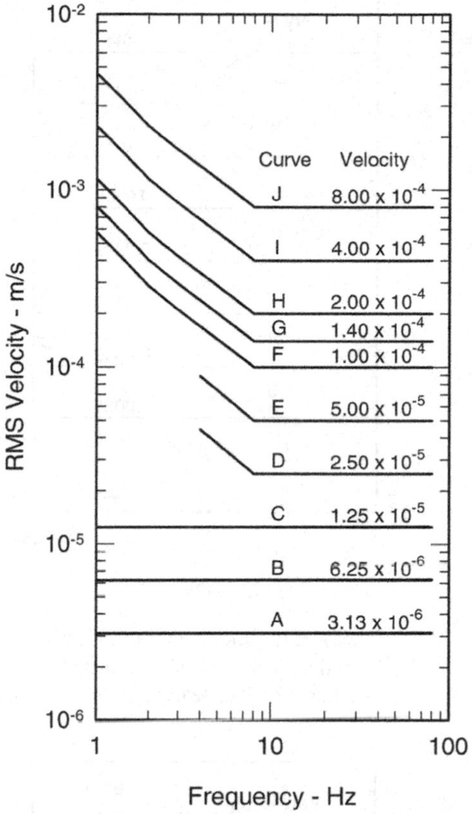

Figure 12.3 (SI) Building Vibration Criteria Measured on the Building Structure[1, 2]

vibration values specified by equipment manufacturers should be used. When this information is not available, the values specified in Figures 12.2 and 12.3 can be used. Figures 12.4 and 12.5 specify mechanical equipment *vibration severity ratings*.[2] These ratings are based on the measured rms velocity values, which are measured with vibration transducers (normally accelerometers) mounted directly on mechanical equipment structures and at bearing shaft locations. The measured vibration severity associated with vibrating mechanical equipment should be in the Good range or lower. When the measured vibration severity is in the Fair range or higher, steps should be initiated to reduce the vibration values.

Vibration present in mechanical equipment and equipment components is usually associated with reciprocating and rotating unbalance, misaligned coupled rotating machine components, and resonance interactions between vibrating mechanical equipment and building structures to which they are attached. When mechanical equipment with rotating or reciprocating components that are balanced within acceptable tolerances has excessive vibration, the equipment installation and support structure should be examined for the possible existence of resonance conditions.

12.3 Vibration Isolation System Design Guidelines

Factors to consider when designing a vibration isolation system include:

- Harmonic vibration can have a single frequency, multiple individual frequencies, or frequencies that continuously vary over a specified frequency range.
- Multiple types of resilient (spring) elements can be used in vibration isolation systems.

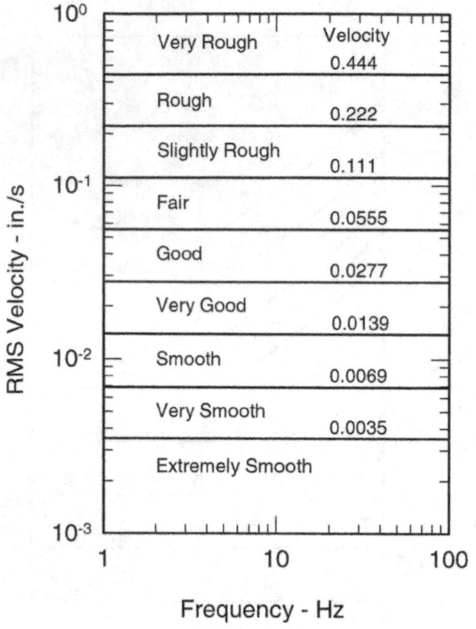

Figure 12.4 (U.S.) Equipment Vibration Severity for Vibration Measured on Mechanical Equipment Structures and at Bearing Shaft Locations[2]

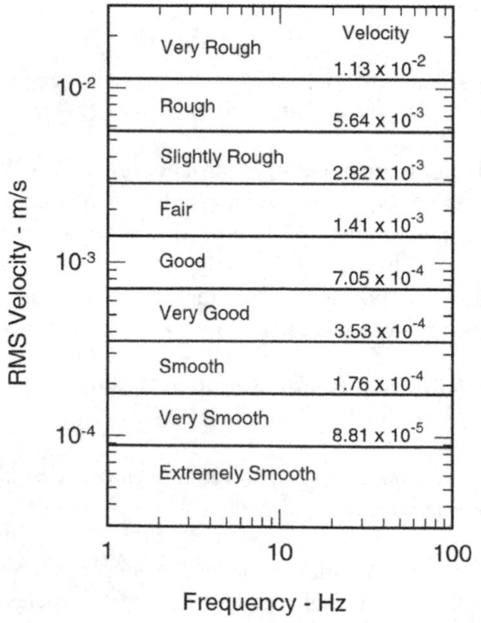

Figure 12.5 (SI) Equipment Vibration Severity for Vibration Measured on Mechanical Equipment Structures and at Bearing Shaft Locations[2]

- Viscous, structural or another type of damping can be present in a vibration system.
- Vibration systems can be mass-excited where vibration is applied directly to or generated within the system or base-excited where vibration is applied to the system from the structure to which the system is attached.
- An excitation force or moment can be externally applied to a vibration system at a single or multiple frequencies. It can also be generated by an unbalanced rotating or reciprocating component within the system.

General Guidelines Associated with Resonance Frequency and Damping

Resonance Frequency: Whenever possible when harmonic excitation is present, the resonance frequency ω_n of a vibration isolation system should be less than one-third of the value of the system forcing frequency ω ($\omega_n < \omega / 3$). If multiple or variable system forcing frequencies are present, the resonance frequency should be less than one-third of the value of the lowest forcing frequency. Depending on the design requirements for a vibration isolation system, it may be necessary to design the system so its resonance frequency ω_n is much less than one-third of the value of the lowest forcing frequency ω. For example, ω_n may have to be less than $\omega/5$ or $\omega/10$.

Damping: Viscous and structural damping were discussed in Chapter 3. When harmonic excitation is present and viscous damping is present in the vibration isolation system, the damping ratio ξ should be less than 0.2 and preferably less than 0.1. If the vibration isolation system is excited with a shock input (normally results from an initial velocity input with zero initial displacement - see Section 2.3 in Chapter 2), the damping ratio ξ should be in the range of 0.5 to 1.0 to cause the system oscillation to cease in a short period of time. Structural damping, which normally has a structural damping factor γ of less than 0.1, has no appreciable effect on the response characteristics of vibration isolation systems. All metal spring elements and many other types of resilient elements have structural damping.

Vibration Systems with Viscous Damping

A summary of the system response equations for mass-excited and base-excited vibration isolation systems with viscous damping is presented in this section for the design region where $\omega > 3\omega_n$ $(r > 3)$. The equations indicate viscus damping has a demonstrable effect in this design region on $|\bar{F}_T|$ in mass-excited systems and on $|\bar{Y}_2|$, $|\ddot{\bar{Y}}_2|$ and $|\bar{F}_T|$ in base-excited systems.

Mass-Excited System with Applied Harmonic Forces at Discrete Frequencies: For mass-excited vibration isolation systems with viscous damping that are excited by harmonic forces at discrete frequencies, the equations for the value of the displacement Y [equation (3.94)] of the system mass and for the value of the force $|\bar{F}_T|$ [equation (3.117)] transmitted to the base of the system at frequencies when $\omega > 3\omega_n$ $(r > 3)$ reduce to:

$$Y \approx \frac{F}{M\omega^2} \tag{12.1}$$

and:

$$|\bar{F}_T| = \frac{F\sqrt{K}\sqrt{\omega_n^2 + (2\xi\omega)^2}}{\sqrt{M}\,\omega^2} \tag{12.2}$$

or:

$$|\bar{F}_T| = \frac{F\omega_n\sqrt{\omega_n^2 + (2\xi\omega)^2}}{\omega^2}. \tag{12.3}$$

The values for Y and $|\bar{F}_T|$ decrease as a function of ω^2. When $\xi \neq 0$, the value of the numerator for $|\bar{F}_T|$ in equation (12.2) increase as ω increases. When $\xi = 0$, $|\bar{F}_T|$ reduces to:

$$|\bar{F}_T| = \frac{FK}{M\omega^2} \quad \text{or} \quad |\bar{F}_T| = F\frac{\omega_n^2}{\omega^2}. \tag{12.4}$$

Mass-Excited System with Applied Harmonic Forces Associated with a Rotating or Reciprocating Unbalance: For mass-excited vibration isolation systems with viscous damping that are excited by a rotating or reciprocating unbalance, the equations for the value of the displacement Y of the system mass and for the value of the force $|\bar{F}_T|$ [equation (3.132)] transmitted to the base of the system at frequencies when $\omega > 3\omega_n$ $(r > 3)$ are:

$$Y = \frac{me}{M} \quad \text{and} \quad |\bar{F}_T| = me\,\omega_n\sqrt{\omega_n^2 + (2\xi\omega)^2}. \tag{12.5}$$

As ω is increased, Y approaches the limiting value specified by equation (3.132). When $\xi \neq 0$, the value for $|\bar{F}_T|$ increase as ω increases. When $\xi = 0$, $|\bar{F}_T|$ reduces to:

$$|\bar{F}_T| \approx \frac{me\,K}{M} \quad \text{or} \quad |\bar{F}_T| \approx me\,\omega_n^2. \tag{12.6}$$

Base Excited System with Applied Harmonic Forces at Discrete Frequencies: For base-excited vibration isolation systems with viscous damping, the equations for $|\bar{Y}_2|$ [equation (3.211)], $|\ddot{\bar{Y}}_2|$ [equation (3.219)], and $|\bar{F}_T|$ [equation (3..225)]at frequencies when $\omega > 3\omega_n$ $(r > 3)$ are:

$$|\bar{Y}_2| \approx \frac{Y_1\,\omega_n\sqrt{\omega_n^2 + (2\xi\omega)^2}}{\omega^2} \tag{12.7}$$

$$|\ddot{\bar{Y}}_2| \approx Y_1\,\omega_n\sqrt{\omega_n^2 + (2\xi\omega)^2} \tag{12.8}$$

and:

$$|\bar{F}_T| \approx Y_1\sqrt{KM}\sqrt{\omega_n^2 + (2\xi\omega)^2}. \tag{12.9}$$

$|\bar{Y}_2|$ decreases as a function of ω^2. When $\xi \neq 0$, the numerator for $|\bar{Y}_2|$ and the values for $|\ddot{\bar{Y}}_2|$ and $|\bar{F}_T|$ increase as ω increases. When $\xi = 0$, these equations reduce to;

$$|\bar{Y}_2| \approx \frac{Y_1 K}{M\omega^2} \quad \text{or} \quad |\bar{Y}_2| \approx Y_1\frac{\omega_n^2}{\omega^2}. \tag{12.10}$$

$$|\ddot{\bar{Y}}_2| = \frac{Y_1 K}{M} \quad \text{or} \quad |\ddot{\bar{Y}}_2| = Y_1\,\omega_n^2. \tag{12.11}$$

and:

$$|\bar{F}_T| = Y_1 K. \tag{12.12}$$

When $\xi = 0, |\ddot{\bar{Y}}_2|$ and $|\bar{F}_T|$ reach the limiting values specified by equations (12.11) and (12.12).

Vibration Systems with Structural Damping

A summary of the system response equations for mass-excited and base-excited vibration isolation systems with structural damping is presented in this section for the design region where $\omega > 3\omega_n$ ($r > 3$). The equations indicated structural damping has no demonstrable effect on the response of mass-excited and base-excited vibration isolation systems.

Mass-Excited System with Applied Harmonic Forces at Discrete Frequencies: For mass-excited vibration isolation systems with structural damping that are excited by harmonic forces at discrete frequencies, the equations for the value of the displacement Y [equation (3.147)] of the system mass and for the value of the force $|\bar{F}_T|$ [equation (3.157)] transmitted to the base of the system at frequencies when $\omega > 3\omega_n$ ($r > 3$) are:

$$\boxed{Y = \frac{F}{M\omega^2}.} \tag{12.13}$$

and:

$$\boxed{|\bar{F}_T| = \frac{FK}{M\omega^2} \quad \text{or} \quad |\bar{F}_T| = F\frac{\omega_n^2}{\omega^2}.} \tag{12.14}$$

The values for Y and $|\bar{F}_T|$ decrease as a function of ω^2.

Mass-Excited System with Applied Harmonic Forces associated with a Rotating or Reciprocating Unbalance: For mass-excited vibration isolation systems with structural damping that are excited by a rotating or reciprocating unbalance, the equations for the value of the displacement Y of the system mass and for the value of the force $|\bar{F}_T|$ [equation (3.160)] transmitted to the base of the system at frequencies when $\omega > 3\omega_n$ ($r > 3$ are:

$$\boxed{Y = \frac{me}{M} \quad \text{and} \quad |\bar{F}_T| = \frac{me\,K}{M} \quad \text{or} \quad |\bar{F}_T| = me\,\omega_n^2.} \tag{12.15}$$

As ω is increased, Y and $|\bar{F}_T|$ approach the limiting values specified by equation (3.160).

Base Excited System with Applied Harmonic Forces at Discrete Frequencies: For base-excited vibration isolation systems with structural damping, the equations for $|\bar{Y}_2|$ [equation (3.233)], $|\ddot{\bar{Y}}_2|$ [equation (3.236)], and $|\bar{F}_T|$ [equation (3.239)]at frequencies when $\omega > 3\omega_n$ ($r > 3$) are:

$$\boxed{|\bar{Y}_2| = \frac{Y_1 K}{M\omega^2} \quad \text{or} \quad |\bar{Y}_2| = Y_1\frac{\omega_n^2}{\omega^2}.} \tag{12.16}$$

$$\boxed{|\ddot{\bar{Y}}_2| = \frac{Y_1 K}{M} \quad \text{or} \quad |\ddot{\bar{Y}}_2| = Y_1\,\omega_n^2.} \tag{12.17}$$

$$\boxed{|\bar{F}_T| = K\,Y_1.} \tag{12.18}$$

As ω is increased, $|\ddot{\bar{Y}}_2|$ and $|\bar{F}_T|$ reach the limiting values specified by equations (12.17) and (12.18). $|\bar{Y}_2|$ decreases as a function of ω^2.

Relationship between F, me, Y_1, M and K

The response associated with all parameters that describe the motions of mass-excited and base-excited vibration isolation systems can be reduced by reducing the system inputs. Y and $|\bar{F}_T|$ for mass-excited systems can be reduced by decrease F and me. me in mass-excited systems with rotation unbalance can be reduced by balancing the rotating components to a closer tolerance. $|\bar{Y}_2|$, $|\ddot{\bar{Y}}_2|$, and $|\bar{F}_T|$ in base-excited systems can be reduced by decreasing Y_1.

The values for Y and $|\bar{F}_T|$ in a mass-excited system can be reduced by decreasing the resonance frequency ω_n of the system. Y can be reduced by increasing M. Noting that $\omega_n = \sqrt{K/M}$, when M is increased, to maintain the same value for ω_n, K must also be increased in the same proportion to the increase in M. However, when Y is reduced by increase M, $|\bar{F}_T|$ is reduced by leaving K unchanged, thereby reducing ω_n, or by increasing K by a proportionately lower amount relative to the increase in M, which will also reduce ω_n. The system resilient element represented by K must be sized to support the weight of the system mass represented by M.

When coil springs are used as the system resilient element, they must be sized to support the weight of the system mass represented by M in a stable manner (See Section 2.7). The working height h of the springs must be greater than their static deflection δ associated with the weight of M. This will allow the dynamic motion of M to occur without the spring coils totally closing.

The response for a base-excited system associated with $|\bar{Y}_2|$, $|\ddot{\bar{Y}}_2|$, and $|\bar{F}_T|$ can also be reduced by decreasing the resonance frequency ω_n for the system. When designing a vibration isolation system for a base-excited system, the relationship requirements between the mass M and the resilient element K are the same as those described for a mass-excited system.

12.4 Resilient Elements and Commercial Vibration Isolation Mounts

Helical (Coil) Springs

The most common vibration isolation element is a helical spring (Figure 12.6). Helical springs can be designed with a maximum static deflection of up to 5 in. (127 mm) However, for many applications, their static deflection is limited to 2 in. (51 mm) They can be used in almost any mechanical environment and, if properly designed, are as permanent as the machines they support. Helical springs are used in critical design applications where resonance frequencies below 5 Hz are required.

The static stiffness K_y of a helical spring in the y-direction is given by:

$$K_y = \frac{G d^4}{8 D^3 n} \tag{12.19}$$

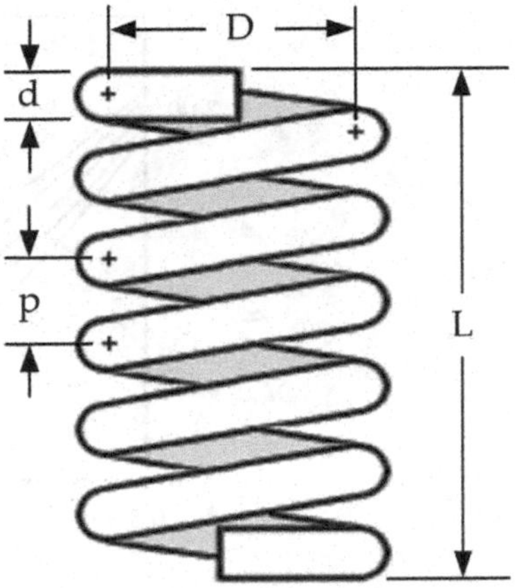

Figure 12.6 Helical Spring with five active coils

where:

G = the shear modulus ($lb_f/in.^2$),
d = the diameter of the spring wire (in.),
D = the mean coil diameter (in.), and
n = the number of active coils.

L in Figure 12.6 is the spring free length, and p is the spring pitch. Table 12.2 list Young's modulus and the shear modulus for some common helical spring materials.

Helical springs that support mechanical equipment are often loaded in compression (y-direction) and in the lateral direction (x-direction). The analysis for determining the spring stiffness K_x in the lateral direction is rather involved. The results for this analysis for a helical spring with a shear modulus of $G = 11.5 \times 10^6$ and a Young's modulus of $E = 30 \times 10^6$ are shown in Figure 12.7.[3] In the figure:

Table 12.2 Young's Modulus and Shear Modulus for Common Helical Spring Metals

Material	Young's Modulus E - $lb_f/in.^2$	Shear Modulus G - $lb_f/in.^2$
Carbon and Steel Alloy	30×10^6	11.5×10^6
Stainless Steel 18-8	30×10^6	10.0×10^6
Spring Brass	15×10^6	5.5×10^6
Phosphor Bronz	15×10^6	6.5×10^6
Berylium Copper	19×10^6	7.0×10^6

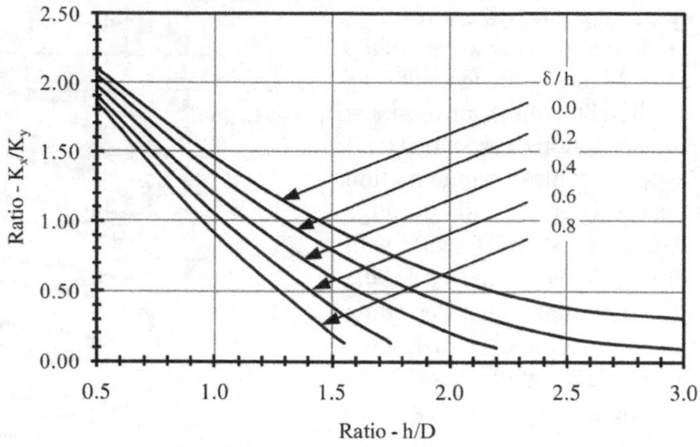

Figure 12.7 Ratio of K_x/K_y for Helical Compression Springs as a Functionof δ/h and h/D [3]

h = the working height, which is the free height of the spring (in.) with no load minus the static deflection (in.) of the spring under load; and
δ = the static deflection of the spring (in.).

Figure 12.7 also applies for other spring materials where the ratio of E/G is around 2.6.

Helical springs are used where static deflections of 2 in. (51 mm) and sometimes higher are required. Mechanical equipment supported by these springs can experience lateral deflections when exposed to a lateral force. To ensure the system will remain stable and not tip to the side, the following condition must be met:

$$\boxed{\frac{K_x}{K_y} > 1.2\frac{\delta}{h}.} \tag{12.20}$$

Pneumatic Springs

For most applications, the maximum practical static deflection for helical springs is around 2 in. (50.8 mm). This results in a limiting resonance frequency of around 2.2 Hz. In very critical situations where very low vibration transmission to a structure frame a vibrating machine or to a very sensitive piece of equipment from a vibrating structure are required, a vibration isolation system with a resonance frequencies of 1 Hz or lower may be required. Pneumatic (air) springs are often used for these situations.

3. Crede, C. E., *Vibration and Shock Isolation*, John Wiley & Sons, 1951.

A pneumatic spring is designed to support the static weight of a piece of equipment by exerting a pressure over a designated surface area (Figure 12.8). To determine the associated stiffness of the pneumatic spring, assume a change in air pressure associated with a corresponding change in air volume is an adiabatic process. For this process:

$$PV^{\gamma} = \text{Constant} \tag{12.21}$$

where;

P = absolute pressure ($lb_f/in.^2$),
V = volume of air in the pneumatic spring ($in.^3$), and
γ = the ratio of specific heats (γ = 1.4 for air).

Taking the derivative of equation (12.21) yields:

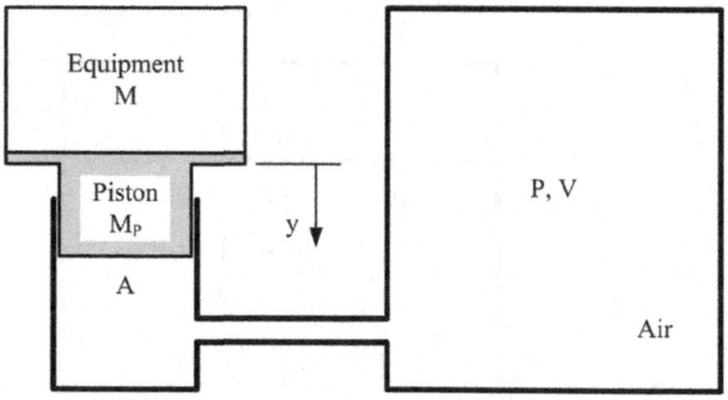

Figure 12.8 Schematic of a Pnuematic Spring

$$dP\,V^{\gamma} + \gamma P\,V^{\gamma-1}\,dV = 0. \tag{12.22}$$

Rearranging equation (12.22) and simplifying gives:

$$dP = -\frac{\gamma P}{V}\,dV. \tag{12.23}$$

If the incremental movement dy of the piston is assumed to be positive downward:

$$dV = -A\,dy \tag{12.24}$$

where A is the cross-section area of the piston (in.^{2}). A positive displacement y of the piston results in a decrease in volume V. Substituting equation (12.24) into (12.22 yields:

$$dP = \frac{\gamma P A}{V}\,dy \quad \text{or} \quad \frac{dP}{dy} = \frac{\gamma P A}{V}. \tag{12.25}$$

Multiplying both sides of equation (12.25) by the surface area of the piston A gives:

$$\frac{dF}{dy} = \frac{\gamma P A^2}{V} \quad \text{where} \quad dF = A\,dP. \tag{12.26}$$

The spring coefficient K is defined as dF/dy. Therefore, the spring coefficient K of the pneumatic spring is:

$$\boxed{K = \frac{\gamma P A^2}{V}.} \tag{12.27}$$

If the mass of the piston is designated M_p and M is the mass of the piece of equipment supported by the pneumatic spring, then the resonance frequency of the system is:

$$\boxed{f_n = \frac{1}{2\pi}\sqrt{\frac{\gamma P A^2}{(M+M_p)V}}.} \tag{12.28}$$

EXAMPLE 12.1

A large machine is to be supported by a pneumatic spring. The platform on which the machine is placed is supported by five 6-in.-diameter pistons that are connected to a common air reservoir. The weight of the machine and platform is 50,000 lb_f. The desired resonance frequency is 0.3 Hz. Determine the required air pressure P and the volume V of the common air reservoir.

SOLUTION

The system air pressure must be sufficient to support the static weight of the machine plus platform. The equation for determining the required air pressure is:

$$W = P_g\,A \tag{12.1a}$$

where W is the weight (lb_f), P_g is gauge pressure (lb_f/in.^{2}), and A_p is the combined cross-section surface area of the five pistons (in.^{2}). A_p is:

$$A_p = 5\pi 3^2 = 141.37 \text{ in.}^2. \tag{12.1b}$$

Therefore, the required air pressure P_g is:

$$P_g = \frac{50{,}000}{141.37} = 353.68\,\frac{lb_f}{in.^2}. \tag{12.1c}$$

The pressure in equation (3.273) is absolute pressure. Therefore,

$$\boxed{P = P_g + 14.7 = 368.38.} \tag{12.1d}$$

Rearranging equation (3.273):

$$V = \frac{\gamma P A^2}{(2\pi f_n)^2\,M}. \tag{12.1e}$$

$$\boxed{V = \frac{1.4\times 368.38\times 141.37^2\times 386}{(2\pi\times 0.3)^2\times 50{,}000} = 22.395 \text{ in.}^2.} \tag{12.1f}$$

Commercial Vibration Isolation Mounts

Commercial vibration isolation mounts are resilient elements used to support mechanical equipment. When properly integrated with the mechanical equipment,

Figure 12.9 Steel Open Coil Spring Mount

they reduce vibration transmitted to the structure that supports the mechanical equipment. Common resilient mounts include steel coil springs, elastomeric mounts, pneumatic isolators and glass fiber pads.

Steel open coil Spring Mount: Steel open coil spring mounts are the most commonly used resilient element in vibration isolation systems for mechanical equipment and for mechanical, optical, electronic, instrument and computer systems. They are available for most required static deflections, and they have a virtually unlimited life. Steel open coil springs are normally used in vibration isolations systems where a static deflection δ in the range of 0.5 in. to 2.0 in. is required.

The steel open coil spring mount shown in Figure 12.9 incorporates a steel open coil spring with a threaded stud at the top to attach the mount to a machine base and a bottom plate with a neoprene pad under the plate. The top threaded stud can be used to level the machine supported by the mount. These spring elements should be selected such that the ratio of the spring coil diameter D divided by the spring working height h is between 0.8 and 1.0. The horizontal stiffness K_x should be equal to or greater than the vertical stiffness K_y to ensure stability. The allowable spring deflection beyond its static deflection δ under load should be at least one-half its working height h before the coils totally close. The neoprene pad should be at least 0.25 in. (6 mm) thick to reduce the transmission of high frequency vibration through the steel coils of the spring to the structure supporting the spring mount and to permit the spring mount to be mounted on a floor without the use of bolts or other fasteners.

Figure 12.10 Restraint Open Coil Spring Mount

Vertical motion restraints are often used with steel open coil spring mounts to limit the upward movement of the springs when weight is temporarily removed or to restrict the free height of the spring coils during mechanical equipment installation. Restraint spring mounts commonly are used with: (a) equipment with large variations in mass (boilers, refrigeration machines, etc.) to restrict vertical movement and prevent strain on pipes when water is removed and (b) outdoor equipment to prevent excessive horizontal movement associated with wind load.

A common restraint configuration used with steel open coil springs is shown in Figure 12.10. It uses rigid sides where threaded studs attached to a top plate pass through holes in restraint plates attached to the tops of the rigid sides. Nuts applied to the studs on the undersides of these restraint plates can be adjusted to limit the vertical movement of the top plate. This configuration also limits the horizontal motion of the spring.

Figure 12.11 shows a housed steel open coil spring mount. These mounts are often used to limit the horizontal motion of the spring. They tend to bind up when the load on the spring causes it to tilt. Therefore, these spring mounts should never be used to support

Figure 12.11 Housed Steel Open Coil Spring Mount

mechanical equipment.

Steel open coil restraint spring mounts are used in outdoor applications (cooling towers, rooftop mounted mechanical equipment, etc.) where wind loads can develop large horizontal forces that act against the sides of mechanical equipment. These forces are then transferred to the vibration isolation mounts. Therefore, rubber sleeves should be placed in the holes of the restraint plates on the rigid sides of the vibration isolation mounts to prevent the threaded studs attached to the top plate from contacting the sides of the holes in the restraint plates.

Spring vibration isolation hangers are used to vibration isolate ducts, pipes, and small pieces of mechanical equipment that are suspended from a ceiling. Vibration isolation hangers may have only a steel open coil spring or a combination of a steel open coil spring plus a neoprene mount (Figure 12.12). Vibration isolation hangers can also have only a neoprene pad. For either type of vibration isolation hanger, the hole at the bottom of the hanger box must be sufficiently large to permit the hanger rod to swing through a 20 to 35 degree arc before the rod contacts the side of the hole. Contact between the rod and the hanger box can reduce the vibration isolation efficiency of the spring hanger.

Elastomeric Mounts: Elastomers are resilient materials such as neoprene, silicone, polyurethane, natural and butyl rubber. Neoprene is most commonly used because it is resistant to oils, acids and alkalis encountered

Figure 12.13 Neoprene Mounts

in machinery environments. Elastomer isolators are available in molded and pad configurations. They can be used when the required static deflection associated with vibration isolation is less than 0.5 in. (13 mm). Normally, elastomer isolators are used when 0.3 in. (8 mm) or less static deflections are required.

Molded elastomeric mounts are inexpensive. They can be molded in many different sizes and shapes and with and without metallic inserts to receive bolts or other fasteners (Figure 12.13). The hardness of the neoprene mounts furnished for mechanical and electrical equipment normally range from 30 to 70 durometer. The durometer of the mount is color coded as follows: (a) 30 durometer - black, (b) 40 durometer - green, (c) 50 durometer - red, (d) 60 durometer - white, and (e) 70 durometer - yellow. Elastomer mounts are normally used to vibration isolate light and low horsepower equipment or for equipment located in basements.

Neoprene is used for the manufacture of vibration isolation pads. Figure 12.14 shows a neoprene waffle

Figure 12.12 Vibration Isolation Hangers

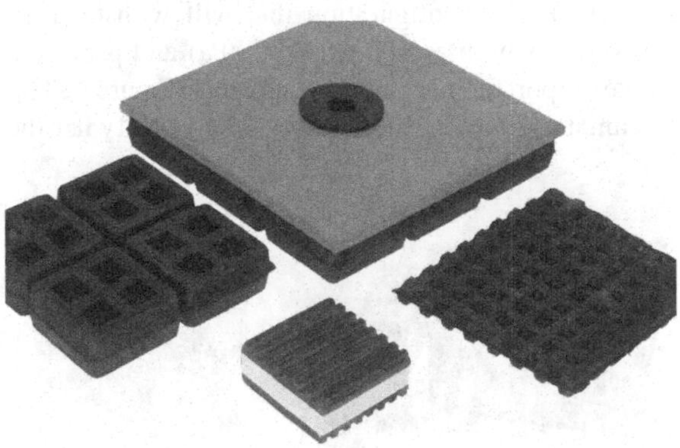

Figure 12.14 Neoprene Pads

pad. This pad can be cut into sections that contain one or more of the square waffle sections. These pads normally do not require fastening to a machine base. When necessary, they can be attached to a machine base or to a surface that supports the base with an adhesive. They can be stacked to provide increased deflection. The durometer of neoprene pads normally range from 30 to 60.

Vibration isolation hangers are also fabricated with the lower steel open coil spring being replaced by a molded neoprene mount similar to the one shown in Figure 12.12. The durometer of the neoprene used in the vibration isolation hangers range from 30 to 70 (color coded as noted for neoprene mounts).

Glass Fiber Pads: Glass fiber pads are made of a high density matrix of pre-compressed, inorganic, inert fiber glass that is molded into specified shapes and coated with a flexible elastomeric moisture barrier (Figure 12.15. Glass fiber pads can support weights that range from 20 lb to 16,000 lb (9 kg to 7,256 kg) per pad. The pads normally come in thicknesses that range from 1 in. to 4 in. (25 mm to 102 mm), and they can have static deflections that range from 0.2 in. to 1.0 in. (5 mm to 25 mm). Glass fiber pads are used for vibration isolation of pumps, chillers, cooling towers and other similar equipment. They are effective in reducing shock vibration produced by punch presses and other impact-producing machinery, and they are commonly used to support floating floors and slabs.

Pneumatic Vibration Isolation Mounts: A pneumatic vibration isolation mount (also called an air spring) consists of a cylindrical shaped air bladder that is fabricated in a configuration that will withstand air pressures as high as 100 psi (690 kPa) and provide a stable support for mechanical equipment (Figure 12.16). Pneumatic vibration isolation mounts are used when the required design resonance frequency is between 1 - 2 Hz. This provides the vibration isolation equivalence of steel open coil springs with static deflections under the weight of supported mechanical equipment of 2.4 to 9.8 in. (61 mm to 245 mm). Pneumatic vibration isolation mounts have the advantage of being able to support a wide range of loads by varying the air pressure in its pneumatic bladder. The resonance frequency of a pneumatic vibration isolation mount is a function of the air pressure and air volume of the pneumatic bladder and the weight of the mechanical equipment supported by the mount. Pneumatic vibration isolation mounts are installed with a replenishing air source since some leakage is inevitable. Makeup air is required for expansion and contraction in cases of wide variations of temperature. Height control valves are provided to maintain the elevation.

Figure 12.15 Glass Fiber Pads

Figure 12.16 Pneumatic Vibration Isolation Mounts

12.5 Structural Bases

Whenever a motor driving a machine is connected to the machine by a belt or other coupling, a structural base must be provided to maintain proper alignment between the motor and machine. Structural rigidity is an essential property of a structural base to ensure proper alignment between a motor and the machine it drives and to resists motor-machine belt forces and torques. A structural base that is mounted on spring or other resilient mounts reduces mechanical equipment vibration energy that is transmitted to the structure that supports the structural base.

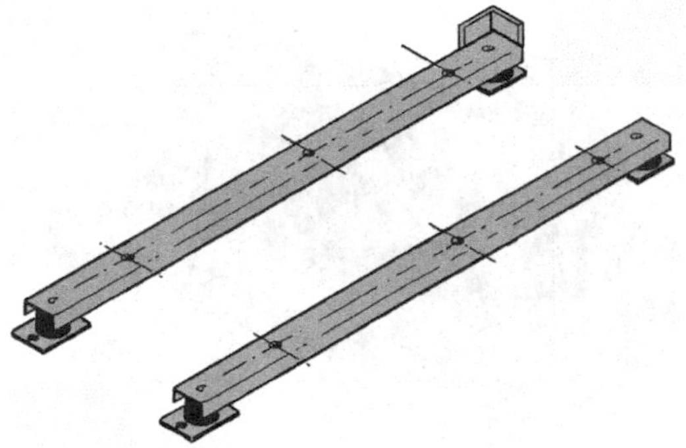

(a) Structural Steel Rails

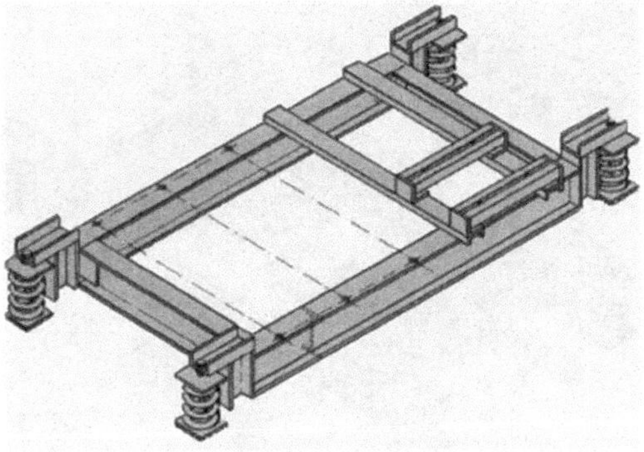

(b) Structural Steel Base

Figure 12.17 Structural Bases

A structural base of substantial mass is sometimes required to provide increased mass-related inertia to resist forces associated with severe unbalance of rotating and reciprocating machine components and to resist large movements of machines during machine start-up. A massive *structural inertia base* requires spring or other resilient mounts with corresponding large stiffness coefficients to support it. This is necessary to support the weight of the structural inertia base plus the equipment that is attached to it and to achieve a required low resonance frequency for the vibration isolation system. Structural inertia bases are normally placed on raised concrete platforms that are poured integrally with a floor slab. These raised platforms are referred to as *housekeeping pads*.

The proper design of a structural base is as important as the proper selection of the resilient mounts that will support it. A structural base can become warped, twisted or distorted if it is not properly designed to support the weight distribution of a machine it will support or to withstand the belt forces and torques produced by the connection between a machine and a motor. Resonances within a structural base itself can result from improper design. Lightweight and overly long structural base members tend to have resonances at low frequencies.

Structural Steel Rails: Structural steel rails with vibration isolation mounts are used to support mechanical equipment that already have a rigid structural frame [Figure 12.17(a)]. The structure steel rails can be I-beams, angle beams or channel beams, and they should have sufficient rigidity to support mechanical equipment without flexing.

Structural Steel Base: Structural steel bases are used when mechanical equipment cannot be supported at individual locations on the equipment and when it is necessary to maintain proper alignment between component parts associated with the equipment. These bases can be used with springs or other resilient mounts. They should have sufficient rigidity to resist all equipment starting and operating forces and torques without supplemental hold-down devices. Structural steel bases are fabricated using structural steel members as shown in Figure 12.17(b). The structure steel members can be I-beams, angle beams or channel beams. Structural steel bases are normally rectangular in shape for most types of mechanical equipment. However, sometimes they are T or L shaped. Figure 12.18 shows a centrifugal

Figure 12.18 Centrifugal Fan Mounted on a Structural Steel Base

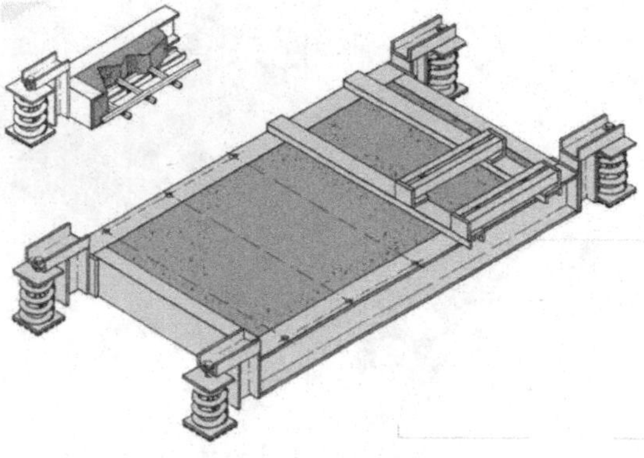

(a) Concrete Inertia Base

(b) Pump Mounted on a Concrete Inertia Base

Figure 12.19 Concrete Inertia BaseStructural Base

fan mounted on a structural steel base.

Concrete Inertia Base: Concrete inertia bases are fabricated with structurally reinforced concrete with reinforcement bars usually placed 6 in. (152 mm) on center in both directions as shown in Figure 12.19(a). The concrete is normally poured on the job when the base is in place. Concrete inertia bases are used to vibration isolate mechanical equipment with unbalanced rotating or reciprocating components and with very low rotational operating speeds. They are often substituted for structural steel bases when the center of gravity of the entire machine-isolation system must be lowered or when an exceedingly firm and rigid base is required. A concrete inertia base also aids in obtaining a more uniform load distribution for the spring or other mounts that support the base. Figure 19(b) shows a motor-pump assembly mounted on a concrete inertia base.

Rooftop Curb Mounted Base: Rooftop curb mounted bases are used to support rooftop mechanical equipment mounted over an open roof curb with penetrations through the roof deck (Figure 12.20). Rooftop equipment typically supported by rooftop curb mounted bases includes unitary packaged air handling and refrigeration units and large rooftop exhaust fans. Motor/air handler assemblies and refrigeration compressors placed in unitary packaged air handling and refrigeration units may or may not be internally mounted on a structural steel base with vibration isolation mounts as shown in Figures 12.17(b) and 12.18. When they are not internally mounted on a structural steel base with vibration isolation mounts, the rooftop curb mounted base must be supported on perimeter vibration isolation mounts as shown in Figure 12.20. Most often these mounts are restraint spring mounts. When they are internally mounted on a structural steel base with vibration isolation mounts, the rooftop curb mounted base must not be placed on perimeter vibration isolation mounts. Rooftop curb mounted bases designed to resist high wind loads against the sides of the enclosured equipment they support and with perimeter water seals that do not short circuit the vibration isolation mounts that support the curb mounted bases.

12.6 Resilient Support Locations for Structural Steel and Concrete Inertia Bases

Figure 12.21 shows a schematic drawing of a structural base with an attached mass that is support by four vibration isolation mounts. The center of gravity is shown such that each mount supports a different fraction

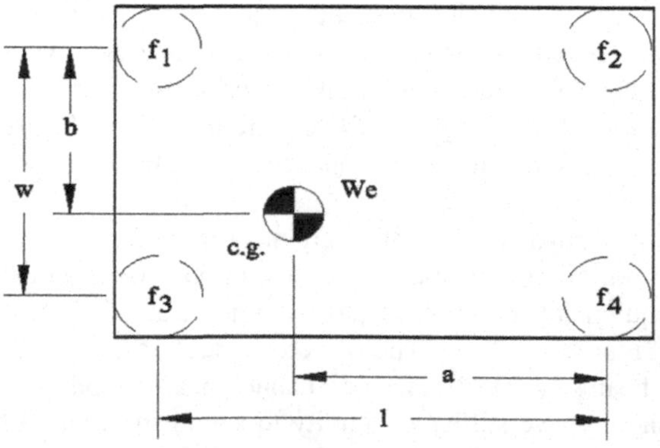

Figure 12.21 Structural Base with Four Springs

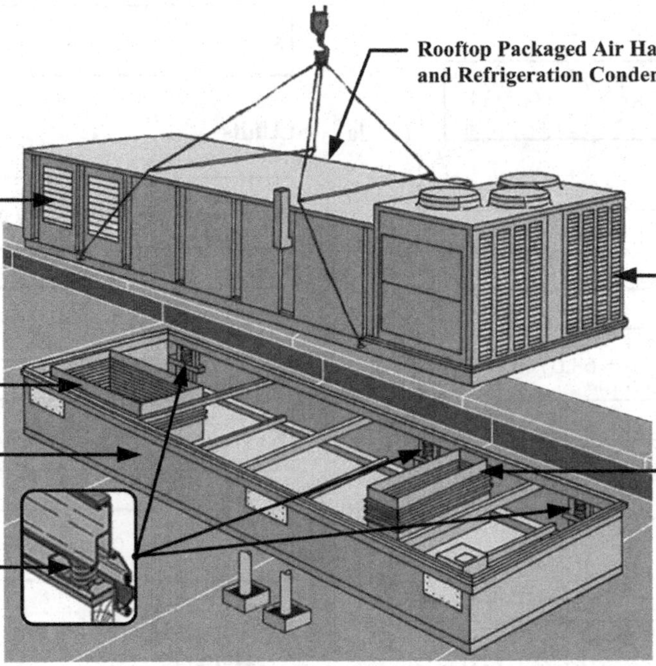

Figure 12.20 Rooftop Curb Mounted Base

of the total weight We. Let f_1, f_2, f_3 and f_4 represent the forces supported by each resilient element. Therefore:

$$f_1 = We\frac{a}{L}\left[1-\frac{b}{W}\right] \tag{12.29}$$

$$f_2 = We\left[1-\frac{a}{L}\right]\left[1-\frac{b}{W}\right] \tag{12.30}$$

$$f_3 = We\frac{a}{L}\frac{b}{W} \tag{12.31}$$

$$f_4 = We\left[1-\frac{a}{L}\right]\frac{b}{W} \tag{12.32}$$

If each resilient mount has the same static deflection δ:

$$K_1 = \frac{f_1}{\delta},\quad K_2 = \frac{f_2}{\delta},\quad K_3 = \frac{f_3}{\delta},\quad \text{and}\quad K_4 = \frac{f_4}{\delta}. \tag{12.33}$$

Resilient mount manufacturers market springs and other types of resilient mounts that have specified stiffness coefficients. When a spring or other type of resilient mount with a specific calculated stiffness coefficient is not available, select one with a stiffness coefficient closest to the value that has been calculated.

EXAMPLE 12.2

Referring to Figure 12.18, a machine mounted on a structural base with four resilient mounts has the following dimensions: L = 25 in. (635 mm), W = 35.5 in. (902 mm), a = 15.3 in. (389 mm), and b = 17.9 in. (455 mm). The weight of the machine plus structural base is We = 220.6 lb_f (981.3 N). The required static deflection δ is 0.25 in. (6.4 mm). Determine the values for f_1, f_2, f_3 and f_4, and the corresponding values for K_1, K_2, K_3 and K_4.

SOLUTION

From equations (12.29) to (12.33):

U.S. Units

$$f_1 = 220.6 \times \frac{15.3}{25} \times \left[1 - \frac{17.9}{35.5}\right] = 66.96\ lb_f \quad (12.2a)$$

(3.11b)

$$f_2 = 220.6 \times \left[1 - \frac{15.3}{25}\right] \times \left[1 - \frac{17.9}{35.5}\right] = 42.45\ lf_f$$

$$f_3 = 220.6 \times \frac{15.3}{25} \times \frac{17.9}{35.5} = 68.07\ lb_f \quad (12.2c)$$

$$f_4 = 220.6 \times \left[1 - \frac{15.3}{25}\right] \times \frac{17.9}{35.5} = 43.16\ lb_f \quad (12.2d)$$

SI Units

$$f_1 = 981.3 \times \frac{389}{635} \times \left[1 - \frac{455}{902}\right] = 297.9\ N \quad (12.2e)$$

$$f_2 = 981.3 \times \left[1 - \frac{389}{635}\right] \times \left[1 - \frac{455}{902}\right] = 188.4\ N \quad (12.2f)$$

$$f_3 = 981.3 \times \frac{389}{635} \times \frac{455}{902} = 303.2\ N \quad (12.2g)$$

$$f_4 = 981.3 \times \left[1 - \frac{389}{635}\right] \times \frac{455}{902} = 191.8\ N \quad (12.2h)$$

The corresponding values for K_1, K_2, K_3 and K_4 are:

U.S. Units

$$K_1 = \frac{66.96}{0.25} = 267.8\ lb_f / in. \quad (12.2i)$$

$$K_2 = \frac{42.45}{0.25} = 169.8\ lb_f / in. \quad (12.2j)$$

$$K_3 = \frac{68.07}{0.25} = 272.3\ lb_f / in. \quad (12.2k)$$

$$K_4 = \frac{43.16}{0.25} = 172.6\ lb_f / in. \quad (12.2l)$$

SI Units

$$K_1 = \frac{297.9}{0.0064} = 46{,}547\ N/m \quad (12.2m)$$

$$K_2 = \frac{188.4}{0.0064} = 29{,}438\ N/m \quad (12.2n)$$

$$K_3 = \frac{303.2}{0.0064} = 47{,}375\ N/m \quad (12.2o)$$

$$K_2 = \frac{188.4}{0.0064} = 29{,}438\ N/m \quad (12.2p)$$

12.7 Thrust Restraints and Vibration Isolation Hangers

Air Handler Flexible Duct Connections and Thrust Restraints

Flexible duct connections should be used where ducts are connected to the intake and discharge sections of fans or fan assemblies. If the fan static pressure is greater than 2.1 in. wg (523 Pa), spring thrust restraints should be used to prevent excessive horizontal displacements across the flexible connection (Figure 2-22).

Vibration Isolation Hangers

Duct Vibration Isolation Hanger: Flexible connections as shown in Figure 12.22 prevent air handler fan vibration from being transmitted to attached air distribution ducts. However, flow pulsations within the duct can cause duct vibration to be transmitted to the building structure through rigid hangers. Spring or combination spring and neoprene (or glass fiber) hangers should be used when ducts are suspended below noise sensitive building spaces (Figure 12.23). These hangers are especially recommended for large ducts with flow velocities that are greater than 2,000 fpm (10 m/s) and for all duct sizes when the duct static pressures are greater than 2.1 in. wg (523 Pa).

When a flexible connection between a fan and a related

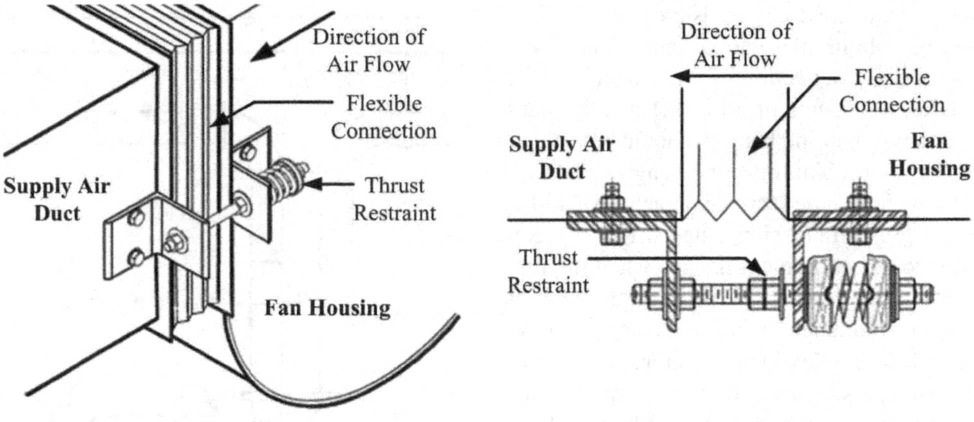

Figure 12.22 Air Handler Spring Thrust Restraint

duct section cannot be used, the duct should be vibration isolated with either resilient floor mounts or hangers to a maximum distance of 50 ft (15.2 m) from the point where the duct connects to the fan or fan assembly. When both a flexible connection between a fan or fan assembly and a duct and resilient floor mounts or hangers to support the duct are used, the first support or hanger at the point where the duct is attached to the fan or fan assembly should be rigid.

Pipe Vibration Isolation Hangers: When it is not possible to use flexible pipe connectors for pips extending from the inlet and discharge ports of a pump, the pipes should be vibration isolated using vibration isolation hangers as shown in Figure 12.24. The static deflections of the pipe vibration isolation hangers should be equal to the static deflection of the equipment vibration isolation mounts to a minimum distance of 30 ft (9.1 m). The pipes should be isolated within the equipment room or to a distance of 50 ft (15.2 m) from equipment to which

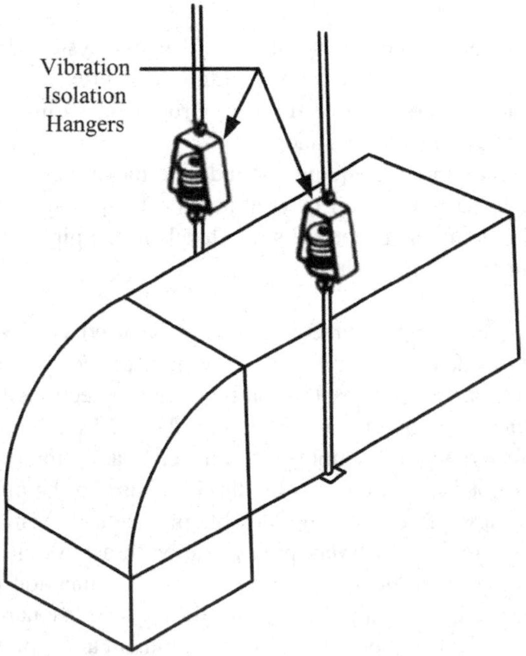

Figure 12.23 Vibration Isolation Hangers Used To Support Air Distribution Ducts

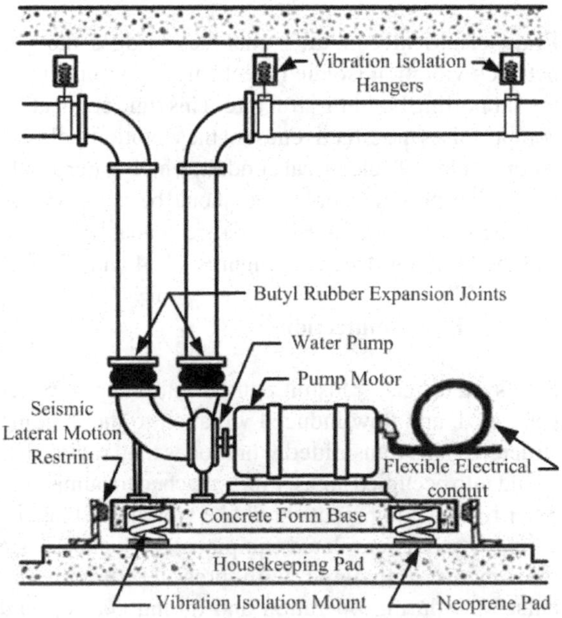

Figure 12.24 Water Pump/Moter Vibration Isolation Assembly with Expansion Joints

the pipes are attached, whichever is greater.

When spring vibration isolation hangers are used, the maximum static deflection of the hangers is 2 in. (51 mm). After a distance of 30 ft (9.1 m), the static deflection of the remaining hangers should be reduced to 0.75 in. (19 mm). The first two hangers from the equipment to which a pipe section is attached should be positioning or pre-compression hangers. This prevents load transfer to the equipment flanges when the pipe is filled. Positioning hangers assist in the installation of large pipes with diameters greater than or equal to 8 in. (203 mm). If both a flexible connector and vibration isolation pipe hangers are used, the first pipe hanger after the flexible connector should be a rigid hanger. Even though pipe isolation hangers are usually not specified for pipes beyond equipment rooms, they should be used for all pipes over 2 in. (51 mm) in diameter and for pipes suspended below noise-sensitive areas.

12.8 Flexible Connections

Care must be exercised to ensure that vibration isolation mounts and structural support bases are not bridged and compromised by electrical conduits and rigid pipes.

Flexible Electrical Conduit Connections

Rigid electrical conduit sometimes bridges the gap between vibration isolated mechanical equipment and the supporting building structure. This rigid connection compromises the effectiveness of the vibration isolation mounts. Flexible electrical conduit, which is purposely longer than necessary or looped, should be used to bridge gaps between vibration isolated mechanical equipment and the building structure (Figures 12.24 and 12.25).

Flexible Pipe Connections

Pipe systems can transmit equipment, pump impeller generated, and flow induced vibration to the building structure. This transmitted vibration is then radiated as sound into occupied areas. Pipes attached to pumps and other related equipment should be vibration isolated.

Pumps and other related equipment often experience movements associated with pressure thrusts and pulsations during operation and during start-up and shutdown when the operating speed of the equipment moves through the system resonance frequency.

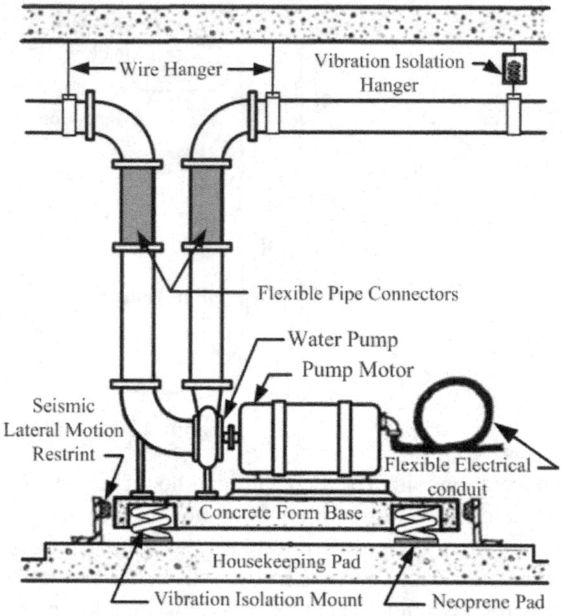

Figure 12.25 Water Pump/Moter Vibration Isolation Assembly with Flexible Pipe Conextors

Therefore, flexibility should be built into pipe systems to:

- Prevent or reduce the vibration in a pipe system that can be transmitted to the building structure;
- Prevent pipe systems from compromising equipment vibration isolation; and
- Accommodate equipment and pipe movements and thermal expansion or contraction of pipes without inducing unacceptable strain levels in the pipes and related equipment.

Flexible pipe connectors should be used between pumps and other related vibration isolated system components at or near the points where pipe sections are attached to the pumps (Figures 12.24 and 12.25). The effectiveness of flexible pipe connectors as a vibration isolator becomes less as the fluid pressure in the pipe increases. Even though a flexible pipe connector may not completely prevent pipe vibration transmission, it does provide flexibility at the point of connection to mechanical equipment. It also allows for reasonable movement of vibration isolated equipment at the point where the pipe is connected to the equipment. Flexible pipe connectors perform other functions, such as,

Figure 12.26 Pipe Expansion Joints and Flexible Pipe Connectors

compensating for minor pipe misalignment and relieving strain at pipe flanges.

Figure 12.24 shows a water pump/motor vibration isolation system assembly with a butyl rubber pipe expansion joint (Figure 12.26). These expansion joints are too short to be effective in reducing vibration transmission. However, they allow for pipe expansion and contraction and for axial, transverse, and angular movement of pipes.

Figure 12.25 shows a water pump/motor vibration isolation system assembly with flexible pipe connectors. These flexible connectors are fabricated using butyl rubber and braided stainless steel (Figure 12.26). Butyl rubber flexible connectors are more effective at reducing pipe vibration than braided stainless steel flexible connectors. The lengths of butyl rubber flexible pipe connectors are normally six to ten times their diameter and do not exceed 36 in. (914 mm). Braided stainless steel flexible pipe connectors are used when the fluid temperatures inside the pipe exceed 220 °F (104 °C) or pressures exceed those recommended for rubber hoses. Braided stainless steel flexible connectors can also be used as pipe expansion joints.

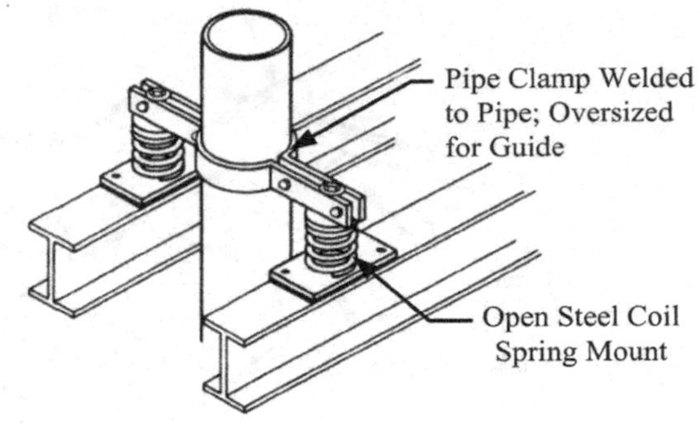

Figure 12.27 Vertical Pipe Riser Vibration Isolation Using Open Coil Steel Spring Mounts

12.9 Vibration Isolation of Vertical Pipe Risers

Vertical pipe risers have supports and guides to facilitate pipe expansion joints, bends and loops to function properly. Supports and guides are designed to control pipe movement and are usually attached rigidly to the building structure. Whenever possible, vertical pipe risers should be placed in non-critical areas adjacent to elevator shafts, stairwells, and other similar areas. When this is not possible and vertical pipe riser must be placed adjacent to noise sensitive occupied areas, the vertical riser supports should be vibration isolated. Figure 12.27shows a vertical pipe riser being supported by a pair of steel open coil spring mounts similar to the one shown in Figure 12.8. Depending on pipe vibration isolation requirements, the steel open coil spring mounts can be replaced by neoprene mounts (Figure 12.11), neoprene pads (Figure 12.12), glass fiber pads (12.14), or other similar vibration isolation mounts or pads.

12.10 Floating Floor Vibration Isolation Systems

Quiet spaces, such as executive offices, conference rooms, living areas, theatres, recording studios, etc. are sometimes placed beneath mechanical equipment rooms, commercial kitchens, athletic areas, and other similar spaces. Floating floor vibration isolation systems can be used in these situations to reduce the transmission of sound associated with structure-borne vibration to acceptable levels in the quiet spaces.

Figure 12.28 shows two form-work floating floor systems. Floating floor system A is placed over a

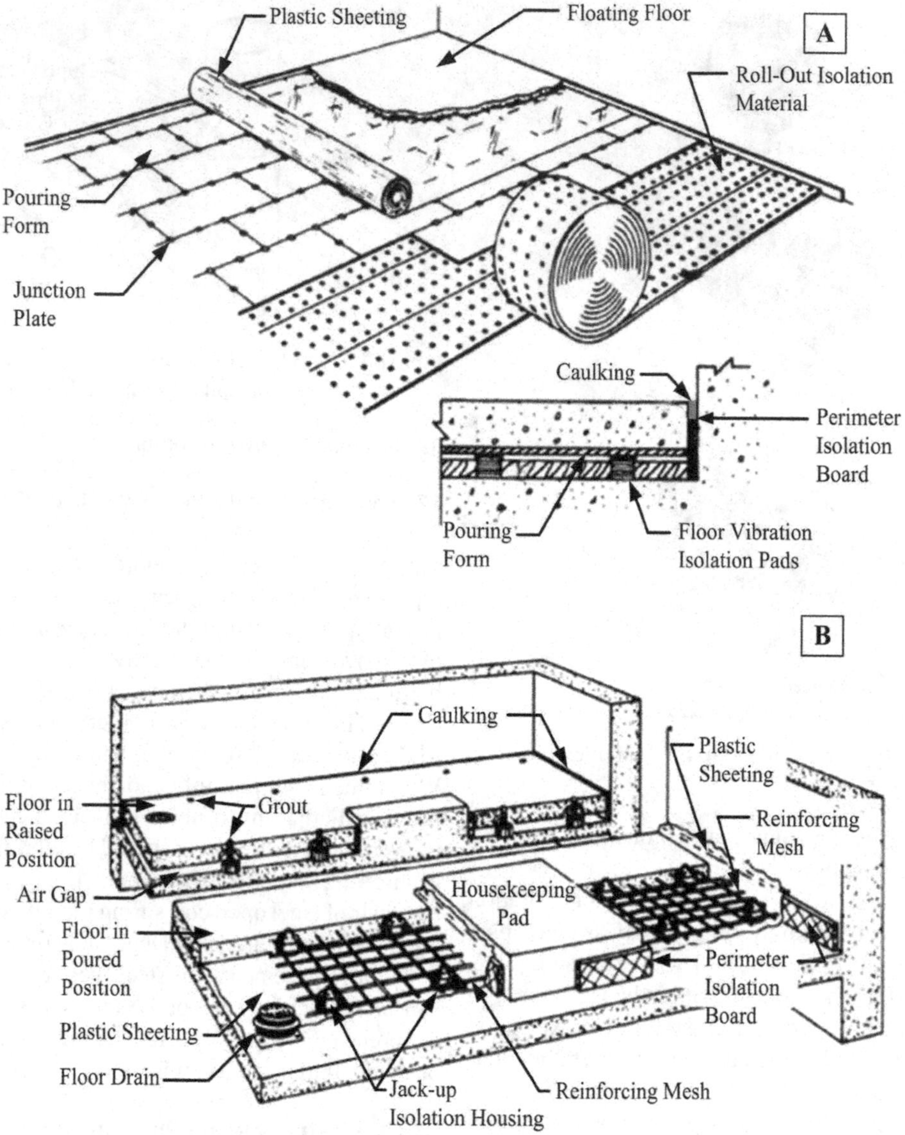

Figure 12.28 Two Floating Floor Vibration Isolation Systems

rectangular grid of glass fiber or neoprene pads that are imbedded in an acoustic fiberglass blanket. Floating floor system B is a jack-up floating floor system that uses steel open coil springs. Figure 12.29 shows a more detailed cross-section view of this floating floor system. It is similar to floating floor system A, except the concrete floor slab is poured with the slab in the lowered position. After the concrete has cured, the slab is raised to its operating height with the built-in jack screws in each of the vibration isolation mounts.

Both floating floor systems consist of a reinforced concrete slab that is supported by neoprene, glass fiber or steel open coil spring vibration mounts. The floating slab in each system must be free to move along their entire perimeters and around columns and protruding housekeeping pads. Perimeter isolation boards, which are around 1 in. (25 mm) thick and wide enough to extend from the supporting structural floor to the top

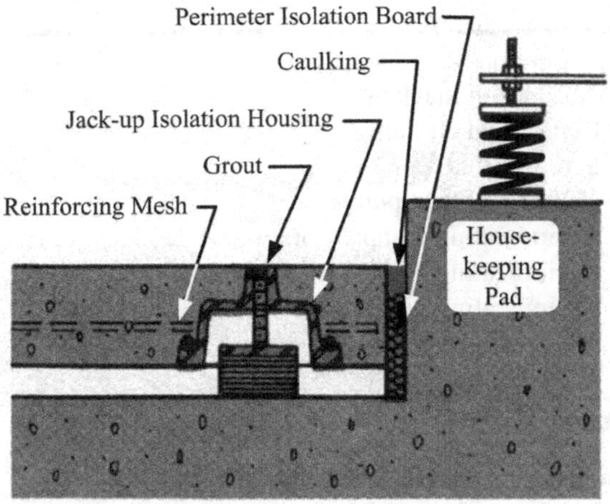

Figure 12.29 Jack-up Floating Floor System

of the floating slab, are installed around the perimeter of the floating slab and the perimeters of columns and housekeeping pads. This ensures the free movement of the floating floor and prevents sound from entering the space between the floating floor and the supporting building structure. The floating slab is caulked around its perimeter and the perimeters of columns and housekeeping pads to ensure an airtight seal.

Floating floor isolation systems are not mechanical equipment vibration isolation structures. Sometimes lightweight machinery may be located and vibration isolated on a floating floor. However, heavy mechanical equipment should be vibration isolated on individual concrete housekeeping pads, which extend from the supporting structural floor up through the floating floor (Figure 12.28 and 12.29.

Figure 12.30 shows a wood floating floor system. This type of floating floor is used for situations where a dance studio, basketball court, track or other similar type of athletic area is located above critical noise areas.

12.11 Floor Constructions Used to Support Mechanical Equipment

Mechanical equipment is supported on several different types of floor systems. These include:

- Concrete floor slabs on grade,
- Concrete floor slabs on upper floors,
- Floating floors, and
- Lightweight floor constructions.

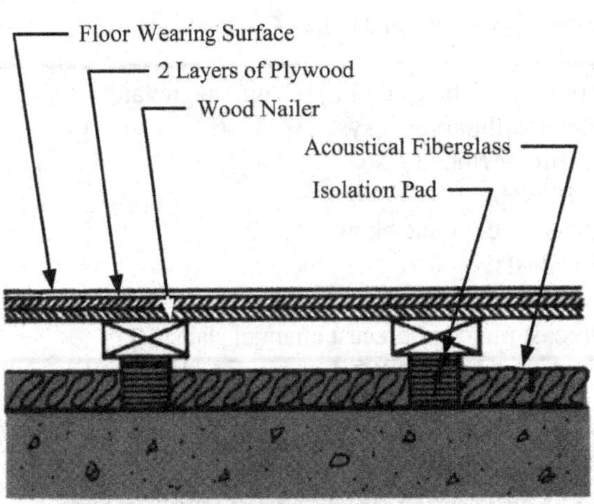

Figure 12.30 Wood Floating Floor System

Concrete Floor Slabs on Grade

Concrete floor slabs on grade are usually very stiff, and they are not deflected by heavy loads. No rooms are below these slabs and only occasionally may a noise-sensitive room be located adjacent to a basement mechanical equipment room (MER).

The theoretical vibration isolation efficiency for vibration isolation mounts associated with mechanical equipment mounted on the floor can be obtained from the 1-DOF analysis procedures outlined in Chapter 3.

Concrete Floor Slabs on Upper Floors

Dense concrete floors [140 to 150 lb_f/ft^3 (2,248 to 2,409 kg/m^3)] 8 to 12 in. (203 and 305 mm) thick should be used for upper-floor mechanical equipment rooms and other rooms that contain vibrating equipment. Even though these floors are massive, they are still flexible. This flexibility must be addressed when selecting vibration isolation systems for mechanical equipment placed on these floors. When occupied spaces below these floors require very low sound levels, it may be necessary to use a floating floor in the mechanical equipment room and gypsum board ceilings attached to resilient channels in the rooms below. When located over noncritical sound areas, light weight mechanical equipment that do not have reciprocating components can be mounted on vibration isolation mounts that are placed on dense concrete slabs 4 to 6 in. (102 to 152 mm) thick. The flexibility of these floors must be addressed when designing vibration isolation systems

Table 12.3 Lightweight Floor Constructions

Concrete slabs using lightweight aggregate
Metal cellular deck systems
Waffle or ribbed slabs
Tube slabs
Gypsum concrete slabs
Bar joists
Wood joist constructions
Precast plank or precast channel slabs

for mechanical equipment.

Floating Floors

Floating floors in a mechanical equipment room are primarily used to increase the airborne sound transmission loss of the mechanical equipment room floor by isolating it from the building structure. Mechanical equipment that does not require the use of a structural or concrete inertia base (Figures 12.16 and 12.18) can be vibration isolated on a mechanical equipment room floating floor. However, when it is necessary to mount mechanical equipment on a structural or concrete inertia base, they should be mounted on a housekeeping pad that extends through the floating floor to the building structure as shown in Figures 12.27 and 12.28.

Lightweight Floor Constructions

Typical lightweight floor systems are listed in Table 12.3. These lightweight floor constructions are not massive or stiff enough to be used as a mechanical equipment room floor where medium or large size mechanical equipment will be mounted on the floors or suspended from the underside of the floors.

Walls

Walls should not be used to support vibrating mechanical equipment. They usually do not have sufficient stiffness and mass for this application. This is particularly true where a wall is adjacent to an occupied area.

12.12 Mechanical Equipment Vibration Isolation Requirements

Table 12.4 lists types of mechanical equipment that normally require vibration isolation. With regard to the design of vibration isolation systems for mechanical equipment, the following factors should always be considered: load distribution and center of gravity.

Table 12.4 Equipment that requires vibration isolation

Refrigeraton Machines
Absorption machines
Centrifugal chillers
Screw chillers
Centrifugal heat pumps
Reciprocating compressors
Reciprocating chillers
Reciprocating heat pumps
Packaged Steam Generators
Pumps
Factory Assembled Air Handling Equpment
Roof mounted units
Suspended units
Floor mounted units
Air Compressors
Tank type
Vertical tank type
Slow speed
Utility Blowers
Floor mounted units
Roof mounted units
Suspended units
Centrifugal Blowers
Floor mounted units
Suspended units
Tubular, Centrifugal, and Axial Fans
Suspended units
Floor mounted units
Cooling Towers and Condensing Units

Load Distribution

Vibration isolation systems that consist of a structural or concrete inertia base and vibration isolation mounts often support mechanical equipment that have unevenly distributed masses. For example, most of the weight a fan and motor mounted on a structural base will normally be at the center of gravity of the fan. The deflection of the vibration isolation mounts, however, must be reasonably equal even though the center of gravity of the system is not located at the horizontal geometric center of the structural base. This is accomplished by selecting spring isolation mounts with different stiffness coefficients so they will have equal deflections associated with the different weights each spring must support. Refer to Example 12.2.

Center of Gravity

The vertical location of the center of gravity of a machine's mass relative to excitation and reaction forces acting on the mass must always be considered with mechanical equipment with rotating and reciprocating components. Excessive rocking motions and instability can occur when the center of gravity of a machine is high relative to the height of its vibration isolation mounts. The center of gravity can be lowered by mounting the machine on a concrete inertia base that is mounted on vibration isolation mounts.

12.13 Vibration Isolation Systems

Vibration isolation systems normally consist of a structural or concrete inertia base, vibration isolation mounts, and the building support structure. The system may also include bridging components, such as pipe, electrical conduit, and air distribution ducts connections. To ensure a vibration isolation system is properly designed and installed, the following information should be obtained from vibration isolation mount and mechanical equipment manufacturers and design engineers:

- Copies of all submittals associated with the vibration isolation system;
- Literature that describes the free, solid and operating heights of the vibration isolation mounts;
- Literature that describes the types, sizes, rated load capacities, and identification system (usually a color code) associated with the vibration isolation mounts; and
- Information on the weight distributions of the mechanical equipment that are to be vibration isolated and drawings that show the locations where the vibration isolation mounts that will support the equipment will be attached.

12.14 Vibration Isolation Mount Selection

Table 12.5 is a guide for selecting vibration isolation mounts for different types of mechanical equipment and floor span lengths.[4] The isolation mount deflections in this table are based on 1-DOF analysis for specifying vibration isolation efficiency. These deflections assume the floors on which mechanical equipment is to be placed are constructed of dense concrete [140 to 150 lb_f/ft^3 (2,248 to 2,409 kg/m^3)] 8 - 12 in. (203 and 305 mm) thick.

Upper level floors in buildings are flexible floors, even though they may be constructed of thick, dense concrete. The flexibility of a floor system increases with the length of its span. The static deflection of a floor associated with a static load placed on the floor increases with increasing floor span. To compensate for this, vibration isolation mounts with larger static deflections are specified for floors with longer spans.

Flexible floor systems often have multiple resonance frequencies that are not affected by the selected static deflection of a vibration isolation mount. Many buildings have structural resonance frequencies that range from 8 to 12 Hz. This corresponds to an equipment operational speed range of 480 to 720 rpm. If a rotating or reciprocating machine has an operating speed within this range, a more detailed vibration analysis should be conducted to ensure that a building resonance frequency will not be excited. Refer to Section 4.6 in Chapter 4.

Problems associated with building structural resonance frequencies and excessive floor vibration become more acute when mechanical equipment are located on lightweight floor systems similar to the ones described in Table 12.3 and on dense concrete floor systems that have span lengths greater than those listed in Table 12.5. Problems often occur when packaged rooftop units are placed on lightweight roof systems. When any of these conditions are encountered, a more detailed vibration analysis should be conducted to ensure that related building vibration problems do not occur.

---

4. Chapter 48, Noise and Vibration Control, *ASHRAE 2015 HVAC Applications Handbood*, American Society for Heating, Refrigerating and Air Conditioning Engineers, 1791 Tullie Circle, N.E.n Atlanta, GA 30329.

Table 12.5 Vibration Isolation Mount Selection Guide[4]

			Equipment Location											
			Slab on Grade			Up to 20 ft Floor Span			20-30 ft Floor Span			30-40 ft Floor Span		
Equipment Type	Horse power & other	RPM	Base Type	Iso-lator Type	Min. Defl. in.	Base Type	Iso-lator Type	Min. Defl. in.	Base Type	Iso-lator Type	Min. Defl. in.	Base Type	Iso-lator Type	Min. Defl. in.
Refrigeration Machines & Chillers														
Bare Compressors	All	All	A	2	0.25	C	3	0.75	C	3	1.75	C	4	2.50
Reciprocating	All	All	A	2	0.25	A	4	0.75	A	3	1.75	A	4	2.50
Centrifugal	All	All	A	1	0.25	A	4	0.75	A	3	1.75	A	3	1.75
Open Centrifugal	All	All	C	1	0.25	C	4	0.75	C	3	1.75	C	3	1.75
Absorption	All	All	A	1	0.25	A	4	0.75	A	3	1.75	A	3	1.75
Air Compressors and Vacuum Pumps														
Tank Mounted	up to 10	All	A	3	0.75	A	3	0.75	A	3	1.75	A	3	1.75
	15 & over	All	C	3	0.75	C	3	0.75	C	3	1.75	C	3	1.75
Base Mounted	All	All	C	3	0.75	C	3	0.75	C	3	1.75	C	3	1.75
Large Reciprocating	All	All	C	3	0.75	C	3	0.75	C	3	1.75	C	3	1.75
Pumps														
Closed Coupled	up to 7.5	All	B	2	0.25	C	3	0.75	C	3	0.75	C	3	0.75
	10 & over	All	C	3	0.75	C	3	0.75	C	3	1.75	C	3	1.75
Large Inline	5 to 25	All	A	3	0.75	A	3	1.75	A	3	1.75	A	3	1.75
	30 & over	All	A	3	1.75	A	3	1.75	A	3	1.75	A	3	2.50
End Suction and Split Case	up to 40	All	C	3	0.75	C	3	0.75	C	3	1.75	C	3	1.75
	50 to 125	All	C	3	0.75	C	3	0.75	C	3	1.75	C	3	2.50
	150 & over	All	C	3	0.75	C	3	1.75	C	3	1.75	C	3	2.50
Cooling Towers - All	All	up to 300	A	1	0.25	A	4	3.50	A	4	3.50	A	4	3.50
		301 to 500	A	1	0.25	A	4	2.50	A	4	2.50	A	4	2.50
		500 & over	A	1	0.25	A	4	0.75	A	4	0.75	A	4	1.75
Boilers - Firetube	All	All	A	1	0.25	B	4	0.75	B	4	1.75	B	4	2.50
Axial Fans, Fan Heads, Cabinet Fans, Fan Sections														
Up to 22 in. dia.	All	All	A	2	0.25	A	3	0.75	A	3	0.75	C	3	0.75
24 in. dia. & Over	[up to 2 in.	up to 300	B	3	2.50	C	3	3.50	C	3	3.50	C	3	3.50
	Static	300 to 500	B	3	0.75	B	3	1.75	C	3	2.50	C	3	2.50
	Pressure]	501 & over	B	3	0.75	B	3	1.75	B	3	1.75	B	3	1.75
	[2.1 in.	up to 300	C	3	2.50	C	3	3.50	C	3	3.50	C	3	3.50
	Static	300 to 500	C	3	1.75	C	3	1.75	C	3	2.50	C	3	2.50
	Pressure]	501 & over	C	3	0.75	C	3	1.75	C	3	1.75	C	3	2.50

12.15 Investigation of Mechanical Equipment Vibration Problems

The investigation of mechanical equipment vibration problems should include:

- Measurement of the imbalance of rotating and reciprocating mechanical equipment components.
- Measurement of vibration amplitudes on vibrating mechanical equipment. Refer to Figures 12.4 and 12.5 for the classification of vibration severity for mechanical equipment.
- Measurement of vibration amplitudes in occupied building spaces that are impacted by mechanical equipment. Refer to Table 12.1 and Figurse 12.2 and 12.3 for recommended building vibration criteria.
- Examination of mechanical equipment vibration generated by system components, such as bearings, drives, etc. Refer to Table 12.6.[5]
- Examination of equipment installation factors, such

Table 12.5 Vibration Isolation Mount Selection Guide[4]

Equipment Type	Horse power & other	RPM	Slab on Grade: Base Type	Slab on Grade: Isolator Type	Slab on Grade: Min. Defl. in.	Up to 20 ft Floor Span: Base Type	Up to 20 ft Floor Span: Isolator Type	Up to 20 ft Floor Span: Min. Defl. in.	20-30 ft Floor Span: Base Type	20-30 ft Floor Span: Isolator Type	20-30 ft Floor Span: Min. Defl. in.	30-40 ft Floor Span: Base Type	30-40 ft Floor Span: Isolator Type	30-40 ft Floor Span: Min. Defl. in.
Centrifugal Fans														
Up to 22 in. dia.	All	All	B	2	0.25	B	3	0.75	B	3	0.75	C	3	1.75
24 in. dia. & Over	up to 40	up to 300	B	3	2.50	B	3	3.50	B	3	3.50	B	3	3.50
		300 to 500	B	3	1.75	B	3	1.75	B	3	2.50	B	3	2.50
		501 & over	B	3	0.75	B	3	0.75	B	3	0.75	B	3	1.75
	50 & over	up to 300	C	3	2.50	C	3	3.50	C	3	3.50	C	3	3.50
		300 to 500	C	3	1.75	C	3	1.75	C	3	2.50	C	3	2.50
		501 & over	C	3	1.00	C	3	1.75	C	3	1.75	C	3	2.50
Propeller Fans														
Wall Mounted	All	All	A	1	0.25	A	1	0.25	A	1	0.25	A	1	0.25
Roof Mounted	All	All	A	1	0.25	A	1	0.25	B	4	1.75	D	4	1.75
Heat Pumps - All	All	All	A	3	0.75	A	3	0.75	A	3	0.75	A/D	3	1.75
Condensing Units - All	All	All	A	1	0.25	A	4	0.75	A	4	1.75	A/D	4	1.75
Packaged AH, AC, H & V Units														
All	up to 10	All	A	3	0.75	A	3	0.75	A	3	0.75	A	3	0.75
	[15 & over	up to 300	A	3	0.75	A	3	3.50	A	3	3.50	C	3	3.50
	up to 4 in.	301 to 500	A	3	0.75	A	3	2.50	A	3	2.50	A	3	2.50
	Static Press.]	501 & over	A	3	0.75	A	3	1.75	A	3	1.75	A	3	1.75
	[15 & over	up to 300	B	3	0.75	C	3	3.50	C	3	3.50	C	3	3.50
	4 in. Static	301 to 500	B	3	0.75	C	3	1.75	C	3	2.50	C	3	2.50
	Press. & over]	501 & over	B	3	0.75	C	3	1.75	C	3	1.75	C	3	2.50
Packaged Rooftop Equipment - AllAll		All	A/D	1	0.25	D	3	0.75	Normally Not Recommended			Normally Not Recommended		
Ducted Rotating Equipment														
Small Fans, Fan-Powered Boxes	Up to 600 cfm	All	A	3	0.50	A	3	0.50	A	3	0.50	A	3	0.50
	601 cfm & over	All	A	3	0.75	A	3	0.75	A	3	0.75	A	3	0.75
Engine-Driven Generators - All	All	All	A	3	0.75	C	3	1.75	C	3	2.50	C	3	3.50

Base Types:
A. No base, vibration mounts attached directly to equipment
B. Structural steel rails or base (Figures 12.17 and 12.18)
C. Concrete inertia base (Figure 12.19)
D. Curb-mounted base (Figure 12.20)

Isolator Types:
1. Rubber or glass fiber pad (Figures 12.14 and12.16)
2. Molded neoprene floor mount or hanger (Figures 12.12 and 12.13)
3. Spring floor mount or hanger (Figures 12.9 and 12.12)
4. Restraint spring floor mount (Figure 12.10)
5. Thrust restraint (Figure 12.21)

as equipment alignment, vibration isolation mount placement, etc.

When a vibration problem exists:

Identify the source of the vibration: The system or equipment that is causing a vibration problem can often be identified without instrumentation. Vibration levels are usually well above the sensory level of perception and readily felt or heard. A simple method for determining the problem source is to turn individual pieces of mechanical equipment on and off to identify the source of vibration. Since the source of a problem is often more than one piece of equipment or the interaction of two or more

5. IRD Mechanalysis, *Vibration Technology-1*, 6150 Huntley Road, Columbus, OH 43229.

Table 12.6 Potential Causes of Machine Equipment Related Vibration Problems[5]

Cause	Amplitude	Frequency	Phase	Remarks
Unbalance	Proportional to unbalance. Largest in radial direction.	1 x RPM	Single reference mark stable repeatable	Most common cause of vibration
Misalignment couplings or bearings and bent shaft	Large in axial direction. 50% or more of radial vibration.	1 x RPM usual 2 & 3 x RPM sometimes	Single, double or triple reference mark	Best found by appearance of large axial vibration. Use dial indicators or other method for positive diagnosis. If sleeve bearing machine and no coupling misalignment, balance the rotor.
Bad bearings anti-friction type	Unsteady. Use velocity, acceleration, and spike energy measurements.	Very high, several times RPM	Erratic Multiple marks	If bearing is responsible, it is most likely the one nearest the point of largest high frequency vibration. Spike energy measurement recommended when analyzing bearing failures.
Eccentric journals	Usually not large.	1 x RPM	Single mark	If on gears, largest vibration is in line with the gear center. If on motor or generator, vibration disappears when power is turned off. If on pump or blower, attempt to balance.
Bad gears or gear noise	Low. Use velocity, acceleration, and spike energy measurements.	Very high, gear teeth x RPM	Erratic Multiple marks	Velocity, acceleration, and spike energy measurements recommended when analyzing gear problems. Analyze higher orders and sideband frequencies.
Mechanical looseness	Sometimes erratic.	2 x RPM	Two reference marks Slightly erratic	Usually accompanied by unbalance and/or misalignment.
Bad drive belts	Erratic or pulsing.	1, 2, 3 & 4 x RPM of belt	One or two depending on frequency, usually unsteady	Strobe light best tool to visually freeze faulty belt.
Electrical	Disappears when power is turned off.	1 x RPM or 1 or 2 x synchronous frequency	Single or rotating double mark	Common electrical problems include broken rotor bars, eccentric rotor, unbalanced phases in poly-phase systems, unequal air gaps.
Torque Pulsation	Disapears when power is turned off.	2 x synchronous frequency	Single or rotating double mark	Rare as a cause of trouble except in cases of resonance problems.
Aerodynamic or hydraulic forces	Can be large in the axial direction.	1 x RPM or number of blades on fan impeller x RPM	Multiple marks	Rare as a cause of trouble except in cases of resonance problems.
Reciprocating forces	Higher in line with motion.	1, 2 & higher orders x RPM	Multiple marks	Inherent in reciprocating machines. Can only be reduced by design changes or vibration isolation.

systems, it is always good practice to double check by shutting off all equipment and operating each piece of equipment individually.

Determine if the problem is related to sound or vibration: When vibration is perceptible, vibration transmission is usually the major cause of a problem. The possibility that lightweight walls or ceiling panels are excited by airborne sound should be considered. If vibration is not perceptible, the problem may still be one of vibration transmission causing structure-borne sound. This sound can be checked by following the procedure below:

- If a sound level meter is available, check the A-scale and the C-scale or linear sound readings. If the difference is greater than 6 dB, or if the slope of the octave band frequency curve is greater than 5 to 6 dB/octave in the low frequencies, vibration is probably the cause.
- When the affected area is remote from the vibration source, there are no sound or vibration problems in intermediary spaces, and sound does not appear to be coming from the duct system or diffusers, structure-borne sound is most likely the cause.

12.16 Causes of Structure-Bourne Vibration Problems

Vibration and structure-borne vibration problems can be caused and made worse by:

- Equipment with excessive dynamic unbalance;
- Excessive vibration of mechanical equipment components;
- Lack of proper vibration isolation mounts;
- Improperly selected or installed vibration isolation mounts;
- Flanking transmission paths, such as rigid pipe connections and obstructions under a structural or concrete inertia base supporting mechanical equipment;
- Floor flexibility;
- Vibration isolation system resonances
- Building structure resonances.

Equipment with Excessive Dynamic Unbalance

Rotating and reciprocating equipment should always be dynamically balanced to within an acceptable limit. The balancing process is usually accomplished in two steps: measure and locate the required mass alterations and apply correction masses at properly designated locations. The equipment used to measure and locate the positions of the corrections may be of two types: non-rotating and rotating. The considerations for balancing rotating and reciprocating machines include number of planes, speed range, and adequacy of balance. ANSI Standard S2.19, *Standard for Balance Quality of Rotating Rigid Bodies*, specifies quality grades of balance for a large number of rotating and reciprocating machines.

Excessive Vibration of Mechanical Equipment Components

Excessive mechanical equipment vibration has several potential causes. Table 12.6 identifies many of these causes. Excessive equipment vibration can be associated with improper equipment installation and signify operational problems associated with equipment components that can ultimately result in the failure of these components. The causes of excessive equipment vibration must always be identified and corrected.

Lack of Vibration isolation

When mechanical equipment lack vibration isolation mounts, they normally can be added by using structural brackets to the equipment base. Related electrical, duct, and pipe connections may have to be isolated by using flexible connectors.

Improperly Selected or Installed Vibration isolation Mounts

Improperly selected or installed vibration isolation mounts can be the cause of field vibration problems. A common problem is inadequate deflection associated with under-loaded isolator isolation mounts. Determine the static deflections of spring isolation mounts by measuring the operating height and comparing it to the free height information available from the manufacturer. Compare the measured deflection with the recommended deflection in Table 12.5. Overloaded isolation mounts are not generally a problem as long as the system is free-floating and there is space between the spring coils.

If the resonance frequency of a vibration isolation system is 25 percent or less than the disturbing frequency (usually considered the operating speed of the machine), the vibration isolation mounts should be reasonably efficient, except in situations where heavy equipment is installed on extremely long floor or roof span or very flexible floor or roof systems. \

Vibration transmission problems can be caused by excessively rough equipment operation, by the system not being free-floating and generating flanking vibration transmission, and by a resonance or low floor stiffness situation. While it may be easy to determine the resonance frequency of vibration isolation system by the measurement of the vibration isolation mount, such measurements are often difficult with neoprene or similar pads. Although these isolation pads can theoretically provide system resonance frequencies as low as 4 Hz, they actually result in resonance frequencies higher than those achievable with spring isolation mounts, and they normally do not provide the desired isolation efficiencies for mechanical equipment located on upper floors.

It normally is not possible to determine if design vibration isolation efficiencies have been achieved in field installations by field vibration measurements. However, vibration measurements can be conducted on mechanical equipment, equipment supports, floors supporting vibration isolated equipment, and floors in adjacent areas to determine if vibration criteria specified in Figures 12.2 and 12.3 have been achieved in field installations.

Flanking vibration transmission

Rigid connections caused by rigid electrical conduits, pipes, and ducts create vibration transmission flanking paths in vibration isolation systems. Ensure that mechanical equipment and vibration isolation mounts are free floating by stepping on the structural base and releasing it. This should cause the equipment to oscillate freely and easily. Check to ensure there are no obstructions between the equipment or structural base and the floor that can short circuit the vibration isolation system on floor mounted equipment. This is best accomplished by passing a rod under the base. A small obstruction might permit the base to move, giving the impression it is free-floating when it is not. On suspended equipment, make sure the rods are not touching the hanger box.

Floor and Roof Deck Flexibility

Floor flexibility usually is not a problem with most mechanical equipment and floor structures. However, problems can occur when heavy equipment is installed on long-span floors and on thin floor slabs and when rooftop packaged units and other mechanical equipment are installed on lightweight roof structures. When floor flexibility is suspected, spring isolation mounts should be used. The design static deflection of the spring isolation mounts should be at least 10 times greater than the corresponding static deflection of the floor or roof system supporting the weight of the equipment. Information related to floor and roof deflections most often must be obtained from a structural engineer. Heavy mechanical equipment with rotating and reciprocating components should not be placed on extremely flexible floor and roof systems.

Vibration Isolation System Resonances

A vibration isolation system resonance is normally characterized by excessive mechanical equipment and structural support or inertia base vibration. It results in objectionable vibration transmission to the building structure. A vibration isolation system resonance is usually visually observable and can be measured using the instrumentation described in Chapter 9. A vibration isolation system resonance is normally addressed by changing the operation speed of the mechanical equipment when possible or by changing the system resonance frequency by increasing or decreasing the static deflection of the spring isolation mounts.

Building Structure Resonances

Building structure resonance problems occur when some part of the building structure has a structural resonance frequency that corresponds to the operating speed of one of the machines mounted on the structure. These problems can exist even when mechanical equipment is properly vibration isolated from the building structure. This resonance problem is often associated with a structural resonance frequency present in the floor on which the equipment is installed. However, it can occur in a remotely located floor, wall or other building component. When a vibration problem has a remote source that is not associated with pipes or ducts, structural resonances must be considered. Some resonance problems can be resolved by the following steps.

Dynamically balance rotating or reciprocating equipment: Reduce the vibration force by more effectively balancing the rotating or reciprocating components of mechanical equipment. This is often not a practical solution for building structure resonant problem. Balancing these components most often must be done by the equipment manufacturer.

Change the deflection of vibration isolation mounts: This approach is sometimes considered. However, in many cases, the resonance frequency of a machine vibration isolation system is significantly lower than the rotational operating speeds of the machine. When the operational frequency of a machine corresponds to a building structural resonance, changing the resonance frequency of the machine vibration isolation frequency will not solve the problem.

Change the structure stiffness: Change the structure stiffness. A change in structure stiffness changes the structure resonance frequency; the greater the stiffness, the higher the structural resonance frequency. While this approach may be difficult and expensive, it may be the only solution in certain cases.

Change the equipment operating speed: Change the disturbing frequency by changing the equipment operating speed. Do not allow mechanical equipment to operate at speeds that are coincident building structural resonances.

12.17 Solutions to Building Structure Borne Vibration Problems - Figure 12.1 Revisited

Figure 12.31 present a schematic representation of solutions to the structure borne vibration problems identified in Figure 12.1 at the beginning of this chapter. Vibration transmitted to the building structure from the cooling tower on the roof is reduced to acceptable amplitudes by mounting the cooling tower on a structural steel base that is support by appropriately sized vibration isolation mounts. Vibration transmitted to the building structure from the pump to the left of the room is reduced to acceptable amplitudes by placing the pump on a concrete inertia base that is supported by appropriately sized vibration isolation mounts. Vibration transmitted to the building structure from the pipe between the discharge section of the pump (supplies water to the cooling tower) and the cooling tower is reduced to acceptable amplitudes by support the pipe between the pump and cooling tower with appropriately sized vibration isolation mounts. Vibration transmitted to the building structure from the pipe that

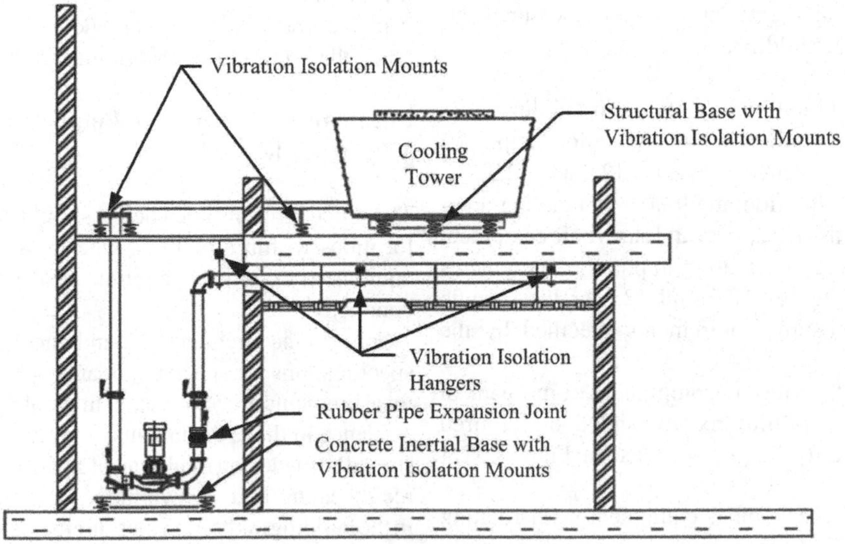

Figure 12.31 Solutions to Structure-Borne Vibration in a Building

directs water to the inlet section of the pump is reduced to acceptable amplitudes by supporting the pipe with appropriately sized vibration isolation hangers and by placing a rubber pipe expansion joint between the pipe and inlet to the pump.

The above solutions to building vibration problems associated with mechanical equipment are specifically tailored to the building mechanical systems shown in Figures 12.1 and 12.30. Mechanical systems and related equipment associated with building heating, ventilating and air conditioning (HVAC) system are designed to meet the specific requirements of buildings in which they are to be installed. Appropriately applying the vibration isolations principles, methods and components presented in this chapter during the design phase of building HVAC systems will significantly reduce the potential for mechanical equipment related vibration problems in building where the equipment will be placed. When mechanical equipment vibration problems do occur in buildings, appropriately applying the vibration isolations principles, methods and components presented in this chapter will aid in trouble shooting this problems and developing appropriate solutions.

12.18 Mechanical Equipment Vibration Measurements

Information presented in this section is an extension of and builds upon the vibration transducer and instrumentation information covered in Chapter 9.

Mechanical equipment related vibration measurements are conducted in buildings to:

- Determine if vibration amplitudes in building areas occupied by people meet the vibration amplitude requirements specified in Figures 12.2 and 12.3.
- Determine if vibration amplitudes in building areas where very sensitive optical and electronic equipment are located meet the vibration amplitude requirements specified in Figures 12.2 and 12.3 or the specific vibration exposure requirements specified by the equipment manufacturers.
- Determine if the vibration amplitudes of mechanical equipment are within recommended or required vibration severity ranges specified in Figures 12.4 and 12.5.
- Determine if mechanical equipment with rotating and reciprocating components are operating properly and obtain vibration signatures necessary to identify potential machine related vibration problems and failure modes outlined in Table 12.6.

The frequency range of interest for vibration measurements in building areas:

- Occupied by people,
- Where vibration sensitive optical and electronic equipment are located, and
- On mechanical equipment that can transmit vibration to occupied building areas

is normally 1 to 100 Hz. The frequency range of interest for mechanical equipment vibration diagnostic investigations associated with Table 12.6 is normally 1 to 1,000 Hz.

Building and mechanical equipment vibration measurements should only be conducted after:

- All building construction equipment has either been removed from the building or turned off, and
- The testing, adjusting and balancing of all mechanical equipment have been completed and the inspection of the mechanical equipment vibration isolation systems has been completed and approved.

If the building is subject to vibration sources other than the mechanical equipment and related systems in the building (e.g., railroad, subway or vehicle traffic), it may be necessary to schedule a measurement time when these effects can be minimized.

Locations for Mechanical Equipment Related Vibration Measurements

The procedures outlined below should be used to select locations within a building where and on mechanical equipment on which vibration measurements will be conducted.

Consult the architectural and mechanical contract specifications for specific guidance on where vibration measurements are to be made. In the absence of specific guidance in these documents, vibration measurements should be made on a minimum of 5 percent of the rotating, reciprocating, and other vibrating mechanical equipment in the building. Vibration measurements should be made on all noticeably vibrating equipment and on equipment

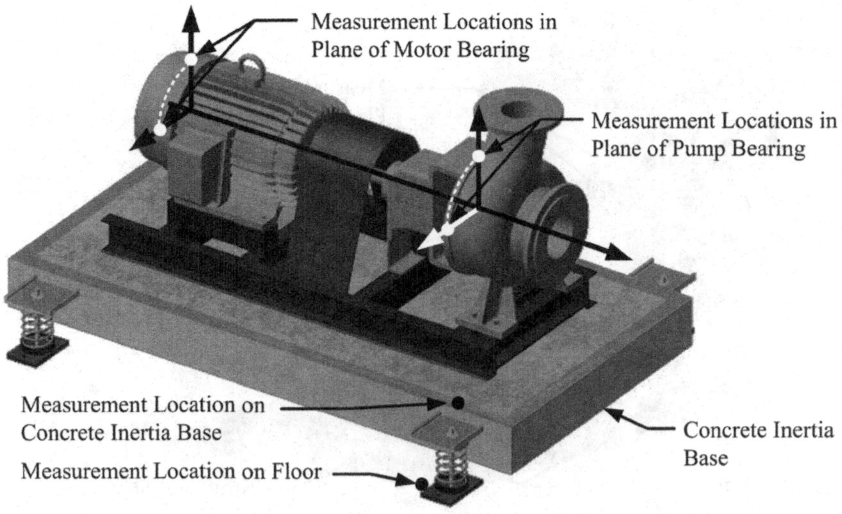

Figure 12.32 Vibration Measurement Locations for Pumps

that is generating noticeable structural vibration. The equipment on which vibration measurements will be made should be evenly distributed between fans, pumps, compressors and other vibrating equipment present in the building. Vibration measurements do not need to be made in rooms occupied by people unless noticeable building structural vibration is observed to exist during a walk-through of the building when all mechanical equipment is operating properly. Vibration measurements should be made in areas where very sensitive optical and electronic equipment will be or are located. Based on a walk-through of the building, select the areas in the building and the equipment on which vibration measurements will be made. Submit a list summarizing the results of this selection process to the building owner, general contractor, or mechanical contractor for review and approval.

Check the architectural and mechanical contract specifications to determine locations within rooms and on mechanical equipment where vibration measurements will be made. If this information is not contained in these documents, use the following locations:

- Vibration measurements in rooms where people will be present and building structural vibration has been observed to exist should be made at the location(s) of the observed vibration.
- In rooms where very sensitive optical and electronic equipment will be or are placed, make vibration measurements at the locations where the equipment will be or are placed.
- Vibration measurements should be made on mechanical equipment motor and shaft bearings locations at the positions and in the directions shown in Figures 12.32 and 12.33 and at other similar locations of other types of mechanical equipment with rotating and reciprocating components.
- Vibration measurements should be made on structural and inertia bases that support rotating and reciprocating mechanical equipment at the positions shown in Figures 12.32 and 12.33.

Vibration Measurements

Vibration measurements should be made with a parallel processing, real time analyzer that has 1/24, 1/12 or 1/3 octave band filters or a Fast Fourier Transform (FFT) analyzer. If a real time analyzer is used, at a minimum, the frequency range of the analyzer should cover the filter band center frequencies of 1 to 100 Hz. If an FFT analyzer is used, the spectral line resolution should not exceed 0.5 Hz/line, and the frequency range of the analyzer, at a minimum, should cover the frequency range of 0 to 100 Hz. Calibrate the real-time or FFT analyzer per the procedures outlined in Chapter 3.

Use the following procedure to make the vibration measurements in the absence of specific instructions in the architectural and mechanical contract specifications:

- Select one of the methods discussed in Chapter to

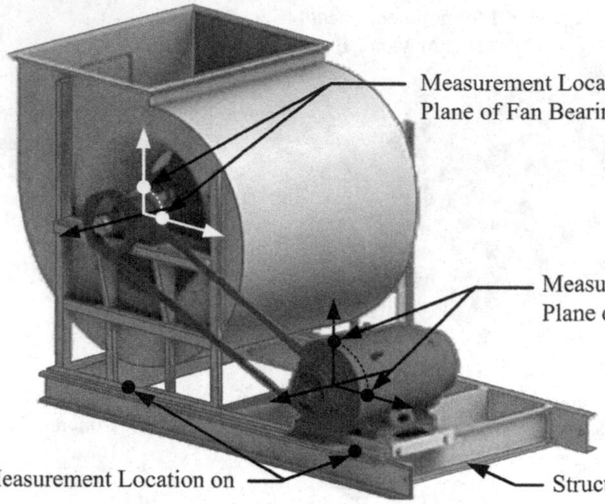

Figure 12.33 Vibration Measurement Locations on Centrifugal Fans

mount the accelerometer onto the building or machine surfaces where vibration measurements will be made. Normally, beeswax or a magnet can be used.

- Building structural vibration measurements should be made in the direction normal to the vibrating surface.
- Initially make vibration measurements only in the vertical direction at one motor or shaft bearing location on each piece of mechanical equipment selected for vibration measurements. If excessive vibration is observed to exist on a piece of equipment, make vibration measurements in the vertical direction at all accessible motor and shaft bearings locations and in the vertical, horizontal, and axial directions at the bearing location with the highest vibration amplitudes.
- Initially make vibration measurements in the vertical direction near one of the structural vibration isolation mount supports on the structural steel and concrete inertia bases selected for vibration measurements and on the floor near the base of the vibration isolation mount. If excessive vibration is observed to exist on a structural or inertia base, repeat the vibration measurements at the other three structural vibration isolation mount supports.

Assessment of Vibration Measurements

Mechanical equipment construction documents should specify acceptable vibration criteria for rooms and mechanical equipment in a building. The manufacturers of very sensitive optical and electronic equipment should specify acceptable vibration criteria for their equipment. When this information is not available, use Table 12.1 and Figures 12.2 and 12.3 to access the acceptability of vibration that may exist in areas where people will be present or very sensitive optical and/or electronic may be present. Use Figures 12.4 and 12.5 and Table 12.6 to assess the acceptability of vibration measured on mechanical equipment.

Vibration Measurement Report

The vibration measurement report should include:

- Building owner, architect, contractor, etc. for whom the vibration measurements were made and to whom the report is being submitted;
- The address of the building where the vibration measurements were made;
- Manufacturer, model, and serial number of the vibration analyzer used for the vibration measurements;
- Manufacturer, model, and serial number of the vibration calibrator;
- Dates of last calibrations of the vibration analyzer and vibration calibrator;
- Plots of the measured vibration data; and
- Assessment(s) of the measured vibration results.

12.19 Mechanical Equipment Vibration Isolation Specifications and Site Inspections

During the building commissioning process, the

architectural and mechanical equipment drawings and specifications should be reviewed and the mechanical and environmental equipment and equipment rooms should be inspected to ensure that the sound and vibration requirements contained in the architectural and mechanical contract specifications have been achieved.

Mechanical Equipment Vibration Isolation Specifications

Architectural and mechanical specifications contain the mechanical equipment vibration and sound specifications for a building. Vibration specifications should include:

- Mechanical equipment and building vibration and sound criteria requirements for building spaces;
- List of building areas and rooms that require low vibration and sound and the vibration and sound criteria that must be achieved;
- Schedule of mechanical equipment to be vibration isolated. The schedule may be in the mechanical specification or on the mechanical drawings; and
- Descriptions, approved manufacturers, suppliers, etc. of mechanical equipment structural and concrete inertia bases and vibration isolation mounts.

The architectural and mechanical drawings should contain information necessary to achieve specified building vibration criteria. The drawings should include:

- Site plan showing building location. This plan should identify critical buildings and all building vibration and sound sources;
- Locations of cooling towers and other outside mechanical equipment on the building site;
- Locations of roof mounted mechanical equipment, such as cooling towers, chillers and packaged air handling and refrigeration units;
- Locations of all mechanical equipment rooms;
- Locations of building duct and pipe shafts;
- Locations of all critical occupied spaces adjacent to mechanical equipment rooms, pipe and duct shafts, and any other spaces containing mechanical equipment;
- Schedules identifying required vibration and sound criteria for all building spaces;
- Schedules identifying required building vibration criteria in critical building spaces;
- Types of walls, partitions, and door constructions that enclose mechanical equipment rooms, duct and pipe shafts, etc.;
- Types of suspended ceilings below mechanical equipment rooms, high-velocity ducts and air terminal units; and
- Locations of exterior wall openings for air intakes and discharges for mechanical systems.

The mechanical drawings and specifications should contain information necessary to achieve specified building vibration and sound criteria in building spaces. With respect to the mechanical drawings and specifications:

- Examine the schedule of all vibrating equipment associated with the building air distribution and environmental systems. This schedule should be contained in the mechanical specifications and drawings and normally includes the vibration isolation system specifications and required static deflections for the vibration isolation mounts.
- Confirm the locations of all mechanical equipment rooms shown on the architectural drawings.
- Confirm the locations of vibrating equipment located outside of mechanical equipment rooms.
- Identify duct and pipe systems that penetrate mechanical equipment room floor slabs, partitions and ceilings.
- Identify partition constructions that separate mechanical equipment rooms and pipe and duct shafts from adjacent occupied building spaces that have low sound and vibration requirements. Check the architectural and mechanical specifications and drawings to identify what these requirements are. Note all penetrations through mechanical equipment room slabs, partitions and ceilings.
- Identify mechanical equipment rooms and pipe and duct shafts that are adjacent to occupied building spaces that have very low vibration and sound requirements. Check the architectural and mechanical specifications and drawings to identify what these requirements are. Note all penetrations through mechanical equipment room slabs, partitions and ceilings.
- Examine all potential vibration and sound transmission paths between mechanical equipment rooms and duct and pipe shafts and adjacent occupied building spaces that have low and very low vibration and sound

requirements. Paths include supply and return air ducts; partition, ceiling and floor penetrations; doors; etc.

Manufacturers' submittals for vibration isolation mounts, mechanical equipment structural and concrete inertia bases, and other vibration control equipment should be reviewed to ensure they comply with the architectural and mechanical drawings and specifications. During field inspections, verify that mechanical equipment installed in the building comply with each manufacturer's submittals.

Mechanical Equipment Vibration Isolation Site Inspections

Mechanical equipment rooms and pipe and duct shafts are often adjacent to occupied building spaces that have low and very low vibration and sound requirements. Properly designed mechanical equipment room floors, partitions and ceilings and pipe and duct shafts reduce vibration and sound transmission to these adjacent building spaces. Properly designed mechanical equipment vibration isolation systems prevent excessive vibration from being structurally transmitting as vibration and sound to adjacent occupied building spaces.

Inspection of vibration isolation mounts: Vibration isolated equipment should be inspected before mechanical equipment start-up. The following should be checked:

- Check structural bases for cracked welds, excessive bending, and torque on steel members.
- Check concrete inertia bases for cracked concrete and their vibration isolation mount retainer brackets for looseness.
- Check concrete inertia base to ensure their mounting surfaces are flat and they are level.
- Check elastomeric vibration isolation mounts and hangers for cracks in the rubber, for loose bonds between the rubber and steel plates and other steel components, and for adequate clearances between support studs and the side of the stud holes in elastomeric hangers.
- Check open coil steel spring mounts for loose or missing bolts, nuts and lock washers; spring overloading; completely collapsed spring coils; and cocked springs.
- Check open coil steel spring mounts for high frequency rubber or glass fiber pads [0.25 in. (6.4 mm) thick, unless greater thickness is specified] between the bottom plate of the mount and the supporting structure.
- Check housed open coil steel spring mounts for proper centering of the springs, clearances between the mount housing and rubber snubbers, and for tilted or cocked springs.
- Check to ensure that bolts used to mount vibration isolation mounts from supporting structures are properly vibration isolated by means of rubber bushings and rubber washers when specifications require that the mounts be bolted to a supporting structure.
- Check vibration isolation mounts with restraint devices to ensure that all shims have been removed and support nuts have been properly adjusted.
- Check pneumatic vibration isolation mounts for potential overload by confirming its air pressure is consistent with the manufacturer's submittal or catalog information. Ensure they are protected from potential damage from vehicle or other traffic. The pneumatic vibration isolation system should include the pneumatic chamber, strainer, oil separator, height regulator, and air pressure gage.
- Check the spaces under all structural and concrete inertia bases to ensure they are clean and free of debris.
- Check to ensure all shipping bolts associated with open coil steel spring mounts have been removed.

Inspection of mechanical equipment structural and concrete inertia bases: Manufacturers of mechanical equipment normally provide the weights of their equipment to manufacturers of vibration isolation mounts and hangers and structural and concrete inertia bases by means of their catalog information or equipment submittals. The vibration isolation mount manufacturers use this information to determine equipment weight distributions and centers of gravity. These data are used to select the spring coefficients or rubber durometers of the vibration isolation mounts and hangers that are listed on the submittal drawings. The weight distribution and vibration isolation mount or hanger loading for all vibration isolated mechanical equipment should be inspected. The following should be checked:

- Check to ensure that vibration isolation mount locations are within ±0.25 in. (6.4 mm) of the indicated locations shown on a structural or concrete inertia base submittal drawings.

- Compare the color codes of vibration isolation mounts with their respective manufacturer's color code chart and check the manufacturer's submittal as to vibration isolation mount locations and related spring or stiffness constants.
- Measure the heights of loaded vibration isolation mounts and subtract the results from the corresponding "free" or unloaded heights shown in the manufacturer's catalog or submittal drawings. The differences between the free and loaded heights are the vibration isolation mounts static deflections under load. Check to ensure the measured vibration isolation mount static deflections under load correspond to their respective static deflections listed in the manufacturer's submittal.
- Check structural and concrete inertia bases for "out of level", which indicates possible incorrectly calculated weight distribution on a base or improper placement of vibration isolation mounts supporting the base. Permissible tolerances are 1/32 in./ft (2.6 mm/m) of base lengthwise, or 1/64 in./ft (1.3 mm/m) of base length parallel to the equipment shaft, or as otherwise specified in the mechanical specifications. When "out of level" base conditions prevail, also check the flatness of the supporting structures.

Inspection of mechanical equipment flexible connections: Ducts, pipes, and electrical conduits are connected to mechanical equipment. If they are not properly connected to the mechanical equipment, they can "short circuit" the equipment vibration isolation mounts and allow unwanted vibration to be transmitted to the building structure.

- Check flexible pipe connectors, hoses, and pipe expansion joints with regard to type, length and location to ensure they are installed as required by the mechanical specifications.
- Check flexible electrical and control connections to ensure they do not restrain the movement of vibration isolated mechanical equipment.
- Check fabric connections between fans and ducts to ensure that a fabric "bellows" exists when the fans are operating.
- Check all vibration isolated mechanical equipment to ensure they are free of structural ties and rigid connections that short circuit their vibration isolation systems.
- Check limit stops and leveling bolts on all vibration isolation mounts to ensure they are not "short circuiting" the vibration isolation system.

Inspection of suspended vibration isolated mechanical equipment, ducts and pipes: Vibration isolation hangers consist of a rubber, glass fiber, or open coil steel spring mount or a combination of these mounts housed in a hanger box. Two major problems in the installation of vibration isolation hangers are misalignment and over/under loading.

- Check to ensure the bottom vibration isolation hanger rod does not contact the sides of the hole in the bottom of the hanger box. This contact can significantly reduce the vibration isolation effectiveness of the vibration isolation hanger. The hanger manufacturer can minimize the potential for this contact by providing a large enough hole in the bottom of the hanger so the hanger rod can swing through a 30 degree arc before contacting the edges of the hole.
- Make appropriate dimensional measurements and load calculations to ensure vibration isolation hangers are loaded within specified limits and to detect overloaded or completely collapsed open coil springs. Problems are most frequently encountered when vibration isolation hanger are used to support piping.

Inspection of duct and pipe partition penetrations: Duct and pipe penetrations through partitions that separate mechanical equipment rooms and adjacent occupied building spaces can result in the transmission of unwanted sound to these areas. Check the following:

- When ducts leaves the mechanical equipment room through a duct shaft, the duct shaft normally acts as an effective barrier that reduces the sound transmission from the mechanical equipment room to building spaces above and below the room. Check to ensure there are no sound leaks or holes in partitions that separate the duct shafts from adjacent spaces.
- When ducts or pipes penetrate a floor slab (either above or below the mechanical equipment room) the effectiveness of the slab as a sound barrier can be diminished. Check to ensure these penetrations are properly sealed.
- Ducts or pipes that penetrate partitions that separate

mechanical equipment rooms from adjacent building spaces can diminish the sound attenuation effectiveness of these partitions. For each penetration, check to ensure the open area around the duct or pipe penetration is filled with compressed fiberglass and the area between the duct or pipe and the partition is tightly sealed.

- Check the entire perimeter of the mechanical equipment room to ensure the partitions are properly sealed at the floor and the underside of the slab above the mechanical equipment room. Check all partition penetrations to ensure they are properly sealed around their complete perimeter.
- When a duct shaft acts as a return air duct or plenum, sound from a mechanical equipment room can be transmitted to adjacent building spaces served by the return air shaft or plenum. Check for and report this arrangement when it is present.

Inspection of mechanical equipment room floating floors: A properly designed and installed mechanical equipment room floating floor will yield a substantial reduction in the transmission of sound from a mechanical equipment room to the building space below. Mechanical equipment supported by structural bases and vibration isolation mounts can be mounted on a floating floor. However, floating floors are not designed to support mechanical equipment that requires concrete inertia bases. For this case, the machine and concrete inertial base must be mounted on a structural pedestal that passes through the floating floor to the supporting structure under the floating floor. When a floating floor has been installed in a mechanical equipment room, check for the following:

- Ensure the floating floor is not used as a vibration isolation base for large mechanical equipment mounted on concrete inertial bases.
- Ensure the floating floor is free of structural ties and rigid connections between mechanical equipment mounted on the floor and the floor's supporting structure, peripheral partitions, columns, protruding machinery pedestals, protruding pipes, drain pipes, and any other part of the building structure that can short circuit the floating floor.
- Ensure gaps between the floating floor and perimeter partitions, structural columns, protruding machinery pedestals, protruding pipes, and any other protruding element are properly sealed. Refer to Figures 12.28 and 12.29.

Inspection of vibration isolated ceiling: Properly designed and installed vibration isolated ceilings will yield a substantial increase in the sound attenuation between two vertically adjacent building spaces. Vibration isolated ceilings can be substituted for a floating floor or used to supplement a floating floor when very high sound isolation requirements exist. Vibration isolated ceilings are usually suspended from vibration isolation hangers. A structural grid is attached to the hangers to support a gypsum board ceiling. T-bar acoustical tile or panel ceiling should not be used as a vibration isolated ceiling. Sound absorbing ceiling tile can be attached on the underside of the gypsum board ceiling. Sound absorption batts are normally placed on top of the isolated ceiling. The periphery of the ceiling should not be attached to vertical partitions. The ceiling should not be rigidly connected to pipes, ducts, structural columns, and other elements that penetrate the ceiling. These penetrations should be caulked or sealed with flexible material. Access doors should be provided in vibration isolated ceilings so the plenum area above the ceiling should be checked:

- Check to ensure the structural and vibration isolation hangers and ceiling suspension structure do not contact ducts, pipes and any building structural components.
- Check to ensure penetrations through the ceiling are properly sealed and vibration isolated from the ceiling.
- Check to ensure the perimeter of the ceiling is properly sealed.
- Check to ensure the vibration isolation hangers are as specified with regard to material, type, loading and spacing;
- Check to ensure the sound absorption batts are as specified with regard to thickness and density.
- Check to ensure the ceiling material is as specified with regard to the type, thickness, and weight per square foot of the material.

QUESTIONS AND PROBLEMS - CHAPTER 12

1. Discuss the importance of appropriately applying the building vibration criteria specified in Figures 12.2 and 12.3 when designing vibration isolation systems to reduce the negative effects of structure borne vibration in buildings.

2. Discuss the importance of appropriately applying the equipment vibration severity specified in Figures 12.4 and 12.5 when designing vibration isolation systems to reduce the negative effects of structure borne vibration in buildings.

3. When designing a vibration isolation system, why is it important for the resonance frequency of the system to be significantly lower than the lowest harmonic excitation frequency applied to the system? For good design, how much lower should the resonance frequency be?

4. When designing a vibration isolation system, why is it important to minimize the damping present in the system.

5. When designing a vibration isolation system, why is it important to examine both the harmonic response function and vibration transmissibility? What additional consideration must given to these function when the vibration excitation is associated with rotating and reciprocating unbalance?

6. Specify the suggested maximum static deflections associated with the different commercial vibration isolation mounts listed on pages 432-434.

7. When is it necessary to use pneumatic vibration isolation mounts?

8. Discuss the design conditions associated with the use of the structural bases presented on pages 434-436.

8. Referring to Figure 12.21, develop equations (12.29) through (12.21).

9. A vibration isolation system similar to the one shown in Figure 12.21 is designed to have a static deflection of 0.5 in. We = 2,200 lb_f, l = 5 ft, a = 3 ft, w = 4 ft, and b = 3 ft. Determine the stiffness coefficients of each of the four springs (K_1, K_2, K_3, and K_4) and the total stiffness coefficient of the vibration isolation system.

10. What is the purpose of thrust restraints?

11. How should pipe expansion joints and flexible pipe couples be used to ensure the proper functioning of concrete form bases?

12. What is the purpose of floating floors. Where are they commonly used?

13. What are the different types of floor constructions used to support mechanical equipment?

14. What two physical requirements must be properly addressed when designing vibration isolation systems for mechanical equipment?

15. What is the purpose of the information presented in Table 12.5? How should this information be used?

16. List and discuss the physical factors that should be addressed when investigation mechanical equipment vibration problems.

17. When conducting mechanical equipment vibration measurements, where should accelerometers be mounted on mechanical equipment and on structural bases?

18. What factors should be considered when examining the results of mechanical equipment vibration measurements?

19. Briefly describe the physical factors that should be examined when reviewing mechanical equipment vibration isolation specifications.

20. Briefly describe the physical installations that should be examined when conducting mechanical equipment site inspections.

General References

1. J. S. Bendat and A. G. Piersol, *Engineering Applications of Correlation and Spectral Analysis*, John Wiley & Sons, Inc., 1993.
2. J. S. Bendat and A. G. Piersol, *Random Data: Analysis and Measurement Procedures*, John Wiley & Sons, Inc., 2000.
3. A. D. D. Dimarogonas, *Vibrations for Engineers*, Prentice Hall, 1996.
4. D. J. Ewins, S. S. Rao, S. G. Braun (Eds.), *Encyclopedia of Vibration*, Three-Volume Set, Elsevier Science & Technology Books, 2001.
5. D. G. Fertis, *Mechanical and Structural Vibrations*, John Wiley & Sons, Inc., 1995.
6. P. L. Gatti and V. Ferrari, *Applied Structural and Mechanical Vibrations: Theory Methods and Measuring Instrumentation*, Taylor & Francis, Inc., 1999.
7. J. H. H. Ginsberg, *Mechanical and Structural Vibrations: Theory and Applications*, John Wiley & Sons, Inc., 2001.
8. C. M. Harris, A. G. Piersol (Eds.), *Harris' Shock and Vibration Handbook*, McGraw-Hill Companies, 2001.
9. J. P. Den Hartog (Ed.), *Mechanical Vibrations*, Dover Publications, 1985.
10. D. J. Inman, Engineering Vibrations, Prentice Hall, 2007.
11. K. G. McConnell and P. S. Varoto, *Vibration Testing: Theory and Practice*, John Wiley & Sons, Inc., 1995.
12. L. Meirovitch, *Analytical Methods in Vibrations*, Prentice Hall, 1997.
13. L. Meirovitch, *Fundamentals of Vibrations*, McGraw-Hill Companies, 2002.
14. P. M. Moretti, *Modern Vibrations Primer*, CRC Press, 1999.
15. A. D. Nashif, J. P. Henderson, D. I. G. Jones , *Vibration Damping*, John Wiley & Sons, Inc., 1985.
16. P. V. O'Neil, *Advanced Engineering Mathematics*, Engineering - Nelson, 2006.
17. W. J. Palm, *Mechanical Vibrations*, John Wiley & Sons, Inc., 2006.
18. M. Petyt, *Introduction to Finite Element Vibration Analysis*, Cambridge University Press, 1998.
19. S. S. Rao, *Mechanical Vibrations* with Disk, Addison-Wesley, 1995.
20. W.Soedel, *Vibrations of Shells and Plates* (Third Edition), CRC Press, 2004.
21. M. R. Spiegel, Schaums, John Liu, *Schaum's Mathematical Handbook of Formulas and Tables*, McGraw-Hill Companies, 1998.
22. R. F. Steidel, *Introduction to Mechanical Vibrations*, John Wiley & Sons, Inc., 2001.
23. C. T. Sun, Y. P. Lu, Y. P. Lu, *Damping of Structural Elements*, Prentice Hall Professional Technical Reference, 1995.
24. W. T. Thomson and M. D. Dahleh, *Theory of Vibrations with Applications*, Prentice Hall, 1997.
25. B. H. Tongue, *Principles of Vibration*, Oxford University Press, USA, 2003.
26. W. Weaver, S. R. Timoshenko, D. H. Young, *Vibration Problems in Engineering*, John Wiley & Sons, Inc., 1990.
27. V. Wowk, *Machinery Vibration: Measurement and Analysis*, McGraw-Hill Companies, 1991.

General References

INDEX

Printed in the United States
By Bookmasters